Marius Grundmann

The Physics of Semiconductors

Marius Grundmann

The Physics of Semiconductors

An Introduction Including Devices and Nanophysics

With 587 Figures, 6 in Color, and 36 Tables

Springer

Marius Grundmann
Institut für Experimentelle Physik II
Universität Leipzig
Linnéstraße 5
04103 Leipzig
e-mail: grundmann@physik.uni-leipzig.de

Library of Congress Control Number: 2006923434

ISBN-10 3-540-25370-X Springer Berlin Heidelberg New York
ISBN-13 978-3-540-25370-9 Springer Berlin Heidelberg New York

Springer is a part of Springer Science+Business Media

springer.com

Printed in Germany

Typesetting: Protago-TEX-Production GmbH, Berlin
Production: LE-TEX Jelonek, Schmidt & Vöckler GbR, Leipzig
Cover design: eStudio Calamar S.L., F. Steinen-Broo, Pau/Girona, Spain

Printed on acid-free paper 57/3100/YL 5 4 3 2 1 0

To Michelle,
Sophia Charlotte
and Isabella Rose

Preface

Semiconductor devices are nowadays commonplace in every household. In the late 1940s the invention of the transistor was the start of a rapid development towards ever faster and smaller electronic components. Complex systems are built with these components. The main driver of this development was the economical benefit from packing more and more wiring, transistors and functionality on a single chip. Now every human is left with about 100 million transistors (on average). Semiconductor devices have also enabled economically reasonable fiber-based optical communication, optical storage and high-frequency amplification and have only recently revolutionized photography, display technology and lighting. Along with these tremendous technological developments, semiconductors have changed the way we work, communicate, entertain and think. The technological sophistication of semiconductor materials and devices is progressing continuously with a large worldwide effort in human and monetary capital, partly evolutionary, partly revolutionary embracing the possibilities of nanotechnology. For students, semiconductors offer a rich, diverse and exciting field with a great tradition and a bright future.

This book is based on the two semester semiconductor physics course taught at Universität Leipzig. The material gives the students an overview of the subject as a whole and brings them to the point where they can specialize and enter supervised laboratory research. For the interested reader some additional topics are included in the book that are taught in subsequent, more specialized courses.

The first semester contains the fundamentals of semiconductor physics (Part I – Chaps. 1–17). Besides important aspects of solid-state physics such as crystal structure, lattice vibrations and band structure, semiconductor specifics such as technologically relevant materials and their properties, electronic defects, recombination, hetero- and nanostructures are discussed. Semiconductors with electric polarization and magnetization are introduced. The emphasis is put on inorganic semiconductors, but a brief introduction to organic semiconductors is given in Chap. 16. In Chap. 17 dielectric structures are treated. Such structures can serve as mirrors, cavities and microcavities and are a vital part of many semiconductor devices.

The second part (Part II – Chaps. 18–21) is dedicated to semiconductor applications and devices that are taught in the second semester of the course. After a general and detailed discussion of various diode types, their applications in electrical circuits, photodetectors, solar cells, light-emitting diodes and lasers are treated. Finally, bipolar and field-effect transistors are discussed.

The course is designed to provide a balance between aspects of solid-state and semiconductor physics and the concepts of various semiconductor devices and their applications in electronic and photonic devices. The book can be followed with little or no pre-existing knowledge in solid-state physics.

I would like to thank several colleagues for their various contributions to this book, in alphabetical order (if no affiliation is given, from Universität Leipzig): Klaus Bente, Rolf Böttcher, Volker Gottschalch, Axel Hoffmann (Technische Universität Berlin), Alois Krost (Otto-von-Guericke Universität Magdeburg), Michael Lorenz, Thomas Nobis, Rainer Pickenhain, Hans-Joachim Queisser (Max-Planck-Institut für Festkörperforschung, Stuttgart), Bernd Rauschenbach (Leibniz-Institut für Oberflächenmodifizierung, Leipzig), Bernd Rheinländer, Heidemarie Schmidt, Rüdiger Schmidt-Grund, Mathias Schubert, Gerald Wagner, Holger von Wenckstern, Michael Ziese, and Gregor Zimmermann. Their comments, proof reading and graphic material improved this work. Also, numerous helpful comments from my students on my lectures and on preliminary versions of the present text are gratefully acknowledged. I am also indebted to many other colleagues, in particular to (in alphabetical order) Gerhard Abstreiter, Zhores Alferov, Levon Asryan, Günther Bauer, Manfred Bayer, Immanuel Broser, Jürgen Christen, Laurence Eaves, Ulrich Gösele, Alfred Forchel, Manus Hayne, Frank Heinrichsdorff, Fritz Henneberger, Detlev Heitmann, Robert Heitz†, Nils Kirstaedter, Fred Koch, Nikolai Ledentsov, Evgeni Kaidashev, Eli Kapon, Claus Klingshirn, Jörg Kotthaus, Axel Lorke, Anupam Madhukar, Bruno Meyer, David Mowbray, Hisao Nakashima, Mats-Erik Pistol, Fred Pollak, Volker Riede, Hiroyuki Sakaki, Lars Samuelson, Vitali Shchukin, Maurice Skolnick, Oliver Stier, Robert Suris, Volker Türck, Konrad Unger, Victor Ustinov, Leonid Vorob'jev, Richard Warburton, Alexander Weber, Eicke Weber, Peter Werner, Ulrike Woggon, Roland Zimmermann and Alex Zunger, with whom I have worked closely, had enjoyable discussions with and who have posed questions that stimulated me. I reserve special thanks for Dieter Bimberg, who supported me throughout my career. I leave an extra niche – as the Romans did, in order not to provoke the anger of a God missed in a row of statues – for those who had an impact on my scientific life and that I have omitted to mention.

Leipzig, January 2006 *Marius Grundmann*

Contents

Part II Applications

Abbreviations

2DEG	two-dimensional electron gas
AAAS	American Association for the Advancement of Science
AB	antibonding (position)
ac	alternating current
AFM	atomic force microscopy
AIP	American Institute of Physics
AM	air mass
APD	antiphase domain
APD	avalanche photodiode
APS	American Physical Society
AR	antireflection
ASE	amplified spontaneous emission
AVS	American Vacuum Society (The Science & Technology Society)
BC	bond center (position)
bcc	body-centered cubic
BEC	Bose–Einstein condensation
BGR	bandgap renormalization
CAS	calorimetric absorption spectroscopy
CCD	charge coupled device
CD	compact disc
CEO	cleaved-edge overgrowth
CIE	Commission Internationale de l'Éclairage
CIS	$CuInSe_2$ material
CL	cathodoluminescence
CMOS	complementary metal–oxide–semiconductor
CMY	cyan-magenta-yellow (color system)
COD	catastrophical optical damage
CPU	central processing unit
CRT	cathode ray tube
CVD	chemical vapor deposition

cw	continuous wave
CZ	Czochralski (growth)
DAP	donor–acceptor pair
DBR	distributed Bragg reflector
dc	direct current
DFB	distributed feedback
DH(S)	double heterostructure
DLTS	deep level transient spectroscopy
DMS	diluted magnetic semiconductor
DOS	density of states
DPSS	diode-pumped solid-state (laser)
DRAM	dynamic random access memory
DVD	digital versatile disc
EEPROM	electrically erasable programmable read-only memory
EHL	electron–hole liquid
EL	electroluminescence
ELO	epitaxial lateral overgrowth
EMA	effective mass approximation
EPROM	erasable programmable read-only memory
ESF	extrinsic stacking fault
EXAFS	extended X-ray absorption fine structure
fcc	face-centered cubic
FeRAM	ferroelectric random access memory
FET	field-effect transistor
FIR	far infrared
FKO	Franz–Keldysh oscillation
FPA	focal plane array
FQHE	fractional quantum Hall effect
FWHM	full width at half-maximum
FZ	float-zone (growth)
GLAD	glancing-angle deposition
GRINSCH	graded-index separate confinement heterostructure
GSMBE	gas-source molecular beam epitaxy
HBT	heterobipolar transistor
hcp	hexagonally close packed
HCSEL	horizontal cavity surface-emitting laser
HEMT	high electron mobility transistor
HIGFET	heterojunction insulating gate FET
HJFET	heterojunction FET
hh	heavy hole

HOMO	highest occupied molecular orbital
HR	high reflection
HRTEM	high-resolution transmission electron microscopy
HWHM	half-width at half-maximum
IC	integrated circuit
IDB	inversion domain boundary
IF	intermediate frequency
IPAP	Institute of Pure and Applied Physics, Tokyo
IQHE	integral quantum Hall effect
IR	infrared
ISF	intrinsic stacking fault
ITO	indium tin oxide
JDOS	joint density of states
JFET	junction field-effect transistor
KKR	Kramers–Kronig relation
KTP	$KTiOPO_4$ material
LA	longitudinal acoustic (phonon)
LCD	liquid crystal display
LDA	local density approximation
LEC	liquid encapsulated Czochralski (growth)
LED	light-emitting diode
lh	light hole
LO	longitudinal optical (phonon), local oscillator
LPE	liquid phase epitaxy
LPCVD	low-pressure chemical vapor deposition
LPP	longitudinal phonon plasmon (mode)
LST	Lyddane–Sachs–Teller (relation)
LT	low temperature
LUMO	lowest unoccupied molecular orbital
LVM	local vibrational mode
MBE	molecular beam epitaxy
MEMS	micro-electro-mechanical system
MESFET	metal–semiconductor field-effect transistor
MIGS	midgap (surface) states
MIOS	metal–insulator–oxide–semiconductor
MIR	mid-infrared
MIS	metal–insulator–semiconductor
MHEMT	metamorphic HEMT
ML	monolayer
MMIC	millimeter-wave integrated circuit

MO	master oscillator
MODFET	modulation-doped FET
MOMBE	metal-organic molecular beam epitaxy
MOPA	master oscillator power amplifier
MOS	metal–oxide–semiconductor
MOSFET	metal–oxide–semiconductor field-effect transistor
MOVPE	metal-organic chemical vapor deposition
MRS	Material Research Society
MS	metal–semiconductor (diode)
MSM	metal–semiconductor–metal (diode)
MWQ	multiple quantum well
NDR	negative differential resistance
NEP	noise equivalent power
NIR	near infrared
NMOS	n-channel metal–oxide–semiconductor (transistor)
NTSC	national television standard colors
OPSL	optically pumped semiconductor laser
PA	power amplifier
PBG	photonic bandgap
pc	primitive cubic
PFM	piezoresponse force microscopy
PHEMT	pseudomorphic HEMT
PL	photoluminescence
PLD	pulsed laser deposition
PLE	photoluminescence excitation (spectroscopy)
PMMA	poly-methyl methacrylate
PMOS	p-channel metal–oxide–semiconductor (transistor)
PPC	persistent photoconductivity
PPLN	perodically poled lithium niobate
PV	photovoltaic
PWM	pulsewidth modulation
PZT	$PbTi_xZr_{1-x}O_3$ material
QCL	quantum cascade laser
QCSE	quantum confined Stark effect
QD	quantum dot
QHE	quantum Hall effect
QW	quantum well
QWIP	quantum-well intersubband photodetector
QWR	quantum wire

RAM	random access memory
RAS	reflection anisotropy spectroscopy
REI	random element isodisplacement
RF	radio frequency
RGB	red-green-blue (color system)
RHEED	reflection high-energy electron diffraction
RKKY	Ruderman–Kittel–Kasuya–Yoshida (interaction)
rms	root mean square
ROM	read-only memory
SAGB	small-angle grain boundary
SAM	separate absorption and amplification (structure)
sc	simple cubic
SCH	separate confinement heterostructure
SEL	surface-emitting laser
SEM	scanning electron microscopy
SET	single-electron transistor
SGDBR	sampled grating distributed Bragg reflector
SHG	second-harmonic generation
si	semi-insulating
SIA	Semiconductor Industry Association
SIMS	secondary ion mass spectroscopy
SL	superlattice
s-o	spin-orbit (or split-off)
SOA	semiconductor optical amplifier
SPD	spectral power distribution
SPIE	International Society for Optical Engineering
SPS	short-period superlattice
sRGB	standard RGB
SRH	Shockley–Read–Hall (kinetics)
SSR	side-mode suppression ratio
STM	scanning tunneling microscopy
TA	transverse acoustic (phonon)
TCO	transparent conductive oxide
TE	transverse electric (polarization)
TEGFET	two-dimensional electron gas FET
TEM	transmission electron microscopy
TES	two-electron satellite
TF	thermionic field emission
TFT	thin-film transistor
TM	transverse magnetic (polarization)
TO	transverse optical (phonon)
TOD	turn-on delay (time)
TPA	two-photon absorption

UHV	ultrahigh vacuum
UV	ultraviolet
VCA	virtual crystal approximation
VCO	voltage-controlled oscillator
VCSEL	vertical-cavity surface-emitting laser
VFF	valence force field
VGF	vertical gradient freeze (growth)
VIS	visible
VLSI	very large scale integration
WGM	whispering gallery mode
WKB	Wentzel–Kramer–Brillouin (approximation or method)
WS	Wigner–Seitz (cell)
XSTM	cross-sectional STM

Symbols

α	Madelung constant, disorder parameter, linewidth enhancement factor
$\alpha(\omega)$	absorption coefficient
α_{m}	mirror loss
α_{n}	electron ionization coefficient
α_{p}	hole ionization coefficient
β	used as abbreviation for $e/(k_{\mathrm{B}}T)$, spontaneous emission coefficient
γ, Γ	broadening parameter
γ_1, γ_2, γ_3	Luttinger parameter
Δ_0	spin-orbit splitting
$\epsilon(\omega)$	dielectric function
ϵ_0	permittivity of vacuum
ϵ_{r}	relative dielectric function
ϵ_{xy}	strain components
η	quantum efficiency
η_{d}	differential quantum efficiency
η_{w}	wall-plug efficiency
Θ_{D}	Debye temperature
κ	imaginary part of index of refraction, heat conductivity
λ	wavelength
μ	mobility
μ_0	magnetic susceptibility of vacuum
μ_{h}	hole mobility
μ_{n}	electron mobility
ν	frequency
Π	Peltier coefficient
ρ	mass density, charge density, resistivity
σ	standard deviation, conductivity
σ_{n}	electron capture cross section
σ_{p}	hole capture cross section
σ_{P}	polarization charge
σ_{xy}	stress components
τ	lifetime, time constant

ϕ	phase
ϕ_{Bn}	Schottky barrier height
χ	electron affinity, electric susceptibility
$\chi(\mathbf{r})$	envelope wavefunction
$\Psi(\mathbf{r})$	wavefunction
ω	angular frequency
Ω	interaction parameter
a	hydrostatic deformation potential
$\mathbf{a}$	accelaration
A	area
$\mathbf{A}$, A	vector potential
A^*	Richardson constant
A^{**}	effective Richardson constant
a_0	(cubic) lattice constant
b	bowing parameter, deformation potential
$\mathbf{b}$	Burger's vector
B	bimolecular recombination coefficient, bandwidth
$\mathbf{B}$, B	magnetic field
c	velocity of light in vacuum, lattice constant (along c-axis)
C	capacity, spring constant
C_{n}, C_{p}	Auger recombination coefficient
C_{ij}	elastic constants
d	distance, shear deformation potential
D	density of states, diffusion coefficient
$\mathbf{D}$, D	displacement field
$D_{\mathrm{e}}(E)$	electron density of states
$D_{\mathrm{h}}(E)$	hole density of states
D_{n}	electron diffusion coefficient
D_{p}	hole diffusion coefficient
e	elementary charge
E	energy
$\mathbf{E}$, E, $\mathcal{E}$	electric field
E_{A}	energy of acceptor level
$E_{\mathrm{A}}^{\mathrm{b}}$	acceptor ionization energy
E_{C}	energy of conduction-band edge
E_{D}	energy of donor level
$E_{\mathrm{D}}^{\mathrm{b}}$	donor ionization energy
E_{F}	Fermi energy
$E_{\mathrm{F_n}}$	electron quasi-Fermi energy
$E_{\mathrm{F_p}}$	hole quasi-Fermi energy
E_{g}	bandgap
E_{P}	energy parameter
E_{V}	energy of valence-band edge
E_{X}	exciton energy

E_X^b	exciton binding energy
f	oscillator strength
F	free energy
$\mathbf{F}$, F	force
$F(M)$	excess noise factor
f_e	Fermi–Dirac distribution function
f_i	ionicity
F_n	electron quasi-Fermi energy
F_p	hole quasi-Fermi energy
g	degeneracy, g-factor, gain
G	free enthalpy, generation rate
$\mathbf{G}$	vector of reciprocal lattice
g_m	transconductance
h	Planck constant
H	enthalpy
$\mathbf{H}$, H	magnetic field
$\mathcal{H}$	Hamiltonian
$\hbar$	$h/(2\pi)$
i	imaginary number
I	current
I_s	saturation current
j	current density, orbital momentum
j_s	saturation current density
k, k_B	Boltzmann constant
$\mathbf{k}$	wavevector
k_F	Fermi wavevector
l	angular orbital momentum
L	length of line element
$\mathbf{L}$	line vector (of dislocation)
L_D	diffusion length
L_z	quantum-well thickness
m	mass
m^*	effective mass
M	mass, multiplication factor
$\mathbf{M}$, M	magnetization
m_e	effective electron mass
m_h	effective hole mass
m_j	magnetic quantum number
n	electron concentration (in conduction band), ideality factor
$\mathbf{n}$	normal vector
$N(E)$	number of states
n^*	complex index of refraction ($= n_r + i\kappa$)
N_A	acceptor concentration
N_c	critical doping concentration

N_{C}	conduction-band edge density of states
N_{D}	donor concentration
n_{i}	intrinsic electron concentration
n_{r}	index of refraction (real part)
n_{s}	sheet electron density
N_{t}	trap concentration
n_{tr}	transparency electron concentration
n_{thr}	threshold electron concentration
N_{V}	valence-band edge density of states
p	pressure, free hole density
$\mathbf{p}$, p	momentum
P	power
$\mathbf{P}$, P	electric polarization
p_{cv}	momentum matrix element
p_{i}	intrinsic hole concentration
q	charge
$\mathbf{q}$, q	heat flow
Q	charge, quality factor
r	radius
$\mathbf{r}$	spatial coordinate
R	resistance, radius, recombination rate
$\mathbf{R}$	vector of direct lattice
r_{H}	Hall factor
R_{H}	Hall coefficient
s	spin
S	entropy, Seebeck coefficient, total spin
S_{ij}	stiffness coefficients
t	time
T	temperature
u	displacement, cell-internal parameter
$u_{n\mathbf{k}}$	Bloch function
U	energy
$\mathbf{v}$, v	velocity
v_{s}	drift-saturation velocity
V	volume, voltage, potential
$V(\lambda)$	(standardized) sensitivity of human eye
V_{a}	unit-cell volume
v_{g}	group velocity
v_{s}	velocity of sound
v_{th}	thermal velocity
w	depletion-layer width
W_{m}	work function
X	electronegativity
Y	Young's module, CIE brightness parameter
Z	partition sum, atomic order number

Physical Constants

constant	symbol	numerical value	unit
speed of light in vacuum	c_0	2.99792458×10^{8}	$\mathrm{m\,s^{-1}}$
permeability of vacuum	μ_0	$4\pi \times 10^{-7}$	$\mathrm{N\,A^{-2}}$
permittivity of vacuum	$\epsilon_0 = (\mu_0 c_0^2)^{-1}$	$8.854187817 \times 10^{-12}$	$\mathrm{F\,m^{-1}}$
		8.617385×10^{-5}	$\mathrm{eV\,K^{-1}}$
elementary charge	e	$1.60217733 \times 10^{-19}$	C
electron mass	m_e	$9.1093897 \times 10^{-31}$	kg
Planck constant	h	$6.6260755 \times 10^{-34}$	J s
	$\hbar = h/(2\pi)$	$1.05457266 \times 10^{-34}$	J s
	$\hbar$	6.582122×10^{-16}	eV s
Boltzmann constant	k_B	1.380658×10^{-23}	$\mathrm{J\,K^{-1}}$
von-Klitzing constant	R_H	25812.8056	Ω
Rydberg constant		13.6056981	eV
Bohr radius	a_B	$5.29177249 \times 10^{-11}$	m

1 Introduction

> The proper conduct of science
> lies in the pursuit of Nature's puzzles,
> wherever they may lead.
> *J.M. Bishop* [1]

The historic development of semiconductor physics and technology began in the second half of the 19th century. In 1947, the realization of the transistor was the impetus to a fast-paced development that created the electronics and photonics industries. Products founded on the basis of semiconductor devices such as computers (CPUs, memories), optical-storage media (CD, DVD), communication infrastructure (optical-fiber technology, mobile communication) and lighting (LEDs) are commonplace. Thus, fundamental research on semiconductors and semiconductor physics and its offspring in the form of devices has contributed largely to the development of modern civilization and culture.

1.1 Timetable

In this section early important milestones in semiconductor physics and technology are listed.

1821
T.J. Seebeck – discovery of semiconductor properties of PbS [2].

1833
M. Faraday – discovery of the temperature dependence of the conductivity of AgS (negative $\mathrm{d}R/\mathrm{d}T$) [3].

1873
W. Smith – discovery of photoconductivity in selenium [4].

1874
F. Braun[1] – discovery of rectification in metal–sulfide semiconductor contacts [6], e.g. for $CuFeS_2$ and PbS. The current through a metal–semiconductor contact is nonlinear (as compared to that through a metal, Fig. 1.1), i.e. a deviation from Ohm's law. Braun's structure is similar to a MSM diode.

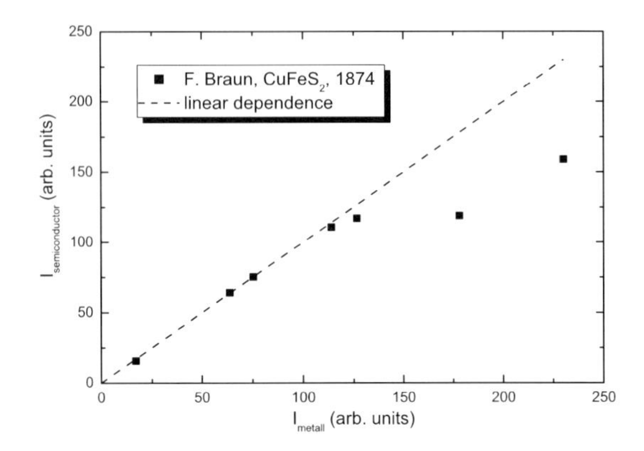

Fig. 1.1. Current through a silver–$CuFeS_2$–silver structure as a function of the current through the metal only, 1874. Data points are for different applied voltages. Experimental data from [6]

1876
W.G. Adams and R.E. Day – discovery of the photovoltaic effect in selenium [7].

1883
Ch. Fritts – first solar cell, based on an Au/selenium rectifier [5]. The efficiency was below 1%.

1907
H.J. Round – discovery of electroluminescence investigating blue light emission from SiC [8].

1911
The term 'Halbleiter' (semiconductor) is introduced for the first time by J. Königsberger and J. Weiss [9].

[1]F. Braun made his discoveries on metal–semiconductor contacts in Leipzig while a teacher at the Thomasschule zu Leipzig. He conducted his famous work on vacuum tubes later as a professor in Strasbourg, France.

1925
J.E. Lilienfeld[2] – proposal of the field-effect transistor (Fig. 1.2) (Method and Apparatus for Controlling Electric Currents, US patent 1,745,175, 1930, filed 1926). J.E. Lilienfeld was also awarded patents for a depletion mode MOSFET (US patent 1,900,018, 1933) and current amplification with nppn- and pnnp-transistors (US patent 1,877,140, 1932).

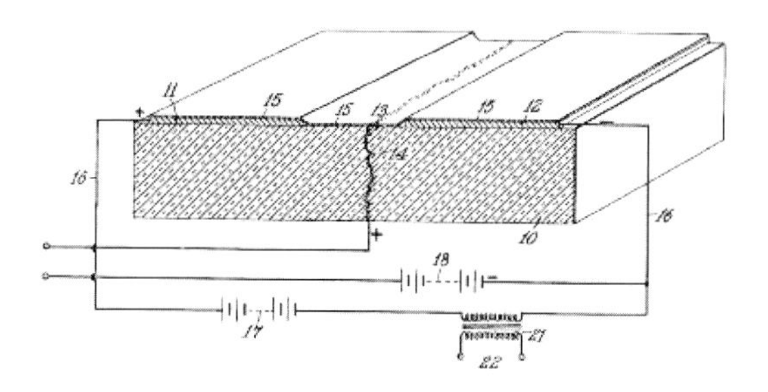

Fig. 1.2. Sketch of a field-effect transistor, 1926. From [12]

1927
A. Schleede and Baggisch – impurities are of decisive importance for conductivity.

1931
R. de L. Kronig and W.G. Penney – properties of periodic potentials in solids [13].

A.H. Wilson – development of band-structure theory [14].

C. Zener – Zener tunneling [15].

1936
J. Frenkel – description of excitons [16].

1938
B. Davydov – theoretical prediction of rectification in Cu_2O [17].

W. Schottky – theory of the boundary layer in metal–semiconductor contacts [18], being the basis for Schottky contacts and field-effect transistors (FETs).

[2]After obtaining his PhD in 1905 from the Friedrich-Wilhelms-Universität Berlin, J.E. Lilienfeld joined the Physics Department of University of Leipzig and worked on gas liquification and with Lord Zeppelin on hydrogen-filled blimps. In 1910 he became professor at the University of Leipzig where he mainly researched on X-rays and vacuum tubes. To the surprise of his colleagues he left in 1926 to join a US industrial laboratory [10, 11].

N.F. Mott – metal–semiconductor rectifier theory [19].

R. Hilsch and R.W. Pohl – proposal of a three-electrode crystal (from NaCl).

1941
R.S. Ohl – Si rectifier with point contact (Fig. 1.3) (US patent 2,402,661).

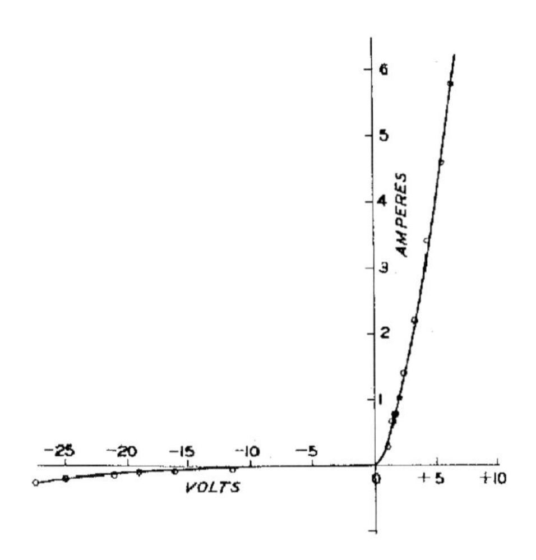

Fig. 1.3. Characteristics of a silicon rectifier, 1941. From [20]

1942
K. Clusius, E. Holz and H. Welker – rectification in germanium (Elektrische Gleichrichteranordnung mit Germanium als Halbleiter und Verfahren zur Herstellung von Germanium für eine solche Gleichrichteranordnung, German patent DBP 966 387, 21g, 11/02)

1945
H. Welker – patents for JFET and MESFET (Halbleiteranordnung zur kapazitiven Steuerung von Strömen in einem Halbleiterkristall, German patent DBP 980 084, 21g, 11/02)

1947
W. Shockley, J. Bardeen and W. Brattain fabricate the first transistor in the AT&T Bell Laboratories, Holmdel, NJ in an effort to improve hearing aids [21].[3] Strictly speaking the structure was a point-contact transistor. A 50-µm wide slit was cut with a razor blade into gold foil over a plastic (insulating) triangle and pressed with a spring on n-type germanium (Fig. 1.4). The one

[3]Subsequently, AT&T, under pressure from the US Justice Department's antitrust division, licensed the transistor for $25,000. This action initiated the rise of companies like Texas Instruments, Sony and Fairchild.

gold contact controls via the field effect (depletion of a surface layer) the current from Ge to the other gold contact. For the first time, amplification was observed [22]. More details about the history and development of the semiconductor transistor can be found in [23], written on the occasion of the 50th anniversary of its invention.

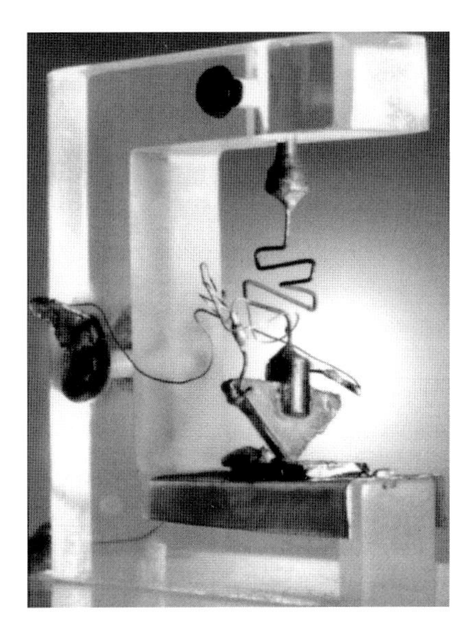

Fig. 1.4. The first transistor, 1947 (length of side of wedge: 32 mm)

1952
H. Welker – fabrication of compound semiconductors [24] (Verfahren zur Herstellung eines Halbleiterkristalls aus einer A III - B V - Verbindung mit Zonen verschiedenen Leitungstyps, German patent DBP 976 791, 12c, 2)

W. Shockley – today's version of the (J)FET [25].

1953
G.C. Dacey and I.M. Ross – first realization of a JFET [26].

D.M. Chapin, C.S. Fuller and G.L. Pearson – invention of the silicon solar cell at Bell Laboratories [27]. A single 2-cm^2 photovoltaic cell from Si, Si:As with an ultrathin layer of Si:B, with about 6% efficiency generated 5 mW of electrical power.[4] Previously existing solar cells based on selenium had very low efficiency ($< 0.5\%$).

[4]A solar cell with 1 W power cost $300 in 1956 ($3 in 2004). Initially, 'solar batteries' were only used for toys and were looking for an application. H. Ziegler proposed the use in satellites in the 'space race' of the late 1950s.

1958
J. Kilby made the first integrated circuit at Texas Instruments. The simple oscillator consisted of one transistor, three resistors and a capacitor on a $11 \times 1.7\,\mathrm{mm}^2$ Ge platelet (Fig. 1.5a). J. Kilby filed in 1959 for US patent 3,138,743 for miniaturized electronic circuits. At practically the same time R. Noyce from Fairchild Semiconductors, the predecessor of INTEL, invented the integrated circuit on silicon using planar technology (US patent 2,981,877, 1959, for a silicon-based integrated circuit).

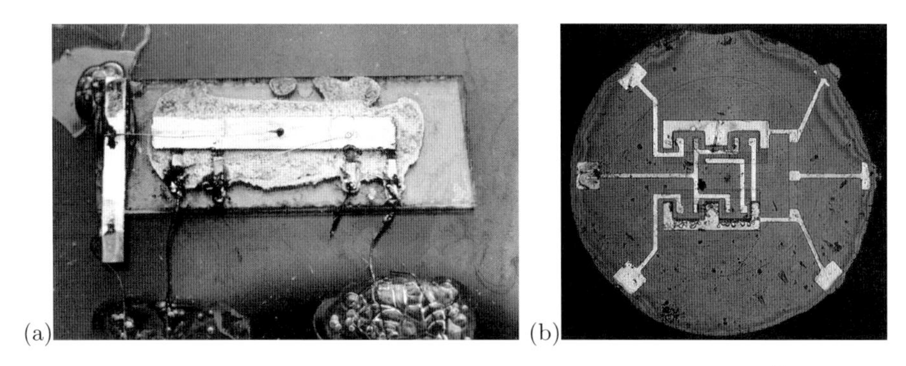

Fig. 1.5. (**a**) The first integrated circuit, 1958 (germanium, $11 \times 1.7\,\mathrm{mm}^2$). (**b**) The first planar integrated circuit, 1959 (silicon, diameter: 1.5 mm)

Figure 1.5b shows a flip-flop with four bipolar transistors and five resistors. Initially, the invention of the integrated circuit[5] met scepticism because of concerns regarding yield and the achievable quality of the transistors and the other components (such as resistors and capacitors).

1959
J. Hoerni and R. Noyce – first realization of a planar transistor (Fig. 1.6) [29].

1960
D. Kahng and M.M. Atalla – first realization of a MOSFET [30].

1962
The first semiconductor laser on GaAs basis at 77 K at GE [28] and at IBM [31].

1963
Proposal of a double heterostructure laser (DH laser) by Zh.I. Alferov [32] and H. Kroemer [33].

[5]The two patents led to a decade-long legal battle between Fairchild Semiconductors and Texas Instruments. Eventually, the US Court of Customs and Patent Appeals upheld R. Noyce's claims on interconnection techniques but gave J. Kilby and Texas Instruments credit for building the first working integrated circuit.

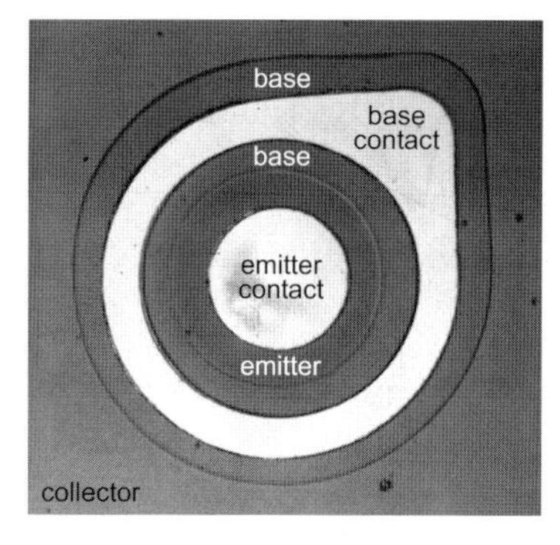

Fig. 1.6. Planar pnp silicon transistor, 1959. The contacts are Al surfaces (not bonded)

1966
Zh.I. Alferov – report of the first DH laser on the basis of GaInP at 77 K [34].

C.A. Mead – proposal of the MESFET ('Schottky Barrier Gate FET') [35].

1967
W.W. Hooper and W.I. Lehrer – first realization of a MESFET [36].

1968
DH laser on the basis of GaAs/AlGaAs at room temperature by Zh.I. Alferov [37] and I. Hayashi [38].

1.2 Nobel Prize Winners

Several Nobel Prizes[6] have been awarded for discoveries and inventions in the field of semiconductor physics (Fig. 1.2).

1909
Karl Ferdinand Braun
'in recognition of his contributions to the development of wireless telegraphy'

1914
Max von Laue 'for his discovery of the diffraction of X-rays by crystals'

1915
Sir William Henry Bragg
William Lawrence Bragg
'for their services in the analysis of crystal structure by means of X-rays'

[6] www.nobel.se

1909
Karl Ferdinand Braun
(1850–1918)

1914
Max von Laue
(1879–1960)

1915
Sir William Henry Bragg
(1862–1942)

1915
William Laurence Bragg
(1890–1971)

1946
Percy Williams Bridgman
(1882–1961)

1953
William B. Shockley
(1910–1989)

1953
John Bardeen
(1908–1991)

1953
Walter Hauser Brattain
(1902–1987)

1973
Leo Esaki
(*1925)

1985
Klaus von Klitzing
(*1943)

1998
Robert B. Laughlin
(*1930)

1998
Horst L. Störmer
(*1949)

1998
Daniel C. Tsui
(*1939)

2000
Zhores I. Alferov
(*1938)

2000
Herbert Kroemer
(*1928)

2000
Jack St. Clair Kilby
(1923–2005)

Fig. 1.7. Winners of Nobel Prize in Physics and year of award with great importance for semiconductor physics

1946
Percy Williams Bridgman
'for the invention of an apparatus to produce extremely high pressures, and for the discoveries he made therewith in the field of high pressure physics'

1953
William Bradford Shockley
John Bardeen
Walter Houser Brattain
'for their researches on semiconductors and their discovery of the transistor effect'

1973
Leo Esaki
'for his experimental discoveries regarding tunneling phenomena in semiconductors'

1985
Klaus von Klitzing
'for the discovery of the quantized Hall effect'

1998
Robert B. Laughlin
Horst L. Störmer
Daniel C. Tsui
'for their discovery of a new form of quantum fluid with fractionally charged excitations'

2000
Zhores I. Alferov
Herbert Kroemer
'for developing semiconductor heterostructures used in high-speed and opto-electronics'
Jack St. Clair Kilby
'for his part in the invention of the integrated circuit'

1.3 General Information

In Fig. 1.8, the periodic table of elements is shown. In Table 1.1 the physical properties of various semiconductors are summarized.

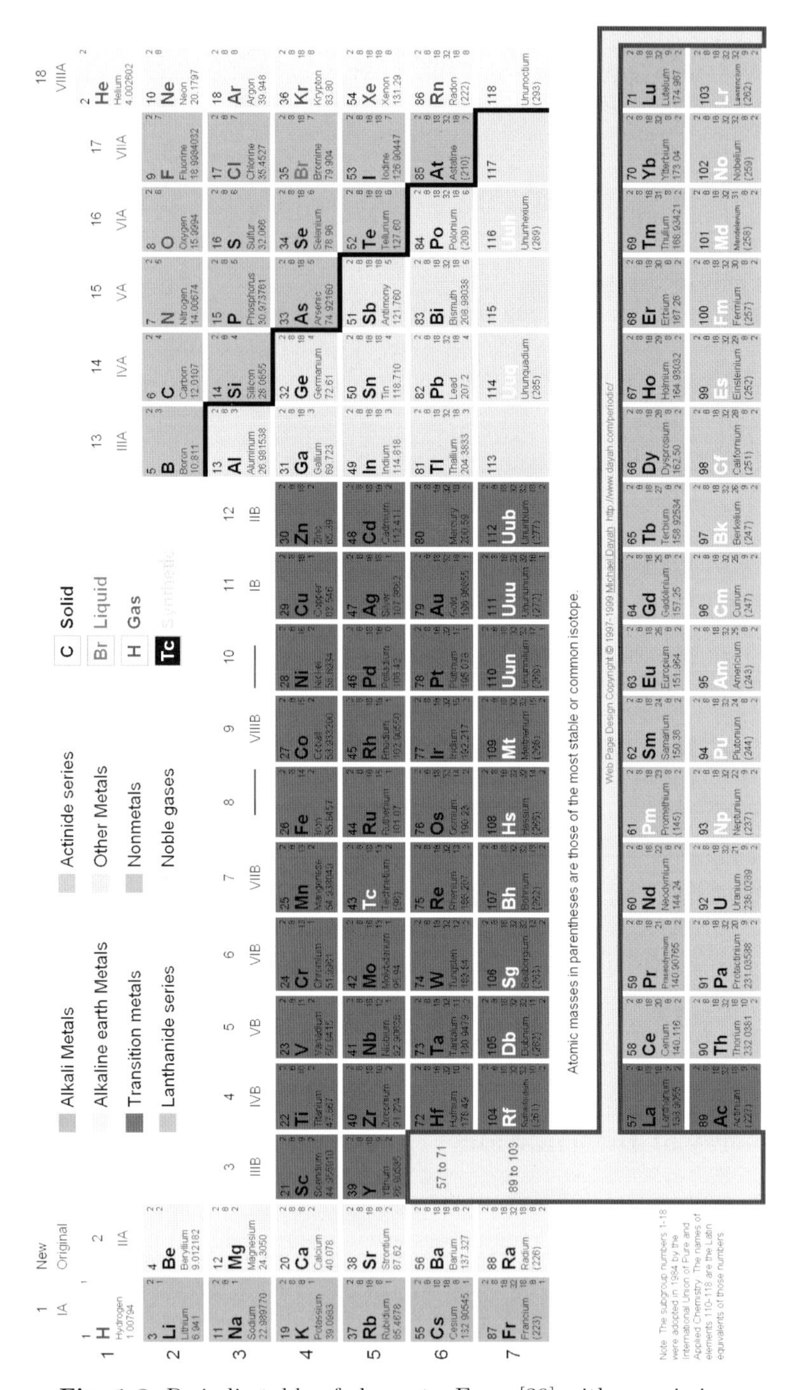

Fig. 1.8. Periodic table of elements. From [39] with permission

Table 1.1. Physical properties of various semiconductors at room temperature. 'S' denotes the crystal structure (d: diamond, w: wurtzite, zb: zincblende, ch: chalcopyrite, rs: rocksalt) ZnS, Cds and CdTe can realize zb and w structures

	S	a_0 (nm)	E_g (eV)	m_e^*	m_h^*	ϵ_0	n_r	μ_e ($cm^2/V\,s$)	μ_h ($cm^2/V\,s$)
C	d	0.3567	5.45 (Γ)			5.5	2.42	2200	1600
Si	d	0.5431	1.124 (X)	0.98 (m_l) 0.19 (m_t)	0.16 (m_{lh}) 0.5 (m_{hh})	11.7	3.44	1350	480
Ge	d	0.6461	0.67 (L)	1.58 (m_l) 0.08 (m_t)	0.04 (m_{lh}) 0.3 (m_{hh})	16.3	4.00	3900	1900
α-Sn	d	0.64892	0.08 (Γ)	0.02				2000	1000
3C-SiC	zb	0.436	2.4			9.7	2.7	1000	50
4H-SiC	w	0.3073 (a) 1.005 (c)	3.26			9.6	2.7		120
6H-SiC	w	0.30806 (a) 1.5117 (c)	3.101			10.2	2.7	1140	850
AlN	w	0.3111 (a) 0.4978 (c)	6.2			8.5	3.32		
AlP	zb	0.54625	2.43 (X)	0.13		9.8	3.0	80	
AlAs	zb	0.56605	2.16 (X)	0.5	0.49 (m_{lh}) 1.06 (m_{hh})	12		1000	80
AlSb	zb	0.61335	1.52 X)	0.11	0.39	11	3.4	200	300
GaN	w	0.3189 (a) 0.5185 (c)	3.4 (Γ)	0.2	0.8	12	2.4	1500	
GaP	zb	0.54506	2.26 (Γ)	0.13	0.67	10	3.37	300	150
GaAs	zb	0.56533	1.43 (Γ)	0.067	0.12 (m_{lh}) 0.5 (m_{hh})	12.5	3.4	8500	400
GaSb	zb	0.60954	0.72 (Γ)	0.045	0.39	15	3.9	5000	1000
InN	w	0.3533 (a) 0.5693 (c)	0.9 (Γ)						
InP	zb	0.58686	1.35 (Γ)	0.07	0.4	12.1	3.37	4000	600
InAs	zb	0.60584	0.36 (Γ)	0.028	0.33	12.5	3.42	22 600	200
InSb	zb	0.64788	0.18 (Γ)	0.013	0.18	18	3.75	100 000	1700
ZnO	w	0.325 (a) 0.5206 (c)	3.4 (Γ)	0.28	0.59	6.5	2.2	220	
ZnS	zb	0.54109	3.6 (Γ)	0.3		8.3	2.4	110	
ZnSe	zb	0.56686	2.58 (Γ)	0.17		8.1	2.89	600	
ZnTe	zb	0.61037	2.25 (Γ)	0.15		9.7	3.56		
CdO	rs	0.47	2.16						
CdS	w	0.416 (a) 0.6756 (c)	2.42 (Γ)	0.2	0.7	8.9	2.5	250	
CdSe	zb	0.6050	1.73 (Γ)	0.13	0.4	10.6		650	
CdTe	zb	0.64816	1.50 (Γ)	0.11	0.35	10.9	2.75	1050	100

Table 1.1. (continued)

	S	a_0 (nm)	E_g (eV)	m_e^*	m_h^*	ϵ_0	n_r	μ_e (cm^2/V s)	μ_h (cm^2/V s)
MgO	rs	0.421	7.3			10.3	3.0		
HgS	zb	0.5852	2.0 (Γ)					50	
HgSe	zb	0.6084	−0.15 (Γ)	0.045		25		18 500	
HgTe	zb	0.64616	−0.15 (Γ)	0.029	0.3	20	3.7	22 000	100
PbS	rs	0.5936	0.37 (L)	0.1	0.1	170	3.7	500	600
PbSe	rs	0.6147	0.26 (L)	0.07 (m_{lh}) 0.039 (m_{hh})	0.06 (m_{lh}) 0.03 (m_{hh})	250		1800	930
PbTe	rs	0.645	0.29 (L)	0.24 (m_{lh}) 0.02 (m_{hh})	0.3 (m_{lh}) 0.02 (m_{hh})	412		1400	1100
$ZnSiP_2$	ch	0.54 (a) 1.0441 (c)	2.96 (Γ)	0.07					
$ZnGeP_2$	ch	0.5465 (a) 1.0771 (c)	2.34 (Γ)		0.5				
$ZnSnP_2$	ch	0.5651 (a) 1.1302 (c)	1.66 (Γ)						
$CuInS_2$	ch	0.523 (a) 1.113 (c)	1.53 (Γ)						
$CuGaS_2$	ch	0.5347 (a) 1.0474 (c)	2.5 (Γ)						
$CuInSe_2$	ch	0.5784 (a) 1.162 (c)	1.0 (Γ)						
$CuGaSe_2$	ch	0.5614 (a) 1.103 (c)	1.7 (Γ)						

Part I

Fundamentals

2 Bonds

2.1 Introduction

The positively charged atomic nuclei and the electrons in the atomic shells of the atoms making up the semiconductor (or any other solid) are in a binding state. Several mechanisms can lead to such cohesiveness. First, we will discuss the homopolar, electron-pair or covalent bond, then the ionic bond and subsequently the mixed bond. We will only briefly touch on the metallic bond and the van-der-Waals bond.

2.2 Covalent Bonds

Covalent bonds are formed due to quantum-mechanical forces. The prototype covalent bond is the bonding of the hydrogen molecule due to overlapping of the atomic shells. If several electron pairs are involved, directional bonds can be formed in various spatial directions, eventually making up a solid.

2.2.1 Electron-Pair Bond

The covalent bond of two hydrogen atoms in a H_2 molecule can lead to a reduction of the total energy of the system, compared to two single (distant) atoms (Fig. 2.1). For fermions (electrons have spin 1/2) the two-particle wavefunction of the two (indistinguishable) electrons A and B must be antisymmetric, i.e. $\Psi(A, B) = -\Psi(B, A)$ (Pauli principle). The wavefunction of each electron has degrees of freedom in real space ($\mathbf{r}$) and spin (σ), $\Psi(A) = \Psi_{\mathbf{r}}(A)\Psi_{\sigma}(A)$. The two-particle wavefunction of the molecule is non-separable and has the form $\Psi(A, B) = \Psi_{\mathbf{r}}(r_A, r_B)\Psi_{\sigma}(\sigma_A, \sigma_B)$. The binding state has a wavefunction with a symmetric orbital and antiparallel spins, i.e. $\Psi_{\mathbf{r}}(r_A, r_B) = \Psi_{\mathbf{r}}(r_B, r_A)$ and $\Psi_{\sigma}(\sigma_A, \sigma_B) = -\Psi_{\sigma}(\sigma_B, \sigma_A)$. The antisymmetric orbital with parallel spins is antibinding for all distances of the nuclei (protons).

2.2.2 sp^3 Bond

Elements from group IV of the periodic system (C, Si, Ge, ...) have 4 electrons on the outer shell. Carbon has the electron configuration $1s^2 2s^2 2p^2$.

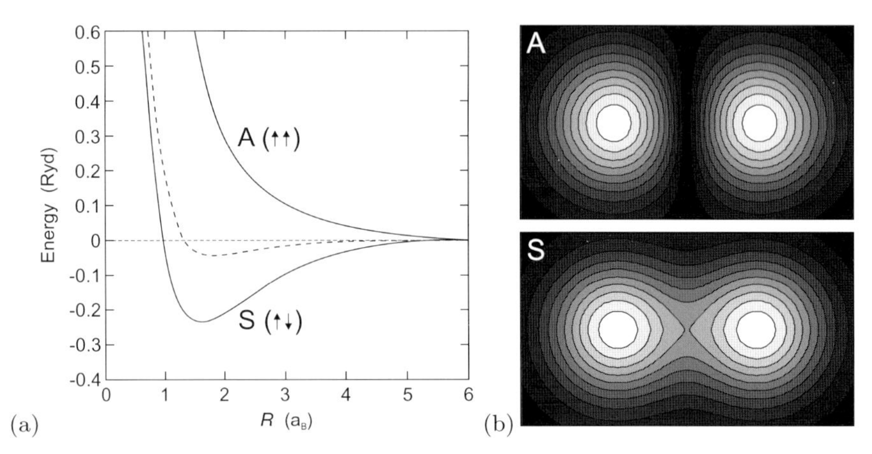

Fig. 2.1. Binding of the hydrogen molecule. (**a**) *Dashed line*: classical calculation (electrostatics), 'S', 'A': quantum-mechanical calculation taking into account Pauli's principle (S: symmetric orbital, antiparallel spins, A: antisymmetric orbital, parallel spins. The distance of the nuclei (protons) is given in units of the Bohr radius $a_{\mathrm{B}} = 0.053\,\mathrm{nm}$, the energy is given in Rydberg units (13.6 eV). (**b**) Schematic contour plots of the probability distribution ($\Psi^* \Psi$) for the S and A states

For an octet configuration bonding to four other electrons would be optimal (Fig. 2.2). This occurs through the mechanism of sp^3 hybridization.[1] First, one S-electron of the ns^2np^2 configuration is brought into a p orbital, such that the outermost shell contains one s, p_x, p_y, and p_z orbital each (Figs. 2.3a–e). The energy necessary for this step is more than regained in the subsequent formation of the covalent bonds. The four orbitals can be reconfigured into four other wavefunctions, the sp^3 hybrids (Figs. 2.3f–i), i.e.

$$\Psi_{++++} = (\mathrm{s} + \mathrm{p}_x + \mathrm{p}_y + \mathrm{p}_z)/2 \tag{2.1a}$$
$$\Psi_{++--} = (\mathrm{s} + \mathrm{p}_x - \mathrm{p}_y - \mathrm{p}_z)/2 \tag{2.1b}$$
$$\Psi_{+-+-} = (\mathrm{s} - \mathrm{p}_x + \mathrm{p}_y - \mathrm{p}_z)/2 \tag{2.1c}$$
$$\Psi_{+--+} = (\mathrm{s} - \mathrm{p}_x - \mathrm{p}_y + \mathrm{p}_z)/2\,. \tag{2.1d}$$

These orbitals have a directed form along tetrahedral directions. The binding energy (per atom) of the covalent bond is large, for H–H 4.5 eV, for C–C 3.6 eV, for Si–Si 1.8 eV, and for Ge–Ge 1.6 eV. Such energy is, for *neutral* atoms, comparable to the ionic bond, discussed in the next section.

In Fig. 2.4a the energy of a crystal made up from silicon atoms is shown for various crystal structures or phases (see Chap. 3). The lattice constant with the lowest total energy determines the lattice spacing for each crystal

[1] It is debated in femtosecond chemistry whether the bond *really* forms in this way. However, it is a picture of overwhelming simplicity.

Die wundersame Welt der Atomis

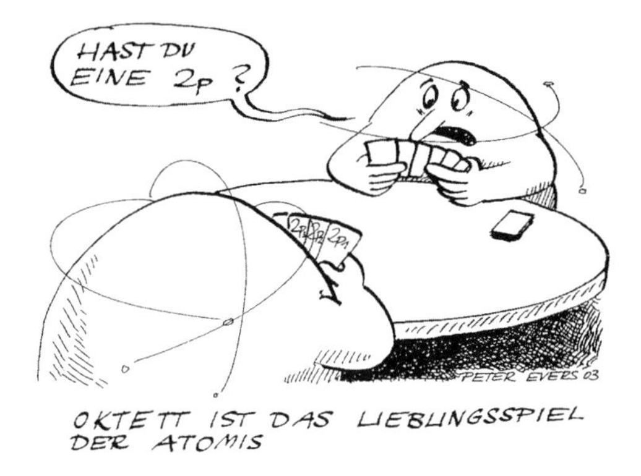

Fig. 2.2. Octet, the favorite card game of the 'Atomis' (trying to reach octet configuration in a bond by swapping wavefunctions). The bubble says: 'Do you have a 2p?'. Reprinted with permission from [40], ©2002 Wiley-VCH

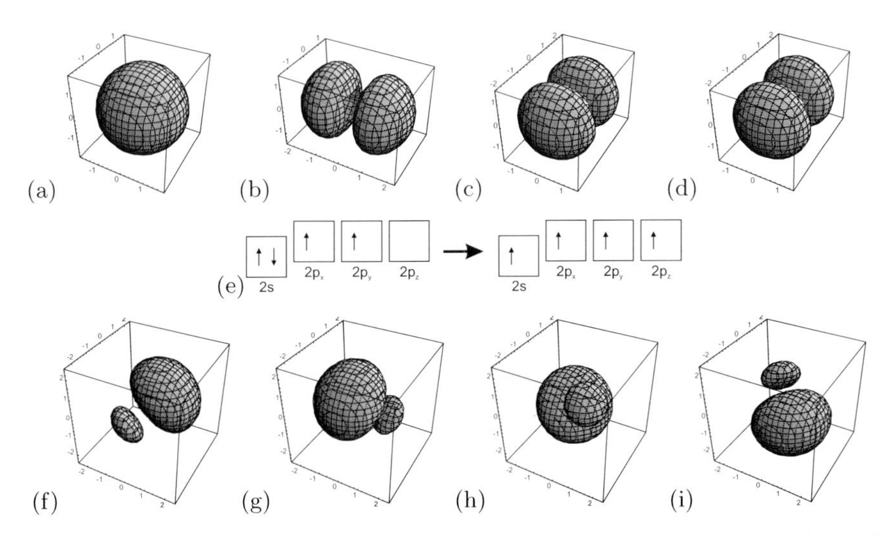

Fig. 2.3. (**a**) s orbital, (**b,c,d**) p_x, p_y and p_z orbital, (**e**) hybridization, (**f,g,h,i**) orbitals of the sp^3 hybridization: (**f**) $(s+p_x+p_y+p_z)/2$, (**g**) $(s+p_x-p_y-p_z)/2$, (**h**) $(s-p_x+p_y-p_z)/2$, (**i**) $(s-p_x-p_y+p_z)/2$

structure. The thermodynamically stable configuration is the phase with the lowest overall energy for given external conditions.

The covalent bond of a group-IV atom to other group-IV atoms has a tetrahedral configuration with electron-pair bonds, similar to the hydrogen molecule bond. In Fig. 2.4b the energy states of the $n = 2$ shell for tetrahedrally bonded carbon (diamond, see Sect. 3.4.3) are shown as a function of the distance from the nuclei. First, the energetically sharp states become a band

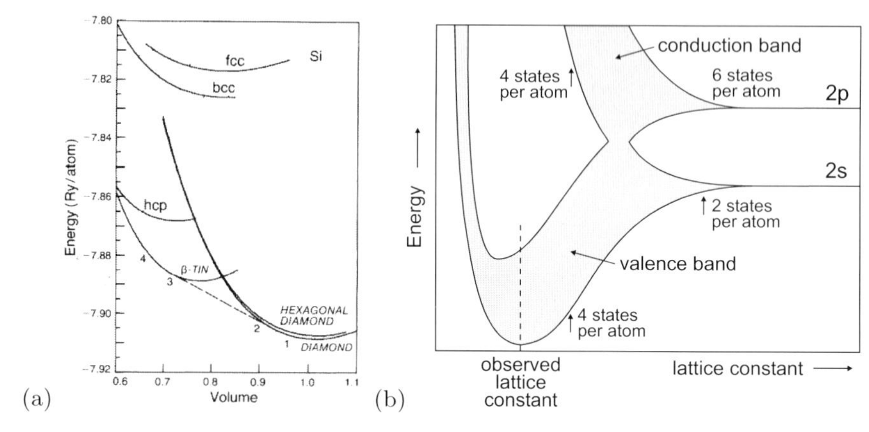

Fig. 2.4. (**a**) Energy per atom in silicon for various crystal structures. Reprinted with permission from [41], ©1980 APS. (**b**) Electron energy levels in (diamond structure) carbon as a function of the distance of the atomic nuclei (schematic). Adapted from [42]

due to the overlap and coupling of the atomic wavefunctions (see Chap. 6). The mixing of the states leads to the formation of the filled lower valence band (binding states) and the empty upper conduction band (antibinding states). This principle is valid for most semiconductors and is shown schematically also in Fig. 2.5. The configuration of bonding and antibinding p orbitals is depicted schematically in Fig. 2.6. The bonding and antibinding sp^3 orbitals are depicted in Figs. 2.7a,b and 2.13. We note that the energy of the crystal does not only depend on the distance of the nuclei but also on their geometric arrangement (crystal structure).

Per carbon atom there are (in the second shell) four electrons and four unoccupied states, altogether eight. These are redistributed into four states (filled) per atoms in the valence band and four states per atom (empty) in the lowest conduction bands. Between the top of the valence band and the bottom of the conduction band there is an energy gap, later called the *bandgap* (see Chap. 6).

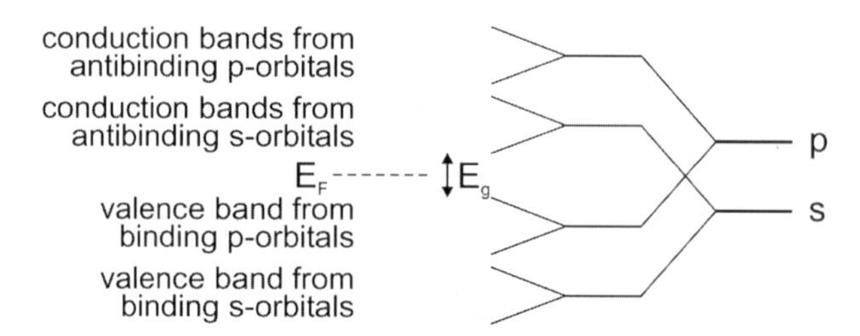

Fig. 2.5. Schematic of the origin of valence and conduction band from the atomic s and p orbitals. The bandgap E_g and the position of the Fermi level E_F are indicated

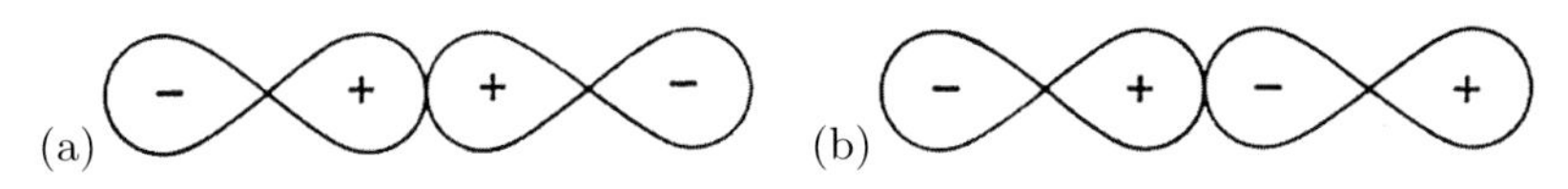

Fig. 2.6. Schematic representation of (**a**) bonding and (**b**) antibinding p orbitals. The signs denote the phase of the wavefunction

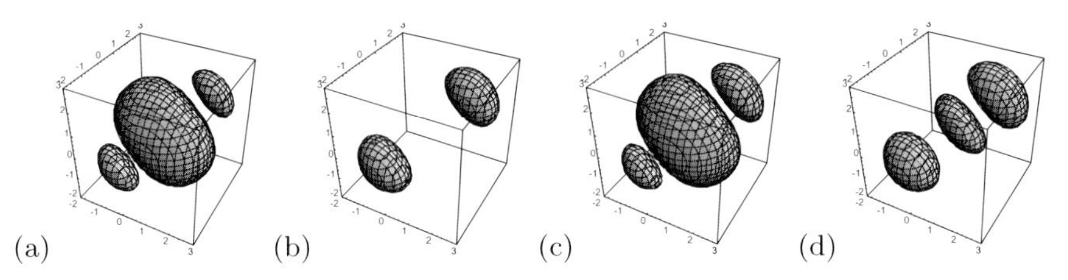

Fig. 2.7. Schematic representation of (**a**,**c**) bonding and (**b**,**d**) antibinding symmetric (**a**,**b**) and nonsymmetric (**c**,**d**) sp^3 orbitals

2.2.3 sp^2 Bond

Organic semiconductors (see Chap. 16) are made up from carbon compounds. While for inorganic semiconductors the covalent (or mixed, see Sect. 2.4) bond with sp^3 hybridization is important, the organic compounds are based on the sp^2 hybridization. This bonding mechanism, which is present in graphite, is stronger than the sp^3-bond present in diamond. The prototype organic molecule is the benzene ring[2] (C_6H_6), shown in Fig. 2.8. The benzene ring is the building block for small organic molecules and polymers.

In the benzene molecule neighboring carbon atoms are bonded within the ring plane via the binding σ states of the sp^2 orbitals (Fig. 2.8a). The wavefunctions (Fig. 2.9) are given by (2.2a–c).

[2]Supposedly, the chemist Friedrich August Kekulé von Stadonitz had a dream about dancing carbon molecules and thus came up with the ring-like molecule structure [43]. Kekulé remembered: 'During my stay in Ghent, I lived in elegant bachelor quarters in the main thoroughfare. My study, however, faced a narrow side-alley and no daylight penetrated it.... I was sitting writing on my textbook, but the work did not progress; my thoughts were elsewhere. I turned my chair to the fire and dozed. Again the atoms were gamboling before my eyes. This time the smaller groups kept modestly in the background. My mental eye, rendered more acute by the repeated visions of the kind, could now distinguish larger structures of manifold conformation; long rows sometimes more closely fitted together all twining and twisting in snake-like motion. But look! What was that? One of the snakes had seized hold of its own tail, and the form whirled mockingly before my eyes. As if by a flash of lightning I awoke; and this time also I spent the rest of the night in working out the consequences of the hypothesis.'

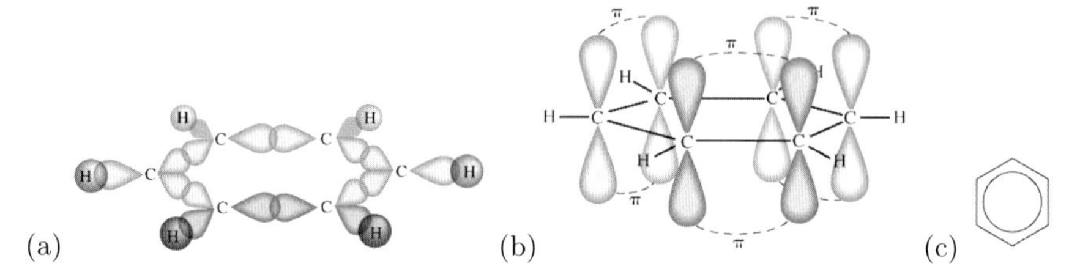

Fig. 2.8. Schematic representation of the (**a**) σ and (**b**) π bonds in benzene, (**c**) schematic symbol for benzene

$$\Psi_1 = (\mathrm{s} + \sqrt{2}\mathrm{p}_x)/\sqrt{3} \tag{2.2a}$$

$$\Psi_2 = (\mathrm{s} - \sqrt{1/2}\mathrm{p}_x + \sqrt{3/2}\mathrm{p}_y)/\sqrt{3} \tag{2.2b}$$

$$\Psi_3 = (\mathrm{s} - \sqrt{1/2}\mathrm{p}_x - \sqrt{3/2}\mathrm{p}_y)/\sqrt{3}\,. \tag{2.2c}$$

The 'remaining' p_z orbitals do not directly take part in the binding (Fig. 2.8b) and form bonding (π, filled) and antibinding (π^*, empty) orbitals (see Fig. 2.10). The π and π^* states are delocalized over the ring. Between the highest populated molecular orbital (HOMO) and the lowest unoccupied molecular orbital (LUMO) is typically an energy gap (Fig. 2.11). The antibinding σ^* orbitals are energetically above the π^* states.

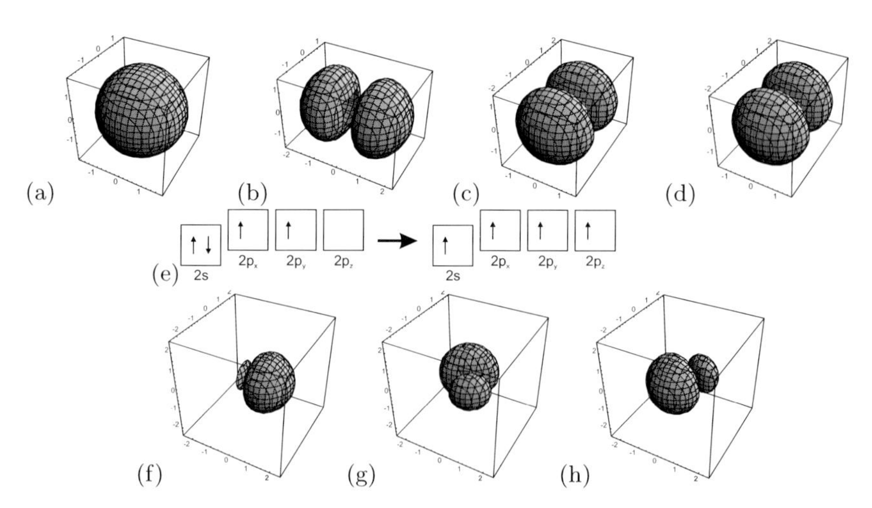

Fig. 2.9. (**a**) s orbital, (**b,c,d**) p_x, p_y and p_z orbital, (**e**) hybridization, (**f,g,h**) orbitals of the sp^2 hybridization: (**f**) $(\mathrm{s}+\sqrt{2}\mathrm{p}_x)/\sqrt{3}$, (**g**) $(\mathrm{s}-\sqrt{1/2}\mathrm{p}_x+\sqrt{3/2}\mathrm{p}_y)/\sqrt{3}$, (**h**) $(\mathrm{s}-\sqrt{1/2}\mathrm{p}_x - \sqrt{3/2}\mathrm{p}_y)/\sqrt{3}$

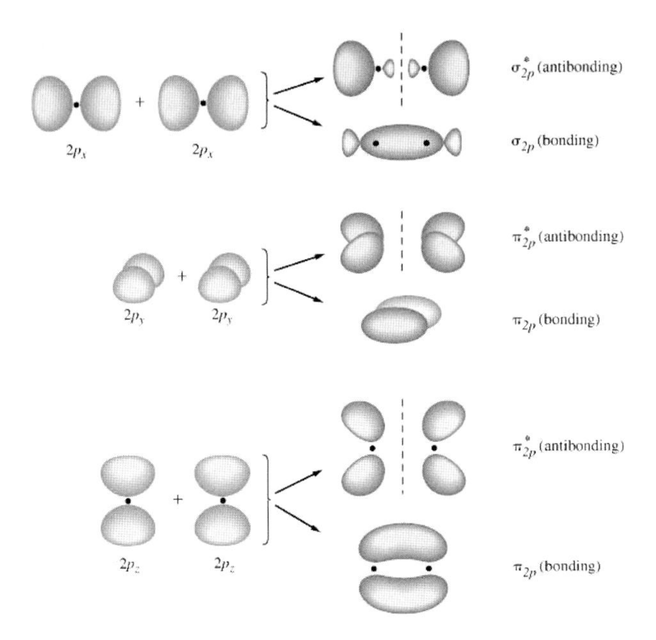

Fig. 2.10. Orbitals due to binding and antibinding configurations of various π orbitals

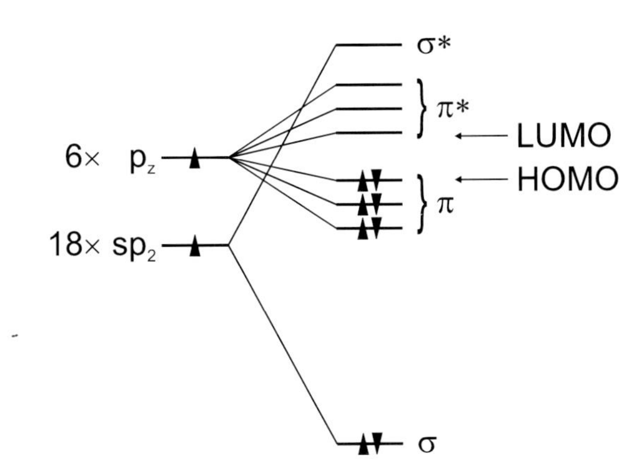

Fig. 2.11. Schematic energy terms of the benzene molecule

2.3 Ionic Bonds

Ionic crystals are made up from positively and negatively charged ions. The heteropolar or ionic bond is the consequence of the electrostatic attraction between the ions. However, the possibly repulsive character of next neighbors has to be considered.

For I–VII compounds, e.g. LiF or NaCl, the shells of the singly charged ions are complete: Li: $1s^2 2s^1 \rightarrow$ Li^+: $1s^2$, F: $1s^2 2s^2 2p^5 \rightarrow$ F^-: $1s^2 2s^2 2p^6$.

Compared to ions in a gas, a Na–Cl pair in the crystal has a binding energy of 7.9 eV that mostly stems from the electrostatic energy (Madelung energy). Van-der-Waals forces (see Sect. 2.6) only contribute 1–2%. The ionization energy of Na is 5.14 eV, the electron affinity of Cl is 3.61 eV. Thus the energy of the NaCl pair in the solid is 6.4 (=7.9−5.1+3.6) eV lower than in a gas of neutral atoms.

The interaction of two ions with distance vector $\mathbf{r}_{ij}$ is due to the Coulomb interaction

$$U_{ij}^{\mathrm{C}} = \frac{q_i q_j}{4\pi\epsilon_0} \frac{1}{r_{ij}} = \pm \frac{e^2}{4\pi\epsilon_0} \frac{1}{r_{ij}} \tag{2.3}$$

and a repulsive contribution due to the overlap of (complete) shells. This contribution is typically approximated by a radially symmetric core potential

$$U_{ij}^{\mathrm{core}} = \lambda \exp(-r/\varrho) \tag{2.4}$$

that only acts on next neighbors. λ describes the strength of this interaction and ρ parameterizes its range.

The distance of ions is denoted as $r_{ij} = p_{ij} R$, where R denotes the distance of next neighbors and the p_{ij} are suitable coefficients. The electrostatic interaction of an ion with *all* its neighbors is then written as

$$U_{ij}^{\mathrm{C}} = -\alpha \frac{e^2}{4\pi\epsilon_0} \frac{1}{R} , \tag{2.5}$$

where α is the Madelung constant. For an attractive interaction (as in a solid), α is positive. It is given (calculated for the i-th ion) as

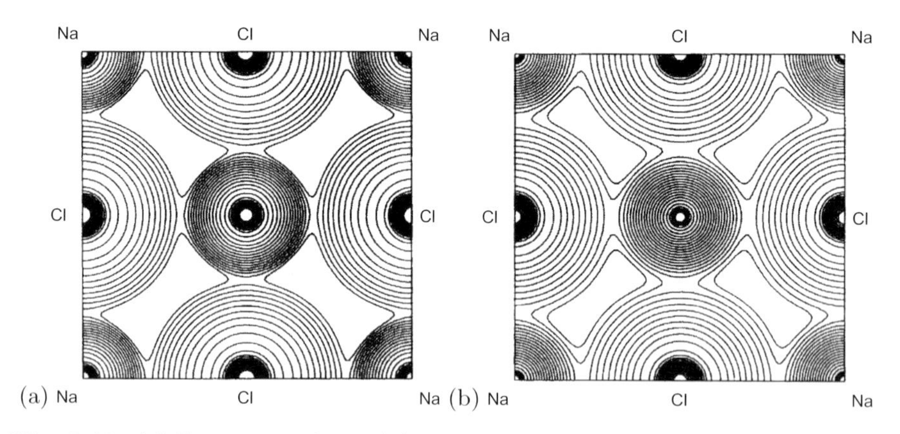

Fig. 2.12. (**a**) Experimental and (**b**) theoretical charge distribution in the (100) plane of NaCl. The lowest contour in the interstitial region corresponds to a charge density of 7 e/nm^3 and adjacent contours differ by a factor of $\sqrt{2}$. Differences are mainly due to the fact that the X-ray experiments have been made at room temperature. Reprinted with permission from [45], ©1986 APS

$$\alpha = \sum_{ij} \frac{\pm 1}{p_{ij}} \,. \tag{2.6}$$

For a one-dimensional chain $\alpha = 2 \ln 2$. For the rocksalt (NaCl) structure (see Sect. 3.4.1) it is $\alpha \approx 1.7476$, for the CsCl structure (see Sect. 3.4.2) it is $\alpha \approx 1.7627$, and for the zincblende structure (see Sect. 3.4.4) it is $\alpha \approx 1.6381$. This shows that ionic compounds prefer the NaCl or CsCl structure. The charge distribution for NaCl is shown in Fig. 2.12. For tetragonal and orthorhombic structures, the Madelung constant has been calculated in [44].

2.4 Mixed Bond

The group-IV crystals are of perfectly covalent nature, the I–VII are almost exclusively ionically bonded. For III–V (e.g. GaAs, InP) and II–VI compounds (e.g. CdS, ZnO) we have a mixed case.

The (screened) Coulomb potentials of the A and B atoms (in the AB compound) shall be denoted V_A and V_B. The origin of the coordinate system is in the center of the A and B atom (i.e. for the zincblende structure (see Sect. 3.4.4) at $(1/8, 1/8, 1/8)a$). The valence electrons then see the potential

$$V_{\mathrm{crystal}} = \sum_{\alpha} V_A(\mathbf{r} - \mathbf{r}_\alpha) + \sum_{\beta} V_B(\mathbf{r} - \mathbf{r}_\beta) \,, \tag{2.7}$$

where the sum α (β) runs over all A (B) atoms. This potential can be split into a symmetric (V_{c}, covalent) and an antisymmetric (V_{i}, ionic) part (2.8b), i.e. $V_{\mathrm{crystal}} = V_{\mathrm{c}} + V_{\mathrm{i}}$

$$\begin{aligned} V_{\mathrm{c}} = \frac{1}{2} \Bigg\{ &\sum_{\alpha} V_A(\mathbf{r} - \mathbf{r}_\alpha) + \sum_{\alpha} V_B(\mathbf{r} - \mathbf{r}_\alpha) \\ &+ \sum_{\beta} V_B(\mathbf{r} - \mathbf{r}_\beta) + \sum_{\beta} V_A(\mathbf{r} - \mathbf{r}_\beta) \Bigg\} \end{aligned} \tag{2.8a}$$

$$\begin{aligned} V_{\mathrm{i}} = \frac{1}{2} \Bigg\{ &\sum_{\alpha} V_A(\mathbf{r} - \mathbf{r}_\alpha) - \sum_{\alpha} V_B(\mathbf{r} - \mathbf{r}_\alpha) \\ &+ \sum_{\beta} V_B(\mathbf{r} - \mathbf{r}_\beta) - \sum_{\beta} V_A(\mathbf{r} - \mathbf{r}_\beta) \Bigg\} \,. \end{aligned} \tag{2.8b}$$

For homopolar bonds $V_{\mathrm{i}} = 0$ and the splitting between bonding and antibinding states is E_{h}, which mainly depends on the bond length l_{AB} (and the related overlap of atomic wavefunctions). In a partially ionic bond the orbitals are not symmetric along A–B, but the center is shifted towards the more electronegative atom (Figs. 2.7c,d and 2.13).

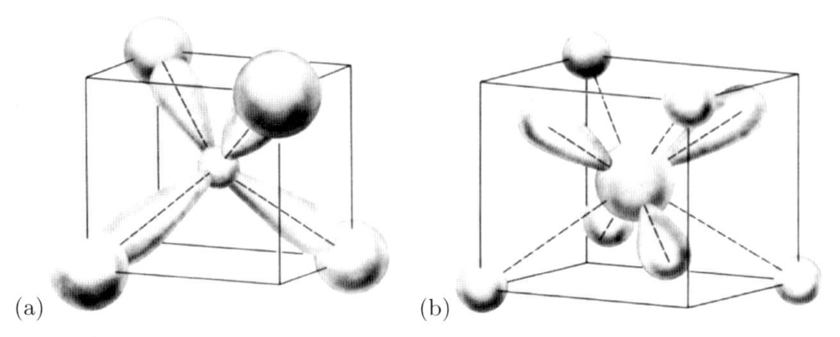

Fig. 2.13. Schematic representation of (**a**) bonding and (**b**) antibinding sp^3 orbitals. From [46]

The band splitting[3] between the (highest) bonding and (lowest) antibinding state E_{ba} is then written as

$$E_{ba} = E_h + iC , \tag{2.9}$$

where C denotes the band splitting due to the ionic part of the potential and depends only on $V_A - V_B$. C is proportional to the difference of the

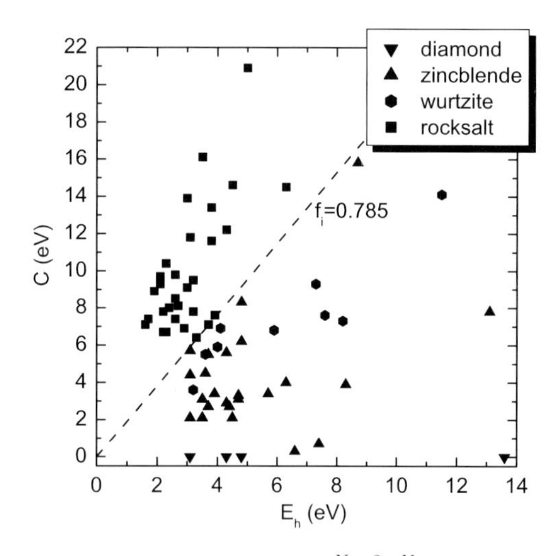

Fig. 2.14. Values of E_h and C for various $A^N B^{8-N}$ compounds. The *dashed line* $f_i = 0.785$ separates 4-fold from 6-fold coordinated structures. Most data taken from [47]

[3]This energy should not be confused with the *bandgap* ΔE_{cv}, the energy separation of the highest valence-band state and the lowest conduction-band state. The energy splitting E_{ba} is the energy separation between the centers of the valence and conduction bands. Mostly, the term E_g is used for ΔE_{cv}.

electronegativities X of the A and B atoms, $C(A,B) = 5.75(X_A - X_B)$. A material thus takes a point in the (E_h,C) plane (Fig. 2.14). The absolute value for the band splitting is given as $E_\mathrm{ba}^2 = E_\mathrm{h}^2 + C^2$.

The ionicity of the bond is described with the ionicity (after Phillips) f_i, defined as [48]

$$f_\mathrm{i} = \frac{C^2}{E_h^2 + C^2} . \tag{2.10}$$

The covalent part is $1 - f_\mathrm{i}$. In Table 2.1 the ionicity is given for a number of binary compounds. The ionicity can also be interpreted as the angle $\tan\phi = C/E_h$ in the (E_h,C) diagram. The critical value of $f_\mathrm{i} = 0.785$ for the ionicity separates quite exactly (for about 70 compounds) the 4-fold (diamond, zincblende and wurtzite) from the 6-fold (rocksalt) coordinated substances ($f_\mathrm{i} = 0.785$ is indicated by a dashed line in Fig. 2.14).

Table 2.1. Ionicity f_i (2.10) for various binary compounds

C	0.0	AlAs	0.27	BeO	0.60	CuCl	0.75
Si	0.0	BeS	0.29	ZnTe	0.61	CuF	0.77
Ge	0.0	AlP	0.31	ZnO	0.62	AgI	0.77
Sn	0.0	GaAs	0.31	ZnS	0.62	MgS	0.79
BAs	0.002	InSb	0.32	ZnSe	0.63	MgSe	0.79
BP	0.006	GaP	0.33	HgTe	0.65	CdO	0.79
BeTe	0.17	InAs	0.36	HgSe	0.68	HgS	0.79
SiC	0.18	InP	0.42	CdS	0.69	MgO	0.84
AlSb	0.25	AlN	0.45	CuI	0.69	AgBr	0.85
BN	0.26	GaN	0.50	CdSe	0.70	LiF	0.92
GaSb	0.26	MgTe	0.55	CdTe	0.72	NaCl	0.94
BeSe	0.26	InN	0.58	CuBr	0.74	RbF	0.96

2.5 Metallic Bond

In a metal, the positively charged atomic cores are embedded in a more or less homogeneous sea of electrons. The valence electrons of the atoms become the conduction electrons of the metal. These are freely moveable and at $T = 0\,\mathrm{K}$ there is no energy gap between filled and empty states. The bonding is mediated by the energy reduction for the conduction electrons in the periodic potential of the solid compared to free atoms. This will be clearer when the band structure is discussed (Chap. 6). In transition metals the overlap of (partially filled) inner shells (d or f) can also contribute to the bonding.

2.6 van-der-Waals Bond

The van-der-Waals bond is a dipole bond that leads to bonding in the noble-gas crystals (at low temperature). Ne, Ar, Kr and Xe crystallize in the densely packed fcc lattice (see Sect. 3.3.5). He^3 and He^4 represent an exception. They do not solidify at zero pressure at $T = 0\,K$ due to the large zero-point energy. This quantum-mechanical effect is especially strong for oscillators with small mass.

When two neutral atoms come near to each other (distance of the nuclei R), an attractive dipole–dipole interaction $-AR^{-6}$ arises (London interaction) the van-der-Waals interaction. The quantum-mechanical overlap of the (filled) shells leads to a strong repulsion $+BR^{-12}$. Altogether, a binding energy minimum results for the Lennard–Jones potential V_{LJ} (see Fig. 2.15)

$$V_{\mathrm{LJ}}(R) = -\frac{A}{R^6} + \frac{B}{R^{12}} . \tag{2.11}$$

The energy minimum $E_{\min} = -A^2/(2B)$ is at $R = (2B/A)^{1/6}$.

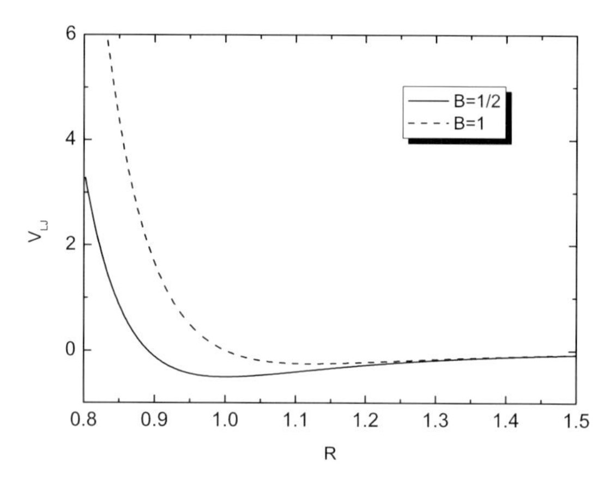

Fig. 2.15. Lennard–Jones potential (2.11) for $A = 1$ and two values of B

The origin of the attractive dipole–dipole interaction can be understood from a one-dimensional (1D) model as follows: Two atoms are modeled by their fixed positively charged nuclei in a distance R and their negatively charged electron shells that are polarizable, i.e. can be displaced along one direction x. Additionally, we assume (two identical) 1D harmonic oscillators for the electron motion at the positions 0 and R. Then, the Hamilton operator H_0 of the system without interaction (R is very large)

$$H_0 = \frac{1}{2m}p_1^2 + Cx_1^2 + \frac{1}{2m}p_2^2 + Cx_2^2 . \tag{2.12}$$

The indices 1 and 2 denote the two electrons of atoms. x_1 and x_2 are the displacements of the electrons. Both harmonic oscillators have a resonance frequency $\omega_0 = \sqrt{C/m}$, and the zero-point energy is $\hbar\omega_0/2$.

Taking into account the Coulomb interaction of the four charges, an additional term H_1 arises

$$H_1 = \frac{e^2}{R} + \frac{e^2}{R + x_1 + x_2} - \frac{e^2}{R + x_1} - \frac{e^2}{R - x_2} \approx -\frac{2e^2}{R^3} x_1 x_2 \,. \tag{2.13}$$

The approximation is valid for small amplitudes $x_i \ll R$. A separation of variables can be achieved by transformation to the normal modes

$$x_s = \frac{x_1 + x_2}{\sqrt{2}}, x_a = \frac{x_1 - x_2}{\sqrt{2}} \,. \tag{2.14}$$

Then we find

$$\begin{aligned} H &= H_0 + H_1 \\ &= \left[\frac{1}{2m} p_s^2 + \frac{1}{2}\left(C - \frac{2e^2}{R^3}\right) x_s^2\right] + \left[\frac{1}{2m} p_a^2 + \frac{1}{2}\left(C - \frac{2e^2}{R^3}\right) x_a^2\right] \,. \end{aligned} \tag{2.15}$$

This equation is the Hamiltonian of two decoupled harmonic oscillators with the normal frequencies

$$\omega_\pm = \sqrt{\left(C \pm \frac{2e^2}{R^3}\right)/m} \approx \omega_0 \left[1 \pm \frac{1}{2}\left(\frac{2e^2}{CR^3}\right) - \frac{1}{8}\left(\frac{2e^2}{CR^3}\right)^2 + \ldots\right] \,. \tag{2.16}$$

The coupled system thus has a lower (zero-point) energy than the uncoupled. The energy difference per atom is (in lowest order) proportional to R^{-6}.

$$\Delta U = \hbar\omega_0 - \frac{1}{2}\left(\omega_+ - \omega_-\right) \approx -\hbar\omega_0 \frac{1}{8}\left(\frac{2e^2}{CR^3}\right)^2 = -\frac{A}{R^6} \,. \tag{2.17}$$

The interaction is a true quantum-mechanical effect, i.e. the reduction of the zero-point energy of coupled oscillators.

2.7 Hamilton Operator of the Solid

The total energy of the solid, including kinetic and potential terms, is

$$\begin{aligned} H = &\sum_i \frac{\mathbf{p}_i^2}{2m_i} + \sum_j \frac{\mathbf{P}_j^2}{2M_j} \\ &+ \frac{1}{2}\sum_{j,j'} \frac{Z_j Z_{j'} e^2}{4\pi\epsilon_0 |\mathbf{R}_j - \mathbf{R}_{j'}|} + \frac{1}{2}\sum_{i,i'} \frac{e^2}{4\pi\epsilon_0 |\mathbf{r}_i - \mathbf{r}_{i'}|} \\ &- \sum_{i,j} \frac{Z_j e^2}{4\pi\epsilon_0 |\mathbf{R}_j - \mathbf{r}_i|} \,, \end{aligned} \tag{2.18}$$

where $\mathbf{r}_i$ and $\mathbf{R}_i$ are the position operators and $\mathbf{p}_i$ and $\mathbf{P}_i$ are the momentum operators of the electrons and nuclei, respectively. The first term is the kinetic energy of the electrons, the second term is the kinetic energy of the nuclei. The third term is the electrostatic interaction of the nuclei, the fourth term is the electrostatic interaction of the electrons. In the third and fourth terms the summation over the same indices is left out. The fifth term is the electrostatic interactions of electrons and nuclei.

In the following, the usual approximations in order to treat (2.18) are discussed. First, the nuclei and the electrons tightly bound to the nuclei (inner shells) are united to form ion cores. The remaining electrons are the valence electrons.

The next approximation is the Born–Oppenheimer (or adiabatic) approximation. Since the ion cores are much heavier than the electrons (factor $\approx 10^3$) they move much slower. The frequencies of the ion vibrations are typically in the region of several tens of meV (phonons, see Sect. 5.2), the energy to excite an electron is typically 1 eV. Thus, the electrons always 'see' the momentary position of the ions; the ions, however, 'see' the electron motions averaged over many periods. Thus, the Hamiltonian (2.18) is split into three parts:

$$H = H_{\text{ions}}(\mathbf{R}_j) + H_{\text{e}}(\mathbf{r}_i, \mathbf{R}_{j0}) + H_{\text{e-ion}}(\mathbf{r}_i, \delta\mathbf{R}_j) \ . \tag{2.19}$$

The first term contains the ion cores with their potential and the time-averaged contribution of the electrons. The second term is the electron motion around the ion cores at their averaged positions $\mathbf{R}_{j0}$. The third term is the Hamiltonian of the electron–phonon interaction that depends on the electron positions and the deviation of the ions from their average position $\delta\mathbf{R}_j = \mathbf{R}_j - \mathbf{R}_{j0}$. The electron–phonon interaction is responsible for such effects as electrical resistance and superconductivity.

3 Crystals

3.1 Introduction

The economically most important semiconductors have a relatively simple atomic arrangement and are highly symmetric. The symmetry of the atomic arrangement is the basis for the classification of the various crystal structures. Using group theory, basic and important conclusions can be drawn about the physical properties of the crystal, such as its elastic and electronic properties. The presence of highly symmetric planes is obvious from the crystal shape of the minerals and their cleavage behavior.

Polycrystalline semiconductors consist of grains of finite size that are structurally perfect. The grain boundaries are a lattice defect (see also Sect. 4.6). Amorphous semiconductors are disordered on the atomic scale, see Sect. 3.3.7.

3.2 Crystal Structure

The crystals are built up by the (quasi-) infinite periodic repetition of identical building blocks. This lattice (Bravais lattice) is generated by the three fundamental *translation* vectors $\mathbf{a}_1$, $\mathbf{a}_2$ and $\mathbf{a}_3$. These three vectors may not lie in a common plane. The lattice (Fig. 3.1) is the set of all points $\mathbf{R}$ (n_i are integer numbers)

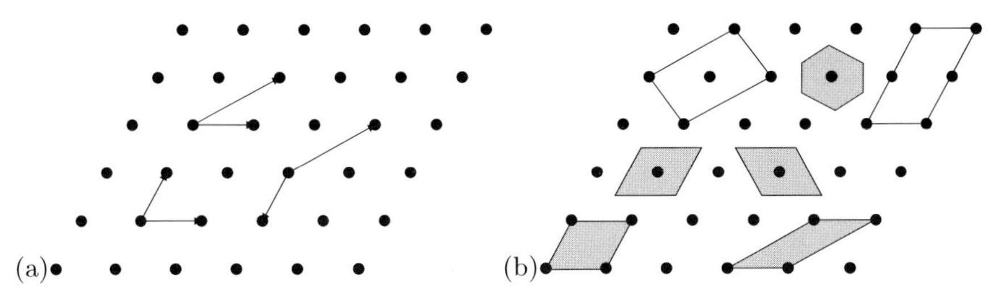

Fig. 3.1. (**a**) Two-dimensional lattice. It can be generated by various pairs of translation vectors. (**b**) Elementary cells of the lattice. Primitive elementary cells are shaded

$$\mathbf{R} = n_1\mathbf{a}_1 + n_2\mathbf{a}_2 + n_3\mathbf{a}_3 \, . \tag{3.1}$$

The crystal structure is made up by the lattice and the building block that is attached to each lattice point. This building block is called the base (Fig. 3.2). In the simplest case, e.g. for simple crystals like Cu, Fe or Al, this is just a single atom (monoatomic base). In the case of C (diamond), Si or Ge, it is a diatomic base with two identical atoms (e.g. Si–Si or Ge–Ge), in the case of compound semiconductors, such as GaAs or InP, it is a diatomic base with nonidentical atoms such as Ga–As or In–P. There exist far more involved structures, e.g. $NaCd_2$ where the smallest cubic cell contains 1192 atoms. In protein crystals, the base of the lattice can contain 10 000 atoms.

In summary: Crystal structure = Lattice × Base.

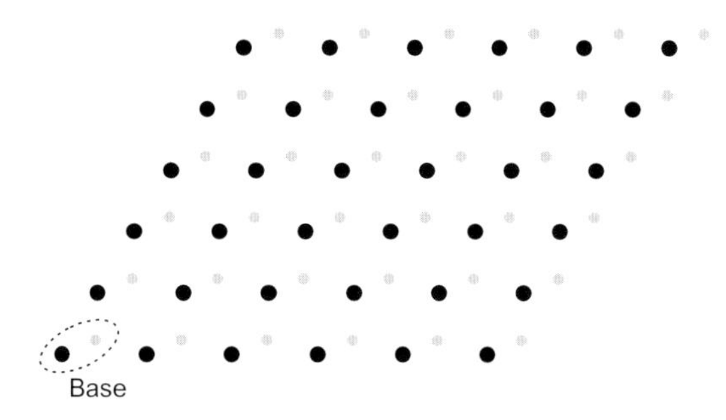

Fig. 3.2. Crystal structure, consisting of a lattice and a base

3.3 Lattice

As described in Sect. 3.2 the lattice is spanned by three translation vectors $\mathbf{a}_i$. The lattice symmetry is decisive for the physical properties of the semiconductor. It is completely described by the appropriate groups of the symmetry operations.

3.3.1 Unit Cell

The choice of the vectors $\mathbf{a}_i$ making up the lattice is not unique (Fig. 3.1). The volume that is enclosed in the parallelepiped spanned by the vectors $\mathbf{a}_1$, $\mathbf{a}_2$ and $\mathbf{a}_3$ is called the *elementary cell*. A *primitive* elementary cell is an elementary cell with the smallest possible volume (Fig. 3.1b). In each primitive elementary cell there is exactly *one* lattice point. The coordination number is the number of next-neighbor lattice points. A primitive cubic (pc) lattice, e.g. has a coordination number of 6.

The typically chosen primitive elementary cell is the *Wigner–Seitz* (WS) cell that reflects the symmetry of the Bravais lattice best. The Wigner–Seitz cell around a lattice point R_0 contains all points that are closer to this lattice point than to any other lattice point. Since all points fulfill such a condition for some lattice point R_i, the Wigner–Seitz cells fill the volume completely. The boundary of the Wigner–Seitz cell is made up by points that have the same distance to R_0 and some other lattice point(s). The Wigner–Seitz cell around R_0 is constructed by drawing lines from R_0 to the next neighbors R_j, taking the point at half distance and erecting a perpendicular plane at $(R_j + R_0)/2$. The WS cell is the smallest polyhedra circumscribed by these planes. A two-dimensional construction is shown in Fig. 3.3.

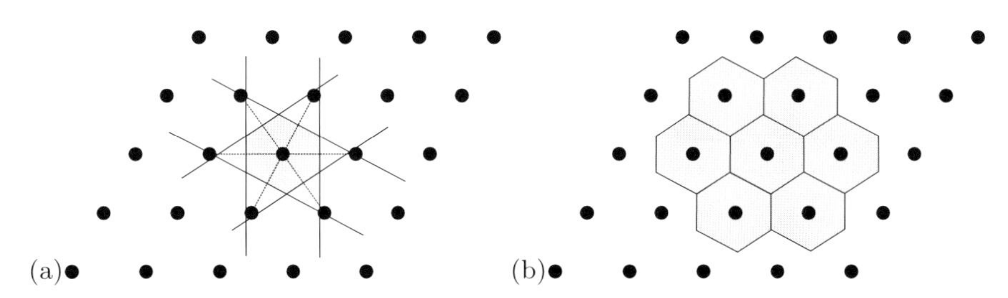

Fig. 3.3. (**a**) Construction of a two-dimensional Wigner–Seitz cell, (**b**) filling of space with WS cells

3.3.2 Point Group

Besides the translation there are other operations under which the lattice is invariant, i.e. the lattice is imaged into itself. These are:

Identity. The neutral element of any point group is the identity that does not change the crystal. It is denoted as 1 (E) in international (Schönfließ) notation.

Rotation. The rotation around an axis may have a rotation angle of 2π, $2\pi/2$, $2\pi/3$, $2\pi/4$ or $2\pi/6$ or their integer multiples. The axis is then called n = 1-, 2-, 3-, 4- or 6-fold, respectively, and denoted as n (international notation) or C_n (Schönfließ). Objects with C_n symmetry are depicted in Fig. 3.4.

Mirror operation with respect to a plane through a lattice point. Different mirror planes are discerned (Fig. 3.5) (after Schönfließ) σ_h: a mirror plane

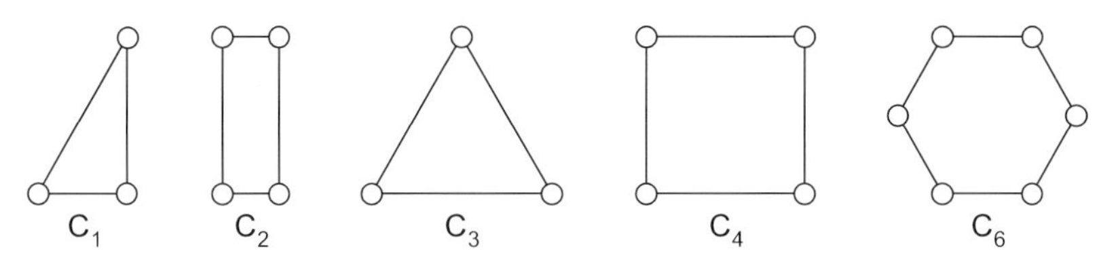

Fig. 3.4. Two-dimensional objects with perpendicular rotation axis C_n. Note that the circles do not exhibit σ_h symmetry with respect to the paper plane, i.e. they are different on the top and bottom side

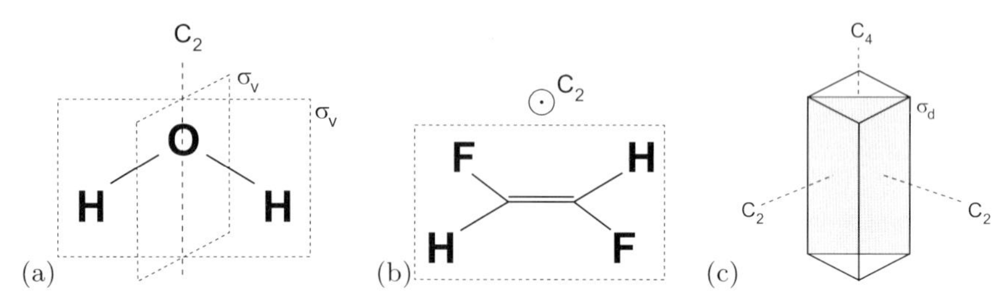

Fig. 3.5. Mirror planes: (**a**) σ_v (at H_2O molecule), (**b**) σ_h (at F_2H_2 molecule) and (**c**) σ_d

perpendicular to a rotational axis, σ_v: a mirror plane that contains a rotational axis, and σ_d: a mirror plane that contains a rotational axis and bisects the angle between two C_2 axes. The international notation is $\bar{2}$.

Inversion. All points $\mathbf{r}$ around the inversion center are replaced by $-\mathbf{r}$. The inversion is denoted $\bar{1}$ (i) in international (Schönfließ) notation.

Improper rotation. The improper rotation S_n is a rotation C_n followed immediately by the inversion operation i denoted as $\bar{n}$ in international notation. There are $\bar{3}$, $\bar{4}$ and $\bar{6}$ and their powers. Only the combined operation $\bar{n}$ is a symmetry operation, while the individual operations C_n and i alone are not symmetry operations. In the Schönfließ notation the improper rotation is defined as $S_n = \sigma_\mathrm{h} C_n$, with σ_h being a mirror operation with a plane perpendicular to the axis of the C_n rotation, denoted as S_n. There are S_3, S_4 and S_6 and $\bar{3} = S_6^5$, $\bar{4} = S_4^3$ and $\bar{6} = S_3^5$. For successive applications, the S_n yield previously known operations, e.g. $S_4^2 = C_2$, $S_4^4 = E$, $S_6^2 = C_3$, $S_6^3 = i$, $S_3^2 = C_3^2$, $S_3^3 = \sigma_\mathrm{h}$, $S_3^4 = C_3$, $S_3^6 = E$. We note that formally S_1 is the inversion i and S_2 is the mirror symmetry σ. Objects with S_n symmetry are schematically shown in Fig. 3.6.

These symmetry operations form 32 *point groups.* These groups are shown (with their different notations and elements) in Table 3.1. The highest symmetry is the cubic symmetry $O_h = O \times i$. The tetraeder group T_d (methane molecule) is a subgroup of O_h.

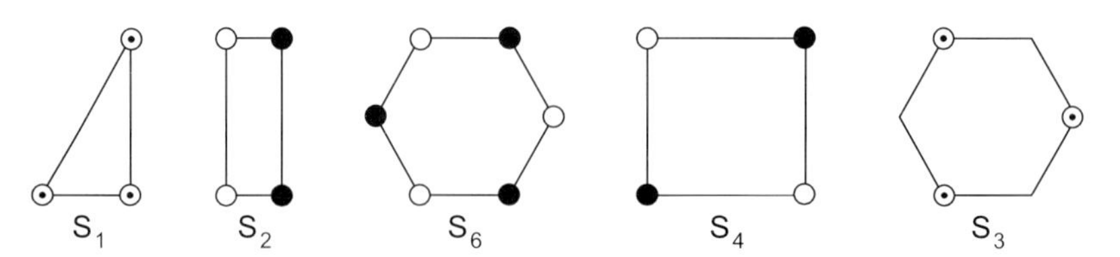

Fig. 3.6. Two-dimensional objects with perpendicular improper rotation axis S_n. Note that the white and black circles do not exhibit σ_h symmetry with respect to the paper plane, i.e. they are white on the top and black on the bottom. The circles with a dot in the center exhibit σ_h symmetry, i.e. they look the same from top and bottom

Table 3.1. The 32 point groups

System	Class		Symmetry elements
	International	Schönfließ	
triclinic	1	C_1	E
	$\bar{1}$	C_i	$E\ i$
monoclinic	m	C_s	$E\ \sigma_h$
	2	C_2	$E\ C_2$
	$2/m$	C_{2h}	$E\ C_2\ i\ \sigma_h$
orthorhombic	$2mm$	C_{2v}	$E\ C_2\ \sigma_v'\ \sigma_v''$
	222	D_2	$E\ C_2\ C_2'\ C_2''$
	mmm	D_{2h}	$E\ C_2\ C_2'\ C_2''\ i\ \sigma_h\ \sigma_v'\ \sigma_v''$
tetragonal	4	C_4	$E\ 2C_4\ C_2$
	$\bar{4}$	S_4	$E\ 2S_4\ C_2$
	$4/m$	C_{4h}	$E\ 2C_4\ C_2\ i\ 2S_4\ \sigma_h$
	$4mm$	C_{4v}	$E\ 2C_4\ C_2\ 2\sigma_v'\ 2\sigma_d$
	$\bar{4}2m$	D_{2d}	$E\ C_2\ C_2'\ C_2''\ 2\sigma_d\ 2S_4$
	422	D_4	$E\ 2C_4\ C_2\ 2C_2'\ 2C_2''$
	$4/mmm$	D_{4h}	$E\ 2C_4\ C_2\ 2C_2'\ 2C_2''\ i\ 2S_4\ \sigma_h\ 2\sigma_v'\ 2\sigma_h$
trigonal (rhombohedral)	3	C_3	$E\ 2C_3$
	$\bar{3}$	S_6	$E\ 2C_3\ i\ 2S_6$
	$3m$	C_{3v}	$E\ 2C_3\ 3\sigma_v$
	32	D_3	$E\ 2C_3\ 3C_2$
	$\bar{3}m$	D_{3d}	$E\ 2C_3\ 3C_2\ i\ 2S_6\ 3\sigma_d$
hexagonal	$\bar{6}$	C_{3h}	$E\ 2C_3\ \sigma_h\ 2S_3$
	6	C_6	$E\ 2C_6\ 2C_3\ C_2$
	$6/m$	C_{6h}	$E\ 2C_6\ 2C_3\ C_2\ i\ 2S_3\ 2S_6\ \sigma_h$
	$\bar{6}m2$	D_{3h}	$E\ 2C_3\ 3C_2\ \sigma_h\ 2S_3\ 3\sigma_v$
	$6mm$	C_{6v}	$E\ 2C_6\ 2C_3\ C_2\ 3\sigma_v\ 3\sigma_d$
	622	D_6	$E\ 2C_6\ 2C_3\ C_2\ 3C_2'\ 3C_2''$
	$6/mmm$	D_{6h}	$E\ 2C_6\ 2C_3\ C_2\ 3C_2'\ 3C_2''\ i\ 2S_3$ $2S_6\ \sigma_h\ 3\sigma_d\ 3\sigma_v$
cubic	23	T	$E\ 4C_3\ 4C_3^2\ 3C_2$
	$m3$	T_h	$E\ 4C_3\ 4C_3^2\ 3C_2\ i\ 8S_6\ 3\sigma_h$
	$\bar{4}3m$	T_d	$E\ 8C_3\ 3C_2\ 6\sigma_d\ 6S_4$
	432	O	$E\ 8C_3\ 3C_2\ 6C_2'\ 6C_4$
	$m3m$	O_h	$E\ 8C_3\ 3C_2\ 6C_2\ 6C_4\ i\ 8S_6\ 3\sigma_h\ 6\sigma_d\ 6S_4$

3.3.3 Space Group

The space group is formed by the combination of the elements of the point group with translations. The combination of a translation along a rotational axis with a rotation around this axis creates a screw axis n_m. In Fig. 3.7a, a so-called 4_1 screw axis is shown. The first index n indicates the rotation angle, i.e. $2\pi/n$, the second index indicates the translation, i.e. c/m, c being the periodicity along the axis. There are eleven crystallographically allowed screw rotations.[1]

[1] 2_1, 3_1, 3_2, 4_1, 4_2, 4_3, 6_1, 6_2, 6_3, 6_4, 6_5.

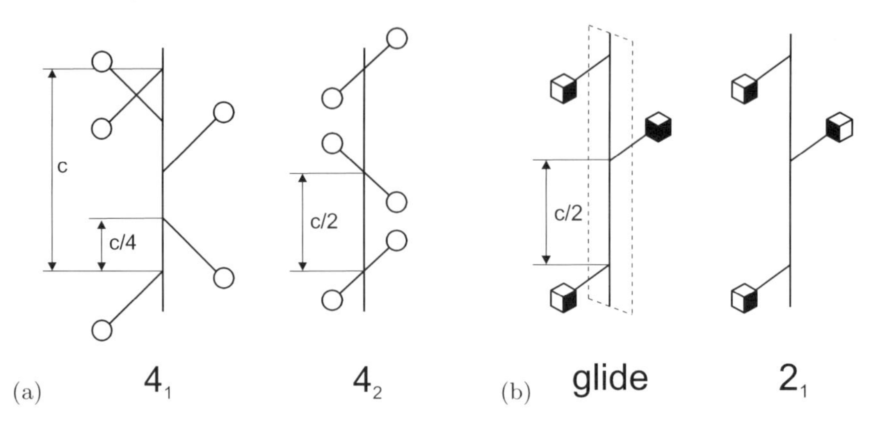

Fig. 3.7. (**a**) Schematic drawing of a 4_1 and 4_2 screw axis. (**b**) Schematic drawing of an axial glide reflection. The mirror plane is shown with dashed outline. Opposite faces of the cube have opposite color. For comparison a 2_1 screw axis is shown

The combination of the mirror operation at a plane that contains a rotational axis with a translation along this axis creates a glide reflection (Fig. 3.7b). For an axial glide (or b-glide) the translation is parallel to the reflection plane. A diagonal glide (or d-glide) involves translation in two or three directions. A third type of glide is the diamond glide (or d-glide). There are 230 different space groups [49].

3.3.4 2D Bravais Lattices

There are five two-dimensional (2D) Bravais lattices (Fig. 3.8). These are very important for the description of symmetries at surfaces. The 2D Bravais lattices are the square, hexagonal, rectangular and centered-rectangular lattice.

3.3.5 3D Bravais Lattices

In three dimensions, the operations of the point group results in fourteen 3D Bravais lattices (Fig. 3.9), that are categorized into seven crystal classes (trigonal, monoclinic, rhombic, tetragonal, cubic, rhombohedral and hexagonal). These classes are discerned by the conditions for the lengths and the mutual angles of the vectors that span the lattice (Table 3.2). Some classes have several members. The cubic crystal can have a simple (sc), face-centered (fcc) or body-centered (bcc) lattice.

In the following, some of the most important lattices, in particular those most relevant to semiconductors, will be treated in some more detail.

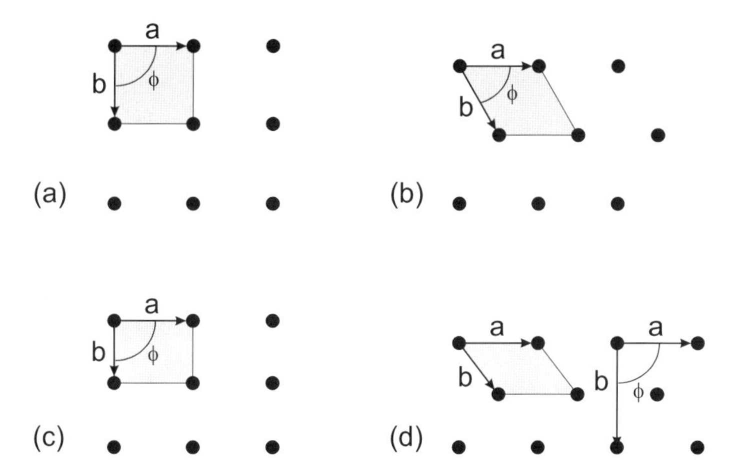

Fig. 3.8. The two-dimensional Bravais lattices with the primitive unit cells: (**a**) square lattice ($a = b$, $\phi = 90°$), (**b**) hexagonal lattice ($a = b$, $\phi = 60°$), (**c**) rectangular lattice ($a \neq b$, $\phi = 90°$), (**d**) centered-rectangular lattice ($a \neq b$, $\phi = 90°$, for the (nonprimitive) rectangular unit cell shown on the right)

Table 3.2. Conditions for lengths and angles for the 7 crystal classes. Note that only the positive conditions are listed. The rhombohedral system is a special case of the trigonal class. Conditions for the trigonal and hexagonal classes are the same, however, trigonal symmetry includes a single C_3 or S_6 axis, while hexagonal symmetry includes a single C_6 or S_6^5 axis

System	#	lattice symbol	conditions for the usual unit cell
triclinic	1		none
monoclinic	2	s, c	$\alpha = \gamma = 90°$ or $\alpha = \beta = 90°$
orthorhombic	4	s, c, bc, fc	$\alpha = \beta = \gamma = 90°$
tetragonal	2	s, bc	$a = b$, $\alpha = \beta = \gamma = 90°$
cubic	3	s, bc, fc	$a = b = c$, $\alpha = \beta = \gamma = 90°$
trigonal	1		$a = b$, $\alpha = \beta = 90°$, $\gamma = 120°$
(rhombohedral)	1		$a = b = c$, $\alpha = \beta = \gamma$
hexagonal	1		$a = b$, $\alpha = \beta = 90°$, $\gamma = 120°$

Cubic fcc and bcc Lattices

The primitive translation vectors for the cubic face-centered (fcc) and the cubic body-centered (bcc) lattice are shown in Fig. 3.10 and Fig. 3.11, respectively. Many metals crystallize in these lattices, e.g. copper (fcc) and tungsten (bcc).

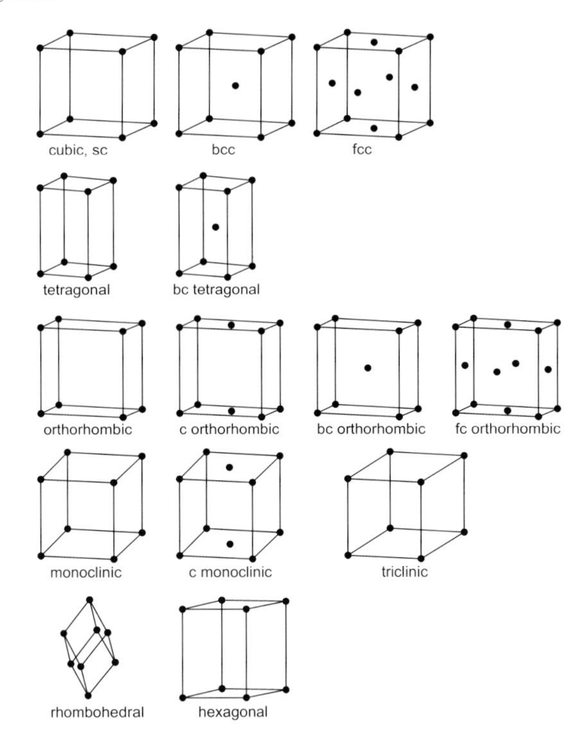

Fig. 3.9. The 14 three-dimensional Bravais lattices: cubic (sc: simple cubic, bcc: body-centered cubic, fcc: face-centered cubic), tetragonal (simple and body-centered), orthorhombic (simple, centered, body-centered and face-centered), monoclinic (simple and centered), triclinic, rhombohedral and hexagonal

In the fcc lattice, one lattice point sits in the center of each of the six faces of the usual cubic unit cell. The vectors spanning the primitive unit cell are

$$\begin{aligned} \mathbf{a}' &= \frac{a}{2}\left(\mathbf{e_x} + \mathbf{e_y} - \mathbf{e_z}\right) \\ \mathbf{b}' &= \frac{a}{2}\left(-\mathbf{e_x} + \mathbf{e_y} + \mathbf{e_z}\right) \\ \mathbf{c}' &= \frac{a}{2}\left(\mathbf{e_x} - \mathbf{e_y} + \mathbf{e_z}\right) . \end{aligned} \tag{3.2}$$

In the bcc lattice, one extra lattice point sits at the intersection of the three body diagonals at $(\mathbf{a}_1 + \mathbf{a}_2 + \mathbf{a}_3)/2$. The vectors spanning the primitive unit cell are

$$\begin{aligned} \mathbf{a}' &= \frac{a}{2}\left(\mathbf{e_x} + \mathbf{e_y} - \mathbf{e_z}\right) \\ \mathbf{b}' &= \frac{a}{2}\left(-\mathbf{e_x} + \mathbf{e_y} + \mathbf{e_z}\right) \\ \mathbf{c}' &= \frac{a}{2}\left(\mathbf{e_x} - \mathbf{e_y} + \mathbf{e_z}\right) . \end{aligned} \tag{3.3}$$

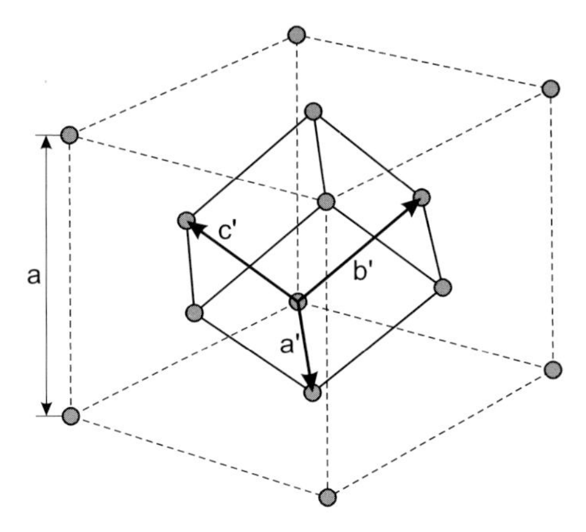

Fig. 3.10. Primitive translations of the fcc lattice. These vectors connect the origin with the face-center points. The primitive unit cell is the rhombohedron spanned by these vectors. The primitive translations $\mathbf{a}'$, $\mathbf{b}'$ and $\mathbf{c}'$ are given in (3.2). The angle between the vectors is 60°

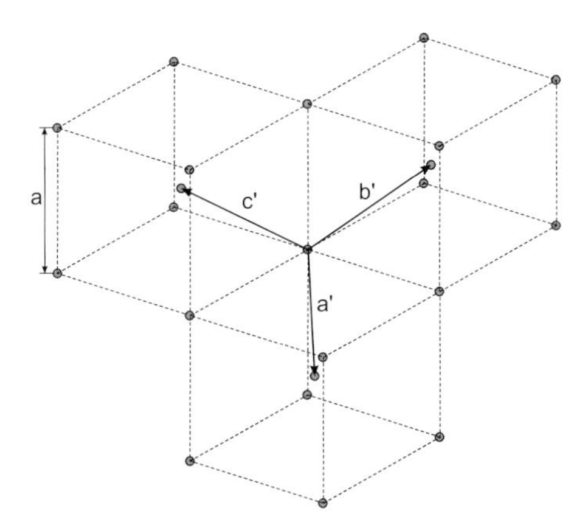

Fig. 3.11. Primitive translations of the bcc lattice. These vectors connect the origin with the lattice points in the cube centers. The primitive unit cell is the rhombohedron spanned by these vectors. The primitive translations $\mathbf{a}'$, $\mathbf{b}'$ and $\mathbf{c}'$ are given in (3.3). The angle between the vectors is $\approx 109.5°$

Hexagonally Close Packed Lattice (hcp)

The 2D hexagonal Bravais lattice fills a plane with spheres (or circles) with maximum filling factor. There are two ways to fill space with spheres and highest filling factor. One is the fcc lattice, the other is the hexagonally close packed (hcp) structure. Both have a filling factor of 74%.

For the hcp, we start with a hexagonally arranged layer of spheres (A), see Fig. 3.12. Each sphere has six next-neighbor spheres. This could, e.g., be a plane in the fcc perpendicular to the body diagonal. The next plane B is

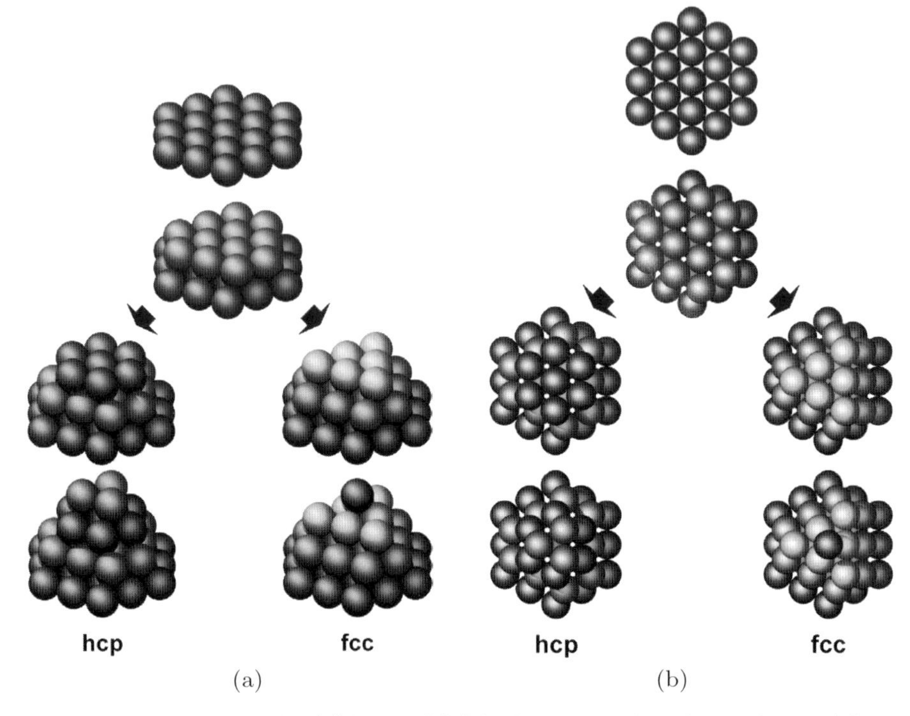

Fig. 3.12. Structure of the (**a**) hcp and (**b**) fcc lattice. For hcp the stacking is (along the c-axis) ABABAB..., for fcc (along the body diagonal) ABCABCABC...

put on top in such a way that each new sphere touches three spheres of the previous layer. The third plane can now be added in two different ways: If the spheres of the third layer are vertically on top of the spheres of layer A, a plane A' identical to A has been created that is shifted from A along the stacking direction (normally called the c-axis) by

$$c_{\text{hcp}} = (8/3)^{1/2} a \approx 1.633a \ . \tag{3.4}$$

The hcp stacking order is ABABAB..., the coordination number is 12. In the fcc structure, the third layer is put on the thus far unfilled positions and forms a new layer C. Only the forth layer is again identical to A and is shifted by

$$c_{\text{fcc}} = \sqrt{6}a \approx 2.45a \ . \tag{3.5}$$

The fcc stacking order is ABCABCABC....

In the hexagonal plane of the fcc lattice (which will later be called a $\{111\}$ plane) the distance between lattice points is $a = a_0/\sqrt{2}$, where a_0 is the cubic lattice constant. Thus $c = \sqrt{3}a_0$, just what is expected for the body diagonal.

For real materials with hexagonal lattice the ratio c/a deviates from the ideal value given in (3.4). Helium comes very close to the ideal value, for Mg it is 1.623, for Zn 1.861. Many metals exhibit a phase transition between hcp and fcc at higher temperatures.

3.3.6 Polycrystalline Semiconductors

A polycrystalline material consists of crystal grains that are randomly oriented with respect to each other. Between two grains a (large-angle) grain boundary (see also Sect. 4.6) exists. An important parameter is the grain size and its statistical distribution. It can be influenced via processing steps such as annealing. Polycrystalline semiconductors are used in cheap, large-area applications such as solar cells (e.g. polysilicon, $CuInSe_2$) or thin-film transistors (poly-Si) or as n-conducting contact material in MOS diodes (poly-Si) as shown in Fig. 3.13 (see also Fig. 18.23).

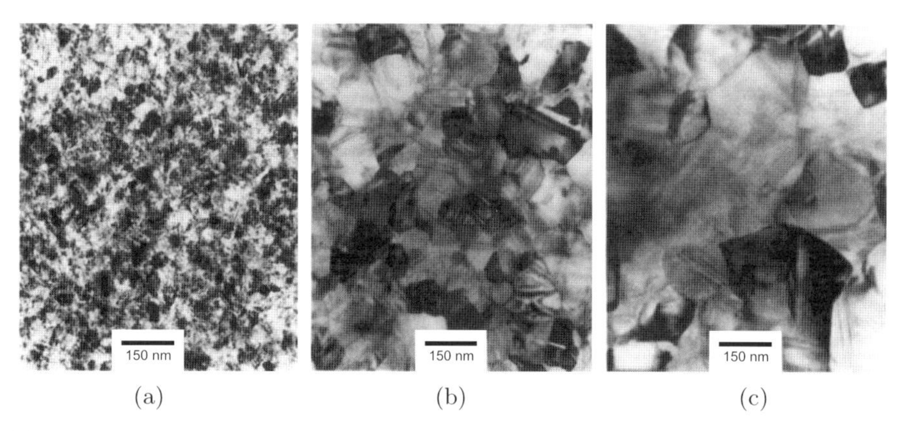

Fig. 3.13. Transmission electron micrographs of polycrystalline silicon (poly-Si). (**a**) As-deposited material from low-pressure chemical vapor deposition (LPCVD) at about 620°C, grain size is about 30 nm. (**b**) After conventional processing (annealing at 1150°C), average grain size is about 100 nm. (**c**) After annealing in HCl that provides enhanced point defect injection (and thus increased possibility to form larger grains), average grain size is about 250 nm. Reprinted with permission from [50], ©1989 AIP

3.3.7 Amorphous Semiconductors

An amorphous material lacks the long-range order of the direct lattice. It is disordered on the atomic scale. Amorphous silicon is, e.g., denoted as 'a-Si'.

The local quantum mechanics provides almost rigorous requirements for the bond length to next neighbors. The constraints for the bond angle are less strict. Covalently bonded atoms arrange in an open network with correlations up to the third and fourth neighbors. The short-range order is responsible

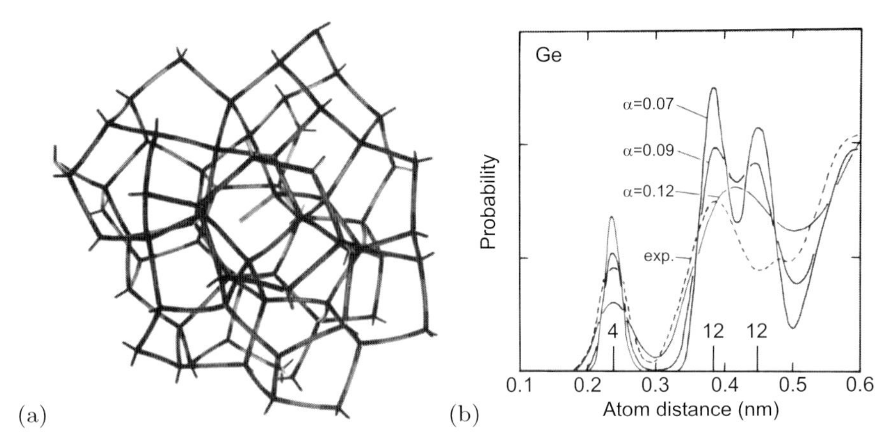

Fig. 3.14. (**a**) A continuous random network model of amorphous silicon containing one dangling bond. Reprinted with permission from [51]. (**b**) Calculated radial atomic distribution functions of amorphous Ge (*solid lines*) for three different values of the disorder parameter α (3.6) as labeled and experimental result (*dashed line*). The positions of next, second-next and third-next neighbors are indicted by *vertical bars* with numbers of their multiplicity (4, 12, and 12). Adapted from [52], reprinted with permission

for the observation of semiconductor properties such as an optical absorption edge and also thermally activated conductivity. In Fig. 3.14a a model of a continuous random network (with a bond-angle distortion of less than about 20%) of a-Si is depicted. The diameter d_{SR} of the short-range order region is related to the disorder parameter α via [52]

$$d_{\mathrm{SR}} = \frac{a}{2\alpha} , \tag{3.6}$$

where a is the next-neighbor interatomic distance. For a diamond structure it is related to the lattice constant by $a = \sqrt{3}a_0/4$.

Typically, a significant number of dangling bonds exists. Bonds try to pair but if an odd number of broken bonds exists locally, an unsaturated, dangling bond remains. This can be passivated by a hydrogen atom. Thus, the hydrogenation of amorphous semiconductors is very important. A hydrogen atom can also break an overlong (and therefore weak) bond, saturate one side and eventually leave a dangling bond.

3.4 Important Crystal Structures

Now the crystal structures that are important for semiconductor physics will be discussed. These are the rocksalt (PbS, MgO, ...), diamond (C, Si, Ge), zincblende (GaAs, InP, ...) and wurtzite (GaN, ZnO, ...) structures.

3.4.1 Rocksalt Structure

The rocksalt (rs, NaCl) structure (Fig. 3.15a) consists of a fcc lattice with the period a and a diatomic base in which the Cl atom is positioned at (0,0,0) and the Na atom at $(1/2,1/2,1/2)a$ with a distance $\sqrt{3}a/2$. Materials that crystallize (under normal conditions) in the rocksalt lattice are, e.g., KCl, KBr, PbS (galena), PbSe, PbTe, AgBr, MgO, CdO, MnO. AlN, GaN and InN undergo, under high pressure, a phase transition from the wurtzite into the rocksalt structure.

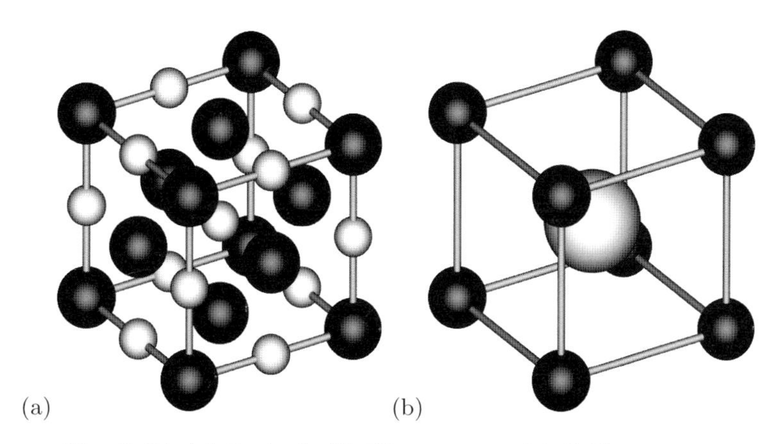

Fig. 3.15. (**a**) Rocksalt (NaCl) structure, (**b**) CsCl structure

3.4.2 CsCl Structure

The CsCl structure (Fig. 3.15b) consists of a simple cubic lattice. Similar as for the rocksalt structure, the base consists of different atoms at (0,0,0) and $(1/2,1/2,1/2)a$. Typical crystals with CsCl-structure are TlBr, TlI, CuZn (β-brass), AlNi.

3.4.3 Diamond Structure

The diamond structure (Fig. 3.16a) has the fcc lattice. The base consists of two identical atoms at (0,0,0) and $(1/4,1/4,1/4)a$. Each atom has a tetrahedral configuration. The packing density is only about 0.34. The materials that crystallize in the diamond lattice are C, Ge, Si and α-Sn. The diamond structure (point group O_h) has an inversion center, located between the two atoms of the base, i.e. at $(1/8,1/8,1/8)a$. The radii of the wavefunctions for various group-IV elements increases with the order number (Table 3.3), and accordingly increases the lattice constant.

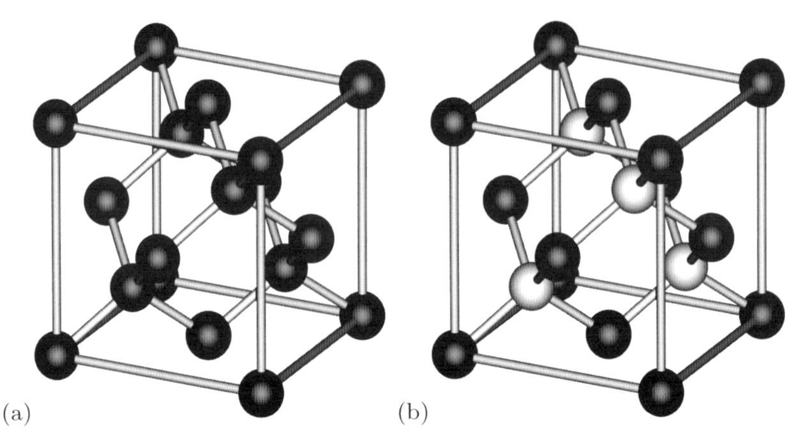

Fig. 3.16. (**a**) Diamond structure and (**b**) zincblende structure (*black spheres*: A atoms, *white spheres*: B atoms). The tetragonal bonds are indicated

Table 3.3. Radii of the wavefunctions in the diamond structure, r_s and r_p are related to s^1p^3, r_d to $s^1p^2d^1$ and lattice constant a_0

	r_s (nm)	r_p (nm)	r_d (nm)	a_0 (nm)
C	0.121	0.121	0.851	0.356683
Si	0.175	0.213	0.489	0.543095
Ge	0.176	0.214	0.625	0.564613

In Fig. 3.17a the unit cell with tetragonal symmetry of three places along the ⟨111⟩ direction is shown. In Fig. 3.17b the arrangement of atoms along ⟨111⟩ is depicted. The symmetry along this line is at least C_{3v}. At the atoms sites it is O_h. The bond center (BC) and the hexagonal (H) position are a center of inversion and have D_{3d} symmetry. The unoccupied 'T' positions have T_d symmetry.

3.4.4 Zincblende Structure

The zincblende (sphalerite, ZnS) structure (Fig. 3.16b) has a fcc lattice with a diatomic base. The metal (A) atom is at $(0, 0, 0)$ and the nonmetal (B) atom is at $(1/4, 1/4, 1/4)a$. Thus the cation and anion sublattices are shifted with respect to each other by a quarter of the body diagonal of the fcc lattice. The atoms are tetrahedrally coordinated, a Zn atom is bonded to four S atoms and vice versa. However, no inversion center is present any longer (point group T_d). In the zincblende structure the stacking order of diatomic planes along the body diagonal is aAbBcCaAbBcC....

Many important compound semiconductors, such as GaAs, InAs, AlAs, InP, GaP and their alloys (see Sect. 3.7), but also the II–VI compounds ZnS, ZnSe, ZnTe, HgTe, CdSe and CdTe crystallize in the zincblende structure.

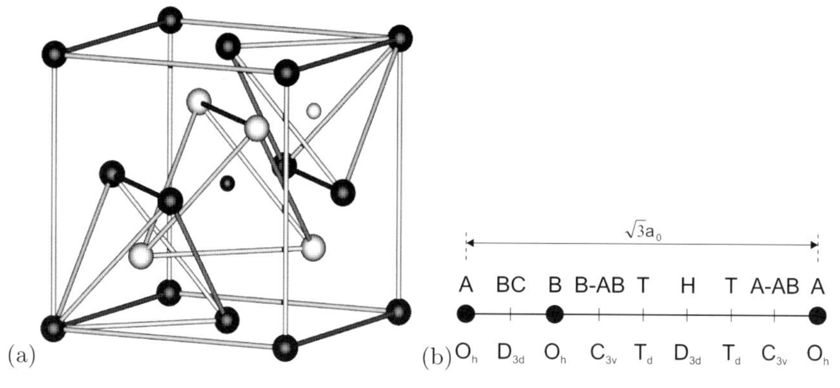

Fig. 3.17. (**a**) Unit cell of the zincblende structure with the indication of tetragonal symmetries. The position of the small *white* (*black*) *sphere* is the tetrahedrally configured unoccupied positions of the A (B) sublattice, denoted with 'T' in part (**b**). (**b**) Line along $\langle 111 \rangle$ in the zincblende structure. *Full circles* denote the positions of the A and B atoms as labeled. Other positions are called the bond center ('BC'), antibonding ('AB') relative to A and B atoms ('A–AB', 'B–AB'), hexagonal ('H') and tetrahedral position ('T'). The point symmetries of the various locations are given in the lower line

Four-fold coordinated materials (zincblende and wurtzite) typically undergo a phase transition into 6-fold coordinated structures upon hydrostatic pressure [53]. For GaAs under pressure see [54].

3.4.5 Wurtzite Structure

The wurtzite structure is also called the hexagonal ZnS structure (because ZnS has both modifications). It consists (Fig. 3.18) of a hcp lattice with a diatomic base. The c/a ratio typically deviates from the ideal value $\zeta_0 = (8/3)^{1/2} \approx 1.633$ (3.4) as listed in Table 3.4.

The Zn atom is located at $(0,0,0)$, the S atom at $(0,0,\sqrt{3/8})a$. This corresponds to a shift of $3/8c$ along the c-axis. This factor is called the cell-internal parameter u that has the value $u_0 = 3/8 = 0.375$ for the ideal wurtzite structure. For real wurtzite crystals u deviates from the ideal value, e.g. for group-III nitrides $u > u_0$. The ZnS diatomic planes have a stacking order of aAbBaAbB... in the wurtzite structure.

In Fig. 3.19 the different local structural environment of the atoms in the zincblende and wurtzite structure is shown.

Many important semiconductors with large bandgap crystallize in the wurtzite structure, such as GaN, AlN, InN, ZnO, SiC, ZnS[2], CdS und CdSe.

[2]For ZnS, CdS and CdTe the high-temperature phase is cubic and the low-temperature phase hexagonal.

Table 3.4. c/a ratio of various wurtzite semiconductors. Listed is $\xi = (c/a - \zeta_0)/\zeta_0$. Data based on [55]

material	ξ (%)	material	ξ(%)	material	ξ (%)	material	ξ (%)
AlN	−2.02	CdS	−0.61	CuBr	0.43	BeO	−0.61
GaN	−0.49	CdSe	−0.18	CuCl	0.55	ZnO	−1.9
InN	−1.35	CdTe	0.25	CuI	0.74	6H-SiC	0.49
ZnS	0.25	MgS	−0.80	AgI	0.12	BN	0.74
ZnSe	0.06	MgSe	−0.67	ZnTe	0.74	MgTe	−0.67

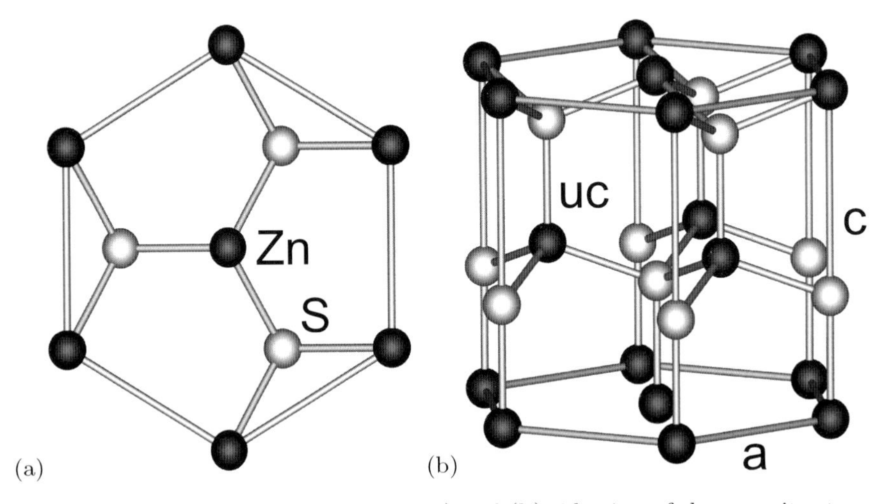

Fig. 3.18. (**a**) Top view (along the c-axis) and (**b**) side view of the wurtzite structure with the tetragonal bonds indicated. The top (bottom) surface of the depicted structure is termed the Zn-face, (00.1) (O-face, $(00.\bar{1})$)

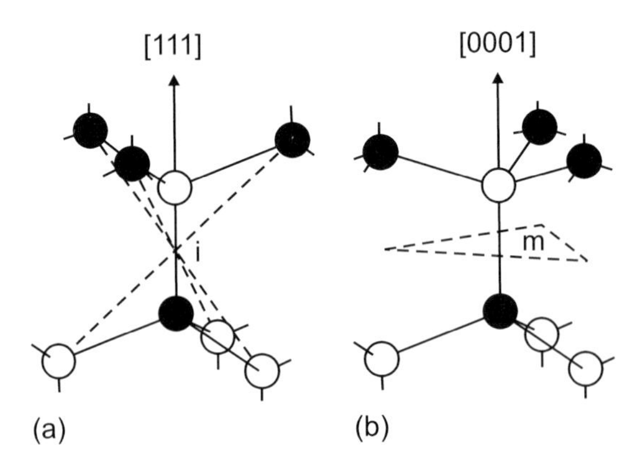

Fig. 3.19. Comparison of the tetragonal bonds in the (**a**) zincblende and (**b**) wurtzite structure (i: inversion center, m: symmetry plane)

3.4.6 Chalcopyrite Structure

The chalcopyrite (ABC_2, named after 'fool's gold' $CuFeS_2$) structure is relevant for I–III–VI_2 (with chalcogenide anions) and II–IV–V_2 (with pnictide anions) semiconductors such as, e.g., (Cu,Ag)(Al,Ga,In)$(S,Se,Te)_2$ and (Mg,Zn,Cd)(Si,Ge,Sn)$(As,P,Sb)_2$. A nonmetallic anion atom ('C') is tetrahedrally bonded to two different types of cation atoms ('A' and 'B') as shown in Fig. 3.20. The local surrounding of each anion is identical, two of both the A and B atoms. The structure is often tetragonal, i.e. $\eta = c/(2a)$ deviates from the ideal value 1; typically $\eta < 1$ [56, 57].

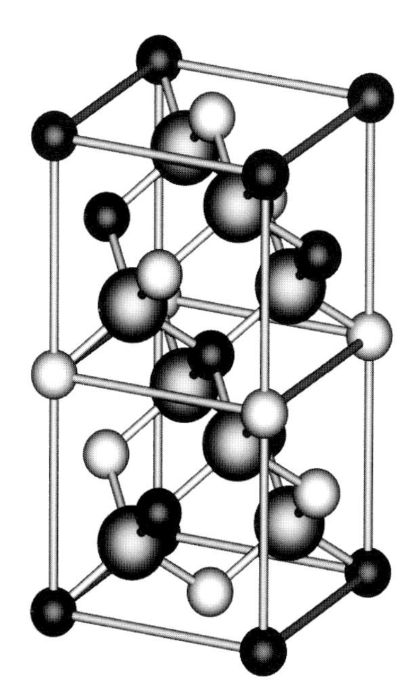

Fig. 3.20. Chalcopyrite structure, *black* and *white spheres* denote the metal species. The bigger *grey spheres* represent the nonmetal anion

If the C atom is in the tetrahedral center of the two A and two B atoms, the bond lengths R_{AC} and R_{BC} of the A–C and B–C bonds, respectively, are equal. Since the ideal A–C and B–C bond lengths d_{AC} and d_{BC} are typically unequal, this structure is strained. The common atom C is therefore displaced along [100] and [010] such that it is closer (if $d_{\mathrm{AC}} < d_{\mathrm{BC}}$) to the pair of A atoms and further away from the B atoms. The displacement parameter is

$$u = \frac{1}{4} + \frac{R_{\mathrm{AC}}^2 - R_{\mathrm{BC}}^2}{a^2} \tag{3.7}$$

and it deviates from the ideal value $u_0 = 1/4$ for the zincblende structure as listed in Table 3.5 for a number of chalcopyrite compounds. In the chalcopyrite structure

$$R_{\rm AC} = \left[u^2 + \frac{1+\eta^2}{16}\right]^{1/2} a \tag{3.8a}$$

$$R_{\rm BC} = \left[\left(u - \frac{1}{2}\right)^2 + \frac{1+\eta^2}{16}\right]^{1/2} a\,. \tag{3.8b}$$

The minimization of the microscopic strain yields (in first order) [58]

$$u \cong \frac{1}{4} + \frac{3}{8}\frac{d_{\rm AC}^2 - d_{\rm BC}^2}{d_{\rm AC}^2 + d_{\rm BC}^2}\,. \tag{3.9}$$

Compounds with $u > u_{\rm c}$, $u_{\rm c} = 0.265$ being a critical displacement parameter, (or $u < 1/2 - u_{\rm c} = 0.235$) are stable with regard to cation disorder [57]. In Fig. 3.21 the correlation of the calculated value for u according to (3.9) and the experimental values is shown.

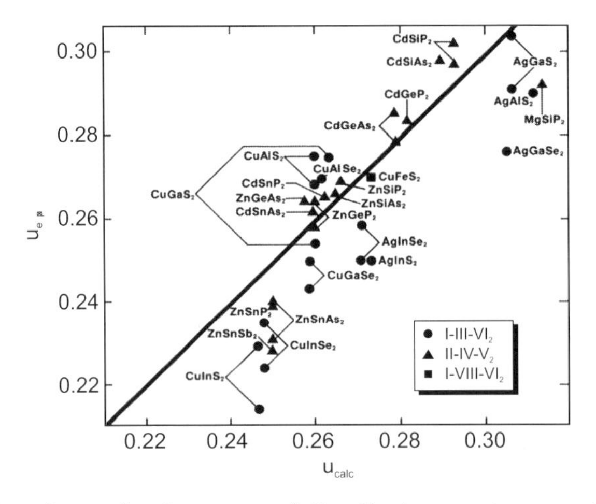

Fig. 3.21. Experimental values $u_{\rm exp}$ of the displacement parameter for various chalcopyrites vs. the calculated value $u_{\rm calc}$ according to (3.9). Adapted from [58], reprinted with permission, ©1984 APS

3.4.7 Delafossite Structure

The I–III–O_2 materials crystallize in the trigonal delafossite ($CuFeO_2$, space group 166, R$\bar{3}$m) structure (Fig. 3.22). This structure is also called caswellsilverite ($NaCrS_2$). In Table 3.6 the lattice parameters of some delafossite compounds are given. The (Cu,Ag) (Al,Ga,In)O_2 materials are transparent conductive oxides (TCO). We note that Pt and Pd as group-I component create metal-like compounds because of the d^9 configuration as opposed to the d^{10} configuration of Cu and Ag.

Table 3.5. Lattice nonideality parameters η and u (from (3.9)) of various chalcopyrite compounds and their experimentally observed disorder stability (+/− indicates compound with/without order–disorder (D–O) transition, respectively). Data from [57]

	η	u	D–O		η	u	D–O
$CuGaSe_2$	0.983	0.264	+	$ZnSiAs_2$	0.97	0.271	−
$CuInSe_2$	1.004	0.237	+	$ZnGeAs_2$	0.983	0.264	+
$AgGaSe_2$	0.897	0.287	−	$CdSiAs_2$	0.92	0.294	−
$AgInSe_2$	0.96	0.261	+	$CdGeAs_2$	0.943	0.287	
$CuGaS_2$	0.98	0.264		$ZnSiP_2$	0.967	0.272	−
$CuInS_2$	1.008	0.236	+	$ZnGeP_2$	0.98	0.264	+
$AgGaS_2$	0.895	0.288	−	$CdSiP_2$	0.92	0.296	−
$AgInS_2$	0.955	0.262		$CdGeP_2$	0.939	0.288	−

Table 3.6. Lattice parameters a, c, and u of some delafossite compounds. Theoretical values are shown with asterisk. Data from [59]

	a (nm)	c (nm)	u (nm)
$CuAlO_2$	0.2858	1.6958	0.1099
$CuGaO_2$	0.2980	1.7100	0.1073*
$CuInO_2$	0.3292	1.7388	0.1056*

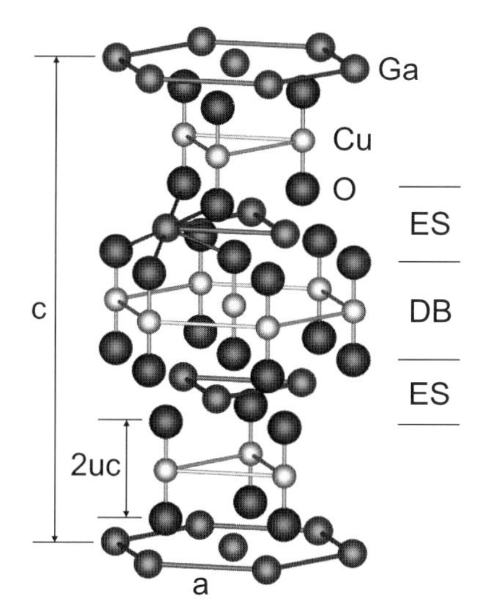

Fig. 3.22. Hexagonal unit cell of delafossite $CuGaO_2$. Oxygen atoms are bonded to the Cu in a dumbbell ('DB') configuration. In the edge-sharing ('ES') layer the Ga atoms are octahedrally configured as GaO_6

3.4.8 Perovskite Structure

The perovskite structure (calcium titanate, $CaTiO_3$) (Fig. 3.23) is relevant for ferroelectric semiconductors (see Sect. 14.3). It is cubic with the Ca (or Ba, Sr) ions (charge state 2+) on the corners of the cube, the O ions (2−) on the face centers and the Ti (4+) in the body center. The lattice is simple cubic, the base is Ca at (0,0,0), O at (1/2,1/2,0), (1/2,0,1/2) and (0,1/2,1/2) and Ti at (1/2,1/2,1/2). The ferroelectric polarization is typically evoked by a shift of the negatively and positively charged ions relative to each other.

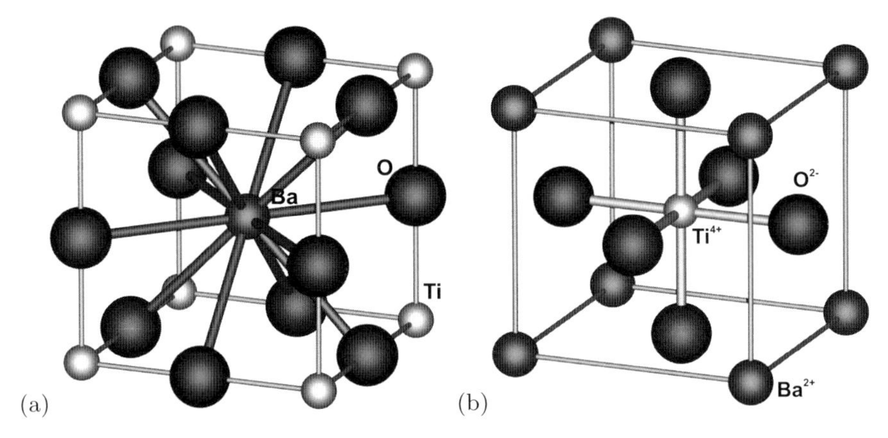

Fig. 3.23. Perovskite structure ($BaTiO_3$), (**a**) A cell with 12-fold (cuboctahedrally) configured Ba, (**b**) B cell with octahedrally configured Ti

3.4.9 NiAs Structure

The NiAs structure (Fig. 3.24) is relevant for magnetic semiconductors, such as MnAs, and also occurs in the formation of Ni/GaAs Schottky contacts [60]. The structure is hexagonal. The arsenic atoms form a hcp structure and are trigonal prismatically configured with six nearest metal atoms. The metal atoms form hcp planes and fill all octahedral holes of the As lattice. For a cubic close packed, i.e. fcc, structure this would correspond to the rocksalt crystal. The stacking is ABACABAC... (A: Ni, B,C: As).

3.5 Polytypism

In polytype materials the stacking order is not only hcp or fcc but takes different sequences, such as, e.g., ACBCABAC as the smallest unit cell along the stacking direction. A typical example is SiC, for which in addition to hcp and fcc 45 other stacking sequences are known. The largest primitive unit

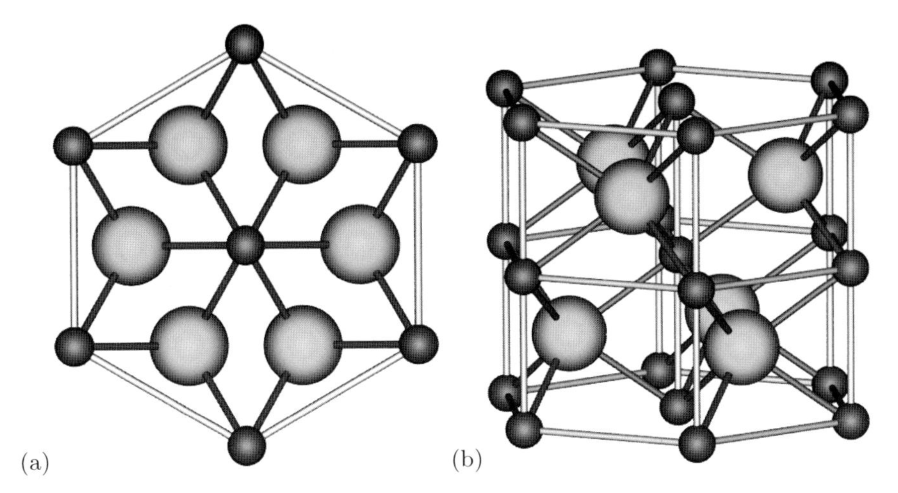

Fig. 3.24. NiAs structure, metal atoms: *dark grey*, chalcogenide atoms: *light grey*

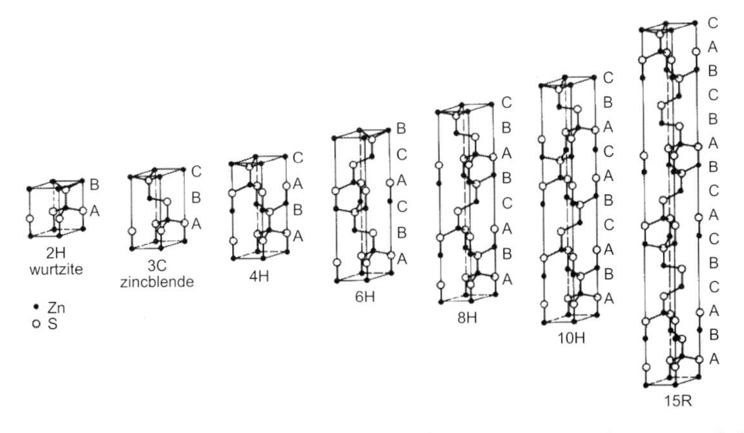

Fig. 3.25. Polytypes of the zincblende and wurtzite lattice (found in SiC), the letters A, B and C denote the three possible positions of the diatomic layers (see Fig. 3.12)

cell of SiC contains 594 layers. Some of the smaller polytypes are shown in Fig. 3.25. In Fig. 3.26 cubic diamond crystallites and metastable hexagonal and orthorhomic phases (in silicon) are shown.

For the ternary alloy (see Sect. 3.7) $Zn_{1-x}Cd_xS$ the numbers n_h of diatomic layers with hexagonal stacking (AB) and n_c of layers with cubic stacking (ABC) have been investigated. CdS has wurtzite structure and ZnS mostly zincblende structure. The hexagonality index α as defined in (3.10) is shown in Fig. 3.27 for $Zn_{1-x}Cd_xS$

$$\alpha = \frac{n_h}{n_h + n_c} . \tag{3.10}$$

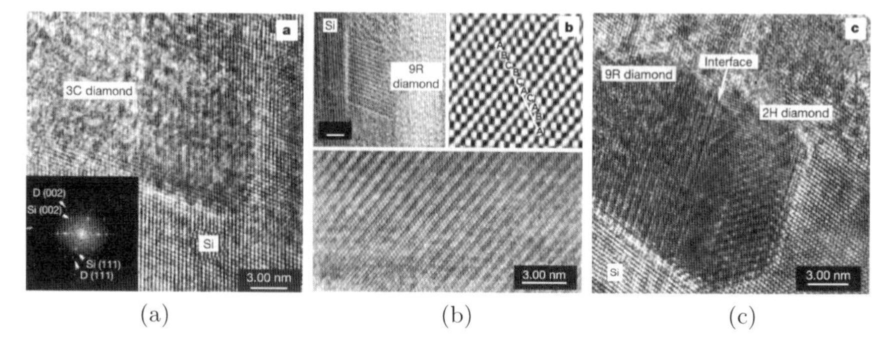

Fig. 3.26. Polytypes of diamond found in crystallites (metastable phases in silicon). (**a**) cubic type (3C) with stacking ABC, *inset* shows a diffractogram and the alignment of the C and Si lattice, (**b**) rhombohedral 9R crystallite with ABCB-CACABA stacking, (**c**) 9R phase with interface to a hexagonal 2H (AB stacking) phase. Reprinted with permission from Nature [61], ©2001 Macmillan Magazines Limited

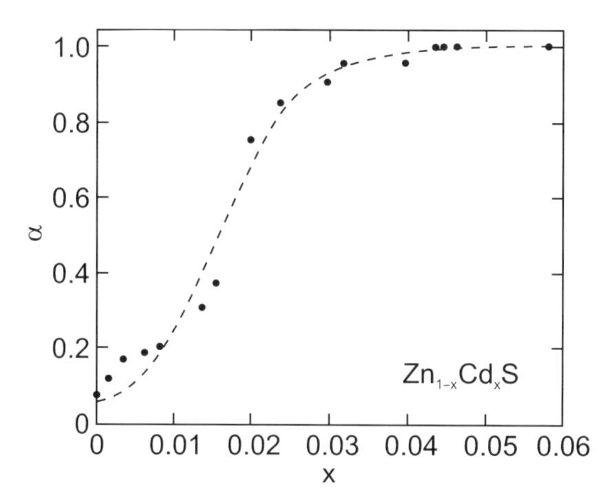

Fig. 3.27. Hexagonality index α (3.10) of $Zn_{1-x}Cd_xS$ for various ternary compositions. *Dashed line* is a guide to the eye

3.6 Reciprocal Lattice

The *reciprocal* lattice is of utmost importance for the description and investigation of periodic structures, in particular for X-ray diffraction, surface electron diffraction, phonons and the band structure. It is the quasi-Fourier transformation of the crystal lattice. The crystal lattice is also called the *direct* lattice, in order to distinguish it from the reciprocal lattice.

3.6.1 Reciprocal Lattice Vectors

When $\mathcal{R}$ denotes the set of vectors of the direct lattice, the set $\mathcal{G}$ of the reciprocal lattice vectors is given by the condition[3]

$$\exp(\mathrm{i}\mathbf{GR}) = 1 \tag{3.11}$$

for all $\mathbf{R} \in \mathcal{R}$ and $\mathbf{G} \in \mathcal{G}$. Therefore, for all vectors $\mathbf{r}$ and a reciprocal lattice vector $\mathbf{G}$

$$\exp(\mathrm{i}\mathbf{G}(\mathbf{r}+\mathbf{R})) = \exp(\mathrm{i}\mathbf{G}\mathbf{r}) \ . \tag{3.12}$$

Each Bravais lattice has a certain reciprocal lattice. The reciprocal lattice is also a Bravais lattice, since when $\mathbf{G}_1$ and $\mathbf{G}_2$ are two reciprocal lattice vectors, then this is obviously true also for $\mathbf{G}_1+\mathbf{G}_2$. For the primitive translation vectors $\mathbf{a}_1$, $\mathbf{a}_2$ and $\mathbf{a}_3$ of the direct lattice, the vectors $\mathbf{b}_1$, $\mathbf{b}_2$ and $\mathbf{b}_3$ that span the reciprocal lattice can be given directly as

$$\mathbf{b}_1 = \frac{2\pi}{V_a}(\mathbf{a}_2 \times \mathbf{a}_3) \tag{3.13a}$$

$$\mathbf{b}_2 = \frac{2\pi}{V_a}(\mathbf{a}_3 \times \mathbf{a}_1) \tag{3.13b}$$

$$\mathbf{b}_3 = \frac{2\pi}{V_a}(\mathbf{a}_1 \times \mathbf{a}_2) \ , \tag{3.13c}$$

where $V_a = \mathbf{a}_1(\mathbf{a}_2 \times \mathbf{a}_3)$ is the volume of the unit cell spanned by the vectors $\mathbf{a}_i$. The volume of the unit cell in the reciprocal space is $V_a^* = (2\pi)^3/V_a$.

The vectors $\mathbf{b}_i$ fulfill the conditions

$$\mathbf{a}_i\mathbf{b}_j = 2\pi\delta_{ij} \ . \tag{3.14}$$

For an arbitrary reciprocal lattice vector $\mathbf{G} = k_1\mathbf{b}_1 + k_2\mathbf{b}_2 + k_3\mathbf{b}_3$ and a vector $\mathbf{R} = n_1\mathbf{a}_1 + n_2\mathbf{a}_2 + n_3\mathbf{a}_3$ in direct space we find

$$\mathbf{GR} = 2\pi\,(n_1k_1 + n_2k_2 + n_3k_3) \ . \tag{3.15}$$

The number in brackets is an integer. Thus, it is clear that (3.11) is fulfilled. Additionally, we note that the reciprocal lattice of the reciprocal lattice is again the direct lattice. The reciprocal lattice of the fcc is bcc and vice versa. The reciprocal lattice of hcp is hcp (rotated by 30° with respect to the direct lattice).

For later, we note two important theorems. A (sufficiently well behaved) function $f(\mathbf{r})$ that is periodic with the lattice, i.e. $f(\mathbf{r}) = f(\mathbf{r}+\mathbf{R})$ can be expanded into a Fourier series with the reciprocal lattice vectors according to

[3] The dot product $\mathbf{a}\cdot\mathbf{b}$ of two vectors shall also be denoted as $\mathbf{ab}$.

$$f(\mathbf{r}) = \sum a_{\mathbf{G}} \exp(\mathrm{i}\mathbf{G}\mathbf{r}) , \tag{3.16}$$

where $a_{\mathbf{G}}$ denotes the Fourier component of the reciprocal lattice vector $\mathbf{G}$, $a_{\mathbf{G}} = \int_V f(\mathbf{r}) \exp(-\mathrm{i}\mathbf{G}\mathbf{r})\, \mathrm{d}^3\mathbf{r}$. If $f(\mathbf{r})$ is lattice periodic, the integral given in (3.17) is zero unless $\mathbf{G}$ is a reciprocal lattice vector.

$$\int_V f(\mathbf{r}) \exp(-\mathrm{i}\mathbf{G}\mathbf{r}) = \begin{cases} a_{\mathbf{G}} \\ 0, \mathbf{G} \notin \mathcal{G} \end{cases} . \tag{3.17}$$

3.6.2 Miller Indices

A lattice plane is the set of all lattice points in a plane spanned by two independent lattice vectors $\mathbf{R_1}$ and $\mathbf{R_2}$. The lattice points on that plane form a two-dimensional Bravais lattice. The entire lattice can be generated by shifting the lattice plane along its normal $\mathbf{n} = (\mathbf{R_1} \times \mathbf{R_2})/|\mathbf{R_1} \times \mathbf{R_2}|$. The plane belongs to the reciprocal lattice vector $\mathbf{G_n} = 2\pi\mathbf{n}/d$, d being the distance between planes.

This correspondence between reciprocal lattice vectors and sets of planes allows the orientation of planes to be described in a simple manner. The shortest reciprocal lattice vector perpendicular to the plane is used. The coordinates with respect to the primitive translation vectors of the reciprocal space $\mathbf{b}_i$ form a triplet of integer numbers and are called Miller indices.

The plane described by $\mathbf{G_n r} = A$ fulfills the condition for a suitable value of A. The plane intersects the axes $\mathbf{a}_i$ at the points $x_1\mathbf{a}_1$, $x_2\mathbf{a}_2$ and $x_3\mathbf{a}_3$. Thus we find $\mathbf{G_n} x_i \mathbf{a}_i = A$ for all i. From (3.15) follows $\mathbf{G_n a}_1 = 2\pi h$, $\mathbf{G_n a}_2 = 2\pi k$ and $\mathbf{G_n a}_3 = 2\pi l$, where h, k and l are integers. The triplet of integer numbers (hkl), the reciprocal values of the axis intersections in the direct lattice, are the Miller indices. An example is shown in Fig. 3.28.

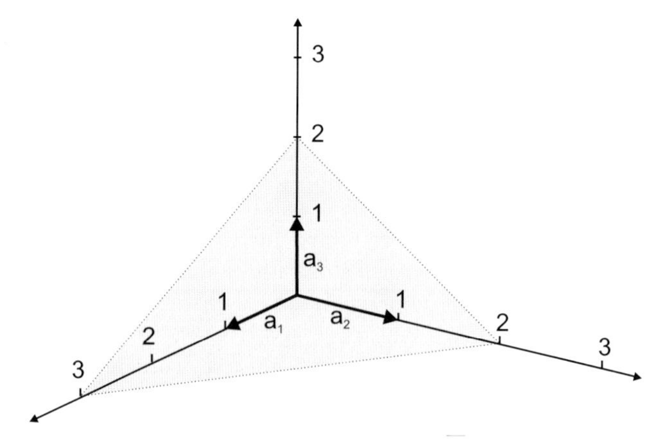

Fig. 3.28. The plane intersects the axes at 3, 2, and 2. The inverse of these numbers is 1/3, 1/2, and 1/2. The smallest integer numbers of this ratio form the Miller indices (233)

Planes are denoted as (hkl) with parentheses. The (outward) normal direction is denoted with $[hkl]$ (square brackets). A set of equivalent planes is denoted with curly brackets as $\{hkl\}$. For example, in the simple cubic lattice (100), (010), (001), (−1 00), (0−1 0) are (00−1) equivalent and are denoted by $\{100\}$. (−1 00) can also be written as $(\bar{1}00)$. A set of equivalent directions is denoted with $\langle hkl \rangle$.

In a cubic lattice the faces of the cubic unit cell are $\{001\}$, the planes perpendicular to the area diagonals are $\{110\}$ and the planes perpendicular to the body diagonals are $\{111\}$ (Fig. 3.29). In the zincblende lattice the $\{111\}$ planes consist of diatomic planes with Zn and S atoms. It depends on the direction whether the metal or the nonmetal is on top. These two cases are denoted by A and B. We follow the convention that the (111) plane is (111)A and the metal is on top (as in Fig. 3.16b). For each change of sign the type changes from A to B and vice versa, e.g. (111)A, (1−1 0)B, (−1 − 11)A and (−1 − 1 − 1)B.

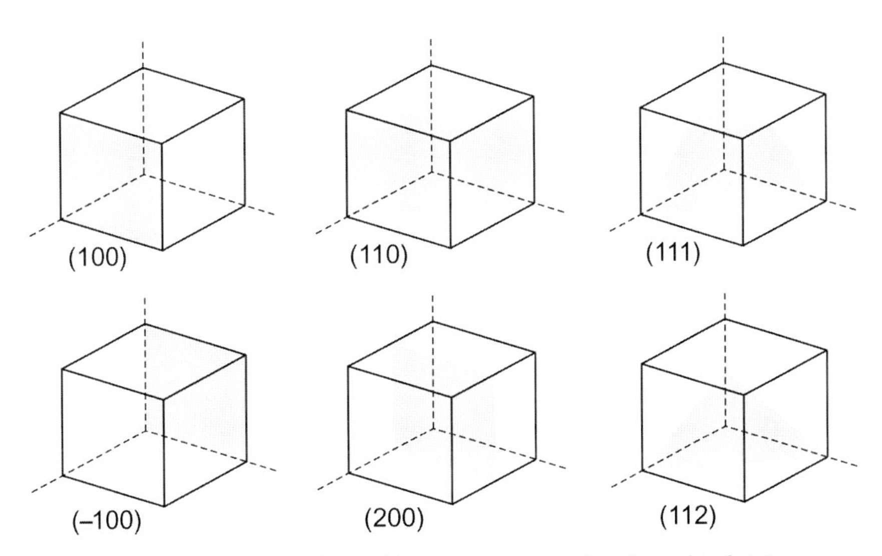

Fig. 3.29. Miller indices of important planes for the cubic lattice

In the wurtzite lattice the Miller indices are denoted as $[hklm]$ (Fig. 3.30). Within the (0001) plane three indices hkl are used that are related to the three vectors $\mathbf{a}_1$, $\mathbf{a}_2$ and $\mathbf{a}_3$ (see Fig. 3.30a) rotated with respect to each other by 120°. Of course, the four indices are not independent and $l = -(h+k)$. The third (redundant) index can therefore be denoted as a dot. The c-axis [0001] is then denoted as [00.1]. Wurtzite (and trigonal, e.g. sapphire) substrates are available typically with a ([11.0]), r ([01.2]), m ([01.0]) and c ([00.1]) surface orientations (Fig. 3.30b).

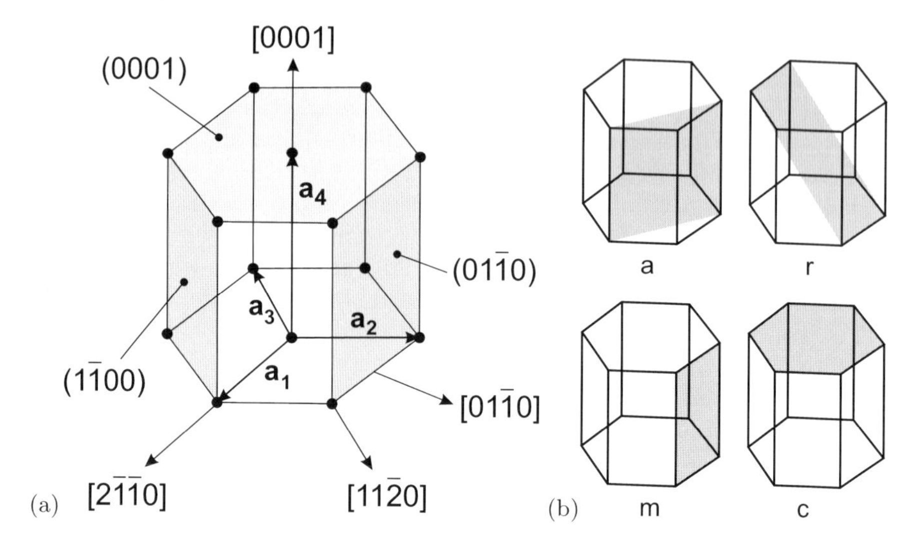

Fig. 3.30. (**a**) Miller indices for the wurtzite (or hcp) structure. (**b**) Orientation of the a-, r-, m-, and c-plane in the wurtzite structure

3.6.3 Brillouin Zone

The Wigner–Seitz cell in reciprocal space is called the (first) Brillouin zone. In Fig. 3.31, the Brillouin zones for the most important lattices are shown. Certain points in the Brillouin zone are labeled with dedicated letters. The Γ point always denotes $\mathbf{k} = 0$ (zone center). Certain paths in the Brillouin zone are labeled with dedicated Greek symbols.

In the Brillouin zone of the fcc lattice (Si, Ge, GaAs, ...) the X point denotes the point at the zone boundary in [001] direction (at a distance π/a from Γ), K for [110] (at a distance $(3\sqrt{2}/4)\pi/a$) and L for the [111] direction (at a distance $\sqrt{3/4}\pi/a \approx 0.87\pi/a$). The straight paths from Γ to X, K, and L are denoted as Δ, Σ, and Λ, respectively.

3.7 Alloys

When different semiconductors are mixed various cases can occur:

- The semiconductors are not miscible and have a so-called miscibility gap. They will tend to form clusters that build up the crystal. The formation of defects is probable.
- They form an ordered (periodic) structure that is called a superlattice.
- They form a random alloy.

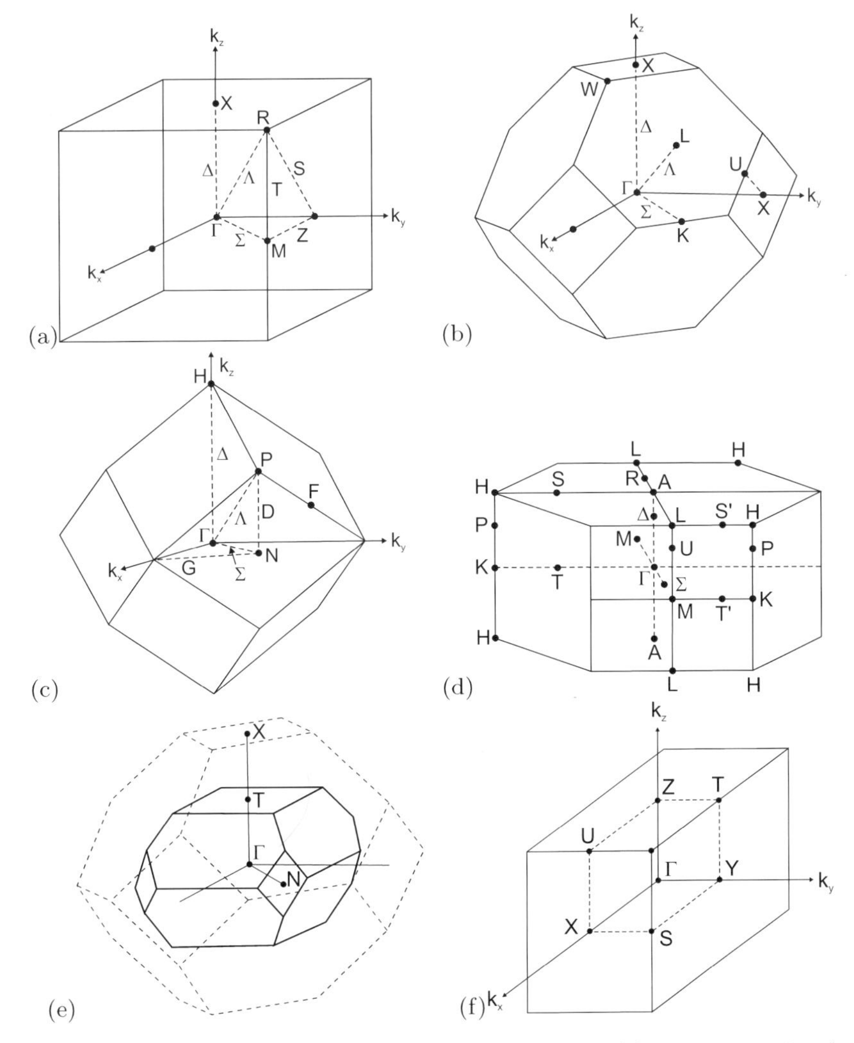

Fig. 3.31. Brillouin zones and special **k** points for the (**a**) primitive cubic (pc), (**b**) fcc, (**c**) bcc, and (**d**) hcp lattice. (**e**) Brillouin zone for chalcopyrite structure with fcc Brillouin zone shown as *dashed outline.* (**f**) Brillouin zone for orthorhombic lattice with one quadrant shown with *dashed lines.*

3.7.1 Random Alloys

Alloys for which the probability to find an atom at a given lattice site is given by the fraction of these atoms (i.e. the stoichiometry) are called *random* alloys. Deviations from the random population of sites is termed *clustering*.

For a Ge_xSi_{1-x} alloy this means that any given atom site has the probability x to have a Ge atom and $1-x$ to have a Si atom. The probability p_n that a Si atom has n next-neighbor Ge atoms is

$$p_n = \binom{4}{n} x^n (1-x)^{4-n} , \tag{3.18}$$

and is depicted in Fig. 3.32 as a function of the alloy composition. The symmetry of the Si atom is listed in Table 3.7 for these five cases. If it is surrounded by four of the same atoms (either Ge or Si), the symmetry is T_d. If one atom is different from the other three next neighbors, the symmetry is reduced to C_{3v} since one bond is singled out. For two atoms each the symmetry is lowest (C_{2v}).

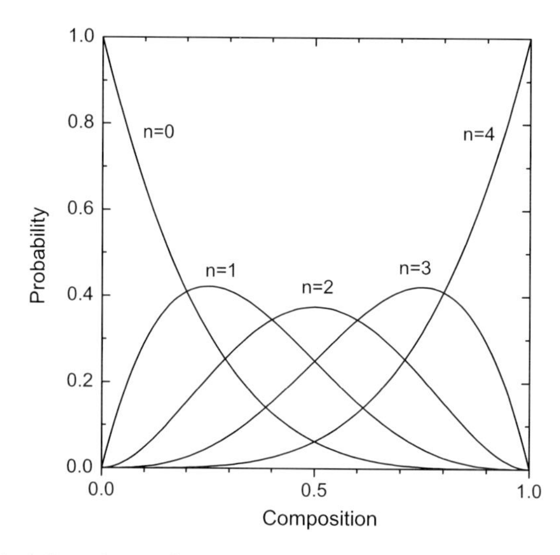

Fig. 3.32. Probability that a Si atom has n next-neighbor Ge atoms in a random Ge_xSi_{1-x} alloy

In an alloy from binary compound semiconductors such as $Al_xGa_{1-x}As$ the mixing of the Al and Ga metal atoms occurs only on the metal (fcc) sublattice. Each As atom is bonded to four metal atoms. The probability

Table 3.7. Probability p (3.18) and symmetry of an A atom being surrounded by n B atoms in a tetrahedrally configured B_xA_{1-x} random alloy

n	p	symmetry
0	p^4	T_d
1	$4p^3(1-p)$	C_{3v}
2	$6p^2(1-p)^2$	C_{2v}
3	$4p(1-p)p^3$	C_{3v}
4	$(1-p)^4$	T_d

that it is surrounded by n Al atoms is given by (3.18). The local symmetry of the As atom is also given by Table 3.7. For $AlAs_xP_{1-x}$ the mixing occurs on the nonmetal (anion) sublattice. If the alloy contains three atom species it is called a *ternary* alloy. If the binary end components have different crystal structure, the alloy shows a transition (or transition range) from one structure to the other at a particular concentration. An example is the alloy between wurtzite ZnO and rocksalt MgO. The $Mg_xZn_{1-x}O$ alloy exhibits wurtzite structure up to about $x = 0.5$ and rocksalt structure for $x > 0.6$ [64].

If the alloy contains four atom species it is called a *quaternary*. A quaternary zincblende alloy can have the mixing of three atom species on one sublattice, such as $Al_xGa_yIn_{1-x-y}As$ or $GaAs_xP_ySb_{1-x-y}$ or the mixing of two atom species on both of the two sublattices, such as $In_xGa_{1-x}As_yN_{1-y}$.

The random placement of different atoms on the (sub)lattice in an alloy represents a perturbation of the ideal lattice and causes additional scattering (alloy scattering). In the context of cluster formation, the probability of an atom having a direct neighbor of the same kind on its sublattice is important. Given a $A_xB_{1-x}C$ alloy, the probability p_S to find a single A atom surrounded by B atoms is given by (3.19a). The probability p_{D^1} to find a cluster of two neighbored A atoms surrounded by B atoms is given by (3.19b).

$$p_S = (1-x)^{12} \tag{3.19a}$$
$$p_{D^1} = 12x(1-x)^{18} \,. \tag{3.19b}$$

These formulas are valid for fcc and hcp lattices. For larger clusters [65, 66], probabilities in fcc and hcp structures differ.

3.7.2 Phase Diagram

The mixture A_xB_{1-x} with average composition x between two materials A and B can result in a single phase (alloy), a two-phase system (phase separation) or a metastable system. The molar free enthalpy ΔG of the mixed system is approximated by

$$\Delta G = \Omega x(1-x) + kT\left[x\ln(x) + (1-x)\ln(1-x)\right] \,. \tag{3.20}$$

The first term on the right-hand side of (3.20) is the (regular solution) enthalpy of mixing with the interaction parameter Ω, which can depend on x. The second term is the ideal configurational entropy based on a random distribution of the atoms. The function is shown for various ratios of kT/Ω in Fig. 3.33a. In an equilibrium phase diagram (see Fig. 3.33b) the system is above the binodal curve in one phase (miscible). On the binodal line $T_b(x)$ in the (x, T) diagram the A- and B-rich disordered phases have equal chemical potentials, i.e. $\partial G/\partial x = 0$. For Ω independent of x the temperature T_b is given by (3.21a). A critical point is at the maximum temperature T_{mg} and concentration x_{mg} of the miscibility gap. For Ω independent of x it is given by

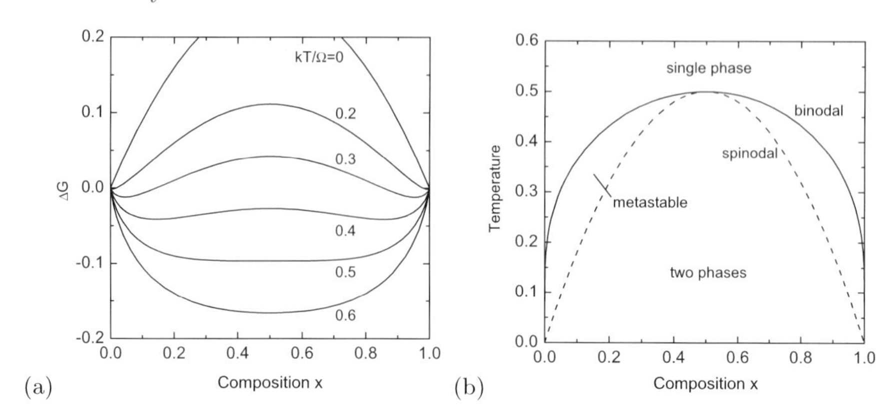

Fig. 3.33. (**a**) Free enthalpy ΔG of mixed binary system (3.20) in units of Ω for Ω =const. and various values of kT/Ω as labeled. (**b**) Schematic phase diagram for binary mixture. The temperature is given in units of Ω/k. The *solid* (*dashed*) denotes the binodal (spinodal) line

$T_{\mathrm{mg}} = \Omega/2$ and $x_{\mathrm{mg}} = 1/2$. In the region below the spinodal boundary, the system is immiscible and phases immediately segregate (by spinodal decomposition). On the spinodal line $T_{\mathrm{sp}}(x)$ the condition $\partial^2 G/\partial x^2 = 0$ is fulfilled. For Ω independent of x the temperature T_{sp} is given by (3.21b). The region between the binodal and spinodal curves is the metastable region, i.e. the system is stable to small fluctuations of concentration or temperature but not for larger ones.

$$kT_{\mathrm{b}}(x) = \Omega \frac{2x - 1}{\ln(x) - \ln(1 - x)} \tag{3.21a}$$

$$kT_{\mathrm{sp}}(x) = 2\Omega x(1 - x) \; . \tag{3.21b}$$

In Fig. 3.34 calculated diagrams for GaAs-AlAs and GaAs-GaP [62] are shown. The arrows denote the critical point. These parameters and the inter-

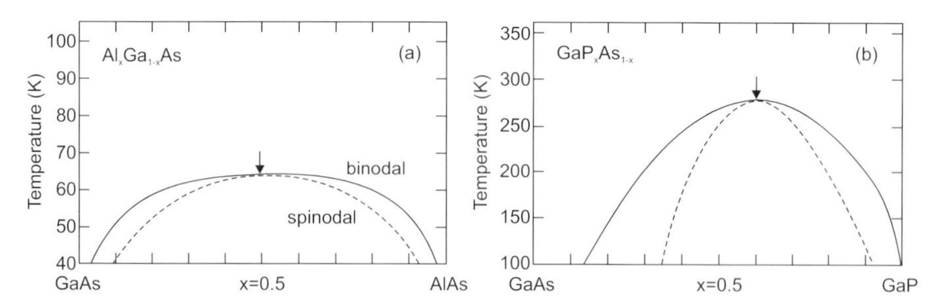

Fig. 3.34. Calculated phase diagrams for (**a**) $\mathrm{Al}_x\mathrm{Ga}_{1-x}\mathrm{As}$ and (**b**) $\mathrm{GaP}_x\mathrm{As}_{1-x}$. The binodal (spinodal) curve is shown as *solid* (*dashed*) line. Adapted from [62]

action parameters for a number of ternary alloys are given in Table 3.8. For example, for $Al_xGa_{1-x}As$ complete miscibility is possible for typical growth temperatures (> 700 K), but for $In_xGa_{1-x}N$ the In solubility at a typical growth temperature of 1100 K is only 6% [63].

Table 3.8. Calculated interaction parameter $\Omega(x)$ (at $T = 800$ K, 1 kcal/mol= 43.39 meV), miscibility-gap temperature T_{mg} and concentration x_{mg} for various ternary alloys. Data for InGaN from [63], other data from [62]

alloy	T_{mg} (K)	x_{mg}	$\Omega(0)$ kcal/mol	$\Omega(0.5)$ kcal/mol	$\Omega(1)$ kcal/mol
$Al_xGa_{1-x}As$	64	0.51	0.30	0.30	0.30
GaP_xAs_{1-x}	277	0.603	0.53	0.86	1.07
$Ga_xIn_{1-x}P$	961	0.676	2.92	3.07	4.60
$GaSb_xAs_{1-x}$	1080	0.405	4.51	3.96	3.78
$Hg_xCd_{1-x}Te$	84	0.40	0.45	0.80	0.31
$Zn_xHg_{1-x}Te$	455	0.56	2.13	1.88	2.15
$Zn_xCd_{1-x}As$	605	0.623	2.24	2.29	2.87
$In_xGa_{1-x}N$	1505	0.50	6.32	5.98	5.63

3.7.3 Virtual Crystal Approximation

In the *virtual crystal approximation* (VCA) the disordered alloy AB_xC_{1-x} is replaced by an ordered binary compound AD with D being a 'pseudoatom' with properties that are configuration averaged over the properties of the B and C atoms, e.g. their masses or charges. Such an average is weighted with the ternary composition, e.g. the mass is $M_D = xM_B + (1-x)M_C$. For example, the A–D force constant would be taken as the weighted average over the A–B and A–C force constants.

3.7.4 Lattice Parameter

In the VCA for an alloy a new sort of effective atom is assumed that has an averaged bond length that depends linearly on the composition. Typically, Vegard's law (3.22), which predicts that the lattice constant of a ternary alloy $A_xB_{1-x}C$ depends linearly on the lattice constants of the binary alloys AC and BC, is indeed fulfilled

$$a_0(A_xB_{1-x}C) = a_0(BC) + x[a_0(AC) - a_0(BC)] \, . \quad (3.22)$$

In reality, the bond length of the AC and BC bonds changes rather little (Fig. 3.35a) such that the atoms in the alloy suffer a displacement from

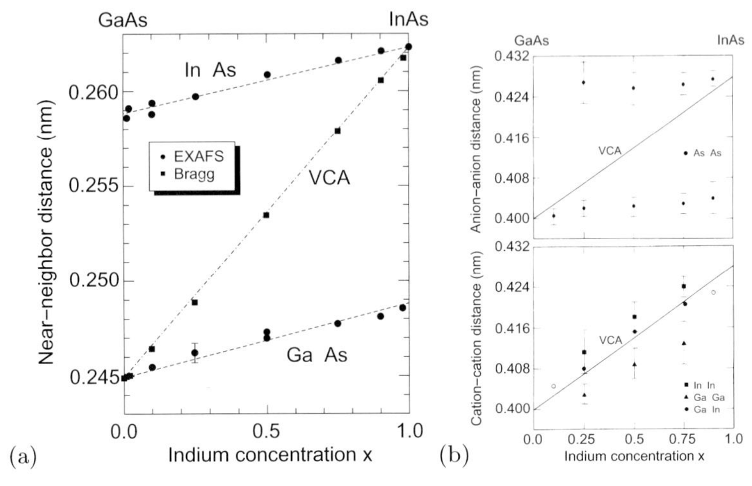

Fig. 3.35. (**a**) Near-neighbor distance ($\sqrt{3}a_0/4$) of $In_xGa_{1-x}As$ as measured by standard X-ray diffraction (Bragg reflection, *solid squares*) and VCA approximation (*dash-dotted line*). Near-neighbor Ga–As and In–As distances as determined by EXAFS (extended X-ray absorption fine structure, *solid squares*). *Dashed lines* are guides to the eye. Data from [67]. (**b**) Second-neighbor distances for $In_xGa_{1-x}As$ as determined from EXAFS, *top*: anion–anion distance (for As–As), *bottom*: cation–cation distance (for In–In, Ga–Ga, and Ga–In). *Solid lines* in both plots are the VCA ($a_0/\sqrt{2}$). Data from [68]

their average position and the lattice is deformed on the nanoscopic scale. In a lattice of the type $In_xGa_{1-x}As$ the anions suffer the largest displacement since their position adjusts to the local cation environment. For $In_xGa_{1-x}As$ a bimodal distribution, according to the As–Ga–As and As–In–As configura-

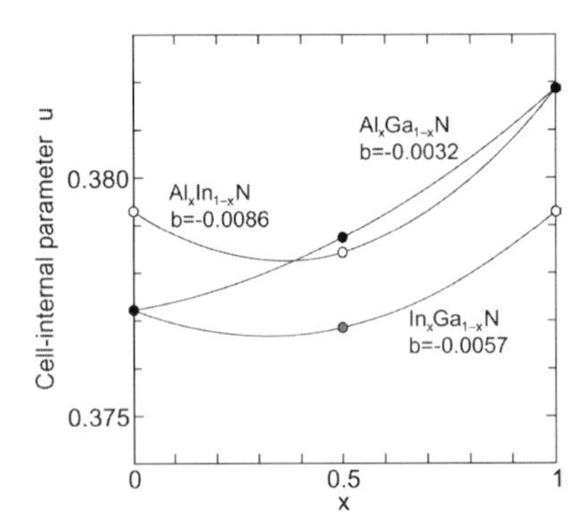

Fig. 3.36. Theoretical values ($T = 0$ K) for the cell-internal parameter u as a function of the composition for group-III nitride alloys. The *solid lines* are quadratic curves (bowing parameter b is shown) through the points for $x = 0$, 0.5, and 1.0. Data from [69]

tions, is observed (Fig. 3.35b). The cation–cation second-neighbor distances are fairly close to the VCA.

While the average lattice parameter in alloys changes linearly with composition, the cell-internal parameter u (for wurtzite structures, see Sect. 3.4.5) exhibits a nonlinear behavior as shown in Fig. 3.36. Therefore physical properties connected to u, such as the spontaneous polarization, will exhibit a bowing.

3.7.5 Ordering

Some alloys have the tendency for the formation of a superstructure [70]. Growth kinetics at the surface can lead to specific adatom incorporation leading to ordering. For example, in $In_{0.5}Ga_{0.5}P$ the In and Ga atoms can be ordered in subsequent (111) planes (CuPt structure) instead of being randomly mixed (Fig. 3.37). This impacts fundamental properties such as the phonon spectrum or the bandgap. CuPt ordering on (111) and $(\bar{1}\bar{1}1)$ planes is called $CuPt_A$, on $(\bar{1}11)$ and $(1\bar{1}1)$ planes $CuPt_B$ ordering. In Fig. 3.38, a TEM image of a $Cd_{0.68}Zn_{0.32}Te$ epilayer is shown with simultaneous ordering in the CuPt structure (doublet periodicity along $[1\bar{1}1]$ and $[\bar{1}11]$) and in the CuAu-I structure[4] (doublet periodicity along [001] and $[\bar{1}10]$).

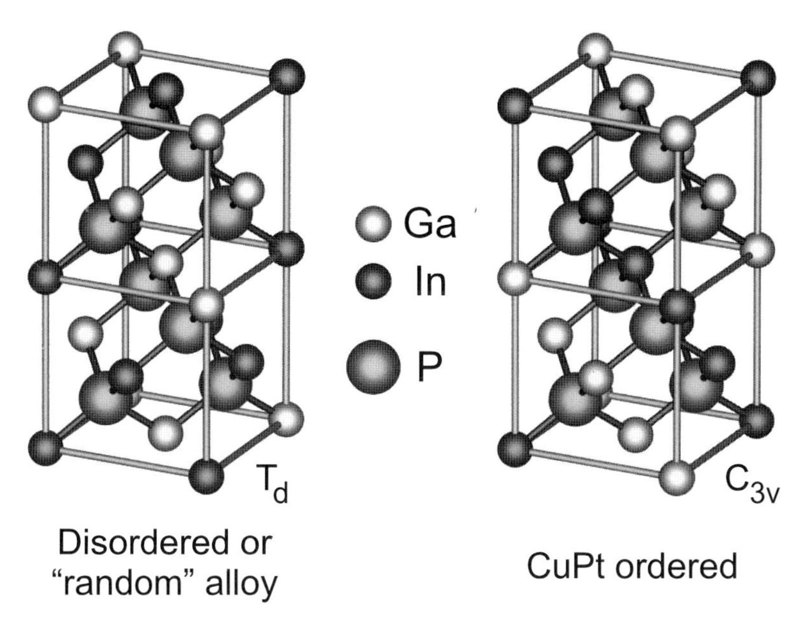

Fig. 3.37. CuPt-ordered ternary alloy $In_{0.5}Ga_{0.5}P$; the lattice symmetry is reduced from T_d to C_{3v}

[4]The CuAu-I structure has tetragonal symmetry. There exists also the CuAu-II structure that is orthorhombic.

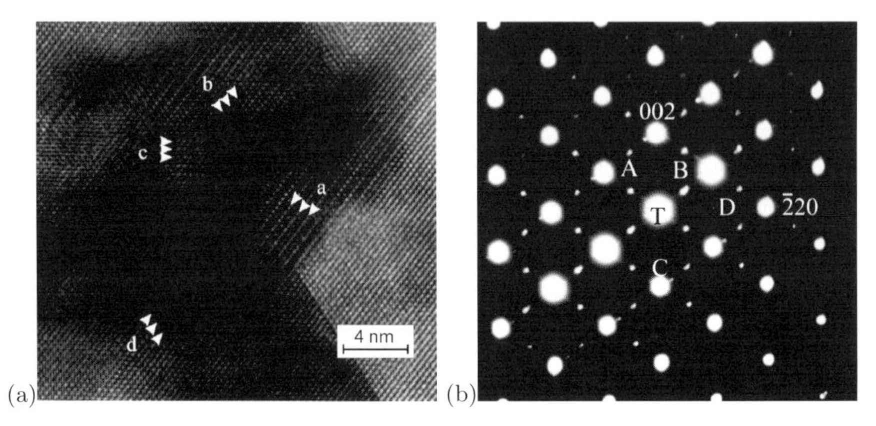

Fig. 3.38. (**a**) Cross-sectional transmission electron microscope image along the [110] zone axis of a $Cd_{0.68}Zn_{0.32}Te$ epilayer on GaAs showing ordered domains having a doublet periodicity on the {111} and {001} lattice planes. Two different {111} variants are labeled 'a' and 'b'. The doublet periodicity in the [001] is seen in the 'c' region. (**b**) Selected-area diffraction pattern along the [110] zone. Strong peaks are fundamental peaks of the zincblende crystal, weak peaks are due to CuPt ordering, labeled A and B, and CuAu-I ordering, labeled C and D. The latter are the weakest due to a small volume fraction of CuAu-ordered domains. Reprinted with permission from [71], ©2001 AIP

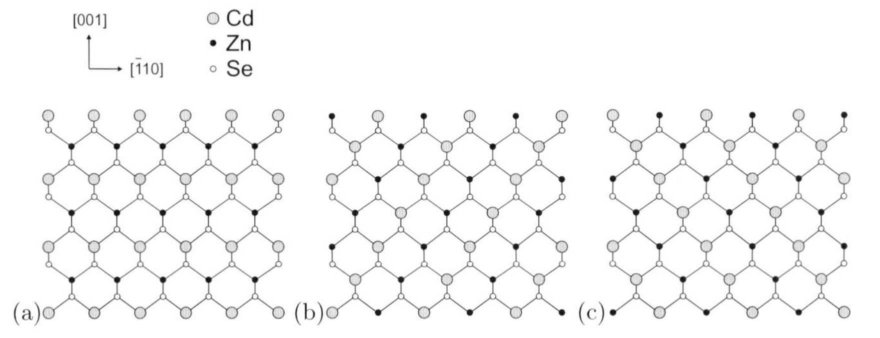

Fig. 3.39. Schematic diagrams of zincblende $Cd_xZn_{1-x}Te$ along [110] with (**a**) CuAu-I type ordering and (**b,c**) two types of the $CuPt_B$ type ordering. Doublet periodicity is along (**a**) [001] and [$\bar{1}$10], (**b**) [1$\bar{1}$1] and (**c**) [$\bar{1}$11]. Reprinted with permission from [71], ©2001 AIP

4 Defects

4.1 Introduction

In an ideal lattice each atom is at its designated position. Deviations from the ideal structure are called defects. In the following, we will briefly discuss the most common defects. The electrical activity of defects will be discussed in Sects. 7.5 and 7.7. For the creation of a defect a certain free energy G_{D} is necessary. At thermodynamical equilibrium a (point) defect density proportional to $e^{-G_{\mathrm{D}}/kT}$ will always be present (see Sect. 4.3).

4.2 Point Defects

Point defects are deviations from the ideal structure involving essentially only one lattice point. The simplest point defect is a vacancy, a missing atom at an atomic position. If an atom is at a position that does not belong to the crystal structure an interstitial (or Frenkel defect) is formed.

The formation energy for line or area defects scales with $N^{1/3}$ and $N^{2/3}$, respectively, N being the number of atoms in the crystal. Therefore, these defects are not expected in thermodynamic equilibrium. However, the path into thermodynamical equilibrium might be so slow that these defects are metastable and must be considered quasi-frozen. There may also exist metastable point defects. By annealing the crystal, the thermodynamic equilibrium concentration might be re-established.

If an atom site is populated with an atom of different order number, an *impurity* is present. If the number of valence electrons is the same as for the original (or correct) atom, then it is an *isovalent* impurity and quasi fits into the bonding scheme. If the valence is different, the impurity adds extra (negative or positive) charge to the crystal bonds, which is compensated by the extra, locally fixed charge in the nucleus. This mechanism will be discussed in detail in the context of doping (Chap. 7). If in an AB compound an A atom sits on the B site, the defect is called an *antisite* defect A_B.

A point defect is typically accompanied by a relaxation of the surrounding host atoms. As an example, we discuss the vacancy in Si (Fig. 4.1a). The missing atom leads to a lattice relaxation with the next neighbors moving some way into the void (Fig. 4.1b). The bond lengths of the next and second-next neighbor Si atoms around the neutral vacancy are shown in Fig. 4.1c. The lattice relaxation depends on the charge state of the point defect (Jahn–Teller effect) that is discussed in more detail in Sect. 7.7. In Fig. 4.1d the situation for the positively charged vacancy with one electron missing is shown. One of the two bonds is weakened since it lacks an electron. The distortion is therefore different from that for V^0.

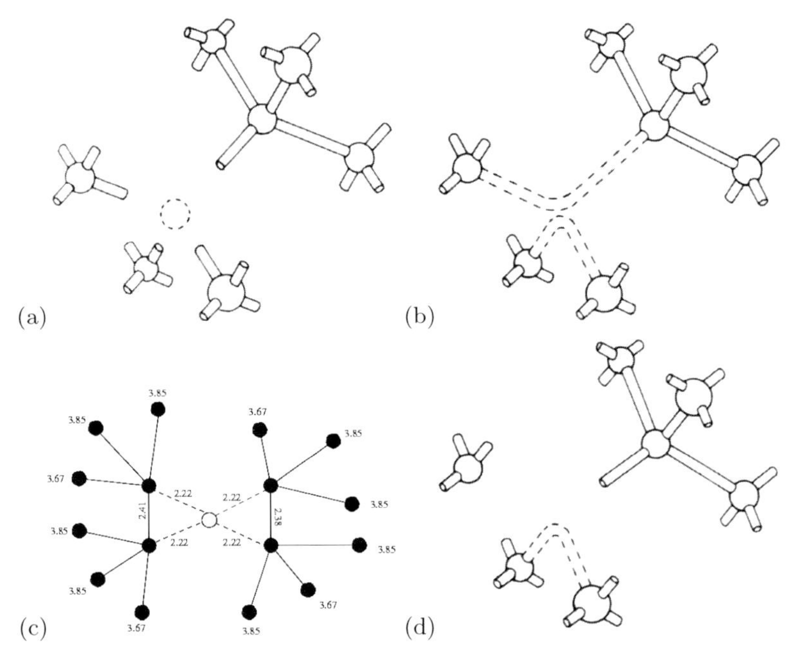

Fig. 4.1. (**a**) Schematic diamond lattice with vacancy, i.e. a missing Si atom without relaxation. (**b**) Si with neutral vacancy (V^0), lattice relaxation and formation of two new bonds. (**c**) Schematic diagram showing the relaxation of the neighbors around the neutral Si vacancy defect site calculated by an ab initio method. The distances (in Å) on the outer shell of atoms show the distance from the vacant site. The lengths of the two new bonds are also indicated. The bond length in bulk Si is 2.352 Å, the second-neighbor distance 3.840 Å. (**d**) Si unit cell with positively charged vacancy (V^+). Parts (**a**,**b**,**d**) reprinted with permission from [72]

As an example, we depict in Fig. 4.2a the lowest-energy configuration of a boron-related defect in silicon. It is B_s–Si_i^T, i.e. boron on a substitutional site and a self-interstitial Si on the 'T' place with highest symmetry[1] (cf. Fig. 3.17). Due to its importance as an acceptor in Si, the configuration and diffusion of B in Si has found great interest [73, 74]. The diffusion of boron

[1]The positive charge state is stable, the neutral charge state is metastable since the defect is a negative-U center (see Sect. 7.7.3).

has been suggested [74] to occur via the following route: The boron leaves its substitutional site and goes to the hexagonal site (‘H’) (Fig. 4.2b) with an activation energy of about 1 eV. It can then relax (~ 0.1 eV) without barrier to the tetrahedral ‘T’ position (Fig. 4.2c). It could then diffuse through the crystal by going from ‘H’ to ‘T’ to ‘H’ and so on. However, long-range diffusion seems to be not possible in this way because the kick-in mechanism will bring back the boron to its stable configuration. The pair diffusion mechanism B_s–$Si_i^T \rightarrow B_i^H \rightarrow B_s$–$Si_i^T$ provides an activation energy of about 1 eV.

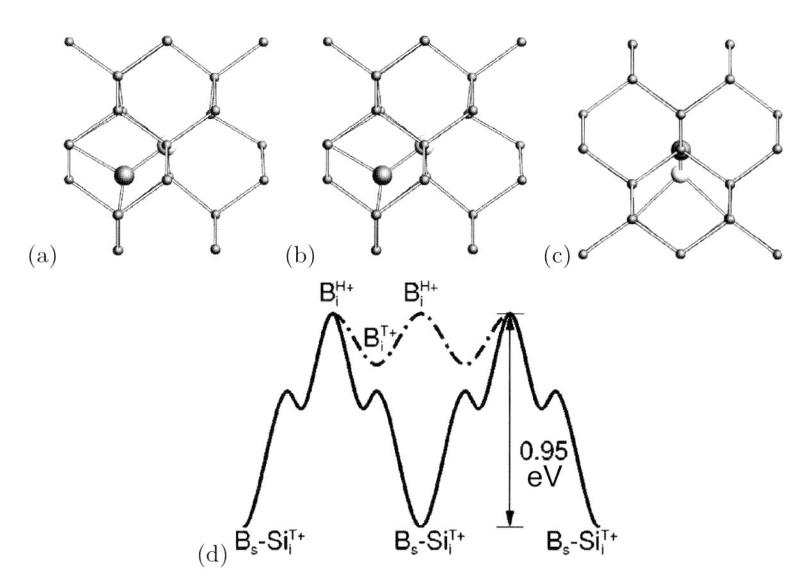

Fig. 4.2. Configurations of boron in Si. (**a**) Substitutional boron and Si self-interstitial at ‘T’ site. (**b**) Boron at ‘H’ and (**c**) boron at ‘T’ site, each with the Si atoms on the Si lattice sites. The *large bright ball* represents the boron atom, *large* and *small dark balls* represent Si atoms. (**d**) Diffusion of the positively charged B–Si state, configuration vs. total energy, with two pathways, kick-out (*dashed-dotted line*) and pair diffusion (*solid line*); the activation energy is labeled. Reprinted with permission from [74], ©2001 APS

4.3 Thermodynamics of Defects

For a given temperature the free enthalpy G of a crystal (a closed system with regard to particle exchange)

$$G = H - TS \tag{4.1}$$

is minimum. H is the enthalpy and S the entropy. The enthalpy $H = E + pV$ is the thermodynamic potential for a system whose only external parameter is the volume V. It is used when the independent variables of the system are the entropy S and pressure p. The free enthalpy is used when the independent parameters are T and p. G_0 (H_0) is the free energy (enthalpy) of the perfect crystal. H_f is the formation enthalpy of an isolated defect. This could be, e.g., the enthalpy of a vacancy, created by bringing an atom from the (later) vacancy site to the surface, or an interstitial, created by bringing an atom from the surface to the interstitial site. In the limit that the n defects do not interact with each other, i.e. their concentration is sufficiently small, they can be considered independent and the enthalpy is given by

$$H = H_0 + nH_\mathrm{f} \; . \tag{4.2}$$

The increase of entropy due to increased disorder is split into the configurational disorder over the possible sites, denoted as S_d, and the formation entropy S_f due to localized vibrational modes. The total change ΔG of the free energy is

$$\Delta G = G - G_0 = n(H_\mathrm{f} - TS_\mathrm{f}) - TS_\mathrm{d} = nG_\mathrm{f} - TS_\mathrm{d} \; , \tag{4.3}$$

where $G_\mathrm{f} = H_\mathrm{f} - TS_\mathrm{f}$ denotes the free enthalpy of formation of a single isolated defect. In Table 4.1 experimental values for the formation entropy and enthalpy are given for several defects. The defect concentration is obtained by minimizing ΔG, i.e.

$$\frac{\partial \Delta G}{\partial n} = G_\mathrm{f} - T\frac{\partial S_\mathrm{d}}{\partial n} = 0 \; . \tag{4.4}$$

Table 4.1. Formation enthalpy H_f and entropy S_f of interstitials (I) and vacancy (V) in Si and the Ga vacancy in GaAs. Data for Si from [75], for GaAs from [76]

Material	defect	H_f (eV)	S_f (k_B)
Si	I	3.2	4.1
Si	V	2.0	~ 1
GaAs	V_Ga	3.2	9.6

The entropy S_d due to disorder is given as

$$S_\mathrm{d} = k_\mathrm{B} \ln W \; , \tag{4.5}$$

where W is the complexion number, usually the number of distinguishable ways to distribute n defects on N lattice sites

$$W = \binom{N}{n} = \frac{N!}{n!(N-n)!} \; . \tag{4.6}$$

With Stirling's formula $\ln x! \approx x(\ln x - 1)$ for large x we obtain

$$\frac{\partial S_{\mathrm{d}}}{\partial n} = k_{\mathrm{B}} \left[\frac{N}{n} \ln \left(\frac{N}{N-n} \right) + \ln \left(\frac{N-n}{n} \right) \right] . \tag{4.7}$$

If $n \ll N$ and $\partial N/\partial n = 0$, the right side of (4.7) reduces to $k_{\mathrm{B}} \ln(N/n)$. The condition (4.4) reads $G_{\mathrm{f}} + k_{\mathrm{B}} T \ln(n/N)$, or

$$\frac{n}{N} = \exp \left(- \frac{G_{\mathrm{f}}}{k_{\mathrm{B}} T} \right) . \tag{4.8}$$

In the case of several different defects i with a degeneracy Z_i, e.g. a spin degree of freedom or several equivalent configurations, (4.8) can be generalized to

$$\frac{n_i}{Z_i N} = \exp \left(- \frac{G_{\mathrm{f}}^i}{k_{\mathrm{B}} T} \right) . \tag{4.9}$$

4.4 Dislocations

Dislocations are line defects along which the crystal lattice is shifted. The vector along the dislocation line is called line vector $\mathbf{L}$. A closed path around the dislocation core differs from that in an ideal crystal. The difference vector is called the Burger's vector $\mathbf{b}$.

Since the energy of a dislocation is proportional to $\mathbf{b}^2$, only dislocations with the shortest Burger's vector are stable. The plane spanned by $\mathbf{L}$ and $\mathbf{b}$ is called the *glide* plane. In Fig. 4.3 a high-resolution image of the atoms around a dislocation and the phase and amplitude of the (111) reflection are shown. The phase corresponds to the atomic columns, the amplitude to the displacement of the atoms at the dislocation core (see also Fig. 5.18).

For an *edge* dislocation (Fig. 4.4a) $\mathbf{b}$ and $\mathbf{L}$ are perpendicular to each other. An extra half-plane spanned by $\mathbf{L}$ and $\mathbf{b} \times \mathbf{L}$ is inserted.

For a *screw* dislocation (Fig. 4.4b) $\mathbf{b}$ and $\mathbf{L}$ are collinear. The solid has been cut along a half-plane up to the dislocation line, shifted along $\mathbf{L}$ by the amount $\mathbf{b}$ and reattached.

The most important dislocations in the zincblende lattice (Fig. 4.5) have the line vector along $\langle 110 \rangle$. With the Burger's vector $1/2a\,\langle 110 \rangle$ three different types of dislocations can be formed: edge, screw and 60° dislocations. The vicinity of the core of the latter is shown in more detail in Fig. 4.5d. We note that the atomistic structure of 60° dislocations is different for $\mathbf{L}$ along [110] and [−1 10]; depending on whether the cation or anion are in the core, they are labeled α or β dislocations.

When materials with different lattice constants are grown on top of each other, the strain can plastically relax via the formation of misfit dislocations. A typical network of such dislocations is shown in Fig. 4.6 for InGaAs on InP.

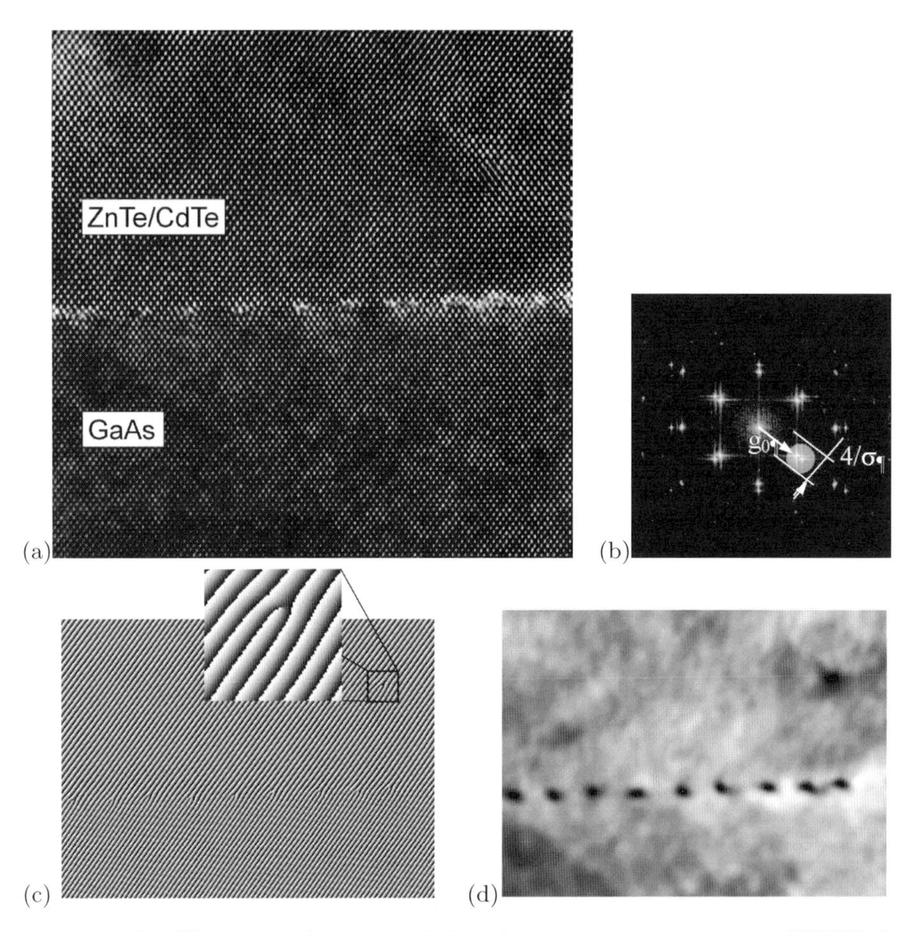

Fig. 4.3. (**a**) High-resolution transmission electron microscopy image (HRTEM) in the ⟨110⟩ projection of a network of misfit dislocations at a GaAs/CdTe/ZnTe interface. Substrate: GaAs (001), 2° off ⟨110⟩, ZnTe buffer layer is 2 monolayers thick. (**b**) Fourier transform with round mask around the (111) Bragg reflection. (**c**) Phase and (**d**) amplitude images for the mask from (**b**). From [77]

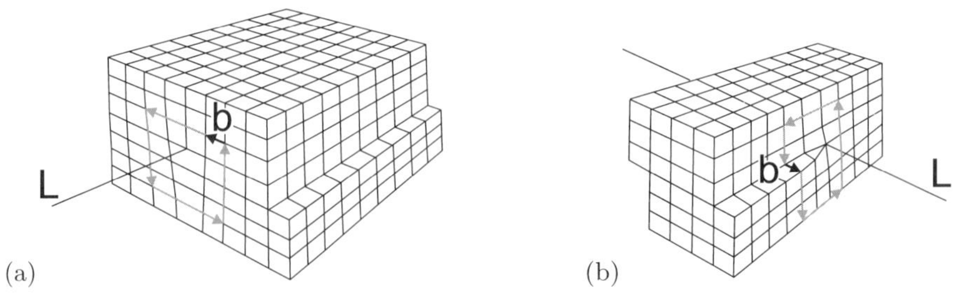

Fig. 4.4. Model of (**a**) an edge and (**b**) a screw dislocation. The Burger's vector **b** is indicated

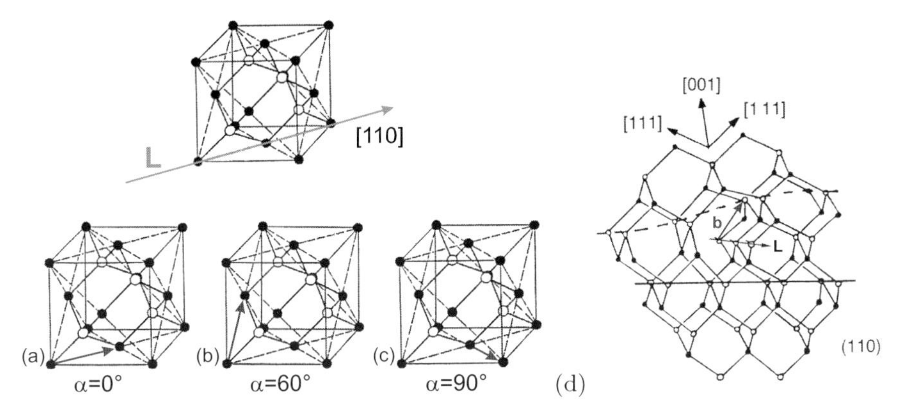

Fig. 4.5. Dislocations in the zincblende structure. The line vector is along [100]. The Burger's vector $1/2a\,\langle 110\rangle$ can create an (**a**) screw dislocation, a (**b**) 60° dislocation, and (**c**) an edge dislocation. (**d**) Atomistic structure of a 60° dislocation

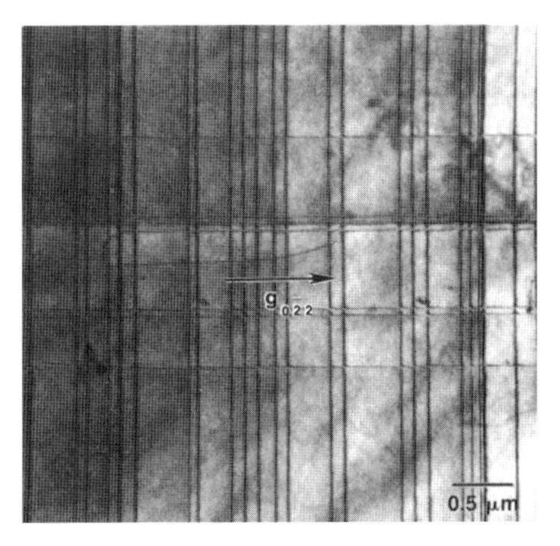

Fig. 4.6. Plan-view transmission electron microscopy image of a network of ⟨110⟩ dislocation lines in InGaAs on InP (001) with a lattice mismatch of about 0.1%. The TEM diffraction vector is $\mathbf{g}$ = [220]. Reprinted with permission from [78], ©1989 AIP

Around the intersection of a screw dislocation with a surface, the epitaxial growth occurs typically in the form of a growth spiral that images the lattice planes around the defect.

It has been found that the addition of impurities can lead to a substantial reduction of the dislocation density. This effect is known as impurity hardening and is caused by a hardening of the lattice due to an increase of the so-called critical resolved shear stress [80]. In Fig. 4.8 the dependence of the dislocation density in GaAs and InP is shown as a function of the carrier density that is induced by the incorporation of (electrically active) group-II

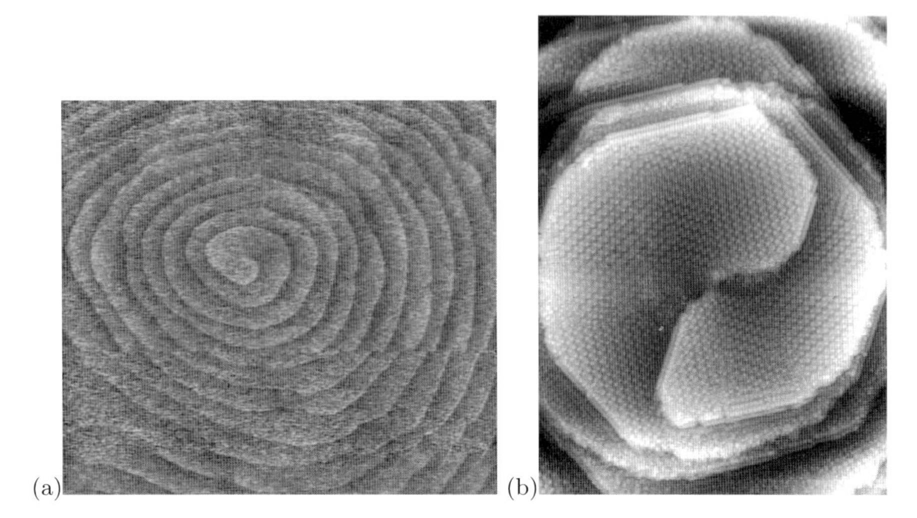

Fig. 4.7. (**a**) Atomic force microscopy image of growth spiral around a screw dislocation on a silicon surface; image width: 4 µm. (**b**) STM image (width: 75 nm) of a screw-type dislocation with a Burgers vector of [000-1] on the N-face. The reconstruction is c(6×12). The c(6×12) row directions correspond to $\langle\bar{1}100\rangle$. Reprinted with permission from [79], ©1998 AVS

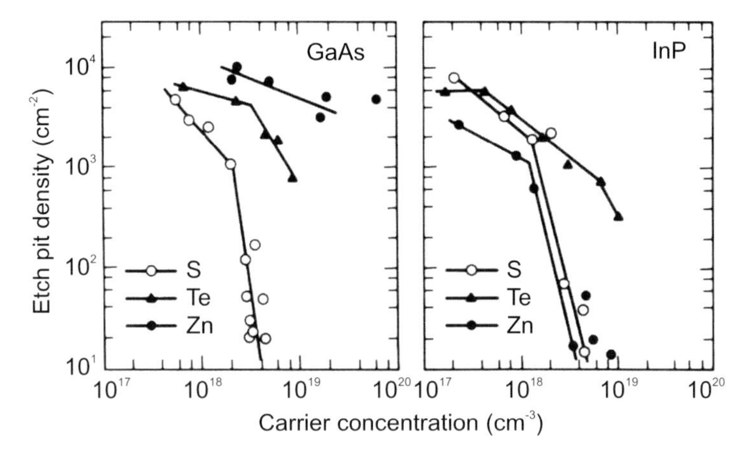

Fig. 4.8. Dislocation density (as revealed by etch pits) for GaAs and InP as a function of the carrier concentration for various concentrations of impurities (S, Se, and Zn). Adapted from [81].

or group-VI atoms (acceptors or donors, cf. Sect. 7.5). The high carrier concentration is unwanted when semi-insulating substrates (cf. Sect. 7.7.6) or low optical absorption (cf. Sect. 9.6) are needed. Thus the incorporation of isovalent impurities, such as In, Ga or Sb in GaAs and Sb, Ga or As in InP, has been investigated and found to be remarkably effective. Material containing such impurities in high concentration ($> 10^{19}\,\mathrm{cm}^{-3}$) must be considered a low-concentration alloy. The lattice constant is thus slightly changed, which can cause problems in the subsequent (lattice-mismatched) epitaxy of pure layers.

4.5 Stacking Faults

The ideal stacking of (111) planes in the zincblende structure, ABCABC..., can be disturbed in various ways and creates area defects. If one plane is missing, i.e. the stacking is ABC*AC*ABC, an *intrinsic* stacking fault is present. If an additional plane is present, the defect is called an *extrinsic* stacking fault, i.e. ABCAB*A*CABC. An extended stacking fault in which the order of stacking is reversed is called a *twin* lamella, e.g. ABCABC*BACB*ABCABC. If two regions have inverted stacking order they are called twins and their joint interface is called a twin boundary, e.g. ...ABCABCAB*C*BACBACBA... The various types of stacking faults are shown in Fig. 4.9. In Fig. 4.10 a cross-sectional image of stacking faults in GaAs on Si is shown. They block each other and thus partially annihilate with increasing thickness.

A stacking fault is bounded by two partial dislocations formed by the dissociation of a perfect dislocation. A perfect dislocation with Burger's vector $a/2[110]$ in a III–V compound is dissociated into two Shockley partials according to [83]

$$\frac{a}{2}[0\,\bar{1}\,\bar{1}] \rightarrow \frac{a}{6}[1\,2\,1] + \frac{a}{6}[\bar{1}\,1\,2]\,. \tag{4.10}$$

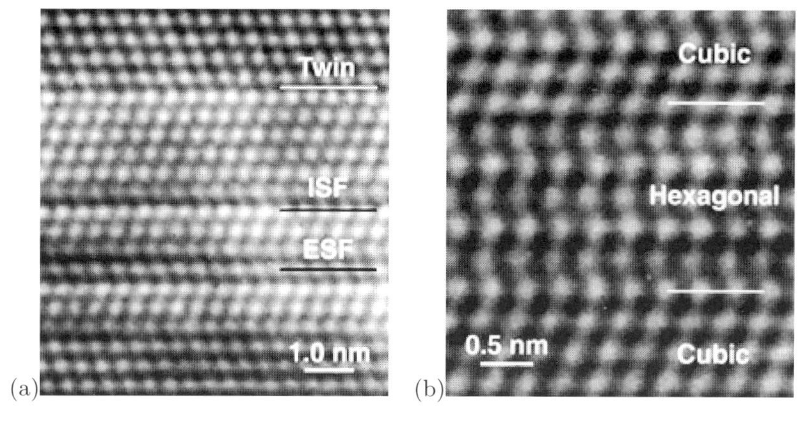

Fig. 4.9. HRTEM images of (**a**) thin-film silicon with intrinsic (labeled 'ISF') and extrinsic ('ESF') stacking faults and twin boundary ('Twin'). (**b**) Six monolayer thick hexagonal (wurtzite) CdTe layer in cubic (zincblende) CdTe. Stacking order (from bottom to top) is: ABCABABABABC... Reprinted with permission from [86]

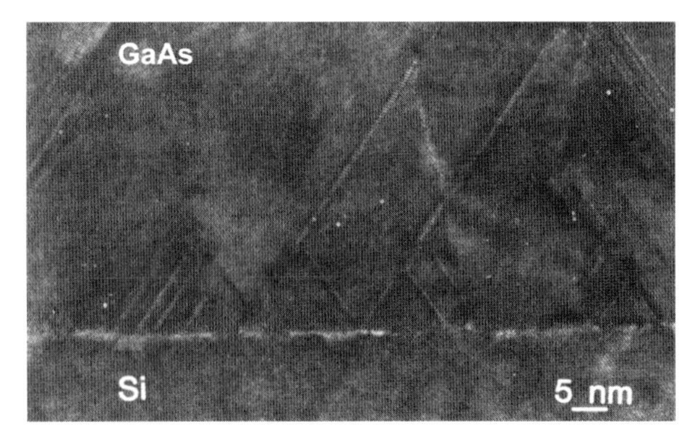

Fig. 4.10. Cross-sectional TEM image showing stacking faults in heteroepitaxial GaAs on Si. Adapted from [82]

Since the dislocation energy is proportional to $|\mathbf{b}|^2$, the dissociation is energetically favored. The stacking-fault energy in undoped GaAs is about $50\,\mathrm{mJ\,m^{-2}}$. It is reduced upon doping, e.g. to $6\,\mathrm{mJ\,m^{-2}}$ for GaAs:Sn with $[\mathrm{Sn}] \approx 10^{17}\,\mathrm{cm^{-3}}$.

4.6 Grain Boundaries

The boundaries of crystal grains are called grain boundaries. Such defects can have a large impact on the electric properties. They can act as barriers for transport or as carrier sinks. Details of their structure and properties can be found in [84, 85]. The two crystal grains meet each other with a relative tilt and/or twist. The situation is shown schematically in Fig. 4.11a for a small angle between the two crystals. A periodic pattern of dislocations forms at the interface that is called a small-angle grain boundary (SAGB) (Fig. 4.11b). In

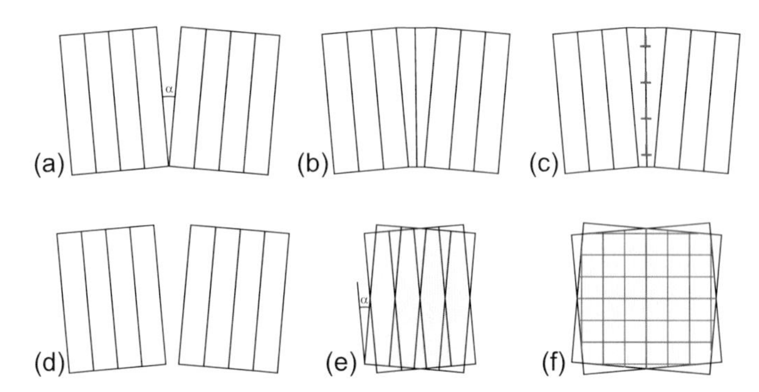

Fig. 4.11. Schemes of (**a**,**b**,**c**) pure tilt and (**d**,**e**,**f**) pure twist boundary, dislocation formation in (**c**) pure tilt and (**f**) twist boundaries.

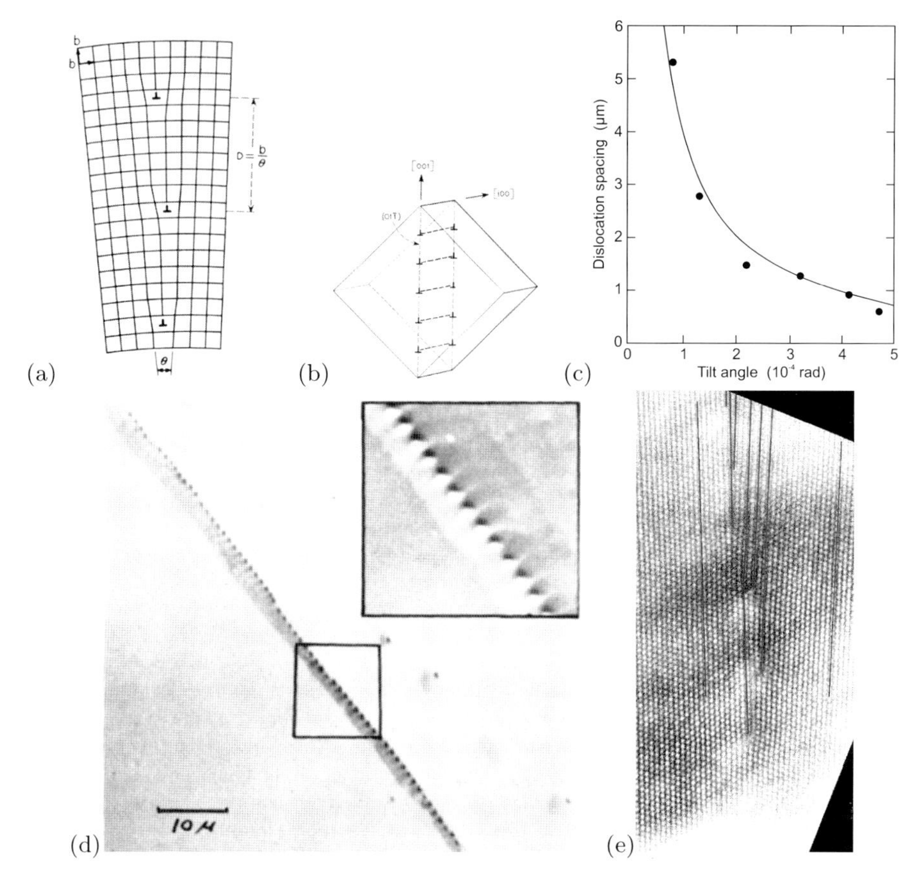

Fig. 4.12. (**a**) Scheme of a small-angle (pure tilt) grain boundary. (**b**) Model of edge dislocations in a {110} plane in Ge. (**c**) Relation of dislocation distance d and tilt angle θ for various small-angle grain boundaries in Ge. *Solid line* is relation $d = 4.0 \times 10^{-8}/\theta$. (**d**) Optical image of an etched (CP–4 etch) Ge sample with a small-angle grain boundary. Reprinted with permission from [87], ©1953 APS. (**e**) HRTEM image of a small-angle grain boundary in Si with dislocations highlighted. From [88]

Fig. 4.12, experimental results for pure tilt SAGB are shown. The dislocation spacing is inversely proportional to the tilt angle θ. An image of a twist SAGB is shown in Fig. 4.13.

4.7 Antiphase and Inversion Domains

Antiphase domains occur when one part of the crystal is shifted with respect to another by an antiphase vector **p**. This does not form a twin. If the polar direction changes between two domains they are called inversion domains.

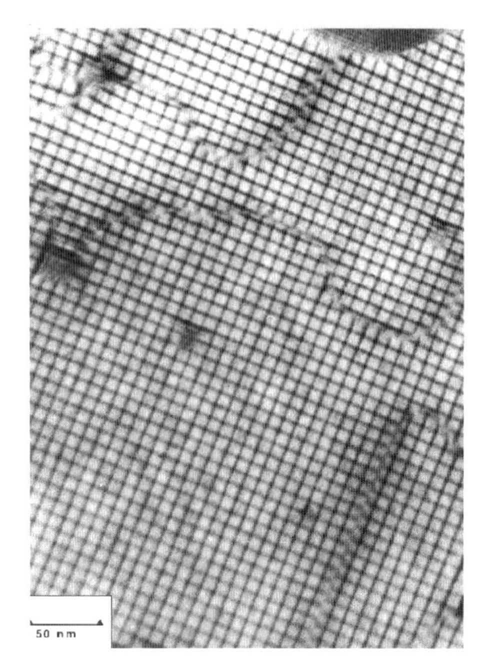

Fig. 4.13. Bright-field TEM image of pure twist boundary with network of pure twist dislocations fabricated by wafer bonding of two Si (001) surfaces with a relative twist. From [89]

In the zincblende structure the [110] and $[\bar{1}10]$ directions are not equivalent. In one case there is a Zn-S lattice and in the other a S-Zn lattice. Both lattices vary by a 90° rotation or an inversion operation (which is not a symmetry operation of the zincblende crystal). If, e.g., a zincblende crystal is grown on a Si surface with monoatomic steps (Fig. 4.14), adjoint regions have a different phase; they are called antiphase domains (APD). The antiphase vector is $(0, 0, 1)a_0/4$. At the boundaries a two-dimensional defect,

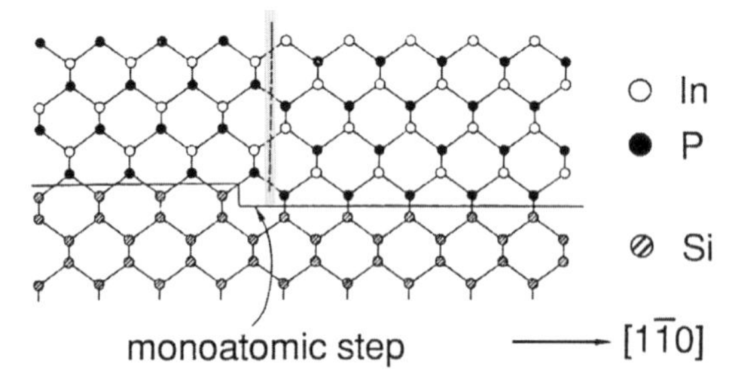

Fig. 4.14. Monoatomic step on the Si (001) surface and subsequent formation of an antiphase boundary in InP (zincblende)

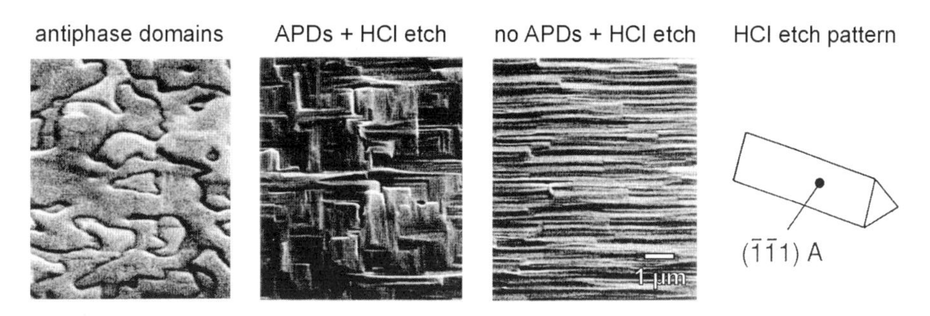

Fig. 4.15. Antiphase domains in InP on Si. HCl etchs InP anisotropically and prepares A planes. The etch patterns of layers with (without) APDs are cross-hatched (linear). Reprinted from [90], ©1991, with permission from Elsevier

an antiphase domain boundary, develops. The APD boundary contains bonds between identical atom species. In Fig. 4.15, intertwinning APD boundaries are shown on the surface of InP layers on Si. The antiphase domains can be visualized with an anisotropic etch.

In Fig. 4.16a, inversion domains in iron-doped ZnO are shown. Between domains the direction of the c-axis is reversed. The iron is found preferentially

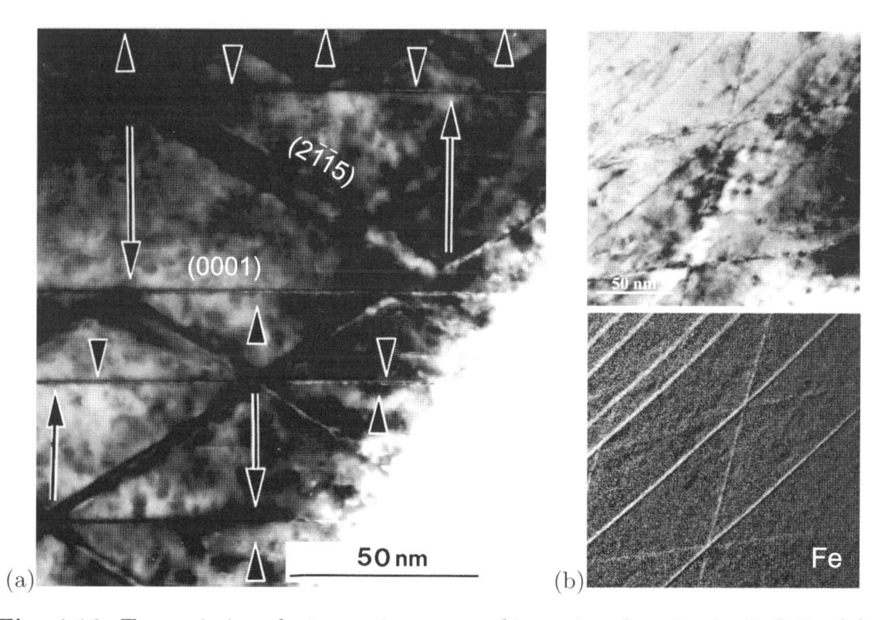

Fig. 4.16. Transmission electron microscopy of inversion domains in ZnO:Fe. (**a**) Inversion domains in iron-doped ZnO (ZnO:Fe_2O_3 =100:1). *Arrows* denote the orientation of the c-axis in the respective domains. (**b**) *Top*: bright field TEM, *bottom* Fe distribution from energy-filtered image. Adapted from [92]

in the inversion domain boundary (IDB) (Fig. 4.16b) and plays an important role in its formation [91].

4.8 Disorder

Disorder is a general term for deviations from the ideal structure on a microscopic scale. Examples are

- The presence of various isotopes of an element. This introduces disorder with regard to the mass of the atoms and impacts mostly phonon properties (see, e.g., Fig. 8.15).
- The occupation of lattice sites in alloys (Sect. 3.7) ranging from a random alloy, clustering to (partially) ordered phases.

5 Mechanical Properties

5.1 Introduction

The atoms making up the solid have an average position from which they can deviate since they are elastically bonded. The typical atomic interaction potential looks like the one shown in Fig. 2.1. The atoms thus perform a vibrational motion and the solid is elastic. The potential is essentially asymmetric, being steeper for small distances due to quantum-mechanical overlap of orbitals. However, for small amplitudes around the minimum a harmonic oscillator is assumed (harmonic approximation).

5.2 Lattice Vibrations

In the following we will discuss the dispersion relations for lattice vibrations, i.e. the connection between the frequency ν (or energy $E = h\nu = \hbar\omega$) of the wave and its wavelength λ (or k-vector $k = 2\pi/\lambda$).

5.2.1 Monoatomic Linear Chain

The essential physics of lattice vibrations can best be seen from a one-dimensional model that is called the linear chain. The mechanical vibrations will also be called *phonons*, although technically this term is reserved for the quantized lattice vibrations resulting from the quantum-mechanical treatment.

In the monoatomic linear chain the atoms of mass M are positioned along a line (x-axis) with a period (lattice constant) a at the positions $x_{n_0} = na$. This represents a one-dimensional Bravais lattice. The Brillouin zone of this system is $[-\pi/a, \pi/a]$.

The atoms will interact with a harmonic potential, i.e. the energy is proportional to the displacement $u_n = x_n - x_{n_0}$ to the second power. The total potential energy of the system is then:

$$U = \frac{1}{2} C \sum_n (u_n - u_{n+1})^2 \ . \tag{5.1}$$

The model assumes that the mass points are connected via massless, ideal springs with a spring constant C. If $\phi(d)$ is the interaction energy between two atoms as a function of their distance d, C is given by $C = \phi''(a)$. Again, the harmonic approximation is only valid for small displacements, i.e. $u_n \ll a$. The displacement of the atoms can be *along* the chain (longitudinal wave) or *perpendicular* to the chain (transverse wave), see Fig. 5.1. We note that for these two types of waves the elastic constant C must not be the same.

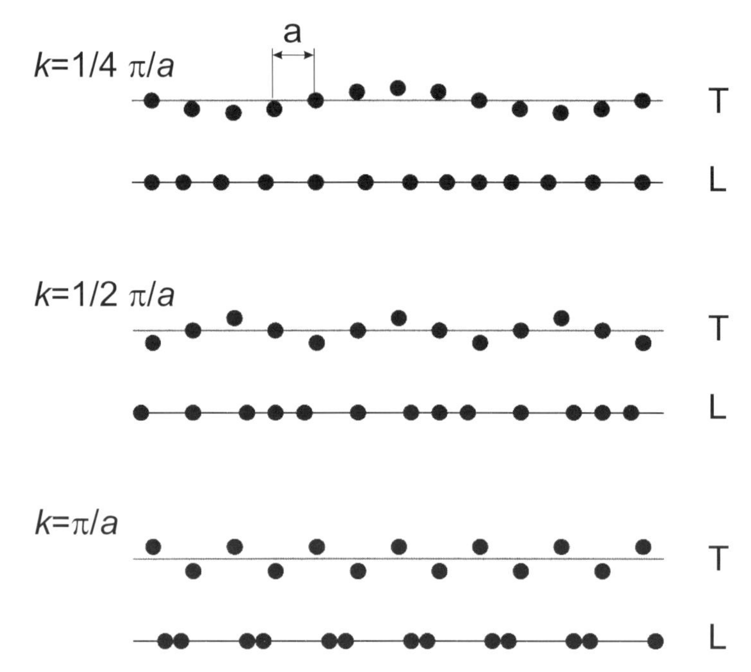

Fig. 5.1. Visualization of transverse ('T') and longitudinal ('L') waves in a linear monoatomic chain at different wavevectors

When the sum in (5.1) has a finite number of terms ($n = 0,\ldots,N-1$), the boundary conditions have to be considered. There are typically two possibilities: The boundary atoms are fixed, i.e. $u_0 = u_{N-1} = 0$, the boundary conditions are periodic (Born–von Karman), i.e. $u_i = u_{N+i}$. If $N \gg 1$, the boundary conditions play no significant role anyway, thus those with the greatest ease for subsequent math are chosen. In solid-state physics typically periodic boundary conditions are used. Boundary phenomena, such as at surfaces, are then treated separately.

The equations of motion derived from (5.1) are

$$M\ddot{u}_n = F_n = -\frac{\partial U}{\partial u_n} = -C(2u_n - u_{n-1} - u_{n+1}) \,. \tag{5.2}$$

We solve for solutions that are periodic in time (harmonic waves), i.e. $u_n(x,t) = u_n \exp(-\mathrm{i}\omega t)$. Then the time derivative can be executed immediately as $\ddot{u}_n = -\omega^2 u_n$ and we obtain:

$$M\omega^2 u_n = C(2u_n - u_{n-1} - u_{n+1}) \,. \tag{5.3}$$

If, also, the solution is periodic in space, i.e. is a (one-dimensional) plane wave, i.e. $u_n(x,t) = v_0 \exp[\mathrm{i}(kx - \omega t)]$ with $x = na$, we find from the periodic boundary condition $\exp(\mathrm{i}kNa) = 1$ and thus

$$k = \frac{2\pi}{a}\frac{n}{N}, \; n \in \mathbf{N} \,. \tag{5.4}$$

It is important that, when k is altered by a reciprocal space vector, i.e. $k' = k + 2\pi n/a$, the displacements u_n are unaffected. This property means that there are only N values for k that generate independent solutions. These can be chosen as $k = -\pi/a, \ldots, \pi/a$, so that k lies in the Brillouin zone of the lattice. In the Brillouin zone there is a total number of N k-values, i.e. one for each lattice point. The distance between adjacent k-values is $\frac{2\pi}{Na} = \frac{2\pi}{L}$, L being the lateral extension of the system.

The displacements at the lattice points n and $n+m$ are now related to each other via

$$\begin{aligned} u_{n+m} &= v_0 \exp(\mathrm{i}k(n+m)a) \\ &= v_0 \exp(\mathrm{i}kna) \exp(\mathrm{i}kma) = \exp(\mathrm{i}kma) u_n \,. \end{aligned} \tag{5.5}$$

Thus, the equation of motion (5.3) reads

$$M\omega^2 u_n = C(2 - \exp(-\mathrm{i}ka) - \exp(\mathrm{i}ka)) u_n \,. \tag{5.6}$$

Using the identity $\exp(\mathrm{i}ka) + \exp(-\mathrm{i}ka) = 2\cos(ka)$, we find the dispersion relation of the monoatomic linear chain (Fig. 5.2):

$$\omega^2(k) = \frac{4C}{M}\frac{1 - \cos(ka)}{2} = \frac{4C}{M}\sin^2\left(\frac{ka}{2}\right) \,. \tag{5.7}$$

The solutions describe plane waves that propagate in the crystal with a phase velocity $c = \omega/k$ and a group velocity $v_g = \mathrm{d}\omega/\mathrm{d}k$

$$v_g = \left[\frac{4C}{M}\right]^{1/2} \frac{a}{2} \cos\left(\frac{ka}{2}\right) \,. \tag{5.8}$$

In the vicinity of the Γ point, i.e. $k \ll \pi/a$ the dispersion relation is linear in k

$$\omega(k) = a\sqrt{C/M}|k| \,. \tag{5.9}$$

We are used to such linear relations for sound (and also light) waves. The phase and group velocity are the same and do not depend on k. Thus, such

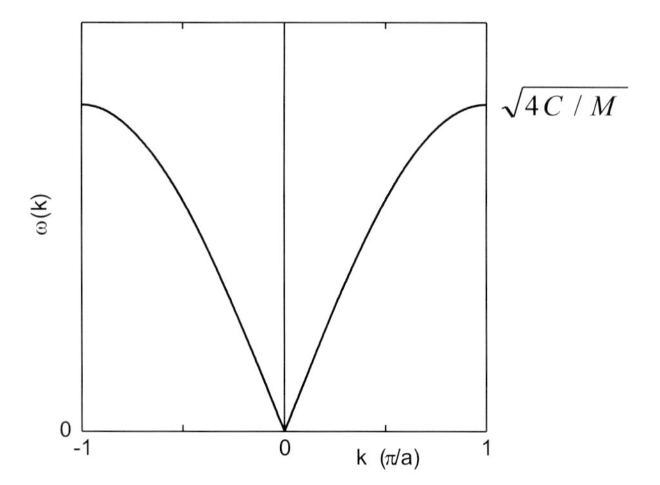

Fig. 5.2. Dispersion relation for a monoatomic linear chain

solutions are called *acoustic*. The sound velocity of the medium is given by $c_\mathrm{s} = a\sqrt{C/M}$.

It is characteristic of the nonhomogeneous medium that, when k approaches the boundary of the Brillouin zone, the behavior of the wave is altered. For $k = \pi/a$ the wavelength is just $\lambda = 2\pi/k = 2a$, and thus samples the granularity of the medium. The maximum phonon frequency ω_m is

$$\omega_\mathrm{m} = \sqrt{\frac{4C}{M}} \,. \tag{5.10}$$

The group velocity is zero at the zone boundary, thus a standing wave is present.

Since the force constants of the longitudinal and transverse waves can be different, the dispersion relations are different. The transverse branch of the dispersion relation is 2-fold degenerate, unless the two directions that are perpendicular to x are not equivalent.

5.2.2 Diatomic Linear Chain

Now we consider the case that the system is made up from two different kinds of atoms (Fig. 5.3). This will be a model for semiconductors with a diatomic base, such as zincblende. We note that the diamond structure also needs to be modeled in this way, although both atoms in the base are the same.

The lattice will be the same and the lattice constant will be a. Alternating atoms of sort 1 and 2 with a relative distance of $a/2$ are on the chain. The displacements of the two atoms are labeled u_n^1 and u_n^2, both belonging to the lattice point n. The atoms have the masses M_1 and M_2. The force constants are C_1 (for the 1–2 bond within the base) and C_2 (for the 2–1 bond between different bases).

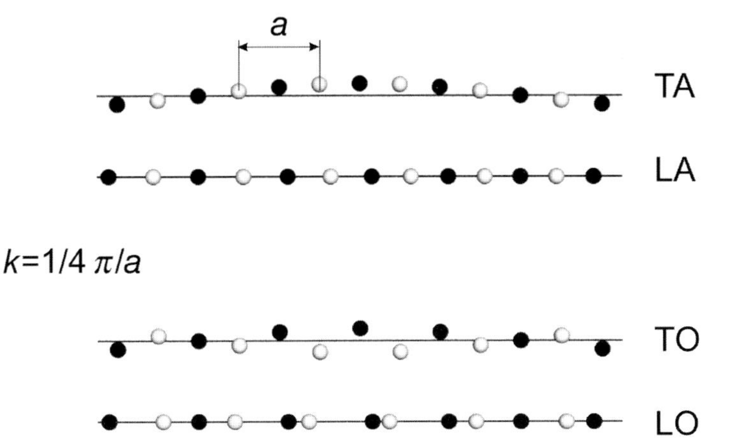

Fig. 5.3. Visualization of acoustic and optical waves in a diatomic linear chain

The total potential energy of the system is then given as

$$U = \frac{1}{2} C_1 \sum_n \left(u_n^1 - u_n^2\right)^2 + \frac{1}{2} C_2 \sum_n \left(u_n^2 - u_{n+1}^1\right)^2 . \tag{5.11}$$

The equations of motion are

$$M_1 \ddot{u}_n^1 = -C_1 \left(u_n^1 - u_n^2\right) - C_2 \left(u_n^1 - u_{n-1}^2\right) \tag{5.12a}$$

$$M_2 \ddot{u}_n^2 = -C_1 \left(u_n^2 - u_n^1\right) - C_2 \left(u_n^2 - u_{n+1}^1\right) . \tag{5.12b}$$

With the plane-wave ansatz $u_n^1(x,t) = v_1 \exp\left[\mathrm{i}(kna - \omega t)\right]$ and $u_n^2(x,t) = v_2 \exp\left[\mathrm{i}(kna - \omega t)\right]$ and periodic boundary conditions we find

$$0 = -M_1 \omega^2 v_1 + C_1(v_1 - v_2) + C_2(v_1 - \exp(-\mathrm{i}ka) v_2) \tag{5.13a}$$

$$0 = -M_2 \omega^2 v_2 + C_1(v_2 - v_1) + C_2(v_2 - \exp(\mathrm{i}ka) v_1) . \tag{5.13b}$$

These equations for v_1 and v_2 can only be solved nontrivially if the determinant vanishes, i.e.

$$0 = \begin{vmatrix} M_1 \omega^2 - (C_1 + C_2) & C_1 + \mathrm{e}^{-\mathrm{i}ka} C_2 \\ C_1 + \mathrm{e}^{\mathrm{i}ka} C_2 & M_2 \omega^2 - (C_1 + C_2) \end{vmatrix}$$
$$= M_1 M_2 \omega^4 - (M_1 + M_2)(C_1 + C_2)\omega^2 + 2 C_1 C_2 (1 - \cos(ka)) . \tag{5.14}$$

Using the substitutions $C_+ = (C_1 + C_2)/2$, $C_\times = \sqrt{C_1 C_2}$, the arithmetic and geometrical averages, and accordingly for M_+ and $M_\times$, the solution is

$$\omega^2(k) = \frac{2 C_\times}{\gamma M_\times} \left[1 \pm \sqrt{1 - \gamma^2 \frac{1 - \cos(ka)}{2}} \right] , \tag{5.15}$$

with

$$\gamma = \frac{C_\times M_\times}{C_+ M_+} \ . \tag{5.16}$$

The dispersion relation, as shown in Fig. 5.4, now has (for each longitudinal and transverse mode) two branches. The lower branch ('−' sign in (5.15)) is related to the acoustic mode; neighboring atoms have similar phase (Fig. 5.3). For the acoustic mode $\omega = 0$ at the Γ point and the frequency increases towards the zone boundary. The maximum phonon frequency ω_{m} is in the upper branch ('+' sign in (5.15)) at the zone center

$$\omega_{\mathrm{m}} = \sqrt{\frac{4C_\times}{\gamma M_\times}} = \sqrt{\frac{4C_+ M_+}{M_\times^2}} \ . \tag{5.17}$$

In the vicinity of the Γ point the dispersion is parabolic with negative curvature:

$$\omega(k) \cong \omega_{\mathrm{m}} \left[1 - \frac{1}{2} \left(\frac{\gamma a}{4} \right)^2 k^2 \right] \ . \tag{5.18}$$

The upper branch is called the optical mode (since it can interact strongly with light, see Sect. 9.7) and neighboring atoms have opposite phase. Thus, four different vibrations exist that are labeled TA, LA, TO, and LO. Both the TA and TO branches are degenerate.

At the zone boundary (X point) a frequency gap exists. The gap center is at

$$\bar{\omega}_{\mathrm{X}} = \frac{\omega_{\mathrm{m}}}{\sqrt{2}} \ , \tag{5.19}$$

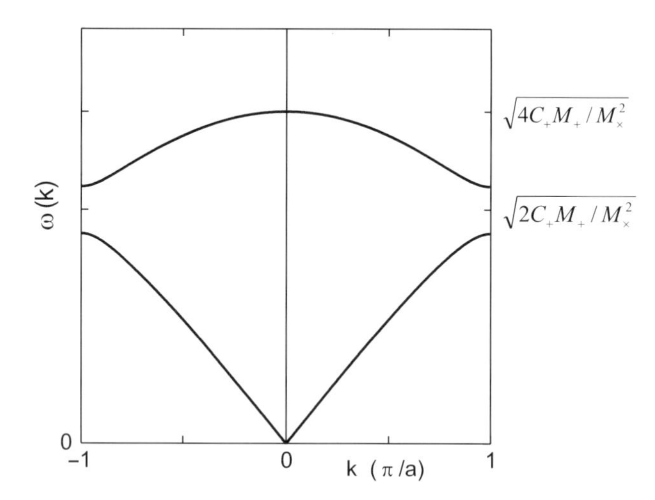

Fig. 5.4. Dispersion relation for a diatomic linear chain.

and the total width is

$$\Delta\omega_X = \omega_{\rm m}\sqrt{1-\gamma} = 2\sqrt{\frac{C_+ M_+ - C_\times M_\times}{M_\times^2}} \ . \tag{5.20}$$

The group velocity is zero for optical and acoustic phonons at $k = \pi/a$ and for optical phonons at the Γ point.

Usually two cases are treated explicitly: (i) atoms with equal mass ($M = M_1 = M_2$) and different force constants or (ii) atoms with unequal mass and identical force constants $C = C_1 = C_2$. For the case $C_1 = C_2$ and $M_1 = M_2$, $\gamma = 1$ and thus $\Delta\omega_X = 0$. Then the dispersion relation is the same as for the monoatomic chain, except that the k space has been folded since the actual lattice constant is now $a/2$.

$M_1 = M_2$

In this case, $M_+ = M_\times = M$ and the dispersion relation is

$$\omega^2 = \frac{2C_+}{M}\left[1 \pm \sqrt{1 - \frac{C_\times^2}{C_+^2}\frac{1-\cos(ka)}{2}}\right] \ . \tag{5.21}$$

At the zone boundary the frequencies for the acoustic and the optical branch are $\omega_{X,1} = \sqrt{2C_1/M}$ with $v_1 = v_2$ and $\omega_{X,2} = \sqrt{2C_2/M}$ with $v_1 = -v_2$, respectively (assuming $C_2 > C_1$). The motion for $k = \pi/a$ is phase shifted by 180° for adjacent bases. Additionally, for the acoustic branch the atoms of the base are in phase, while for the optical branch the atoms of the base are 180° out of phase. The vibration looks as if only one of the springs is strained.

$C_1 = C_2$

In this case, $C_+ = C_\times = C$ and the dispersion relation is

$$\omega^2 = \frac{2CM_+}{M_\times^2}\left[1 \pm \sqrt{1 - \frac{M_\times^2}{M_+^2}\frac{1-\cos(ka)}{2}}\right] \ . \tag{5.22}$$

At the zone boundary the frequencies for the acoustic and the optical branch are $\omega_{X,1} = \sqrt{2C/M_1}$ with $v_2 = 0$ and $\omega_{X,2} = \sqrt{2C/M_2}$ with $v_1 = 0$, respectively (assuming $M_2 < M_1$). In the vibration for $k = \pi/a$ thus only one atom species oscillates, the other does not move. Close to the Γ point the atoms are in phase in the acoustic branch, i.e. $v_1 = v_2$. For the optical branch, the frequency at the Γ point is given by $\omega = \sqrt{2C/M_{\rm r}}$ (with the reduced mass $M_{\rm r}^{-1} = M_1^{-1} + M_2^{-1} = 2M_+/M_\times^2$) and the amplitude ratio is given by the mass ratio: $v_2 = -(M_1/M_2)v_1$, i.e. the heavier atom has the smaller amplitude.

5.2.3 Lattice Vibrations of a Three-Dimensional Crystal

When calculations are executed for a three-dimensional crystal with a monoatomic base, there are $3N$ equations of motion. These are transformed to normal coordinates and represent 3 acoustic branches (1 LA phonon mode and 2 TA phonon modes) of the dispersion relation. In a crystal with a base with p atoms, there are also 3 acoustic branches and $3(p-1)$ optical branches. For a diatomic base (as in the zincblende structure) there are 3 optical phonon branches (1 LO phonon mode and 2 TO phonon modes). The total number of modes is $3p$. The dispersion $\omega(\mathbf{k})$ now has to be calculated for all *directions* of $\mathbf{k}$.

In Figs. 5.5 and 5.6, the phonon dispersion in silicon and GaAs is shown along particular lines in the Brillouin zone (cf. Fig. 3.31b). The main differences are: (i) the degeneracy of the acoustic and optical branch at the X point for the group-IV semiconductor is lifted for the III–V semiconductor due to the different mass of the constituents, (ii) the degeneracy of the LO and TO energies at the Γ point for the group-IV semiconductor is lifted for the III–V semiconductor due to the ionic character of the bond and the macroscopic electric field connected with the long-wavelength LO phonon (see Sect. 5.2.7).

We note that the degeneracy of the TA phonon is lifted for propagation along the $\langle 110 \rangle$ directions (Σ) because the two transverse directions $\langle 001 \rangle$ and $\langle 1\bar{1}0 \rangle$ are not equivalent.

The dependence of the phonon frequency on the mass of the atoms ($\propto M^{-1/2}$) can be demonstrated with the isotope effect, visualized for GaAs

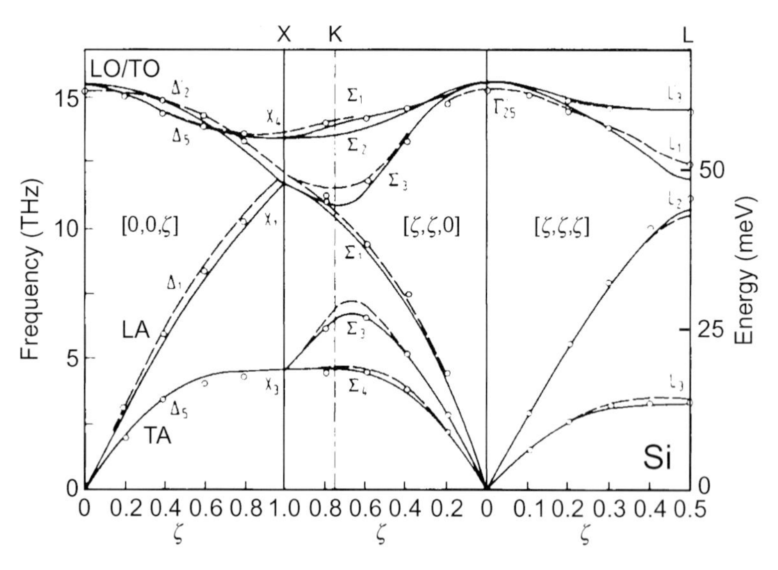

Fig. 5.5. Phonon dispersion in Si, experimental data and theory (*solid lines*: bond charge model, *dashed lines*: valence force field model). Adapted from [93]

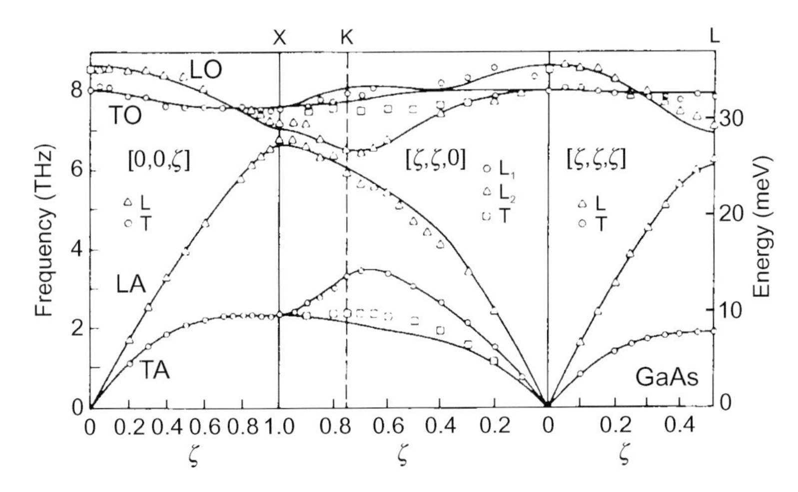

Fig. 5.6. Phonon dispersion in GaAs, experimental data and theory (*solid lines*, 14-parameter shell model). 'L' and 'T' refer to longitudinal and transverse modes, respectively. 'I' and 'II' (along $[\zeta,\zeta,0]$) are modes whose polarization is in the $(1,\bar{1},0)$ plane. Adapted from [93], based on [94]

in Fig. 5.7. The dependence of the phonon frequencies on the stiffness of the spring can be seen from Fig. 5.8; the smaller lattice constant provides the stiffer spring.

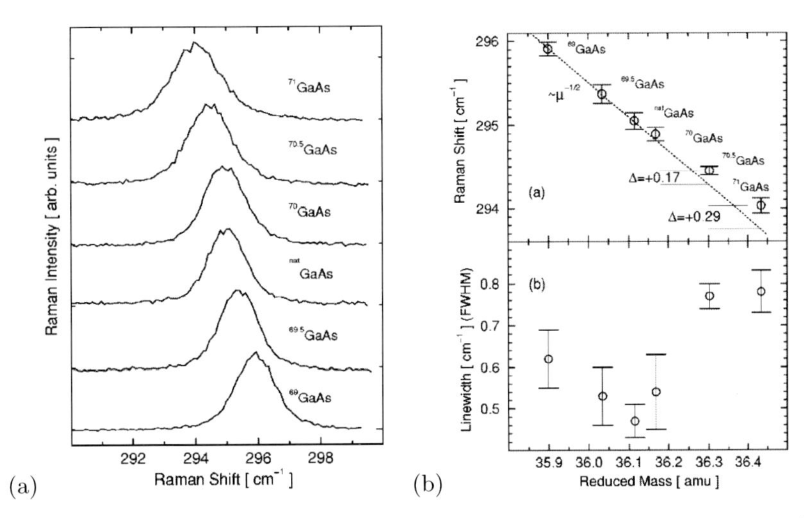

Fig. 5.7. (**b**) Energy of optical phonons in GaAs with different isotope content [using the Raman spectra shown in (**a**)]. Reprinted with permission from [95], ©1999 APS

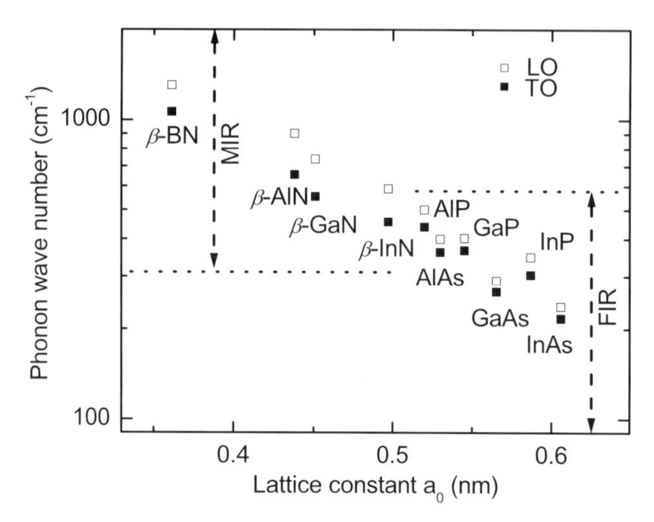

Fig. 5.8. Optical phonon frequencies (TO: *filled squares*, LO: *empty squares*) for a number of III–V compounds with different lattice constant a_0. 1 meV corresponds to 8.065 wave numbers (or cm^{-1}). From [100]

5.2.4 Phonons

Phonons are the quantized quasi-particles of the lattice vibrations (normal modes). The energy of a phonon can take the discrete values of a harmonic oscillator

$$E_{ph} = \left(n + \frac{1}{2}\right) \hbar\omega \,, \tag{5.23}$$

where n denotes the quantum number of the state, which corresponds to the number of energy quanta $\hbar\omega$ in the vibration. The amplitude of the vibration can be connected to n via the following discussion. For the classical oscillation $u = u \exp[\mathrm{i}(kx - \omega t)]$ the space and time average for the kinetic energy yields

$$E_{\text{kin}} = \frac{1}{2}\rho V \overline{\left(\frac{\partial u}{\partial t}\right)^2} = \frac{1}{8}\rho V \omega^2 u_0^2 \,, \tag{5.24}$$

where ρ is the density and V the volume of the (homogeneous) solid. The energy of the oscillation is split in half between kinetic and potential energy. From $2E_{\text{kin}} = E_{\text{ph}}$ we find

$$u_0^2 = \left(n + \frac{1}{2}\right) \frac{4\hbar}{\rho V \omega} \,. \tag{5.25}$$

The number of phonons with which a vibrational mode is populated is thus directly related to the classical amplitude square.

Phonons act with a momentum $\hbar\mathbf{k}$, the so-called crystal momentum. When phonons are created, destroyed or scattered the crystal momentum is conserved, except for an arbitrary reciprocal-space vector $\mathbf{G}$. Scattering with $\mathbf{G} = 0$ is called a normal process, otherwise (for $\mathbf{G} \neq 0$) it is called an *umklapp* process.

5.2.5 Localized Vibrational Modes

A defect in the crystal can induce localized vibrational modes (LVM). The defect can be a mass defect, i.e. one of the masses M is replaced by M_{d}, or the force constants in the neighborhood are modified to C_{d}. A detailed treatment can be found in [101]. LVM are discussed, e.g., in [102–104].

First we consider the LVM for the one-dimensional, monoatomic chain. If the mass at lattice point $i = 0$ is replaced by $M_{\mathrm{d}} = M + \Delta M$ ($\epsilon_{\mathrm{M}} = \Delta M/M$), the displacements are given by $u_i = AK^{|i|}$, A being an amplitude, with

$$K = -\frac{1+\epsilon_{\mathrm{M}}}{1-\epsilon_{\mathrm{M}}} , \tag{5.26}$$

and the defect phonon frequency ω_{d} is

$$\omega_{\mathrm{d}} = \omega_{\mathrm{m}} \sqrt{\frac{1}{1-\epsilon_{\mathrm{M}}^2}} . \tag{5.27}$$

A real frequency is obtained for $|\epsilon_{\mathrm{M}}| < 1$. ω_{d} is then higher than the highest frequency of the bulk modes $\omega_{\mathrm{m}} = \sqrt{4C/M}$ (5.10). For $\epsilon_{\mathrm{M}} < 0$, i.e. the mass of the defect is smaller than the mass of the host atoms, K is negative and $|K| < 1$. Thus, the displacement can be written as

$$u_i \propto (-|K|)^{|i|} = (-1)^{|i|} \exp\left(+|i| \log|K|\right) . \tag{5.28}$$

The numerical value of the exponent is negative, thus the amplitude decreases exponentially from the defect and indeed makes a localized vibrational mode. For small mass $M_{\mathrm{d}} \ll M$ (5.27) yields approximately $\omega_{\mathrm{d}} = \sqrt{2C/M_{\mathrm{d}}}$. This approximation is the so-called one-oscillator model. Since typically the extension of the localized mode is only a few lattice constants, the picture of LVM remains correct for impurity concentrations up to $\sim 10^{18} - 10^{20}\,\mathrm{cm}^{-3}$. For higher concentrations the concept of alloy modes has to be invoked (see Sect. 5.2.6).

For the case of group-III or -V substitutional impurities in group-IV semiconductors the change in force constants (treated below) can be neglected to some extent. For silicon ($M = 28$) and germanium ($M = 73$) the effect of various substitutions is shown in Fig. 5.9.

Now, additionally the force constants left and right of the defect are replaced by $C_{\mathrm{d}} = C + \Delta C$ ($\epsilon_{\mathrm{C}} = \Delta C/C$). The displacements are still given by $u_i = AK^{|i|}$, now with

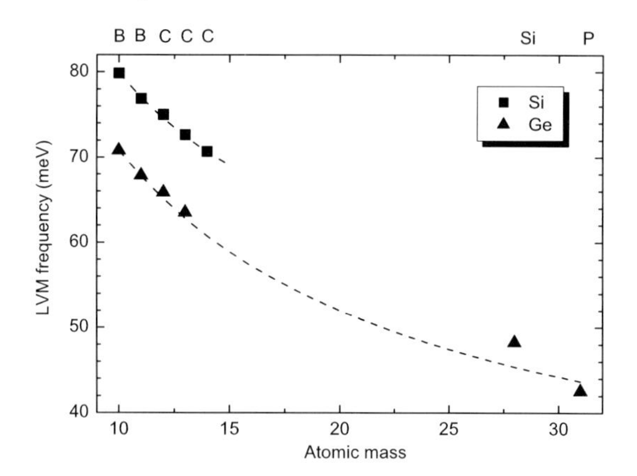

Fig. 5.9. Energy of local vibrational modes in Si and Ge. Experimental values at $T = 300\,\mathrm{K}$ (B in Ge: $T = 80\,\mathrm{K}$) taken from [101] and references therein and from [105] (C in Ge). The *dashed lines* are the mass dependence according to (5.27) scaled to the experimental frequency of the ^{10}B LVM

$$K = -\frac{(1+\epsilon_\mathrm{M})(1+\epsilon_\mathrm{C})}{1-\epsilon_\mathrm{M}-2\epsilon_\mathrm{C}} \ . \tag{5.29}$$

An exponential decrease of the LVM amplitude occurs for negative K that is ensured for $\epsilon_\mathrm{M}+2\epsilon_\mathrm{C}<0$ (and $\epsilon_\mathrm{M}>-1$ and $\epsilon_\mathrm{C}>-1$). The defect frequency is given by

$$\omega_\mathrm{d} = \omega_\mathrm{m}\sqrt{\frac{(1+\epsilon_\mathrm{C})(2+\epsilon_\mathrm{C}(3+\epsilon_\mathrm{M}))}{2(1+\epsilon_\mathrm{M})(2\epsilon_\mathrm{C}+1-\epsilon_\mathrm{M})}} \ . \tag{5.30}$$

We note that for $\epsilon_\mathrm{C}=0$ (5.26) and (5.27) are recovered.

For a given mass defect, the change of frequency with ΔC is (in linear order, i.e. for $\epsilon_\mathrm{C} \ll 1$)

$$\frac{\partial\omega_\mathrm{d}(\epsilon_\mathrm{M},\epsilon_\mathrm{C})}{\partial\epsilon_\mathrm{C}} = \frac{1-4\epsilon_\mathrm{M}-\epsilon_\mathrm{M}^2}{4(1-\epsilon_\mathrm{M})\sqrt{1-\epsilon_\mathrm{M}^2}}\,\epsilon_\mathrm{C} \ . \tag{5.31}$$

The linear coefficient diverges for $\epsilon_\mathrm{M}\to -1$. For ϵ_M between -0.968 and 0 the linear coefficient varies between 2 and 1/4. Therefore, a larger force constant ($\epsilon_\mathrm{C}>0$) increases the LVM frequency of the defect, as expected for a stiffer spring.

In a binary compound the situation is more complicated. We assume here that the force constants remain the same and only the mass of the substitution atom M_d is different from the host. The host has the atom masses M_1 and M_2 with $M_1<M_2$. Substitution of the heavy atom with a lighter one creates a LVM above the optical branch for $M_\mathrm{d}<M_2$. Additionally, a level in the

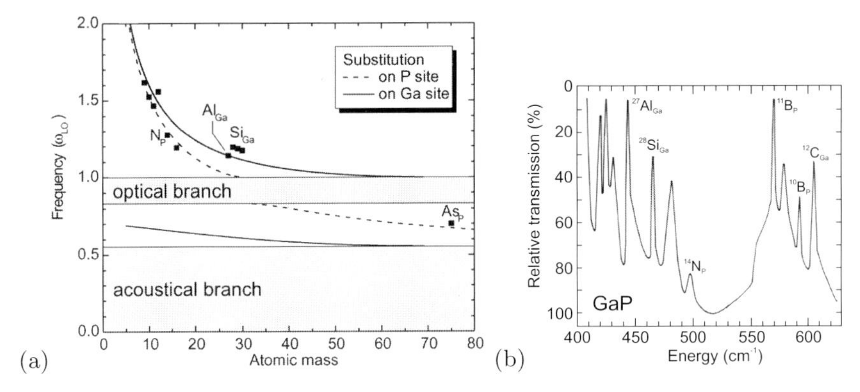

Fig. 5.10. (**a**) Numerical simulation of a linear chain model for GaP ($M_1 = 31$, $M_2 = 70$). Energy of local vibrational modes (*dashed* (*solid*) *line*): substitution on P (Ga) site) in units of the optical phonon frequency at Γ ($\omega_{\mathrm{m}} = 45.4\,\mathrm{meV}$). The *grey areas* indicate the acoustic and optical phonon bands. *Solid squares* are experimental data (from [101]), scaled to the theoretical curve for the $^{27}\mathrm{Al_{Ga}}$ LVM frequency. (**b**) Differential transmission spectrum of GaP structure (nitrogen-doped layer on zinc-doped compensated substrate) against pure crystal ($T = 77\,\mathrm{K}$). Data from [106]

gap between the optical and acoustic branch is induced. Such LVM is called a *gap mode*. Substitution of the lighter atom of the binary compound induces a LVM above the optical branch for $M_{\mathrm{d}} < M_1$. A gap mode is induced for $M_{\mathrm{d}} > M_1$. The situation for GaP is depicted in Fig. 5.10.

The energy position of a local vibrational mode is sensitive to the isotope mass of the surrounding atoms. In Fig. 5.11, a high-resolution ($0.03\,\mathrm{cm}^{-1}$) spectrum of the $^{12}\mathrm{C_{As}}$ LVM in GaAs is shown together with a theoretical simulation. The various theoretical peak positions are given as vertical bars, their height indicating the oscillator strength. Five experimental peaks are obvious that are due to a total of nine different transitions. The C atom can experience five different surroundings (see Table 3.7) with the four neighbors being ^{69}Ga or ^{71}Ga. The natural isotope mix is an 'alloy' $^{69}\mathrm{Ga}_x{}^{71}\mathrm{Ga}_{1-x}\mathrm{As}$ with $x = 0.605$. The configurations with T_d symmetry contribute one peak each, the lowest (^{71}Ga surrounding) and highest (^{69}Ga surrounding) energy transitions. The configurations with C_{3v} and C_{2v} symmetry contribute each with 2 and 3 nondegenerate modes, respectively.

5.2.6 Phonons in Alloys

In an alloy of the type $\mathrm{AB}_{1-x}\mathrm{C}_x$ the phonon frequencies will depend on the ternary composition. For the binary end materials AB and AC clearly TO and LO frequencies exist. The simplest behavior of the alloy is the one-mode behavior (Fig. 5.12d) where the mode frequencies vary continuously (and approximately linearly) with the composition. The oscillator strength (LO–

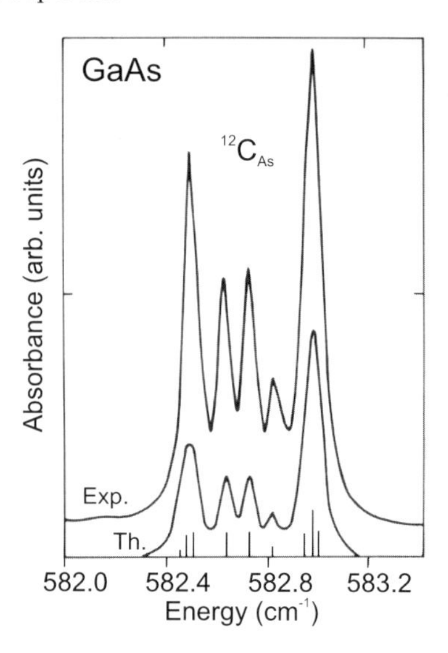

Fig. 5.11. Experimental (Exp., $T = 4.2$ K, resolution $0.03\,\mathrm{cm}^{-1}$) and theoretical (Th., artificial Lorentzian broadening) infrared spectra of LVM of $^{12}C_{As}$ in GaAs. The positions and oscillator strengths of the theoretical transitions involving different configurations with ^{69}Ga and ^{71}Ga isotopes are shown as *vertical bars*. Data from [102]

TO splitting, (9.61)) remains approximately constant. In many cases, the two-mode behavior is observed where the LO–TO gap closes (accompanied by decreasing oscillator strength) and a localized vibrational mode and a gap mode occur for the binary end materials (Fig. 5.12a). Also, a mixed-mode behavior (Fig. 5.12b,c) can occur.

The masses of the three constituent atoms will be M_A, M_B, and M_C. Without limiting the generality of our treatment, we assume $M_\mathrm{B} < M_\mathrm{C}$. From the considerations in Sect. 5.2.5 on LVM and gap modes, the condition

$$M_\mathrm{B} < M_\mathrm{A}, M_\mathrm{C} \tag{5.32}$$

for two-mode behavior can be deduced. This ensures a LVM of atom B in the compound AC and a gap mode of atom C in the compound AB. However, it turns out that this condition is not sufficient, e.g. $\mathrm{Na}_{1-x}\mathrm{K}_x\mathrm{Cl}$ fulfills (5.32) but exhibits one-mode behavior. From a modified REI model (for $\mathbf{k} \approx 0$ modes) it has been deduced that

$$M_\mathrm{B} < \mu_\mathrm{AC} = \frac{M_\mathrm{A} M_\mathrm{C}}{M_\mathrm{A} + M_\mathrm{C}} < M_\mathrm{A}, M_\mathrm{C} \tag{5.33}$$

is a necessary and sufficient condition (unless the force constants between A–B and A–C are significantly different) for two-mode behavior [107]. A detailed discussion is given in [108]. Equation (5.33) is a stronger condition than the

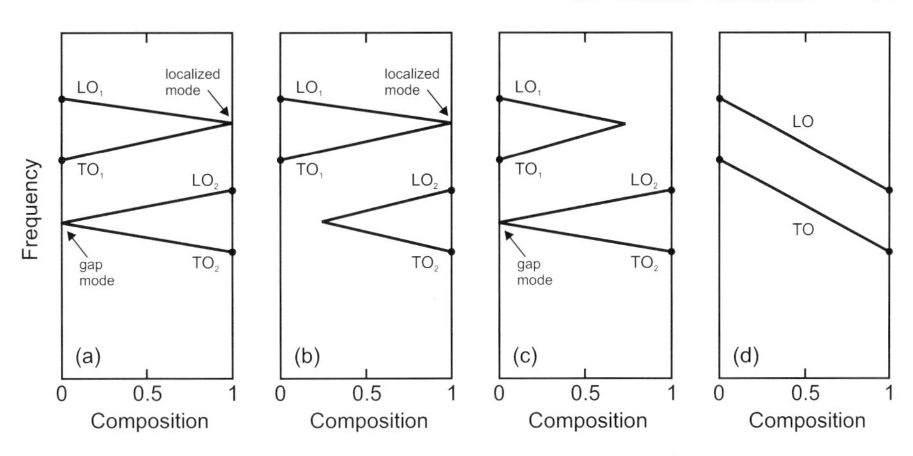

Fig. 5.12. Schematic behavior of phonon modes in an alloy. (**a**) Two-mode behavior with gap mode and localized mode, (**b**,**c**) mixed-mode behavior, (**b**) only localized mode allowed, (**c**) only gap mode allowed, (**d**) one-mode behavior with neither localized mode nor gap mode allowed

previous one (5.32). If (5.33) is not fulfilled the compound exhibits one-mode behavior. As an example, we show the mass relations for $CdS_{1-x}Se_x$ and $Cd_xZn_{1-x}S$ in Table 5.1 and the experimental phonon energies in Fig. 5.13. Also in Table 5.1 the masses for $GaP_{1-x}As_x$ ($GaAs_{1-x}Sb_x$) exhibiting two- (one-) mode behavior are shown.

Table 5.1. Atomic masses of the constituents of various ternary compounds, reduced mass μ_{AC} (5.33), fulfillment of the relation from (5.33) ('+': fulfilled, '−': not fulfilled) and experimental mode behavior ('2': two-mode, '1': one-mode)

Alloy	A	B	C	M_A	M_B	M_C	μ_{AC}	Rel.	Modes
$GaP_{1-x}As_x$	Ga	P	As	69.7	31.0	74.9	36.1	+	2
$GaAs_{1-x}Sb_x$	Ga	As	Sb	69.7	74.9	121.8	44.3	−	1
$CdS_{1-x}Se_x$	Cd	S	Se	112.4	32.1	79.0	46.4	+	2
$Cd_xZn_{1-x}S$	S	Zn	Cd	32.1	65.4	112.4	25.0	−	1

5.2.7 Electric Field Created by Optical Phonons

Adjacent atoms oscillate with opposite phase in an optical phonon. If the bond has (partial) ionic character, this leads to a time-dependent polarization and subsequently to a macroscopic electric field. This additional field will influence the phonon frequencies obtained from a purely mechanical approach. We consider in the following the case $\mathbf{k} \approx 0$. The phonon frequency for TO and LO vibrations is given by

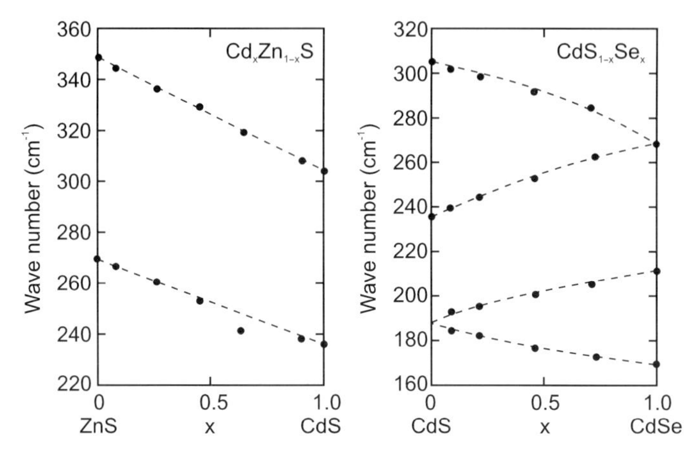

Fig. 5.13. Phonon energies of $Cd_xZn_{1-x}S$ and $CdS_{1-x}Se_x$ as a function of the ternary composition. Experimental data (*solid circles*) are from [107], *dashed lines* are guides to the eye

$$\omega_0 = \left(\frac{2C}{M_r}\right)^{1/2}, \tag{5.34}$$

where M_r is the reduced mass of the two different atoms (cf. Sect. 5.2.2). $\mathbf{u}$ is the relative displacement $\mathbf{u}_1 - \mathbf{u}_2$ of the two atoms in a diatomic base. When the interaction with the electric field $\mathbf{E}$ (which will be calculated self-consistently in the following) is considered, the Hamiltonian for the long-wavelength limit is given by [109]:

$$\hat{H}(\mathbf{p}, \mathbf{u}) = \frac{1}{2}\left(\frac{1}{M_r}\mathbf{p}^2 + b_{11}\mathbf{u}^2 + 2b_{12}\mathbf{u}\cdot\mathbf{E} + b_{22}\mathbf{E}^2\right). \tag{5.35}$$

The first term is the kinetic energy ($\mathbf{p}$ stands for the momentum of the relative motion of the atoms 1 and 2 in the base, $\mathbf{p} = M_r\dot{\mathbf{u}}$), the second the potential energy, the third the dipole interaction and the fourth the electric-field energy. The equation of motion for a plane wave $\mathbf{u} = \mathbf{u}_0 \exp\left[-\mathrm{i}(\omega t - \mathbf{k}\cdot\mathbf{r})\right]$ ($\ddot{\mathbf{u}} = -\omega^2\mathbf{u}$) yields

$$M_r\omega^2\mathbf{u} = b_{11}\mathbf{u} + b_{12}\mathbf{E}. \tag{5.36}$$

Thus, the electric field is

$$\mathbf{E} = (\omega^2 - \omega_{\mathrm{TO}}^2)\frac{M_r}{b_{12}}\mathbf{u}. \tag{5.37}$$

Here, the substitution $\omega_{\mathrm{TO}}^2 = b_{11}/M_r$ was introduced that is consistent with (5.34) and $b_{11} = 2C$. ω_{TO} represents the mechanical oscillation frequency of the atoms undisturbed by any electromagnetic effects. Already now the important point is visible: If ω approaches ω_{TO}, the system plus electric field oscillates with the frequency it has without an electric field. Therefore the electric field must be zero. Since the polarization $\mathbf{P} = (\epsilon - 1)\epsilon_0 \mathbf{E}$ is finite, the dielectric constant ϵ thus diverges.

The polarization is

$$\mathbf{P} = -\nabla_{\mathbf{E}} \hat{H} = -\left(b_{12}\mathbf{u} + b_{22}\mathbf{E}\right) . \tag{5.38}$$

The displacement field is

$$\mathbf{D} = \epsilon_0 \mathbf{E} + \mathbf{P} = \epsilon_0 \mathbf{E} - \left(b_{22} - \frac{b_{12}^2/M_{\mathrm{r}}}{\omega_{\mathrm{TO}}^2 - \omega^2}\right) \mathbf{E} = \epsilon_0 \epsilon(\omega) \mathbf{E} . \tag{5.39}$$

Therefore, the dielectric constant is

$$\epsilon(\omega) = \epsilon(\infty) + \frac{\epsilon(0) - \epsilon(\infty)}{1 - \frac{\omega^2}{\omega_{\mathrm{TO}}^2}} . \tag{5.40}$$

Here, $\epsilon(\infty) = 1 - b_{22}/\epsilon_0$ is the high-frequency dielectric constant and $\epsilon(0) = \epsilon(\infty) + b_{12}^2/(b_{11}\epsilon_0)$ the static dielectric constant. The relation (5.40) is shown in Fig. 5.14.

From the Maxwell equation $\nabla \cdot \mathbf{D} = 0$ for zero free charge we obtain the relation

$$\epsilon_0 \epsilon(\omega) \nabla \cdot \mathbf{E} = 0 . \tag{5.41}$$

Thus, either $\epsilon(\omega) = 0$ or $\nabla \cdot \mathbf{E} = 0$, i.e. $\mathbf{u}$ is perpendicular to $\mathbf{k}$. In the latter case we have a TO phonon and, neglecting retardation effects, using

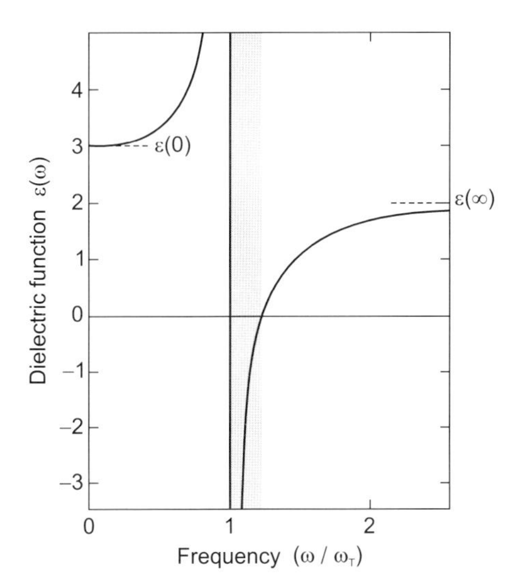

Fig. 5.14. Dielectric function according to (5.40) with $\epsilon(0) = 3$ and $\epsilon(\infty) = 2$ (without damping). *Grey area* denotes the region of negative ϵ

$\nabla \times \mathbf{E} = 0$ we find $\mathbf{E} = 0$ and therefore $\omega = \omega_{\mathrm{TO}}$, justifying our notation. In the case of $\epsilon(\omega) = 0$, we call the related frequency ω_{LO} and find the so-called Lyddane–Sachs–Teller (LST) relation

$$\frac{\omega_{\mathrm{LO}}^2}{\omega_{\mathrm{TO}}^2} = \frac{\epsilon(0)}{\epsilon(\infty)} \, . \tag{5.42}$$

This relation holds reasonably well for optically isotropic, heteropolar materials with two atoms in the basis, such as NaI and also GaAs. Since at high frequencies, i.e. $\omega \gg \omega_{\mathrm{TO}}$, only the individual atoms can be polarized, while for low frequencies the atoms can also be polarized against each other, $\epsilon(0) > \epsilon(\infty)$ and therefore also $\omega_{\mathrm{LO}} > \omega_{\mathrm{TO}}$. For GaAs, the quotient of the two phonon energies is 1.07. Using the LST relation (5.42), we can write for the dielectric constant

$$\epsilon(\omega) = \epsilon(\infty) \left(\frac{\omega_{\mathrm{LO}}^2 - \omega^2}{\omega_{\mathrm{TO}}^2 - \omega^2} \right) \, . \tag{5.43}$$

The (long-wavelength) TO-phonon does not create a long-range electric field. Using $\nabla \cdot \mathbf{D} = 0$ and (5.39) and looking at the longitudinal fields, we have

$$\epsilon_0 \mathbf{E} = b_{12} \mathbf{u} + b_{22} \mathbf{E} \, . \tag{5.44}$$

This can be rewritten as

$$\mathbf{E} = -\omega_{\mathrm{LO}} \left(\frac{M_{\mathrm{r}}}{\epsilon_0} \right)^{1/2} \left(\frac{1}{\epsilon(\infty)} - \frac{1}{\epsilon(0)} \right)^{1/2} \mathbf{u} \propto -\mathbf{u} \, . \tag{5.45}$$

The (long-wavelength) LO-phonon thus creates a long-range electric field acting *against* the ion displacement and represents an additional restoring force. This effect is also consistent with the fact that $\omega_{\mathrm{LO}} > \omega_{\mathrm{TO}}$.

5.3 Elasticity

The elastic properties of the semiconductor are important if the semiconductor is subjected to external forces (pressure, temperature) or to lattice mismatch during heteroepitaxy.

5.3.1 Stress–Strain Relation

In this section, we recall the classical theory of elasticity [110]. The solid is treated as a continuous medium (piecewise homogeneous) and the displacement vector is thus a continuous function $\mathbf{u}(\mathbf{r})$ of the spatial coordinates. When the spatial variation $\nabla \mathbf{u}$ of $\mathbf{u}$ is small, the elastic energy can be written as

$$U = \frac{1}{2} \int \frac{\partial u_l}{\partial x_k} C_{klmn} \frac{\partial u_n}{\partial x_m} \mathrm{d}^3\mathbf{r} , \tag{5.46}$$

where C_{klmn} is the (macroscopic) tensor of the elastic coefficients. 21 components of this tensor can be independent. For crystals with cubic symmetry the number of independent constants is reduced to 3. An exchange $k \leftrightarrow l$ and $m \leftrightarrow n$ does not matter, only six indices have to be considered (xx, yy, zz, yz, xz, and xy). The strain components ϵ_{ij} are symmetrized according to

$$\epsilon_{ij} = \frac{1}{2} \left(\frac{\partial u_j}{\partial x_i} + \frac{\partial u_i}{\partial x_j} \right) . \tag{5.47}$$

The strains ϵ_{xx} are along the main axes of the crystal as visualized in Fig. 5.15.

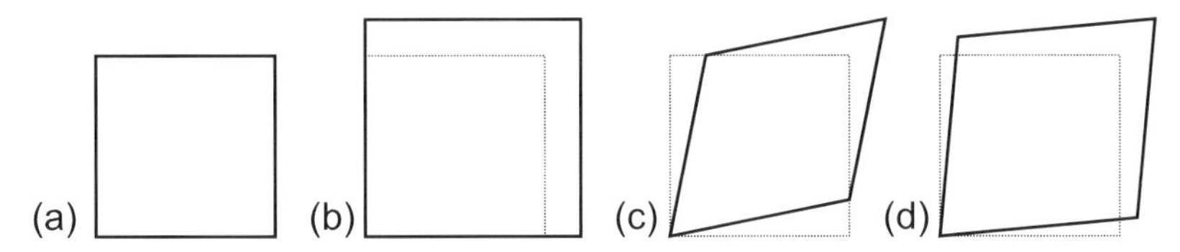

Fig. 5.15. Deformation of a square (**a**). (**b**) Pure hydrostatic deformation ($\epsilon_{xx} = \epsilon_{yy} = 0.2$, $\epsilon_{xy} = 0$), (**c**) pure shear deformation ($\epsilon_{xx} = \epsilon_{yy} = 0$, $\epsilon_{xy} = 0.2$), and (**d**) mixed deformation ($\epsilon_{xx} = \epsilon_{yy} = 0.1$, $\epsilon_{xy} = 0.1$)

The stresses[1] σ_{kl} are then given by

$$\sigma_{kl} = C_{klmn} \epsilon_{mn} . \tag{5.48}$$

The inverse relation is mediated by the stiffness tensor $\mathbf{S}$.

$$\epsilon_{kl} = S_{klmn} \sigma_{mn} . \tag{5.49}$$

Typically, the strain components e_{ij} or e_i are used with the convention $xx \rightarrow 1$, $yy \rightarrow 2$, $zz \rightarrow 3$, $yz \rightarrow 4$, $xz \rightarrow 5$, and $xy \rightarrow 6$:

$$e_{ij} = \epsilon_{ij}(2 - \delta_{ij}) . \tag{5.50}$$

Then, $\sigma_m = C_{mn} e_n$ with the C_{ij} being the elastic constants, i.e. $C_{11} = C_{1111}$, $C_{12} = C_{1122}$ and $C_{44} = C_{1212}$. The x, y, and z directions are the main axes of the cubic solid, i.e. the $\langle 100 \rangle$ directions. Then, we have explicitly

$$\begin{pmatrix} \sigma_1 \\ \sigma_2 \\ \sigma_3 \\ \sigma_4 \\ \sigma_5 \\ \sigma_6 \end{pmatrix} = \begin{pmatrix} C_{11} & C_{12} & C_{12} & 0 & 0 & 0 \\ C_{12} & C_{11} & C_{12} & 0 & 0 & 0 \\ C_{12} & C_{12} & C_{11} & 0 & 0 & 0 \\ 0 & 0 & 0 & C_{44} & 0 & 0 \\ 0 & 0 & 0 & 0 & C_{44} & 0 \\ 0 & 0 & 0 & 0 & 0 & C_{44} \end{pmatrix} \begin{pmatrix} e_1 \\ e_2 \\ e_3 \\ e_4 \\ e_5 \\ e_6 \end{pmatrix} . \tag{5.51}$$

[1]The stress is a force per unit area and has the dimensions of a pressure or energy density.

Values for some semiconductors are given in Table. 5.2. The inverse relation is given by the matrix

$$\begin{pmatrix} S_{11} & S_{12} & S_{12} & 0 & 0 & 0 \\ S_{12} & S_{11} & S_{12} & 0 & 0 & 0 \\ S_{12} & S_{12} & S_{11} & 0 & 0 & 0 \\ 0 & 0 & 0 & S_{44} & 0 & 0 \\ 0 & 0 & 0 & 0 & S_{44} & 0 \\ 0 & 0 & 0 & 0 & 0 & S_{44} \end{pmatrix}, \tag{5.52}$$

with the stiffness coefficients in this notation given by

$$S_{11} = \frac{C_{11} + C_{12}}{(C_{11} - C_{12})(C_{11} + 2C_{12})} \tag{5.53a}$$

$$S_{12} = \frac{C_{12}}{-C_{11}^2 - C_{11}C_{12} + 2C_{12}^2} \tag{5.53b}$$

$$S_{44} = \frac{1}{C_{44}} \,. \tag{5.53c}$$

We emphasize that in this convention (also called the *engineering* convention), e.g. $e_1 = \epsilon_{xx}$ and $e_4 = 2\epsilon_{yz}$. There is also another convention (the *physical* convention) without this factor of two; in this case the matrix in (5.51) contains the elements $2C_{44}$. We note that for an isotropic material the isotropy parameter ξ is zero

$$\xi = \frac{C_0}{C_{44}} = \frac{2C_{44} + C_{12} - C_{11}}{C_{44}} = 0 \,. \tag{5.54}$$

This isotropy condition is not obeyed by real semiconductors as shown in Table 5.2. Another relation, known as the Keating criterion [111, 112]

Table 5.2. Elastic constants of some cubic semiconductors at room temperature. ξ refers to (5.54) and is a measure for the isotropy, I_K refers to the Keating criterion (5.55). For MgO, the Keating criterion is not fulfilled because it has (six-fold coordinated) rocksalt structure and is thus not tetrahedrally bonded

Material	C_{11} (10^{10} Pa)	C_{12} (10^{10} Pa)	$2C_{44}$ (10^{10} Pa)	ξ	I_K
C	107.64	12.52	57.74	−1.3	1.005
Si	16.58	6.39	7.96	−0.56	1.004
Ge	12.85	4.83	6.68	−0.40	1.08
BN	82.0	19.0	48.0	−0.63	1.11
GaAs	11.9	5.34	5.96	−0.20	1.12
InAs	8.33	4.53	3.96	0.08	1.22
AlAs	12.05	4.686	5.94	−0.48	1.03
ZnS	10.46	6.53	4.63	0.30	1.33
MgO	29.7	15.6	9.53	−0.96	0.80

$$I_\mathrm{K} = \frac{2C_{44}(C_{11}+C_{12})}{(C_{11}-C_{12})(C_{11}+3C_{12})} = 1 \tag{5.55}$$

stems from the consideration of bending and stretching of the tetrahedral bonds in the valence force field (VFF) model. It is closely fulfilled (Table 5.2), in particular for the covalent semiconductors.

If $C_0 \neq 0$, the Young's modulus ($\sigma_{\mathbf{nn}} = Y(\mathbf{n})\epsilon_{\mathbf{nn}}$) is nonspherical (see Fig. 5.16a for GaAs). Typically, the $\langle 100 \rangle$ directions are the softest. In terms of the Euler angles (Fig. A.1) the angular dependence of the stiffness constant S_{11} is (Fig. 5.16b,c)

$$\begin{aligned}\frac{S_{11}(\phi,\theta)}{S_{11}^0} = &\frac{C_{11}-C_{12}}{C_0+C_{11}-C_{12}} + \frac{C_0(21C_{11}+10C_{12})}{32(C_0+C_{11}-C_{12})(C_{11}+C_{12})} \\ &+\frac{C_0(C_{11}+2C_{12})}{32(C_0+C_{11}-C_{12})(C_{11}+C_{12})} \times \\ &(4\cos 2\theta + 7\cos 4\theta + 8\cos 4\phi \sin^4\theta) ,\end{aligned} \tag{5.56}$$

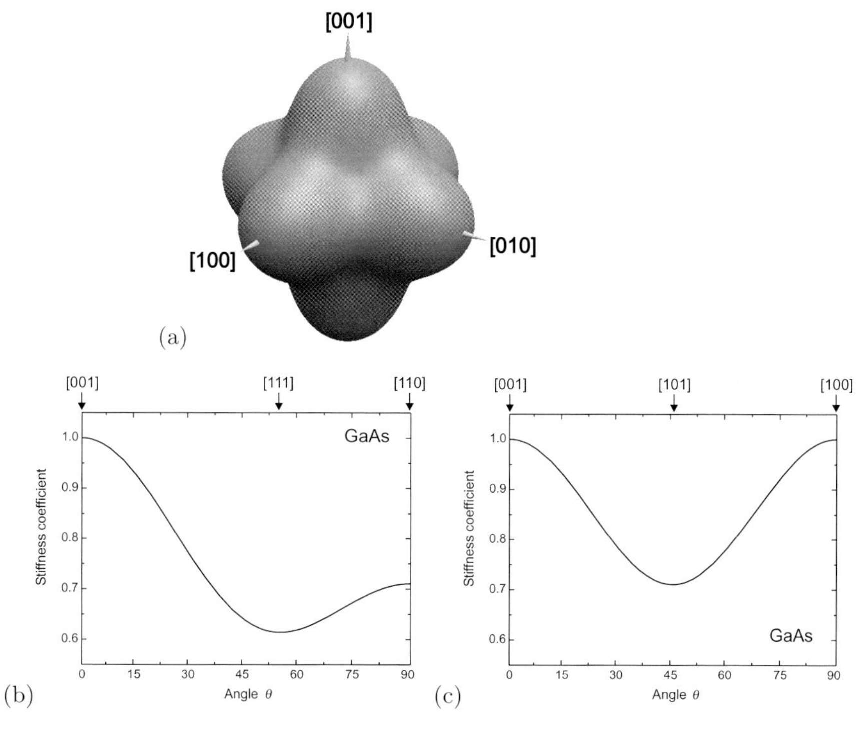

Fig. 5.16. The directional dependence of the inverse of the Young's modulus ($1/Y(\mathbf{n})$) for uniaxial stress (along the direction $\mathbf{n}$) in GaAs. (**a**) 3D view, arbitrary units, (**b**) dependence on polar angle θ in the (01–1) plane (for $\phi = \pi/4$, normalized to S_{11}), (**c**) dependence on polar angle θ in the (010) plane (for $\phi = 0$, normalized to S_{11})

where S^0_{11} is the stiffness along the $\langle 100 \rangle$ directions as given in (5.53a). We note that the Young's modulus Y (Y of (5.57a) is equivalent to $1/S_{11}$ of (5.53a)) and the Poisson ratio ν used for isotropic materials are related to the constants of cubic material by

$$Y = C_{11} - \frac{2C_{12}^2}{C_{11} + C_{12}} \tag{5.57a}$$

$$\nu = \frac{C_{12}}{C_{11} + C_{12}} \,. \tag{5.57b}$$

For isotropic materials Lamé's constants λ and μ are also used. They are given by $C_{11} = \lambda + 2\mu$, $C_{12} = \lambda$ and $C_{44} = \mu$.

Beyond the dependence of the elastic constants on the bond length (as materialized in the phonon frequencies in Fig. 5.8), they depend on the ionicity. In Fig. 5.17, the elastic constants of various zincblende semiconductors are shown as a function of the ionicity f_{i}. The values for the elastic constants are normalized by e^2/a^4, a being the average nearest-neighbor distance.

For wurtzite crystals, five elastic constant are necessary for the stress–strain relation that reads

$$C_{ij} = \begin{pmatrix} C_{11} & C_{12} & C_{13} & 0 & 0 & 0 \\ C_{12} & C_{11} & C_{13} & 0 & 0 & 0 \\ C_{13} & C_{13} & C_{33} & 0 & 0 & 0 \\ 0 & 0 & 0 & C_{44} & 0 & 0 \\ 0 & 0 & 0 & 0 & C_{44} & 0 \\ 0 & 0 & 0 & 0 & 0 & \frac{1}{2}(C_{11} - C_{12}) \end{pmatrix} . \tag{5.58}$$

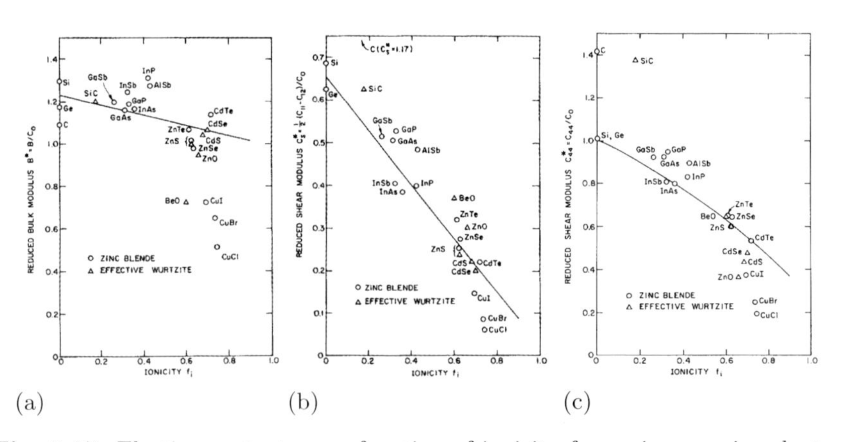

Fig. 5.17. Elastic constants as a function of ionicity for various semiconductors, normalized by e^2/a^4, a being the average nearest-neighbor distance. (**a**) Bulk modulus, $B = (C_{11} + 2C_{12})/3$, (**b,c**) shear moduli, (**b**) $C_{\mathrm{s}} = (C_{11} - C_{12})/2$. *Solid lines* are a simple model as discussed in [113]. Reprinted with permission from [114], ©1972 APS

5.3.2 Biaxial Strain

In heteroepitaxy (cf. Sect. 11.3.1), a biaxial strain situation arises, i.e. layered material is compressed (or expanded in the case of tensile strain) in the interface plane and is expanded (compressed) in the perpendicular direction. Here, we assume that the substrate is infinitely thick, i.e. that the interface remains planar. Substrate bending is discussed in Sect. 5.3.4.

The simplest case is epitaxy on the (001) surface, i.e. $e_1 = e_2 = \epsilon_\parallel$. The component e_3 is found from the condition $\sigma_3 = 0$ (no forces in the z direction). All shear strains are zero. For zincblende material it follows

$$\epsilon_\perp^{100} = e_3 = -\frac{C_{12}}{C_{11}}(e_1 + e_2) = -\frac{2C_{12}}{C_{11}}\,\epsilon_\parallel\,. \tag{5.59}$$

For other crystallographic directions the formula is more involved [115]:

$$\epsilon_\perp^{110} = -\frac{2C_{12} - C_0/2}{C_{11} + C_0/2}\,\epsilon_\parallel \tag{5.60}$$

$$\epsilon_\perp^{111} = -\frac{2C_{12} - 2C_0/3}{C_{11} + 2C_0/3}\,\epsilon_\parallel\,. \tag{5.61}$$

In Fig. 5.18 the strain around misfit dislocations at a GaAs/CdTe heterointerface, as calculated from a TEM image (Fig. 4.3), is shown.

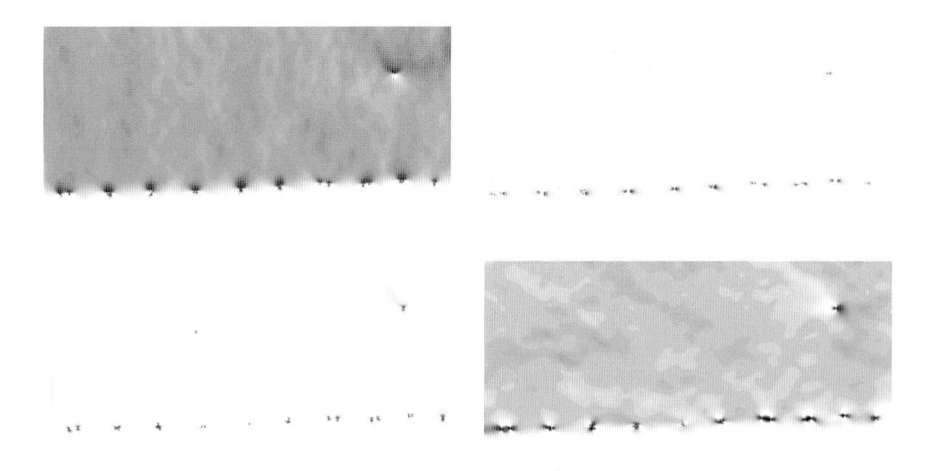

Fig. 5.18. Components $\begin{pmatrix} \epsilon_{xx} & \epsilon_{xz} \\ \epsilon_{zx} & \epsilon_{zz} \end{pmatrix}$ of the strain tensor (with respect to the GaAs lattice constant) of the dislocation array shown in Fig. 4.3, *red/blue*: positive/negative value, *white*: zero. From [77]

For wurtzite crystals and pseudomorphic growth along [00.1] the strain along the c-axis is given by

$$\epsilon_{\perp}^{0001} = -\frac{C_{13}}{C_{33}}(e_1 + e_2) = -\frac{2C_{13}}{C_{33}}\,\epsilon_{\parallel}\;, \tag{5.62}$$

where $\epsilon_{\perp} = (c - c_0)/c_0$ and $\epsilon_{\parallel} = (a - a_0)/a_0$.

5.3.3 Three-Dimensional Strain

The strain distribution in two-dimensional or three-dimensional objects such as quantum wires and dots (see also Chap. 13) is more complicated.

A simple analytical solution for the problem of a strained inclusion is only possible for isotropic material parameters [116].

The solution for a sphere can be extended to yield the strain distribution of an inclusion of arbitrary shape. This scheme applies only for isotropic materials and identical elastic properties of the inclusion and the surrounding matrix. The solution will be given in terms of a surface integral of the boundary of the inclusion, which is fairly easy to handle. Several disconnected inclusions can be treated by a sequence of surface integrals.

The strain distribution for the inner and outer parts of a sphere with radius ρ_0 is given (in spherical coordinates) by

$$\epsilon_{\rho\rho}^{\text{in}} = \frac{2}{3}\epsilon_0\frac{1-2\nu}{1-\nu} = \epsilon_{\theta\theta}^{\text{in}} = \epsilon_{\phi\phi}^{\text{in}} \tag{5.63}$$

$$\epsilon_{\rho\rho}^{\text{out}} = \frac{2}{3}\epsilon_0\frac{1+\nu}{1-\nu}\left(\frac{\rho_0}{\rho}\right)^3 = -2\epsilon_{\theta\theta}^{\text{out}} = -2\epsilon_{\phi\phi}^{\text{out}}\;, \tag{5.64}$$

where ρ denotes the radius, ν the Poisson ratio, and ϵ_0 the relative lattice mismatch of the inclusion and the matrix. The radial displacements are

$$u_{\rho}^{\text{in}} = \frac{2}{3}\epsilon_0\frac{1-2\nu}{1-\nu}\rho \tag{5.65}$$

$$u_{\rho}^{\text{out}} = \frac{2}{3}\epsilon_0\frac{1-2\nu}{1-\nu}\rho_0^3\frac{1}{\rho^2}\;. \tag{5.66}$$

Dividing the displacement by the sphere's volume, we obtain the displacement per unit volume of the inclusion. From the displacement we can derive the stress σ_{ij}^0 per unit volume.

$$\sigma_{ii}^0 = \frac{1}{4\pi}\frac{Y\epsilon_0}{1-\nu}\frac{2x_i^2 - x_j - x_k}{\rho^5} \tag{5.67}$$

$$\sigma_{ij}^0 = \frac{3}{2}\frac{1}{4\pi}\frac{Y\epsilon_0}{1-\nu}\frac{x_i x_j}{\rho^5}\;, \tag{5.68}$$

where i, j and k are pairwise unequal indices. Due to the linear superposition of stresses, the stress distribution σ_{ij}^V for the arbitrary inclusion of volume V can be obtained by integrating over V

$$\sigma_{ij}^V = \int_V \sigma_{ij}^0(\mathbf{r} - \mathbf{r_0}) d^3\mathbf{r} \, . \tag{5.69}$$

The strains can be calculated from the stresses.

When ϵ_0 is constant within V, the volume integral can be readily transformed into an integral over the surface ∂V of V using Gauss' theorem. With the 'vector potentials' $\mathbf{A}_{ij}$ we fulfill div$\mathbf{A}_{ij} = \sigma_{ij}$.

$$\mathbf{A}_{ii} = -\frac{1}{4\pi} \frac{Y\epsilon_0}{1-\nu} \frac{x_i \mathbf{e}_i}{\rho^3} \tag{5.70}$$

$$\mathbf{A}_{ij} = -\frac{1}{2} \frac{1}{4\pi} \frac{Y\epsilon_0}{1-\nu} \frac{x_i \mathbf{e}_j + x_j \mathbf{e}_i}{\rho^3} \, . \tag{5.71}$$

Equation (5.71) is valid for the case $i \neq j$. $\mathbf{e}_i$ is the unit vector in the i-th direction. However, special care must be taken at the singularity $\mathbf{r} = \mathbf{r_0}$ if $\mathbf{r_0}$ lies within V because the stress within the 'δ-inclusion' is not singular (in contrast to the electrostatic analog of a δ-charge). Thus, we find

$$\sigma_{ij}^V(\mathbf{r}_0) = \oint_{\partial V} \mathbf{A}_{ij} d\mathbf{S} + \delta_{ij} \frac{Y\epsilon_0}{1-\nu} \int_V \delta(\mathbf{r} - \mathbf{r_0}) d^3\mathbf{r} \, . \tag{5.72}$$

As an example, we show in Fig. 5.19 the numerically calculated strain components [117] (taking into account the different elastic properties of the dot and matrix materials) in the cross section of a pyramidal InAs quantum dot in a GaAs matrix on top of a two-dimensional InAs layer. The strain component ϵ_{zz} is positive in the 2D layer, as expected from (5.59). However, in the pyramid ϵ_{zz} exhibits a complicated dependence and even takes negative values at the apex.

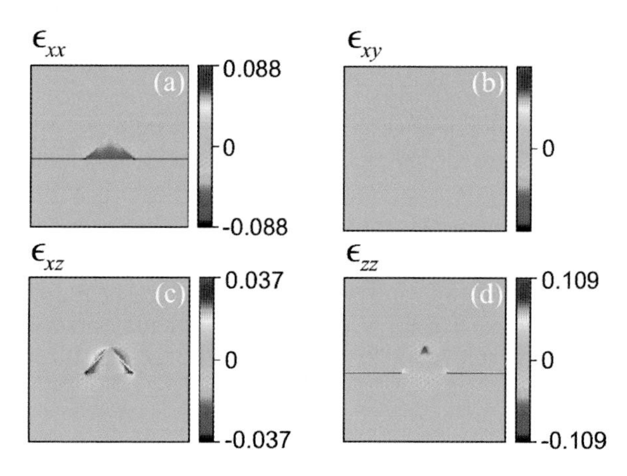

Fig. 5.19. Strain components in an InAs pyramid (quantum dot with {101} faces), embedded in GaAs. The cross section is through the center of the pyramid. The lattice mismatch between InAs and GaAs amounts to ≈ −7%. Reprinted with permission from [117], ©1995 APS

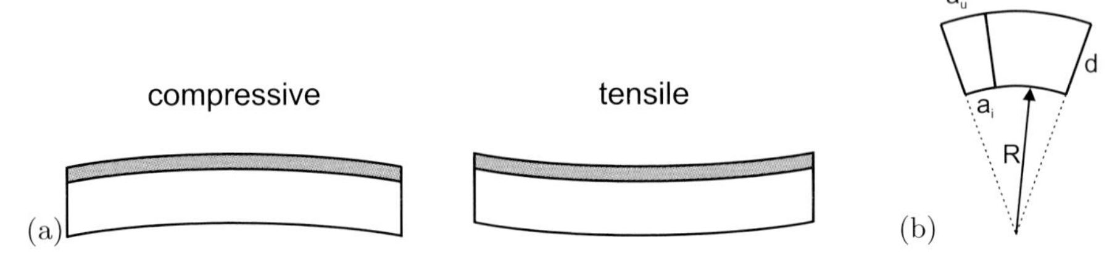

Fig. 5.20. (**a**) Schematic bending of a film/substrate system for compressive (left) and tensile (right) film strain. (**b**) Schematic deformation of curved film of thickness d. The lattice constants at the inner and outer surface are a_{i} and a_{u}, respectively.

5.3.4 Substrate Bending

If a lattice-mismatched layer is pseudomorphically grown on top of a substrate it suffers biaxial strain. For finite substrate thickness part of the strain will relax via substrate bending. If the substrate is circular, a spherical cap is formed. If the lattice constant of the film is larger (smaller) than that of the substrate, the film is under compressive (tensile) strain and the curvature is convex (concave) with respect to the outward normal given by the growth direction (Fig. 5.20a). Substrate bending can also be induced by a mismatch of the thermal expansion coefficients $\alpha^{\mathrm{f}}_{\mathrm{th}}$ and $\alpha^{\mathrm{s}}_{\mathrm{th}}$ of the film and substrate, respectively. If a film/substrate system is flat at a given temperature, e.g. growth temperature, a decrease of temperature, e.g. during cooling, will lead to compressive (tensile) strain if $\alpha^{\mathrm{f}}_{\mathrm{th}}$ is smaller (larger) than $\alpha^{\mathrm{s}}_{\mathrm{th}}$.

In a curved structure, the lattice constant in the tangential direction increases from $a^{\mathrm{t}}_{\mathrm{i}}$ at the inner surface ($r = R = \kappa^{-1}$) to $a^{\mathrm{t}}_{\mathrm{u}}$ at the outer surface ($r = R + d$). Thus, the tangential lattice constant varies with the radial position

$$a^{\mathrm{t}}(r) = a^{\mathrm{t}}_{\mathrm{i}}\,(1 + r\kappa) \ , \tag{5.73}$$

where d is the layer thickness (Fig. 5.20b). Therefore $a_{\mathrm{u}} = a_{\mathrm{i}}(1 + d/R)$. We note that (5.73) holds in all layers of a heterostructure, i.e. the film *and* the substrate.

The lattice constant in the radial direction a^{r}, however, depends on the lattice constant a_0 of the local material and is calculated from the biaxial strain condition, such as (5.59). The in-plane strain is $\epsilon_{\|} = (a^{\mathrm{t}} - a_0)/a_0$ (we assume a spherical cap with $\epsilon_{\|} = \epsilon_{\theta\theta} = \epsilon_{\phi\phi}$). For an isotropic material we find $a^{\mathrm{r}} = a_0\,(1+\epsilon_{\perp})$ with $\epsilon_{\perp} = -2\nu\epsilon_{\|}/(1-\nu)$. The local strain energy density U is given by

$$U = \frac{Y}{1-\nu}\,\epsilon_{\|}^2 \ . \tag{5.74}$$

The total strain energy per unit area U' of a system of two layers with lattice constants a_1, a_2, Young's moduli Y_1, Y_2 and thickness d_1, d_2 (we assume the same Poisson constant ν in both layers) is

$$U' = \int_0^{d_1} U_1 \mathrm{d}r + \int_{d_1}^{d_2} U_2 \mathrm{d}r \,. \tag{5.75}$$

The total strain energy needs to be minimized with respect to a_i and R in order to find the equilibrium curvature κ. We find

$$\kappa = \frac{6a_1a_2(a_2-a_1)d_1d_2(d_1+d_2)Y_1Y_2}{a_2^3d_1^4Y_1^2+\alpha Y_1Y_2+a_1^3d_2^4Y_2^4} \tag{5.76}$$
$$\alpha = a_1a_2d_1d_2\left[-a_2d_1(2d_1+3d_2)+a_1(6d_1^2+9d_1d_2+4d_2^2)\right] \,.$$

For $a_2 = a_1\ (1+\epsilon)$ we develop κ to first order of ϵ and find ($\chi = Y_2/Y_1$) [118, 119]

$$\kappa = \frac{6\chi d_1d_2(d_1+d_2)}{d_1^4+4\chi d_1^3d_2+6\chi d_1^2d_2^2+4\chi d_1d_2^3+\chi^2d_2^4}\,\epsilon \,. \tag{5.77}$$

In the case of a substrate (d_s) with a thin epitaxial layer ($d_\mathrm{f} \ll d_\mathrm{s}$), the radius of curvature is approximately (Stoney's formula [120])

$$\kappa = 6\epsilon \frac{d_\mathrm{f}}{d_\mathrm{s}^2}\frac{Y_\mathrm{f}}{Y_\mathrm{s}} \,. \tag{5.78}$$

Conversely, if the radius of curvature is measured [121], e.g. optically, the film strain can be determined during epitaxy.

5.3.5 Scrolling

In some cases cylindrically scrolled structures are important, e.g. for thin-film flexible electronics, nanotubes and nanoscrolls. The scrolling of thin layers must be avoided by suitable strain management for thin layers that are lifted off from their substrate for transfer to somewhere else. If the film remains attached to its substrate, a scroll can be fabricated as schematically shown in Fig. 5.21. Such structures were first reported in [122].

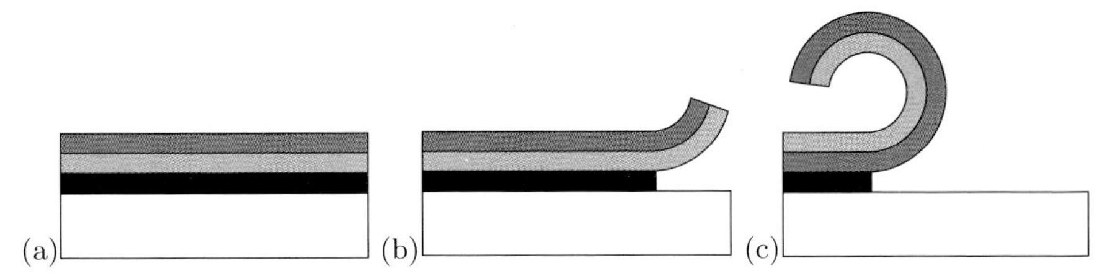

Fig. 5.21. Schematic representation of nanoscroll formation. (**a**) Strained heterostructure (*grey*) that is planar due to large substrate thickness, (**b**) starting removal of sacrificial layer (*black*), (**c**) release of thin film into nanoscroll geometry

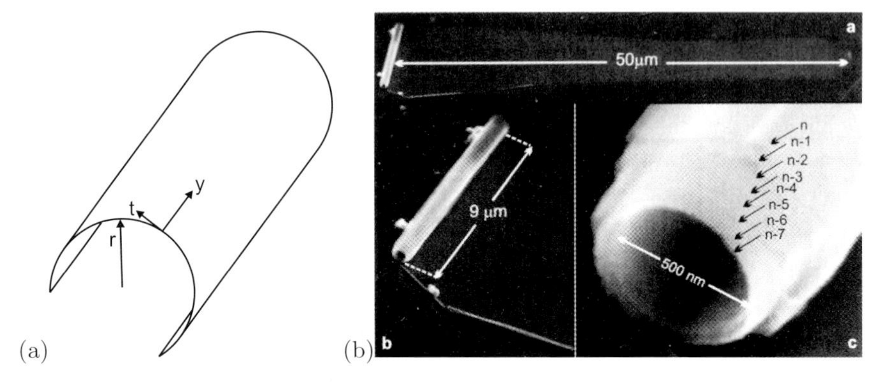

Fig. 5.22. (**a**) Schematic representation of a cylindrically rolled sheet with radial direction r, tangential direction t and direction along the cylinder axis y. (**b**) SEM images of multiwall InGaAs/GaAs nanoscroll rolled up over about 50 μm. Part (**b**) from [123]

If bending strain occurs only in *one* of the tangential directions, the energy density is given by

$$U = \frac{Y}{2(1-\nu^2)} \left(\epsilon_\mathrm{t}^2 + \epsilon_y^2 + 2\nu\epsilon_\mathrm{t}\epsilon_y\right) , \tag{5.79}$$

where ϵ_y is the strain in the unbent direction (cylinder axis) as shown in Fig. 5.22a. For a strained heterostructure made up from two layers the curvature is given by (calculated analogous to (5.77), $\chi = Y_2/Y_1$ [119])

$$\kappa = \frac{6(1+\nu)\chi d_1 d_2 (d_1 + d_2)}{d_1^4 + 4\chi d_1^3 d_2 + 6\chi d_1^2 d_2^2 + 4\chi d_1 d_2^3 + \chi^2 d_2^4}\, \epsilon , \tag{5.80}$$

which differs from (5.77) only by the factor $1+\nu$ in the denominator.

For cubic material and a (001) surface the energy is given as

$$U_{100} = \frac{C_{11} - C_{12}}{2C_{11}} \left[C_{12}(\epsilon_\mathrm{t} + \epsilon_y)^2 + C_{11}(\epsilon_\mathrm{t}^2 + \epsilon_y^2)\right] \tag{5.81}$$

for a scrolling direction along $\langle 100 \rangle$. When the (001)-oriented film winds up along a direction $\langle hk0 \rangle$ having an angle ϕ with the [100] direction ($\phi = 45°$ for $\langle 110 \rangle$), the strain energy is given by (C_0 is given by (5.54))

$$U_\phi = U_{100} + C_0 \left(\frac{\epsilon_\mathrm{t} - \epsilon_y}{2}\right)^2 \sin^2(2\phi) . \tag{5.82}$$

The strain energy vs. bending radius ($= \kappa^{-1}$) is shown for a SiGe nanoscroll in Fig. 5.23. First, the relaxation along the cylinder axis plays a minor role. The smallest strain energy is reached for scrolling along $\langle 100 \rangle$, also yielding the smaller bending radius (larger curvature). Therefore, the film preferentially scrolls along $\langle 100 \rangle$. This explains the observed 'curl' behavior of scrolls winding up for $\phi \neq 0$ (Fig. 5.23b).

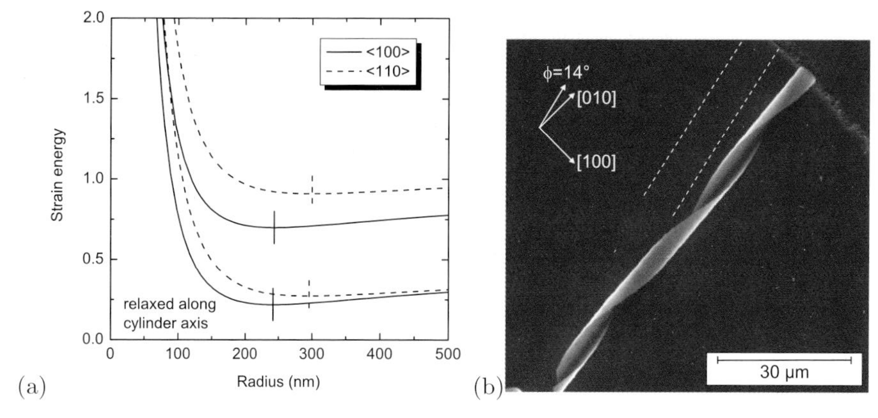

Fig. 5.23. (**a**) Strain energy (in units of the strain energy of the flat pseudomorphic layers) of a scroll of a 4-layer SiGe structure ($Si_{0.3}Ge_{0.7}$, $Si_{0.6}Ge_{0.4}$ and $Si_{0.8}Ge_{0.2}$, each 3 nm thick and a 1-nm Si cap) as a function of radius for winding directions along $\langle 100 \rangle$ and $\langle 110 \rangle$. *Top* (*bottom*) curves without (with complete) strain relaxation along the cylinder axis. *Vertical lines* indicate the positions of the respective energy minima [119]. (**b**) SEM image of curled InGaAs/GaAs nanoscroll rolled $\phi = 14°$ off $\langle 100 \rangle$. The stripe from which the film was rolled off is indicated by *white dashed lines*. Part (**b**) from [125]

5.3.6 Critical Thickness

Strained epitaxial films are called *pseudomorphic* when they do not contain defects and the strain relaxes elastically, e.g. by tetragonal distortion. When the layer thickness increases, however, strain energy is accumulated that will lead at some point to plastic relaxation via the formation of defects. In many cases, a grid of misfit dislocations forms at the interface (Figs. 4.6 and 5.24). The average distance p of the dislocations is related to the misfit $f = (a_1 - a_2)/a_2$ and the edge component $b_\perp$ of the Burger's vector (for a 60° dislocation $b_\perp = a_0/\sqrt{8}$)

$$p = \frac{b_\perp}{f} . \tag{5.83}$$

Two mechanisms have been proposed for the formation of misfit dislocations (Fig. 5.25), the elongation of a grown-in threading dislocation [127] and the nucleation and growth of dislocation half-loops [128]. For the modeling of such systems a mechanical approach based on the forces on dislocations [127] or an energy consideration based on the minimum strain energy necessary for defect formation [128–130] can be followed. Both approaches have been shown to be equivalent [131] (if a periodic array of dislocations is considered). In [132] it was pointed out that the finite speed of plastic flow also has to be considered to explain experimental data. Temperature affects the observed critical thickness and a kinetic model is needed.

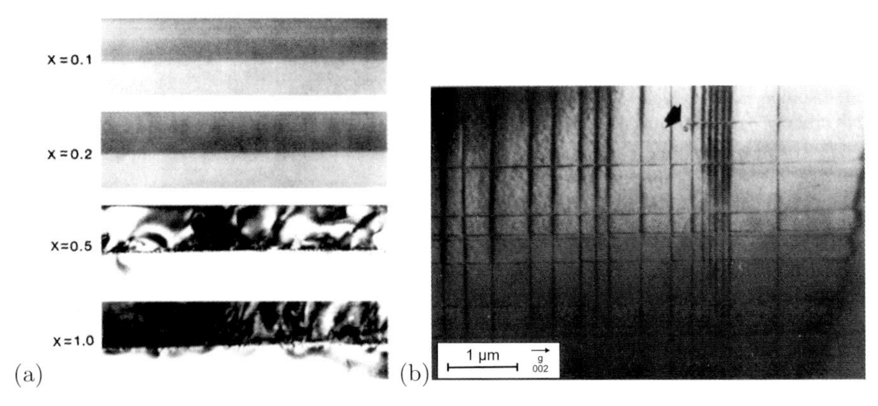

Fig. 5.24. (**a**) Series of cross-sectional TEM images of 100-nm thick Ge_xSi_{1-x} layers on Si(001) with different ternary compositions $x = 0.1, 0.2, 0.5$, and 1.0. The growth temperature was 550°C. The transition from commensurate to incommensurate growth is obvious. From [124]. (**b**) Plan view ⟨022⟩ TEM bright field image of a 250-nm $Ge_{0.15}Si_{0.85}$ layer on Si (001), annealed at about 700°C. The arrow denotes the position of a dislocation loop. Reprinted with permission from [126], ©1989 AVS

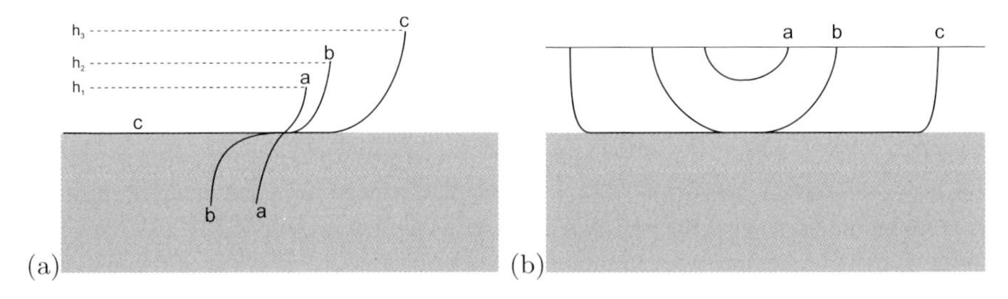

Fig. 5.25. Schematic formation of misfit dislocations by (**a**) elongation of a grown-in threading dislocation and (**b**) by the nucleation and growth of dislocation half-loops. (**a**) depicts a threading dislocation. Initially, for thickness h_1 the interface is coherent 'a', for larger thickness h_2 the interface is critical and the force of the interface on the dislocation is equal to the tension in the dislocation line, 'b'. For larger thickness, e.g. h_3, the dislocation line is elongated in the plane of the interface, 'c'. In (**b**) 'a' denotes a subcritical dislocation half-loop, 'b' depicts a half-loop being stable under the misfit stress and for 'c' the loop has grown under the misfit stress into a misfit dislocation line along the interface

In the following, isotropic materials and identical elastic constants of substrate and thin film are assumed, following [131]. The interface plane is the (x,y)-plane, the growth direction is z. The energy E_d of a periodic dislocation array with period p and Burgers vector $\mathbf{b} = (b_1, b_2, b_3)$ is

$$
\begin{aligned}
E_{\mathrm{d}} &= \frac{Y}{8\pi(1-\nu^2)}\beta^2 \qquad (5.84)\\
\beta^2 &= \left[b_1^2 + (1-\nu)b_2^2 + b_3^2\right] \ln\left(\frac{p\left[1-\exp(-4\pi h/p)\right]}{2\pi q}\right)\\
&+ \left(b_1^2 - b_3^2\right)\frac{4\pi h}{p}\frac{\exp(-4\pi h/p)}{1-\exp(-4\pi h/p)}\\
&- \frac{1}{2}\left(b_1^2 + b_3^2\right)\left(\frac{4\pi h}{p}\right)^2 \frac{\exp(-4\pi h/p)}{\left[1-\exp(-4\pi h/p)\right]^2}\\
&+ b_3^2 \frac{2\pi h}{p}\frac{\exp(-2\pi h/p)}{1-\exp(-2\pi h/p)}\,,
\end{aligned}
$$

where h is the film thickness and q denotes the cutoff length for the dislocation core, taken as $q = b$. The misfit strain including the relaxation due to dislocations with Burger's vectors $\mathbf{b}$ and $\hat{\mathbf{b}}$ in the two orthogonal interface $\langle 110 \rangle$ directions $\mathbf{n}$ and $\hat{\mathbf{n}}$. We chose the coordinate system such that $\mathbf{n} = (1,0,0)$ and $\hat{\mathbf{n}} = (0,1,0)$ (the z direction remains). With respect to these axes the Burger's vectors are $\left(\pm\frac{1}{2}, \frac{1}{2}, \frac{1}{\sqrt{2}}\right) a_0/\sqrt{2}$. The misfit strain $\epsilon_{ij}^{\mathrm{m}}$ is reduced due to the dislocation formation to the 'relaxed' misfit strain $\epsilon_{ij}^{\mathrm{r}}$ with

$$
\epsilon_{ij}^{\mathrm{r}} = \epsilon_{ij}^{\mathrm{m}} + \frac{b_i n_j + b_j n_i}{2p} + \frac{\hat{b}_i \hat{n}_j + \hat{b}_j \hat{n}_i}{2p}\,, \qquad (5.85)
$$

with an associated stress σ_{ij}. The strain energy E_{s} of the layer due to the relaxed misfit is then

$$
\begin{aligned}
E_{\mathrm{s}} &= \frac{1}{2} h \sigma_{ij} \epsilon_{ij}^{\mathrm{r}} \qquad (5.86)\\
\lim_{p\to\infty} E_{\mathrm{s}} &= 2h \frac{Y(1+\nu)}{1-\nu} f^2\,. \qquad (5.87)
\end{aligned}
$$

The total strain energy E is given by

$$
\begin{aligned}
pE &= 2E_{\mathrm{d}} + 2E_{\mathrm{c}} + pE_{\mathrm{s}} \qquad (5.88)\\
E_{\infty} &= \lim_{p\to\infty} E\,, \qquad (5.89)
\end{aligned}
$$

with the core energy E_{c} of the dislocation that needs to be calculated with an atomistic model (not considered further here). This energy is shown in Fig. 5.26a for the material parameters of $Ge_{0.1}Si_{0.9}/Si(001)$ (misfit -0.4%) for various layer thicknesses as a function of $1/p$. This plot looks similar to that for a first-order phase transition (with $1/p$ as the order parameter, cmp. Fig. 14.7a). For a certain critical thickness h_{c1} the energy of the layer without any dislocation and the layer with a particular dislocation density p_1 are identical ($E - E_{\infty} = 0$) and additionally $\partial E/\partial p|_{p=p_1} = 0$. However, between $p \to \infty$ and $p = p_0$ there is an energy barrier. The critical thickness h_{c2} is reached when

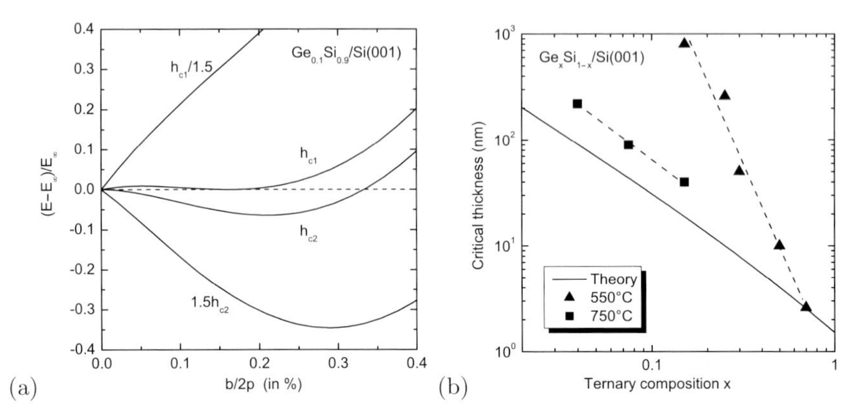

Fig. 5.26. (**a**) Theoretical calculation for the strain energy vs. inverse dislocation density for various thicknesses of $Ge_{0.1}Si_{0.9}$ layers on Si (001). The ordinate is $b/2p$, $b/2$ being the edge component of the Burgers vector and p being the dislocation spacing. The abscissa is the strain energy E scaled with E_∞ (5.89). (**b**) Critical thickness for Ge_xSi_{1-x} layers on Si (001). The *solid line* is theory (h_{c2}) according to (5.91). Data points are from [135] (*squares*, growth temperature of 750°C) and from [124] (*triangles* for growth temperature of 550°C)

$$\partial E/\partial p\,|_{p\to\infty} = 0\;, \tag{5.90}$$

i.e. the energy decreases monotonically for increasing dislocation density up to the global energy minimum at a certain equilibrium dislocation spacing p_2. Equation (5.90) leads to the following implicit equation for the determination of h_{c2}:

$$h_{c2} = \frac{b\left[-16 + 3b^2 + 8(-4+\nu)\ln\left(2h_{c2}/q\right)\right]}{128 f\pi(1+\nu)}\;, \tag{5.91}$$

with the length of the Burgers vector $b = a_0/\sqrt{2}$.

The theoretical dependence of h_{c2} for Ge_xSi_{1-x}/Si(001) with varying composition is shown in Fig. 5.26b together with experimental data. The critical thickness for a fairly high growth temperature is much closer to the energetic equilibrium than that deposited at lower temperature. This shows that there are kinetic limitations for the system to reach the mechanical equilibrium state. Also, the experimental determination of the critical thickness is affected by finite resolution for large dislocation spacing, leading generally to an overestimate of h_c.

In zincblende materials two types of dislocations are possible, α and β, with Ga- and As-based cores, respectively. They have $[\bar{1}10]$ and $[110]$ line directions for a compressively strained interface. The α dislocation has the larger glide velocity. Therefore, strain relaxation can be anisotropic with regard to the $\langle 110\rangle$ directions for zincblende material, e.g. InGaAs/GaAs [133, 134].

5.4 Cleaving

The cleavage planes of the diamond structure are $\{111\}$ planes. It is easiest to break the bonds connecting the double layers in the $\langle 111 \rangle$ directions.

The cleavage planes of the zincblende structure are $\{110\}$ planes. Due to the ionic character, breaking the bonds connecting the double layers in the $\langle 111 \rangle$ directions would leave charged surfaces, which is energetically unfavorable. The $\{100\}$ planes contain only one sort of atom and would also leave highly charged surfaces. The $\{110\}$ planes contain equal amounts of A and B atoms and are neutral.

Ideally, the cleaving plane is atomically flat or exhibits large monoatomically flat terraces. However, certain dopants in high concentrations, e.g. GaAs:Te, can induce a rough surface due to lattice distortion [136].

6 Band Structure

Silicon is a metal.
A.H. Wilson, 1931 [14]

6.1 Introduction

Valence electrons that move in the crystals feel a periodic potential

$$U(\mathbf{r}) = U(\mathbf{r} + \mathbf{R}) \tag{6.1}$$

for all vectors $\mathbf{R}$ of the direct lattice. The potential[1] is due to the effect of the ion cores and all other electrons. Thus a serious many-body problem is present. In principle, the band structure can be calculated from the periodic arrangements of the atoms and their atomic order number. We note that for some problems, e.g. the design of optimal solar cells, a certain band structure is known to be ideal and a periodic atomic arrangement, i.e. a material, needs to be found that generates the optimal band structure. This problem is called the *inverse band structure problem.*

6.2 Bloch's Theorem

First, we will deduce some general conclusions about the structure of the solution as a consequence of the periodicity of the potential. We first investigate the solution of a Schrödinger equation of the type

$$H\Psi(\mathbf{r}) = \left(-\frac{\hbar^2}{2m}\nabla^2 + U(\mathbf{r}) \right) \Psi(\mathbf{r}) = E\Psi(\mathbf{r}) \tag{6.2}$$

for an electron. U will be periodic with the lattice, i.e. it will obey (6.1).

[1] In this book the form of the potential will never be explicitly given.

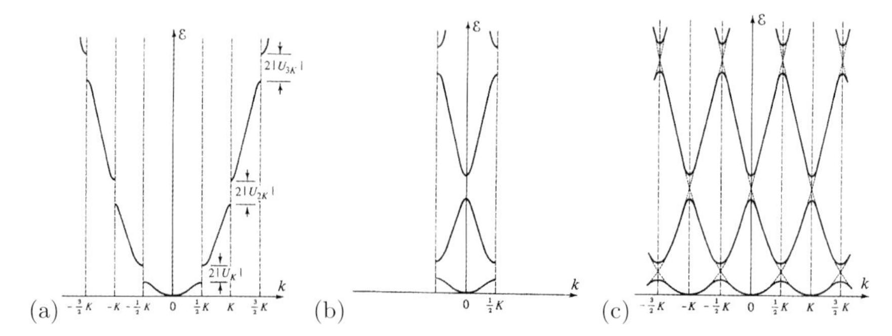

Fig. 6.1. Zone schemes for a band structure: (**a**) extended, (**b**) reduced and (**c**) repetitive zone scheme

Bloch's theorem says that the eigenstates Ψ of a one-particle Hamiltonian as in (6.2) can be written as the product of plane waves and a lattice-periodic function, i.e.

$$\Psi_{n\mathbf{k}}(\mathbf{r}) = A e^{\mathrm{i}\mathbf{k}\mathbf{r}} u_{n\mathbf{k}}(\mathbf{r}) \ . \tag{6.3}$$

The normalization constant A is often omitted. If $u_{n\mathbf{k}}(\mathbf{r})$ is normalized, $A = 1/\sqrt{V}$, where V is the integration volume. The wavefunction is indexed with a quantum number n and the wavevector $\mathbf{k}$. The key is that the function $u_{n\mathbf{k}}(\mathbf{r})$, the so-called Bloch function, is periodic with the lattice, i.e.

$$u_{n\mathbf{k}}(\mathbf{r}) = u_{n\mathbf{k}}(\mathbf{r} + \mathbf{R}) \tag{6.4}$$

for all vectors $\mathbf{R}$ of the direct lattice. The proof is simple in one dimension and more involved in three dimensions with possibly degenerate wavefunctions, see [137].

If $E_{n\mathbf{k}}$ is an energy eigenvalue, then $E_{n\mathbf{k}+\mathbf{G}}$ is also an eigenvalue for all vectors $\mathbf{G}$ of the reciprocal lattice, i.e.

$$E_n(\mathbf{k}) = E_n(\mathbf{k} + \mathbf{G}) \ . \tag{6.5}$$

Thus the energy values are periodic in reciprocal space. The proof is simple, since the wavefunction (for $\mathbf{k} + \mathbf{G}$) $\exp\left(\mathrm{i}(\mathbf{k} + \mathbf{G})\mathbf{r}\right) u_{n(\mathbf{k}+\mathbf{G})}(\mathbf{r})$ is for $u_{n(\mathbf{k}+\mathbf{G})}(\mathbf{r}) = \exp\left(-\mathrm{i}\mathbf{G}\mathbf{r}\right) u_{n\mathbf{k}}(\mathbf{r})$ obviously an eigenfunction to $\mathbf{k}$.

A band structure along one $\mathbf{k}$-direction can be displayed in various zone schemes as depicted in Fig. 6.1. The most frequently used scheme is the *reduced* zone scheme. In three dimensions, the band structure is typically shown along particular paths in the Brillouin zone, as depicted, e.g., in Fig. 6.2c.

6.3 Free-Electron Dispersion

If the entire wavefunction (from (6.3)) obeys the Schrödinger equation (6.2), the Bloch function $u_{n\mathbf{k}}$ fulfills the equation

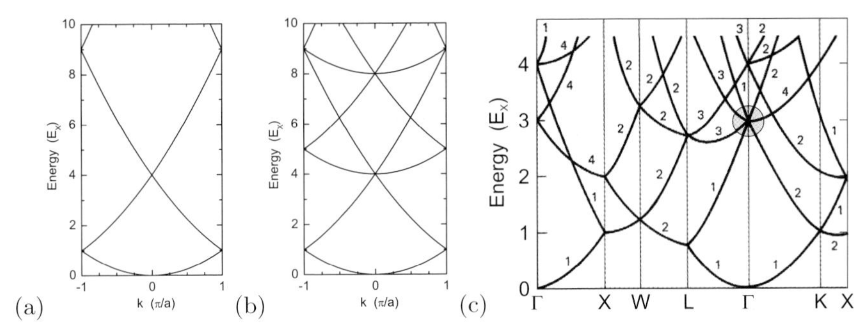

Fig. 6.2. Dispersion of free electrons (empty lattice calculation, $U = 0$, shown in the first Brillouin zone) in (**a**) a one-dimensional lattice ($\mathbf{G} = n\, 2\pi/a$), (**b**) a simple cubic lattice ($\mathbf{G} = (h, k, l)\, 2\pi/a$) and (**c**) in a fcc lattice. The energy is measured in units of the energy at the X-point, $E_{\mathrm{X}} = (\hbar^2/2m)(\pi/a)^2$. The *shaded circle* in (**c**) represents the region where the bandgap develops for finite periodic potential $U \neq 0$

$$\left(-\frac{1}{2m}(\mathbf{p}+\mathbf{k})^2 + U(\mathbf{r})\right) u_{n\mathbf{k}}(\mathbf{r}) = E_{n\mathbf{k}} u_{n\mathbf{k}}(\mathbf{r}) \,, \tag{6.6}$$

which is easy to see from $\mathbf{p} = -i\hbar\nabla$.

First, we discuss the simplest case of a periodic potential, $U \equiv 0$. This calculation is also called the *empty lattice* calculation. The solution of (6.6) is then just constant, i.e. $u_{\mathbf{k}} = c$ and $\Psi_{\mathbf{k}}(\mathbf{r}) = c \exp(i\mathbf{k}\mathbf{r})$. The dispersion of the free electron is then given by

$$E(\mathbf{k}) = \frac{\hbar^2}{2m}\mathbf{k}^2 \,, \tag{6.7}$$

where $\mathbf{k}$ is an arbitrary vector in the reciprocal space. $\mathbf{k}'$ is a vector from the Brillouin zone such that $\mathbf{k} = \mathbf{k}' + \mathbf{G}$ with a suitable reciprocal lattice vector $\mathbf{G}$. Because of (6.5) the dispersion relation can be written also as

$$E(\mathbf{k}) = \frac{\hbar^2}{2m}(\mathbf{k}' + \mathbf{G})^2 \,, \tag{6.8}$$

where $\mathbf{k}'$ denotes a vector from the Brillouin zone. Thus, many branches of the dispersion relation arise from using various reciprocal lattice vectors in (6.8).

The resulting dispersion relation for the free electron is shown in Fig. 6.2a for a one-dimensional system ($\mathbf{k}'$ and $\mathbf{G}$ are parallel) and in Fig. 6.2b for the simple cubic lattice (in the so-called reduced zone scheme). In Fig. 6.2c, the (same) dispersion of the free electron is shown for the fcc lattice.

6.4 Kronig–Penney Model

The Kronig–Penney model [13] is a simple, analytically solvable model that visualizes the effect of the periodic potential on the dispersion relation of the electrons, i.e. the formation of a band structure.

A one-dimensional periodic hard-wall potential of finite height is assumed (Fig. 6.3a). The well width is a, the barrier width b and thus the period $P = a + b$. The potential is zero in the well (regions $(0, a) + nP$) and $+U_0$ in the barrier. The Schrödinger equation

$$-\frac{\hbar^2}{2m}\frac{\partial^2 \Psi}{\partial x^2} + U(x)\Psi(x) = E\Psi(x) \tag{6.9}$$

has to be solved. The solutions for a single hard-wall potential well are well known. In the well, they have oscillatory character, i.e. $\Psi \propto \exp(\pm \mathrm{i}kx)$ with real k. In the barrier, they have exponential character, i.e. $\Psi \propto \exp(\pm kx)$ with real k. Thus we chose

$$\Psi(x) = A \exp(\mathrm{i}Kx) + B \exp(-\mathrm{i}Kx) \tag{6.10a}$$

$$\Psi(x) = C \exp(\kappa x) + D \exp(-\kappa x) \ . \tag{6.10b}$$

The wavefunction from (6.10a) is for the well between 0 and a with $E = \hbar^2 K^2/2m$. The wavefunction from (6.10b) is for the barrier between a and $a + b$ with $U_0 - E = \hbar^2\kappa^2/2m$. From the periodicity and Bloch's theorem the wavefunction at $x = -b$ must have the form $\Psi(-b) = \exp(-\mathrm{i}kP)\Psi(a)$, i.e. between the two wavefunctions is only a phase factor. The wavevector k of the Bloch function (plane-wave part of the solution) is a new quantity and must be carefully distinguished from K and κ.

Both K and κ are real numbers. As boundary conditions, the continuity of Ψ and Ψ' are used.[2] At $x = 0$ and $x = a$ this yields

$$A + B = C + D \tag{6.11a}$$

$$\mathrm{i}KA - \mathrm{i}KB = \kappa C - \kappa D \tag{6.11b}$$

$$A\exp(\mathrm{i}Ka) + B\exp(-\mathrm{i}Ka) = C\exp(\kappa a) + D\exp(-\kappa a) \tag{6.11c}$$

$$\mathrm{i}KA\exp(\mathrm{i}Ka) - \mathrm{i}KB\exp(-\mathrm{i}Ka) = \kappa C\exp(\kappa a) - \kappa D\exp(-\kappa a) . \tag{6.11d}$$

The continuity of Ψ and Ψ' at $x = -b$ is used in the left sides of (6.11c,d).

A nontrivial solution arises only if the determinant of the coefficient matrix is zero. This leads (after some tedious algebra) to

$$\cos(kP) = \left[\frac{\kappa^2 - K^2}{2\kappa K}\right] \sinh(\kappa b)\sin(Ka) + \cosh(\kappa b)\cos(Ka) \ . \tag{6.12}$$

---

[2]Generally, Ψ'/m should be continuous, however, in the present example the mass is assumed constant throughout the structure.

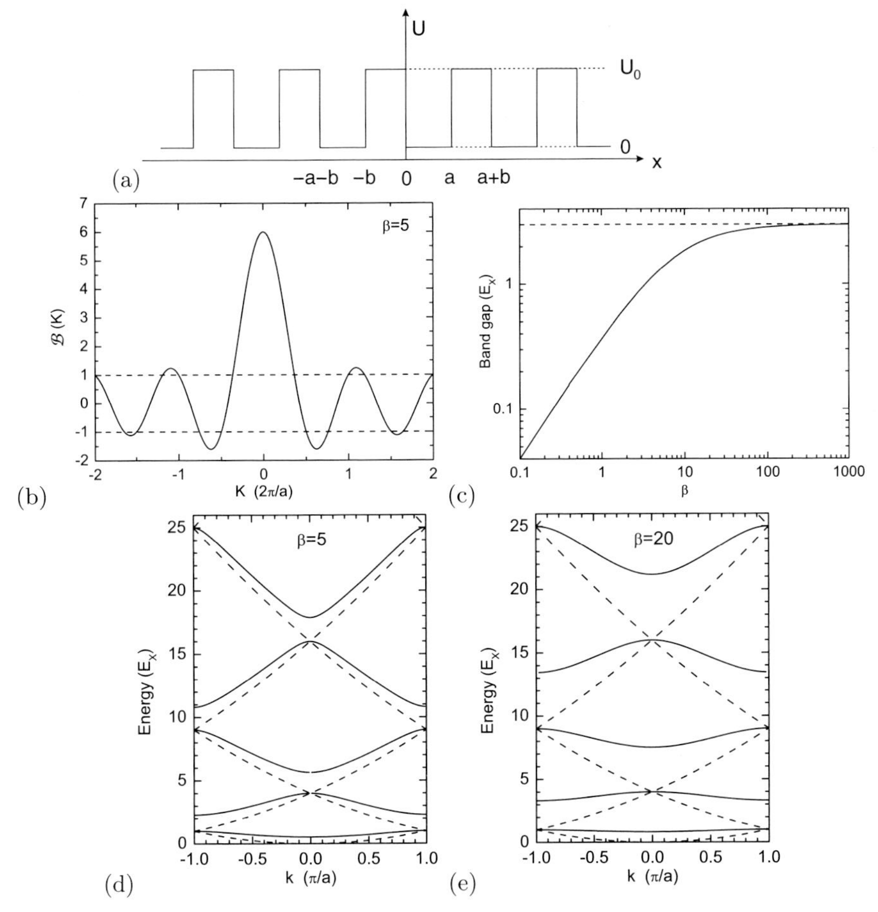

Fig. 6.3. (**a**) One-dimensional periodic hard-wall potential (Kronig–Penney model). (**b**) Transcendental function $\mathcal{B}(K)$ from (6.13) for $\beta = 5$. The *dashed lines* indicate the $[-1, 1]$ interval for which solutions exist for (6.13). (**c**) Bandgap between first and second subband (in units of $E_X = (\hbar^2/2m)(\pi/a)^2$) as a function of β. For smaller β the bandgap is $\propto \beta$. For thick barriers ($\beta \to \infty$) the bandgap saturates towards $3E_X$ as expected for uncoupled wells. (**d**,**e**) The resulting energy dispersion (in units of E_X) as a function of the superlattice wavevector k for (**d**) $\beta = 5$ and (**e**) $\beta = 20$ in (6.13). The *dashed lines* are the free-electron dispersion (for $\beta = 0$) (cf. Fig. 6.2a)

Further simplification can be reached by letting the barrier thickness $b \to 0$ and $U_0 \to \infty$. Then $P \to a$. The limit, however, is performed in such a way that the barrier 'strength' $U_0 b \propto \kappa^2 b$ remains constant and finite. Equation (6.12) then reads (for $\kappa b \to 0$: $\sinh(\kappa b) \to \kappa b$ and $\coth(\kappa b) \to 1$):

$$\cos(ka) = \beta \frac{\sin(Ka)}{Ka} + \cos(Ka) = \mathcal{B}(K) \,. \tag{6.13}$$

The coupling strength $\beta = \kappa^2 ba/2$ represents the strength of the barrier. Equation (6.13) only has a solution if the right side is in the interval $[-1, 1]$ (Fig. 6.3b). The function $\sin(x)/x$ oscillates with decreasing amplitude such that for sufficiently high values of Ka a solution can always be found. The resulting dispersion is shown in Fig. 6.3c. The dispersion is different from the free-electron dispersion and has several separated bands. The bandgaps are related to the K values, i.e. energies for which (6.13) cannot be fulfilled. At the zone boundary ($k = \pi/a$) the bands are split and the tangent is horizontal ($\mathrm{d}E/\mathrm{d}k = 0$). The form of the dispersion is similar to the arccos function.

For large coupling between the potential wells (small β, $\beta \lesssim 1$) the bandgap E_{12} between the first and the second subband at the X-point is $E_{12} = \beta(4/\pi^2)E_{\mathrm{X}}$ with $E_{\mathrm{X}} = (\hbar^2/2m)(\pi/a)^2$. In this case, the width of the subbands is wide. For small coupling (large β) the bandgap E_{12} converges towards $3E_{\mathrm{X}}$ as expected (Fig. 6.3c) for decoupled potential wells with energy levels $E_n = E_{\mathrm{X}} n^2$ and the width of the bands is small.

6.5 Electrons in a Periodic Potential

In this section, we will discuss the solution of a general wave equation for electrons in a periodic potential. The solution is investigated particularly at the zone boundary. The potential U is periodic with the lattice (6.1). It can be represented as a Fourier series with the reciprocal lattice vectors (cf. (3.16)):

$$U(\mathbf{r}) = \sum_{\mathbf{G}} U_{\mathbf{G}} \exp(\mathrm{i}\mathbf{G}\mathbf{r}) \ . \tag{6.14}$$

Since U is a real function, $U_{-\mathbf{G}} = U^*_{\mathbf{G}}$. The deeper reason for the success of such an approach is that for typical crystal potentials, the Fourier coefficients decrease rapidly with increasing $\mathbf{G}$, e.g. for the unscreened Coulomb potential $U_{\mathbf{G}} \propto 1/G^2$. The wavefunction is expressed as a Fourier series (or integral) over all allowed (Bloch) wavevectors $\mathbf{K}$,

$$\Psi(\mathbf{r}) = \sum_{\mathbf{K}} C_{\mathbf{K}} \exp(\mathrm{i}\mathbf{K}\mathbf{r}) \ . \tag{6.15}$$

The kinetic and potential energy terms in the Schrödinger equation (6.6) are

$$\nabla^2 \Psi = -\sum_{\mathbf{K}} \mathbf{K}^2 C_{\mathbf{K}} \exp(\mathrm{i}\mathbf{K}\mathbf{r}) \tag{6.16a}$$

$$U\Psi = \sum_{\mathbf{G}} \sum_{\mathbf{K}} U_{\mathbf{G}} C_{\mathbf{K}} \exp[\mathrm{i}(\mathbf{G}+\mathbf{K})\mathbf{r}] \ . \tag{6.16b}$$

With $\mathbf{K}' = \mathbf{K} + \mathbf{G}$, (6.16b) can be rewritten as

$$U\Psi = \sum_{\mathbf{G}} \sum_{\mathbf{K}'} U_{\mathbf{G}} C_{\mathbf{K}'-\mathbf{G}} \exp(\mathrm{i}\mathbf{K}'\mathbf{r}) . \tag{6.17}$$

Now, the Schrödinger equation can be written as an (infinite) system of algebraic equations:

$$(\lambda_{\mathbf{K}} - E)C_{\mathbf{K}} + \sum_{\mathbf{G}} U_{\mathbf{G}} C_{\mathbf{K}-\mathbf{G}} = 0 , \tag{6.18}$$

with $\lambda_{\mathbf{K}} = \hbar^2 \mathbf{K}^2/(2m)$.

6.5.1 Approximate Solution at the Zone Boundary

We assume that the potential energy has only one important Fourier coefficient U for the shortest reciprocal lattice vector $\mathbf{G}$. Also, we have $U_{-\mathbf{G}} = U_{\mathbf{G}}$. Thus, the (one-dimensional) potential has the form $U(x) = 2U\cos(Gx)$. We consider the solution at the zone boundary, i.e. in the vicinity of $\mathbf{K} = \mathbf{G}/2$. The kinetic energy is then the same for $\mathbf{K} = \pm\mathbf{G}/2$, i.e. $\lambda_{\mathbf{K}} = \lambda_{\mathbf{K}-\mathbf{G}} = (\hbar^2/2m)\,(G^2/4)$. $C_{\mathbf{G}/2}$ is an important coefficient of the wavefunction, thus $C_{-\mathbf{G}/2}$ will also be important. We will now limit the $\mathbf{K}$ vectors for the series of (6.15) to just these two. For the coefficient vector $\begin{pmatrix} C_{\mathbf{G}/2} \\ C_{-\mathbf{G}/2} \end{pmatrix}$ the determinant is

$$\begin{vmatrix} \lambda - E & U \\ U & \lambda - E \end{vmatrix} = (\lambda - E)^2 - U^2 = 0 , \tag{6.19}$$

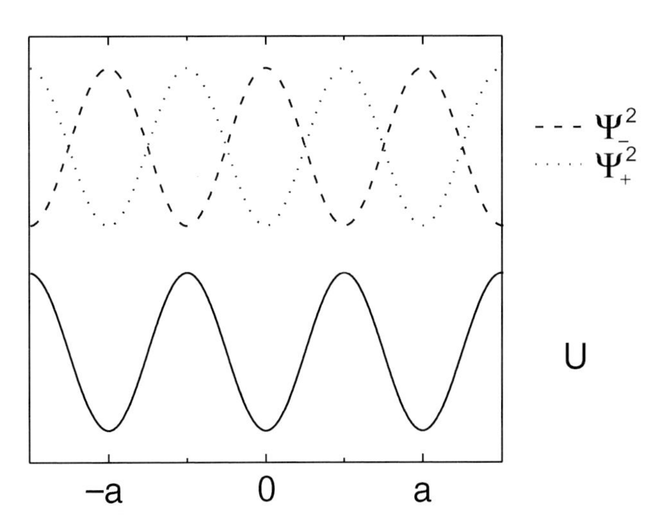

Fig. 6.4. Periodic potential U (one-dimensional cosine, *solid line*) and the squares of the wavefunctions Ψ_+ (*dashed line*) and Ψ_- (*dotted line*) for the wavevector at the zone boundary, $K = G/2 = \pi/a$

with $\lambda = \lambda_{\mathbf{K}} = \lambda_{-\mathbf{K}} = (\hbar^2/2m)\,(G^2/4)$ for $\mathbf{K} = \mathbf{G}/2$. Thus the energy values are

$$E_{\pm} = \lambda \pm U = \frac{\hbar^2}{2m}\frac{G^2}{4} \pm U\ . \tag{6.20}$$

At the zone boundary, a splitting of the size $E_+ - E_- = 2U$ occurs. The center of the energy gap is given by the energy $\lambda_{\mathbf{K}}$ of the free-electron dispersion. The ratio of the coefficients is $C_{\mathbf{G}/2}/C_{-\mathbf{G}/2} = \mp 1$. The '$-$' solution of (6.20) (lower energy) is a standing cosine wave (Ψ_-), the '+' solution (Ψ_+) is a standing sine wave as visualized in Fig. 6.4. For the lower-energy (binding) state the electrons are localized at the potential minima, i.e. at the atoms, for the upper state (antibinding) the electrons are localized between the atoms. Both wavefunctions have the same periodicity since they belong to the same wavevector $\mathbf{K} = \mathbf{G}/2$. We note that the periodicity of Ψ is $2a$, while the periodicity of Ψ^2 is equal to the lattice constant a.

6.5.2 Solution in the Vicinity of the Zone Boundary

In the vicinity of the zone boundary we rewrite (6.19) as

$$\begin{vmatrix} \lambda_{\mathbf{K}} - E & U \\ U & \lambda_{\mathbf{K}-\mathbf{G}} - E \end{vmatrix} = 0\ . \tag{6.21}$$

We find two solutions

$$E_{\pm} = \frac{1}{2}\left(\lambda_{\mathbf{K}} + \lambda_{\mathbf{K}-\mathbf{G}}\right) \pm \left[\frac{1}{4}\left(\lambda_{\mathbf{K}} - \lambda_{\mathbf{K}-\mathbf{G}}\right)^2 + U^2\right]^{1/2}\ . \tag{6.22}$$

For $\mathbf{K}$ in the vicinity of the zone boundary these solutions can be developed. Therefore, we use the (small) distance from the zone boundary $\widetilde{\mathbf{K}} = \mathbf{K} - \mathbf{G}/2$. With $\lambda = (\hbar^2/2m)\,(G^2/4)$ we rewrite still exactly (6.22):

$$E_{\pm}\left(\widetilde{\mathbf{K}}\right) = \frac{\hbar^2}{2m}\left(\frac{1}{4}\mathbf{G}^2 + \widetilde{\mathbf{K}}^2\right) \pm \left[4\lambda\frac{\hbar^2\widetilde{\mathbf{K}}^2}{2m} + U^2\right]^{1/2}\ . \tag{6.23}$$

For small $\widetilde{\mathbf{K}}$ with $\hbar^2\mathbf{G}\widetilde{\mathbf{K}}/(2m) \ll |U|$, the energy is then approximately given by

$$E_{\pm}\left(\widetilde{\mathbf{K}}\right) \approx \lambda \pm U + \frac{\hbar^2\widetilde{\mathbf{K}}^2}{2m}\left(1 \pm \frac{2\lambda}{U}\right)\ . \tag{6.24}$$

Thus the energy dispersion in the vicinity of the zone boundary is parabolic. The lower state has a negative curvature, the upper state a positive curvature. The curvature is

$$m^* = m\frac{1}{1 \pm 2\lambda/U} \approx \pm m\frac{U}{2\lambda}\ , \tag{6.25}$$

and will be later related to the *effective mass*. The approximation in (6.25) is valid for $|U| \ll 2\lambda$. We note that in our simple model m^* increases linearly with increasing bandgap $2U$ (see Fig. 6.20 for experimental data).

6.5.3 Kramer's degeneracy

$E_n(\mathbf{k})$ is the dispersion in a band. The time-reversal symmetry (Kramer's degeneracy) implies

$$E_{n\uparrow}(\mathbf{k}) = E_{n\downarrow}(-\mathbf{k}) \ , \tag{6.26}$$

where the arrow refers to the direction of the electron spin. If the crystal is symmetric under inversion, we have additionally

$$E_{n\uparrow}(\mathbf{k}) = E_{n\uparrow}(-\mathbf{k}) \ . \tag{6.27}$$

With both time reversal and inversion symmetry the band structure fulfills

$$E_{n\uparrow}(\mathbf{k}) = E_{n\downarrow}(\mathbf{k}) \ . \tag{6.28}$$

6.6 Band Structure of Selected Semiconductors

In the following, the band structures of various important and prototype semiconductors are discussed. The band below the fundamental energy gap is called the valence band, the band above it the conduction band. The bandgap ΔE_{cv}, mostly denoted as E_{g}, is the energy separation between the highest valence-band state and the lowest conduction-band state. The maximum of the valence band is for most semiconductors at the Γ point.

6.6.1 Silicon

For silicon, an elemental semiconductor, (Fig. 6.5a) the minimum of the conduction band is located close to the X point at $0.85\pi/a$ in the $\langle 100\rangle$ direction. Thus, it is not at the same point in $\mathbf{k}$-space as the top of the valence band. Such a band structure is called *indirect*. Since there are six equivalent $\langle 100\rangle$ directions, there are six equivalent minima of the conduction band.

6.6.2 Germanium

Germanium, another elemental semiconductor, (Fig. 6.5b) also has an indirect band structure. The conduction minima are at the L point in the $\langle 111\rangle$ direction. Due to symmetry there are eight equivalent conduction-band minima.

6.6.3 GaAs

GaAs (Fig. 6.6a) is a compound semiconductor with a *direct* bandgap since the top of the valence band and the bottom of the conduction band are at the same position in $\mathbf{k}$-space (at the Γ point). The next highest (local) minimum in the conduction band is close to the L point.

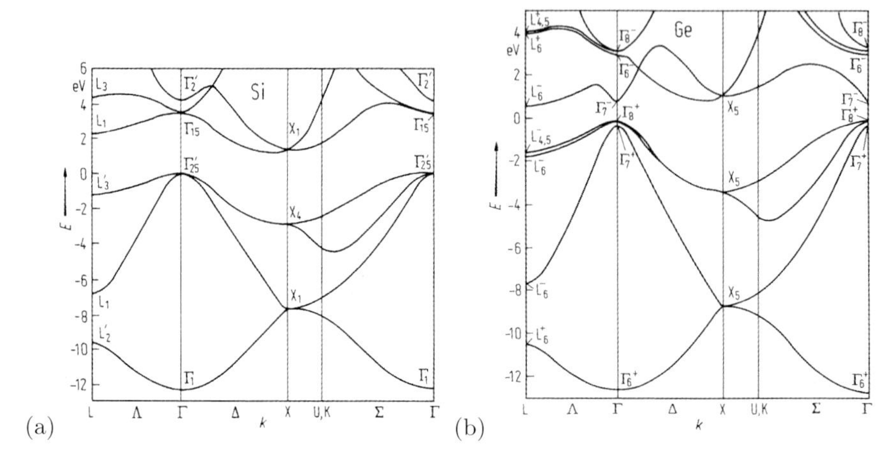

Fig. 6.5. Band structure of (**a**) silicon (indirect) and (**b**) germanium (indirect). In Si, the minima of the conduction band are in the ⟨100⟩ direction, for germanium in the ⟨111⟩ direction. From [93], based on [138]

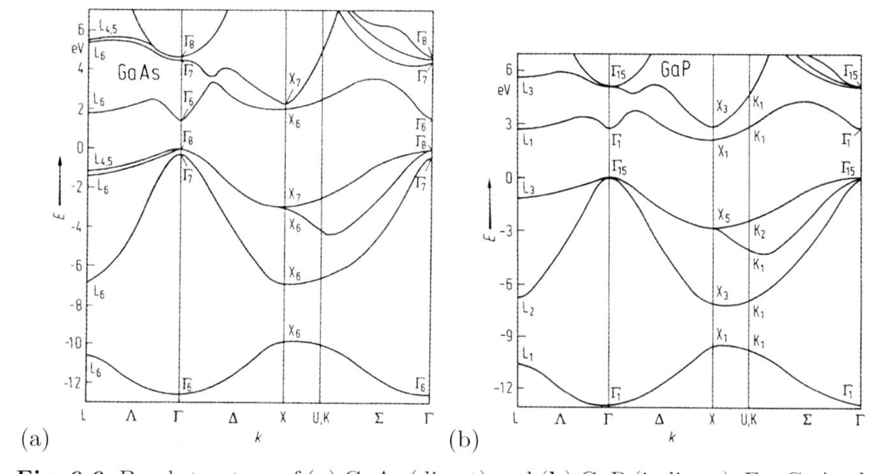

Fig. 6.6. Band structure of (**a**) GaAs (direct) and (**b**) GaP (indirect). For GaAs the minimum of the conduction band is at Γ, for GaP in the ⟨100⟩ direction. From [93], based on [138]

6.6.4 GaP

GaP (Fig. 6.6b) is an indirect compound semiconductor. The conduction-band minima are along the ⟨100⟩ directions.

6.6.5 GaN

GaN (Fig. 6.7) is a direct semiconductor that has wurtzite structure but can also occur in the metastable cubic (zincblende) phase.

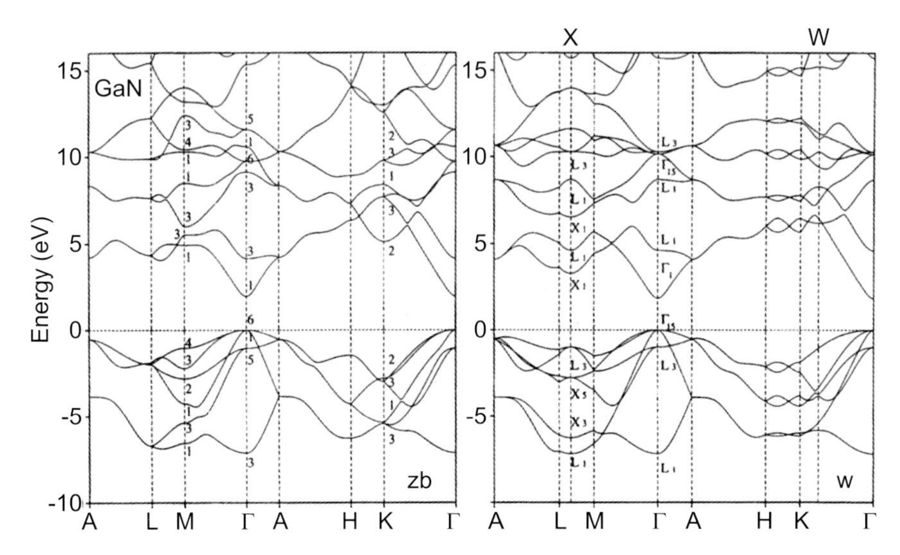

Fig. 6.7. Band structure of GaN (direct) in zincblende modification (*left*) and wurtzite modification (*right*), both displayed in the wurtzite Brillouin zone to facilitate comparison

6.6.6 Lead Salts

The band structures of PbS, PbSe and PbTe are shown in Fig. 6.8.

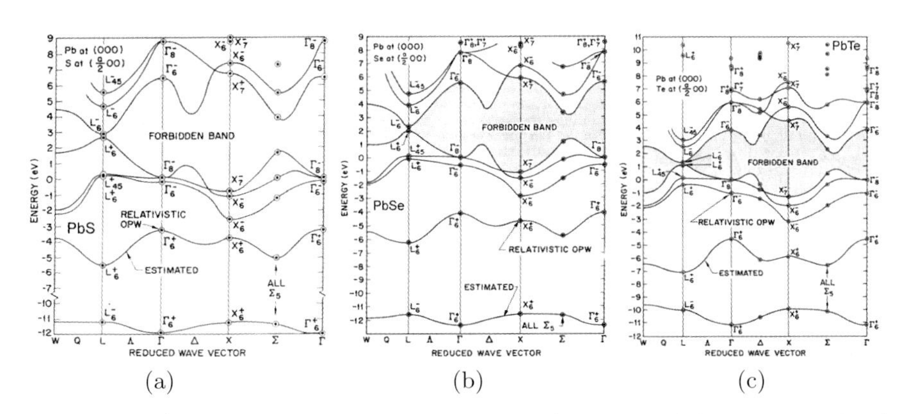

Fig. 6.8. Calculated band structures of (**a**) PbS, (**b**) PbSe and (**c**) PbTe. From [139]

The bandgap is located at the L point. The lead chalcogenide system shows the anomaly that with increasing atomic weight the bandgap does not decrease monotonically. At 300 K, the bandgaps are 0.41, 0.27 and 0.31 eV for PbS, PbSe and PbTe, respectively.

6.6.7 Chalcopyrites

The experimental bandgaps of a number of chalcopyrite semiconductors are listed in Table 6.1. The band structures of $CuAlS_2$, $CuAlSe_2$, and $CuGaSe_2$ are compared in Fig. 6.9.

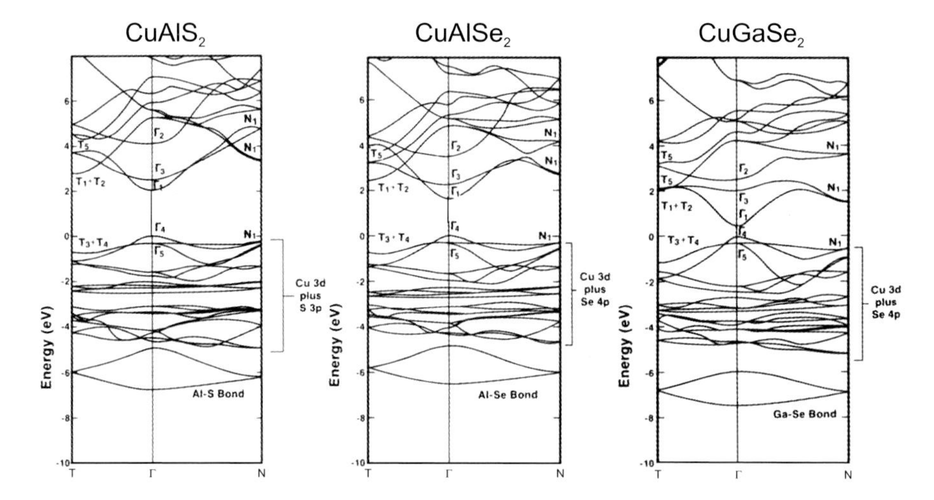

Fig. 6.9. Calculated band structures of $CuAlS_2$, $CuAlSe_2$, and $CuGaSe_2$. Absolute values of gaps are incorrect due to LDA. Adapted from [140], reprinted with permission, ©1983 APS

Table 6.1. Bandgaps of various chalcopyrite semiconductors

	E_g (eV)		E_g (eV)		E_g (eV)
$CuAlS_2$	3.5	$CuGaS_2$	2.5	$CuInS_2$	1.53
$CuAlSe_2$	2.71	$CuGaSe_2$	1.7	$CuInSe_2$	1.0
$CuAlTe_2$	2.06	$CuGaTe_2$	1.23	$CuInTe_2$	1.0-1.15
$AgAlS_2$	3.13	$AgGaS_2$	2.55	$AgInS_2$	1.87
$AgAlSe_2$	2.55	$AgGaSe_2$	1.83	$AgInSe_2$	1.24
$AgAlTe_2$	2.2	$AgGaTe_2$	1.1-1.3	$AgInTe_2$	1.0
$ZnSiP_2$	2.96	$ZnGeP_2$	2.34	$ZnSnP_2$	1.66
$ZnSiAs_2$	2.12	$ZnGeAs_2$	1.15	$ZnSnAs_2$	0.73
$CdSiP_2$	2.45	$CdGeP_2$	1.72	$CdSnP_2$	1.17
$CdSiAs_2$	1.55	$CdGeAs_2$	0.57	$CdSnAs_2$	0.26

In Fig. 6.10, the theoretical band structure of GaN and its closest related chalcopyrite $ZnGeN_2$ are compared, both shown in the chalcopyrite (orthorhombic) Brillouin zone. The bandgap of $ZnGeN_2$ is smaller than that of GaN and the difference of 0.4 eV is fairly well reproduced by the calculation[3] (giving 0.5 eV).

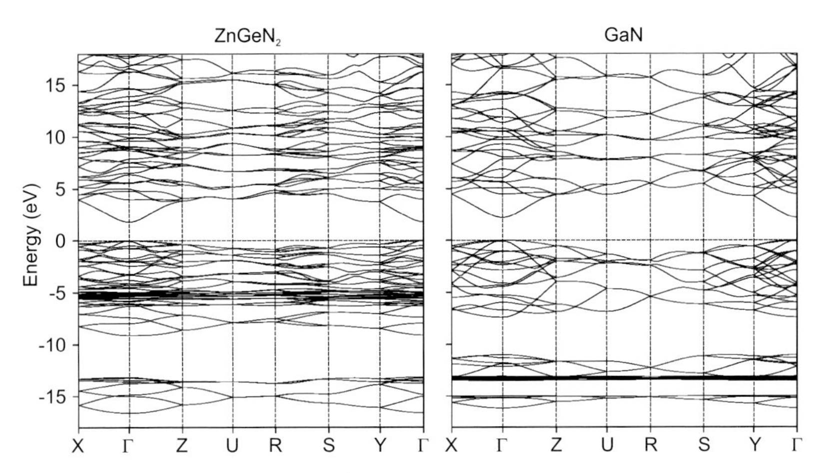

Fig. 6.10. Calculated (within LDA) band structures of $ZnGeN_2$ and its related III–V compound GaN, both displayed in the chalcopyrite (orthorhombic) Brillouin zone to facilitate comparison. Adapted from [141], reprinted with permission, ©1999 MRS

6.6.8 Delafossites

In Fig. 6.11, the theoretical band structures of the delafossites $CuAlO_2$, $CuGaO_2$, and $CuInO_2$ are shown. The maximum of the valence band is not at Γ but near the F point. The direct bandgap at Γ decreases for the sequence Al $\rightarrow$ Ga $\rightarrow$ In, similar to the trend for AlAs, GaAs and InAs. The direct bandgap at F and L, causing the optical absorption edge, increases, however (experimental values are 3.5, 3.6, and 3.9 eV).

6.6.9 Perovskites

The calculated band structure of $BaTiO_3$ in the tetragonal phase is shown in Fig. 6.12. The minimum of the conduction band is at the Γ point. The maximum of the valence band is not at the Γ point but at the M point. The bandgap of the LDA calculation is too small (2.2 eV) compared to the experimental value ~ 3.2 eV.

[3]Due to the local density approximation (LDA) the absolute values of the bandgaps are too small by about 1 eV.

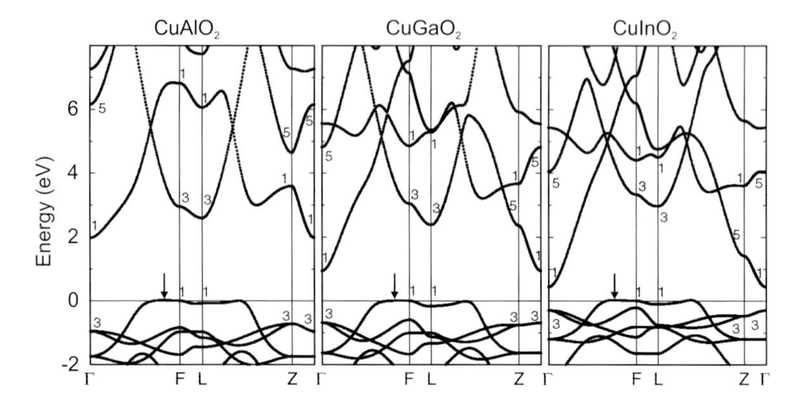

Fig. 6.11. Band structures of $CuAlO_2$, $CuGaO_2$, and $CuInO_2$, calculated with LDA (underestimating the absolute value of the bandgaps). The *arrows* denote the maximum of the valence band that has been set to zero energy for each material. Adapted from [59], reprinted with permission, ©2002 APS

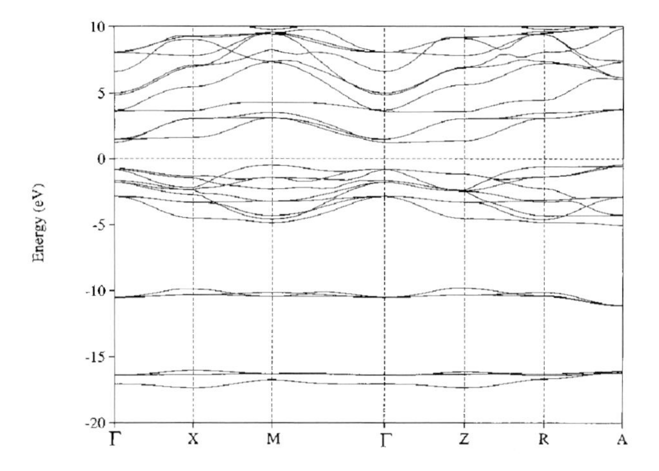

Fig. 6.12. Calculated energy band structure of $BaTiO_3$ along the major symmetry directions. The Fermi level (E_F) is set at zero energy. Reprinted with permission from [142], ©2001 AIP

6.7 Alloy Semiconductors

In alloy semiconductors, the size of the bandgap and the character of the bandgap will depend on the composition. The dependence of the bandgap on the ternary composition is mostly nonlinear and can usually be expressed with a bowing parameter b that is mostly positive. For a compound $A_xB_{1-x}C$ the bandgap is written as

$$E_g(A_xB_{1-x}C) = E_g(BC) + x\left[E_g(AC) - E_g(BC)\right] - bx(1-x)\ . \qquad (6.29)$$

Even on the virtual crystal approximation (VCA) level (Sect. 3.7.3) a nonzero bowing parameter b is predicted. However, a more thorough analysis shows that the bowing cannot be treated adequately within VCA and is due to the combined effects of volume deformation of the band structure with the alloy lattice constant, charge exchange in the alloy with respect to the binary end components, a structural contribution due to the relaxation of the cation–anion bond lengths in the alloy and a small contribution due to disorder [143].

The Si_xGe_{1-x} alloy has diamond structure for all concentrations and the position of the conduction-band minimum in **k**-space switches from L to X at about $x = 0.15$ (Fig. 6.13a). However, for all concentrations the band structure is indirect. The $In_xGa_{1-x}As$ alloy has zincblende structure for all compositions. The bandgap is direct and decreases with a bowing parameter of $b = 0.6\,\text{eV}$ [144] (Fig. 6.13b). This means that for $x = 0.5$ the bandgap is 0.15 eV smaller than expected from a linear interpolation between GaAs and InAs, as reported by various authors [145].

If one binary end component has a direct band structure and the other is indirect, a transition occurs from direct to indirect at a certain composition. An example is $Al_xGa_{1-x}As$ where GaAs is direct and AlAs is indirect. For all concentrations the crystal has zincblende structure. In Fig. 6.13c, the Γ, L and X conduction-band minima for ternary $Al_xGa_{1-x}As$ are shown. Up to an aluminum concentration of $x = 0.4$ the band structure is direct. Above this value the band structure is indirect with the conduction-band minimum being at the X point. The particularity of $Al_xGa_{1-x}As$ is that the lattice constant is almost independent of x. For other alloys lattice match to GaAs or InP substrates is only obtained for specific compositions, as shown in Fig. 6.14.

If the two binary end components have different crystal structure, a phase transition occurs at a certain composition (range). An example is $Mg_xZn_{1-x}O$, where ZnO has wurtzite structure and MgO has rocksalt structure. The bandgap is shown in Fig. 6.13d. In this case, each phase has its own bowing parameter. A discontinuity arises when the crystal changes from fourfold to sixfold coordination.

6.8 Amorphous Semiconductors

In a perfectly crystalline semiconductor the eigenenergies of the states in the bands are real. An amorphous semiconductor can be modelled using a spectrum of complex energies [150]. In Fig. 6.15 the band structure of crystalline silicon is shown next to that calculated for amorphous silicon with $\alpha = 0.05$.

6.9 Systematics of Semiconductor Bandgaps

The trends with regard to the size of the bandgap for elemental, III–V and II–VI semiconductors can essentially be understood in terms of the bond

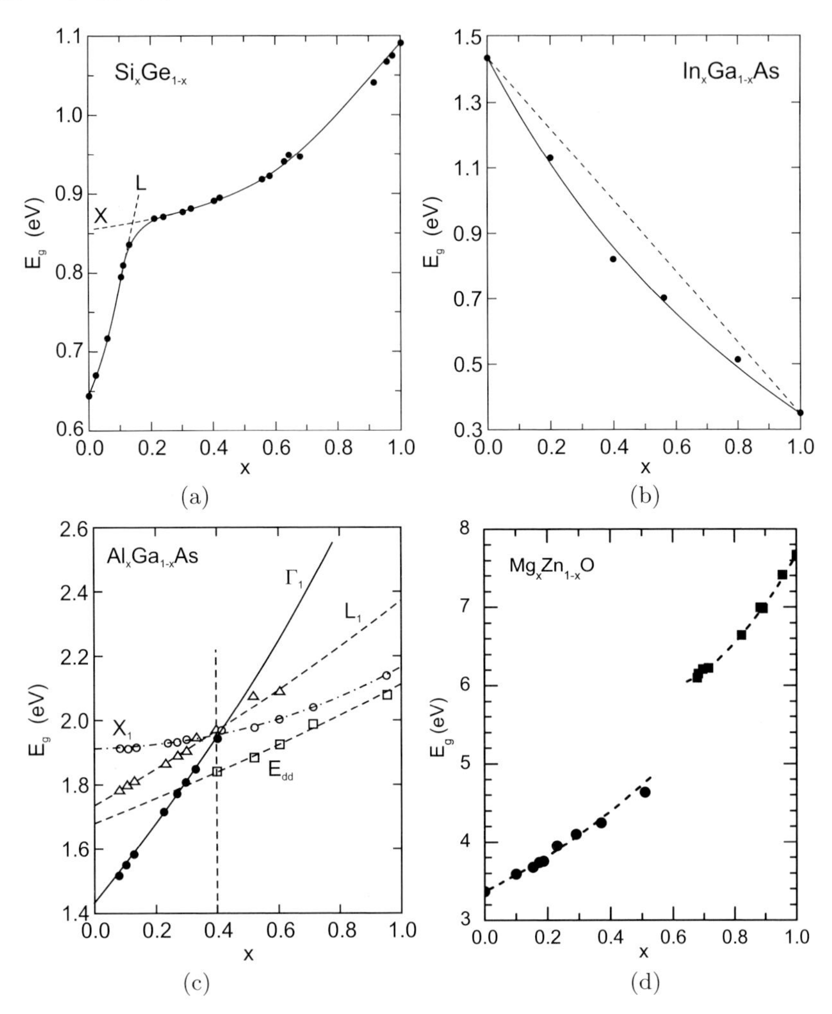

Fig. 6.13. (**a**) Bandgap of Si_xGe_{1-x} alloy ($T = 296$ K) with a change from the conduction-band minimum at L (Ge-rich) to X. Adapted from [146]. (**b**) Bandgap (at room temperature) of $In_xGa_{1-x}As$. The *solid line* is an interpolation with bowing ($b = 0.6$ eV) and the *dashed line* is the linear interpolation. Data from [144]. (**c**) Bandgap (at room temperature) in the ternary system $Al_xGa_{1-x}As$. For $x < 0.4$ the alloy is a direct, for $x > 0.4$ an indirect, semiconductor. E_{dd} denotes the energy position of a deep donor (see Sect. 7.7.4). Adapted from [147]. (**d**) Bandgap (at room temperature) in the ternary system $Mg_xZn_{1-x}O$. Data (from spectroscopic ellipsometry [148,149]) are for hexagonal wurtzite phase (*circles*), and Mg-rich cubic rocksalt phase *(squares)*. *Dashed lines* are fits to data with a bowing parameter for each phase

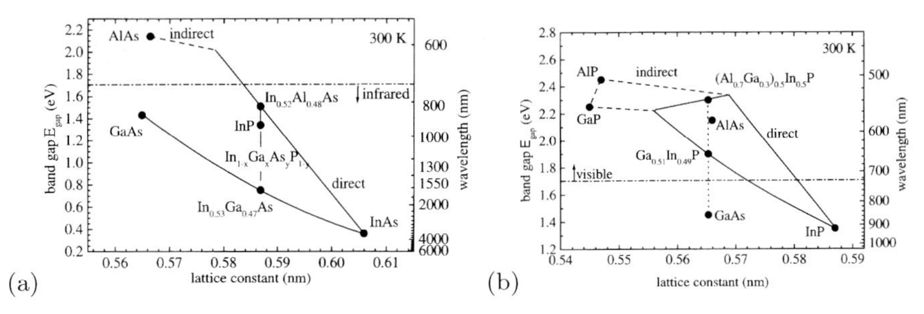

Fig. 6.14. Bandgap vs. lattice constant for (**a**) $Ga_xIn_{1-x}P$ and $Al_xIn_{1-x}P$ (lattice matched to GaAs) as well as for (**b**) $In_xAl_{1-x}As$ and $In_xGa_{1-x}As$ alloys (lattice matched to InP)

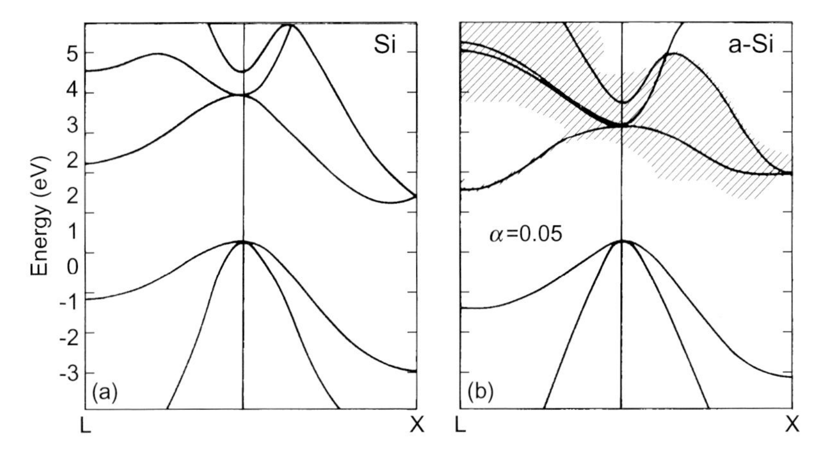

Fig. 6.15. (**a**) Calculated band structure of crystalline silicon. (**b**) Calculated band structure of amorphous silicon with $\alpha = 0.05$ (cf. (3.6)). The *solid lines* denote the real part of the energy, the *shaded areas* denote the regions with a width of twice the imaginary part of the energies centered around the real part. Adapted from [151]

strength and ionicity. In Fig. 6.16, the bandgaps of many important semiconductors are shown as a function of the lattice constant. For elemental semiconductors, the bandgap decreases with reduced bond strength, i.e. lattice constant (C→Si→Ge). A similar trend exists both for the III–V and the II–VI semiconductors.

For the same lattice constant, the bandgap increases with increasing ionicity, i.e. IV–IV→III–V→II–VI. The best example is the sequence Ge→GaAs→ZnSe for which all materials have almost the same lattice constant.

This behavior can be understood within the framework of a modified Kronig–Penney model [152] (Sect. 6.4). Double potential wells ($b/a = 3$) are chosen to mimic the diatomic planes along the $\langle 111 \rangle$ direction in the zincblende structure (Fig. 6.17a). Symmetric wells (depth P_0) are chosen to model covalent semiconductors and asymmetric wells with depths $P_0 \pm \Delta P$

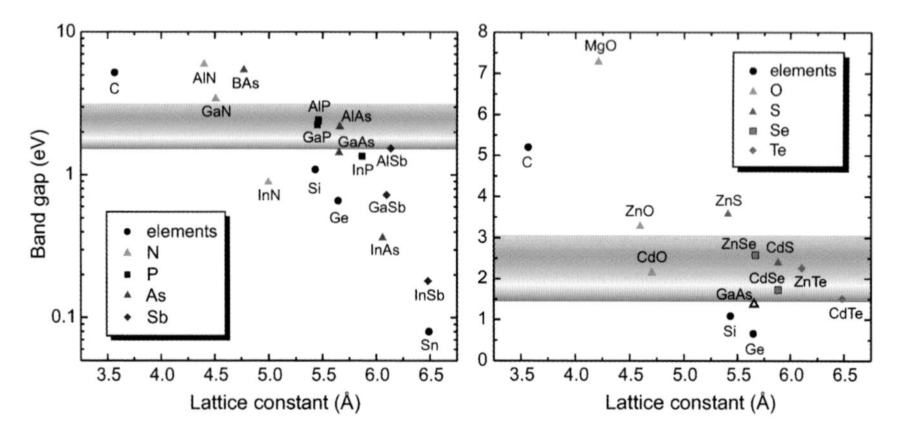

Fig. 6.16. Bandgaps as a function of the lattice constant for various IV–IV, III–V and II–VI semiconductors. (lattice constant of wurtzite semiconductors has been recalculated for a cubic cell)

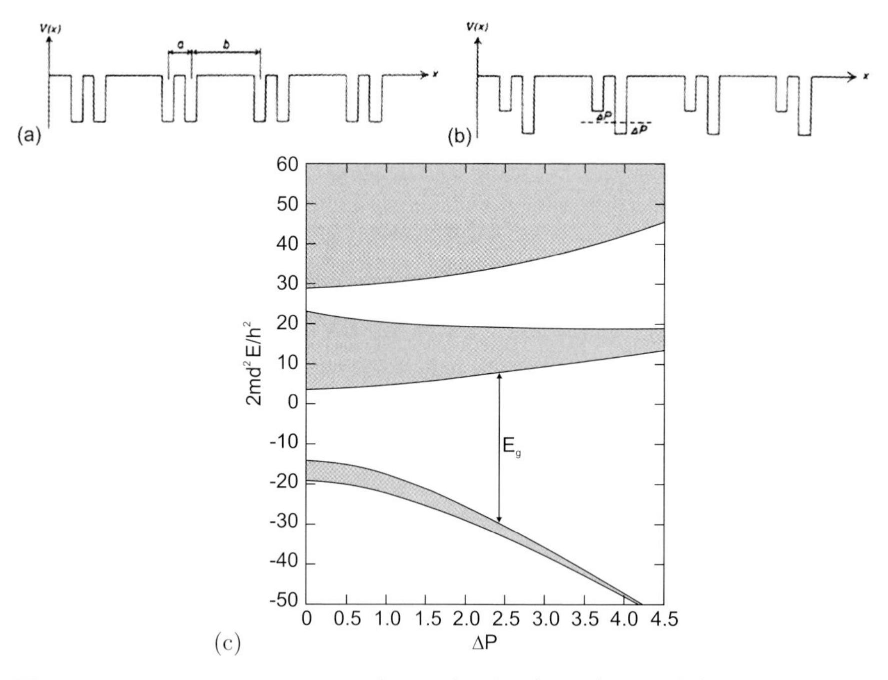

Fig. 6.17. Kronig–Penney model (along $\langle 111 \rangle$, $b/a = 3$) for *a* (**a**) IV–IV semiconductor and (**b**) for a III–V (or II–VI) semiconductor, (**c**) resulting band structure ($P_0 = -3$). d is the lattice constant ($b + a$). Adapted from [152]

to model partially ionic semiconductors. Results are shown in Fig. 6.17a for $P_0 = -3$. With increasing asymmetry, i.e. ionicity, the bandgap increases, mostly due to a downward shift of the valence band. The case of III–V (II–VI) semiconductors is reached for $\Delta P \approx 2$ (4).

6.10 Temperature Dependence of the Bandgap

The bandgap typically decreases with increasing temperature (see Fig. 6.18 for Si and ZnO). The reasons for this are the change of electron–phonon interaction and the expansion of the lattice. The temperature coefficient may be written as

$$\left(\frac{\partial E_\mathrm{g}}{\partial T}\right)_p = \left(\frac{\partial E_\mathrm{g}}{\partial T}\right)_V - \frac{\alpha}{\beta}\left(\frac{\partial E_\mathrm{g}}{\partial p}\right)_T , \tag{6.30}$$

where α is the volume coefficient of thermal expansion and β is the volume compressibility. An anomaly is present for the lead salts (PbS, PbSe, PbTe) for which the temperature coefficient is positive (Fig. 6.19a). Theoretical calculations [153] show that both terms in (6.30) are positive for the lead salts. The L_6^+ and L_6^- levels (see Fig. 6.8) shift as a function of temperature in such a way that their separation increases (Fig. 6.19b).

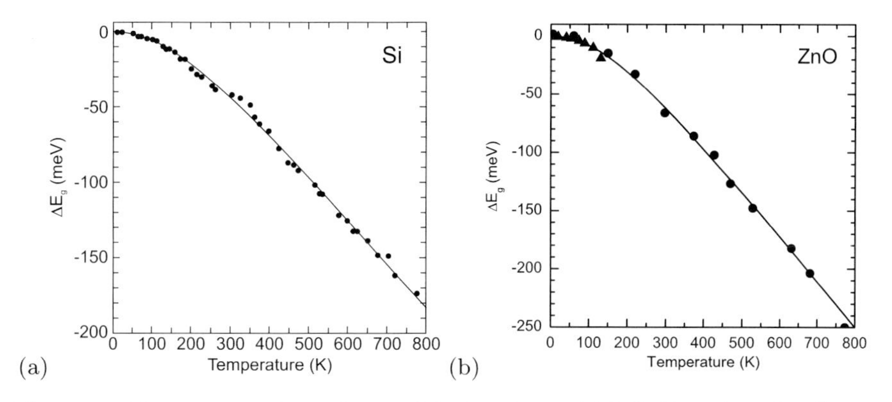

Fig. 6.18. Temperature dependence of the bandgap of (**a**) Si (data from [154]) and (**b**) ZnO (experimental data from photoluminescence (*triangles*) and ellipsometry (*circles*)). The *solid lines* are fits with (6.33) and the parameters given in Table 6.2

For many semiconductors the temperature dependence can be described with the empirical, three-parameter Varshni formula [155],

$$E_\mathrm{g}(T) = E_\mathrm{g}(0) - \frac{\alpha T^2}{T + \beta} , \tag{6.31}$$

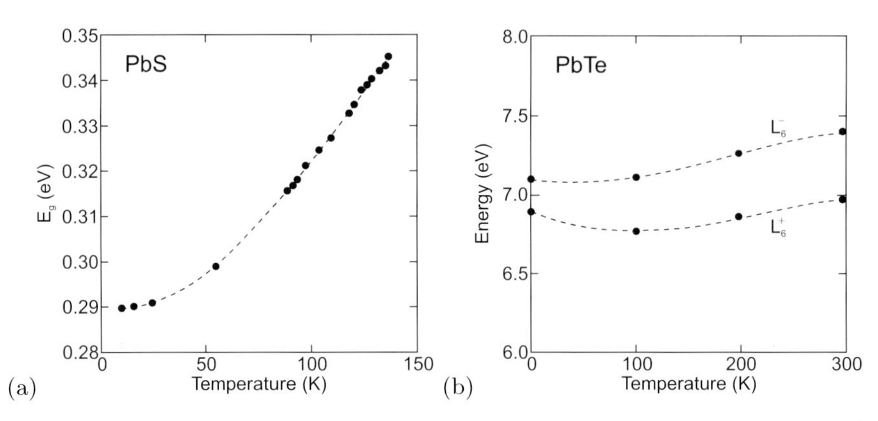

Fig. 6.19. (**a**) Bandgap vs. temperature for PbS. (**b**) Theoretical position of L_6^+ and L_6^- as a function of temperature for PbTe. Adapted from [139]

where $E_g(0)$ is the bandgap at zero temperature. A more precise and physically motivated formula (based on a Bose–Einstein phonon model) has been given in [156]

$$E_g(T) = E_g(0) - 2\alpha_B \Theta_B \left[\coth\left(\frac{\Theta_B}{2T}\right) - 1\right] , \tag{6.32}$$

where α_B is a coupling constant and $k\Theta_B$ is a typical phonon energy; typical values are given in Table 6.2. This model reaches a better description of the fairly flat dependence at low temperatures. However, experimentally the dependence at low temperatures is rather quadratic.

Table 6.2. Parameters for the temperature dependence of the bandgap (6.32) and (6.33) for various semiconductors

	α (10^{-4} eV/K)	Θ (K)	Δ	α_B (10^{-4} eV/K)	Θ_B (K)
Si	3.23	446	0.51	2.82	351
Ge	4.13	253	0.49		
GaAs	4.77	252	0.43	5.12	313
InP	3.96	274	0.48		
InAs	2.82	147	0.68		
ZnSe	5.00	218	0.36		
ZnO	3.8	659	0.54		

The more elaborate model of [157] takes into account a more variable phonon dispersion, including optical phonons, and proposes the four-parameter formula

$$E_{\mathrm{g}}(T) = E_{\mathrm{g}}(0) - \alpha\Theta \left[\frac{1-3\Delta^2}{\exp{(2/\gamma)} - 1} + \frac{3\Delta^2}{2} \left(\sqrt[6]{1+\beta} - 1 \right) \right] \quad (6.33)$$

$$\beta = \frac{\pi^2}{3(1+\Delta^2)}\gamma^2 + \frac{3\Delta^2 - 1}{4}\gamma^3 + \frac{8}{3}\gamma^4 + \gamma^6$$

$$\gamma = 2T/\Theta \; ,$$

where α is the high-temperature limiting magnitude of the slope (of the order of several $10^{-4}\,\mathrm{eV/K}$), Θ is an effective average phonon temperature and Δ is related to the phonon dispersion (typically between zero (Bose–Einstein model) and 3/4).

6.11 Equation of Electron Motion

The equation of motion for the electron in the band structure is no longer given by Netwon's law $\mathbf{F} = \mathrm{d}(m\mathbf{v})/\mathrm{d}t$ as in vacuum. Instead, the propagation of quantum-mechanical electron wave packets has to be considered. Their group velocity is given by

$$v_g = \frac{\mathrm{d}\omega}{\mathrm{d}k} = \frac{1}{\hbar}\frac{\mathrm{d}E}{\mathrm{d}k} \; . \quad (6.34)$$

Through the dispersion relation the influence of the crystal and its periodic potential on the motion enters the equation.

An electric field $\mathcal{E}$ acts on an electron during the time δt the work $\delta E = -e\mathcal{E}v_g\delta t$. This change in energy is related to a change in k via $\delta E = \mathrm{d}E/\mathrm{d}k\delta k = \hbar v_g \delta k$. Thus, we arrive at $\hbar \mathrm{d}k/\mathrm{d}t = -e\mathcal{E}$. For an external force we thus have

$$\hbar\frac{\mathrm{d}\mathbf{k}}{\mathrm{d}t} = -e\mathbf{E} = \mathbf{F} \; . \quad (6.35)$$

Thus, the crystal momentum $\mathbf{p} = \hbar\mathbf{k}$ takes the role of the momentum. A more rigorous derivation can be found in [137].

In the presence of a magnetic field $\mathbf{B}$ the equation of motion is:

$$\hbar\frac{\mathrm{d}\mathbf{k}}{\mathrm{d}t} = -e\mathbf{v}\times\mathbf{B} = -\frac{e}{\hbar}\nabla_{\mathbf{k}}E \times \mathbf{B} \; . \quad (6.36)$$

The motion in a magnetic field is thus perpendicular to the gradient of the energy, i.e. the energy of the electron does not change. It oscillates therefore on a surface of constant energy perpendicular to $\mathbf{B}$.

6.12 Electron Mass

6.12.1 Effective Mass

From the free-electron dispersion $E = \hbar^2 k^2/(2m)$ the mass of the particle is inversely proportional to the curvature of the dispersion relation, i.e. $1/m = (\mathrm{d}^2 E/\mathrm{d}k^2)/\hbar^2$. This relation will now be generalized for arbitrary dispersion relations. The tensor of the effective mass is defined as

$$(m^{*-1})_{ij} = \frac{1}{\hbar^2} \frac{\partial^2 E}{\partial k_i \partial k_j} . \tag{6.37}$$

The equation $\mathbf{F} = m^* \dot{\mathbf{v}}$ must be understood as a tensor equation, i.e. for the components of the force $F_i = m^*_{ij} a_j$. Force and acceleration must no longer be collinear. In order to find the acceleration from the force, the inverse of the effective-mass tensor must be used, $\mathbf{a} = (\mathbf{m}^*)^{-1}\mathbf{F}$.

In (6.25) the ratio of the effective mass and the free-electron mass is of the order of $m^*/m \approx U/\lambda$, the ratio of the free particle energy and the bandgap. For typical semiconductors, the width of the (valence) band is of the order of 20 eV, and the gap is about 0.2–2 eV. Thus, the effective mass is expected to be 10–100 times smaller than the free-electron mass. Additionally, the relation $m^* \propto E_\mathrm{g}$ is roughly fulfilled (Fig. 6.20).

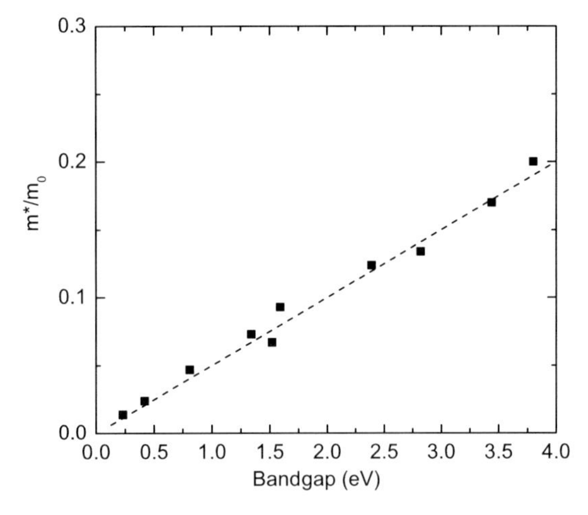

Fig. 6.20. Effective mass (in units of the free-electron mass m_0) as a function of the (low-temperature) bandgap for several semiconductors. The *dashed line* fulfills $m^*/m_0 = E_\mathrm{g}/20\,\mathrm{eV}$

From so-called $\mathbf{k} \cdot \mathbf{p}$ theory [158] (see Appendix E) the effective electron mass is predicted to be related to the momentum matrix element $\mathbf{p}_{\mathrm{cv}}$

$$\mathbf{p}_{\mathrm{cv}} = \langle c|\mathbf{p}|v\rangle , \tag{6.38}$$

with the Bloch functions $|c\rangle$ and $|v\rangle$ of the conduction and valence band, respectively, given as

$$|c\rangle = u_{\mathrm{c},\mathbf{k}_\mathrm{c}}(\mathbf{r}) \exp\left(\mathrm{i}\mathbf{k}_\mathrm{c}\mathbf{r}\right) \tag{6.39a}$$

$$|v\rangle = u_{\mathrm{v},\mathbf{k}_\mathrm{v}}(\mathbf{r}) \exp\left(\mathrm{i}\mathbf{k}_\mathrm{v}\mathbf{r}\right) \; . \tag{6.39b}$$

Typically, the **k**-dependence of the matrix element is small and neglected. The momentum matrix element will also be important for optical transitions between the valence and conduction bands (Sect. 9.3). Other related quantities that are often used are the energy parameter E_P

$$E_\mathrm{P} = \frac{2\left|\mathbf{p}_\mathrm{cv}\right|^2}{m_0} \; , \tag{6.40}$$

and the bulk momentum matrix element M_b^2 that is given by

$$M_\mathrm{b}^2 = \frac{1}{3}\left|\mathbf{p}_\mathrm{cv}\right|^2 = \frac{m_0}{6}E_\mathrm{P} \; . \tag{6.41}$$

The electron mass is given by[4]

$$\begin{aligned} \frac{m_0}{m_\mathrm{e}^*} &= 1 + \frac{E_\mathrm{P}}{3}\left(\frac{2}{E_\mathrm{g}} + \frac{1}{E_\mathrm{g}+\Delta_0}\right) \\ &= 1 + E_\mathrm{P}\frac{E_\mathrm{g}+2\Delta_0/3}{E_\mathrm{g}\left(E_\mathrm{g}+\Delta_0\right)} \approx 1 + \frac{E_\mathrm{P}}{E_\mathrm{g}+\Delta_0/3} \approx \frac{E_\mathrm{P}}{E_\mathrm{g}} \; . \end{aligned} \tag{6.42}$$

Comparison with the fit from Fig. 6.20 yields that E_P is similar for all semiconductors [159] and of the order of 20 eV (InAs: 22.2 eV, GaAs: 25.7 eV, InP: 20.4 eV, ZnSe: 23 eV, CdS: 21 eV).

In silicon there are six conduction-band minima. The surfaces of equal energy are schematically shown in Fig. 6.21c. The ellipsoids are extended along the $\langle 100\rangle$ direction because the longitudinal mass (along the Δ path) is larger than the transverse mass in the two perpendicular directions. For example, the dispersion relation around the [100] minimum is then given as (k_x^0 denotes the position of the conduction-band minimum)

$$E(\mathbf{k}) = \hbar^2\left(\frac{(k_x-k_x^0)^2}{2m_\mathrm{l}} + \frac{k_y^2+k_z^2}{2m_\mathrm{t}}\right) \; . \tag{6.43}$$

For germanium surfaces of constant energy around the eight conduction-band minima in the $\langle 111\rangle$ directions are depicted in Fig. 6.21d. The longitudinal and the transverse masses are again different. For GaAs, the conduction-band dispersion around the Γ point is isotropic, thus the surface of constant energy is simply a sphere (Fig. 6.21a). In wurtzite semiconductors the conduction-band minimum is typically at the Γ point. However, the mass

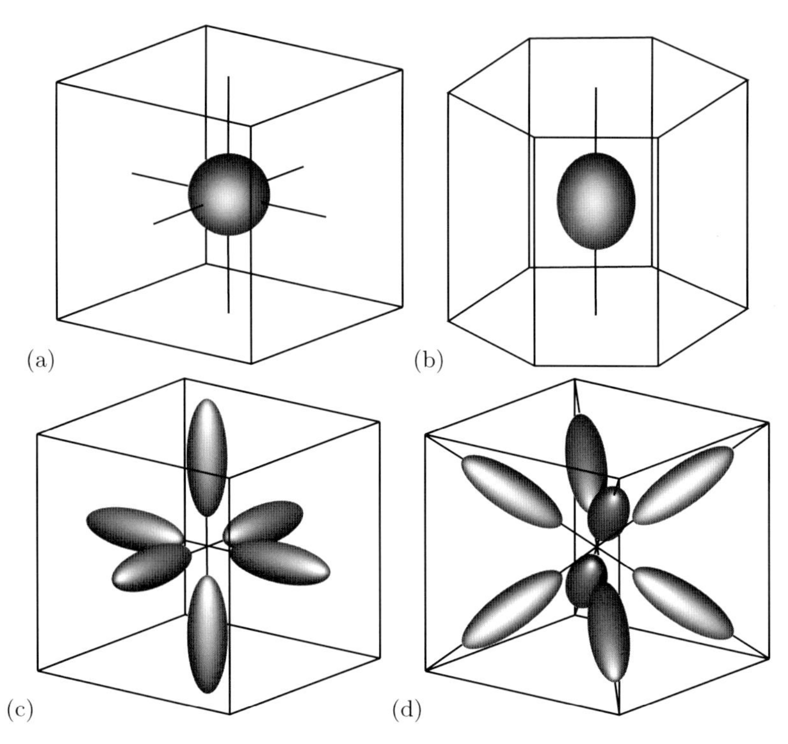

Fig. 6.21. Ellipsoids of constant energy in the vicinity of the conduction-band minima for (**a**) GaAs with isotropic minimum at Γ point, (**b**) ZnO with anisotropic minimum at Γ point (anisotropy exaggerated), (**c**) silicon with six equivalent anisotropic minima ($m_l/m_t = 5$ not to scale) along $\langle 100 \rangle$ and (**d**) germanium with eight equivalent anisotropic minima along $\langle 111 \rangle$. The cube indicates the $\langle 100 \rangle$ directions for the cubic materials. For the wurtzite material (part (**b**)) the vertical direction is along [00.1]

along the c-axis is slightly different from the in-plane mass ($m_l/m_t = 1.08$ for ZnO), see Fig. 6.21b.

The directional dependence of the mass can be measured with cyclotron resonance experiments with varying direction of the magnetic field. In Fig. 6.22, the field $\mathbf{B}$ is in the (110) plane with different azimuthal directions. When the (static) magnetic field makes an angle θ with the longitudinal axis of the energy surface, the effective mass is given as [160]

$$\left(\frac{1}{m^*}\right)^2 = \frac{\cos^2\theta}{m_t^2} + \frac{\sin^2\theta}{m_t m_l} \, . \tag{6.44}$$

[4] Δ_0 is the spin-orbit splitting discussed in Sect. 6.13.2.

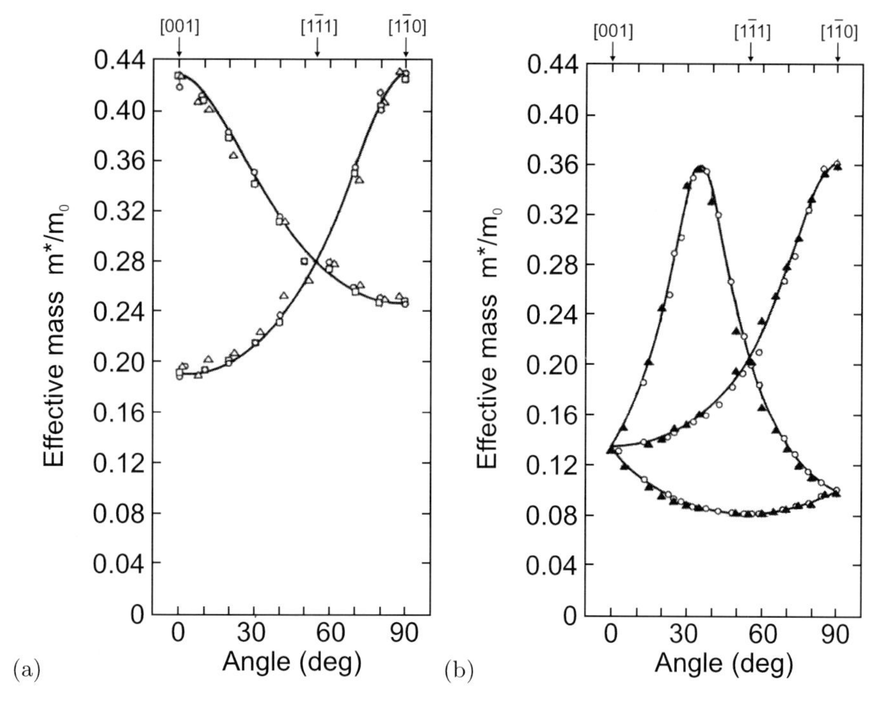

Fig. 6.22. Cyclotron mass (at $T = 4\,\mathrm{K}$) in (**a**) Si and (**b**) Ge for the magnetic field in the (110) plane and various azimuthal directions θ. Experimental data and fits (*solid lines*) using (6.44) with (**a**) $m_l = 0.98$, $m_t = 0.19$ and (**b**) $m_l = 1.58$, $m_t = 0.082$. Data from [161].

6.12.2 Polaron Mass

In an ionic lattice, the electron polarizes the ions and causes a change of their equilibrium position. When the electron moves, it must drag this ion displacement with it. This process it called the *polaronic* effect and requires additional energy. It leads to an increase of the electron mass to the 'polaron mass' m_p,[5]

$$m_\mathrm{p} = m^* \left(1 + \alpha/6 + 0.025\alpha^2\right) , \tag{6.45}$$

with m^* being the band mass as defined in Sect. 6.12.1 and α the Fröhlich coupling constant[6]

[5]For the calculation, many-particle theory and techniques are needed; the best solution is still given by Feynman's calculation [162, 163].

[6]This constant is part of the matrix element in the Hamiltonian of the electron–phonon interaction and is related to the electric field created by LO phonons, as given in (5.45).

$$\alpha = \frac{e^2}{\hbar}\left(\frac{m^*}{2\hbar\omega_{\mathrm{LO}}}\right)^{1/2}\left(\frac{1}{\epsilon(\infty)} - \frac{1}{\epsilon(0)}\right) . \tag{6.46}$$

Often, the polaron mass is given as $m_{\mathrm{p}} = m^*/(1-\alpha/6)$. Polarons in semiconductors are typically 'large' or Fröhlich-type polarons, i.e. the dressing with phonons (as the ion displacement is called in a quantum-mechanical picture) is a perturbative effect and the number of phonons per electron ($\approx \alpha/2$) is small. If α becomes large ($\alpha > 1$, $\alpha \sim 6$), as is the case for strongly ionic crystals such as alkali halides, the polaron becomes localized (self-trapped) by the electron–phonon interaction[7] and hopping occurs infrequently from site to site.

6.12.3 Nonparabolicity of Electron Mass

The dispersion around the conduction-band minimum is only parabolic for small **k**. The further away the wavevector is from the extremum, the more the actual dispersion deviates from the ideal parabola (see, e.g., Fig. 6.6). This effect is termed *nonparabolicity.* Typically, the energy increases less quickly with k than in the parabolic model. This can be described in a so-called two-level model with the dispersion relation

$$\frac{\hbar^2 k^2}{2m_0^*} = E\left(1 + \frac{E}{E_0^*}\right) , \tag{6.47}$$

where $E_0^* > 0$ parameterizes the amount of nonparabolicity (a parabolic band corresponds to $E_0^* = \infty$). The nonparabolic dispersion for GaAs is shown in Fig. 6.23a. The curvature is reduced for larger **k** and thus the effective mass is energy dependent and increases with the energy. Equation (6.47) leads to the energy-dependent effective mass

$$m^*(E) = m_0^*\left(1 + \frac{2E}{E_0^*}\right) , \tag{6.48}$$

where m_0^* denotes here the effective mass at $\mathbf{k} = 0$. Theory and experimental data for the effective electron mass of GaAs are shown in Fig. 6.23b.

6.13 Holes

6.13.1 Hole Concept

Holes are missing electrons in an otherwise filled band. This concept is useful to describe the properties of charge carriers at the top of the valence band. The hole is a new quasi-particle whose dispersion relation is schematically

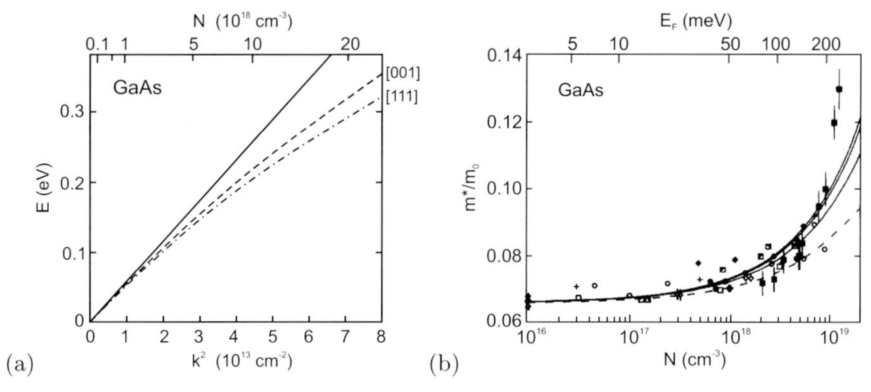

Fig. 6.23. (**a**) Dispersion relations for the conduction band of GaAs. The *solid line* is parabolic dispersion (constant effective mass). The *dashed* (*dash-dotted*) *line* denotes the dispersion for $\mathbf{k}$ along [001] ([111]) from a five-level $\mathbf{k} \cdot \mathbf{p}$ model (5LM). (**b**) Cyclotron resonance effective mass of electrons in GaAs as a function of the Fermi level (upper abscissa) and the corresponding electron concentration (lower abscissa). The *dashed line* is from a 2LM according to (6.48) with $E_0^* = 1.52\,\mathrm{eV}$. The *solid lines* are for a 5LM for the three principal directions of the magnetic field. The *symbols* represent experimental data from different sources. Data from [164]

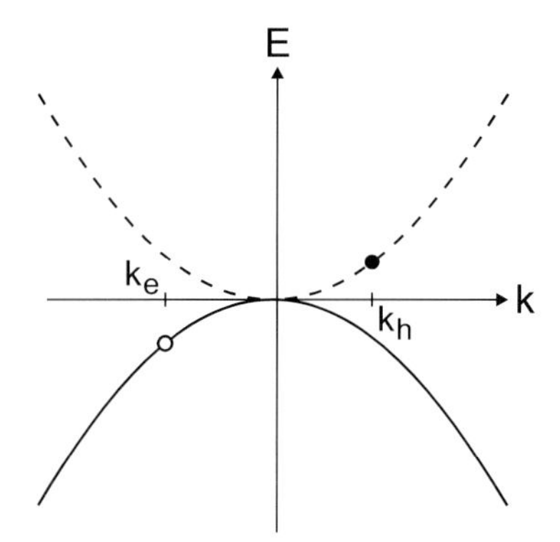

Fig. 6.24. Hole dispersion (*dashed line*) in relation to the electron dispersion in the valence band (*solid line*)

shown in Fig. 6.24 in relation to the dispersion of electrons in the valence band.

The wavevector of the hole (filled circle in Fig. 6.24) is related to that of the 'missing' electron (empty circle in Fig. 6.24) by $\mathbf{k}_\mathrm{h} = -\mathbf{k}_\mathrm{e}$. The energy is $E_\mathrm{h}(\mathbf{k}_\mathrm{h}) = -E_\mathrm{e}(\mathbf{k}_\mathrm{e})$, assuming that $E_\mathrm{V} = 0$, otherwise $E_\mathrm{h}(\mathbf{k}_\mathrm{h}) = -E_\mathrm{e}(\mathbf{k}_\mathrm{e}) + 2E_\mathrm{V}$. The hole energy is larger for holes that are further away from the top of

[7] One can think about it in the way that the electron strongly polarizes the lattice and digs itself a potential hole out of which it can no longer move.

the valence band, i.e. the lower the energy state of the missing electron. The velocity of the hole, $\mathbf{v}_\mathrm{h} = \hbar^{-1}\nabla_{\mathbf{k}_\mathrm{h}} E_\mathrm{h}$, is the same, $\mathbf{v}_\mathrm{h} = \mathbf{v}_\mathrm{e}$, and the charge is positive, $+e$. The effective mass of the hole is positive at the top of the valence band, $m_\mathrm{h}^* = -m_\mathrm{e}^*$. Therefore, the drift velocities of an electron and hole are opposite to each other. The resulting current, however, is the same.

6.13.2 Hole Dispersion Relation

The valence band at the Γ point is 3-fold degenerate. The band developed from the atomic (bonding) p states; the coupling of the spin $s = 1/2$ electrons with the orbital angular momentum $l = 1$ leads to a total angular momentum $j = 1/2$ and $j = 3/2$. The latter states are degenerate at Γ in zincblende bulk material and are called *heavy* holes (hh) for $m_j = \pm 3/2$ and *light* holes (lh) for $m_j = \pm 1/2$ due to their different dispersion (Fig. 6.25a). The two ($m_j = \pm 1/2$) states of the $j = 1/2$ state are split-off from these states by an energy Δ_0 due to spin-orbit interaction and are called *split-off* (s-o) holes.

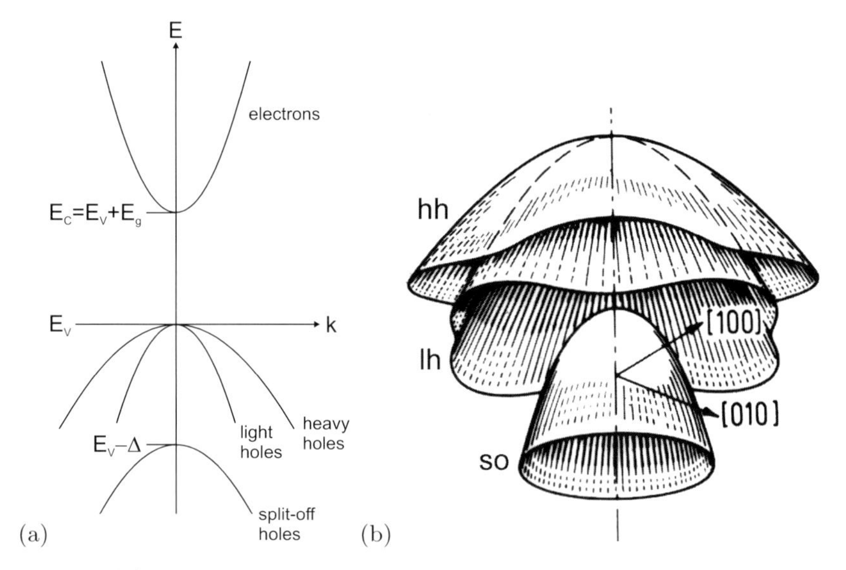

Fig. 6.25. (**a**) Simplified band structure with conduction band and three valence bands and (**b**) three-dimensional visualization (E vs. $(k_\mathrm{x}, k_\mathrm{y})$) of the valence bands of Ge (including warping). Part (**b**) from [165]

All holes have different mass. In the vicinity of the Γ point the dispersion for heavy and light holes can be described with (+:hh, −:lh)

$$E(\mathbf{k}) = Ak^2 \pm \left[B^2k^4 + C^2\left(k_\mathrm{x}^2k_\mathrm{y}^2 + k_\mathrm{y}^2k_\mathrm{z}^2 + k_\mathrm{x}^2k_\mathrm{z}^2\right)\right]^{1/2} . \tag{6.49}$$

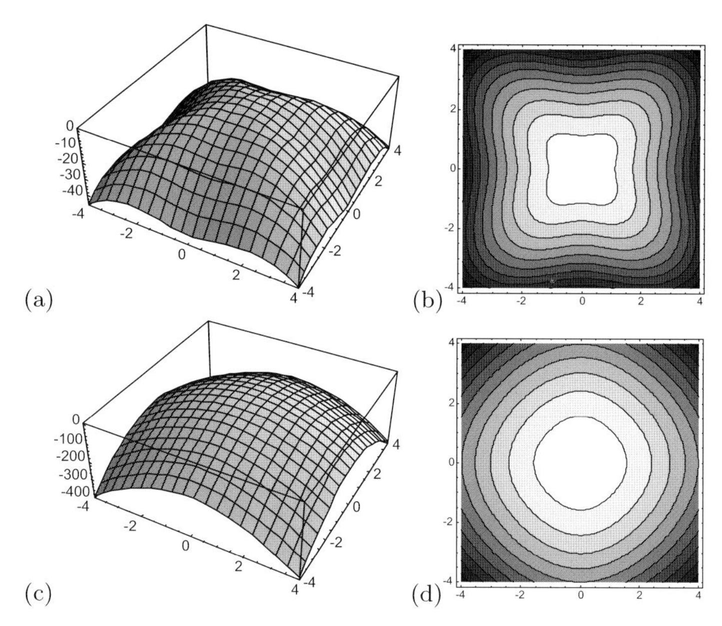

Fig. 6.26. Dispersion in the (k_x, k_y) plane at the valence band edge of GaAs. (**a**,**b**): hh, (**c**,**d**): lh. Dispersion as (**a**,**c**): 3D plot and (**b**,**d**) as isolines

For heavy and light holes there is a dependence of the dispersion, i.e. the mass, in the (001) plane. This effect, sketched in Fig. 6.25b, is called *warping*. The warping at the GaAs valence-band edge is shown in Fig. 6.26.

The s-o holes have the dispersion

$$E(\mathbf{k}) = -\Delta_0 + Ak^2 . \tag{6.50}$$

Values for A, B, C^2 and Δ_0 for a number of semiconductors are given in Table 6.3. The valence-band structure is often described with the Luttinger parameters γ_1, γ_2, and γ_3 that can be represented through A, B, and C via

Table 6.3. Valence-band parameters (for (6.49)) A and B in units of $(\hbar^2/2m_0)$, C^2 in units of $(\hbar^2/2m_0)^2$, and Δ_0 in eV. From [93, 166]

Material	A	B	C^2	Δ_0
Si	−4.28	−0.68	24	0.044
Ge	−13.38	−8.5	173	0.295
GaAs	−6.9	−4.4	43	0.341
InP	−5.15	−1.9	21	0.11
InAs	−20.4	−16.6	167	0.38
ZnSe	−2.75	−1.0	7.5	0.43

$$\frac{\hbar^2}{2m_0}\gamma_1 = A \tag{6.51a}$$

$$\frac{\hbar^2}{2m_0}\gamma_2 = -\frac{B}{2} \tag{6.51b}$$

$$\frac{\hbar^2}{2m_0}\gamma_3 = \frac{\sqrt{B^2 + C^2/3}}{2} . \tag{6.51c}$$

The mass of holes in various directions can be derived from (6.49). The mass along the [001] direction, i.e. $\hbar^2 / \left(\partial^2 E(\mathbf{k})/\partial k_x^2\right)$ for $k_y = 0$ and $k_z = 0$, is

$$\frac{1}{m_{\mathrm{hh}}^{100}} = \frac{2}{\hbar^2}(A + B) \tag{6.52a}$$

$$\frac{1}{m_{\mathrm{lh}}^{100}} = \frac{2}{\hbar^2}(A - B) . \tag{6.52b}$$

In [110] and [111] directions the masses are

$$\frac{1}{m_{\mathrm{hh}}^{110}} = \frac{2}{\hbar^2}\left(A + B\left[1 + \frac{C^2}{4B^2}\right]^{\frac{1}{2}}\right) \tag{6.53a}$$

$$\frac{1}{m_{\mathrm{lh}}^{110}} = \frac{2}{\hbar^2}\left(A - B\left[1 + \frac{C^2}{4B^2}\right]^{\frac{1}{2}}\right) \tag{6.53b}$$

$$\frac{1}{m_{\mathrm{hh}}^{111}} = \frac{2}{\hbar^2}\left(A + B\left[1 + \frac{C^2}{3B^2}\right]^{\frac{1}{2}}\right) \tag{6.54a}$$

$$\frac{1}{m_{\mathrm{lh}}^{111}} = \frac{2}{\hbar^2}\left(A - B\left[1 + \frac{C^2}{3B^2}\right]^{\frac{1}{2}}\right) . \tag{6.54b}$$

For $C^2 = 0$ the hole bands are isotropic, as is obvious from (6.49). In this case $\gamma_2 = \gamma_3$, the so-called spherical approximation. The average of the hole masses over all directions is

$$\frac{1}{m_{\mathrm{hh}}^{\mathrm{av}}} = \frac{2}{\hbar^2}\left(A + B\left[1 + \frac{2C^2}{15B^2}\right]\right) \tag{6.55a}$$

$$\frac{1}{m_{\mathrm{lh}}^{\mathrm{av}}} = \frac{2}{\hbar^2}\left(A - B\left[1 + \frac{2C^2}{15B^2}\right]\right) . \tag{6.55b}$$

6.13.3 Valence-Band Fine Structure

In Fig. 6.27, the schematic structure of the band edges for zincblende structure semiconductors is shown. The s-o holes in the zincblende structure are split-off due to the spin-orbit interaction Δ_{so}, the Γ_8 band is degenerate

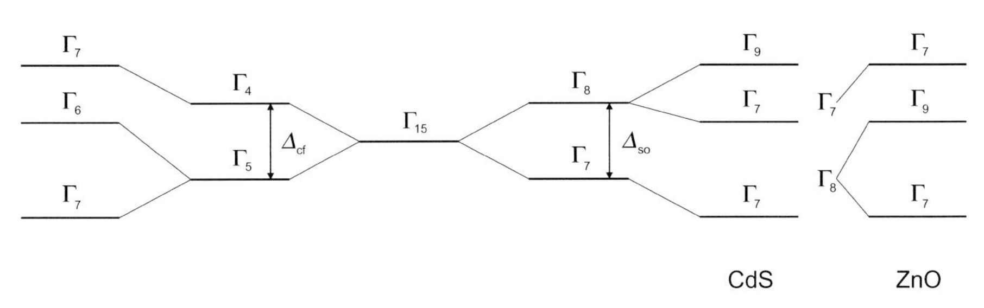

Fig. 6.27. Schematic band structure of zincblende and the valence-band splitting due to spin-orbit interaction Δ_{so} and crystal field splitting Δ_{cf} for chalcopyrites (typically $\Delta_{\mathrm{cf}} < 0$, see Fig. 6.28) and wurtzites. For the wurtzites the situation is schematically shown for CdS ($\Delta_{\mathrm{so}} = 67\,\mathrm{meV}$, $\Delta_{\mathrm{cf}} = 27\,\mathrm{meV}$) (or GaN) and ZnO ($\Delta_{\mathrm{so}} = -8.7\,\mathrm{meV}$, $\Delta_{\mathrm{cf}} = 41\,\mathrm{meV}$)

(heavy and light holes). Degeneracies for the holes are removed in the wurtzite and chalcopyrite structures by the additional crystal field splitting Δ_{cf} due to the anisotropy between the a- and c-axes. Typically, e.g. for CdS, the topmost valence band in the wurtzite structure has Γ_9 symmetry (allowed optical transitions only for $\mathbf{E} \perp \mathbf{c}$); an exception is ZnO for which the two upper bands are believed to be reversed. In the chalcopyrite structure optical transitions involving the Γ_6 band are only allowed for $\mathbf{E} \perp \mathbf{c}$. The three hole bands are usually labeled A, B, and C from the top of the valence band.

The energy positions of the three bands (with respect to the position of the Γ_{15} band) in the presence of spin-orbit interaction and crystal field splitting are given within the quasi-cubic approximation [167] by

$$E_1 = \frac{\Delta_{\mathrm{so}} + \Delta_{\mathrm{cf}}}{2} \tag{6.56a}$$

$$E_{2,3} = \pm \left[\left(\frac{\Delta_{\mathrm{so}} + \Delta_{\mathrm{cf}}}{2} \right)^2 - \frac{2}{3} \Delta_{\mathrm{so}} \Delta_{\mathrm{cf}} \right]^{1/2} . \tag{6.56b}$$

In chalcopyrites the crystal field splitting is typically negative (Fig. 6.28). It is approximately linearly related to $1 - \eta$ (for $\eta = c/2a$ see Sect. 3.4.6).

In certain compounds the bandgap can shrink to zero and even become negative in the sense that the Γ_6 symmetry (conduction) band is below the valence-band edge (Fig. 6.29).

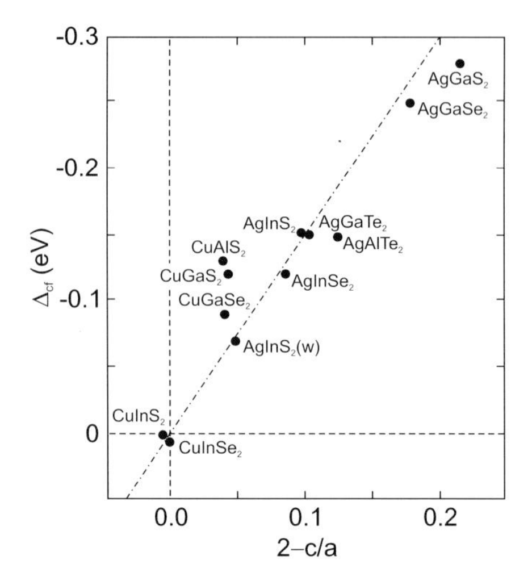

Fig. 6.28. Crystal field splitting Δ_{cf} for various chalcopyrite compounds vs. the tetragonal distortion $2 - c/a = 2(1 - \eta)$. *Dashed line* represents $\Delta_{\mathrm{cf}} = 1.5b(2 - c/a)$ for $b = 1\,\mathrm{eV}$. Data from [168]

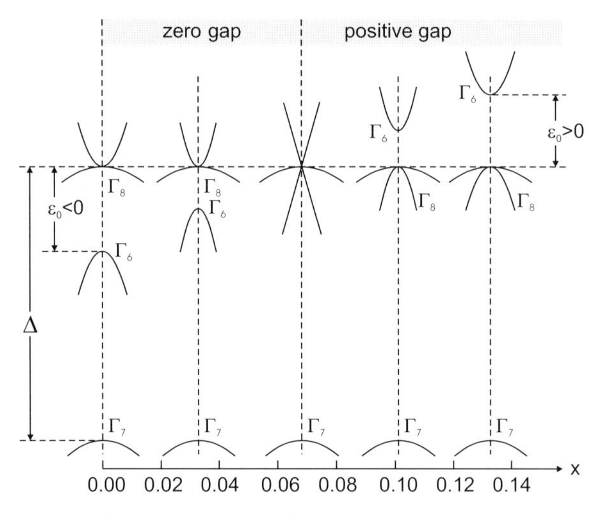

Fig. 6.29. Schematic band structure of zincblende with vanishing energy gap for the ternary compounds $\mathrm{Mn}_x\mathrm{Hg}_{1-x}\mathrm{Te}$

6.14 Strain Effect on the Band Structure

A mechanical strain (or equivalently stress) causes changes in the bond lengths. Accordingly, the band structure is affected. These effects have been exhaustively treated in [169,170]. For small strain, typically $\epsilon \lesssim 0.01$ the shift of the band edges is linear with the strain, for large strain it becomes nonlinear [171]. Often homogeneous strain is assumed, the effect of inhomogeneous strain is discussed in [172].

6.14.1 Strain effect on Band Edges

In a direct-gap zincblende material the position of the conduction-band edge is only affected by the hydrostatic component of the strain

$$E_{\mathrm{C}} = E_{\mathrm{C}}^{0} + a_{\mathrm{c}} \left(\epsilon_{\mathrm{xx}} + \epsilon_{\mathrm{yy}} + \epsilon_{\mathrm{zz}}\right) = E_{\mathrm{C}}^{0} + a_{\mathrm{c}} \mathrm{Tr}(\epsilon) \ , \tag{6.57}$$

where $a_{\mathrm{c}} < 0$ is the conduction-band hydrostatic deformation potential and E_{C}^{0} is the conduction-band edge of the unstrained material. Similarly, the valence-band edge is

$$E_{\mathrm{V}} = E_{\mathrm{V}}^{0} + a_{\mathrm{v}} \mathrm{Tr}(\epsilon) \ , \tag{6.58}$$

where $a_{\mathrm{v}} > 0$ is the valence-band hydrostatic deformation potential. Therefore the bandgap increases by

$$\Delta E_{\mathrm{g}} = a \mathrm{Tr}(\epsilon) = a \left(\epsilon_{\mathrm{xx}} + \epsilon_{\mathrm{yy}} + \epsilon_{\mathrm{zz}}\right) \ , \tag{6.59}$$

with $a = a_{\mathrm{c}} - a_{\mathrm{v}}$.

Biaxial and shear strains affect the valence bands and lead to shifts and splitting of the heavy and light holes at the Γ point:

$$\begin{aligned} E_{\mathrm{v,hh/lh}} &= E_{\mathrm{V}}^{0} \pm E_{\epsilon\epsilon}^{1/2} \qquad (6.60\mathrm{a}) \\ E_{\epsilon\epsilon}^{2} &= b^2/2 \left[\left(\epsilon_{\mathrm{xx}} - \epsilon_{\mathrm{yy}}\right)^2 + \left(\epsilon_{\mathrm{yy}} - \epsilon_{\mathrm{zz}}\right)^2 + \left(\epsilon_{\mathrm{xx}} - \epsilon_{\mathrm{zz}}\right)^2 \right] \\ &\quad + d^2 \left[\epsilon_{\mathrm{xy}}^2 + \epsilon_{\mathrm{yz}}^2 + \epsilon_{\mathrm{xz}}^2 \right] \ , \end{aligned}$$

where E_{V}^{0} denotes the bulk valence-band edge. b and d are the optical deformation potentials. For compressive strain the heavy-hole band is above the light-hole band. For tensile strain there is strong mixing of the bands (Fig. 6.30). In Table 6.4 the deformation potentials for some III–V semiconductors are listed. Typical values are in the eV regime. In a wurtzite crystal, five deformation potentials are needed that are termed a and D_1–D_4 [173, 174].

In Si and Ge four deformation potentials are needed that are termed a, b, d, Ξ_{u} and Ξ_{d} [175]. The energy position of the i-th conduction-band edge (with unit vector $\mathbf{a}_i$ pointing to the valley, e.g. [111]) is

$$E_{\mathrm{C},i} = E_{\mathrm{C},i}^{0} + \Xi_{\mathrm{d}} \mathrm{Tr}(\epsilon) + \Xi_{\mathrm{u}} \mathbf{a}_i \epsilon \mathbf{a}_i \ , \tag{6.61}$$

where $E_{\mathrm{C},i}^{0}$ denotes the energy of the unstrained conduction-band edge.

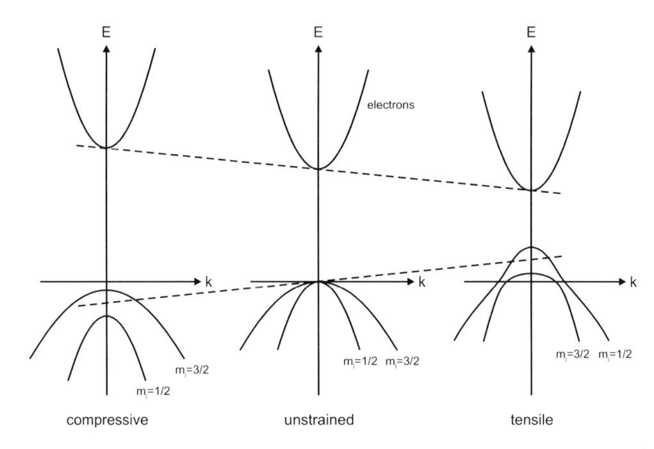

Fig. 6.30. Schematic band structure of GaAs in unstrained state (*center*) and under compressive and tensile biaxial strain as labeled. *Dashed lines* indicate shift of band edges due to hydrostatic part of strain

Table 6.4. Deformation potentials (in eV) for some III–V semiconductors

material	a	b	d
GaAs	−9.8	−1.7	−4.6
InAs	−6.0	−1.8	−3.6

6.14.2 Strain Effect on Effective Masses

In the presence of strain the band edges are shifted (see Sect. 6.14). Since the electron mass is related to the bandgap, it is expected that the mass will also be effected. In the presence of hydrostatic strain ϵ_{H} the electron mass is [176] (cf. to (6.42) for $\epsilon_{\mathrm{H}} \rightarrow 0$)

$$\frac{m_0}{m_{\mathrm{e}}^*} = 1 + \frac{E_{\mathrm{P}}}{E_{\mathrm{g}} + \Delta_0/3}\left[1 - \epsilon_{\mathrm{H}}\left(2 + \frac{3a}{E_{\mathrm{g}} + \Delta_0/3}\right)\right] , \tag{6.62}$$

with a being the hydrostatic deformation potential and $\epsilon_{\mathrm{H}} = \mathrm{Tr}(\epsilon)$. In [176], formulas are also given for biaxial and shear strain and also for hole masses.

6.15 Density of States

6.15.1 General Band Structure

The dispersion relation yields how the energy of a (quasi-) particle depends on the **k** vector. Now we want to know how many states are at a given energy. This quantity is called the *density of states* (DOS) and is written as $D(E)$. It is defined in an infinitesimal sense such that the number of states between E and $E + \delta E$ is $D(E)\delta E$. In the vicinity of the extrema of the band structure many states are at the same energy such that the density of states is high.

The dispersion relation of a band will be given as $E = E(\mathbf{k})$. If several bands overlap, the densities of state of all bands need to be summed up. The density of states at the energy $\tilde{E}$ for the given band is

$$D\left(\tilde{E}\right) \mathrm{d}E = 2 \int \frac{\mathrm{d}^3\mathbf{k}}{(2\pi/L)^3} \, \delta(\tilde{E} - E(\mathbf{k})) \,, \tag{6.63}$$

where $(2\pi/L)^3$ is the $\mathbf{k}$-space volume for one state. The factor 2 is for spin degeneracy. The integral runs over the entire $\mathbf{k}$-space and selects only those states that are at $\tilde{E}$. The volume integral can be converted to a surface integral over the isoenergy surface $S(\tilde{E})$ with $E(\mathbf{k}) = \tilde{E}$. The volume element $\mathrm{d}^3\mathbf{k}$ is written as $\mathrm{d}^2S\mathrm{d}\mathbf{k}_\perp$. The vector $\mathbf{k}_\perp$ is perpendicular to $S(\tilde{E})$ and proportional to $\nabla_\mathbf{k}E(\mathbf{k})$, i.e. $\mathrm{d}E = |\nabla_\mathbf{k}E(\mathbf{k})|\mathrm{d}\mathbf{k}_\perp$.

$$D\left(\tilde{E}\right) = 2 \int_{S(\tilde{E})} \frac{\mathrm{d}^2S}{(2\pi/L)^3} \frac{1}{|\nabla_\mathbf{k}E(\mathbf{k})|} \,. \tag{6.64}$$

In this equation, the dispersion relation is explicitly contained. At band extrema the gradient diverges, however, in three dimensions the singularities are integrable and the density of states takes a finite value. The corresponding peak is named a van-Hove singularity. The concept of the density of states is valid for all possible dispersion relations, e.g. for electrons, phonons or photons.

The density of states for the silicon band structure (cf. Fig. 6.5a) is shown in Fig. 6.31. If disorder is introduced, the density of states is modified as shown in Fig. 6.32 for amorphous germanium. The defects, as compared to the perfect lattice, introduce states in the bandgap and generally wash out the sharp features from the crystalline DOS.

6.15.2 Free-Electron Gas

In M dimensions, the energy states of a free-electron gas are given as

$$E(\mathbf{k}) = \frac{\hbar^2}{2m^*} \sum_{i=1}^{M} k_i^2 \,. \tag{6.65}$$

The k_i can take the values $\pm\pi n/L$ (in the first Brillouin zone) with $n \leq N$, N being the number of unit cells in one dimension. These values are equidistant in $\mathbf{k}$-space. Each M-dimensional $\mathbf{k}$-point takes a volume of $(2\pi/L)^M$. The number of states $N(E_\mathrm{F})$ up to the energy $E_\mathrm{F} = \frac{\hbar^2}{2m}k_\mathrm{F}^2$ (later used as Fermi energy E_F and Fermi vector k_F) is

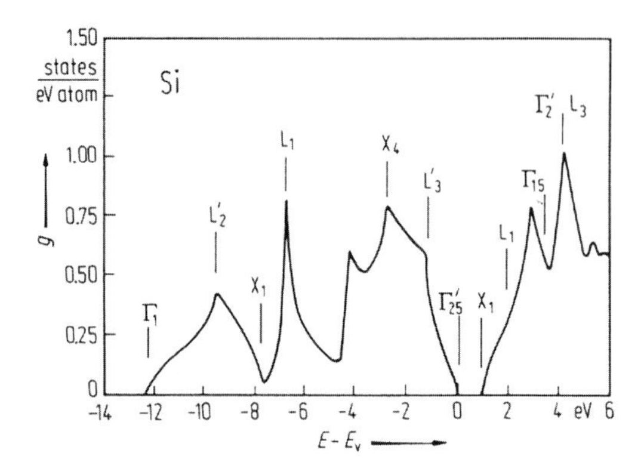

Fig. 6.31. Density of states in the silicon valence- and conduction-band (theoretical calculation using empirical pseudopotentials). From [93], based on [177]

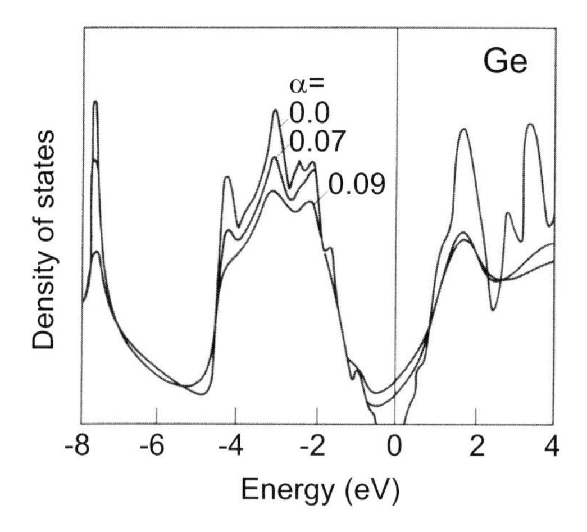

Fig. 6.32. Theoretical calculation for the density of states of amorphous Ge models as obtained for various degrees of disorder α (3.6). $\alpha = 0.09$ corresponds to a mean short-range order distance of about 2.4 lattice constants. From [52]

$$N(E_{\mathrm{F}}) = \frac{2}{(2\pi/L)^M} \int_{\mathbf{k}=0}^{|\mathbf{k}|=k_{\mathrm{F}}} \mathrm{d}^M k \, . \tag{6.66}$$

The factor 2 is for spin degeneracy, the integration runs over M dimensions. The density of states is the derivative $D(E) = \mathrm{d}N/\mathrm{d}E$.

In the following, the density of states for $M = 3, 2, 1$ and zero dimensions is derived. A visualization is given in Fig. 13.1.

$M = 3$

This case relates to bulk material in which electrons are free to move in all three dimensions. Performing the integral (6.66) for $M = 3$ yields

$$N^{3D} = \frac{V}{3\pi^2} k_F^3 = \frac{V}{3\pi^2} \left(\frac{2mE_F}{\hbar^2} \right)^{3/2} . \tag{6.67}$$

Therefore, k_F and E_F are given by

$$k_F = \left(\frac{3\pi^2 N}{V} \right)^{1/3} \tag{6.68}$$

$$E_F = \frac{\hbar^2}{2m^*} \left(\frac{3\pi^2 N}{V} \right)^{2/3} , \tag{6.69}$$

and the density of states in three dimensions is

$$D^{3D}(E) = \frac{V}{2\pi^2} \left(\frac{2m^*}{\hbar^2} \right)^{3/2} \sqrt{E} . \tag{6.70}$$

$M = 2$

This case is important for thin layers in which the electron motion is confined in one direction and free in a plane. Such structures are called quantum wells (see Sect. 11.5). We find for the 2D density of states (for each subband over which it is not summed here, including spin degeneracy)

$$N^{2D} = \frac{A}{2\pi} k_F^2 = \frac{A}{\pi} \frac{m^*}{\hbar^2} E , \tag{6.71}$$

where A is the area of the layer. The density of states is thus constant and given by

$$D^{2D}(E) = \frac{A}{\pi} \frac{m^*}{\hbar^2} . \tag{6.72}$$

$M = 1$

The case $M = 1$ describes a quantum wire in which the electron motion is confined in two dimensions and free in only one dimension (see Sect. 13.2). For this case, we find for a wire of length L

$$N^{1D} = \frac{2L}{\pi} k_F = \frac{2L}{\pi} \left(\frac{2m^* E}{\hbar^2} \right)^{1/2} . \tag{6.73}$$

The density of states becomes singular at $E = 0$ and is given by (for one subband)

$$D^{1D}(E) = \frac{L}{\pi} \left(\frac{2m^*}{\hbar^2} \right)^{1/2} \frac{1}{\sqrt{E}} . \tag{6.74}$$

$M = 0$

In this case electrons have no degrees of freedom, as, e.g., in a quantum dot (see Sect. 13.3), and each state has a δ-like density of states at each of the quantized levels.

7 Electronic Defect States

Über Halbleiter sollte man nicht arbeiten,
das ist eine Schweinerei,
wer weiß ob es überhaupt Halbleiter gibt.[1]
W. Pauli, 1931 [178]

7.1 Introduction

One cm^3 of a semiconductor contains about 5×10^{22} atoms. It is practically impossible to achieve perfect purity. Typical low concentrations of impurity atoms are in the $10^{12} - 10^{13}\,\mathrm{cm}^{-3}$ regime. Such a concentration corresponds to a purity of 10^{-10}, corresponding to about one alien in the world's human population. In the beginning of semiconductor research the semiconductors were so impure that the actual semiconducting properties could only be used inefficiently. Nowadays, thanks to large improvements in high-purity chemistry, the most common semiconductors, in particular silicon, can be made so pure that the residual impurity concentration plays no role in the physical properties. However, the most important technological step for semiconductors is *doping*, the controlled incorporation of impurities, in order to manage the semiconductor's conductivity. Typical impurity concentrations used in doping are $10^{15} - 10^{20}\,\mathrm{cm}^{-3}$.

7.2 Fermi Distribution

In thermodynamic equilibrium, the distribution function for electrons is given by the Fermi–Dirac distribution (Fermi function) $f_\mathrm{e}(E)$ (see Appendix D)

$$f_\mathrm{e}(E) = \frac{1}{\exp\left(\frac{E-E_\mathrm{F}}{kT}\right) + 1}\,, \tag{7.1}$$

[1]One should not work on semiconductors. They are a mess. Who knows whether semiconductors exist at all.

where k (or k_B) denotes the Boltzmann constant, T is the temperature, and E_F is the Fermi level, which is called the chemical potential μ in thermodynamics. The Fermi distribution is shown in Fig. 7.1. The distribution function gives the probability that a state at energy E is populated in thermodynamic equilibrium. For $E = E_F$ the population is 1/2 for all temperatures. At (the unrealistic case of) $T = 0$, the function makes a step from 1 (for $E < E_F$) to 0.

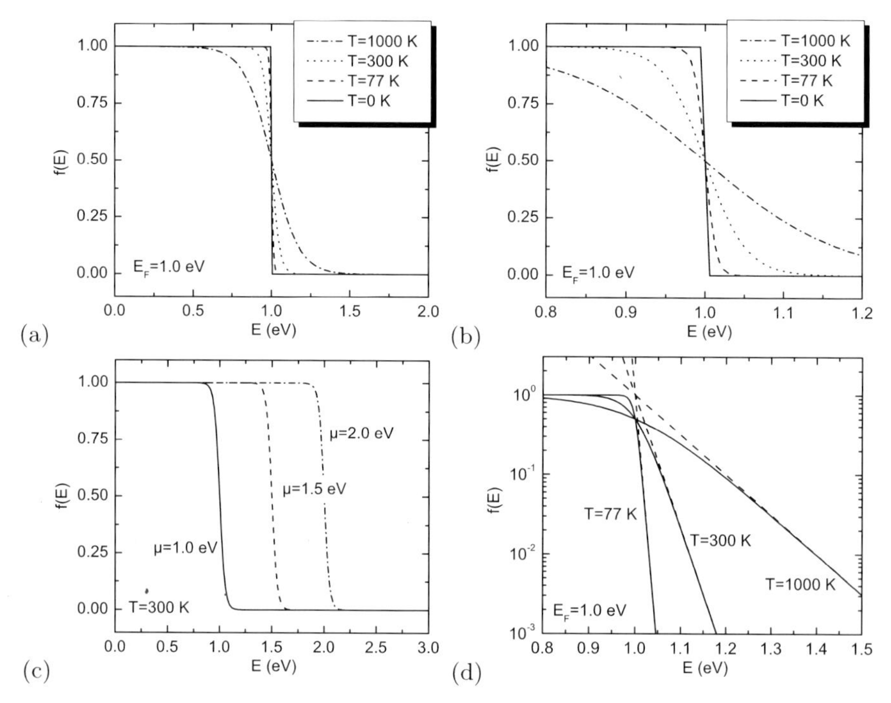

Fig. 7.1. Fermi function for (**a**,**b**) different temperatures (for $E_F = 1.0\,\text{eV}$) and (**c**) for different chemical potentials (for $T = 300\,\text{K}$). (**d**) Fermi function (*solid lines*) compared with Boltzmann approximation (*dashed lines*) for various temperatures and $E_F = 1.0\,\text{eV}$ on semilogarithmic plot

The high-energy tail of the Fermi distribution, i.e. for $E - E_F \gg kT$, can be approximated by the Boltzmann distribution:

$$f_e(E) \approx \exp\left(-\frac{E - E_F}{kT}\right) . \tag{7.2}$$

If the Boltzmann distribution is a good approximation, the carrier distribution is called *nondegenerate*. If the Fermi distribution needs to be invoked, the carrier ensemble is called *degenerate*. If the Fermi level is within the band the ensemble is highly degenerate.

7.3 Carrier Concentration

Generally, the density of electrons in the conduction band is given by

$$n = \int_{E_\mathrm{C}}^{\infty} D_\mathrm{e}(E) f_\mathrm{e}(E) \mathrm{d}E \,, \tag{7.3}$$

and accordingly the density of holes in the valence band is

$$p = \int_{-\infty}^{E_\mathrm{V}} D_\mathrm{h}(E) f_\mathrm{h}(E) \mathrm{d}E \,. \tag{7.4}$$

The energy of the top of the valence band is denoted by E_V, the bottom of the conduction band as E_C. The distribution function for holes is $f_\mathrm{h} = 1 - f_\mathrm{e}$. Thus,

$$f_\mathrm{h}(E) = 1 - \frac{1}{\exp\left(\frac{E-E_\mathrm{F}}{kT}\right)+1} = \frac{1}{\exp\left(-\frac{E-E_\mathrm{F}}{kT}\right)+1} \,. \tag{7.5}$$

We assume parabolic band edges, i.e. effective masses m_e and m_h for electrons and holes, respectively. The density of states in the conduction band (per unit volume) is given (for $E > E_\mathrm{C}$) as (cf. (6.70))

$$D_\mathrm{e}(E) = \frac{1}{2\pi^2} \left(\frac{2m_\mathrm{e}}{\hbar^2}\right)^{3/2} (E - E_\mathrm{C})^{1/2} \,. \tag{7.6}$$

If, in the previous consideration the Boltzmann approximation cannot be applied, i.e. at high temperatures or for very small bandgaps, the integral over Df cannot be explicitly (or analytically) evaluated. In this case the Fermi integral is needed that is defined[2] as

$$F_n(x) = \frac{2}{\sqrt{\pi}} \int_0^{\infty} \frac{y^n}{1+\exp(y-x)} \mathrm{d}y \,. \tag{7.7}$$

In the present case of bulk materials $n = 1/2$. For large negative argument, i.e. $x < 0$ and $|x| \gg 1$, $F_{1/2}(x) \approx \exp(x)$, which is the Boltzmann approximation. $F_{1/2}(0) \approx 0.765 \approx 3/4$. For large argument, i.e. $x \gg 1$, $F_{1/2}(x) \approx (2\sqrt{\pi})(2/3)x^{3/2}$. Such fairly simple approximations are plotted in Fig. 7.2 in comparison with the Fermi integral. For computations, numerical approximations are used [179, 180].

The derivative of the Fermi integral is given by $F_n'(x) = nF_{n-1}(x)$, $n > 0$. For $n = 0$, i.e. a two-dimensional system, the integral can be executed explicitly, $F_0(x) = \ln\left[1 + \exp(x)\right]$.

With the Fermi integral $F_{1/2}$ (7.15) and (7.16) then have the following expressions for the free-carrier densities:

---

[2] Equation (7.7) is restricted to $n > -1$. A form without restriction is $\mathcal{F}_n(x) = \frac{1}{\Gamma(n+1)} \int_0^{\infty} \frac{y^n}{1+\exp(y-x)} \mathrm{d}y$.

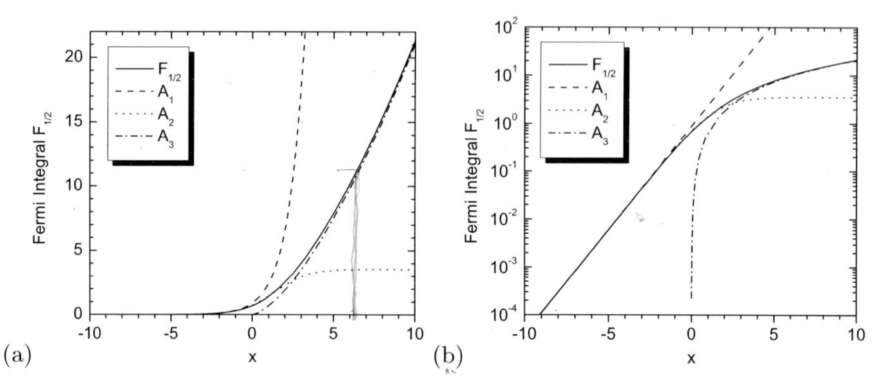

Fig. 7.2. Fermi integral $F_{1/2}$ with approximations in three regions of the argument: $A_1(x) = \exp(x)$ for $x < 2$, $A_2(x) = (1/4 + \exp(-x))^{-1}$ for $-2 < x < 2$, $A_3(x) = (2/\sqrt{\pi})(2/3)x^{3/2}$ for $x > 2$. (**a**) linear, (**b**) semilogarithmic plot

$$n = N_{\mathrm{C}} F_{1/2}\left(\frac{E_{\mathrm{F}} - E_{\mathrm{C}}}{kT}\right) \tag{7.8}$$

$$p = N_{\mathrm{V}} F_{1/2}\left(-\frac{E_{\mathrm{F}} - E_{\mathrm{V}}}{kT}\right) , \tag{7.9}$$

with

$$N_{\mathrm{C}} = 2\left(\frac{m_{\mathrm{e}} kT}{2\pi\hbar^2}\right)^{3/2} \tag{7.10}$$

$$N_{\mathrm{V}} = 2\left(\frac{m_{\mathrm{h}} kT}{2\pi\hbar^2}\right)^{3/2} , \tag{7.11}$$

where N_{C} (N_{V}) is called the conduction-band (valence-band) edge density of states. If the conduction-band minimum is degenerate, a factor g_{v} (valley degeneracy) must be included, i.e. $g_{\mathrm{v}} = 6$ for Si and $g_{\mathrm{v}} = 8$ for Ge ($g_{\mathrm{v}} = 1$ for GaAs). This factor is typically included in the mass used in (7.10) that then becomes the density of states mass $m_{\mathrm{d,e}}$. If the conduction-band minimum has cylindrical symmetry in **k**-space, such as for Si and Ge, the mass that has to be used is

$$m_{\mathrm{d,e}} = g_{\mathrm{v}}^{2/3}\left(m_{\mathrm{t}}^2 m_{\mathrm{l}}\right)^{1/3} . \tag{7.12}$$

In the case of a degeneracy of the valence band, the states of several bands need to be summed for N_{V}. In bulk material, typically the heavy and light hole bands are degenerate at the Γ point. If the split-off band is not populated because of insufficient temperature, the valence-band edge density of states is given by

$$N_{\mathrm{V}} = 2\left(\frac{kT}{2\pi\hbar^2}\right)^{3/2}\left(m_{\mathrm{hh}}^{3/2} + m_{\mathrm{lh}}^{3/2}\right) . \tag{7.13}$$

Alternatively, the mass m_h in (7.11) can be taken as the density of states hole mass

$$m_{\mathrm{d,h}} = (m_{\mathrm{hh}}^{3/2} + m_{\mathrm{lh}}^{3/2})^{2/3} \,. \tag{7.14}$$

Values of $N_{\mathrm{C,V}}$ for Si, Ge and GaAs are given in Table 7.1.

Now, we assume that the Boltzmann approximation (7.2) can be used, i.e. the probability that a band state is populated is $\ll 1$. Then, the integral (7.3) can be executed analytically and the concentration n of electrons in the conduction band is given as

$$n = 2\left(\frac{m_\mathrm{e} kT}{2\pi\hbar^2}\right)^{3/2} \exp\left(\frac{E_\mathrm{F} - E_\mathrm{C}}{kT}\right) = N_\mathrm{C} \exp\left(\frac{E_\mathrm{F} - E_\mathrm{C}}{kT}\right) \,. \tag{7.15}$$

For the Boltzmann approximation and a parabolic valence band, the density of holes is given by

$$p = 2\left(\frac{m_\mathrm{h} kT}{2\pi\hbar^2}\right)^{3/2} \exp\left(-\frac{E_\mathrm{F} - E_\mathrm{V}}{kT}\right) = N_\mathrm{V} \exp\left(-\frac{E_\mathrm{F} - E_\mathrm{V}}{kT}\right) \,. \tag{7.16}$$

The product of the electron and hole density is

$$\begin{aligned} n\,p &= N_\mathrm{V} N_\mathrm{C} \exp\left(-\frac{E_\mathrm{C} - E_\mathrm{V}}{kT}\right) = N_\mathrm{V} N_\mathrm{C} \exp\left(-\frac{E_\mathrm{g}}{kT}\right) \\ &= 4\left(\frac{kT}{2\pi\hbar^2}\right)^3 (m_{\mathrm{d,e}}\, m_{\mathrm{d,h}})^{3/2} \exp\left(-\frac{E_\mathrm{g}}{kT}\right) \,. \end{aligned} \tag{7.17}$$

Thus, the product np is *independent* of the position of the Fermi level, as long as the Boltzmann approximation is fulfilled, i.e. the Fermi level is not in the vicinity of one of the band edges within several kT:

$$E_\mathrm{V} + 4kT < E_\mathrm{F} < E_\mathrm{C} - 4kT \,. \tag{7.18}$$

The relation (7.17) is called the mass–action law.

In Fig. 7.3, the product np is shown for silicon over a wide range of Fermi energies. If E_F is within the bandgap, np is essentially constant. If the Fermi level is in the valence or conduction band, np decreases exponentially.

7.4 Intrinsic Conduction

First, we consider the conductivity of the intrinsic, i.e. an ideally pure, semiconductor. At $T = 0$ all electrons are in the valence band, the conduction band is empty and thus the conductivity is zero (a completely filled band cannot conduct current). Only at finite temperatures the electrons have a finite probability to be in a conduction-band state and contribute to the conductivity. Due to neutrality, the electron and hole concentrations in the intrinsic

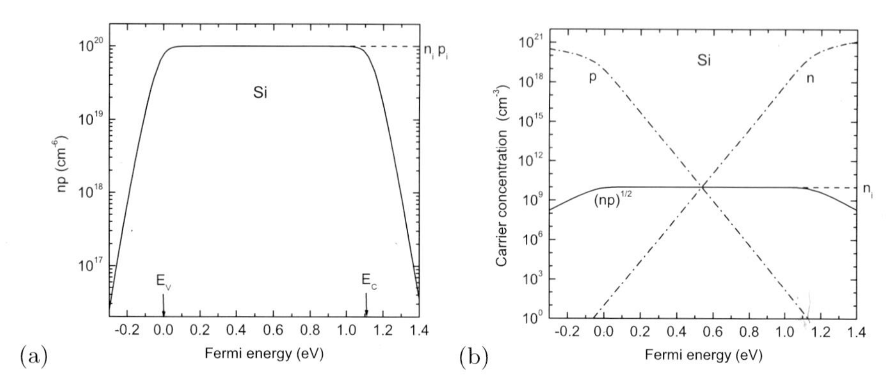

Fig. 7.3. (**a**) np for silicon at $T = 300\,\mathrm{K}$ as a function of the position of the Fermi level. The valence-band edge E_V is chosen as $E = 0$. np is constant for the range of Fermi energies given by (7.18) ($4kT \approx 0.1\,\mathrm{eV}$). (**b**) n, p and $\sqrt{np}$ as a function of the Fermi level

semiconductors are the same, i.e. each electron in the conduction band comes from the valence band,

$$-n + p = 0 \,, \tag{7.19}$$

or $n_i = p_i$. Therefore

$$\begin{aligned} n_i = p_i &= \sqrt{N_V N_C} \exp\left(-\frac{E_g}{2kT}\right) \\ &= 2\left(\frac{kT}{2\pi\hbar^2}\right)^{3/2} (m_e m_h)^{3/4} \exp\left(-\frac{E_g}{2kT}\right) . \end{aligned} \tag{7.20}$$

The mass-action law

$$n\,p = n_i p_i = n_i^2 = p_i^2 \tag{7.21}$$

will be essential also for doped semiconductors. The intrinsic carrier concentration is exponentially dependent on the bandgap. Thus, in thermodynamic equilibrium intrinsic wide-gap semiconductors have much smaller electron concentrations than intrinsic small-gap semiconductors (see Table 7.1). The intrinsic carrier concentration of Si has been determined to be (within 1%, T in K)

$$n_i^{\mathrm{Si}} = 1.640 \times 10^{15}\, T^{1.706} \exp\left(-\frac{E_g(T)}{2kT}\right) \tag{7.22}$$

for temperatures between 77 and 400 K [181, 182].

As we will see later in Part II, many semiconductor devices rely on regions of low conductivity (depletion layers) in which the carrier concentration is

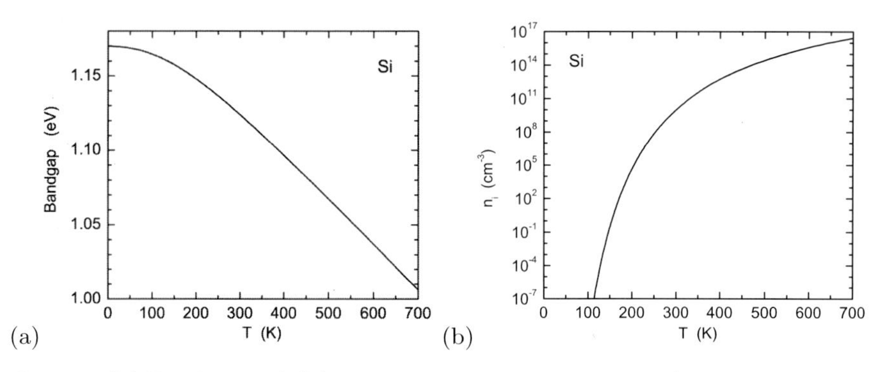

Fig. 7.4. (**a**) Bandgap and (**b**) intrinsic carrier concentration of silicon vs. temperature

Table 7.1. Bandgap, intrinsic carrier concentration, conduction band and valence-band edge density of states at $T = 300$ K for various semiconductors

	E_g (eV)	n_i (cm^{-3})	N_C (cm^{-3})	N_V (cm^{-3})
InSb	0.18	1.6×10^{16}		
InAs	0.36	8.6×10^{14}		
Ge	0.67	2.4×10^{13}	1.04×10^{19}	6.0×10^{18}
Si	1.124	1.0×10^{10}	7.28×10^{19}	1.05×10^{19}
GaAs	1.43	1.8×10^{6}	4.35×10^{17}	5.33×10^{18}
GaP	2.26	2.7×10^{0}		
GaN	3.3	$\ll 1$		

small. Since the carrier concentration cannot be smaller than the intrinsic concentration ($n + p \geq 2n_i$), an increase of temperature leads to increasing ohmic conduction in the depletion layers and thus to a reduction or failure of device performance. The small bandgap of Ge leads to degradation of bipolar device performance already shortly above room temperature. For silicon, intrinsic conduction limits operation typically to temperatures below about 300°C. For higher temperatures, as required for devices in harsh environments, such as close to motors or turbines, other semiconductors with wider bandgaps need to be used, such as GaN, SiC or even diamond.

From the neutrality condition for the intrinsic semiconductor (7.19) and (7.15) and (7.16), the Fermi level of the intrinsic semiconductor can be determined as

$$E_F = \frac{E_V + E_C}{2} + \frac{kT}{2} \ln\left(\frac{N_V}{N_C}\right) = \frac{E_V + E_C}{2} + \frac{3}{4} kT \ln\left(\frac{m_h}{m_e}\right) . \quad (7.23)$$

Since the hole mass is perhaps a factor of ten larger than the electron mass, the second term has the order of kT. Thus, for typical semiconductors

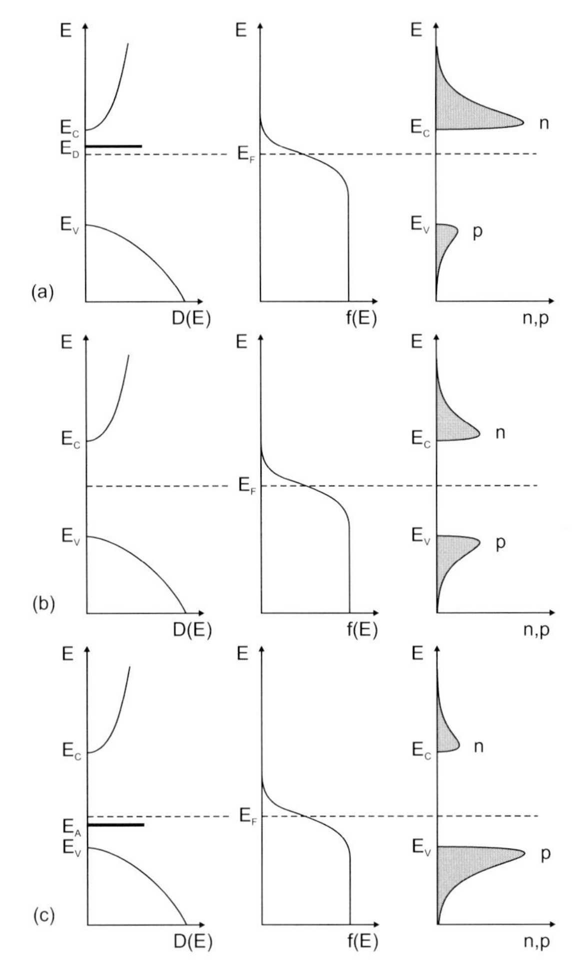

Fig. 7.5. Density of states (*left column*), Fermi distribution (*center column*) and carrier concentration (*right column*) for **(a)** n-type, **(b)** intrinsic and **(c)** p-type semiconductors in thermal equilibrium

where $E_g \gg kT$, the intrinsic Fermi level, denoted by E_i, is close to the middle of the bandgap, i.e. $E_F \approx (E_C + E_V)/2$.

The situation for an intrinsic semiconductor is schematically shown in Fig. 7.5a.

7.5 Shallow Impurities, Doping

The electronic levels of an impurity can be within the forbidden gap of the bulk host material. These levels can be close to the band edges or rather in the middle of the bandgap. In Fig. 7.6, the positions of the energy levels of a variety of impurities are shown for Ge, Si and GaAs. An impurity for which the long-range Coulomb part of the ion-core potential determines the energetic level is termed a *shallow* impurity. The extension of the wavefunction is given by the Bohr radius. This situation is in contrast to a *deep* level where the short-range part of the potential determines the energy level. The extension of the wavefunction is then of the order of the lattice constant.

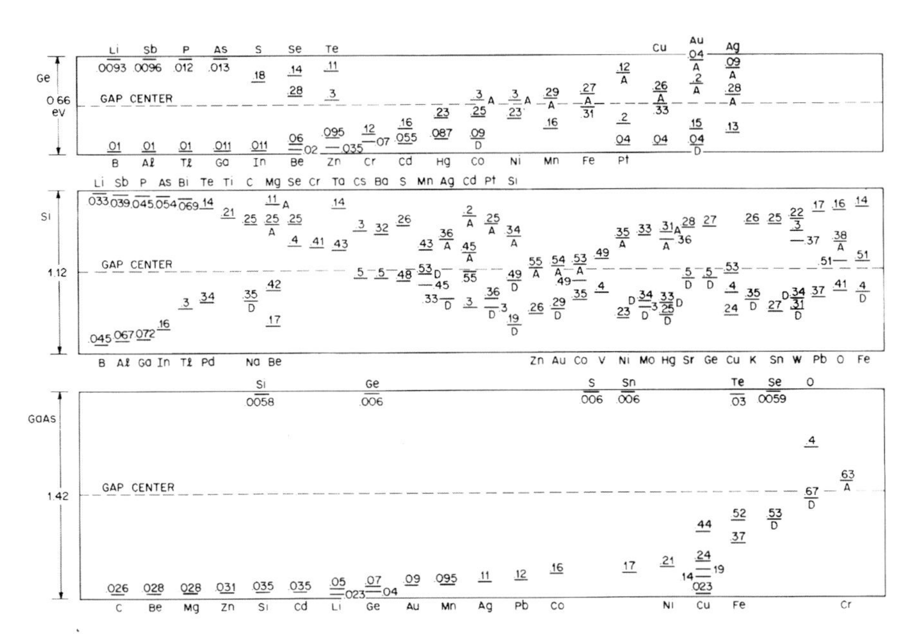

Fig. 7.6. Energetic position (ionization energy) of various impurities in Ge, Si and GaAs. Reprinted with permission from [183], ©1981 Wiley

We will consider first a group-IV semiconductor, Si, and (impurities) dopants from the groups III and V of the periodic system. When these are incorporated on a lattice site (with tetrahedral bonds), there is one electron too few (group III, e.g. B) or one electron too many (group V, e.g. As). The first case is called an *acceptor*, the latter a *donor*. The doping of III–V semiconductors is detailed in [184].

7.5.1 Donors

Silicon doped with arsenic is denoted as Si:As. The situation is schematically shown in Fig. 7.7. The arsenic atom has, after satisfying the tetrahedral bonds, an extra electron. This electron is bound to the arsenic atom via the Coulomb interaction since the ion core is positively charged compared to the silicon cores. If the electron is ionized, a fixed positive charge remains at the As site.

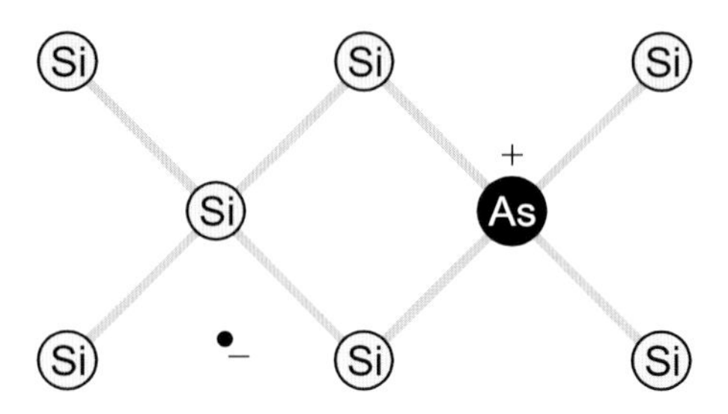

Fig. 7.7. Arsenic impurity in silicon. Arsenic donates one electron, and a fixed positive charge remains

A free arsenic atom has an ionization energy of 9.81 eV. However, in the solid the Coulomb interaction is screened by the (static) dielectric constant ϵ_{r} of the material, which is of the order of 10 for typical semiconductors. Additionally, the mass is renormalized (effective mass) by the periodic potential to a value that is smaller than the free-electron mass. Within effective mass theory (Appendix F) the hydrogen problem is scaled with the (isotropic) effective mass m_{e}^* and the dielectric constant ϵ_{r}, the binding energy (ionization energy) E_{D}^b of the electron to the shallow donor is (relative to the continuum given by the conduction-band edge E_{C})

$$E_{\mathrm{D}}^b = \frac{m_{\mathrm{e}}^*}{m_0} \frac{1}{\epsilon_{\mathrm{r}}^2} \frac{m_0 e^4}{2(4\pi\epsilon_0\hbar)^2} \ . \tag{7.24}$$

The absolute energy position of the level is $E_{\mathrm{D}} = E_{\mathrm{C}} - E_{\mathrm{D}}^b$. The first factor in the right side of (7.24) is the ratio of effective and free-electron mass, typically 1/10, the second factor is typically 1/100. The third factor is the ionization energy of the hydrogen atom, i.e. the Rydberg energy of 13.6 eV. Thus, the binding energy in the solid is drastically reduced by about 10^{-3} to the 10 meV regime. The extension of the wavefunction of the electron bound to the fixed ion is given by the Bohr radius

$$a_{\mathrm{D}} = \frac{m_0}{m_{\mathrm{e}}^*} \epsilon_{\mathrm{r}} \, a_{\mathrm{B}} \ , \tag{7.25}$$

where $a_{\mathrm{B}} = 0.053$ nm denotes the hydrogen Bohr radius. For GaAs $a_{\mathrm{D}} =$ 10.3 nm. For semiconductors with a nonisotropic band minimum, such as Si, Ge or GaP, an 'elliptically deformed' hydrogen problem with the masses m_{l} and m_{t} has to be treated [185].

An impurity that fulfills (7.24) is called an *effective-mass* impurity. For silicon we find, from (7.24), a donor binding energy of 6 meV. Using the correct tensor of the effective masses, the result for the effective-mass donor binding energy is 9.05 meV. Some experimentally observed values are summarized in Table 7.2. For GaAs, the effective-mass donor has a binding energy of 5.715 meV, which is fulfilled for several chemical species (Table 7.3). In

GaP, experimental values deviate considerably from the effective mass donor (59 meV).

Table 7.2. Binding energies $E_{\mathrm{D}}^{\mathrm{b}}$ of group-V donors in Si and Ge. All values in meV

	P	As	Sb
Si	45	49	39
Ge	12.0	12.7	9.6

Table 7.3. Binding energies $E_{\mathrm{D}}^{\mathrm{b}}$ of donors in GaAs (data from [186]), GaP (data from [187]) and GaN (low concentration limits, data from [188, 189]). All values in meV

	V site		III site	
GaAs	S	5.854	C	5.913
	Se	5.816	Si	5.801
	Te	5.786	Ge	5.937
GaP	O	897	Si	85
	S	107	Ge	204
	Se	105	Sn	72
	Te	93		
GaN	O	39	Si	22

The donors are typically distributed statistically (randomly) in the solid. Otherwise their distribution is called clustered. The concentration of donors is labeled N_{D} and usually given in cm^{-3}.

The concentration of donors populated with an electron (neutral donors) is denoted by N_{D}^{0}, the concentration of ionized donors (positively charged) is N_{D}^{+}. Other conventions in the literature label the concentrations N_1 and N_0, respectively:

$$N_1 = N_{\mathrm{D}}^{0} = N_{\mathrm{D}} f_{\mathrm{e}}(E_{\mathrm{D}}) \tag{7.26a}$$

$$N_0 = N_{\mathrm{D}}^{+} = N_{\mathrm{D}}(1 - f_{\mathrm{e}}(E_{\mathrm{D}}))\;, \tag{7.26b}$$

with $f_{\mathrm{e}}(E_{\mathrm{D}}) = [1 + \exp(E_{\mathrm{D}} - E_{\mathrm{F}})]^{-1}$. For the sum of these quantities the condition

$$N_{\mathrm{D}} = N_{\mathrm{D}}^{+} + N_{\mathrm{D}}^{0} \tag{7.27}$$

holds.

The ratio of the two concentrations is first given as (caveat: this formula will be modified below)

$$\frac{N_D^0}{N_D^+} = \frac{N_1}{N_0} = \frac{f}{1-f} = \exp\left(\frac{E_F - E_D}{kT}\right) . \tag{7.28}$$

Now, the degeneracy of the states has to be considered. The donor charged with one electron has a 2-fold degeneracy $g_1 = 2$ since the electron can take the spin up and down states. The degeneracy of the ionized (empty) donor is $g_0 = 1$. Additionally, we assume here that the donor cannot be charged with a second electron. Due to Coulomb interaction the energy level of the possible N_D^- state is in the conduction band. Otherwise, a multiply charged center would be present. We also do not consider excited states of N_D^0 that might be in the bandgap as well. In the following, we will continue with $\hat{g} = g_1/g_0 = 2$. Considering the degeneracy, (7.28) needs to be modified to

$$\frac{N_D^0}{N_D^+} = \frac{N_1}{N_0} = \hat{g} \exp\left(\frac{E_F - E_D}{kT}\right) . \tag{7.29}$$

This can be understood from thermodynamics (see Sect. 4.3), a rate analysis or simply the limit $T \to \infty$.

The probabilities f^1 and f^0 for a populated or empty donor, respectively, are

$$f^1 = \frac{N_1}{N_D} = \frac{1}{\hat{g}^{-1} \exp\left(\frac{E_D - E_F}{kT}\right) + 1} \tag{7.30a}$$

$$f^0 = \frac{N_0}{N_D} = \frac{1}{\hat{g} \exp\left(-\frac{E_D - E_F}{kT}\right) + 1} . \tag{7.30b}$$

First, we assume that no carriers in the conduction band stem from the valence band (no intrinsic conduction). This will be the case at sufficiently low temperatures when $N_D \gg n_i$. Then the number of electrons in the conduction band is equal to the number of ionized donors, i.e.

$$\begin{aligned} n = f^0 N_D = N_0 &= \frac{N_D}{1 + \hat{g} \exp\left(\frac{E_F - E_D}{kT}\right)} \\ &= \frac{n}{n + n_1} N_D = \frac{1}{1 + n_1/n} N_D , \end{aligned} \tag{7.31}$$

with $n_1 = (N_C/\hat{g}) \exp\left(-E_D^b/kT\right)$. The neutrality condition is

$$-n + N_D^+ = -n + N_0 = 0 \,. \tag{7.32}$$

Thus the equation (n is given by (7.15))

$$N_C \exp\left(\frac{E_F - E_C}{kT}\right) - \frac{N_D}{1 + \hat{g}\exp\left(\frac{E_F - E_D}{kT}\right)} = 0 \tag{7.33}$$

needs to be solved to obtain the Fermi level.[3] The result is

$$E_F = E_C - E_D^b + kT \ln \left[\frac{\left[1 + 4\hat{g}\frac{N_D}{N_C}\exp\left(\frac{E_D^b}{kT}\right)\right]^{1/2} - 1}{2\hat{g}} \right] . \tag{7.34}$$

For $T \to 0$ the Fermi level is, as expected, in the middle between the populated and unpopulated states, i.e. at $E_F = E_C - E_D^b/2$. In Fig. 7.8a the position of the Fermi level is shown for a donor with 45 meV binding energy in Si. For low temperatures the solution can be approximated as (dashed curve in Fig. 7.8b)

$$E_F \cong E_C - \frac{1}{2}E_D^b + \frac{1}{2}kT \ln\left(\frac{N_D}{\hat{g}N_C}\right) . \tag{7.35}$$

The freeze-out of carriers in n-type silicon has been discussed in detail in [191], taking into account the effects of the fine structure of the donor states. We note that the fairly high donor binding energy in silicon leads to freeze-out of carriers at about 40 K and is thus limiting for the low-temperature performance of devices. Ge has smaller donor ionization energies and subsequently a lower freeze-out temperature of 20 K. For n-type GaAs, conductivity is preserved down to even lower temperatures.

For higher temperatures, when the electron density saturates towards N_D, the approximate solution is (dash-dotted curve in Fig. 7.8a)

$$E_F \cong E_C + kT \ln\left(\frac{N_D}{N_C}\right) . \tag{7.36}$$

The electron density n is given (still in the Boltzmann approximation) by

$$\begin{aligned} n &= N_C \exp\left(-\frac{E_D^b}{kT}\right) \frac{\left[1 + 4\hat{g}\frac{N_D}{N_C}\exp\left(\frac{E_D^b}{kT}\right)\right]^{1/2} - 1}{2\hat{g}} \\ &= \frac{2N_D}{1 + \left[1 + 4\hat{g}\frac{N_D}{N_C}\exp\left(\frac{E_D^b}{kT}\right)\right]^{1/2}} . \end{aligned} \tag{7.37}$$

---

[3] As usual, the Fermi level is determined by the global charge neutrality, see also Sect. 4.3.

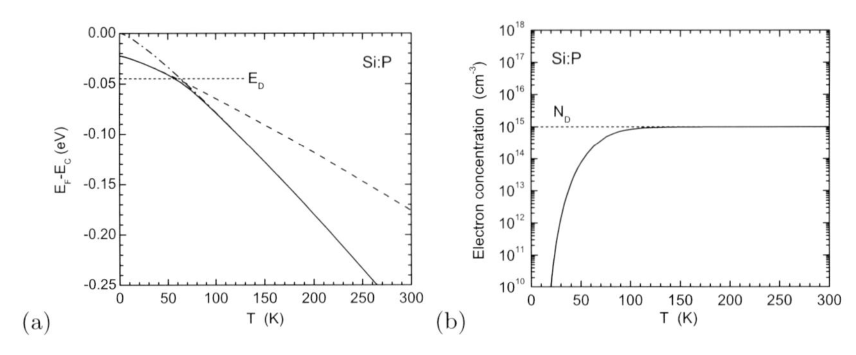

Fig. 7.8. (**a**) Position of the Fermi level in Si:P ($N_\mathrm{D} = 10^{15}\,\mathrm{cm}^{-3}$, $E^b_\mathrm{D} = 45\,\mathrm{meV}$, no acceptors) as a function of temperature *without* consideration of intrinsic carriers. Zero energy refers to the (temperature-dependent, Table 6.2) conduction-band edge E_C with approximative solutions for low (*dashed line*, (7.35)) and high (*dash-dotted line*, (7.36)) temperatures. (**b**) Corresponding density of conduction-band electrons as a function of temperature

The electron density as a function of temperature is shown in Fig. 7.8b. For low temperatures, the solution is close to

$$n \cong \left(\frac{N_\mathrm{D} N_\mathrm{C}}{\hat{g}}\right)^{1/2} \exp\left(-\frac{E^b_\mathrm{D}}{2kT}\right) = \sqrt{n_1 N_\mathrm{D}}\ . \tag{7.38}$$

For high temperatures, $n \cong N_\mathrm{D}$. This regime is called *exhaustion* or *saturation* since all possible electrons have been ionized from their donors. We note that even in this case $np = n_\mathrm{i} p_\mathrm{i}$ holds, however, $n \gg p$.

While the characteristic energy for the ionization of electrons from donors is E^b_D, at high enough temperatures electrons are transferred also from the valence band into the conduction band. Thus, in order to make the above consideration valid for all temperatures the intrinsic conduction also has to be considered. The neutrality condition (still in the absence of any acceptors) is

$$-n + p + N^+_\mathrm{D} = 0\ . \tag{7.39}$$

Using (7.15) and $p = n_\mathrm{i}^2/n$, the equation

$$N_\mathrm{C} \exp\left(\frac{E_\mathrm{F} - E_\mathrm{C}}{kT}\right) - \frac{n_\mathrm{i}^2}{N_\mathrm{C} \exp(\frac{E_\mathrm{F} - E_\mathrm{C}}{kT})} - \frac{N_\mathrm{D}}{1 + \hat{g} \exp(\frac{E_\mathrm{F} - E_\mathrm{D}}{kT})} = 0 \tag{7.40}$$

needs to be solved. The result is

$$E_\mathrm{F} = E_\mathrm{C} - E^b_\mathrm{D} + kT \ln\left[\frac{\frac{\beta}{\gamma} + \frac{\gamma}{N_\mathrm{C}^2} - 1}{3\hat{g}}\right]\ , \tag{7.41}$$

with

$$\gamma = \left(\frac{-N_{\mathrm{C}}^4 \alpha + \sqrt{(N_{\mathrm{C}}^4 \alpha)^2 - 4(N_{\mathrm{C}}^2 \beta)^3}}{2} \right)^{1/3} \tag{7.42}$$

$$\beta = N_{\mathrm{C}}^2 + 3\hat{g} N_{\mathrm{C}} N_{\mathrm{D}} \exp\left(\frac{E_{\mathrm{D}}}{kT}\right) + 3\hat{g}^2 n_{\mathrm{i}}^2 \exp\left(\frac{2E_{\mathrm{D}}}{kT}\right)$$

$$\alpha = 2N_{\mathrm{C}}^2 + 9\hat{g} N_{\mathrm{C}} N_{\mathrm{D}} \exp\left(\frac{E_{\mathrm{D}}}{kT}\right) - 18\hat{g}^2 n_{\mathrm{i}}^2 \exp\left(\frac{2E_{\mathrm{D}}}{kT}\right) .$$

The temperature-dependent position of the Fermi level is shown in Fig. 7.9.

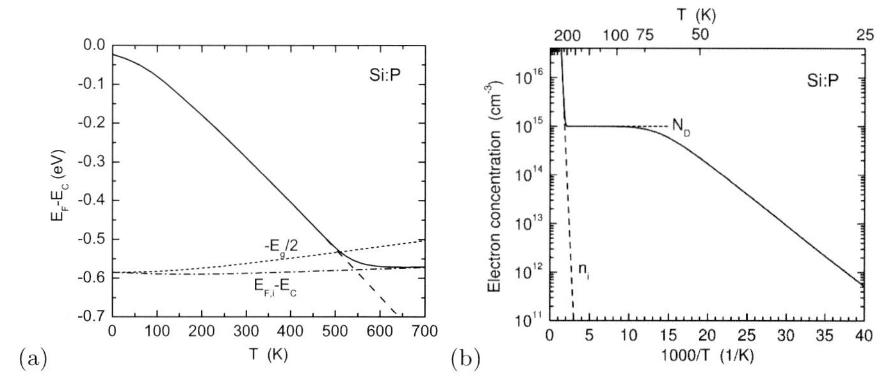

Fig. 7.9. (**a**) Position of the Fermi level in Si:P ($N_{\mathrm{D}} = 10^{15}\,\mathrm{cm}^{-3}$, $E_{\mathrm{D}}^b = 45\,\mathrm{meV}$, no acceptors) as a function of temperature. The temperature dependence of the bandgap (as given in Table 6.2) has been taken into account. Zero energy refers to the conduction-band edge for all temperatures. The *dotted curve* shows $E_{\mathrm{g}}/2$. The *dashed* (*dash-dotted*) *line* shows the low- (high-) temperature limit according to (7.35) and (7.23), respectively. (**b**) Corresponding electron concentration as a function of temperature. The *dashed line* shows the intrinsic carrier density

The carrier concentration is given by

$$n = N_{\mathrm{C}} \exp\left(-\frac{E_{\mathrm{D}}^b}{kT}\right) \frac{\frac{\beta}{\gamma} + \frac{\gamma}{N_{\mathrm{C}}^2} - 1}{3\hat{g}} , \tag{7.43}$$

with α, β and γ having the same meaning as in (7.41). The three important regimes are the intrinsic conduction at high temperatures when $n_{\mathrm{i}} \gg N_{\mathrm{D}}$, the exhaustion at intermediate temperatures when $n_{\mathrm{i}} \ll N_{\mathrm{D}}$ and $kT > E_{\mathrm{D}}$, and finally the freeze-out regime for $kT \ll E_{\mathrm{D}}$ at low temperatures when the electrons condense back into the donors. The rates for these processes will be discussed in Sect. 10.10.

A similar plot as in Fig. 7.9b is shown in Fig. 7.10. With increasing temperature the Fermi level shifts from close to the band edge towards the band center. At higher doping the shift begins at higher temperatures.

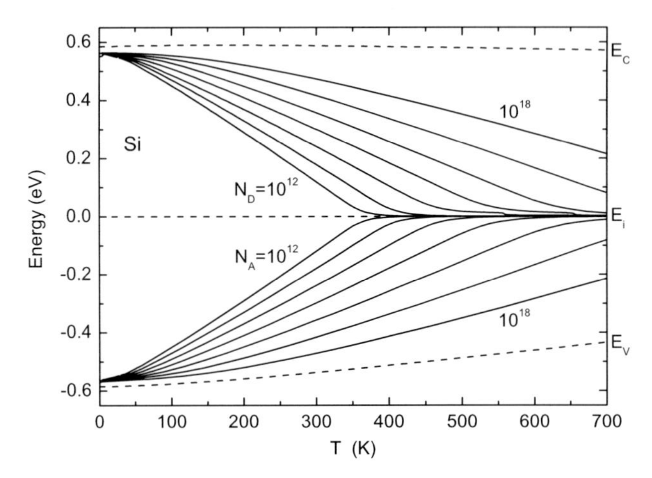

Fig. 7.10. Fermi level in silicon as a function of temperature for various doping levels (n-type and p-type) of $10^{12}, 10^{13}, \ldots, 10^{18}\,\mathrm{cm}^{-3}$. The intrinsic Fermi level is chosen as zero energy for all temperatures. The (temperature-dependent) conduction and valence band edges are shown as *dashed lines*

The electronic states of individual donors can be directly visualized by scanning tunneling microscopy (STM) as shown in Fig. 7.11 for Si:P. For small negative bias, tunneling occurs through the charged dopant that is located within the first three monolayers. At high negative bias the large contribution from the filled valence band masks the effect of the donor. This image, however, shows that the contrast attributed to the dopant atom is not due to surface defects or absorbates.

7.5.2 Acceptors

A group-III atom in Si has one electron too few for the tetrahedral bond. Thus, it 'borrows' an electron from the electron gas (in the valence band) and thus leaves a missing electron (termed hole) in the valence band (Fig. 7.12). The energy level is in the gap close to the valence-band edge. The latter consideration is made in the electron picture. In the hole picture, the acceptor ion has a hole and the hole ionizes (at sufficient temperature) into the valence band. After ionization the acceptor is charged negatively. Also, for this system a hydrogen-like situation arises that is, however, more complicated than for donors because of the degeneracy of the valence bands and their warping.

In Table 7.4 the acceptor binding energies E_{A}^{b} for group-III atoms in Ge and Si are listed. The absolute acceptor energy is given as $E_{\mathrm{A}} = E_{\mathrm{V}} + E_{\mathrm{A}}^{b}$. In Table 7.5 acceptor binding energies are listed for GaAs and GaP. While in GaAs some acceptors are close to the effective mass value of 27 meV, in GaP the deviation from the effective mass value ≈50 meV is large.

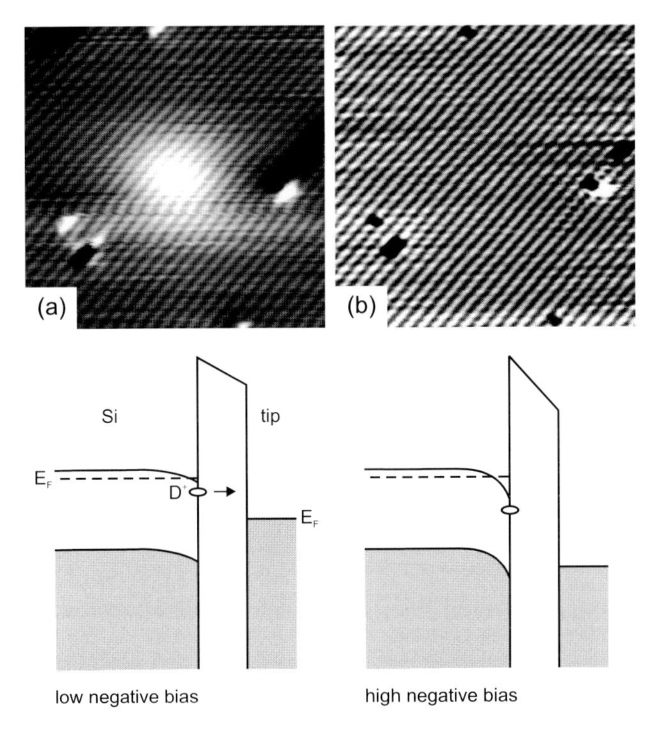

Fig. 7.11. Filled-state image of a phosphorus atom underneath a Si (001) surface at a tunneling current of 110 pA. The doping level is $5 \times 10^{17}\,\mathrm{cm}^{-3}$. (**a**) Sample bias $-0.6\,\mathrm{V}$, (**b**) sample bias $-1.5\,\mathrm{V}$ between Si:P and tip. Image sizes are $22 \times 22\,\mathrm{nm}^2$. Reprinted with permission from [192], ©2004 APS. *Lower row*: Schematic band diagrams for the two bias situations

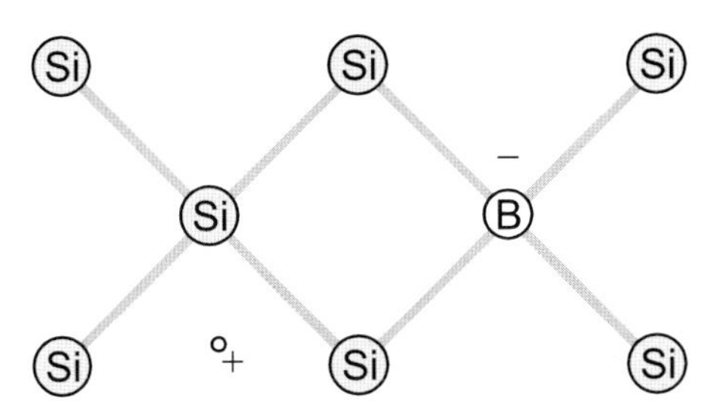

Fig. 7.12. Boron impurity in silicon. Boron accepts one electron and a fixed negative charge remains

When the conductivity is determined by holes or electrons, the material is called p-type or n-type, respectively. We note that some metals also show hole conduction (e.g. Zn). However, for metals the conductivity type is fixed, while the same semiconductor can be made n- or p-type with the appropriate doping.

Table 7.4. Binding energies E_A^b of group-III acceptors in Si and Ge. All values in meV

	B	Al	Ga	In
Si	45	57	65	16
Ge	10.4	10.2	10.8	11.2

Table 7.5. Binding energies E_A^b of acceptors in GaAs, GaP and GaN (low concentration value, data from [193]). All values in meV

	V site		III site	
GaAs	C	27	Be	28
	Si	34.8	Mg	28.8
	Ge	40.4	Zn	30.7
	Sn	167	Cd	34.7
GaP	C	54	Be	57
	Si	210	Mg	60
	Ge	265	Zn	70
			Cd	102
GaN			Mg	220

The acceptor concentration is denoted by N_A. The concentration of neutral acceptors is N_A^0, the concentration of charged acceptors is N_A^-. Of course

$$N_A = N_A^0 + N_A^- \ . \tag{7.44}$$

The ratio of the degeneracy of the (singly) filled and empty acceptor level is $\hat{g}_A$. $\hat{g}_A = 1/2$ for the spin degeneracy of the states in the electron picture.[4] Thus, similar to the considerations for electrons and donors,

$$\frac{N_A^0}{N_A^-} = \hat{g}_A \exp\left(-\frac{E_F - E_A}{kT}\right) \ . \tag{7.45}$$

The population of the acceptor levels is given by

$$N_A^- = \frac{N_A}{1 + \hat{g}_A \exp\left(-\frac{E_F - E_A}{kT}\right)} \ . \tag{7.46}$$

The formulas for the position of the Fermi level and the hole density are analogous to those obtained for electrons and donors and will not be

[4]For donors the neutral state is degenerate, for acceptors the charged state is degenerate.

explicitly given here. In Fig. 7.10, the temperature dependence of the Fermi level is included for p-type Si. With increasing temperature the Fermi level shifts from the valence-band edge (For $T = 0$, $E_{\mathrm{F}} = E_{\mathrm{V}} + E_{\mathrm{A}}^{b}/2$) towards the middle of the bandgap (intrinsic Fermi level).

Also, the wavefunction at acceptors can be imaged using scanning tunneling microscopy [194]. In [195] images of ionized and neutral Mn in GaAs have been reported (Fig. 7.13b). The tunneling I–V characteristics are shown in Fig. 7.13a. At negative bias, the acceptor is ionized and appears spherically symmetric due to the effect of the A^- ion Coulomb potential on the valence-band states. At intermediate positive voltages, tunneling is through the neutral state. The wavefunction of A^0 looks like a bow-tie due to the admixture of d-wavefunctions [196]. The Mn atom is presumably in the third subsurface atomic layer. At even higher positive bias the contrast due to the dopant is lost because the image is dominated by a large tunneling current from the tip to the empty conduction band.

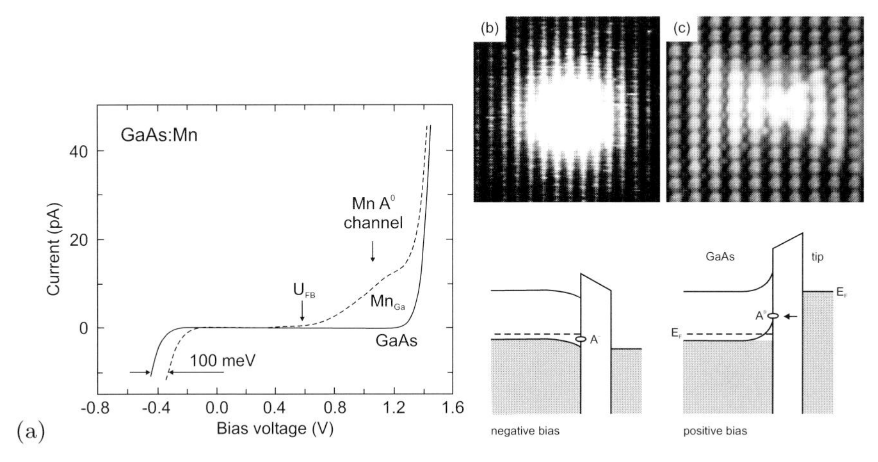

Fig. 7.13. (**a**) Tunneling I–V characteristic of GaAs:Mn sample. *Solid* (*dashed*) line is for pure GaAs (subsurface Mn on Ga site). U_{FB} denotes the simulated flat-band voltage. Adapted from [195]. (**b,c**) STM images of a Mn atom underneath a GaAs (110) surface. The doping level is $3 \times 10^{18}\,\mathrm{cm}^{-3}$. (**b**) Sample bias $-0.7\,\mathrm{V}$, (**c**) sample bias $+0.6\,\mathrm{V}$. Below the images are schematic band diagrams of GaAs:Mn and tip. Image sizes are (**b**) $8\times8\,\mathrm{nm}^2$ and (**c**) $5.6\times5\,\mathrm{nm}^2$. Reprinted with permission from [195], ©2004 APS. *Lower row* under parts (**b,c**): Schematic band diagrams for the two bias situations

7.5.3 Compensation

When donors *and* acceptors are simultaneously present, some of the impurities will compensate each other. Electrons from donors will recombine with holes on the acceptors. Depending on the quantitative situation the semiconductor can be n- or p-type. This situation can be invoked by intentional doping with donors or acceptors or by the unintentional background of donors (acceptors) in p-doped (n-doped) material. The charge-neutrality condition (now finally in its most general form) reads

$$-n + p - N_{\mathrm{A}}^{-} + N_{\mathrm{D}}^{+} = 0 \,. \tag{7.47}$$

We will now investigate the case of the presence of donors and acceptors, but limit ourselves to sufficiently low temperatures (or wide bandgaps) such that the intrinsic carrier density can be neglected. We assume Boltzmann statistics and assume here $N_{\mathrm{D}} > N_{\mathrm{A}}$. Then it is a very good approximation to use $N_{\mathrm{A}}^{-} = N_{\mathrm{A}}$ since there are enough electrons from the donors to recombine with (and thus compensate) all acceptors. Under the given assumptions regarding the temperature $p = 0$ and the material is n-type. Thus, in order to determine the position of the Fermi level, the following equation (charge-neutrality condition $n + N_{\mathrm{A}} - N_{\mathrm{D}}^{+} = 0$) must be solved (compare to (7.33))

$$N_{\mathrm{C}} \exp\left(\frac{E_{\mathrm{F}} - E_{\mathrm{C}}}{kT}\right) + N_{\mathrm{A}} - \frac{N_{\mathrm{D}}}{1 + \hat{g}\exp(\frac{E_{\mathrm{F}} - E_{\mathrm{D}}}{kT})} = 0 \,. \tag{7.48}$$

The solution is

$$E_{\mathrm{F}} = E_{\mathrm{C}} - E_{\mathrm{D}}^{b} + kT \ln\left[\frac{\left[\alpha^2 + 4\hat{g}\frac{N_{\mathrm{D}} - N_{\mathrm{A}}}{N_{\mathrm{C}}}\exp\left(\frac{E_{\mathrm{D}}^{b}}{kT}\right)\right]^{\frac{1}{2}} - \alpha}{2\hat{g}}\right] \tag{7.49}$$

$$\alpha = 1 + \hat{g}\frac{N_{\mathrm{A}}}{N_{\mathrm{C}}}\exp\left(\frac{E_{\mathrm{D}}^{b}}{kT}\right) .$$

For $N_{\mathrm{A}} = 0$ we have $\alpha = 1$ and (7.34) is reproduced, as expected. For $T = 0$ the Fermi energy lies at $E_{\mathrm{F}} = E_{\mathrm{D}}$ since the donor level is *partially* filled ($N_{\mathrm{D}}^{0} = N_{\mathrm{D}} - N_{\mathrm{A}}$). For low temperatures the Fermi level is approximated by

$$E_{\mathrm{F}} \cong E_{\mathrm{C}} - E_{\mathrm{D}}^{b} + kT \ln\left(\frac{N_{\mathrm{D}}/N_{\mathrm{A}} - 1}{\hat{g}}\right) . \tag{7.50}$$

The corresponding carrier density is (α has the same meaning as in (7.49))

$$n = N_{\mathrm{C}} \exp\left(-\frac{E_{\mathrm{D}}^{b}}{kT}\right) \frac{\left[\alpha^2 + 4\hat{g}\frac{N_{\mathrm{D}} - N_{\mathrm{A}}}{N_{\mathrm{C}}}\exp\left(\frac{E_{\mathrm{D}}^{b}}{kT}\right)\right]^{\frac{1}{2}} - \alpha}{2\hat{g}} . \tag{7.51}$$

For very low temperatures ($n < N_{\mathrm{A}}$), the exponential slope of $n(T)$ is given by E_{D}^{b}:

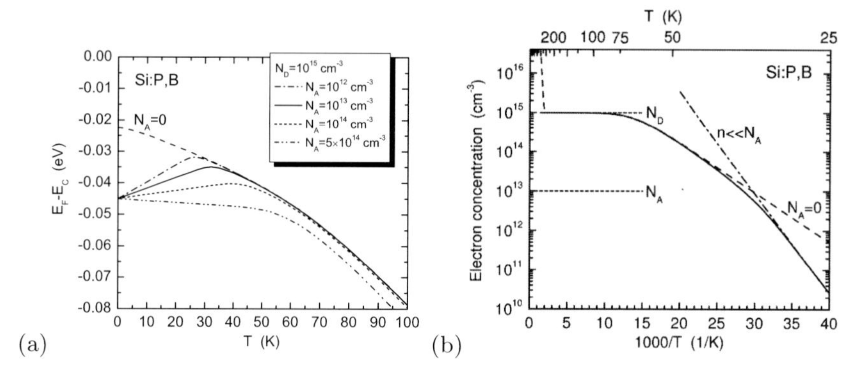

Fig. 7.14. (**a**) Position of Fermi level in partially compensated Si:P,B ($N_{\mathrm{D}} = 10^{15}\,\mathrm{cm}^{-3}$, $E_{\mathrm{D}}^{b} = 45\,\mathrm{meV}$, $E_{\mathrm{A}}^{b} = 45\,\mathrm{meV}$, *solid line*: $N_{\mathrm{A}} = 10^{13}\,\mathrm{cm}^{-3}$, *dashed line*: $N_{\mathrm{A}} = 0$, *dash-dotted line*: $N_{\mathrm{A}} = 10^{12}\,\mathrm{cm}^{-3}$, *short-dashed line*: $N_{\mathrm{A}} = 10^{14}\,\mathrm{cm}^{-3}$, *dash-double dotted line*: $N_{\mathrm{A}} = 5 \times 10^{14}\,\mathrm{cm}^{-3}$) as a function of temperature. (**b**) Corresponding electron concentration for $N_{\mathrm{A}} = 10^{13}\,\mathrm{cm}^{-3}$ as a function of temperature (neglecting intrinsic carriers), *dashed line* for $N_{\mathrm{A}} = 0$ according to (7.38), *dash-dotted line* approximation for $n \ll N_{\mathrm{A}}$ as in (7.52)

$$n \cong \frac{N_{\mathrm{D}} - N_{\mathrm{A}}}{N_{\mathrm{A}}} \frac{N_{\mathrm{C}}}{\hat{g}} \exp\left(-\frac{E_{\mathrm{D}}^{b}}{kT}\right) . \tag{7.52}$$

For higher temperatures ($n > N_{\mathrm{A}}$) the slope is given by $E_{\mathrm{D}}^{b}/2$ as in the uncompensated case (7.38). For sufficiently high temperatures in the exhaustion regime (but still $n_{\mathrm{i}} < n$) the electron density is given by

$$n \approx N_{\mathrm{D}} - N_{\mathrm{A}} . \tag{7.53}$$

At even higher temperatures the electron density will be determined by the intrinsic carrier concentration.

An experimental example is shown in Fig. 7.15 for partially compensated p-Si (with $N_{\mathrm{D}} \ll N_{\mathrm{A}}$).

If donors are added to a p-type semiconductor, first the semiconductor remains p-conducting as long as $N_{\mathrm{D}} \ll N_{\mathrm{A}}$. If the donor concentration becomes larger than the acceptor concentration, the conductivity type switches from p- to n-conduction. If the impurities are exhausted at room temperature, the lowest carrier concentration is reached for $N_{\mathrm{D}} = N_{\mathrm{A}}$. Such a scenario is shown for p-type $\mathrm{In}_x\mathrm{Ga}_{1-x}\mathrm{As}_{1-y}\mathrm{N}_y$ doped with various concentrations of Si in Fig. 7.16. At high Si incorporation, the number of charge carriers saturates due to autocompensation (see next section) and the formation of Si precipitates. Since the ionization energies of donors and acceptors are typically different, the situation for $N_{\mathrm{D}} \approx N_{\mathrm{A}}$ needs, in general, to be investigated carefully and will depend on the temperature.

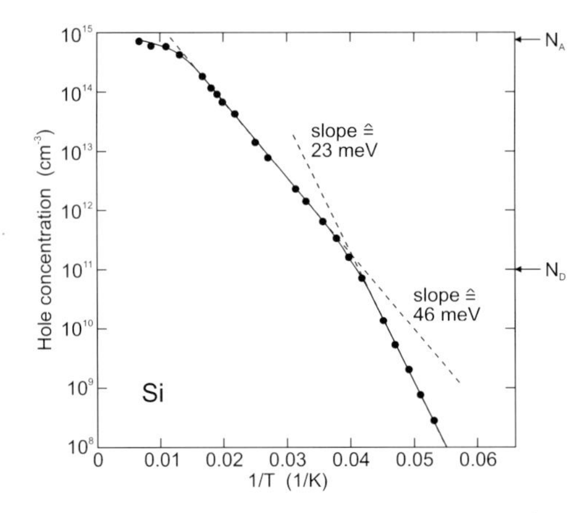

Fig. 7.15. Hole density in p-type silicon ($N_{\mathrm{A}} = 7.4 \times 10^{14}\,\mathrm{cm}^{-3}$, $E_{\mathrm{A}}^{b} = 46\,\mathrm{meV}$ (probably boron) and partial compensation with $N_{\mathrm{D}} = 1.0 \times 10^{11}\,\mathrm{cm}^{-3}$). Adapted from [197]

7.5.4 Amphoteric Impurities

If an impurity atom can act as a donor and acceptor it is called amphoteric. This can occur if the impurity has several levels in the bandgap (such as Au in Ge or Si). In this case, the nature of the impurity depends on the position of the Fermi level. Another possibility is the incorporation on different lattice sites. For example, carbon in GaAs is a donor if incorporated on the Ga-site. On the As-site carbon acts as an acceptor.

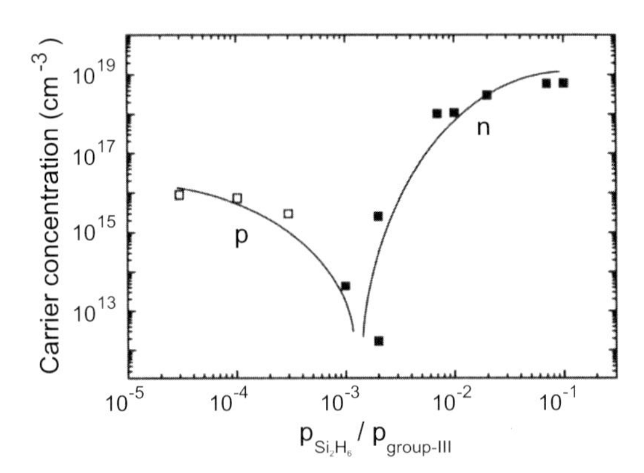

Fig. 7.16. Carrier concentration and conductivity type (*empty squares*: p, *solid squares*: n) for MOVPE-grown $\mathrm{In}_x\mathrm{Ga}_{1-x}\mathrm{As}_{1-y}\mathrm{N}_y$ layers on GaAs (001) (layer thickness $\approx 1\,\mu\mathrm{m}$, $x \approx 5\%$, $y \approx 1.6\%$) doped with different amounts of silicon. The ordinate is the ratio of the partial pressures of disilane and the group-III precursors (TMIn and TMGa) in the gas phase entering the MOVPE reactor. *Solid lines* are guides to the eye. Experimental data from [198]

Thus, e.g., crystal growth kinetics can determine the conductivity type. In Fig. 7.17 the conductivity due to carbon background is shown for GaAs grown using MOVPE under various growth conditions. At high (low) arsine partial pressure incorporation of carbon on As-sites is less (more) probable, thus the conductivity is n-type (p-type). Also, growth on different surfaces can evoke different impurity incorporation, e.g., n-type on (001) GaAs and p-type on (311)A GaAs, since the latter is Ga-stabilized.

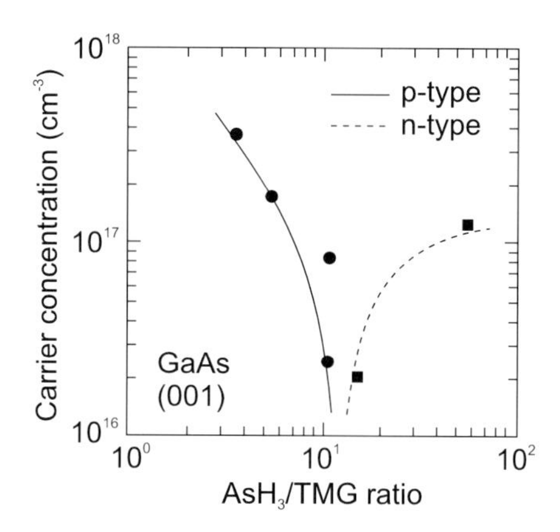

Fig. 7.17. Background doping of GaAs due to carbon in MOVPE for different ratios of the partial pressures of AsH_3 and TMG (trimethylgallium). The conductivity type (*squares*: n-type, *circles*: p-type) depends on the incorporation of C from CH_3 radicals on Ga- or As-site. *Lines* are guides to the eye. Experimental data from [199]

Deviation from the ideal stoichiometry introduces point defects that can be electrically active and change conductivity type and carrier concentration. In the case of $CuInSe_2$, excess Cu could go on interstitial positions or promote selenium vacancies, both leading to n-type behavior. This material is particularly sensitive to deviations from ideal stoichiometry for both Cu/In ratio (Fig. 7.18) and Se deficiency [200].

7.5.5 High Doping

For low doping concentrations, the impurity atoms can be considered to be decoupled. At low temperature, only hopping from one impurity to the next is possible due to thermal emission or tunneling and the semiconductor becomes an insulator.

With increasing concentration, the distance between impurities decreases and their wavefunctions can overlap. Then, an impurity band develops (Fig. 7.19). A periodic arrangement of impurity atoms would result in well-defined band edges as found in the Kronig–Penney model. Since the impurity atoms are randomly distributed, the band edges exhibit tails. For high doping, the impurity band overlaps with the conduction band. In the case of

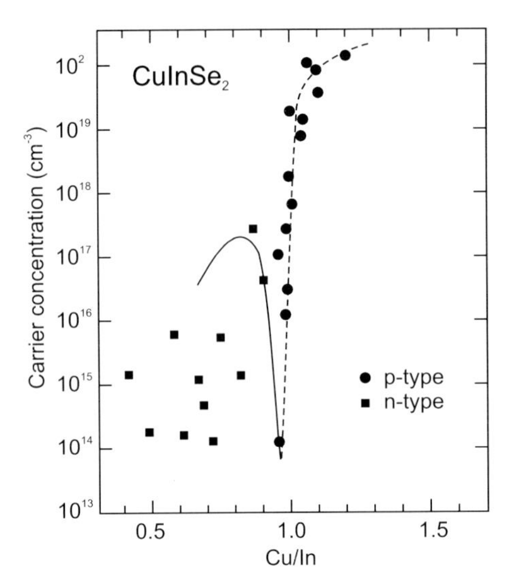

Fig. 7.18. Carrier concentration and conductivity type (*squares*: n-type, *circles*: p-type) as a function of stoichiometry (Cu/In ratio) for $CuInSe_2$ thin films. *Lines* are guides to the eye. Experimental data from [200]

compensation, the impurity band is not completely filled and contains (a new type of) holes. In this case, conduction can take place within the impurity band even at low temperature, making the semiconductor a metal. This metal–insulator transition has been discussed by Mott [201].

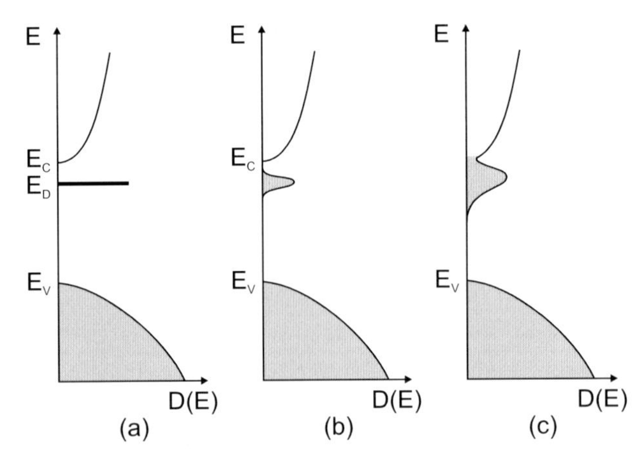

Fig. 7.19. Principle of the formation of a (donor) impurity band. (**a**) Small doping concentration and sharply defined impurity state at E_D, (**b**) increasing doping and development of an impurity band that (**c**) widens further and eventually overlaps with the conduction band for high impurity concentration. The *shaded areas* indicate populated states at $T = 0\,K$

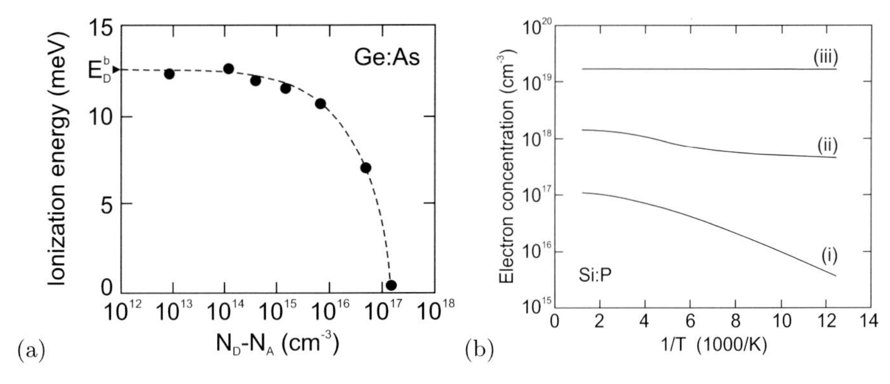

Fig. 7.20. (**a**) Donor ionization energy in n-type Ge for various doping concentrations. *Dashed line* is a guide to the eye. The arrow labeled E_D^b denotes the low-concentration limit (cf. Table 7.2). Experimental data from [202]. (**b**) Electron concentration vs. inverse temperature for Si:P for three different doping concentrations ((i): 1.2×10^{17}, (ii): 1.25×10^{18}, (iii): $1.8 \times 10^{19}\,\mathrm{cm}^{-3}$). Experimental data from [203]

The formation of the impurity band leads to a reduction of the impurity ionization energy as known from (7.24). A typical result is shown in Fig. 7.20a for n-type Ge [202]. At the critical doping concentration of $N_c = 1.5 \times 10^{17}$, the activation energy for the carrier concentration disappears. The freeze-out of the carrier concentration (cf. Fig. 7.8) disappears as shown in Fig. 7.20b. Similar effects have been observed for Si [203] and GaAs [204]. The critical doping concentrations are listed in Table 7.6. The decrease of the ionization energy E^b (donor or acceptor) follows the dependence [202, 203]

$$E^b = E_0^b - \alpha N_i^{1/3} = E_0^b \left[1 - \left(\frac{N_i}{N_c} \right)^{1/3} \right] , \tag{7.54}$$

where N_i is the concentration of ionized dopants.

The critical density can be estimated from the Mott criterion when the distance of the impurities becomes comparable to their Bohr radius (7.25)

$$2a_D = \frac{3}{2\pi} N_c^{1/3} . \tag{7.55}$$

The prefactor $3/(2\pi)$ stems from the random distribution of impurities and disappears for a periodic arrangement. The Mott criterion is (rewriting (7.55))

$$a_D N_c^{1/3} \approx 0.24 . \tag{7.56}$$

For GaAs with $a_D = 10.3\,\mathrm{nm}$, the criterion yields $N_c = 1.2 \times 10^{16}\,\mathrm{cm}^{-3}$, in agreement with experiment.

Table 7.6. Critical doping concentration for various semiconductors (at room temperature)

material	type	$N_\mathrm{c}\,(\mathrm{cm}^{-3})$	Ref.
Ge:As	n	1.5×10^{17}	[202]
Si:P	n	1.3×10^{18}	[203]
Si:B	p	6.2×10^{18}	[203]
GaAs	n	1×10^{16}	[204]
GaP:Zn	p	2×10^{19}	[205]
GaN:Si	n	2×10^{18}	[206]
GaN:Mg	p	4×10^{20}	[193]

The achievable maximum concentration of electrically active dopants is limited by the concentration dependence of the diffusion coefficient, Coulomb repulsion, autocompensation and the solubility limit [184]. In Table 7.7 the maximum carrier concentrations for GaAs with various dopants are listed.

Table 7.7. Maximum electrically active doping concentration for GaAs

material	type	$N_\mathrm{c}\,(\mathrm{cm}^{-3})$	Ref.
GaAs:Te	n	2.6×10^{19}	[207]
GaAs:Si	n	1.8×10^{19}	[208]
GaAs:C	p	1.5×10^{21}	[209]
GaAs:Be	p	2×10^{20}	[210]

7.6 Quasi-Fermi Levels

The carrier concentrations were given by (7.8) and (7.9). So far, we have only considered semiconductors in thermodynamic equilibrium for which $np = n_\mathrm{i}^2$. In a nonequilibrium situation, e.g. for external excitation or carrier injection in a diode, the electron and hole densities can each take arbitrary values, in principle. In particular, np will no longer be equal to n_i^2 and there is no Fermi level constant throughout the structure. In this case, however, quasi-Fermi levels F_n and F_p for electrons and holes, respectively, are defined via

$$n(\mathbf{r}) = N_\mathrm{C} F_{1/2}\left(\frac{F_\mathrm{n}(\mathbf{r}) - E_\mathrm{C}}{kT}\right) \tag{7.57a}$$

$$p(\mathbf{r}) = N_\mathrm{V} F_{1/2}\left(-\frac{F_\mathrm{p}(\mathbf{r}) - E_\mathrm{V}}{kT}\right) . \tag{7.57b}$$

A quasi-Fermi level is sometimes called imref[5] and can also be denoted as E_{F_n} or E_{F_p}. We emphasize that the quasi-Fermi levels are *only* a means to describe the local carrier density in a logarithmical way. The quasi-Fermi levels can be obtained from the density via (for Boltzmann approximation)

$$F_n = E_C + kT \ln\left(\frac{n}{N_C}\right) \tag{7.58a}$$

$$F_p = E_V - kT \ln\left(\frac{p}{N_V}\right) . \tag{7.58b}$$

The quasi-Fermi levels do not imply that the carrier distribution is actually a Fermi distribution. This is generally no longer the case in thermodynamical nonequilibrium. However, in 'well-behaved' cases the carrier distribution in nonequilibrium can be approximated locally as a Fermi distribution using a local quasi-Fermi level and a local temperature, i.e.

$$f_e(\mathbf{r}, E) \cong \frac{1}{\exp\left(\frac{E - F_n(\mathbf{r})}{kT(\mathbf{r})}\right) + 1} . \tag{7.59}$$

Using the quasi-Fermi levels, np is given by (for Boltzmann approximation)

$$n(\mathbf{r})\, p(\mathbf{r}) = n_i^2 \exp\left(\frac{F_n(\mathbf{r}) - F_p(\mathbf{r})}{kT}\right) . \tag{7.60}$$

We note that for an inhomogeneous semiconductor or a heterostructure (see Chap. 11), n_i may also depend on the spatial position. In the case of thermodynamic equilibrium the difference of the quasi-Fermi levels is zero, i.e. $F_n - F_p = 0$ and $F_n = F_p = E_F$.

7.7 Deep Levels

For deep levels the short-range part of the potential determines the energy level. The long-range Coulomb part will only lead to a correction. The term 'deep level' implies that the level is within the bandgap and not close to the band edges. However, some deep levels (in the sense of the potential being determined by the ion core) have energy levels close to the band edges or even within a band. Details can be found in [72, 211, 212].

The wavefunction is strongly localized. Thus, it cannot be composed of Bloch functions, as has been done for the shallow levels for the effective-mass impurity. The localization in $\mathbf{r}$ space leads to a delocalization in $\mathbf{k}$ space. The wavefunction might thus have components far away from the Γ point.

[5] W. Shockley had asked E. Fermi for permission to use his name reversed. Fermi was not too enthusiastic but granted permission.

Examples are Si:S, Si:Cu or InP:Fe, GaP:N, ZnTe:O. Deep levels can also be due to defects such as vacancies or antisite defects.

Due to the larger distance to the band edges, deep levels are not efficient at providing free electrons or holes. Quite the opposite, they rather capture free carriers and thus lead to a reduction of conductivity. Centers that can capture electrons and holes lead to nonradiative recombination of electrons through the deep level into the valence band (see also Chap. 10). This can be useful for the fabrication of semi-insulating layers with low carrier concentration and fast time response of, e.g., switches and photodetectors.

7.7.1 Charge States

The level can have different charge states depending on the occupancy of the levels with electrons. The energy position within the gap varies with the charge state due to the Coulomb interaction. Also, the lattice relaxation around the defect depends on the charge state and modifies the energy level.

The localized charge q_{d} at the defect is the integral over the change $\Delta\rho$ of the charge density compared to the perfect crystal over a sufficiently large volume V_∞ around the defect

$$q_{\mathrm{d}} = \int_{V_\infty} \Delta\rho(\mathbf{r})\mathrm{d}^3\mathbf{r} = \frac{ne}{\epsilon_{\mathrm{r}}} \,. \tag{7.61}$$

In semiconductors, the charge $q_{\mathrm{d}}\epsilon_{\mathrm{r}}$ is an integer multiple of the elementary charge. The defect is said to be in the n-th charge state. Each charge state has a certain stable atomic configuration $\mathbf{R}_n$. Each charge state has a ground state and excited states that can each have different stable atomic configurations.

Now, we discuss how the concentration of the various charge states depends on the position of the Fermi level. The overall constraint of global charge neutrality determines the chemical potential of the electron, i.e. the Fermi level in Fermi–Dirac statistics. We use the approximation that the defect concentration is small and their interactions become negligible.

As an example, we treat the possible reaction $V^0 \rightleftharpoons V^+ + e^-$, where V^0 denotes a neutral vacancy and V^+ is a positively charged vacancy, created by the ionization of an electron from the vacancy into the conduction band. The free energy G depends on the numbers n_0 of neutral and n_+ of positively charged vacancies. The minimum condition is met by

$$\mathrm{d}G = \frac{\partial G}{\partial n_0}\mathrm{d}n_0 + \frac{\partial G}{\partial n_+}\mathrm{d}n_+ = 0 \,. \tag{7.62}$$

The neutrality constraint is $\mathrm{d}n_0 + \mathrm{d}n_+ = 0$ and therefore the minimum condition reads

$$\frac{\partial G}{\partial n_0} = \frac{\partial G}{\partial n_+} \,. \tag{7.63}$$

For noninteracting defects and using (4.9) we write

$$\frac{\partial G}{\partial n_0} = G_{\rm f}(V^0) + kT \ln\left(\frac{n_0}{N_0}\right) \tag{7.64a}$$

$$\frac{\partial G}{\partial n_+} = \frac{\partial G(V^+)}{\partial n_+} + \frac{\partial G(e^-)}{\partial n_+} = G_{\rm f}^{V^+} + kT \ln\left(\frac{n_+}{N_+}\right) + \mu_{\rm e^-}\,, \tag{7.64b}$$

where $N_0 = NZ_0$ and $N_+ = NZ_+$ are the number of available sites, given by the number N of atomic sites and including possible internal degeneracies Z_0 and Z_+, respectively. $G_{\rm f}$ denotes the free enthalpy of formation of the respective defect, as in (4.3). We have written the free enthalpy of the separated pair V^+ and e^- as the sum $G(V^+) + G(e^-)$. $\mu_{\rm e^-} = \frac{\partial G(e^-)}{\partial n_+}$ is (by definition) the chemical potential of the electron, i.e. the Fermi energy $E_{\rm F}$ of Fermi–Dirac statistics.[6] From (7.64a,b) we find for the ratio of the concentrations of defects $c_0 = n_0/N$ and $c_+ = n_+/N$

$$\frac{c_0}{c_+} = \frac{Z_+}{Z_0} \exp\left(-\frac{G_{\rm f}^{V^+} - G_{\rm f}^{V^0} + E_{\rm F}}{kT}\right) = \frac{Z_+}{Z_0} \exp\left(\frac{E_{\rm t}(V^0) - E_{\rm F}}{kT}\right)\,, \tag{7.65}$$

where $-E_{\rm t}(V^0) = G_{\rm f}^{V^+} - G_{\rm f}^{V^0}$ is the negative free enthalpy of ionization of V^0. We note that c_0 can be obtained from (4.9) and $E_{\rm F}$ is determined by the charge-neutrality condition.

7.7.2 Jahn–Teller Effect

The lattice relaxation can reduce the symmetry of the defect. Many defects, such as a vacancy, a tetrahedral interstitial or an impurity, occupy initially tetrahedral sites in the zincblende structure. The lattice relaxation reduces the symmetry, e.g. to tetragonal or trigonal, and therefore causes initially degenerate levels to split. Such splitting is called the static Jahn–Teller effect [211,213]. The energy change in terms of the atomic displacement Q can be denoted (using perturbation theory for the simplest, nondegenerate case) as $-IQ$ ($I > 0$). Including the elastic contribution with a force constant C, the energy of a configuration Q is

$$E = -IQ + \frac{1}{2}CQ^2\,. \tag{7.66}$$

The stable configuration $Q_{\rm min}$, for which the energy is minimal ($E_{\rm min}$), is therefore given by

$$Q_{\rm min} = \frac{I}{C} \tag{7.67a}$$

$$E_{\rm min} = -\frac{I^2}{2C}\,. \tag{7.67b}$$

[6] The chemical potential in a one-component system is $\mu = \partial G/\partial n = G/n$. In a multicomponent system it is, for the i-th component, $\mu_i = \partial G/\partial n_i \neq G/n_i$.

Several equivalent lattice relaxations may exist, e.g. a 3-fold minimum for remaining C_{3v} symmetry. The energy barrier between them has a finite height. Therefore, e.g. at sufficient temperature, the defect can switch between different configurations and eventually again becomes isotropic (dynamic Jahn–Teller effect). The experimental observation depends on the relation between the characteristic time of the experiment and the reorientation time constant of the defect.

7.7.3 Negative-U Center

We explain the principle of a so-called negative-U center for the Si vacancy (cf. Fig. 4.1). Coulomb energy and the Jahn–Teller effect compete for the position of the occupancy level for different charge states. U refers to the additional energy upon charging of the defect with an additional electron. The Coulomb repulsion of electrons leads to an *increase* of the energy, i.e. positive U, which has been calculated to be 0.25 eV for the Si vacancy [214] for all charge states. The occupation level (see Sect. 4.3) $E_0(1,2)$ (the index 0 indicates effects only due to many-electron Coulomb interaction), separating the domination of V^{++} and V^+ (Fig. 7.21) is 0.32 eV above the valance-band edge. Therefore, the occupation level $E_0(0,1)$ is expected to lie at about 0.57 eV above E_V.

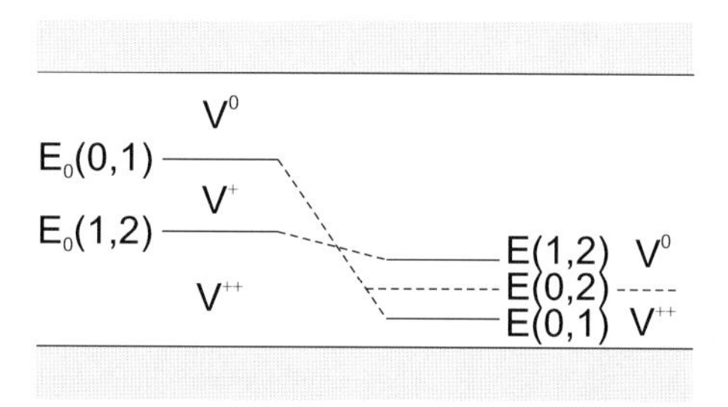

Fig. 7.21. Charge states of the vacancy in silicon. *Left*: level scheme without lattice relaxation, *right*: level scheme including the Jahn–Teller effect. For a Fermi level below (above) $E(0,2)$ the charge state V^{++} (V^0) is dominant

The Jahn–Teller effect may lead to a splitting of the otherwise 4-fold degenerate states of the vacancy. The schematic level diagram for the Jahn–Teller splitting is shown in Fig. 7.22. The V^{++} state (A_1 is always populated with two electrons) is resonant with the valence band. The T_2 state lies in the bandgap. When the Jahn–Teller effect (now on the T_2 state) is included, the energies of the different charge states depend on the configuration coordinate (a mostly tetragonal distortion in the case of the Si vacancy).

$$E_{V^0} = E(0,Q) = E(0,Q=0) - 2IQ + \frac{1}{2}CQ^2 \tag{7.68a}$$

$$E_{V^+} = E(1,Q) = E(1,Q=0) - IQ + \frac{1}{2}CQ^2 \tag{7.68b}$$

$$E_{V^{++}} = E(2,Q) = E(2,Q=0) + \frac{1}{2}CQ^2 \,. \tag{7.68c}$$

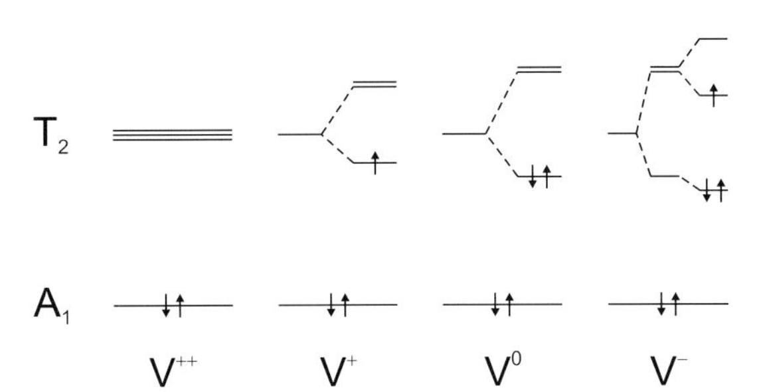

Fig. 7.22. Jahn–Teller splitting for different charge states of the vacancy. A_1 and T_2 refer to irreducible representations of the T_d point symmetry group. A_1 is nondegenerate and therefore does not exhibit a Jahn–Teller effect. T_2 is triply degenerate. The arrows represent electrons and their spin orientation

For the $n = 2$ state the T_2 gap state is empty and thus no degeneracy and Jahn–Teller term arises. For $n = 1$ there is a linear Jahn–Teller term. The occupation with two electrons (V^0) causes an approximately twice as large Jahn–Teller splitting for the $n = 0$ state. The force constant is assumed to be independent of the charge state. The energies for the minimum configurations $Q^n_{\min}$ are therefore

$$E(0,Q^0_{\min}) = E(0,Q=0) - 4\frac{I^2}{2C} \tag{7.69a}$$

$$E(1,Q^1_{\min}) = E(1,Q=0) - \frac{I^2}{2C} \tag{7.69b}$$

$$E(2,Q^2_{\min}) = E(2,Q=0) \,. \tag{7.69c}$$

The Jahn–Teller energy $E_{\mathrm{JT}} = I^2/2C$ lowers the position of the occupancy levels E_0 calculated with Coulomb terms only. The occupancy levels *including* the Jahn–Teller contribution are therefore given as

$$E(1,2) = E_0(1,2) - E_{\mathrm{JT}} \tag{7.70a}$$

$$E(0,1) = E_0(0,1) - 3E_{\mathrm{JT}} \,. \tag{7.70b}$$

For the vacancy in silicon the Jahn–Teller energy E_{JT} is about 0.19 eV. Thus the $E(1,2)$ level is lowered from 0.32 eV to 0.13 eV. The E(0,1) occupancy level, however, is reduced from 0.57 eV to 0 eV, i.e. the valence-band edge below E(1,2) (see Fig. 7.21). The occupancy level $E(0,2)$ is in the middle between $E(0,1)$ and $E(1,2)$. It lies experimentally between 0.16 eV (Si:In) and 0.065 eV (Si:Ga).

The relative concentrations of the three charge states are determined by (7.65) (degeneracy and entropy terms have been neglected)

$$\frac{c(V^{++})}{c(V^{+})} = \exp\left(\frac{E(1,2)-E_\mathrm{F}}{kT}\right) \tag{7.71a}$$

$$\frac{c(V^{+})}{c(V^{0})} = \exp\left(\frac{E(0,1)-E_\mathrm{F}}{kT}\right) . \tag{7.71b}$$

Therefore, V^{++} dominates if $E_\mathrm{F} < E(0,1)$ and V^0 dominates for $E_\mathrm{F} > E(1,2)$. In the intermediate range $E(0,1) < E_\mathrm{F} < E(1,2)$ we know from (7.71a,b) that V^+ is dominated by V^0 and V^{++}. However, at this point it is not clear whether V^{++} or V^0 dominates overall. The ratio of the concentrations of V^{++} and V^0 is given by

$$\frac{c(V^{++})}{c(V^{0})} = \exp\left(\frac{E(1,2)+E(0,1)-2E_\mathrm{F}}{kT}\right) = \left[\exp\left(\frac{E(0,2)-E_\mathrm{F}}{kT}\right)\right]^2 . \tag{7.72}$$

The occupancy level $E(0,2)$ is thus given as

$$E(0,2) = \frac{1}{2}\left[E(0,1)+E(1,2)\right] , \tag{7.73}$$

and is shown in Fig. 7.21. V^{++} dominates if $E_\mathrm{F} < E(0,2)$ and V^0 dominates for $E_\mathrm{F} > E(0,2)$. V^+ is, for no position of the Fermi level, the dominating charge state of the Si vacancy. We note that for n-doped Si the V^- and V^{--} can also be populated. The population of the V^0 state with an extra electron introduces another Jahn–Teller splitting (Fig. 7.22) that has trigonal symmetry.

Generally, the Jahn–Teller effect can make the addition of an electron cause an effectively negative charging energy; in this case the center is termed a negative-U center.

7.7.4 DX Center

The DX center is a deep level that was first investigated for n-doped (e.g. Si-doped) $\mathrm{Al}_x\mathrm{Ga}_{1-x}\mathrm{As}$. It dominates the transport properties of the alloy for $x > 0.22$. For smaller Al concentrations and GaAs the DX level lies in the conduction band. DX-type deep levels have also been found for other alloys and dopants, e.g. GaAsP:S.

It is experimentally found that the capture process of electrons into the DX center is thermally activated. The capture energy E_c depends on the AlAs mole fraction (Fig. 7.23). The (average) barrier for electron capture has a minimum of 0.21 eV for $x \approx 0.35$, near the crossover point between direct and indirect bandgap (cf. Fig. 6.13). For lower Al concentrations, the capture barrier increases to 0.4 eV for $x = 0.27$; for $x > 0.35$ the capture barrier increases to about 0.3 eV for x around 0.7 [215]. The barrier for thermally releasing carriers from the DX center has been determined to be about 0.43 eV, independent of the Al mole fraction [215].

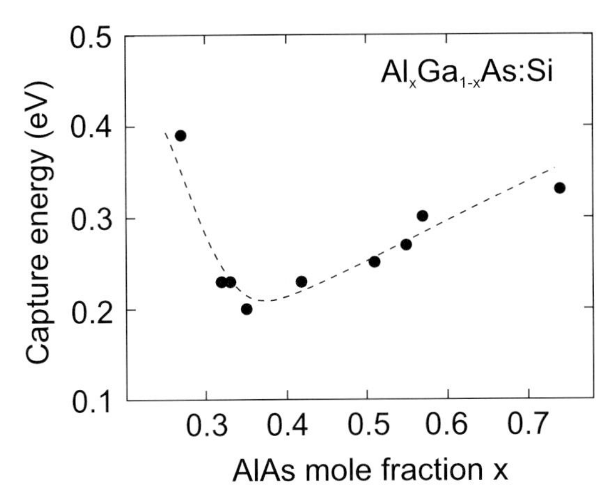

Fig. 7.23. Energy barrier for electron capture E_c at the Si-DX center in $Al_xGa_{1-x}As$ for various compositions. Experimental data from [215]

Carriers can be removed from the DX center by optical absorption of photons with energy larger than about 1.2 eV. If carriers are removed by optical excitation at low temperatures the (re-)capture is so slow ($\sigma < 10^{-30}\,\mathrm{cm}^2$) that the carriers remain in the conduction band and cause persistent photoconductivity (PPC). The PPC is only reduced upon increasing the sample temperature. The concentration of the DX center is about the same as the net doping concentration.

The properties of the DX center are reviewed in [216,217]. So far, no definite microscopic model of the DX center has been agreed on. Lang [218] proposed that the DX center involves a donor and an unknown defect (probably a vacancy). It probably involves large lattice relaxation as in the configuration coordinates model of Fig. 7.24 where the donor binding energy E_D^b with respect to the conduction-band minimum, the barrier for electron capture E_c, the barrier for electron emission E_e and the optical ionization energy E_o are labeled. The donor binding energy is measured with Hall effect (see Sect. 8.4) at temperatures sufficient to overcome the capture and emission barriers, the emission barrier is measured with deep level transient spectroscopy (DLTS). The capture barrier manifests itself in PPC experiments. We note that the DX center is related to the L-conduction band. For small Al mole fraction, the DX level is degenerate with the Γ-related conduction band (see Fig. 7.24b).

Theoretical models and experimental evidence hint at a vacancy-interstitial model for the Si-DX center [219]. The donor (Si) is displaced along the ⟨111⟩ direction from the Ga substitution site. The displacement is predicted to be 0.117 nm and the distorted geometry can be viewed as a Ga vacancy and a Si interstitial. The charge state of the (filled) DX center is proposed to be a two-electron negative-U state.

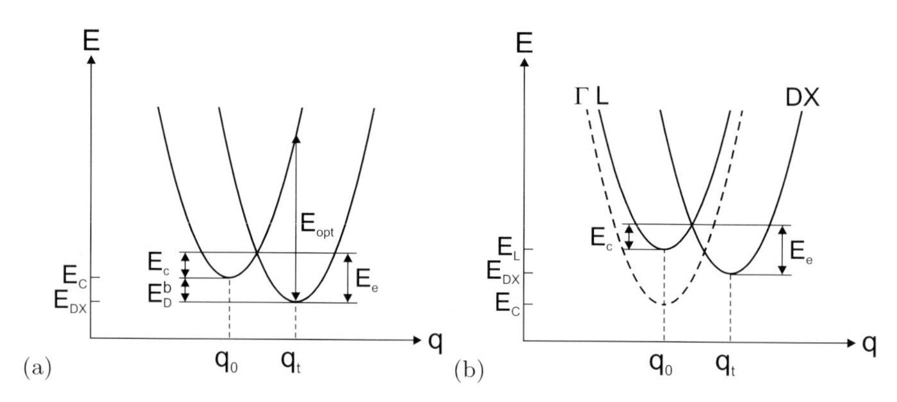

Fig. 7.24. (**a**) Schematic configuration coordinate diagram for the DX level with large lattice relaxation. q_0 is the configuration of the empty defect, q_t is the configuration of the filled defect. The donor binding energy E_D^b, the barrier for electron capture E_c, the barrier for electron emission E_e and the optical ionization energy E_o are labeled. E_C denotes the conduction-band edge. We note that in AlGaAs the DX level is associated with the L conduction band (see Fig. 6.13). (**b**) Schematic configuration coordinate diagram for the DX level in $Al_{0.14}Ga_{0.86}As$ with the DX level being degenerate with the (Γ-related) conduction band

7.7.5 EL2 Defect

The EL2 defect is a deep donor in GaAs. It is not related to impurities but occurs for intrinsic material, in particular grown under As-rich conditions. It has similar physical properties to the DX center. The bleaching of absorption due to EL2, i.e. the optical removal of electrons from the defect at low temperatures, is shown in Fig. 7.25. The microscopic model [221] describes the EL2 defect as an arsenic antisite defect, i.e. an arsenic atom on a Ga site, As_{Ga}. In the charged state the arsenic atom is displaced from the lattice position and a complex of a Ga vacancy (symmetry T_{3d}) and an interstitial As (symmetry C_{3v}) with 0.14 nm displacement along ⟨111⟩ forms (V_{Ga}-As_i). The charged state is filled with two electrons.

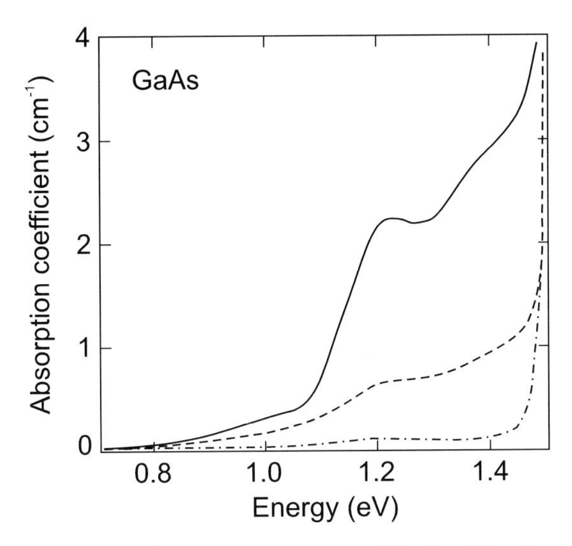

Fig. 7.25. Below bandgap absorption spectrum of GaAs at low temperatures ($T = 10\,\mathrm{K}$) when cooled in the dark (*solid line*). The *dashed* (*dash-dotted*) *line* is the absorption after illuminating the sample for 1 min (10 min) with white light, leading to quenching of the EL2-related absorption. Adapted from [220]

7.7.6 Semi-insulating Semiconductors

Semiconductors with high resistivity (10^7–$10^9\,\Omega\,\mathrm{cm}$) are called semi-insulating or 'si' for short. Semi-insulating substrates are needed for high-speed devices. The high resistivity should stem from a small free-carrier density at finite temperature.[7] For sufficiently wide bandgap, the intrinsic carrier concentration is small and such pure material is semi-insulating, e.g. GaAs with $n_i = 1.47 \times 10^6\,\mathrm{cm}^{-3}$ and $5.05 \times 10^8\,\Omega\,\mathrm{cm}$ [222]. Since shallow impurities are hard to avoid, another route is used technologically. Impurities that form deep levels are incorporated in the semiconductor in order to compensate free carriers. For example, a deep acceptor compensates all electrons if $N_A > N_D$. Since the acceptor is deep ($E_A^b \gg kT$), it does not release holes for reasonable temperatures. Examples of suitable impurities for compensation of electrons are Si:Au [223], GaAs:Cr [224] and InP:Fe [225]. A deep donor, e.g. InP:Cr [226], is necessary to compensate p-type conductivity.

Figure 7.26a shows the terms of Fe in InP [227,228]. An overview of transition metals in III–V semiconductors can be found in [230]. The electron configuration of neutral Fe atoms is $3d^6 4s^2$ (cf. Table 15.2). The Fe is incorporated on the In site and thus has a Fe^{3+} state as a neutral acceptor (A^0). The Fe^{3+} state has the electron configuration $3d^5$. The arrow in Fig. 7.26a

[7] Not from a small mobility due to poor crystal quality.

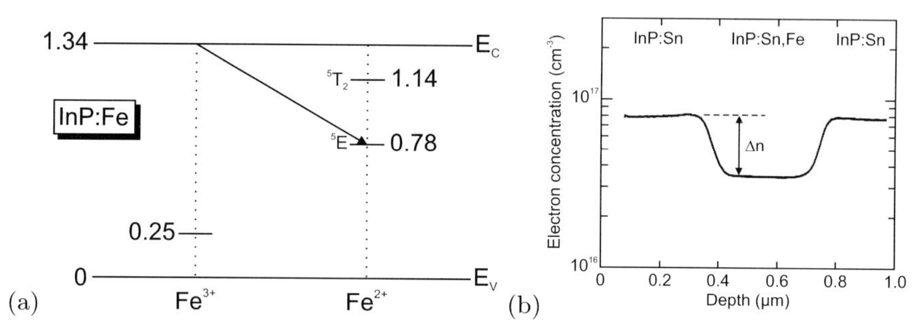

Fig. 7.26. (**a**) Schematic band diagram of InP with levels of Fe impurities in the 3+ and 2+ charge states at low temperature. All energies are given in eV. The *arrow* denotes capture of an electron (from the conduction band or a shallow donor) on the deep acceptor. Compare this figure also with Figs. 9.32 and 10.15. (**b**) Depth profile of electron concentration in an InP:Sn/InP:Fe,Sn/InP:Sn structure. The change $\Delta n \approx 4.5 \times 10^{16}\,\mathrm{cm}^{-3}$ of electron concentration is due to the compensation by Fe and corresponds to the chemical iron concentration determined by SIMS, [Fe]=$4.9 \times 10^{19}\,\mathrm{cm}^{-3}$. Part (**b**) adapted from [235]

represents the capture of an electron from the conduction band or from a shallow donor. The charge state of the Fe becomes Fe^{2+} (charged acceptor, A^-) with the electron configuration $3d^6$. The cubic crystal field (T_d symmetry) splits this ^5D Fe state[8] into two terms [231] that exhibit further fine structure [228]. The large thermal activation energy of 0.64 eV found in the Hall effect on semi-insulating InP:Fe [225] corresponds to the energy separation of the ^5E level and the conduction band.

The maximum electron concentration that can be compensated in this way is limited by the solubility of Fe in InP [232], about $1 \times 10^{17}\,\mathrm{cm}^3$. Higher Fe incorporation leads to the formation of Fe (or FeP) precipitates and degrades the crystal quality. Only a fraction of the incorporated Fe may then be electrically active and contribute to the compensation. The maximum electrically active Fe concentration is found to be 5–$6 \times 10^{16}\,\mathrm{cm}^{-3}$ [233]. The compensation can be directly visualized via the depth profile of the electron concentration in a n-si-n structure (Fig. 7.26b). The poor thermal stability of Fe, i.e. high diffusion coefficient, has evoked proposals for more stable dopants such as InP:Ru [234].

7.7.7 Surface States

The investigation of (semiconductor) surfaces is a large field with sophisticated methods that allow real-space imaging with atomic resolution by scanning probe microscopy and highly depth resolved electronic studies. The surface represents first of all a break in the periodic crystal potential and

[8]The notation is ^{2S+1}J (multiplicity), with S being the total spin and J being the total angular momentum.

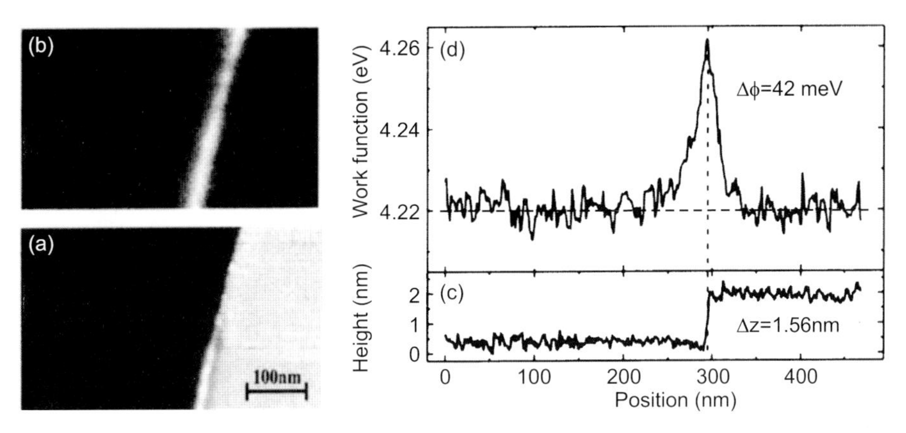

Fig. 7.27. Image of (**a**) topography ($\Delta z = 2.8\,\mathrm{nm}$) and (**b**) work function ($\Delta\phi =$ 4.21–4.26 eV) of a surface step along [111] on a n-GaP(110) surface cleaved in UHV. Reprinted with permission from [237], ©2004 APS. (**c**) and (**d**) show the corresponding linescans. Adapted from [237]

thus a defect of the bulk crystal. The unsatisfied bonds partly rearrange, e.g. by building dimers, forming a surface reconstruction or remain as dangling bonds. The surface exhibits a surface density of states. Such states can lie in the bandgap and capture electrons, leading to recombination and a depletion layer. In this book, we will not get into details of semiconductor surface physics and refer the reader to [236].

As an example of the formation of electronic states at surface defects we show in Fig. 7.27 the comparison of topography and work function (measured by Kelvin probe force microscopy [237]) at a surface step on a GaP(110) surface that has been prepared by cleaving in-situ in ultrahigh vacuum (UHV). The depletion-type band bending of the surface is about 0.4 eV. The further increase of the position of the vacuum level at the step edge shows the presence of trap states in the bandgap causing the conduction band to bend upwards (cf. Sect. 18.2.1). Modeling of the effect shows that the charge density at the surface is $6 \times 10^{11}\,\mathrm{cm}^{-2}$ and at the step edge $1.2 \times 10^{6}\,\mathrm{cm}^{-1}$.

7.8 Hydrogen in Semiconductors

The role of hydrogen in semiconductors was first recognized in studies of ZnO [238]. It is now clear that hydrogen plays an important role in the passivation of defects[9]. As a 'small' atom, it can attach easily to dangling bonds and form an electron-pair bond. Thus, surfaces, grain boundaries, dislocations and shallow (donor and acceptor) and deep impurity levels become passivated. A good overview and many details of the physics and technological

[9]Just not in ZnO where hydrogen is a shallow doner.

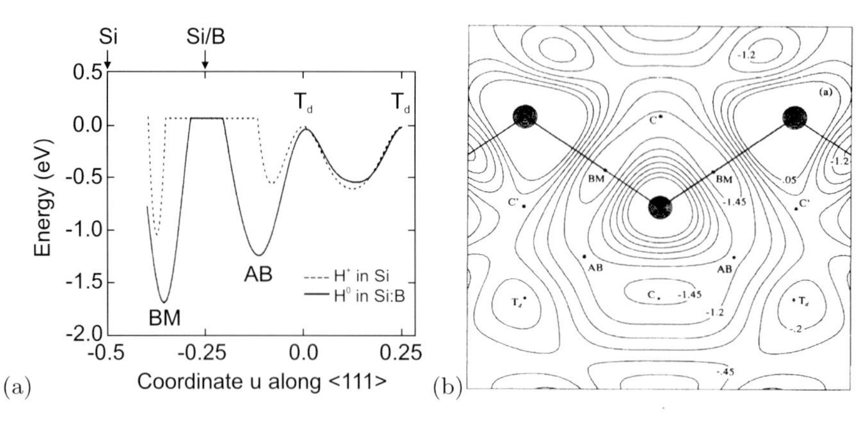

Fig. 7.28. (**a**) Energy for positions u of the hydrogen atom along the $\langle 111 \rangle$ direction for H^+ in pure Si (Si atom at $u = -0.25$) and neutral hydrogen (B atom at $u = -0.25$). u is measured in units of $\sqrt{3}a_0$. For all positions of the hydrogen atom the positions of the other atoms have been relaxed in the calculation. Data from [246]. (**b**) Adiabatic potential energy in the (110) plane for hydrogen in Si:B. 'BM' denotes the bond minimum site (high valence electron density), C and C' are equivalent for pure Si. Reprinted with permission from [246], ©1989 APS

use of hydrogen in semiconductors can be found in [239, 240]. The hydrogen must be typically introduced as atomic species into semiconductors, e.g. from a plasma in the vicinity of the surface or by ion irradiation.

With regard to silicon it is important to note that the Si–H bond is stronger than the Si–Si bond. Thus a silicon surface under atomic hydrogen exhibits Si–H termination rather than Si–Si dimers [241]. Due to the stronger bond, the hydrogenation leads to an increase of the silicon bandgap, which can be used for surface passivation [243], leading to reduced reverse diode current.

The hydrogen concentration in amorphous Si (a-Si) can be as high as 50% [242]. Electronic grade a-Si contains typically 10–30 atomic % hydrogen and is thus rather a silicon–hydrogen alloy.

Hydrogen in crystalline silicon occupies the bond-center interstitial position (cf. Fig. 3.17b) as shown in Fig. 7.28a. The complexes formed by hydrogen with shallow acceptors and donors have been studied in detail. It is now generally accepted that for acceptors (e.g. boron) in silicon the hydrogen is located close to the bond-center position of the Si–B pair (BM, bond minimum) as sketched in Fig. 7.29a. The boron atom forms an electron-pair bond with three silicon atoms of the tetrahedra, the fourth silicon bonds to the hydrogen atom. The complex therefore no longer acts as an acceptor. The silicon atoms and the acceptor relax their positions. The adiabatic potential energy surface of hydrogen in Si:B is shown in Fig. 7.28b. The hydrogen can sit on four equivalent sites (BM) along the $\langle 111 \rangle$ directions of the initial B–Si_4 tetrahedron. This reduces the symmetry, e.g. of H–B vibrations [244]. The energetic barrier for the hydrogen orientation has been determined to be

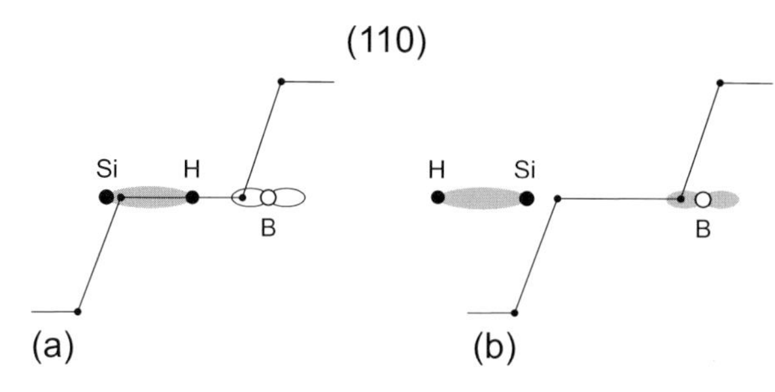

Fig. 7.29. Schematic model for hydrogen in silicon forming a complex with (**a**) a shallow acceptor (boron, empty orbital) and (**b**) a shallow donor (phosphorus, double-filled orbital)

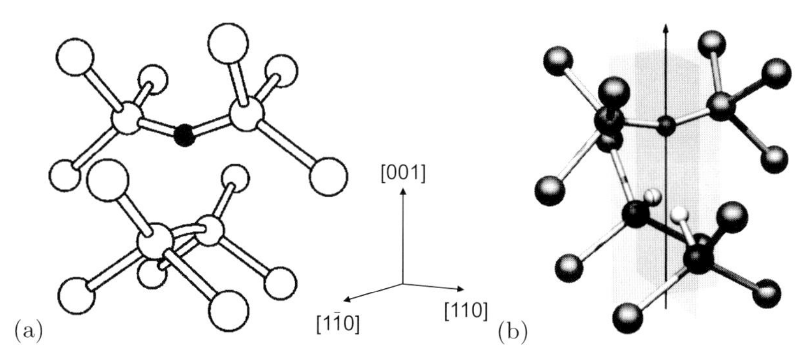

Fig. 7.30. (**a**) Structure of the V–O complex (A center) in silicon. The *black sphere* represents the oxygen atom. Reprinted with permission from [249], ©2004 APS. (**b**) Calculated ground-state structure for the V–O–H_2 center in silicon. Oxygen is over the C_2 axis, and the two *white* spheres represent hydrogen. Reprinted with permission from [247], ©2000 APS

0.2 eV theoretically [246] for a hydrogen motion along the path BM–C–BM in Fig. 7.28b. Stress (along [100] and [112]) reduces the symmetry and leads to splitting of the local vibrational modes, now showing axial symmetry [245]. However, this preferential orientation disappears with an activation energy of 0.19 eV, close to the theoretical value.

Hydrogen has experimentally been found to also passivate shallow donors. The microscopic configuration is sketched in Fig. 7.29b. The hydrogen atom sits on the Si–AB (antibonding) position and forms an electron-pair bond with the silicon atom. The donor, e.g. phosphorus, is left with a double-filled p-orbital (lone pair) whose level is in the valence band and thus no longer contributes to conductivity. Molecular hydrogen can passivate the so-called A center in Si, an oxygen–vacancy complex [247]. The atomistic configuration of the V–O–H_2 complex is shown in Fig. 7.30. The deep double donor S in Si with a level at 0.3 eV below the conduction-band edge can also be passivated by two hydrogen atoms [248].

8 Transport

8.1 Introduction

Charge and heat energy can be transported through the semiconductor in the presence of appropriate (generalized) forces. Such a force can be an electric field or a temperature gradient. Both transport phenomena are coupled since electrons can transport energy and charge through the crystal. First, we will treat the charge transport as a consequence of a gradient in the Fermi level, then the heat transport upon a temperature gradient and finally the coupled system, i.e. the Peltier and Seebeck effects.

Practically all important semiconductor devices are based on the transport of charge, such as diode, transistor, photodetector, solar cell and laser.

Carriers move in the semiconductor driven by a gradient in the Fermi energy. We distinguish

- drift, as a consequence of an electrical field $\mathbf{E}$,
- diffusion, as a consequence of a concentration gradient ∇n or ∇p.

In inhomogeneous semiconductors for which the position of the band edges is a function of position, another force occurs. This will not be treated here, since later (see Chap. 11) it will be included as an additional, internal electrical field.

Many semiconductor properties, such as the carrier concentration and the bandgap, depend on the temperature. Thus, device properties will also depend on temperature. During operation of a device typically heat is generated, e.g. by Joule heating due to finite resistivity. This heat leads to an increase of the device temperature that subsequently alters the device performance, mostly for the worse. Ultimately, the device can be destroyed. Thus cooling of the device, in particular of the active area of the device, is essential. Mostly the thermal management of device heating limits the achievable performance (and lifetime) of the device. In high-power devices quite high energy densities can occur, e.g. the facet of a high-power semiconductor laser has to withstand an energy density beyond $10\,\mathrm{MW\,cm^{-2}}$.

8.2 Conductivity

Under the influence of an electric field the electrons accelerate according to (cf. (6.35))

$$\mathbf{F} = m^* \frac{\mathrm{d}\mathbf{v}}{\mathrm{d}t} = \hbar \frac{\mathrm{d}\mathbf{k}}{\mathrm{d}t} = q\mathbf{E} = -e\mathbf{E} \, . \tag{8.1}$$

In the following, q denotes a general charge, while e is the (positive) elementary charge. After the time δt the $\mathbf{k}$ vector of all conduction electrons (and the center of the Fermi sphere) has been shifted by $\delta\mathbf{k}$

$$\delta\mathbf{k} = -e\mathbf{E}\, \delta t/\hbar \, . \tag{8.2}$$

In the absence of scattering processes this goes on further (similar to an electron in vacuum). This regime is called *ballistic* transport. In a (periodic) band structure, the electron will perform a closed cycle as indicated in Fig. 8.1. Such motion is called a Bloch oscillation. However, in a bulk crystal the period T of such an oscillation $e\mathcal{E}T/\hbar = 2\pi/a_0$ is of the order of 10^{-10} s for $\mathcal{E} = 10^4$ V/cm. This time is much longer than a typical scattering time of 10^{-14} s. Thus, in bulk material the Bloch electron cannot reach the zone boundary. However, in artificial superlattices (see Chap. 11) with larger periodicity (≈ 10 nm), high electric fields ($\approx \mathcal{E} = 10^6$ V/cm) and high quality (reduced collision time) such motion is possible. We note that in the absence of scattering, electrons also perform a periodic oscillation in a magnetic field (cyclotron motion).

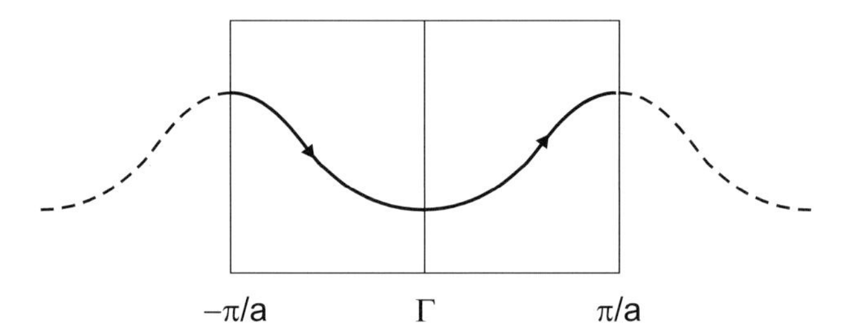

Fig. 8.1. Schematic representation of a Bloch oscillation

In a real semiconductor, at finite temperatures, impurities, phonons and defects (finally also the surface) will contribute to scattering. In the relaxation-time approximation it is assumed that the probability for a scattering event, similar to friction, is proportional to the (average) carrier velocity. The average relaxation time τ is introduced via an additional term $\dot{\mathbf{v}} = -\mathbf{v}/\tau$ that sums up all scattering events. Thus, the maximum velocity that can be reached in a static electric field is given by (steady-state velocity)

$$\mathbf{v} = -e\mathbf{E}\tau/m^* \, . \tag{8.3}$$

The current density per unit area is then linear in the field, i.e. fulfills Ohm's law

$$\mathbf{j} = nq\mathbf{v} = ne^2\mathbf{E}\tau/m^* = \sigma\mathbf{E} . \tag{8.4}$$

The conductivity σ in the relaxation-time approximation is given by

$$\sigma = \frac{1}{\rho} = \frac{ne^2\tau}{m^*} . \tag{8.5}$$

The specific resistivity is the inverse of the conductivity. Metals have a high conductivity, e.g. for Cu at room temperature $\sigma = 5.88 \times 10^5\,\Omega^{-1}\,\mathrm{cm}^{-1}$. At low temperatures (4 K) the conductivity is even a factor of 10^5 higher. The mean free path is then 3 mm (and thus susceptible to the sample geometry) while at room temperature the mean free path is only 30 nm. In semiconductors, the carrier concentration depends strongly on the temperature. At zero temperature the conductivity is zero. Also, the scattering processes and thus the relaxation time constant exhibit a temperature dependence.

8.3 Low-Field Transport

First we consider *small* electric fields. The real meaning of this will only become clear in Sect. 8.5 on high-field transport. In the low-field regime the velocity is proportional to the electric field.

8.3.1 Mobility

The mobility is defined (scalar terms) as

$$\mu = \frac{v}{\mathcal{E}} . \tag{8.6}$$

By definition, it is a negative number for electrons and positive for holes. However, the numerical value is usually given as a positive number for both carrier types. In an intrinsic semiconductor the mobility is determined by scattering with phonons. Further scattering is introduced by impurities, defects or alloy disorder. The conductivity is (8.4)

$$\sigma = qn\mu \tag{8.7}$$

for each carrier type. The mobility in the relaxation time approximation is

$$\mu = \frac{q\tau}{m^*} . \tag{8.8}$$

Thus, in the presence of both electrons and holes,

$$\sigma = -en\mu_\mathrm{n} + ep\mu_\mathrm{p} \;, \tag{8.9}$$

where μ_n and μ_p are the mobilities for electrons and holes, respectively. In the relaxation-time approximation these are given by $\mu_\mathrm{n} = -e\tau_\mathrm{n}/m_\mathrm{e}^*$ and $\mu_\mathrm{p} = e\tau_\mathrm{p}/m_\mathrm{p}^*$.

As a unit, usually $\mathrm{cm}^2/\mathrm{V\,s}$ is used. While Cu at room temperature has a mobility of $35\,\mathrm{cm}^2/\mathrm{V\,s}$, semiconductors can have much higher values. In two-dimensional electron gases (see Chap. 11), the mobility can reach several $10^7\,\mathrm{cm}^2/\mathrm{V\,s}$ at low temperature (Fig. 11.25). In bulk semiconductors with small bandgap, a high electron mobility is caused by its small effective mass. Some typical values are given in Table 8.1.

Table 8.1. Mobilities (in $\mathrm{cm}^2/\mathrm{V\,s}$) for electrons and holes at room temperature for various semiconductors

	μ_n	μ_p
Si	1300	500
Ge	4500	3500
GaAs	8800	400
GaN	300	180
InSb	77 000	750
InAs	33 000	460
InP	4600	150
ZnO	230	8

8.3.2 Microscopic Scattering Processes

The relaxation time constant summarizes all scattering mechanisms. If the relaxation times τ_i of various processes are independent, the Matthiesen rule can be used to obtain the mobility ($\mu_i = q\tau_i/m^*$)

$$\frac{1}{\mu} = \sum_i \frac{1}{\mu_i} \;. \tag{8.10}$$

The different scattering mechanisms have quite different temperature dependences such that the mobility is a rather complicated function of temperature.

8.3.3 Ionized Impurity Scattering

Theoretically, this problem is treated similar to Rutherford scattering. A screened Coulomb potential is assumed, as the scattering potential

$$V(r) = -\frac{Ze}{4\pi\epsilon_0\epsilon_\mathrm{r}}\frac{1}{r}\exp\left(-\frac{r}{L_\mathrm{D}}\right) , \tag{8.11}$$

where L_D is the screening length. The problem has been treated classically by Conwell and Weisskopf [250] and quantum mechanically by Brooks [251] and Herring. An expression for the mobility that encompasses the Conwell–Weisskopf and Brooks–Herring results is derived in [252]. Further details can be found in [253, 254]. For the mobility it is found that

$$\mu_\mathrm{ion.imp.} = \frac{2^{7/2}(4\pi\epsilon_0\epsilon_\mathrm{r})^2}{\pi^{3/2}Z^2e^3\sqrt{m^*}}\frac{(kT)^{3/2}}{N_\mathrm{ion}}\frac{1}{\ln(1+\beta^2)-\beta^2(1+\beta^2)} . \tag{8.12}$$

8.3.4 Deformation Potential Scattering

Acoustic phonons with small wavevector, i.e. a wavelength large compared to the unit cell, can have TA or LA character. The TA phonons represent a shear wave (with zero divergence), the LA phonons are a compression wave (with zero rotation). The LA phonon is a plane wave of displacement $\delta\mathbf{R}$ parallel to the k-vector $\mathbf{q}$,

$$\delta\mathbf{R} = \mathbf{A}\sin\left(\mathbf{q}\cdot\mathbf{R} - \omega t\right) . \tag{8.13}$$

The strain tensor is given by

$$\epsilon_{ij} = \frac{1}{2}\left(\mathbf{q}_i\mathbf{A}_j + \mathbf{q}_j\mathbf{A}_i\right)\cos\left(\mathbf{q}\cdot\mathbf{R} - \omega t\right) . \tag{8.14}$$

It has a diagonal form $\epsilon_{ij} = \mathbf{q}_i\mathbf{A}_j$ for $\mathbf{q}$ and $\omega \to 0$. Therefore, the LA phonon creates an oscillatory volume dilatation (and compression) with amplitude $\mathbf{q}\cdot\mathbf{A}$. This volume modulation affects the position of the band edges (Sect. 6.14). For the conduction-band edge the energy change is related to the volume change by the hydrostatic deformation potential $E_\mathrm{ac.def.} = V\partial E_\mathrm{C}/\partial V$. Since the modulation is small compared to the energy of the charge carriers, it is mostly an elastic scattering process. The Hamilton operator for the LA scattering is

$$\hat{H} = E_\mathrm{ac.def.}(\mathbf{q}\cdot\mathbf{A}) . \tag{8.15}$$

The size of the LA amplitude is given by the number of phonons in the mode that is given by the Bose–Einstein distribution, $N_\mathrm{ph}(\hbar\omega) = \left[\exp\left(\frac{\hbar\omega}{kT}\right)\right]^{-1}$. The mobility due to acoustic deformation potential scattering is found to be

$$\mu_{\mathrm{ac.def.}} = \frac{2\sqrt{2\pi}e\hbar^4 c_\mathrm{l}}{3m^{*5/2}E^2_{\mathrm{ac.def.}}}\left(\frac{1}{kT}\right)^{3/2} , \tag{8.16}$$

where $c_\mathrm{l} = \rho c_\mathrm{s}^\mathrm{LA}$, ρ being the density and c_s being the sound velocity. The acoustical deformation potential scattering is therefore important at high temperatures. It is dominating in nonpolar semiconductors (Ge, Si) at high temperatures (typically at and above room temperature).

8.3.5 Piezoelectric Potential Scattering

In piezoelectric crystals, i.e. crystals that show an electric polarization upon strain, certain acoustic phonons lead to piezoelectric fields. In GaAs this is the case for shear waves. In strongly ionic crystals, e.g. II–VI semiconductors, the piezoelectric scattering can be stronger than the deformation potential scattering. The mobility due to piezoelectric potential scattering is

$$\mu_{\mathrm{pz.el.}} = \frac{16\sqrt{2\pi}}{3}\frac{\hbar\epsilon_0\epsilon_\mathrm{r}}{m^{*3/2}eK^2}\left(\frac{1}{kT}\right)^{1/2} , \tag{8.17}$$

with $K = \frac{e_\mathrm{p}^2/c_\mathrm{l}}{\epsilon_0\epsilon_\mathrm{r}+e_\mathrm{p}^2/c_\mathrm{l}}$, e_p being the piezoelectric coefficient.

8.3.6 Polar Optical Scattering

LO phonons are connected with an electric field antiparallel to the displacement (5.45). In the scattering mechanism the absorbed or emitted phonon energy $\hbar\omega_0$ is comparable to the thermal energy of the carriers. Therefore, the scattering is inelastic and the relaxation-time approximation does not work. The general transport theory is complicated. If the temperature is low compared to the Debye temperature, $T \ll \Theta_\mathrm{D}$

$$\mu_{\mathrm{pol.opt.}} = \frac{e}{2m^*\alpha\omega_0}\exp\left(\frac{\Theta_\mathrm{D}}{T}\right) , \tag{8.18}$$

where $\alpha = \frac{1}{137}\sqrt{\frac{m^*c^2}{2k\Theta_\mathrm{D}}}\left(\frac{1}{\epsilon(\infty)} - \frac{1}{\epsilon(0)}\right)$ is the dimensionless polar constant.

8.3.7 Temperature Dependence

The sum of all scattering processes leads to a fairly complicated temperature dependence of the mobility $\mu(T)$. In covalent semiconductors (Si, Ge) the most important processes are the ionized impurity scattering ($\mu \propto T^{3/2}$) at low temperatures and the deformation potential scattering ($\mu \propto T^{-3/2}$) at high temperatures (Fig. 8.2a). In polar crystals (e.g. GaAs) at high temperatures the polar optical scattering is dominant (Fig. 8.2b).

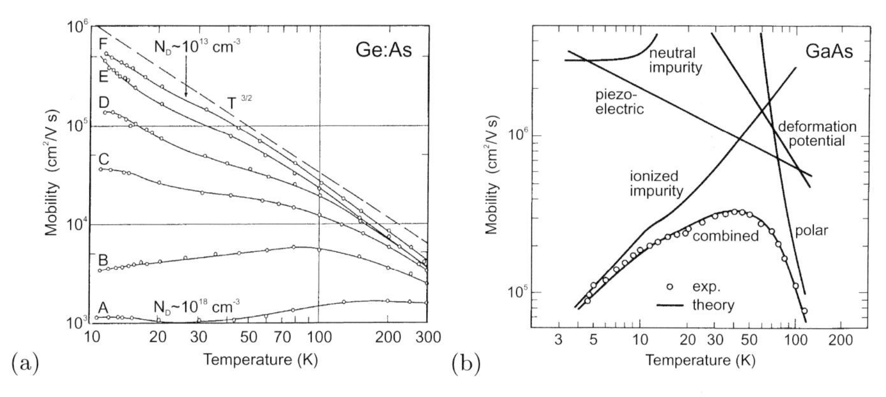

Fig. 8.2. Temperature dependence of the electron mobility in n-doped (**a**) Ge (for various doping levels from $N_\mathrm{D} \approx 10^{18}$ for sample A to $10^{13}\,\mathrm{cm}^{-3}$ for sample F in steps of a factor of ten) and (**b**) GaAs ($N_\mathrm{D} \approx 5\times10^{13}\,\mathrm{cm}^{-3}$, $N_\mathrm{A} \approx 2\times10^{13}\,\mathrm{cm}^{-3}$). Part (**a**) adapted from [202], part (**b**) adapted from [255]

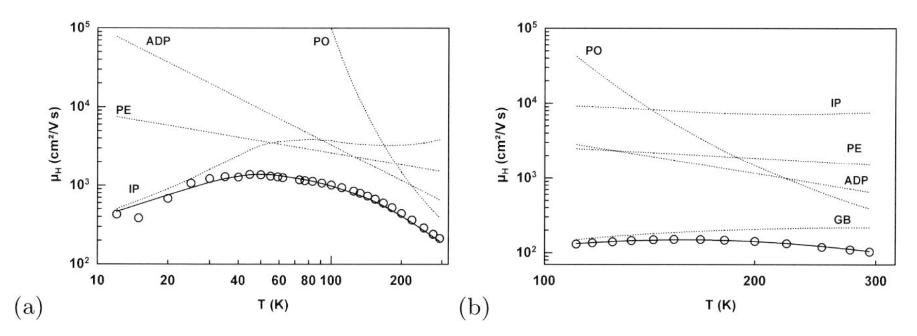

Fig. 8.3. Temperature dependence of the electron mobility in n-type (**a**) bulk ZnO and (**b**) ZnO thin film. In the latter, grain-boundary scattering is limiting the mobility

In Fig. 8.3 the electron mobility of bulk and thin-film ZnO is compared. Since ZnO is polar the mobility at room temperature is limited by polar optical phonon scattering. In the thin film, grain-boundary scattering [256] additionally occurs.

Since the carrier concentration increases with increasing temperature and the mobility decreases, the conductivity has a maximum, typically around 70 K (see Fig. 8.4). At very high temperature, when intrinsic conduction starts, σ shows a strong increase due to the increase in n.

The mobility decreases with increasing dopant concentration as shown in Figs. 8.2a and 8.5. Thus, for bulk material high carrier density and high mobility are contrary targets. A solution will be provided with the concept of *modulation* doping where the dopants and the (two-dimensional) carrier gas will be spatially separated (see Sect. 11.5.3).

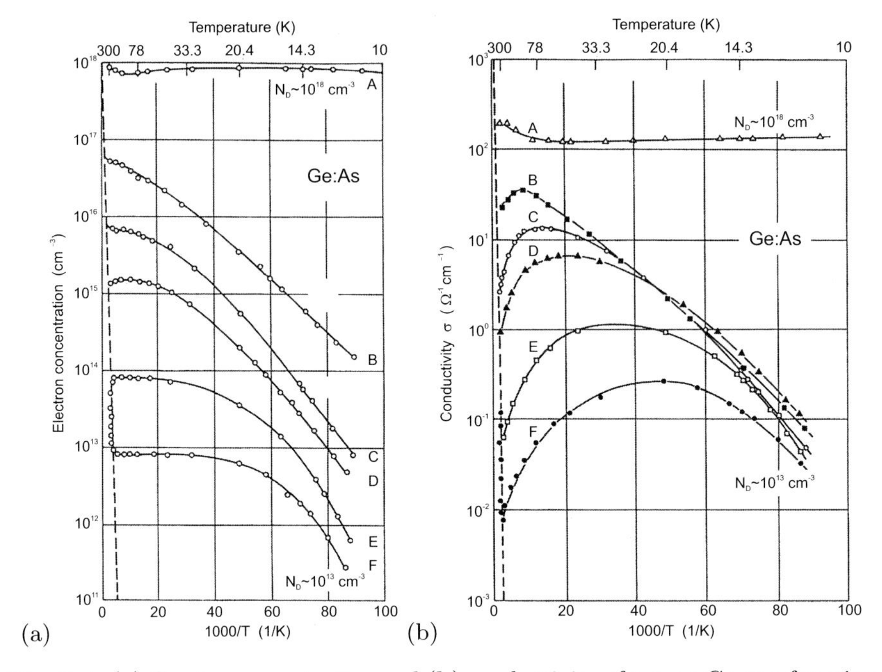

Fig. 8.4. (**a**) Carrier concentration and (**b**) conductivity of n-type Ge as a function of temperature. The doping level varies from $N_D \approx 10^{13}$ to 10^{18} (samples A–F as in Fig. 8.2a where the mobility of the samples is shown). Adapted from [202]

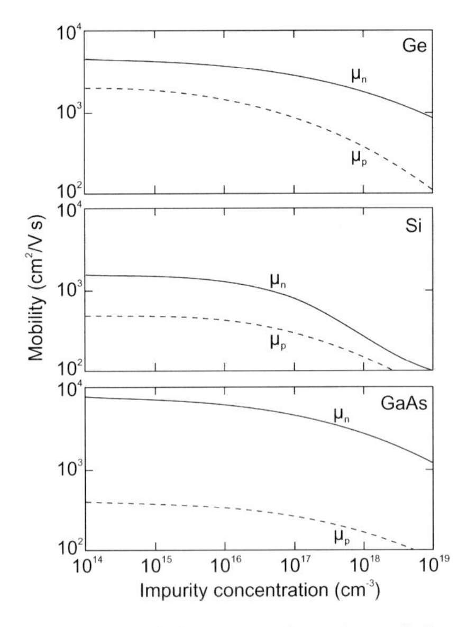

Fig. 8.5. Room-temperature mobility as a function of dopant concentration for Ge, Si, and GaAs. *Solid lines* (*dashed lines*) represent the electron (hole) mobility. Adapted from [183]

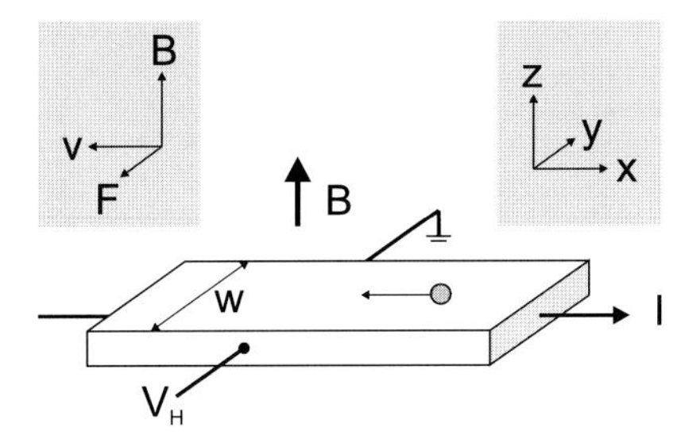

Fig. 8.6. Scheme of the Hall-effect geometry. The movement of one electron in the longitudinal electric current I is shown schematically. The coordinate system (x, y, z) and the directions of the magnetic field $\mathbf{B}$, the drift velocity of an electron $\mathbf{v}$ and the resulting Lorentz force $\mathbf{F}$ are given. The transverse field $\mathcal{E}_y$ is given by V_{H}/w

8.4 Hall Effect

An electrical current along the x (longitudinal) direction in a perpendicular magnetic field $\mathbf{B} = (0, 0, B)$ along z induces an electric field $\mathcal{E}_y$ along the transverse (y) direction. The charge accumulation is due to the Lorentz force. The related transverse voltage is called the Hall voltage and the resistivity $\rho_{xy} = \mathcal{E}_{\mathrm{y}}/j_x$ the Hall resistivity [257]. For the Hall geometry see Fig. 8.6. For thin-film samples typically the van-der-Pauw geometry (Fig. 8.7) and method is used [258–260].

The steady-state equation of motion is

$$m^* \frac{\mathbf{v}}{\tau} = q\,(\mathbf{E} + \mathbf{v} \times \mathbf{B}) \ . \tag{8.19}$$

We note that this equation of motion is also valid for holes, given the convention of Sect. 6.13.1, i.e. positive charge. With the cyclotron frequency $\omega_{\mathrm{c}} = qB/m^*$ the conductivity tensor is ($\mathbf{j} = \sigma\mathbf{E} = qn\mathbf{v}$)

$$\sigma = \begin{pmatrix} \sigma_{xx} & \sigma_{xy} & 0 \\ \sigma_{yx} & \sigma_{yy} & 0 \\ 0 & 0 & \sigma_{zz} \end{pmatrix} \tag{8.20a}$$

$$\sigma_{xx} = \sigma_{yy} = \sigma_0 \frac{1}{1 + \omega_{\mathrm{c}}^2 \tau^2} = \sigma_0 \frac{1}{1 + \mu^2 B^2} \tag{8.20b}$$

$$\sigma_{xy} = -\sigma_{yx} = \sigma_0 \frac{\omega_{\mathrm{c}} \tau}{1 + \omega_{\mathrm{c}}^2 \tau^2} = \sigma_0 \frac{\mu B}{1 + \mu^2 B^2} \tag{8.20c}$$

$$\sigma_{zz} = \sigma_0 = \frac{q^2 n \tau}{m^*} = qn\mu \ . \tag{8.20d}$$

If only one type of carrier (charge q, density n) is considered, the condition $j_y = 0$ leads to $\mathcal{E}_y = \mu B \mathcal{E}_x$ and $j_x = \sigma_0 \mathcal{E}_x$. The Hall coefficient $R_{\mathrm{H}} = \mathcal{E}_y/(j_x\, B)$ is therefore given by

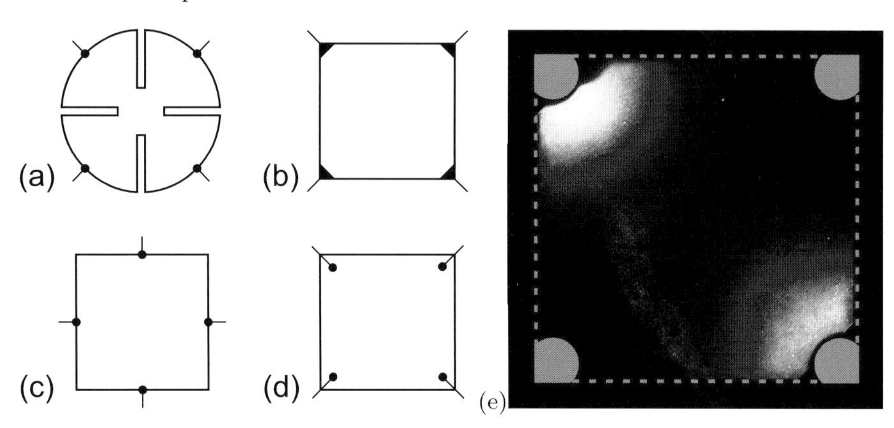

Fig. 8.7. (**a**–**d**) Geometry for van-der-Pauw Hall measurements. (**a**) Best geometry (cloverleaf), (**b**) acceptable square geometry with small contacts on the corners, (**c**,**d**) not recommended geometries with contacts on the edge centers or inside the square, respectively. (**e**) Current distribution, as visualized by lock-in thermography [190], in epitaxial ZnO layer on sapphire with Hall geometry as in part (**b**). *Grey dashed line* indicates the outline of the $10 \times 10\,\mathrm{mm}^2$ substrate, *grey areas* indicate gold ohmic contacts

$$R_{\mathrm{H}} = \frac{1}{qn} \,. \tag{8.21}$$

If both types of carriers are present, again the condition $j_y = 0$ yields the Hall constant

$$R_{\mathrm{H}} = \frac{1}{e} \frac{-n\mu_{\mathrm{e}}^2(1+\mu_{\mathrm{h}}^2 B^2) + p\mu_{\mathrm{h}}^2(1+\mu_{\mathrm{e}}^2 B^2)}{n^2\mu_{\mathrm{e}}^2(1+\mu_{\mathrm{h}}^2 B^2) - 2np\mu_{\mathrm{e}}\mu_{\mathrm{h}}(1+\mu_{\mathrm{e}}\mu_{\mathrm{h}} B^2) + p^2\mu_{\mathrm{h}}^2(1+\mu_{\mathrm{e}}^2 B^2)} \,. \tag{8.22}$$

Under the assumption of small magnetic fields,[1] i.e. $\mu B \ll 1$, the Hall coefficient is

$$R_{\mathrm{H}} = \frac{1}{e} \left[\frac{-n\mu_{\mathrm{e}}^2 + p\mu_{\mathrm{h}}^2}{(-n\mu_{\mathrm{e}} + p\mu_{\mathrm{h}})^2} + \frac{np(-n+p)\mu_{\mathrm{e}}^2\mu_{\mathrm{h}}^2(\mu_{\mathrm{e}} - \mu_{\mathrm{h}})^2}{(-n\mu_{\mathrm{e}} + p\mu_{\mathrm{h}})^4} B^2 + \ldots \right] . \tag{8.23}$$

For small magnetic field this can be written as

$$R_{\mathrm{H}} = \frac{1}{e} \frac{-n\beta^2 + p}{(-n\beta + p)^2} \,, \tag{8.24}$$

with $\beta = \mu_{\mathrm{e}}/\mu_{\mathrm{h}}$. For large magnetic fields, i.e. $\mu B \gg 1$, the Hall coefficient is given by

$$R_{\mathrm{H}} = \frac{1}{e} \frac{-n + p}{(n+p)^2} \,. \tag{8.25}$$

[1] We note that for a mobility of $10^4\,\mathrm{cm}^2/\mathrm{V\,s}$, μ^{-1} is a field of $B = 1\,\mathrm{T}$.

In Fig. 8.8, the absolute value of the Hall coefficient for InSb samples with different doping concentrations is shown. The p-doped samples exhibit a reverse of the sign of the Hall coefficient upon increase of temperature when intrinsic electrons contribute to the conductivity. The zero in R_{H} occurs for $n = p\mu_{\mathrm{h}}^2/\mu_{\mathrm{e}}^2$. For high temperatures, the Hall coefficient for n- and p-doped samples is dominated by the electrons that have much higher mobility (Table 8.1).

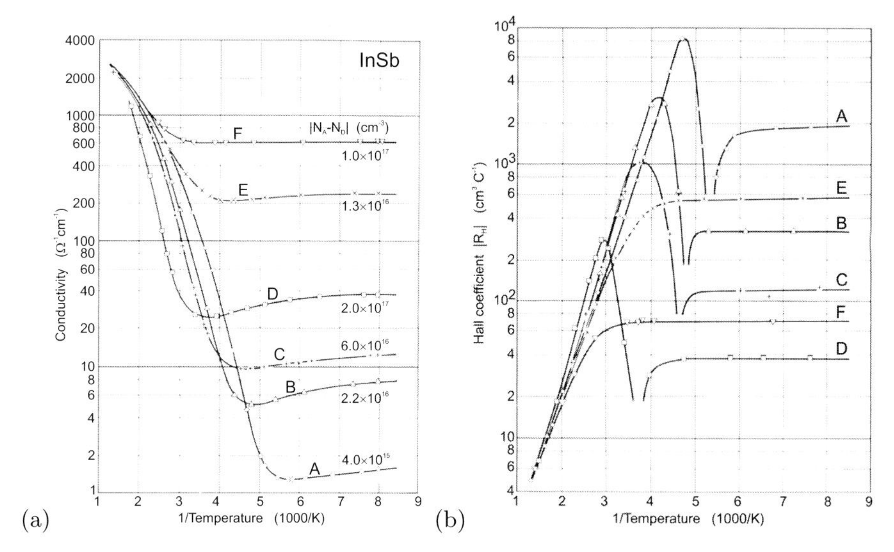

Fig. 8.8. (**a**) Conductivity and (**b**) absolute value of the Hall coefficient vs. inverse temperature for four p-doped (A–D) and two n-doped (E,F) InSb samples. The doping levels are given in (**a**). Adapted from [261]

In the derivation of the Hall coefficient we had assumed that all carriers involved in the transport have the same properties, in particular that they are subject to the same scattering time. This assumption is generally not the case and we need to operate with the ensemble average of the discussed quantities. The ensemble average of an energy-dependent quantity $\zeta(E)$ over the (electron) distribution function $f(E)$ is denoted as $\langle\zeta\rangle$ and is given as[2]

$$\langle\zeta\rangle = \frac{\int \zeta(E) f(E) \mathrm{d}E}{\int f(E) \mathrm{d}E} \ . \tag{8.26}$$

In particular $\langle\tau\rangle^2$ is now different from $\langle\tau^2\rangle$. Considering the equation $\langle\mathbf{j}\rangle = \langle\sigma\rangle\mathbf{E}$ for the ensemble averaged current densities we find (for one type

[2]For this consideration it is assumed that the energy dependence is the decisive one. Generally, averaging may have to be performed over other degrees of freedom as well, such as the spin or, in the case of anisotropic bands, the orbital direction.

of carrier, cf. (8.21))

$$R_{\rm H} = \frac{1}{qn} r_{\rm H} \,, \tag{8.27}$$

with the so-called *Hall factor* $r_{\rm H}$ given by

$$\begin{aligned} r_{\rm H} &= \frac{\gamma}{\alpha^2 + \omega_{\rm c}^2 \gamma^2} \\ \alpha &= \left\langle \frac{\tau}{1+\omega_{\rm c}^2\tau^2} \right\rangle \\ \gamma &= \left\langle \frac{\tau^2}{1+\omega_{\rm c}^2\tau^2} \right\rangle \,. \end{aligned} \tag{8.28}$$

The Hall factor depends on the scattering mechanisms and is of the order of 1. For large magnetic fields the Hall factor approaches 1. For small magnetic fields we have

$$R_{\rm H} = \frac{1}{qn} \frac{\langle \tau^2 \rangle}{\langle \tau \rangle^2} \,. \tag{8.29}$$

The mobility calculated from (cf. (8.20d)) $\sigma_0 R_{\rm H}$ is called the *Hall mobility* $\mu_{\rm H}$ and is related to the mobility via

$$\mu_{\rm H} = r_{\rm H}\, \mu \,. \tag{8.30}$$

8.5 High-Field Transport

8.5.1 Drift-Saturation Velocity

In the case of small electric fields the scattering events are elastic. The drift velocity is linearly proportional to the electric field. The average thermal energy is close to its thermal value $3kT/2$.

The scattering efficiency, however, is reduced already at moderate fields. Then, the electron temperature becomes larger than the lattice temperature. If the carrier energy is large enough it can transfer energy to the lattice by the emission of an optical phonon. This mechanism is very efficient and limits the maximum drift velocity. Such behavior is nonohmic. The limiting value for the drift velocity is termed the *drift-saturation velocity*. It is given by [262]

$$v_{\rm s} = \left(\frac{8}{3\pi}\right)^{1/2} \left(\frac{\hbar\omega_{\rm LO}}{m^*}\right)^{1/2} \,. \tag{8.31}$$

This relation can be obtained from an energy-balance consideration. The energy gain per unit time in the electric field is equal to the energy loss by the emission of an optical phonon.

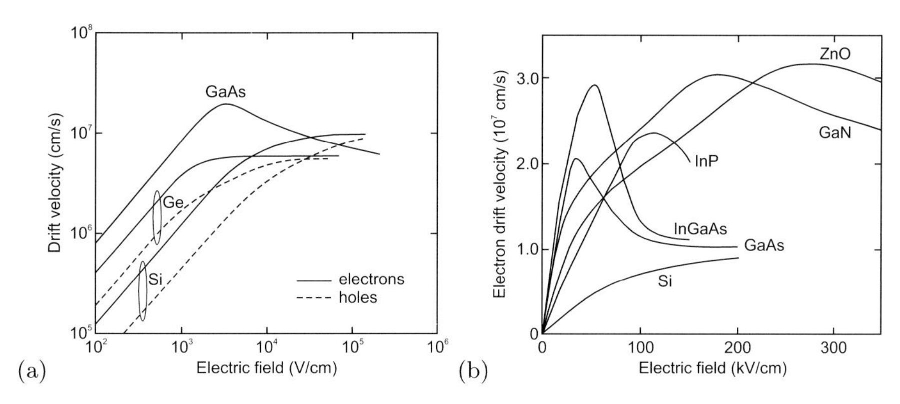

Fig. 8.9. Drift velocity at room temperature as a function of applied electric field for (**a**) high-purity Si, Ge, and GaAs on a double-logarithmic plot and (**b**) on linear plots for Si [263], Ge [264], GaAs [222], InP [265], InGaAs [266], GaN and ZnO [267]

$$q\mathbf{v}\cdot\mathbf{E} = \frac{\hbar\omega_{\mathrm{LO}}}{\tau}\,, \tag{8.32}$$

where τ is the typical relaxation time constant for LO phonon emission. Together with (8.3) we find (8.32) except for the prefactor, which is close to 1. The exact prefactor results from a more exact quantum-mechanical treatment. For Ge the drift-saturation velocity at room temperature is 6×10^6 cm/s, for Si it is 1×10^7 cm/s (Fig. 8.9a).

In GaAs there is a maximum drift velocity of about 2×10^7 cm/s and following a reduction in velocity with increasing field (1.2×10^7 cm/s at 10 kV/cm, 0.6×10^7 cm/s at 200 kV/cm) as shown in Fig. 8.9a. This regime is called *negative differential resistivity* (NDR). This phenomenon can be used in oscillators, e.g. the Gunn diode. The effect occurs in a multivalley band structure (see Fig. 8.10), e.g. in GaAs or InP, when the carrier energy is high enough to scatter (Fig. 8.10c,d) from the Γ minimum (small mass, $m^* = 0.065\,m_0$, high mobility $\mu \approx 8000\,\mathrm{cm^2/V\,s}$) into the L minimum (large mass, $m^* = 1.2\,m_0$, low mobility $\mu \approx 100\,\mathrm{cm^2/V\,s}$).

The temperature dependence of the saturation velocity is shown in Fig. 8.11. With increasing temperature the saturation velocity decreases since the coupling with the lattice becomes stronger.

8.5.2 Velocity Overshoot

When the electric field is switched on, the carriers are at first in the Γ minimum (Fig. 8.10a). Only after a few scattering processes are they scattered into the L minimum. This means that in the first moments transport occurs with the higher mobility of the lowest minimum (Fig. 8.10e). The velocity is then larger than the (steady-state) saturation velocity in a dc field. This phenomenon is called *velocity overshoot* and is a purely dynamic effect (Fig. 8.12) on a sub-ps time-scale.

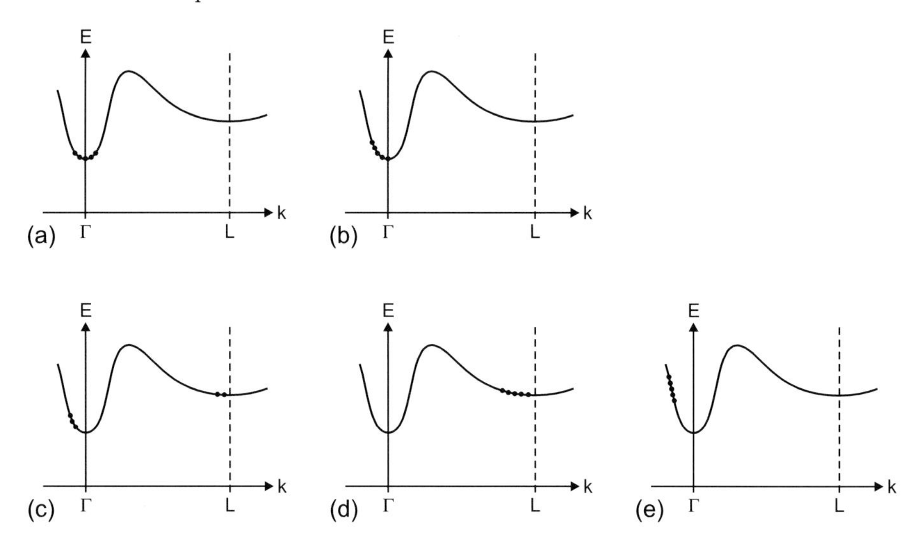

Fig. 8.10. Charge-carrier distribution in a multivalley band structure (e.g. GaAs, InP) for (**a**) zero, (**b**) small ($E < E_a$), (**c**) intermediate and (**d**) large ($E > E_b$) field strength. The situation shown in (**e**) is reached temporarily during velocity overshoot (see also Fig. 8.12)

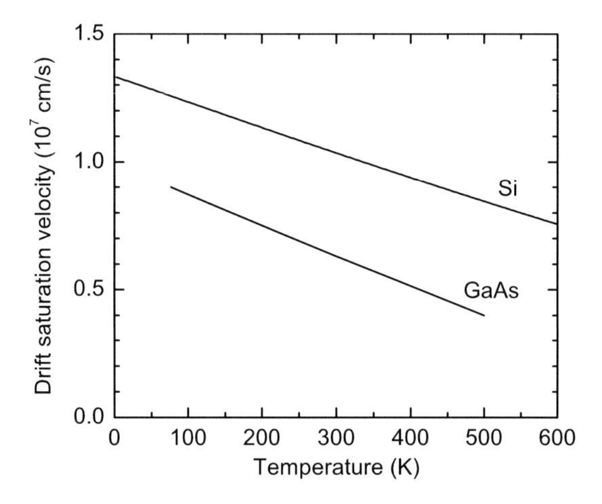

Fig. 8.11. Temperature dependence of the saturation velocity for Si (following $v_s = v_{s0}(1 + 0.8\exp(T/600\,K))^{-1}$ with $v_{s0} = 2.4 \times 10^7$ cm/s from [263]) and GaAs [222, 268, 269]

8.5.3 Impact Ionization

If the energy gain in the field is large enough to generate an electron–hole pair, the phenomenon of impact ionization occurs. The energy is $\propto v^2$. Momentum and energy conservation apply. Thus, at small energies (close to the threshold for impact ionization) the vectors are short and collinear to fulfill momentum

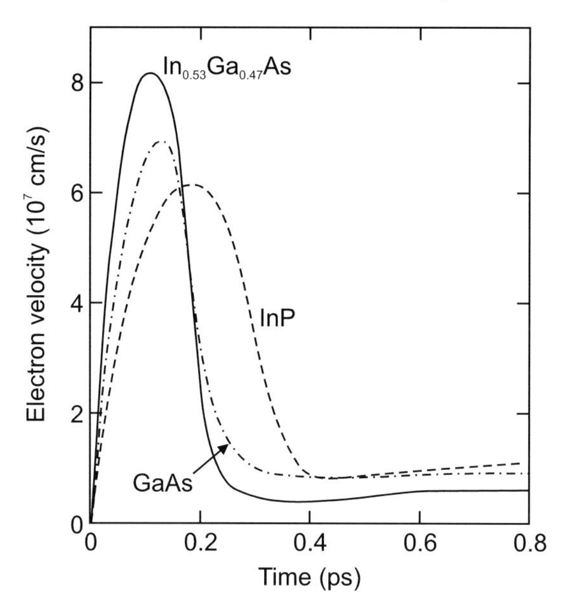

Fig. 8.12. Time dependence of the electron velocity at room temperature upon a step-like electric field (40 kV/cm) for GaAs (*dash-dotted line*), InP (*dashed line*) and $In_{0.53}Ga_{0.47}As$ (*solid line*)

conservation. At higher energy, larger angles between the velocity vectors of the impact partners can also occur. If the process is started by an electron (Fig. 8.13a) the threshold energy is given by

$$E_{\mathrm{e}}^{\mathrm{thr}} = \frac{2m_{\mathrm{e}} + m_{\mathrm{hh}}}{m_{\mathrm{e}} + m_{\mathrm{hh}}} E_{\mathrm{g}} \; . \tag{8.33}$$

The threshold for impact ionization triggered by a s-o hole (shown schematically in Fig. 8.13b) is

$$E_{\mathrm{h}}^{\mathrm{thr}} = \left[1 + \frac{m_{\mathrm{so}}(1 - \Delta_0/E_{\mathrm{g}})}{2m_{\mathrm{hh}} + m_{\mathrm{e}} - m_{\mathrm{so}}} \right] E_{\mathrm{g}} \; . \tag{8.34}$$

The generation rate G of electron–hole pairs during impact ionization is given by

$$G = \alpha_{\mathrm{n}} n v_{\mathrm{n}} + \alpha_{\mathrm{p}} p v_{\mathrm{p}} \; , \tag{8.35}$$

where α_{n} is the electron ionization coefficient. It describes the generation of electron–hole pairs per incoming electron per unit length. α_{p} denotes the hole ionization coefficient. The coefficients depend strongly on the applied electric field. They are shown in Fig. 8.14. They also depend on the crystallographic direction. The ionization rate increases with decreasing bandgap.

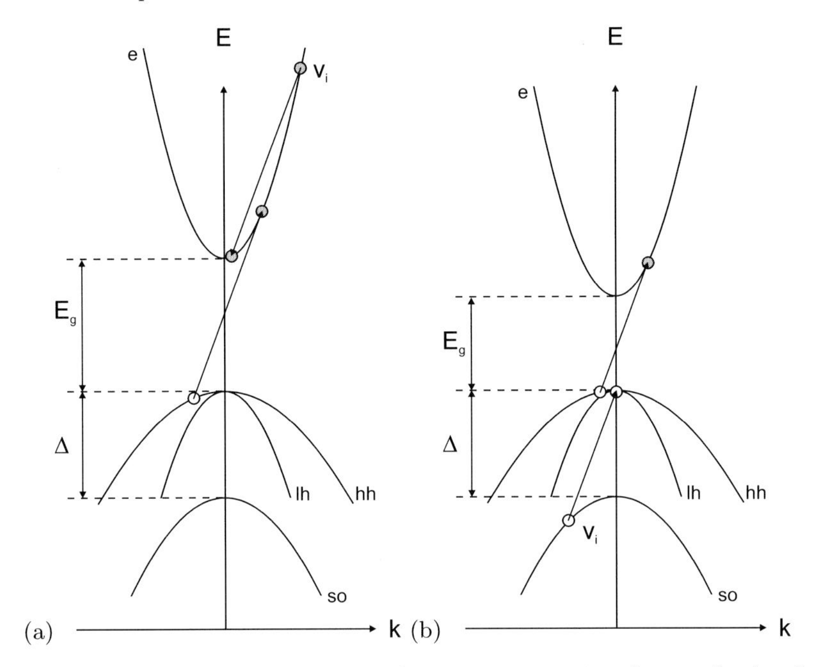

Fig. 8.13. Electron and hole transitions for impact ionization close to the threshold energy. Ionization is triggered by (**a**) an electron and (**b**) a split-off hole of velocity v_i

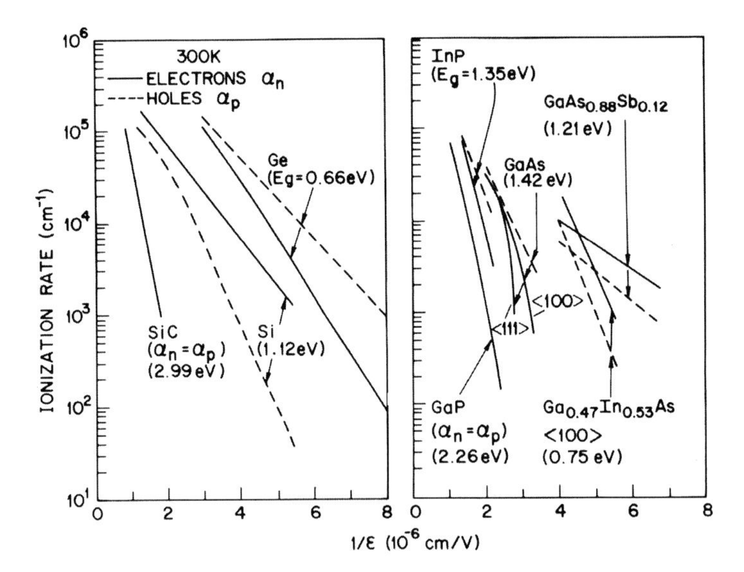

Fig. 8.14. Impact ionization rates for electrons and holes as a function of the inverse electric field for Si, Ge, GaAs and some other semiconductors at 300 K. Reprinted with permission from [183], ©1981 Wiley

8.6 High-Frequency Transport

The above consideration pertained to dc (or slowly varying) fields. Now, we consider an ac field. It accelerates the carriers but at the same time the dissipative force in the relaxation-time approximation is present, i.e.

$$m^* \dot{\mathbf{v}} = -e\mathbf{E} - m^* \frac{\mathbf{v}}{\tau} \,. \tag{8.36}$$

For a harmonic field $\mathcal{E} \propto \exp(-\mathrm{i}\omega t)$ the complex conductivity ($\mathbf{j} = \sigma \mathbf{E} = nq\mathbf{v}$) is

$$\sigma = \frac{ne^2\tau}{m^*} \frac{1}{1 - \mathrm{i}\omega\tau} = \frac{ne^2}{m^*} \frac{\mathrm{i}}{\omega + \mathrm{i}\gamma} \,, \tag{8.37}$$

with $\gamma = 1/\tau$ being the damping constant. Splitting into real and imaginary parts yields

$$\sigma = \frac{ne^2\tau}{m^*} \left(\frac{1}{1+\omega^2\tau^2} + \mathrm{i} \frac{\omega\tau}{1+\omega^2\tau^2} \right) \,. \tag{8.38}$$

For small frequencies ($\omega \to 0$) the dc conductivity from (8.5) is recovered, i.e. $\sigma = ne^2\tau/m^*$. For high frequencies ($\omega\tau \gg 1$)

$$\sigma = \frac{ne^2\tau}{m^*} \left(\frac{1}{\omega^2\tau^2} + \mathrm{i} \frac{1}{\omega\tau} \right) \,. \tag{8.39}$$

8.7 Diffusion

A gradient of a particle concentration n leads to a particle current proportional to $-\nabla n$. This diffusion law (Fick's law) corresponds microscopically to a random walk. The gradients of the semiconductor carrier densities ∇n or ∇p thus lead to electron and hole currents, respectively:

$$\mathbf{j}_\mathrm{n} = eD_\mathrm{n} \nabla n \tag{8.40a}$$

$$\mathbf{j}_\mathrm{p} = -eD_\mathrm{p} \nabla p \,. \tag{8.40b}$$

The coefficients D_n and D_p are called the electron and hole diffusion coefficient, respectively. Thus the total electron and hole currents in the presence of an electric field $\mathbf{E}$ and diffusion are

$$\mathbf{j}_\mathrm{n} = -e\mu_\mathrm{n} n\mathbf{E} + eD_\mathrm{n} \nabla n \tag{8.41a}$$

$$\mathbf{j}_\mathrm{p} = e\mu_\mathrm{p} p\mathbf{E} - eD_\mathrm{p} \nabla p \,. \tag{8.41b}$$

This relation can also be deduced more generally from the gradient of the Fermi level as

$$\mathbf{j}_{\mathrm{n}} = -e\mu_{\mathrm{n}} n\mathbf{E} - n\mu_{\mathrm{n}} \nabla E_{\mathrm{F}} \tag{8.42a}$$

$$\mathbf{j}_{\mathrm{p}} = e\mu_{\mathrm{p}} p\mathbf{E} - p\mu_{\mathrm{p}} \nabla E_{\mathrm{F}} \,. \tag{8.42b}$$

Using (7.8) and (7.9) for the concentrations (valid also in the case of degeneracy) and using $\mathrm{d}F_j(x)/\mathrm{d}x = F_{j-1}(x)$ we obtain

$$\mathbf{j}_{\mathrm{n}} = -e\mu_{\mathrm{n}} n\mathbf{E} - kT\mu_{\mathrm{n}} \frac{F_{1/2}(\eta)}{F_{-1/2}(\eta)} \nabla n \tag{8.43a}$$

$$\mathbf{j}_{\mathrm{p}} = e\mu_{\mathrm{p}} p\mathbf{E} - kT\mu_{\mathrm{p}} \frac{F_{1/2}(\zeta)}{F_{-1/2}(\zeta)} \nabla p \,, \tag{8.43b}$$

with $\eta = (E_{\mathrm{F}} - E_{\mathrm{C}})/kT$ and $\zeta = -(E_{\mathrm{F}} - E_{\mathrm{V}})/kT$. If the prefactor of the density gradient is identified as the diffusion coefficient we find the so-called 'Einstein relations':

$$D_{\mathrm{n}} = -\frac{kT}{e} \mu_{\mathrm{n}} \frac{F_{1/2}(\eta)}{F_{-1/2}(\eta)} \tag{8.44a}$$

$$D_{\mathrm{p}} = \frac{kT}{e} \mu_{\mathrm{p}} \frac{F_{1/2}(\zeta)}{F_{-1/2}(\zeta)} \,. \tag{8.44b}$$

In the case of nondegeneracy, i.e. when the Fermi level is within the bandgap and not closer than about $4kT$ to the band edges, the equation simplifies to $D = \frac{kT}{q}\mu$, i.e.

$$D_{\mathrm{n}} = -\frac{kT}{e} \mu_{\mathrm{n}} \tag{8.45a}$$

$$D_{\mathrm{p}} = \frac{kT}{e} \mu_{\mathrm{p}} \,. \tag{8.45b}$$

In this case, (8.41a,b) read

$$\mathbf{j}_{\mathrm{n}} = -e\mu_{\mathrm{n}} n\mathbf{E} - kT\mu_n \nabla n \tag{8.46a}$$

$$\mathbf{j}_{\mathrm{p}} = e\mu_{\mathrm{p}} p\mathbf{E} - kT\mu_p \nabla p \,. \tag{8.46b}$$

We recall that both diffusion coefficients are positive numbers, since μ_{n} is negative. Generally, the diffusion coefficient depends on the density. A Taylor series of the Fermi integral yields

$$D_{\mathrm{n}} = -\frac{kT}{e} \mu_{\mathrm{n}} \left[1 + 0.35355 \left(\frac{n}{N_{\mathrm{C}}} \right) - 9.9 \times 10^{-3} \left(\frac{n}{N_{\mathrm{C}}} \right)^2 + \ldots \right] \,. \tag{8.47}$$

8.8 Continuity Equation

The balance equation for the charge is called the continuity equation. The temporal change of the charge in a volume element is given by the divergence of the current and any source (generation rate G), e.g. an external excitation, or drain (recombination rate U). Details about recombination mechanisms are discussed in Chap. 10. Thus, we have

$$\frac{\partial n}{\partial t} = G_{\mathrm{n}} - U_{\mathrm{n}} - \frac{1}{q}\nabla \cdot \mathbf{j}_{\mathrm{n}} = G_{\mathrm{n}} - U_{\mathrm{n}} + \frac{1}{e}\nabla \cdot \mathbf{j}_{\mathrm{n}} \tag{8.48a}$$

$$\frac{\partial p}{\partial t} = G_{\mathrm{p}} - U_{\mathrm{p}} - \frac{1}{e}\nabla \cdot \mathbf{j}_{\mathrm{p}} \,. \tag{8.48b}$$

In the case of nondegeneracy we find, using (8.41ab)

$$\frac{\partial n}{\partial t} = G_{\mathrm{n}} - U_{\mathrm{n}} - \mu_{\mathrm{n}} n \nabla \cdot \mathbf{E} - \mu_{\mathrm{n}} \mathcal{E} \nabla n + D_{\mathrm{n}} \Delta n \tag{8.49a}$$

$$\frac{\partial p}{\partial t} = G_{\mathrm{p}} - U_{\mathrm{p}} - \mu_{\mathrm{p}} p \nabla \cdot \mathbf{E} - \mu_{\mathrm{p}} \mathcal{E} \nabla p + D_{\mathrm{p}} \Delta p \,. \tag{8.49b}$$

In the case of zero electric field these read

$$\frac{\partial n}{\partial t} = G_{\mathrm{n}} - U_{\mathrm{n}} + D_{\mathrm{n}} \Delta n \tag{8.50a}$$

$$\frac{\partial p}{\partial t} = G_{\mathrm{p}} - U_{\mathrm{p}} + D_{\mathrm{p}} \Delta p \,, \tag{8.50b}$$

and if the stationary case also applies:

$$D_{\mathrm{n}} \Delta n = -G_{\mathrm{n}} + U_{\mathrm{n}} \tag{8.51a}$$

$$D_{\mathrm{p}} \Delta p = -G_{\mathrm{p}} + U_{\mathrm{p}} \,. \tag{8.51b}$$

8.9 Heat Conduction

We consider here the heat transport [270] due to a temperature gradient. The heat flow $\mathbf{q}$, i.e. energy per unit area per time in the direction $\hat{\mathbf{q}}$, is proportional to the local gradient of temperature. The proportionality constant κ is called the heat conductivity,

$$\mathbf{q} = -\kappa \nabla T \,. \tag{8.52}$$

In crystals, the heat conductivity can depend on the direction and thus κ is generally a tensor of rank 2. In the following, κ will be considered as a scalar quantity. The quite generally valid Wiedemann–Franz law connects the thermal and electrical conductivities

$$\kappa = \frac{\pi^2}{3} \left(\frac{k}{e}\right)^2 T \sigma \,. \tag{8.53}$$

The balance (continuity) equation for the heat energy Q is

$$\nabla \cdot \mathbf{q} = -\frac{\partial Q}{\partial t} = -\rho C \frac{\partial T}{\partial t} + A \, , \tag{8.54}$$

where ρ denotes the density of the solid and C the heat capacity. A denotes a source or drain of heat, e.g. an external excitation. Combining (8.52) and (8.54), we obtain the equation for heat conductivity

$$\Delta T = \frac{\rho C}{\kappa} \frac{\partial T}{\partial t} - \frac{A}{\kappa} \, , \tag{8.55}$$

which simply reads $\Delta T = 0$ for a stationary situation without sources.

The random mixture of various isotopes in natural elements represents a perturbation of the perfectly periodic lattice. Since the mass of the nuclei varies, in particular lattice vibrations will be perturbed. Thus we expect an effect on the heat conductivity. In Fig. 8.15, the thermal conductivity of crystals from natural Ge and enriched ^{74}Ge are compared [272], the latter having, as expected, the higher heat conductivity, i.e. less scattering. The thermal conductivity of isotopically pure ^{28}Si thin films has been measured to be 60% greater than natural silicon at room temperature and at least 40% greater at 100°C, a typical chip operating temperature [273, 274].

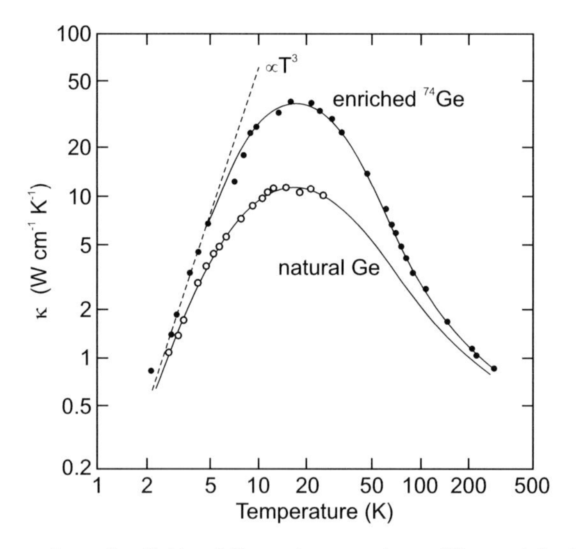

Fig. 8.15. Thermal conductivity of Ge vs. temperature. The enriched Ge consists of 96% ^{74}Ge while the natural isotope mix is 20% ^{70}Ge, 27% ^{72}Ge, 8% ^{73}Ge, 27% ^{74}Ge and 8% ^{76}Ge. The *dashed line* shows the $\kappa \propto T^3$ dependence at low temperatures (Debye's law). Adapted from [272]

8.10 Coupled Heat and Charge Transport

The standard effect of coupled charge and heat transport is that a current heats its conductor via Joule heating. However, more intricate use of thermoelectric effects can also be employed to cool certain areas of a device. For further details see [271].

For the analysis of coupled charge and heat transport we first sum the electric field and the concentration gradient to a new field $\hat{\mathbf{E}} = \mathbf{E} + \nabla E_{\mathrm{F}}/e$. Then, the heat flow and charge current are

$$\mathbf{j} = \sigma \hat{\mathbf{E}} + L \nabla T \tag{8.56}$$

$$\mathbf{q} = M \hat{\mathbf{E}} + N \nabla T , \tag{8.57}$$

where $\hat{\mathbf{E}}$ and ∇T are the stimulators for the currents. From the experimental point of view there is interest to express the equations in $\mathbf{j}$ and ∇T since these quantities are measurable. With new coefficients they read

$$\hat{\mathbf{E}} = \rho \mathbf{j} + S \nabla T \tag{8.58}$$

$$\mathbf{q} = \Pi \mathbf{j} - \kappa \nabla T , \tag{8.59}$$

where ρ, S and Π are the specific resistance, thermoelectric power and Peltier coefficient (transported energy per unit charge), respectively. The relations with the coefficients σ, L, M, and N are given by

$$\rho = \frac{1}{\sigma} \tag{8.60a}$$

$$S = -\frac{L}{\sigma} \tag{8.60b}$$

$$\Pi = \frac{M}{\sigma} \tag{8.60c}$$

$$\kappa = \frac{ML}{\sigma} - N . \tag{8.60d}$$

8.10.1 Seebeck Effect

Assume a semiconductor with a temperature gradient and in an open circuit, i.e. $\mathbf{j} = 0$. Then a field $\hat{\mathbf{E}} = S \nabla T$ will arise. This effect is called the thermoelectric or Seebeck effect. The voltage can be measured and used to determine the temperature at one end if the temperature at the other end is known. Electrons and holes have different sign of the thermoelectric coefficient (see Fig. 8.16). A famous relation from irreversible thermodynamics connects it to the Peltier coefficient via

$$S = \frac{\Pi}{T} . \tag{8.61}$$

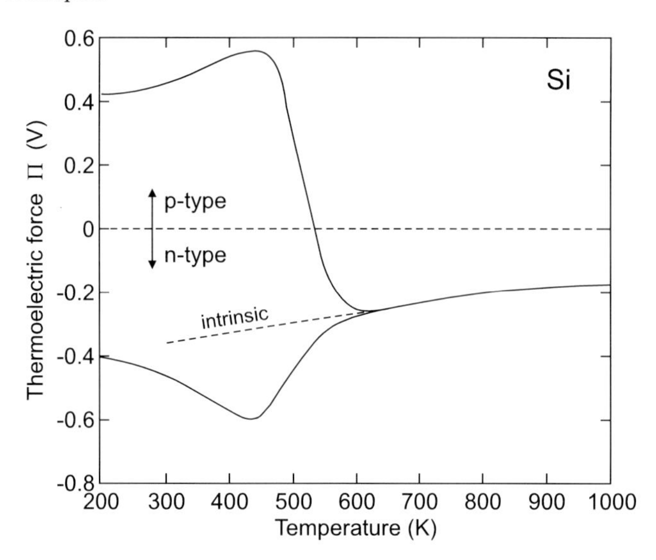

Fig. 8.16. Thermoelectric force Π of n- and p-silicon as a function of temperature. Adapted from [275]

With the Seebeck effect the conductivity type of a semiconductor can be determined. If the cold substrate is grounded, the sign of the voltage at a hot solder tip pressed (carefully) on the surface of the semiconductor yields the conductivity type, n-type (p-type) for a negative (positive) voltage. However, the semiconductor should not be heated so strongly that intrinsic conduction arises.

8.10.2 Peltier Effect

In a semiconductor with a temperature gradient a current flow will be allowed now (short circuit). The current leads via the charge transport also to a heat (or energy) transport. This effect is called the Peltier effect. The Peltier coefficient is negative (positive) for electrons (holes). The total amount of energy P that is transported consists of the generation term and the loss due to transport:

$$P = \mathbf{j} \cdot \hat{\mathbf{E}} - \nabla \cdot \mathbf{q} \,. \tag{8.62}$$

With (8.58) and (8.59) we find

$$P = \frac{\mathbf{j} \cdot \mathbf{j}}{\sigma} + S\mathbf{j} \cdot \nabla T - \Pi \nabla \cdot \mathbf{q} + \kappa \Delta T \,. \tag{8.63}$$

The first term is Joule heating, the second term is Thomson heating. The third exists only when carriers are generated or when they recombine. The fourth term is the heat conduction. In the Thomson term $S\mathbf{j} \cdot \nabla T$ heat is generated in an n-type semiconductor if $\mathbf{j}$ and ∇T are in the same direction. This means that electrons that move from the hotter to the colder part transfer energy to the lattice. The effect can be used to construct a thermoelectric cooler, as shown in Fig. 8.17, that generates a temperature difference due to a current flow. For optimal performance σ should be large to prevent excess Joule heating and κ should be small such that the generated temperature difference is not rapidly equalized.

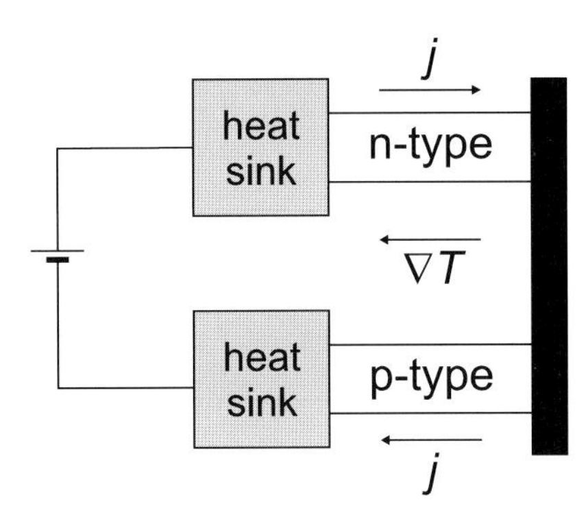

Fig. 8.17. Schematic Peltier cooler. The heat sinks (*grey*) and the cold junction (*black*) on the right are metals that make ohmic contacts with the semiconductors. The current flow is such that electrons move through the n-type semiconductor from right to left

9 Optical Properties

The interaction of semiconductors with light is of decisive importance for photonic and optoelectronic devices as well as for the characterization of semiconductor properties.

9.1 Spectral Regions and Overview

When light hits a semiconductor first reflection, transmission and absorption are considered, as for any dielectric material. The response of the semiconductor largely depends on the photon energy (or wavelength) of the light. In Table 9.1 an overview of the electromagnetic spectrum in the optical range is shown. The energy and wavelength of a photon are related by $E = h\nu = hc/\lambda$, i.e.

$$E = \frac{1240\,\mathrm{eV\,nm}}{\lambda} \tag{9.1}$$

Table 9.1. Spectral ranges with relevance to semiconductor optical properties

Range		wavelengths	energies
deep ultraviolet	DUV	< 250 nm	> 5 eV
ultraviolet	UV	250–400 nm	3–5 eV
visible	VIS	400–800 nm	1.6–3 eV
near infrared	NIR	800 nm–2 µm	0.6–1.6 eV
mid-infrared	MIR	2–20 µm	60 meV–0.6 eV
far infrared	FIR	20–80 µm	1.6–60 meV
THz region	THz	> 80 µm	< 1.6 meV

In Fig. 9.1 a schematic absorption spectrum of a semiconductor is depicted. The transition of electrons from the valence to the conduction band is at the bandgap energy. The bandgaps of Si, Ge, GaAs, InP, InAs, InSb are in the IR, those of AlAs, GaP, AlP, InN in the VIS, those of GaN and ZnO in the UV, MgO and AlN are in the deep UV. The Coulomb correlation of

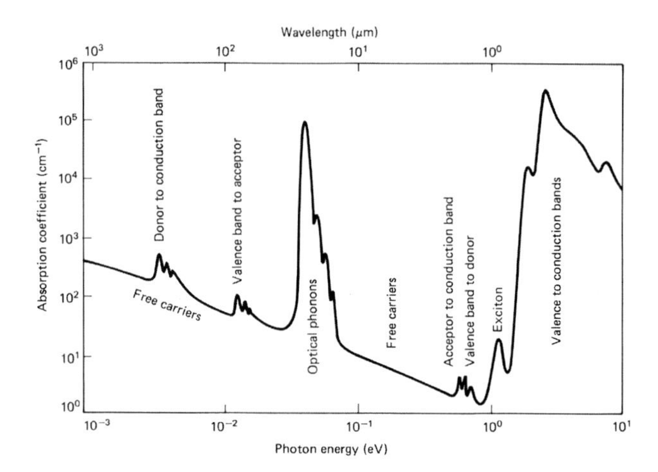

Fig. 9.1. Schematic absorption spectrum of a typical semiconductor. With permission from [278], ©1989 Prentice-Hall

electrons and holes leads to the formation of excitons that leads to absorption below the bandgap. The typical exction binding energy is in the range of 1–100 meV (see Fig. 9.15). Optical transitions from valence-band electrons into donors and from electrons on acceptors into the conduction band lead to band–impurity absorption. In the region from 10–100 meV the interaction with lattice vibrations (phonons) leads to absorption if the phonons are infrared active. Further in the FIR are transitions from impurities to the closest band edge (donor to conduction and acceptor to valence band). A continuous background is due to free-carrier absorption.

9.2 Reflection and Diffraction

From Maxwell's equations and the boundary conditions at a planar interface between two media with different index of refraction for the components of the electric and magnetic fields the laws for reflection and diffraction are derived. We denote the index of refraction as n and also n_{r} in the following. The interface between two media with refractive indices n_1 and n_2 is depicted in Fig. 9.2. In the following we assume first that no absorption occurs.

Snellius' law for the angle of diffraction is

$$n_1 \sin\phi = n_2 \sin\psi \ . \tag{9.2}$$

When the wave enters the denser medium, it is diffracted towards the normal. If the wave propagates into the less-dense medium (reversely to the situation shown in Fig. 9.2), a diffracted wave occurs only up to a critical angle of incidence

$$\sin\phi_{\mathrm{TR}} = \frac{n_2}{n_1} \ . \tag{9.3}$$

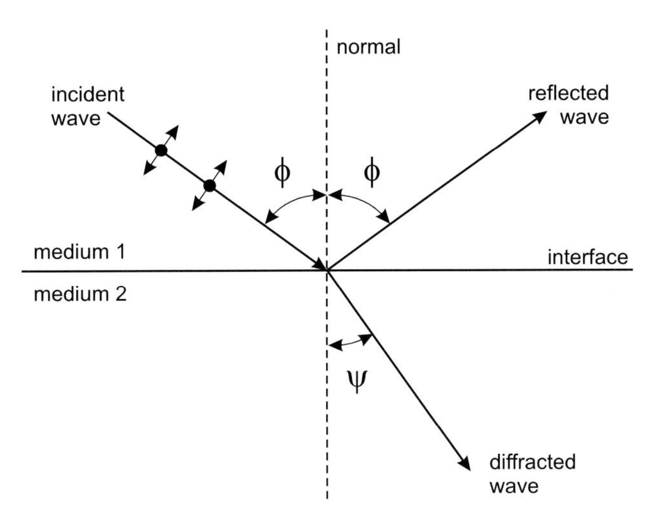

Fig. 9.2. Reflection and diffraction of an electromagnetic wave at the transition between medium '1' and '2', $n_2 > n_1$. The polarization plane is defined by the surface normal and the k-vector of the light (plane of incidence). The parallel ('p') polarized wave (TM-wave, electric field vector oscillates in the plane) is shown as '↔'; perpendicular ('s') polarization (TE-wave, electric field vector is perpendicular to plane) is depicted as '·'

For larger angles of incidence, total internal reflection occurs and the wave remains in the denser medium. Thus, the angle in (9.3) is called the critical angle for total reflection. For GaAs and air the critical angle is rather small, $\phi_{\mathrm{TR}} = 17.4°$.

The reflectivity depends on the polarization (Fresnel formulas). The index 'p' ('s') denotes parallel polarized/TM (perpendicular polarized/TE) waves.

$$R_{\mathrm{p}} = \left(\frac{\tan(\phi - \psi)}{\tan(\phi + \psi)} \right)^2 \tag{9.4}$$

$$R_{\mathrm{s}} = \left(\frac{\sin(\phi - \psi)}{\sin(\phi + \psi)} \right)^2 . \tag{9.5}$$

The situation for GaAs and air is shown for both polarization directions and unpolarized radiation in Fig. 9.3 for a wave going into and out of the GaAs.

When the reflected and the diffracted wave are perpendicular to each other, the reflectivity of the p-polarized wave is zero. This angle is the Brewster angle ϕ_{B},

$$\tan \phi_{\mathrm{B}} = \frac{n_2}{n_1} . \tag{9.6}$$

If a wave has vertical incidence from vacuum on a medium with index of refraction n_{r}, the reflectivity is given (both polarizations are degenerate) as

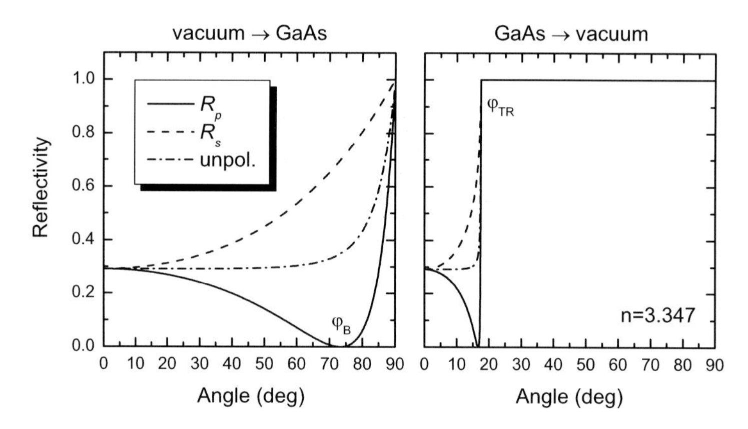

Fig. 9.3. Reflectivity of the GaAs/vacuum interface (close to the bandgap, $n_r = 3.347$) for radiation from vacuum/air and from the GaAs, respectively, as a function of incidence angle and polarization

$$R = \left(\frac{1 - n_r}{1 + n_r}\right)^2 . \tag{9.7}$$

For GaAs, the reflectivity for vertical incidence is 29.2%. If absorption is considered additionally, the reflectivity is (with the complex index of refraction $n^* = n_r + i\kappa$)

$$R = \left(\frac{1 - n^*}{1 + n^*}\right)^2 = \frac{(n_r - 1)^2 + \kappa^2}{(n_r + 1)^2 + \kappa^2} . \tag{9.8}$$

9.3 Electron–Photon Interaction

The absorption process is quantum mechanically described by the coupling of electrons and photons. The process is described with time-dependent perturbation theory. If $H' = \mathbf{H}_{em}$ is the perturbation operator (electromagnetic field), the transition probability per time w_{fi} for electrons from (unperturbed) state 'i' (initial) to state 'f' (final) is given (with certain approximations) by Fermi's golden rule

$$w_{fi}(\hbar\omega) = \frac{2\pi}{\hbar} \left|H'_{fi}\right|^2 \delta(E_f - E_i - \hbar\omega) , \tag{9.9}$$

where $\hbar\omega$ is the photon energy, E_i (E_f) is the energy of the initial (final) state. H'_{fi} is the matrix element

$$H'_{fi} = \langle \Psi_f \left|\mathbf{H}'\right| \Psi_i \rangle , \tag{9.10}$$

where Ψ_i (Ψ_f) are the wavefunctions of the unperturbed initial (final) state.

$\mathbf{A}$ is the vector potential for the electromagnetic field, i.e. $\mathbf{E} = -\dot{\mathbf{A}}$, $\mu\mathbf{H} = \nabla \times \mathbf{A}$, $\nabla \cdot \mathbf{A} = 0$ (Coulomb gange). The Hamiltonian of an electron in the electromagnetic field is

$$\mathbf{H} = \frac{1}{2m} \left(\hbar\mathbf{k} - q\mathbf{A}\right)^2 . \tag{9.11}$$

When terms in $\mathbf{A}^2$ are neglected (i.e. two-photon processes), the perturbation Hamiltonian is thus

$$\mathbf{H}_{\mathrm{em}} = -\frac{q}{m}\mathbf{A} \cdot \mathbf{p} = \frac{\mathrm{i}q\hbar}{m}\mathbf{A} \cdot \nabla \approx q\mathbf{r} \cdot \mathbf{E} . \tag{9.12}$$

The latter approximation is valid for small wavevectors of the electromagnetic wave and is termed the *electric dipole approximation*.

In order to calculate the dielectric function of the semiconductor from its band structure we assume that $\mathbf{A}$ is weak and we can apply (9.9). The transition probability R for the photon absorption rate at photon energy $\hbar\omega$ is then given by[1]

$$R(\hbar\omega) = \frac{2\pi}{\hbar} \int_{\mathbf{k}_{\mathrm{c}}} \int_{\mathbf{k}_{\mathrm{v}}} \left|\langle c|\mathbf{H}_{\mathrm{em}}|v\rangle\right|^2 \delta\left(E_{\mathrm{c}}(\mathbf{k}_{\mathrm{c}}) - E_{\mathrm{v}}(\mathbf{k}_{\mathrm{v}}) - \hbar\omega\right) \mathrm{d}\mathbf{k}_{\mathrm{c}} \mathrm{d}\mathbf{k}_{\mathrm{v}} , \tag{9.13}$$

with the Bloch functions $|c\rangle$ and $|v\rangle$ of the conduction and valence band, respectively, as given in (6.39b).

The vector potential is written as $\mathbf{A} = A\hat{\mathbf{e}}$ with a unit vector $\hat{\mathbf{e}}$ parallel to $\mathbf{A}$. The amplitude is connected to the electric-field amplitude $\mathcal{E}$ via

$$A = -\frac{\mathcal{E}}{2\omega} \left[\exp\left(\mathrm{i}(\mathbf{qr} - \omega t)\right) + \exp\left(-\mathrm{i}(\mathbf{qr} - \omega t)\right)\right] . \tag{9.14}$$

In the electric-dipole approximation the momentum conservation $\mathbf{q}+\mathbf{k}_{\mathrm{v}} = \mathbf{k}_{\mathrm{c}}$, $\mathbf{q}$ being the momentum of the light wave is approximated by $\mathbf{k}_{\mathrm{v}} = \mathbf{k}_{\mathrm{c}}$. The matrix element is then given by

$$\left|\langle c|\mathbf{H}_{\mathrm{em}}|v\rangle\right|^2 = \frac{e^2 \left|A\right|^2}{m^2} \left|\langle c|\hat{\mathbf{e}} \cdot \mathbf{p}|v\rangle\right|^2 , \tag{9.15}$$

with

$$\langle c\,|\hat{\mathbf{e}} \cdot \mathbf{p}|v\rangle|^2 = \frac{1}{3}|\mathbf{p}_{\mathrm{cv}}|^2 = M_{\mathrm{b}}^2 , \tag{9.16}$$

and the momentum matrix element $\mathbf{p}_{\mathrm{cv}}$ given in (6.38). A $\mathbf{k}$-independent matrix element $|\mathbf{p}_{\mathrm{cv}}|^2$ is often used as an approximation. In Fig. 9.4 the matrix elements for valence to conduction band transitions in GaN are shown as a function of $\mathbf{k}$.

[1] Here we assume that the valence-band states are filled and the conduction-band states are empty. If the conduction-band states are filled and the valence-band states are empty, the rate is that of stimulated emission.

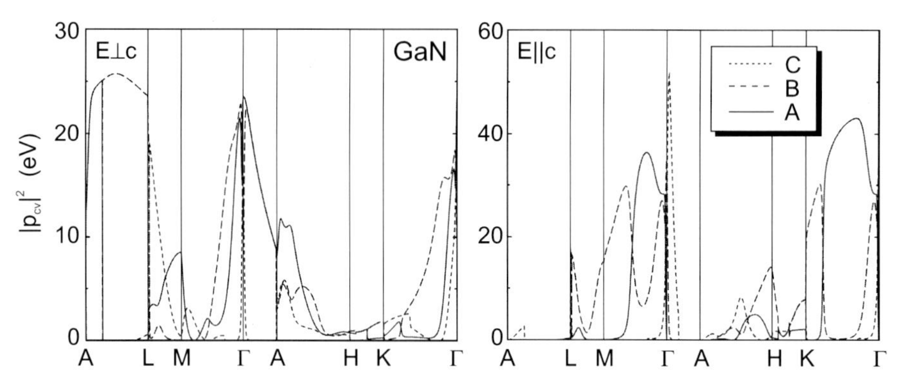

Fig. 9.4. Theoretical momentum matrix elements $|p_{\mathrm{cv}}|^2$ along high-symmetry directions in the Brillouin zone (cf. Fig. 3.31d) for transitions between valence and conduction bands in GaN and light polarized perpendicular (*left panel*) and parallel (*right panel*) to the *c*-axis. The transitions are A: Γ_9(A)→ $\Gamma_{7\mathrm{c}}$, B: Γ_7(B)→ $\Gamma_{7\mathrm{c}}$, C: Γ_7(C)→ $\Gamma_{7\mathrm{c}}$ (cf. Fig. 6.27). Adapted from [276]

In terms of the electric-field amplitude $\mathcal{E}(\omega)$ the transition probability is

$$R(\hbar\omega) = \frac{2\pi}{\hbar}\left(\frac{e}{m\omega}\right)^2 \left|\frac{\mathcal{E}(\omega)}{2}\right|^2 |\mathbf{p}_{\mathrm{cv}}|^2 \int_{\mathbf{k}} \delta\left(E_{\mathrm{c}}(\mathbf{k}) - E_{\mathrm{v}}(\mathbf{k}) - \hbar\omega\right) \mathrm{d}\mathbf{k}\,. \quad (9.17)$$

If the integration over $\mathbf{k}$ is restricted to those values allowed in unit volume, the power that is lost from the field in unit volume is given by $R\hbar\omega$. The dielectric function $\epsilon = \epsilon_{\mathrm{r}} + \mathrm{i}\epsilon_{\mathrm{i}}$ is given by

$$\epsilon_{\mathrm{i}} = \frac{1}{4\pi\epsilon_0}\left(\frac{2\pi e}{m\omega}\right)^2 |\mathbf{p}_{\mathrm{cv}}|^2 \int_{\mathbf{k}} \delta\left(E_{\mathrm{c}}(\mathbf{k}) - E_{\mathrm{v}}(\mathbf{k}) - \hbar\omega\right) \mathrm{d}\mathbf{k} \quad (9.18a)$$

$$\epsilon_{\mathrm{r}} = 1 + \int_{\mathbf{k}} \frac{e^2}{\epsilon_0 m\omega_{\mathrm{cv}}^2} \frac{2|\mathbf{p}_{\mathrm{cv}}|^2}{m\hbar\omega_{\mathrm{cv}}} \frac{1}{1-\omega^2/\omega_{\mathrm{cv}}^2} \mathrm{d}\mathbf{k}\,, \quad (9.18b)$$

with $\hbar\omega_{\mathrm{cv}} = E_{\mathrm{c}}(\mathbf{k}) - E_{\mathrm{v}}(\mathbf{k})$. The second equation has been obtained via the Kramers–Kronig relations[2] (see Appendix B).

Comparison with (C.7) yields that the oscillator strength of the band–band absorption is given by

$$f = \frac{e^2}{\epsilon_0 m\omega_{\mathrm{cv}}^2} \frac{2|\mathbf{p}_{\mathrm{cv}}|^2}{m\hbar\omega_{\mathrm{cv}}}\,, \quad (9.19)$$

with

$$N_{\mathrm{cv}} = \frac{2|\mathbf{p}_{\mathrm{cv}}|^2}{m\hbar\omega_{\mathrm{cv}}} \quad (9.20)$$

being the classical 'number' of oscillators with the frequency ω_{cv}.

[2] The real and imaginary parts of the dielectric function are generally related to each other via the Kramers–Kronig relations.

9.4 Band–Band Transitions

9.4.1 Joint Density of States

The strength of an allowed optical transitions between conduction and valence bands is, apart from the matrix element (9.15), proportional to the joint density of states (JDOS) $D_{\mathrm{j}}(E_{\mathrm{cv}})$ (cf. (6.63), (6.64) and (9.18a))

$$D_{\mathrm{j}}(E_{\mathrm{cv}}) = 2 \int_{S(\tilde{E})} \frac{\mathrm{d}^2 S}{(2\pi/L)^3} \frac{1}{|\nabla_{\mathbf{k}} E_{\mathrm{cv}}|} , \tag{9.21}$$

where E_{cv} is an abbreviation for $E_{\mathrm{c}}(\mathbf{k}) - E_{\mathrm{v}}(\mathbf{k})$ and $\mathrm{d}^2 S$ is a surface element of the constant energy surface with $\tilde{E} = E_{\mathrm{cv}}$. The spin is assumed to generate doubly degenerate bands and accounts for the prefactor 2. Singularities of the JDOS (*van-Hove singularities* or *critical points*) appear where $\nabla_{\mathbf{k}} E_{\mathrm{cv}}$ vanishes. This occurs when the gradient for both bands is zero or when both bands are parallel. The latter generates particularly large JDOS because the condition is valid at many points in **k**-space.

Generally, the (three-dimensional) energy dispersion $E(\mathbf{k})$ around a three-dimensional critical point (here developed at $\mathbf{k} = 0$) can be written as

$$E(\mathbf{k}) = E(0) + \frac{\hbar^2 k_{\mathrm{x}}^2}{2m_{\mathrm{x}}} + \frac{\hbar^2 k_{\mathrm{y}}^2}{2m_{\mathrm{y}}} + \frac{\hbar^2 k_{\mathrm{z}}^2}{2m_{\mathrm{z}}} . \tag{9.22}$$

The singularities are classified as M_0, M_1, M_2 and M_3 with the index being the number of masses m_i in (9.22) that are negative. M_0 (M_3) describes a minimum (maximum) of the band separation. M_1 and M_2 are saddle points. For a two-dimensional **k**-space there exist M_0, M_1 and M_2 points (minimum, saddle point and maximum, respectively). For a one-dimensional **k**-space there exist M_0 and M_1 points (minimum and maximum, respectively). The functional dependence of the JDOS at the critical points is summarized in Table 9.2. The resulting shape of the dielectric function is visualized in Fig. 9.5.

9.4.2 Direct Transitions

Transitions between states at the band edges at the Γ point are possible (Fig. 9.6). The **k** conservation requires (almost) vertical transitions in the $E(\mathbf{k})$ diagram because the length of the light **k** vector, $k = 2\pi/\lambda$, is much smaller than the size of the Brillouin zone $|k| \leq \pi/a_0$. The ratio of the lengths of the **k** vectors is $\sim 2a_0/\lambda$ and typically about 10^{-3} for NIR wavelengths.

For parabolic bands, i.e. bands described with an energy-independent effective mass, the absorption coefficient is (M_0 critical point)

$$\alpha \propto (E - E_{\mathrm{g}})^{1/2} . \tag{9.23}$$

Table 9.2. Functional dependence of the joint density of states for critical points in 3, 2 and 1 dimensions. E_0 denotes the energy (band separation) at the critical point, C stands for a constant value. The type of critical point is given (min.: minimum, saddle: saddle point, max.: maximum)

Dim.	label	type	D_j for $E < E_0$	D_j for $E > E_0$
3D	M_0	min.	0	$\sqrt{E - E_0}$
	M_1	saddle	$C - \sqrt{E_0 - E}$	C
	M_2	saddle	C	$C - \sqrt{E - E_0}$
	M_3	max.	$\sqrt{E_0 - E}$	0
2D	M_0	min.	0	C
	M_1	saddle	$-\ln(E_0 - E)$	$-\ln(E - E_0)$
	M_2	max.	C	0
1D	M_0	min.	0	$\sqrt{E - E_0}$
	M_1	max.	$\sqrt{E_0 - E}$	0

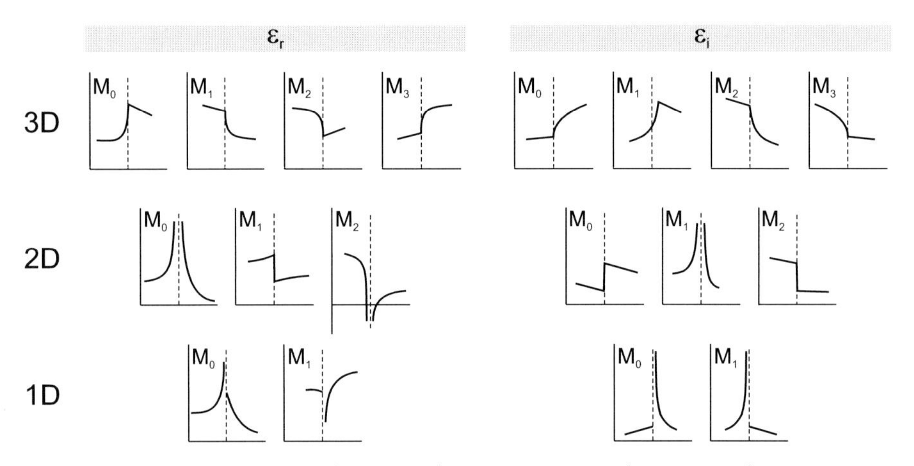

Fig. 9.5. Shape of the real (*left panel*) and imaginary (*right panel*) parts of the dielectric function in the vicinity of critical points in 3, 2 and 1 dimensions (for labels see Table 9.2). The *dashed line* in each graph indicates the energy position of the critical point E_0. Adapted from [277].

The absorption spectrum of GaAs is shown in Fig. 9.7a for photon energies close to the bandgap. The rapid increase, typical for direct semiconductors, is obvious. At low temperatures, however, the absorption lineshape close to the bandgap is dominated by an excitonic feature, discussed in Sect. 9.4.7.

Due to the increasing density of states, the absorption increases with the photon energy (Fig. 9.7c). At 1.85 eV there is a step in the absorption spectrum of GaAs due to the beginning of the contribution of transitions

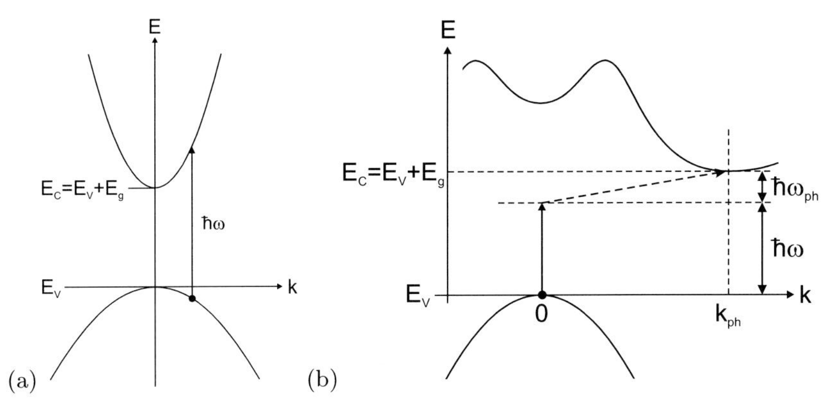

Fig. 9.6. (**a**) Direct optical transition and (**b**) indirect optical transitions between valence and conduction bands. The photon energy is $\hbar\omega$. The indirect transition involves a phonon with energy $\hbar\omega_{\mathrm{ph}}$ and wavevector k_{ph}.

between the s-o hole band and the conduction band (see $E_0 + \Delta_0$ transition in Fig. 9.7e). When bands go parallel, i.e. with the same separation, in the $E(\mathbf{k})$ diagram, the absorption processes contribute at the same transition energy. In this way higher peaks in the absorption spectrum due to the E_1 or E_0' transitions originate as indicated in the band structure in Fig. 9.7e.

The selection rules for transitions from valence to conduction band must take into account the angular momentum and spin states of the wavefunctions. The optical transitions are circularly polarized as shown in Fig. 9.8. A lifting of the energetic degeneracies of these states occurs, e.g. by magnetic fields or spatial confinement.

9.4.3 Indirect Transitions

In an indirect band structure the missing k difference (across the Brillouin zone) between valence- and conduction-band state needs to be provided by a second particle. A phonon can provide this and additionally contributes a small amount of energy $\hbar\omega_{\mathrm{ph}}$. Due to energy conservation this absorption starts already at an energy $E_{\mathrm{g}} - \hbar\omega_{\mathrm{ph}}$ below the bandgap. But the two-particle process is less probable than the direct absorption that only involves the photon. Also there are several processes involving various phonons (or combinations of them). The perturbation calculation yields an absorption coefficient with a quadratic dependence on energy (9.24). However, the strength close to the bandgap is about 10^{-3} smaller than for the direct transition.

$$\alpha \propto (E - E_{\mathrm{g}} - \hbar\omega_{\mathrm{ph}})^2 \,. \tag{9.24}$$

The dependence is shown for Si and GaP in Fig. 9.9. The complicated form is due to the contribution of different phonons.

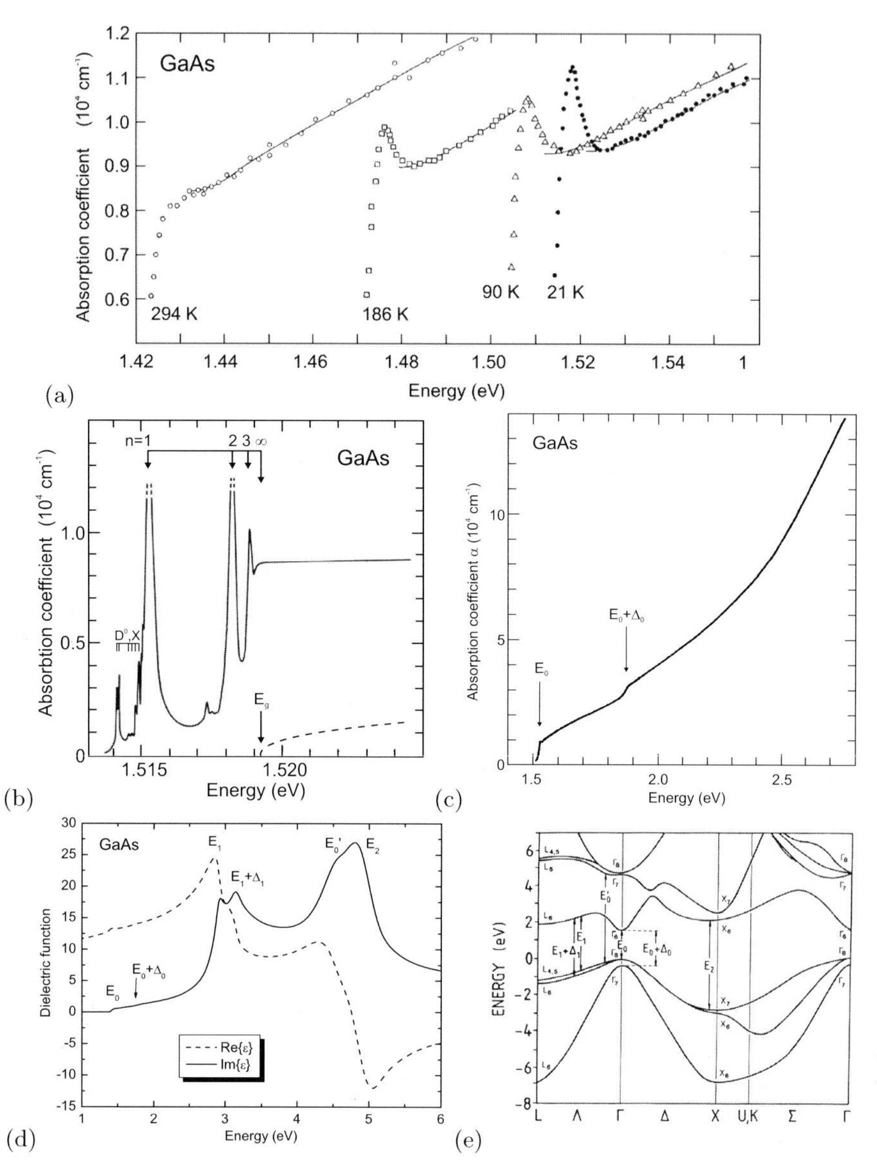

Fig. 9.7. (**a**) Absorption of GaAs close to the bandgap at different temperatures. Adapted from [279]. (**b**) High-resolution absorption spectrum of highly pure GaAs at $T = 1.2\,\mathrm{K}$ in the exciton region. *Dashed line* is theory without excitonic correlation. Adapted from [280]. (**c**) Absorption spectrum of GaAs at $T = 21\,\mathrm{K}$ in the vicinity of the bandgap. Adapted from [279]. (**d**) Complex dielectric function of GaAs at $T = 300\,\mathrm{K}$, *dashed* (*solid*) *line*: real (imaginary) part of dielectric constant. Peak labels relate to transitions shown in part (**e**). (**e**) Band structure of GaAs with bandgap transition (E_0) and higher transitions ($E_0 + \Delta_0$, E_1, $E_1 + \Delta_1$, E_0', and E_2) indicated

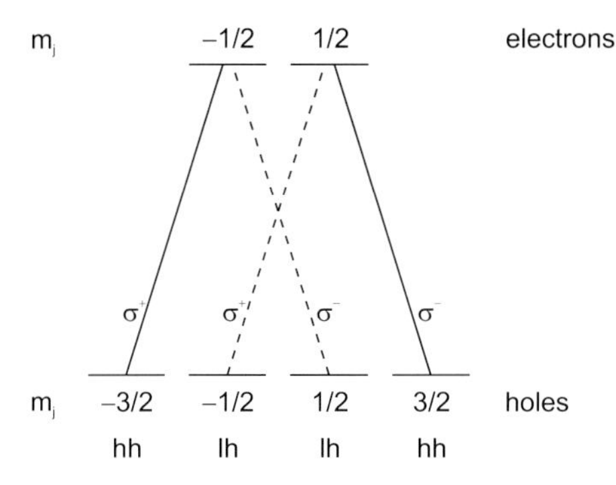

Fig. 9.8. Optical selection rules for band–band transitions in bulk material

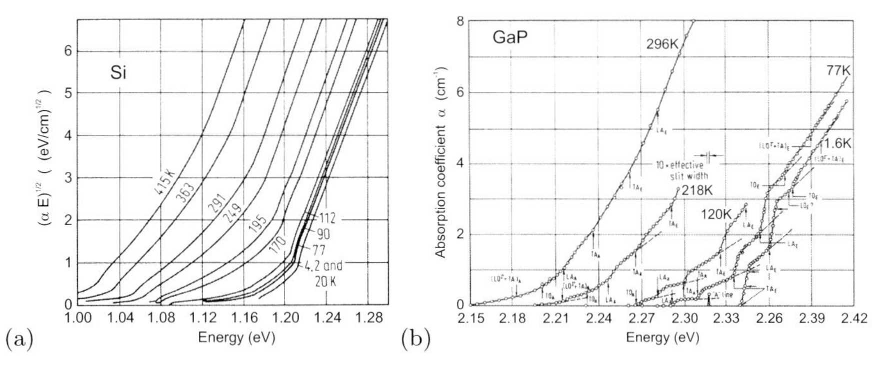

Fig. 9.9. Absorption edge of (**a**) Si and (**b**) GaP at various temperatures. Part (**a**) adapted from [93], based on [281], part (**b**) adapted from [93] based on [282]

We note also that the indirect semiconductors have an optical transition between Γ valence- and conduction-band states. However, this transition is at higher energies than the fundamental bandgap, e.g. for Si ($E_\mathrm{g} = 1.12\,\mathrm{eV}$) at 3.4 eV (see Fig. 6.5a). In Fig. 9.10, the absorption scheme for indirect and direct absorbtion processes starting with an electron at the top of the valence band is shown together with an experimental absorption spectrum for Ge with the direct transition ($\Gamma_8 \rightarrow \Gamma_7$) at 0.89 eV.

In Fig. 9.11, the absorption edge of $BaTiO_3$ is shown. An indirect transition with an increase of (weak) absorption $\propto E^2$ and an indirect gap of $E_\mathrm{i} = 2.66\,\mathrm{eV}$ and a direct transition with an increase of (strong) absorption $\propto E^{1/2}$ and a direct gap of $E_\mathrm{d} = 3.05\,\mathrm{eV}$ are observed. These transitions could be due to holes at the M (indirect gap) and Γ (direct gap) points (see Sect. 6.6.9), respectively.

9.4.4 Urbach Tail

Instead of the ideal $(E - E_\mathrm{g})^{1/2}$ dependence of the direct band-edge absorption, often an exponential tail is observed (see Fig. 9.12). This tail is called

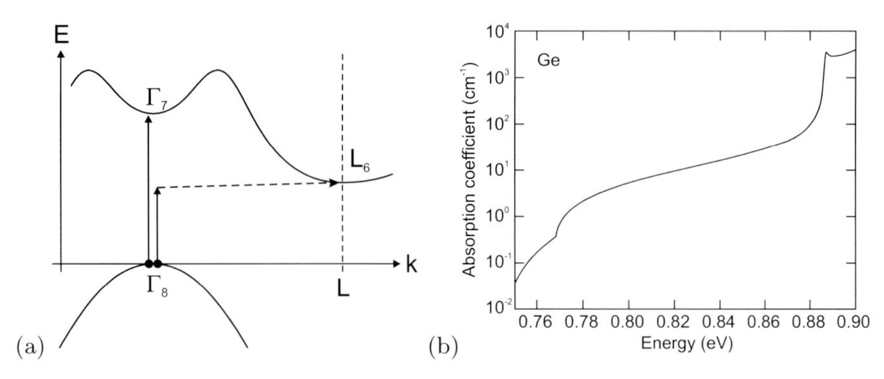

Fig. 9.10. (**a**) Scheme of indirect and direct optical transitions starting at the top of the valence band in Ge. Vertical *solid lines* represent the involved photon, the horizontal *dashed line* the involved phonon. (**b**) Experimental absorption spectrum of Ge ($T = 20\,\mathrm{K}$). Adapted from [281]

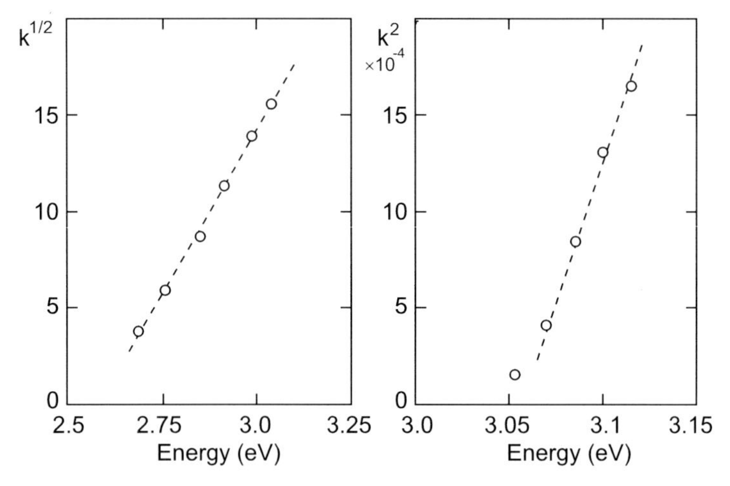

Fig. 9.11. Absorption of $BaTiO_3$ at room temperature. Experimental data (*circles*) from [283] with fits (*dashed lines*) $\propto E^2$ and $\propto E^{1/2}$, respectively

the Urbach tail [284] and follows the functional dependence

$$\alpha(E) = \alpha_{\mathrm{g}} \exp\left(\frac{E - E_{\mathrm{g}}}{E_0}\right) , \tag{9.25}$$

where E_0 is the characteristic width of the absorption edge, the so-called Urbach parameter.

The Urbach tail is attributed to transitions between band tails below the band edges. Such tails can originate from disorder of the perfect crystal, e.g. from defects or doping, and the fluctuation of electronic energy bands due to

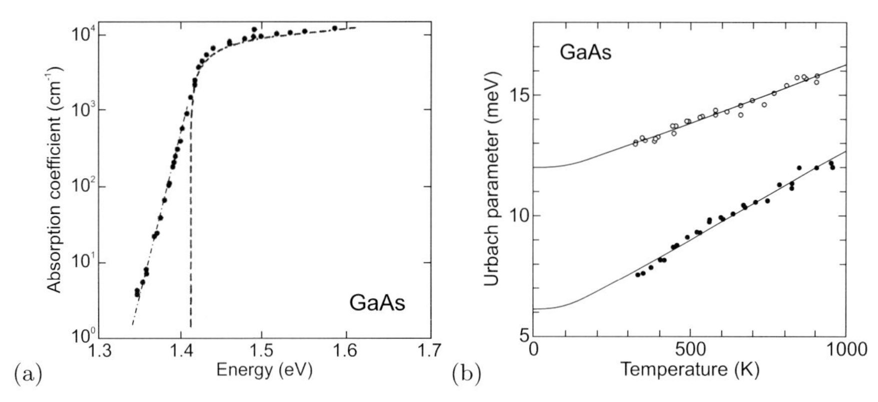

Fig. 9.12. (**a**) Experimental absorption spectrum (*circles*) of GaAs at room temperature on a semilogarithmic plot. The exponential tail below the bandgap is called the Urbach tail (the *dash-dotted* line corresponds to $E_0 = 10.3\,\text{meV}$). The *dashed line* is the theoretical dependence from (9.23). Adapted from [287]. (**b**) Temperature dependence of Urbach parameter E_0 for two GaAs samples. Experimental data for undoped (*solid circles*) and Si-doped ($n = 2 \times 10^{18}\,\text{cm}^{-3}$, *empty circles*) GaAs and theoretical fits (*solid lines*) with one-phonon model. Adapted from [285]

lattice vibrations. The temperature dependence of the Urbach parameter E_0 is thus related to that of the bandgap as discussed in [285, 286].

9.4.5 Intravalence-Band Absorption

Besides optical interband transitions between the valence and the conduction band there are also optical transitions within the valence band. In p-type material holes from the valence-band edge can undergo absorption and end up in the split-off hole band (Fig. 9.13a). Such intravalence-band absorption occurs at photon energies close to Δ_0 as shown in Fig. 9.13b for p-type GaAs:Zn. Also, hh $\rightarrow$ lh transitions are possible.

9.4.6 Amorphous Semiconductors

The sharp features in the dielectric function due to critical points in the band structure of crystalline semiconductors are washed out in amorphous material. As an example the spectra of the imaginary part of the dielectric function for crystalline (trigonal) and amorphous selenium are shown in Fig. 9.14.

9.4.7 Excitons

An electron in the conduction band and a hole in the valence band form a hydrogen-like state due to the mutual Coulomb interaction. Such a state is called an exciton. The center-of-mass motion is separated and has a dispersion

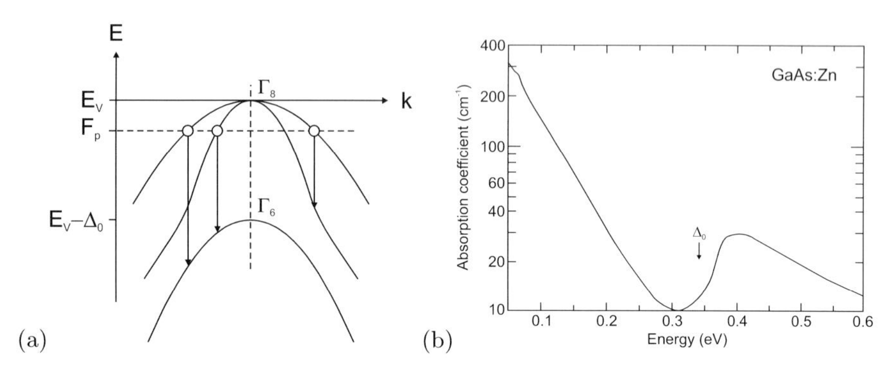

Fig. 9.13. (**a**) Schematic optical transitions within the valence band. Holes (*empty circles*) are shown to start at the hole quasi-Fermi level, transitions are from left to right: 'hh → s-o', 'lh → s-o' and 'hh → lh'. (**b**) Experimental absorption spectrum of GaAs:Zn with $p = 2.7 \times 10^{17}\,\mathrm{cm}^{-3}$ at $T = 84\,\mathrm{K}$. The absorption above the split-off energy Δ_0 is due to the hh/lh → s-o process. Adapted from [288]

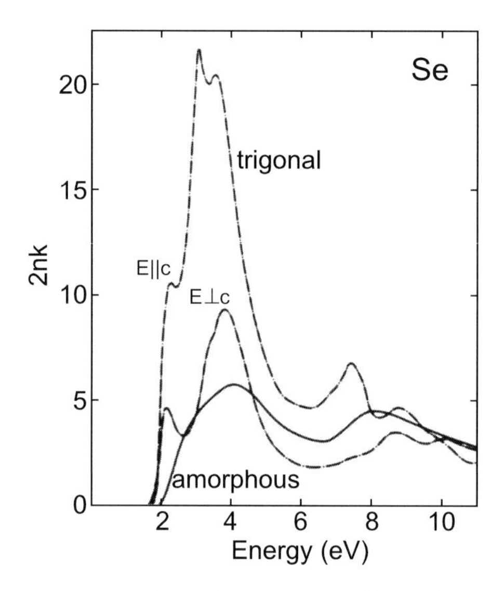

Fig. 9.14. Imaginary part of the dielectric function of amorphous (*solid line*) and crystalline (trigonal) selenium (*dash-dotted lines* for two different polarization directions). From [289]

$E = \frac{\hbar^2}{2M}\mathbf{K}^2$, where $M = m_\mathrm{e} + m_\mathrm{h}$ is the total mass and $\hbar\mathbf{K}$ is the center-of-mass momentum

$$\mathbf{K} = \mathbf{k}_\mathrm{e} + \mathbf{k}_\mathrm{h} \; . \tag{9.26}$$

The relative motion yields hydrogen-like quantized states $E_n \propto n^{-2}$ $(n \geq 1)$:

$$E_X^n = -\frac{m_r^*}{m_0}\frac{1}{\epsilon_r^2}\frac{m_0 e^4}{2(4\pi\epsilon_0\hbar)^2}\frac{1}{n^2}\,, \tag{9.27}$$

where m_r^* denotes the reduced effective mass $m_r^{*-1} = m_e^{*-1} + m_h^{*-1}$. The third factor is the atomic Rydberg energy (13.6 eV). The exciton binding energy $E_X^b = -E_X^1$ is scaled by $\frac{m^*}{m_0}\frac{1}{\epsilon_r^2} \approx 10^{-3}$. A more detailed theory of excitons beyond the simple hydrogen model presented here, taking into account the valence-band structure, can be found in [290] for direct and [291] for indirect cubic and in [292] for wurtzite semiconductors. The exciton binding energies for various semiconductor are shown in Fig. 9.15a.

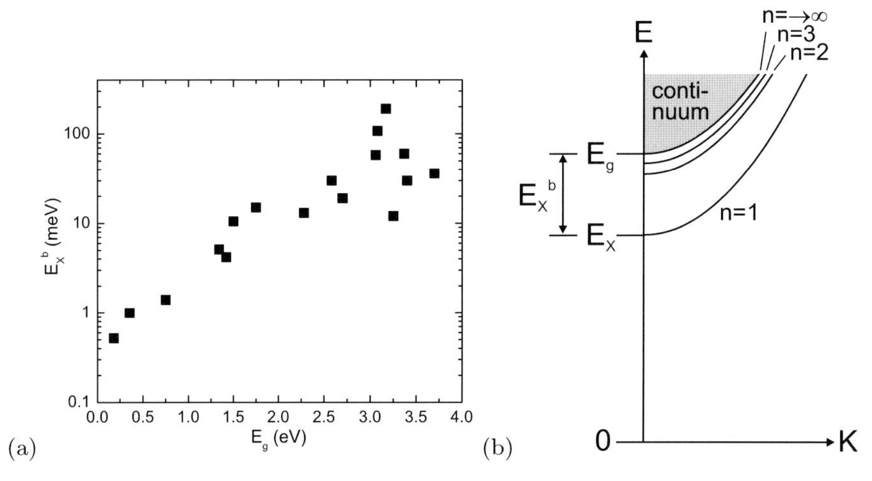

Fig. 9.15. (**a**) Exciton binding energy vs. bandgap for various semiconductors. (**b**) Schematic dispersion of excitonic levels. The K-vector refers to the center-of-mass motion

The radius of the exciton is

$$r_X^n = n^2 \frac{m_0}{m_r^*}\epsilon_r\, a_B\,, \tag{9.28}$$

where $a_B = 0.053$ nm denotes the hydrogen Bohr radius.[3] The Bohr radius of the exciton is $a_X = r_X^1$ (14.6 nm for GaAs). The exciton moves with the center-of-mass **K**-vector through the crystal. The complete dispersion is (see Fig. 9.15b)

$$E = E_g - E_X^n + \frac{\hbar^2}{2M}\mathbf{K}^2\,. \tag{9.29}$$

[3] Cf. (7.25); an electron bound to a donor can be considered as an exciton with an infinite hole mass.

The oscillator strength of the exciton states decays $\propto n^{-3}$. The absorption due to excitons is visible in Fig. 9.7a for GaAs at low temperatures. If inhomogeneities are present, typically only the $n = 1$ transition is seen. However, under special conditions also higher transitions of the exciton Rydberg series are seen (e.g. $n = 2$ and 3 in Fig. 9.7b).

The exciton concept was introduced first for absorption in Cu_2O [293]. The $J = 1/2$ absorption spectrum ('yellow series') is shown in Fig. 9.16. In this particular material both the valence and conduction bands have s character, thus the 1s transition of the exciton is forbidden and the np transitions are observed in normal (one-photon) absorption. Only with two-photon absorption can the s (and d) transitions also be excited.

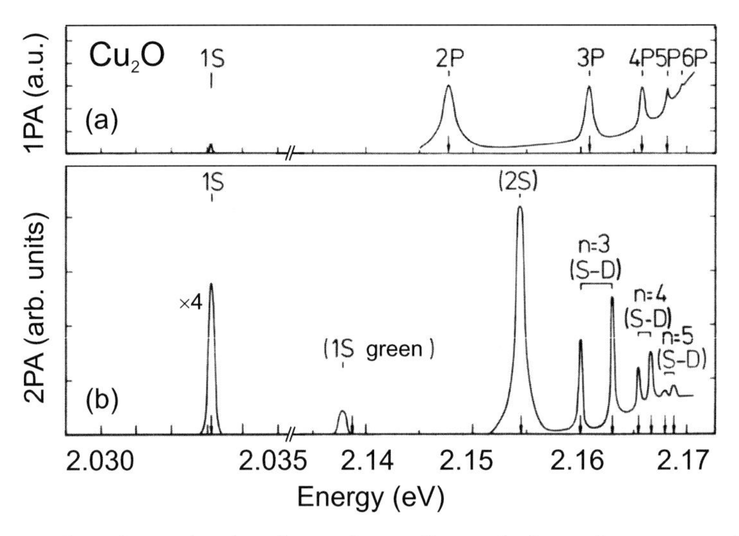

Fig. 9.16. One-photon (*top*) and two-photon (*bottom*) absorption spectra of Cu_2O at $T = 4.2\,\mathrm{K}$. *Arrows* denote theoretical peak positions. Adapted from [294]

The scattering (unbound) states of the exciton [295] for $E > E_g$ contribute to absorption above the bandgap. The factor by which the absorption spectrum is changed is called the Sommerfeld factor. For bulk material it is

$$S(\eta) = \eta \frac{\exp(\eta)}{\sinh(\eta)} , \tag{9.30}$$

with $\eta = \pi \left(\frac{E_X^b}{E - E_g} \right)^{1/2}$. The change of the absorption spectrum due to the Coulomb correlation is shown in Fig. 9.17. There is a continuous absorption between the bound and unbound states. At the bandgap there is a finite absorption ($S(E \to E_g) \to \infty$). The detail to which exciton peaks can be resolved depends on the spectral broadening.

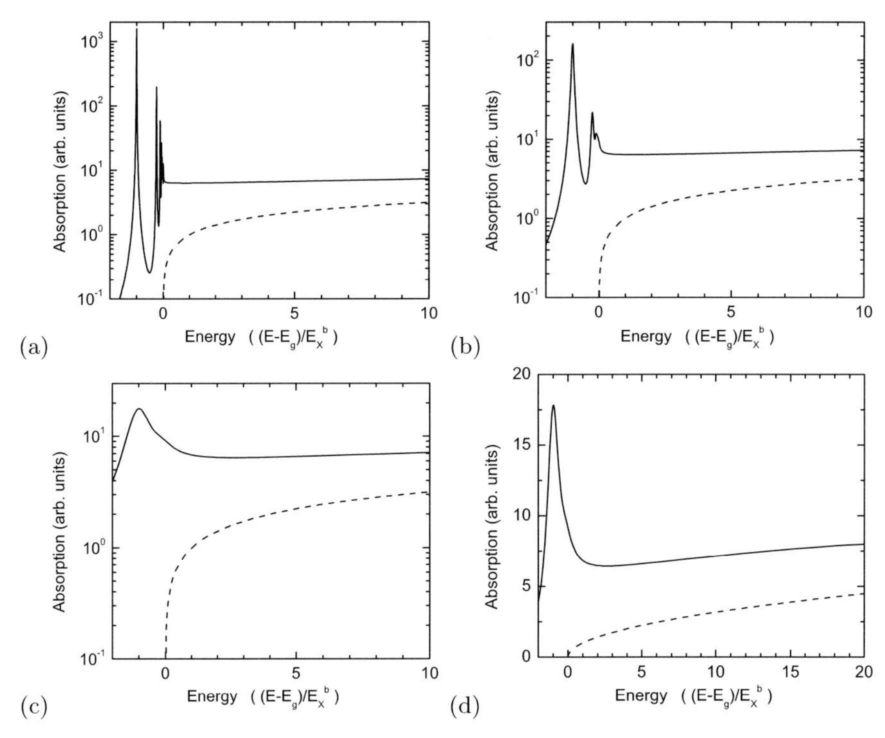

Fig. 9.17. Modification of the absorption edge of a direct transition by excitonic effects for different spectral (Lorentzian) broadening ($\propto (E^2 + \Gamma^2/4)^{-1}$), (**a**) $\Gamma = 0.01 E_X^b$, (**b**) $\Gamma = 0.1 E_X^b$, (**c**) $\Gamma = E_X^b$. (**d**) is (**c**) in linear scale. *Dashed lines* are electron–hole plasma absorption according to (9.23)

In Fig. 9.18 the energy separations of the A-, B-, and C-excitons in GaN are shown [174]. Thus, the ordering of the valence bands depends on the strain state of the semiconductor.

9.4.8 Exciton Polariton

Electrons and holes are particles with spin 1/2. Thus, the exciton can form states with total spin $S = 0$ (para-exciton, singlet) and $S = 1$ (ortho-exciton, triplet). The exchange interaction leads to a splitting of these states, the singlet being the energetically higher. The singlet state splits into the longitudinal and transverse exciton with respect to the orientation of the polarization carried by the Bloch functions and the center-of-mass motion $\mathbf{K}$ of the exciton. Dipole transitions are only possible for singlet excitons (bright excitons). The triplet excitons couple only weakly to the electromagnetic field and are thus also called dark excitons.

The coupling of these states to the electromagnetic field creates new quasiparticles, the exciton polaritons [297]. The dielectric function of the exciton

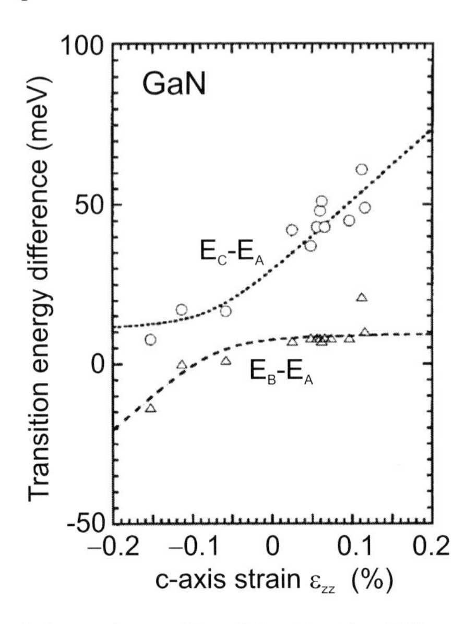

Fig. 9.18. Theoretical dependency (*lines*) for the the differences of the C-line and A-line as well as B-line and A-line exciton transition energies in GaN as a function of the c-axis strain. *Symbols* are experimental data from [296]. Adapted from [174]

(with background dielectric constant ϵ_b) is

$$\epsilon(\omega) = \epsilon_\mathrm{b}\left[1 + \frac{\beta}{1 - \frac{\omega^2}{\omega_\mathrm{X}^2}}\right] \cong \epsilon_\mathrm{b}\left[1 + \frac{\beta}{1 - \frac{\omega^2}{\omega_\mathrm{T}^2} + \frac{\hbar}{M\,\omega_\mathrm{T}}K^2}\right] , \tag{9.31}$$

where β is the oscillator strength and the energy is $\hbar\omega_\mathrm{X} = \hbar\omega_\mathrm{T} + \frac{\hbar^2}{2M}K^2$. $\hbar\omega_\mathrm{T}$ is the energy of the transverse exciton at $K = 0$. With this dispersion the wave dispersion must be fulfilled, i.e.

$$c^2k^2 = \omega^2\epsilon(\omega) , \tag{9.32}$$

where k is the k vector of the light that needs to be $k = K$ due to momentum conservation. The dependence of the dielectric function on k is called *spatial dispersion* [303]. Generally, up to terms in k^2 it is written as

$$\epsilon(\omega) = \epsilon_\mathrm{b}\left[1 + \frac{\beta}{1 - \frac{\omega^2}{\omega_0^2} + Dk^2}\right] . \tag{9.33}$$

The term k^2 with curvature D (for the exciton polariton $D = \hbar/(M\,\omega_\mathrm{T})$) plays a role in particular when $\omega_\mathrm{T}^2 - \omega^2 = 0$. For $\mathbf{k} \neq 0$ even a cubic material is anisotropic. The dimensionless curvature $\hat{D} = Dk'^2$ should fulfill $\hat{D} =$

$(\hbar/(Mc) \ll 1$ in order to make k^4 terms unimportant. For exciton polaritons[4] typically $\hat{D} = \frac{\hbar}{c^2}\frac{\omega_T}{m} \approx 2 \times 10^{-5}$ for $\hbar\omega_T = 1\,\text{eV}$ and $m^* = 0.1$.

From (9.32) together with (9.33) two solutions result:

$$\begin{aligned}2\omega^2 &= c^2k^2 + (1+\beta+Dk^2)\omega_0^2 \\ &\pm \left[-4c^2k^2(1+Dk^2)\omega_0^2 + (c^2k^2 + (1+\beta+Dk^2)\omega_0^2)^2\right]^{1/2} .\end{aligned} \tag{9.34}$$

The two branches are shown schematically in Fig. 9.19a. Depending on the k value they have a photonic (linear dispersion) or excitonic (quadratic dispersion) character. The anticrossing behavior at $k' \approx \omega_T/c$ (for $\hbar\omega_T = 1\,\text{eV}$ $k' \approx 0.5 \times 10^{-5}\,\text{cm}^{-1}$) creates a bottleneck region in the lower polariton branch. This name stems from the small emission rate of acoustic phonons (i.e. cooling) in that region, as predicted in [298] and experimentally found, e.g. in CdS [299]. The polaritons decay into a photon when they hit the surface. The effect of the oscillator strength of the dispersion is shown in Fig. 9.20 for two-exciton resonance. In the case of several excitons (9.33) reads

$$\epsilon(\omega) = \epsilon_b \left[1 + \sum_{i=1}^{n} \frac{\beta_i}{1 - \frac{\omega^2}{\omega_{0,i}^2} + D_i k^2}\right] . \tag{9.35}$$

For $k = 0$ either $\omega = 0$ (lower polariton branch) or $\epsilon(\omega_L) = 0$. We find from (9.33)

$$\omega_L = (1+\beta)^{1/2}\,\omega_T . \tag{9.36}$$

Therefore, the energy splitting ΔE_{LT}, mostly denoted as Δ_{LT}, between the L- and T-exciton energy given by

$$\Delta E_{LT} = \hbar(\omega_L - \omega_T) = \left[(1+\beta)^{1/2} - 1\right]\,\hbar\omega_T \approx \beta\hbar\omega_T/2 \tag{9.37}$$

is proportional to the exciton oscillator strength (for experimental values see Table 9.3). We note that if (C.9) is used for the dielectric function, β in (9.37) needs to be replaced by β/ϵ_b.

The effect of spatial dispersion on the reflection at the fundamental exciton resonance is depicted in Fig. 9.19b. For non-normal incidence an additional feature due to the longitudinal wave is observed for p-polarization [303]. For a detailed discussion additional effects due to anisotropy in wurtzite crystals, an exciton free layer at the semiconductor surface, additional boundary conditions and damping need to be considered [304, 305]. The polariton dispersions of ZnO and GaN are shown in Fig. 9.21.

[4] The dependence of the optical-phonon energies on k is typically too small to make spatial dispersion effects important. According to (5.18) $\hat{D} = -\left(\frac{a_0\omega_{TO}}{4c}\right)^2 \approx 4 \times 10^{-11}$ for typical material parameters (lattice constant $a_0 = 0.5\,\text{nm}$, TO phonon frequency $\omega_{TO} = 15\,\text{THz}$).

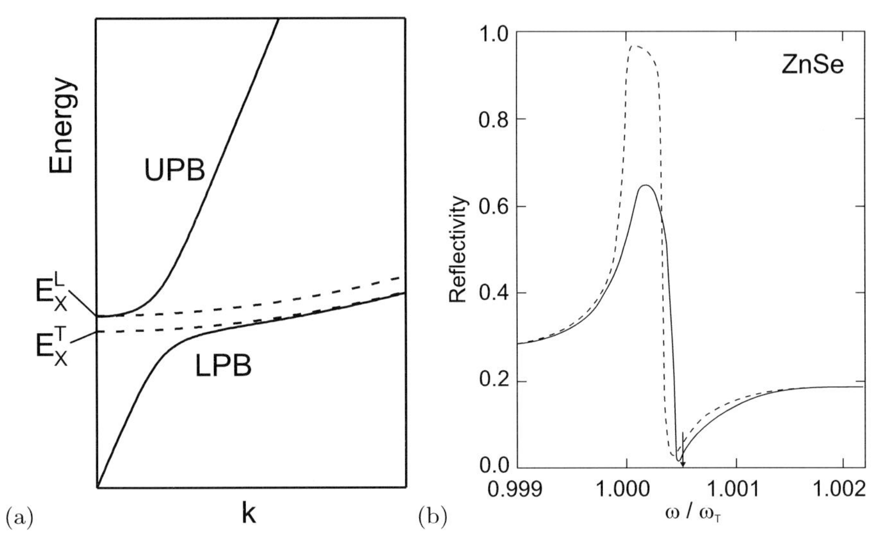

Fig. 9.19. (**a**) Schematic dispersion of exciton polaritons. The lower polariton branch ('LPB') is at small k photon-like, at large k exciton-like. The upper branch ('UPB') is exciton-like at small k and photon-like at larger k. The limit of the UPB for $k \to 0$ is the energy of the longitudinal exciton. The *dashed lines* represent the pure exciton dispersions. (**b**) Theoretical effect of spatial dispersion on the reflectivity at the fundamental exciton resonance at normal incidence for ZnSe material parameters ($\hbar\omega_T$ =2.8 eV, $\beta = 1.0 \times 10^{-3}$ and a background dielectric constant of $\epsilon_b = 8.1$, damping was set to $\Gamma = 10^{-5}\omega_T$). The arrow denotes the position of ω_L. The *solid* (*dashed*) *line* is with (without) spatial dispersion for $\hat{D} = 0.6 \times 10^{-5}$ ($\hat{D} = 0$). Data from [303]

Table 9.3. Exciton energy (low temperature), LT splitting and exciton polariton oscillator strength for various semiconductors. Values for ZnO from [300], values for GaAs from [301], all other values from [302]

	CdS A	CdS B	ZnO A	ZnO B	ZnSe	GaN A	GaN B	GaAs
$\hbar\omega_T$ (eV)	2.5528	2.5681	3.3776	3.3856	2.8019	3.4771	3.4816	1.5153
Δ_{LT} (meV)	2.2	1.4	1.45	5	1.45	1.06	0.94	0.08
β (10^{-3})	1.7	1.1	0.9	3.0	1.0	0.6	0.5	0.11

9.4.9 Bound-Exciton Absorption

Excitons can localize at impurities or inhomogeneities. Such excitons are called *bound excitons*. Here, the absorption due to such complexes is discussed. The recombination is discussed in Sect. 10.4. In GaP:N excitons are bound to the isoelectronic N impurities (substituting P) that results in the 'A' line at 2.3171 eV (at T = 4.2 K). At sufficiently high nitrogen doping, there exist nitrogen pairs, i.e. a complex where a nitrogen impurity has an-

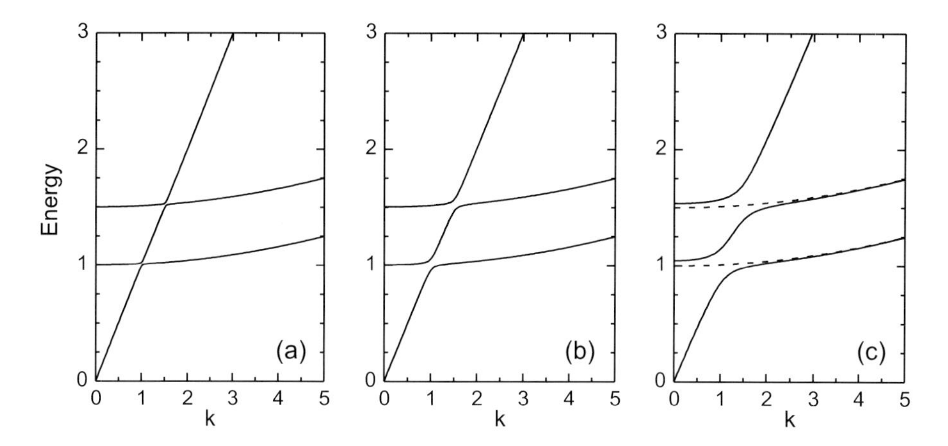

Fig. 9.20. Schematic polariton dispersion for a two-exciton resonance (curvature of exciton dispersion greatly exaggerated, $\hat{D} = 10^{-2}$) at $\omega_{\mathrm{T},1} = 1$ and $\omega_{\mathrm{T},2} = 1.5$ for three different oscillator strengths (**a**) $f = 10^{-3}$, (**b**) $f = 10^{-2}$, (**c**) $f = 10^{-1}$. The *dashed lines* in (**c**) represent the pure exciton dispersions

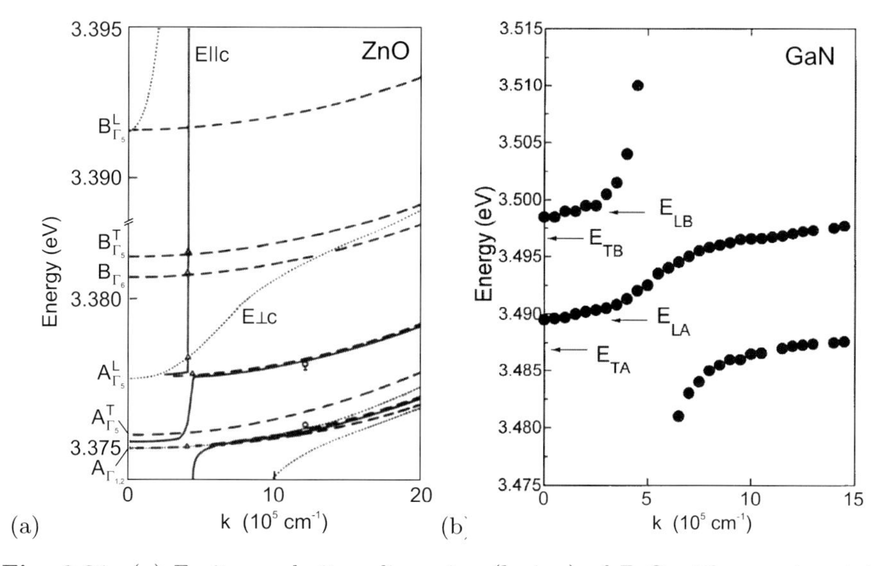

Fig. 9.21. (**a**) Exciton polariton dispersion ($\mathbf{k} \perp c$) of ZnO with experimental data ($T = 1.8\,\mathrm{K}$). *Solid* (*dotted*) *lines* are for polaritons with $\mathbf{E} \parallel c$ ($\mathbf{E} \perp c$). The *dashed lines* refer to excitons. Adapted from [306]. (**b**) Exciton polariton dispersion ($T = 2\,\mathrm{K}$) in GaN (on sapphire) for $\mathbf{E} \perp c$. Reprinted from [307], ©1998, with permission from Elsevier, originally in [308]

other nitrogen impurity in the vicinity. The pair is labeled NN_n if the second nitrogen atom is in the n-th shell around the first. For example, NN_1 relates to a N–Ga–N complex having 12 equivalent sites for the second N atom. The transitions due to excitons bound to NN_n, as shown in Fig. 9.22 give a series of lines (see Table 9.4) that fulfill $\lim_{n\to\infty} NN_n = A$.

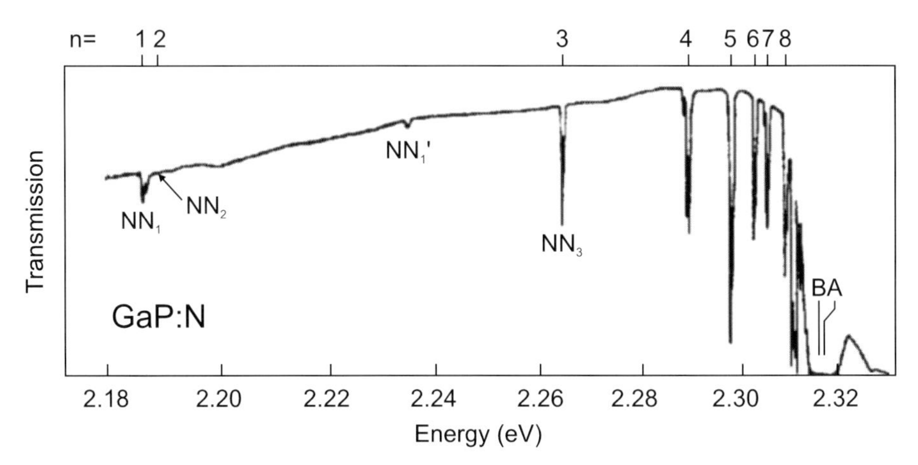

Fig. 9.22. Transmission spectrum of GaP:N with a nitrogen concentration of about $10^{19}\,\mathrm{cm}^{-3}$ at 1.6 K (thickness: 1.1 mm). n is indicated for the first eight transitions due to excitons bound to nitrogen pairs. NN_n' indicate phonon replica. The 'A' line denotes the position of the transition due to excitons bound to a single nitrogen atom (observable for samples with low N doping). The 'B' line is forbidden and due to the $J = 2$ exciton. Adapted from [309]

Table 9.4. Spatial separation d of nitrogen pairs NN_n (in nm) and energy separation ΔE of bound-exciton transitions from the free-exciton line (in meV) for $n = 1 \ldots 10$ and the 'A' line

n	1	2	3	4	5	6	7	8	9	10	A
d	0.385	0.545	0.667	0.771	0.862	0.945	10.2	10.9	11.56	12.19	∞
ΔE	143	138	64	39	31	25	22	20	18	17	11

9.4.10 Biexcitons

Similar to two hydrogen atoms forming a hydrogen molecule two excitons can also form a bound complex, the *biexciton* involving two electrons and two holes. The energy of the biexciton is smaller than $2E_X$. While the exciton density increases linearly with external excitation, the density of biexcitons increases quadratically.

9.4.11 Trions

The complexes 'eeh' and 'ehh' are called trions. Also, the notation X^- and X^+ is common. X^- is typically stable in bulk material but hard to observe. In quantum wells or dots, trions are easier to observe. In quantum dots excitons with higher charge, e.g. X^{2-}, have also been observed (see Fig. 13.29).

9.4.12 Burstein–Moss Shift

In the discussion so far it has been assumed that all target states in the conduction band are empty. In the presence of free carriers the absorption is modified by the

- change of the distribution function
- many-body effects (bandgap renormalization)

The latter is discussed in the next section. For a degenerate electron distribution all states close to the conduction-band edge are populated. Thus a transition from the valence band cannot take place into such states. This shift of the absorption edge to higher energies is called the Burstein–Moss shift [310,311].

k-conserving optical transitions between parabolic hole and electron bands have the dependence

$$E = E_{\mathrm{g}} + \frac{\hbar^2 k^2}{2m_{\mathrm{e}}^*} + \frac{\hbar^2 k^2}{2m_{\mathrm{h}}^*} = E_{\mathrm{g}} + \frac{\hbar^2 k^2}{2m_{\mathrm{r}}}\,, \tag{9.38}$$

where m_{r} is the reduced mass of electron and hole. About $4kT$ below the Fermi level all levels in the conduction band are populated (Fig. 9.23). Thus the k value at which the absorption starts is given as

$$\hat{k} = \left[\frac{2m_{\mathrm{r}}}{\hbar^2}(E_{\mathrm{F}} - E - 4kT)\right]^{1/2}\,. \tag{9.39}$$

Besides the energy shift in the conduction band, the corresponding energy shift in the valence band $\hbar k^2/(2m_{\mathrm{h}})$ must be considered. Thus, the Burstein–Moss shift of the absorption edge is

$$\Delta E = \hbar\omega - E_{\mathrm{g}} = (E_{\mathrm{F}} - 4kT - E_{\mathrm{C}})\left(1 + \frac{m_{\mathrm{e}}}{m_{\mathrm{h}}}\right)\,. \tag{9.40}$$

The relation between n and the Fermi level is given by (7.8). If $E_{\mathrm{F}} - E_{\mathrm{C}} \gg kT$ the Fermi integral can be approximated by $\frac{2}{\sqrt{\pi}}\frac{2}{3}\left(\frac{E_{\mathrm{F}}-E_{\mathrm{C}}}{kT}\right)^{3/2}$ (Sect. 7.3). Using (7.10) for N_{C}, the Burstein–Moss shift can be written for this case (low T) as

$$\Delta E = n^{2/3}\frac{h^2}{8m_{\mathrm{e}}}\left(\frac{3}{\pi}\right)^{2/3}\left(1 + \frac{m_{\mathrm{e}}}{m_{\mathrm{h}}}\right) \approx 0.97\frac{h^2}{8m_{\mathrm{r}}}\,n^{2/3}\,. \tag{9.41}$$

Originally, the Burstein–Moss shift was evoked to explain the absorption shift in InSb with varying carrier concentration (Fig. 9.24).

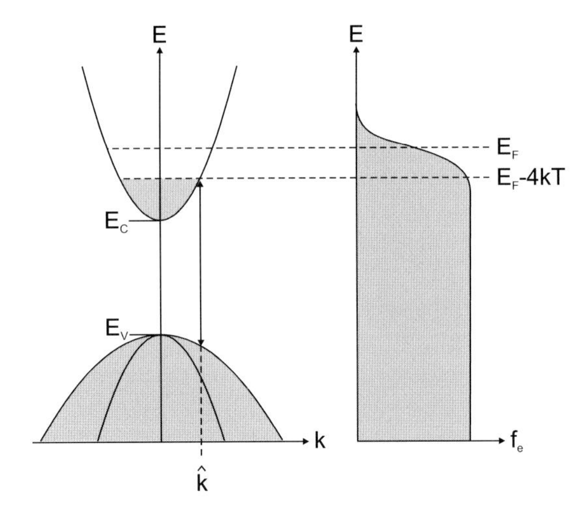

Fig. 9.23. Principle of Burstein–Moss shift. *Left panel*: Schematic band structure with completely filled electron states shown in *grey*. The k-vector for the lowest photon energy optical absorption process is indicated as $\hat{k}$. *Right panel*: Electron distribution function for a degenerate electron gas with Fermi level in the conduction band

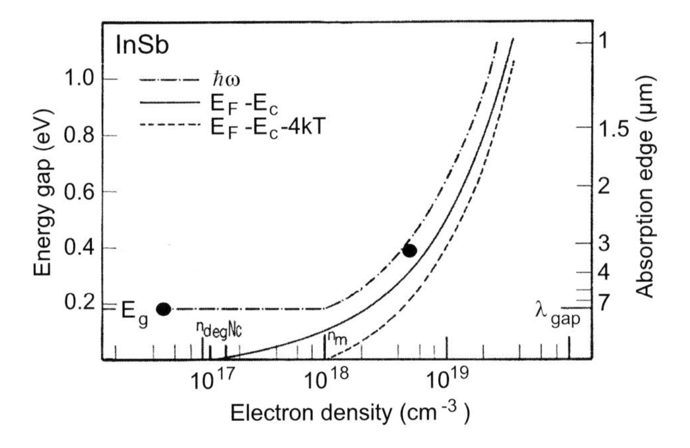

Fig. 9.24. Burstein–Moss effect at InSb ($E_g = 0.18$ eV) at room temperature. Theoretical dependence and data points for intrinsic InSb and $5 \times 10^{18}\,\mathrm{cm}^{-3}$ n-type. Data from [310]

9.4.13 Bandgap Renormalization

The band structure theory has been developed so far for small carrier densities. If the carrier density is large the interaction of free carriers has to be considered. The first step was exciton formation. However, at high temperatures (ionization) and at large carrier density (screening) the exciton is not stable. Exchange and correlation energy leads to a decrease of the optical absorption edge that is called *bandgap renormalization* (BGR).

An effect due to significant carrier density is to be expected when the density is of the order of the exciton volume, i.e. $n \sim a_B^{-3}$ (Mott criterion). For $a_B \sim 15\,\text{nm}$ (GaAs) this means $n \sim 3 \times 10^{17}\text{cm}^{-3}$. The dimensionless radius r_s is defined via

$$\frac{4\pi}{3} r_s^3 = \frac{1}{n a_B^3} . \tag{9.42}$$

The sum of exchange and correlation energies E_{xc} is found to be mostly independent of material parameters [312] (Fig. 9.25a) and follows the form

$$E_{xc} = \frac{a + b r_s}{c + d r_s + r_s^2} , \tag{9.43}$$

with $a = -4.8316$, $b = -5.0879$, $c = 0.0152$ and $d = 3.0426$. Thus the density dependence of the bandgap at small carrier density is $\propto n^{1/3}$. Experimental data for a number of II–VI semiconductors roughly follow such a dependence (Fig. 9.25b).

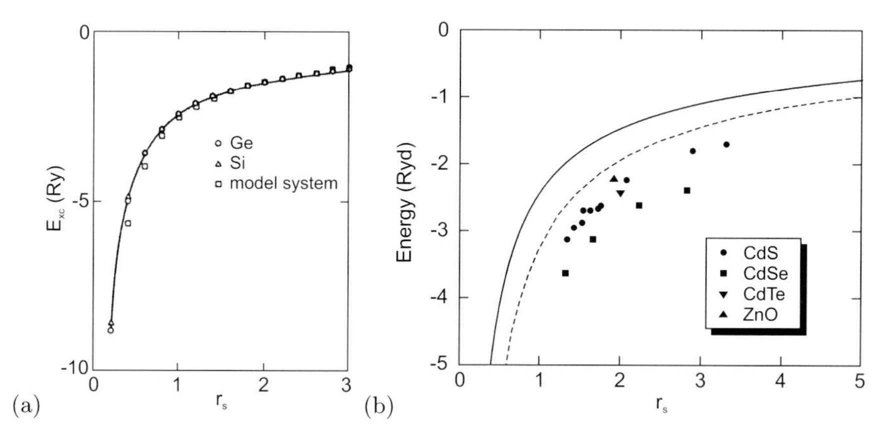

Fig. 9.25. (**a**) Theoretical exchange and correlation energies in units of the exciton Rydberg energy as a function of the dimensionless variable r_s for Ge, Si and a model system (with one isotropic conduction and valence band each). The *solid line* is a fit according to (9.43). Adapted from [312]. (**b**) Bandgap renormalization in terms of the excitonic Rydberg for various II–VI semiconductors. *Solid line* is the relation according to (9.43), *dashed line* is the dependence predicted in [313] for $T = 30\,\text{K}$. Data are compiled in [314]

In Fig. 9.26, a theoretical calculation of the absorption spectrum of bulk GaAs for various carrier densities [315] is shown. With increasing density the excitonic resonance broadens and vanishes. The shape approaches the electron–hole plasma shape. The absorption edge shifts to smaller energies. At high density the absorption becomes negative, i.e. amplification.

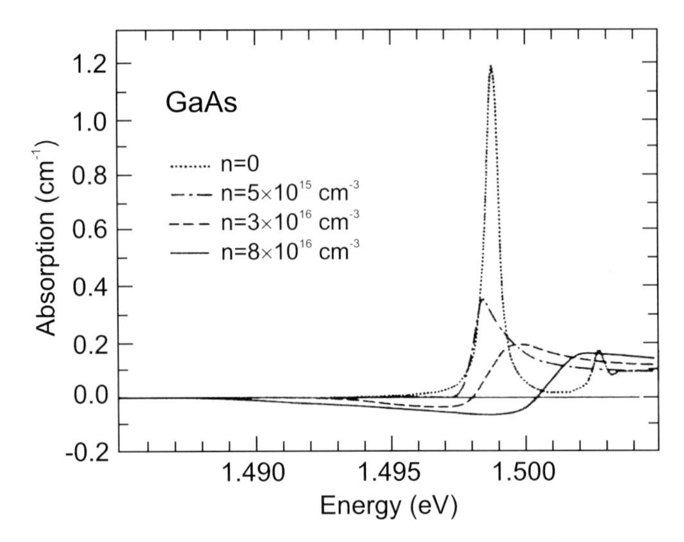

Fig. 9.26. Absorption of GaAs (low temperature, $T = 10\,\mathrm{K}$) as a function of the electron–hole density n (theory). Adapted from [315]

9.4.14 Electron–Hole Droplets

At low temperature and high density, electron–hole pairs in Ge and Si can undergo a phase transition into a liquid state. This *electron–hole liquid* (EHL) was suggested in [316] and is a Fermi liquid exhibiting the high conductivity of a metal and the surface and density of a liquid. The condensation is due to exchange interaction and correlation. The formation is fostered by the band structure of Ge [317] and the long lifetime of carriers in the indirect band structure. In unstressed Ge typically a cloud of electron–hole droplets with diameter in the μm range exists. The phase diagram is shown in Fig. 9.27a. In suitably stressed Ge electron–hole droplets with several hundred μm diameter form around the point of maximum shear strain in inhomogeneously strained crystals, as shown in Fig. 9.27b. The pair density in such a liquid is of the order of $10^{17}\,\mathrm{cm}^{-3}$.

We note that the metallic EHL state hinders observation of the Bose–Einstein condensation (BEC) of (bosonic) excitons. The light-exciton mass offers a high condensation temperature in the 1 K range (compared to the mK range for atoms). Recent experiments with spatially indirect excitons in coupled quantum wells lead towards BEC [321, 322]. A sufficiently long lifetime ensures cooling of the excitons close to the lattice temperature. Another potential candidate for BEC are long-living excitons (ms-range) in Cu_2O [323]. The condensation of polaritons (see Sect. 9.4.8) in microcavities to well-defined regions of **k**-space has been discussed in [324] and compared to bosonic condensation in bulk.

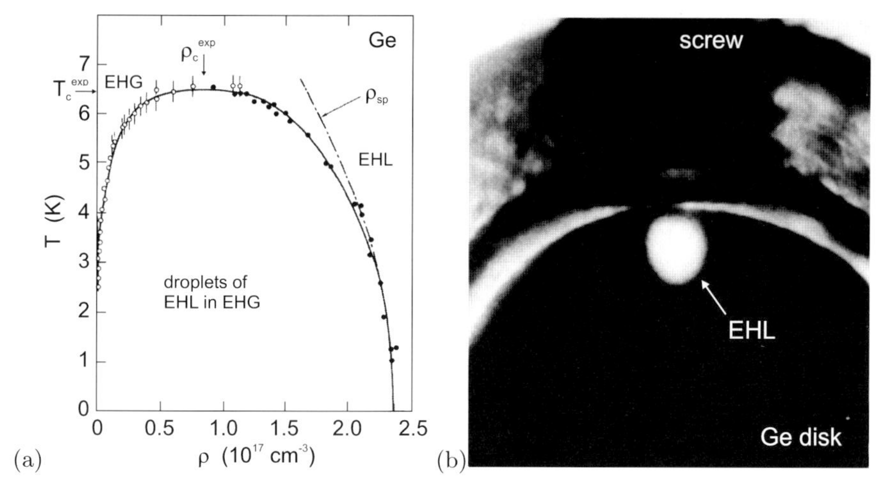

Fig. 9.27. (**a**) Temperature–density phase diagram of electrons and holes in Ge. The regions of electron–hole gas (EHG) and liquid (EHL) and the droplet phase are labeled. *Solid line* is theoretical calculation, *symbols* are experimental data from [318]. The *dash-dotted line* denoted $\rho_{\rm sp}$ is the experimentally obtained temperature dependence of the liquid density due to single-particle excitations. $\rho_{\rm c}^{\rm exp}$ and $T_{\rm c}^{\rm exp}$ denote the experimental critical density and temperature, respectively. Adapted from [319]. (**b**) Photographic image of radiative recombination (at 1.75 µm wavelength) from a 300-µm diameter droplet of electron–hole liquid (EHL) in a stressed (001) Ge disk (diameter 4 mm, thickness 1.8 mm) at $T = 2$ K. The stress is applied from the top by a nylon screw along a ⟨110⟩ direction. Adapted from [320], reprinted with permission, ©1977 APS

9.4.15 Two-Photon Absorption

So far, only absorption processes that involve *one* photon have been considered. The attenuation of the intensity I of a light beam (of frequency ω_0) along the z direction can be written as

$$\frac{\mathrm{d}I}{\mathrm{d}z} = -\alpha I - \beta I^2 \, , \tag{9.44}$$

where α is due to the (linear) absorption coefficient (and possibly scattering) and β is the two-photon absorption coefficient. A two-photon process can occur in two steps, e.g. via a midgap level, which is not considered any further here. Here, we consider two-photon absorption (TPA) via the population of a state at $2\hbar\omega_0$ higher energy than the initial state with a nonlinear optical process. The TPA coefficient is related to the nonlinear third-order electric dipole susceptibility tensor [325] χ_{ijkl}. Within the two-band approximation theory predicts [326]

$$\beta \propto (2\hbar\omega_0 - E_{\rm g})^{3/2} \, . \tag{9.45}$$

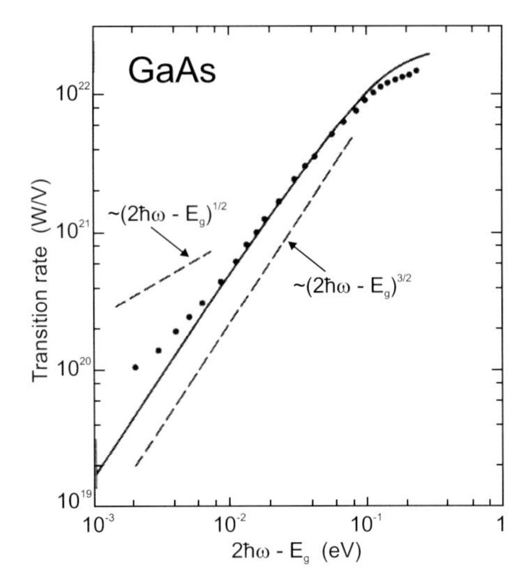

Fig. 9.28. Experimental (*dots*) two-photon absorption of GaAs ($T = 4$ K) as a function of the difference of the double-photon energy $2\hbar\omega$ from the GaAs band edge E_{g}. The *solid line* is a theoretical calculation, the *dashed lines* represent slopes with exponent 1/2 and 3/2, respectively. Adapted from [327]

The exponent 3/2 is indeed found experimentally, as shown in Fig. 9.28 for GaAs. The strength of absorption depends on the relative orientation of the light polarization with respect to the main crystallographic directions, e.g. TPA for polarization along $\langle 110 \rangle$ is about 20% larger than for the $\langle 100 \rangle$ orientation.

9.5 Impurity Absorption

Charge carriers bound to shallow impurities exhibit a hydrogen-like term scheme

$$E_n = \frac{m^*}{m_0} \frac{1}{\epsilon_{\mathrm{r}}^2} \frac{1}{n^2} \times 13.6\,\mathrm{eV} , \tag{9.46}$$

with the ionization limit E_∞ being the conduction (valence) band edge for donors (acceptors), respectively. They can be excited by light to the nearest band edge. Such absorption is typically in the FIR region and can be used for photodetectors in this wavelength regime. The actual energies can deviate from (9.46) due to deviation of the potential close to the impurity from the pure Coulomb potential. Such an effect is known as the chemical shift or central cell effect and is characteristic of the particular impurity. In GaAs such shifts are small ($\sim 100\,\mu\mathrm{eV}$) [328].

The term scheme for P in Si is shown in Fig. 9.29a. The ground state (1s) is split because of a reduction of the tetrahedral symmetry due to intervalley

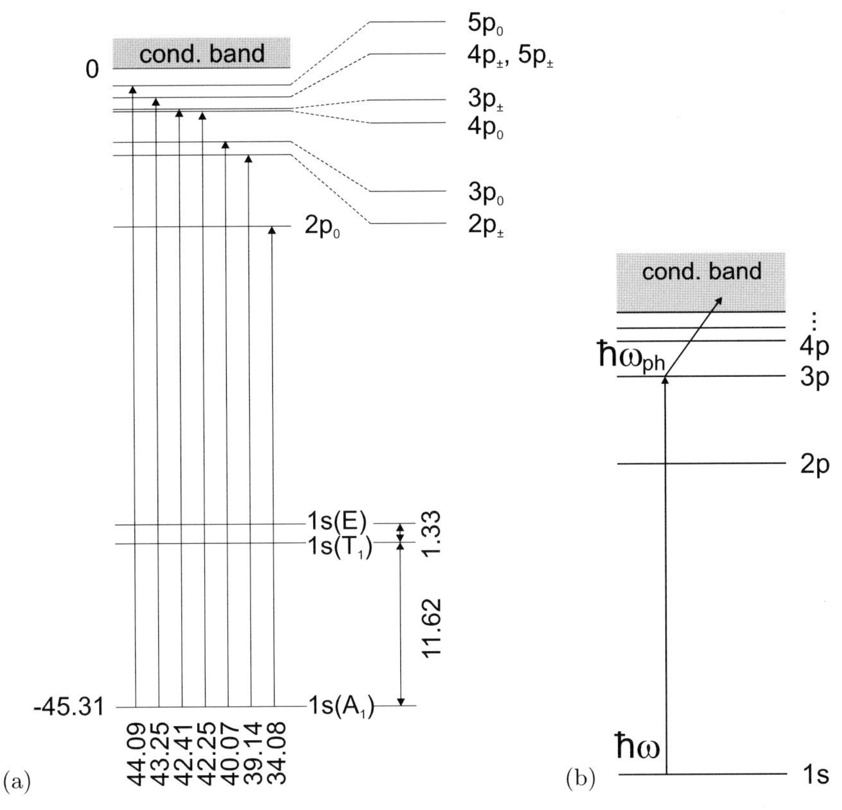

Fig. 9.29. (**a**) Term scheme of phosphorus donor in silicon, all energies in meV. After [329]. (**b**) Schematic sequence for photothermal ionization, here absorption of a photon with $\hbar\omega = E_{3\mathrm{p}} - E_{1\mathrm{s}}$ and subsequent absorption of a phonon with energy $\hbar\omega_{\mathrm{ph}} \geq E_{\infty} - E_{3\mathrm{p}}$

coupling. The anisotropic mass at the X-valley in Si causes the p states (and states with higher orbital momentum) to split into p_0 and $\mathrm{p}_{\pm}$ states. Such an effect is absent in a direct semiconductor with an isotropic conduction-band minimum such as GaAs (Fig. 9.30). Optical transitions between the 1s and various p states are observed because the missing energy to the ionization into the continuum is supplied by a phonon at finite temperature (photothermal ionization) (Fig. 9.29b). The splitting of the 2p transition in Fig. 9.30a is the chemical shift due to different donors incorporated in the GaAs (Si, Sn, and Pb). Peak broadening is mostly Stark broadening due to neighboring charged impurities. The application of a magnetic field induces Zeeman-like splittings and increases the sharpness of the peaks. The peak width can be further increased by illuminating the sample with light having a higher energy than the bandgap. The additional charge carriers neutralize charged impurities and allow higher resolution (Fig. 9.30b).

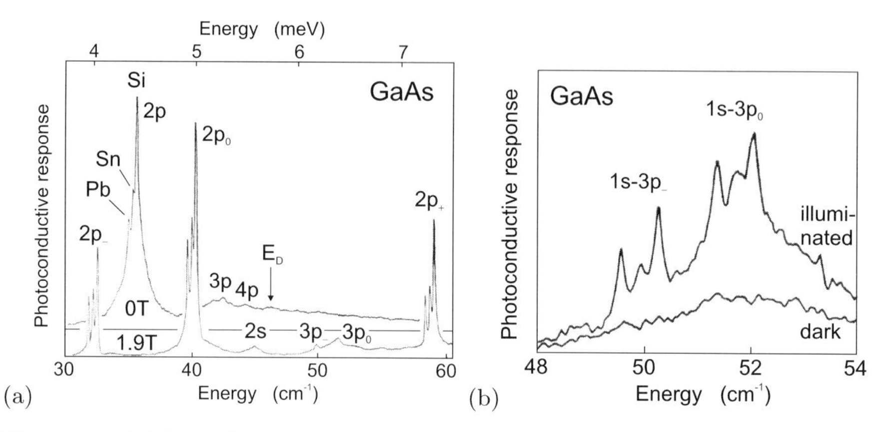

Fig. 9.30. (**a**) Far-infrared photoconductivity response (Lyman-type s→p series) of not intentionally doped GaAs with residual donors Pb, Sn, and Si, $N_{\mathrm{A}} = 2.6 \times 10^{13}\,\mathrm{cm}^{-3}$, $N_{\mathrm{D}} - N_{\mathrm{A}} = 8 \times 10^{12}\,\mathrm{cm}^{-3}$. The *upper* (*lower*) curve is for a magnetic field of 0 (1.9) T. Measurement temperature is 4.2 K. (**b**) Photoconductive response of a (different) GaAs sample with the same impurities ($N_{\mathrm{D}} = 1 \times 10^{13}\,\mathrm{cm}^{-3}$) with (*upper curve*) and without (*lower curve*) illumination with above-bandgap light ($B = 1.9\,\mathrm{T}$, $T = 4.2\,\mathrm{K}$). Adapted from [330]

In Fig. 9.31 absorption spectra of highly doped n-type GaAs are shown. For doping concentrations larger than the critical concentration of $\sim 1 \times 10^{16}\,\mathrm{cm}^{-3}$ (cf. Table 7.6) significant broadening is observed due to the formation of an impurity band (Sect. 7.5.5).

The absorption of deep levels is typically in the infrared. In Fig. 9.32a the possible optical absorption processes involving the Fe levels in InP (cf. Sect 7.7.6) during the charge transfer $\mathrm{Fe}^{3+} \rightarrow \mathrm{Fe}^{2+}$ are shown. These transitions and their fine structure (Fig. 9.32b) have been observed in calorimetric absorption spectroscopy (CAS) experiments [228].

9.6 Free-Carrier Absorption

A time-dependent electric field accelerates the charge carriers. The excess energy is subsequently transferred to the lattice via scattering with phonons. In the relaxation-time approximation energy is relaxed with a time constant τ. Thus energy is absorbed from the electromagnetic wave and dissipated.

The complex conductivity (8.38) is given by

$$\sigma = \sigma_{\mathrm{r}} + \mathrm{i}\sigma_{\mathrm{i}} = \frac{ne^2\tau}{m^*}\left(\frac{1}{1+\omega^2\tau^2} + \mathrm{i}\frac{\omega\tau}{1+\omega^2\tau^2}\right) . \tag{9.47}$$

We note that a static magnetic field introduces birefringence as discussed in more detail in Sect. 12.2.1. The wave equation for the electric field is

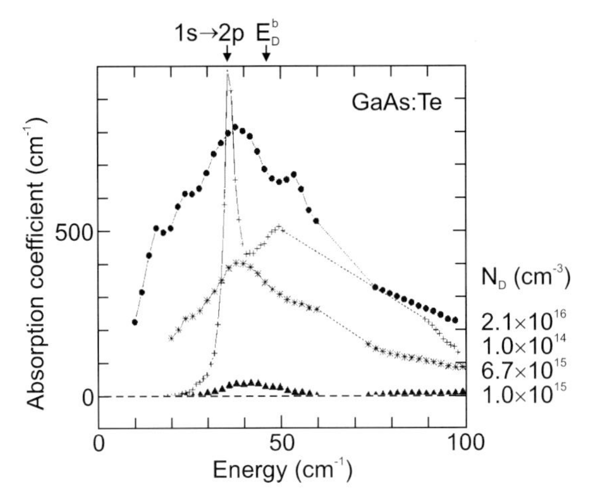

Fig. 9.31. Low-temperature ($T = 1.35\,\mathrm{K}$) absorption spectra of highly doped n-type GaAs:Te with doping concentrations as labeled (*circles*: $N_\mathrm{D} = 2.1 \times 10^{16}\,\mathrm{cm}^{-3}$, *stars*: 6.7×10^{14}, *triangles*: 1.0×10^{15}). A sharp photoconductivity spectrum (in arbitrary units) from low-doped GaAs:Te (*crosses*, $N_\mathrm{D} = 1.0 \times 10^{14}\,\mathrm{cm}^{-3}$) is shown for comparison (cf. Fig. 9.30a). The energy of the 1s→2p transition and the donor binding energy (onset of continuum absorption) are indicated. Adapted from [331]

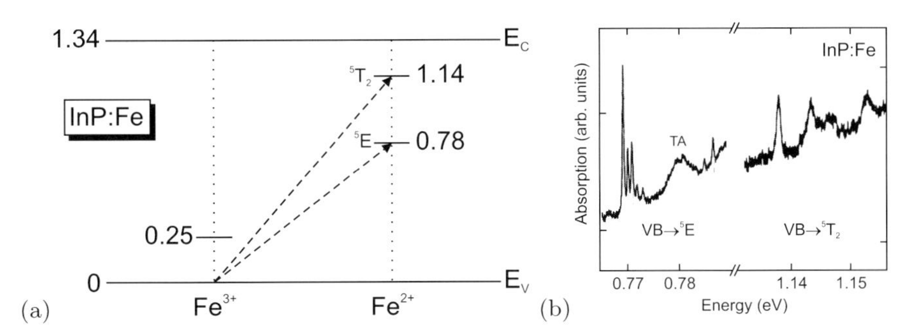

Fig. 9.32. (**a**) Schematic band diagram of InP with levels of Fe impurities in the 3+ and 2+ charge states at low temperature. All energies are given in eV. The *arrows* denote the optical transition of a valence-band electron to the Fe center, $\mathrm{Fe}^{3+} + \hbar\omega \rightarrow \mathrm{Fe}^{2+} + h$. (**b**) Calorimetric absorption spectra (at $T = 1.3\,\mathrm{K}$) of InP:Fe, [Fe]=$5 \times 10^{16}\,\mathrm{cm}^{-3}$. Part (**b**) adapted from [228]

$$\nabla^2 \mathbf{E} = \epsilon \mu_0 \ddot{\mathbf{E}} + \sigma \mu_0 \dot{\mathbf{E}} \,. \tag{9.48}$$

For a plane wave $\propto \exp\left[\mathrm{i}(\mathbf{kr} - \omega t)\right]$ the wavevector obeys

$$k = \frac{\omega}{c} \left(\epsilon_\mathrm{r} + \mathrm{i} \frac{\sigma^*}{\epsilon_0 \omega} \right)^{1/2} , \tag{9.49}$$

where $c = (\epsilon_0 \mu_0)^{-1/2}$ is the velocity of light in vacuum, ϵ_r is the background dielectric constant (for large ω).

The tensor of the dielectric function from free carriers is

$$\epsilon = \frac{\mathrm{i}}{\epsilon_0 \omega} \sigma^* . \tag{9.50}$$

The complex index of refraction is

$$n^* = n_\mathrm{r} + \mathrm{i}\kappa = \left(\epsilon_\mathrm{r} + \mathrm{i}\frac{\sigma^*}{\epsilon_0 \omega}\right)^{1/2} . \tag{9.51}$$

Taking the square of this equation yields

$$n_\mathrm{r}^2 - \kappa^2 = \epsilon_\mathrm{r} + \mathrm{i}\frac{\sigma_\mathrm{i}}{\epsilon_0 \omega} = \epsilon_\mathrm{r} - \frac{ne^2}{\epsilon_0 m^*}\frac{\tau^2}{1+\omega^2\tau^2} \tag{9.52a}$$

$$2n_\mathrm{r}\kappa = \frac{\sigma_\mathrm{r}}{\epsilon_0 \omega} = \frac{ne^2}{\epsilon_0 \omega m^*}\frac{\tau}{1+\omega^2\tau^2} . \tag{9.52b}$$

The absorption coefficient (damping of $\mathbf{E}^2$) of the plane wave is

$$\alpha = 2\frac{\omega}{c}\kappa . \tag{9.53}$$

For the case of higher frequencies, i.e. $\omega\tau \gg 1$, the absorption is

$$\alpha = \frac{ne^2}{\epsilon_0 c n_\mathrm{r} m^* \tau}\frac{1}{\omega^2} . \tag{9.54}$$

The absorption decreases with increasing frequency. For semiconductors it is particularly important in the mid- and far-infrared regions when carriers are present due to doping or thermal excitation. The index of refraction is given by (also for $\omega\tau \gg 1$)

$$\begin{aligned} n_\mathrm{r}^2 &= \epsilon_\mathrm{r} - \frac{ne^2}{\epsilon_0 m^* \omega^2} + \kappa^2 = \epsilon_\mathrm{r}\left[1 - \left(\frac{\omega_\mathrm{p}}{\omega}\right)^2\right] + \frac{\epsilon_\mathrm{r}^2}{4n_\mathrm{r}^2}\left(\frac{\omega_\mathrm{p}}{\omega}\right)^4 \frac{1}{\omega^2\tau^2} \\ &\approx \epsilon_\mathrm{r}\left[1 - \left(\frac{\omega_\mathrm{p}}{\omega}\right)^2\right] , \end{aligned} \tag{9.55}$$

where

$$\omega_\mathrm{p} = \left(\frac{ne^2}{\epsilon_\mathrm{r}\epsilon_0 m^*}\right)^{1/2} \tag{9.56}$$

is the plasma frequency. The approximation is valid for small absorption and when $(\omega\tau)^{-2}$ can be neglected. A graphical representation is given in Fig. 9.33a. For coupling to electromagnetic waves (still $\omega\tau \gg 1$)

$$\epsilon(\omega) = \epsilon_\mathrm{r}\left[1 - \left(\frac{\omega_\mathrm{p}}{\omega}\right)^2\right] = \frac{c^2k^2}{\omega^2} \tag{9.57}$$

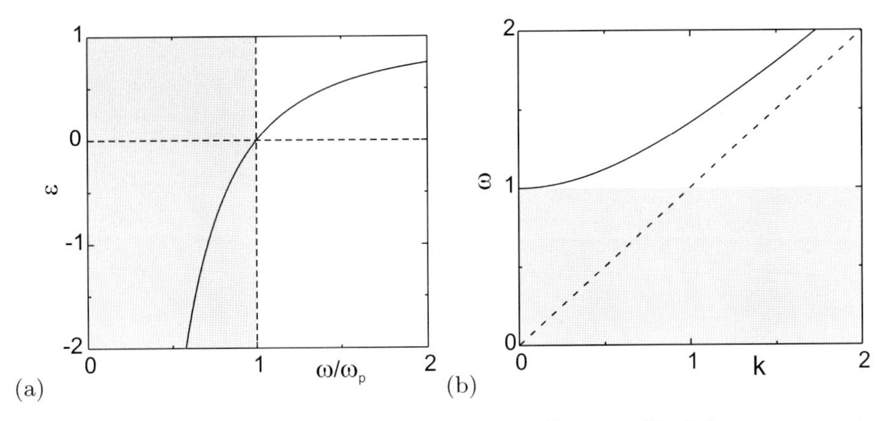

Fig. 9.33. (**a**) Dielectric constant for plasmon oscillations. *Shaded area* represents region of attenuation (negative ϵ). (**b**) Dispersion relation (k in units of ω_p/c, ω in units of ω_p) in the presence of free carriers (9.58, for $\epsilon_\mathrm{r} = 1$). *Shaded area* represents forbidden frequency range for propagating solutions. *Dashed line* is photon dispersion $\omega = ck$

must be fulfilled. It follows that the dispersion relation in the presence of free carriers (Fig. 9.33b) is

$$\omega^2 = \omega_\mathrm{p}^2 + \frac{c^2 k^2}{\epsilon_\mathrm{r}^2} \,. \tag{9.58}$$

For $\omega > \omega_\mathrm{p}$, $\epsilon > 0$, thus waves can propagate. For $\omega < \omega_\mathrm{p}$, however, the dielectric constant is negative, i.e. $\epsilon < 0$. For such frequencies waves are exponentially damped and cannot propagate or penetrate a layer. This effect can be used in a plasmon waveguide. The expected dependence of the plasmon wavelength on the carrier density $\lambda_\mathrm{p} \propto n^{-1/2}$ is depicted in Fig. 9.34 for GaAs. For semiconductors the plasmon frequency is in the mid-or far-infrared spectral region.[5]

9.7 Lattice Absorption

9.7.1 Dielectric Constant

The dielectric constant (with damping parameter $\varGamma$) in the vicinity of the optical phonon energies is given by (cf. (5.43))

$$\epsilon(\omega) = \epsilon(\infty) \left(\frac{\omega_\mathrm{LO}^2 - \omega^2 - \mathrm{i}\omega\varGamma}{\omega_\mathrm{TO}^2 - \omega^2 - \mathrm{i}\omega\varGamma} \right) . \tag{9.59}$$

[5]The much higher free-electron density in metals shifts the plasma frequency to the UV, explaining the reflectivity of metals in the visible and their UV transparency.

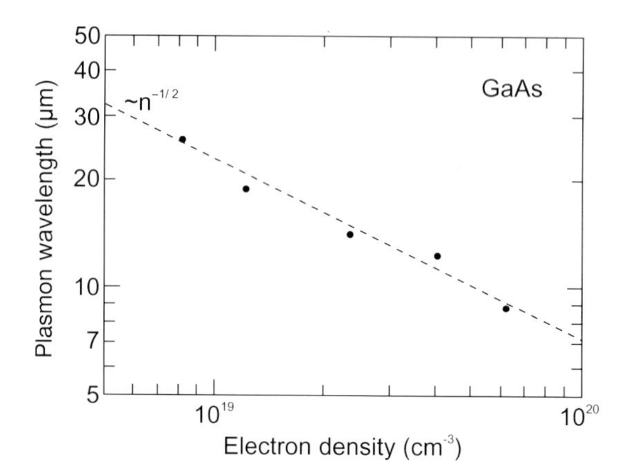

Fig. 9.34. Plasmon wavelength for GaAs with various electron concentrations due to different doping levels. *Filled circles*: experimental values, *dashed line*: $n^{-1/2}$ dependence

The dispersion relation (without damping) can be written as

$$\epsilon(\omega) = \epsilon(\infty) + \frac{\epsilon(0) - \epsilon(\infty)}{1 - \omega^2/\omega_{\mathrm{LO}}^2} = \epsilon(\infty)\left[1 + \frac{f}{1 - \omega^2/\omega_{\mathrm{LO}}^2}\right] . \tag{9.60}$$

Thus the oscillator strength (compare with (C.10)) is $f = \frac{\epsilon(0)-\epsilon(\infty)}{\epsilon(\infty)}$. With the LST relation (5.42) the oscillator strength is

$$f = \frac{\omega_{\mathrm{LO}}^2 - \omega_{\mathrm{TO}}^2}{\omega_{\mathrm{TO}}^2} \approx 2\frac{\omega_{\mathrm{LO}} - \omega_{\mathrm{TO}}}{\omega_{\mathrm{TO}}} , \tag{9.61}$$

and thus proportional to the splitting $\Delta_{\mathrm{LT}} = \omega_{\mathrm{LO}} - \omega_{\mathrm{TO}}$ between the longitudinal and transverse optical phonon frequency. The approximation in (9.61) is valid for $\Delta_{\mathrm{LT}} \ll \omega_{\mathrm{TO}}$.

The oscillator strength increases with the ionicity, i.e. the electronegativity difference of the atoms in the base (Fig. 9.35). Additionally, the oscillator strength depends on the reduced mass and the high-frequency polarizability; this can be seen, e.g., for the series of the Zn compounds that all have similar ionicity. For the series of the nitrides, the mass effect is small since the reduced mass is dominated by the light N mass.

9.7.2 Reststrahlenbande

The absorption of electromagnetic radiation by optical phonons is governed by the dielectric function that has been derived in (9.59). For small damping, i.e. $\Gamma \ll \Delta_{\mathrm{LT}}$, the dielectric constant is negative between ω_{TO} and ω_{LO}. From $\epsilon_{\mathrm{r}} = n_{\mathrm{r}}^2 - \kappa^2$ it follows that κ^2 is much larger than n_{r}^2. Therefore, the reflectivity (9.8) will be close to 1. This energy range is the so-called

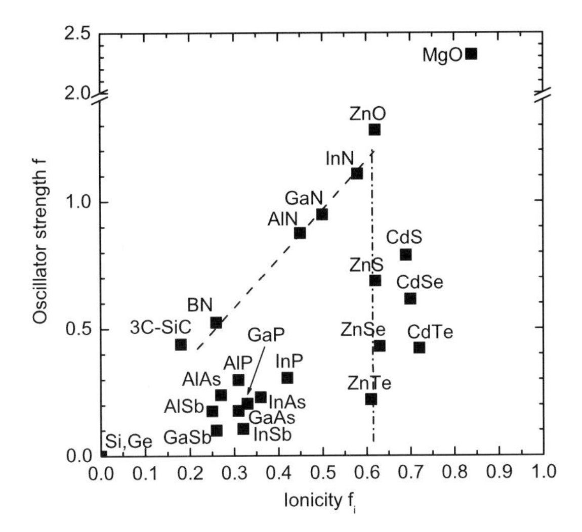

Fig. 9.35. Lattice absorption oscillator strength f from (9.61) for various IV–IV, III–V and II–VI compounds as a function of their ionicity f_i. *Dashed line* is linear dependence on ionicity for similar (reduced) mass, *dash-dotted line* is a guide to the eye for similar ionicity and varying mass

reststrahlenbande. This term stems from multiple reflections in this wavelength regime that suppresses neighboring spectral regions and thus achieves a certain monochromatization in the far-infrared spectral region. Within the semiconductor the absorption is large in the reststrahlenbande (Fig. 9.36).

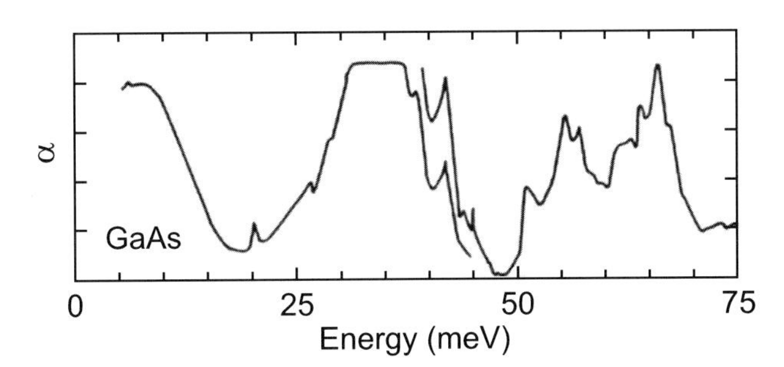

Fig. 9.36. Far-infrared absorption of GaAs. In the region around 35 meV is the reststrahlenbande with high absorption due to optical phonons. Adapted from [93], based on [229]

9.7.3 Polaritons

The coupled propagation of phonons and electromagnetic radiation is related to the equation (without phonon damping)

$$\epsilon(\omega) = \epsilon(\infty)\left(\frac{\omega_{\mathrm{LO}}^2 - \omega^2}{\omega_{\mathrm{TO}}^2 - \omega^2}\right) = \frac{c^2k^2}{\omega^2}\,. \tag{9.62}$$

There are two branches of propagating waves (real k):

$$\omega^2 = \frac{1}{2}\left(\omega_{\mathrm{LO}}^2 + \frac{c^2k^2}{\epsilon(\infty)}\right) \pm \left[\frac{1}{4}\left(\omega_{\mathrm{LO}}^2 + \frac{c^2k^2}{\epsilon(\infty)}\right)^2 - \left(\frac{c^2k^2\omega_{\mathrm{TO}}^2}{\epsilon(\infty)}\right)^2\right]^{1/2}. \tag{9.63}$$

For $k = 0$ we find the solutions $\omega = \omega_{\mathrm{LO}}$ and $\omega = kc/\sqrt{\epsilon(0)}$. For large k we find $\omega = \omega_{\mathrm{TO}}$ and $\omega = kc/\sqrt{\epsilon(\infty)}$. These solutions are shown in Fig. 9.37. Both branches have a phonon- and a photon-like part. The coupled state between the phonon and the photon field is called the (phonon-) polariton.

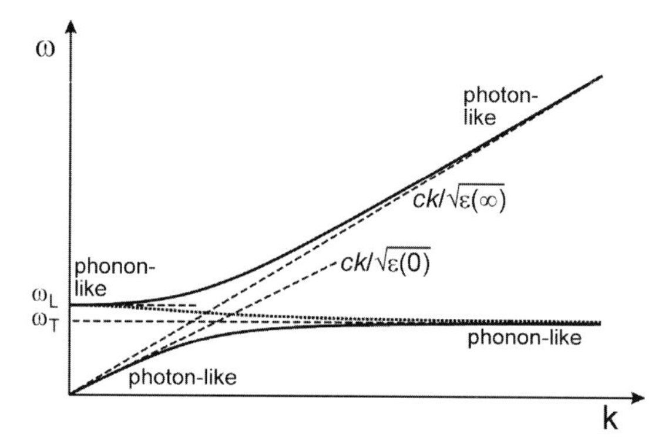

Fig. 9.37. Dispersion of the polariton. The *dotted line* displays the dispersion for a purely imaginary wavevector with the absolute value k

In the interval $[\omega_{\mathrm{TO}}, \omega_{\mathrm{LO}}]$ the wavevector is purely imaginary, i.e. $k = \mathrm{i}\tilde{k}$ with real $\tilde{k}$. For this case there is only one solution that is depicted in Fig. 9.37 as dotted line,

$$\omega^2 = \frac{1}{2}\left(\omega_{\mathrm{LO}}^2 + \frac{c^2\tilde{k}^2}{\epsilon(\infty)}\right) + \left[\frac{1}{4}\left(\omega_{\mathrm{LO}}^2 + \frac{c^2\tilde{k}^2}{\epsilon(\infty)}\right)^2 + \left(\frac{c^2\tilde{k}^2\omega_{\mathrm{TO}}^2}{\epsilon(\infty)}\right)^2\right]^{1/2}. \tag{9.64}$$

9.7.4 Phonon–Plasmon Coupling

The coupling of phonons and plasmons in the spectral region of the reststrahlenbande leads to the development of two new branches, the longitudinal phonon plasmon modes (LPP+ and LPP−), in the common dispersion. The dielectric function is

$$\epsilon(\omega) = \epsilon(\infty)\left(1 + \frac{\omega_{\mathrm{LO}}^2 - \omega^2}{\omega_{\mathrm{TO}}^2 - \omega^2} - \frac{\omega_{\mathrm{p}}^2}{\omega^2}\right) . \tag{9.65}$$

For $\epsilon(\omega) = 0$ for $k = 0$ (coupling to photons) the two solutions $\omega_{\mathrm{LPP+}}$ and $\omega_{\mathrm{LPP-}}$ do not cross as a function of ω_{p} (Fig. 9.38),

$$\omega_{\mathrm{LPP\pm}} = \frac{1}{2}\left[\omega_{\mathrm{LO}}^2 + \omega_{\mathrm{p}}^2 \pm \left((\omega_{\mathrm{LO}}^2 + \omega_{\mathrm{p}}^2)^2 - 4\omega_{\mathrm{TO}}^2\omega_{\mathrm{p}}^2\right)^{1/2}\right] . \tag{9.66}$$

For small plasma frequencies $\omega_{\mathrm{LPP+}} = \omega_{\mathrm{LO}}$, i.e. the optical phonons couple to the electromagnetic field without change. Also $\omega_{\mathrm{LPP-}} = \omega_{\mathrm{p}}$. For large carrier density, i.e. $\omega_{\mathrm{p}} \gg \omega_{\mathrm{LO}}$, we find $\omega_{\mathrm{LPP-}} = \omega_{\mathrm{TO}}$ and $\omega_{\mathrm{LPP+}} = \omega_{\mathrm{p}}$. Thus, the carriers have effectively screened the electric field of the phonon that had led to the increase of the TO to the LO frequency.

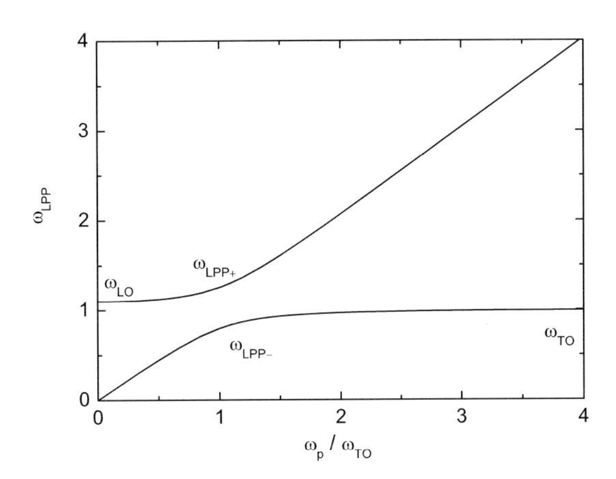

Fig. 9.38. Frequency of the coupled longitudinal-phonon plasmon (LPP) modes as a function of the plasma frequency

10 Recombination

10.1 Introduction

In thermodynamic nonequilibrium excess charges can be present in the semiconductor. They can be created by carrier injection through contacts, an electron beam or the absorption of light with wavelength smaller than the bandgap. After the external excitation is turned off, the semiconductor will return to the equilibrium state. The relaxation of carriers into energetically lower states (and energy release) is called recombination. The term stems from the electron recombining with the hole created after absorption of a photon. However, there are other recombination mechanisms.

In the simplest picture an excitation generates carriers with a rate G (carriers per unit volume and unit time). In the steady state (after all turn-on effects) a constant excess charge n carrier density is present. Then the generation exactly compensates the recombination processes. The principle of detailed balance even says that each microscopic process is balanced by its reverse process. If the time constant of the latter is τ, n is given by $n = G\tau$. This follows from the steady-state solution $\dot{n} = 0$ of

$$\frac{\mathrm{d}n}{\mathrm{d}t} = G - \frac{n}{\tau} \, . \tag{10.1}$$

10.2 Band–Band Recombination

The band–band recombination is the relaxation from an electron in the conduction band into the valence band (the empty state there is the hole). In a direct semiconductor, electrons can make an optical transition between the bottom of the conduction band to the top of the valence band. In an indirect semiconductor, this process is only possible with the assistance of a phonon and is thus much less probable.

10.2.1 Spontaneous Emission

We consider the spontaneous recombination of an electron of energy E_e and a hole of energy E_h (Fig. 10.1a). $C(E_\mathrm{e}, E_\mathrm{h})$ is a constant proportional to the matrix element of the optical transition (see Sect. 9.3). The spontaneous

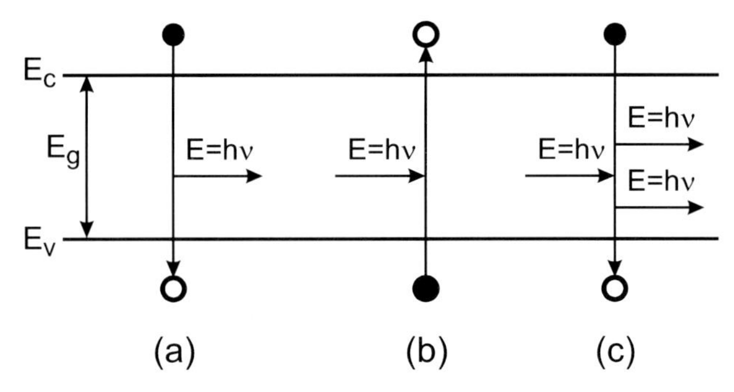

Fig. 10.1. Processes of band–band recombination: (**a**) spontaneous emission, (**b**) absorption and (**c**) stimulated emission. A *full* (*empty*) *circle* represents an occupied (unoccupied) electron state

recombination rate r_{sp} at photon energy $E \geq E_{\mathrm{C}} - E_{\mathrm{V}} = E_{\mathrm{g}}$ is (assuming energy conservation, i.e. $E = E_{\mathrm{e}} - E_{\mathrm{h}}$, but without k-conservation in a dense plasma [332]),

$$\begin{aligned} r_{\mathrm{sp}}(E) = & \int_{E_{\mathrm{C}}}^{\infty} \mathrm{d}E_{\mathrm{e}} \int_{-\infty}^{E_{\mathrm{V}}} \mathrm{d}E_{\mathrm{h}}\; C(E_{\mathrm{e}}, E_{\mathrm{h}}) \\ & \times D_{\mathrm{e}}(E_{\mathrm{e}}) f_{\mathrm{e}}(E_{\mathrm{e}}) D_{\mathrm{h}}(E_{\mathrm{h}}) f_{\mathrm{h}}(E_{\mathrm{h}}) \delta(E - E_{\mathrm{e}} + E_{\mathrm{h}}) \\ = & \int_{E_{\mathrm{C}}}^{E+E_{\mathrm{V}}} \mathrm{d}E_{\mathrm{e}}\; C(E_{\mathrm{e}}, E_{\mathrm{e}} - E) \\ & \times D_{\mathrm{e}}(E_{\mathrm{e}}) f_{\mathrm{e}}(E_{\mathrm{e}}) D_{\mathrm{h}}(E_{\mathrm{e}} - E) f_{\mathrm{h}}(E_{\mathrm{e}} - E)\;, \end{aligned} \tag{10.2}$$

where f_{h} denotes the hole occupation $f_{\mathrm{h}} = 1 - f_{\mathrm{e}}$.

The lineshape of the band–band recombination with k-conservation[1] is proportional to the joint density of states (9.21) and the Fermi distribution function. At small excitation and at low doping it can be approximated by the Boltzmann distribution function and the lineshape is given as

$$I(E) \propto (E - E_{\mathrm{g}})^{1/2} \exp\left(-\frac{E}{kT}\right)\;. \tag{10.3}$$

An experimental spectrum is shown in Fig. 10.2 together with a fit according to (10.3). The expected FWHM of the peak is $1.7954\,kT$, which is about 46 meV at $T = 300$ K.

10.2.2 Absorption

A similar consideration is made for the absorption process (Fig. 10.1b). An electron is transferred upon light absorption from a valence-band state (occupied) to a conduction-band state that must be empty. The coefficient is

[1] Excitonic effects are neglected here, e.g. for temperatures $kT \gg E_{\mathrm{X}}^{\mathrm{b}}$.

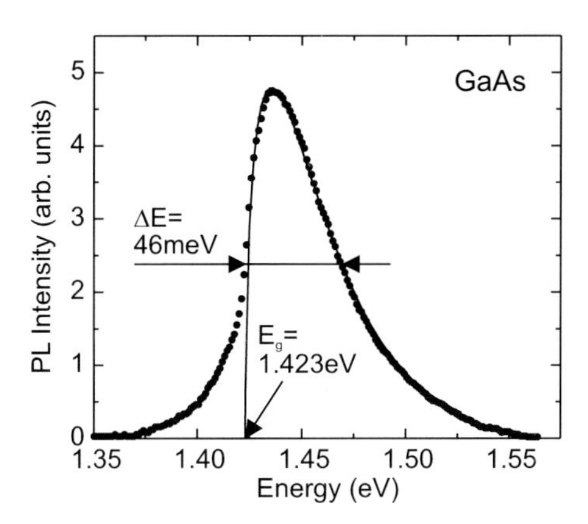

Fig. 10.2. Photoluminescence spectrum of an undoped LPE-grown epitaxial GaAs layer at room temperature and low cw ($\lambda = 647$ nm) excitation density (10 W/cm^2). The *solid line* is a lineshape fit with (10.3) and E_g =1.423 eV and $T = 293$ K

B_1. Also, the process is proportional to the light intensity, represented by the density of occupied photon states $N_\mathrm{ph}(E)$,

$$r_\mathrm{abs}(E) = \int_{E_\mathrm{C}}^{E+E_\mathrm{V}} \mathrm{d}E_\mathrm{e}\; B_1(E_\mathrm{e}, E_\mathrm{e} - E) \\ D_\mathrm{e}(E_\mathrm{e})(1 - f_\mathrm{e}(E_\mathrm{e}))D_\mathrm{h}(E_\mathrm{e} - E)(1 - f_\mathrm{h}(E_\mathrm{e} - E))N_\mathrm{ph}(E) \;. \tag{10.4}$$

10.2.3 Stimulated Emission

In this case, an incoming photon 'triggers' the transition of an electron in the conduction band into an empty state in the valence band. The emitted photon is in phase with the initial photon (Fig. 10.1c). The rate is (with coefficient B_2):

$$r_\mathrm{st}(E) = \int_{E_\mathrm{C}}^{E+E_\mathrm{V}} \mathrm{d}E_\mathrm{e}\; B_2(E_\mathrm{e}, E_\mathrm{e} - E) \\ D_\mathrm{e}(E_\mathrm{e})f_\mathrm{n}(E_\mathrm{e})D_\mathrm{h}(E_\mathrm{e} - E)f_\mathrm{p}(E_\mathrm{e} - E)N_\mathrm{ph}(E) \;. \tag{10.5}$$

10.2.4 Net Recombination Rate

In thermodynamical equilibrium the rates fulfill

$$r_\mathrm{sp}(E) + r_\mathrm{st}(E) = r_\mathrm{abs}(E) \;. \tag{10.6}$$

The population functions are Fermi functions with quasi-Fermi levels F_n and F_p. The photon density is given by Planck's law and the Bose–Einstein distribution (Appendix D)

$$N_{\mathrm{ph}} = N_0 \frac{1}{\exp\left(\frac{E}{kT}\right) - 1} \,. \tag{10.7}$$

The prefactor is $N_0 = \hbar\omega^3 n_r^2/(\pi^2 c^3)$. Since for absorption and stimulated emission the same quantum-mechanical matrix element is responsible, $B_1 = B_2$. The detailed balance (10.6) yields

$$C(E_1, E_2) = B_1(E_1, E_2) N_{\mathrm{ph}} \left[\exp\left(\frac{E - (F_{\mathrm{n}} - F_{\mathrm{p}})}{kT}\right) - 1\right] \,. \tag{10.8}$$

In thermodynamic equilibrium, i.e. $F_{\mathrm{n}} = F_{\mathrm{p}}$,

$$C(E_1, E_2) = N_0 B_1(E_1, E_2) = B \,. \tag{10.9}$$

If the constant B, the bimolecular recombination coefficient, is independent of the energy E, the integration for the net recombination rate R can be executed analytically and we find

$$\begin{aligned} r_{\mathrm{B}} &= \int_{E_{\mathrm{g}}}^{\infty} \left[r_{\mathrm{sp}}(E) + r_{\mathrm{st}}(E) - r_{\mathrm{abs}}(E)\right] \mathrm{d}E \\ &= Bnp \left[1 - \exp\left(-\frac{F_{\mathrm{n}} - F_{\mathrm{p}}}{kT}\right)\right] \,. \end{aligned} \tag{10.10}$$

In thermodynamic equilibrium, of course, $r_{\mathrm{B}} = 0$. The recombination rate Bnp is then equal to the thermal generation rate G_{th}

$$G_{\mathrm{th}} = B n_0 p_0 \,. \tag{10.11}$$

The bimolecular recombination rate typically used in Shockley–Read–Hall (SRH) [333, 334] kinetics is

$$r_{\mathrm{B}} = B(np - n_0 p_0) \,. \tag{10.12}$$

Values for the coefficient B are given in Table 10.1. In the case of carrier injection, np is larger than in thermodynamical equilibrium, i.e. $np > n_0 p_0$, and the recombination rate is positive, i.e. light is emitted. If the carrier density is smaller than in thermodynamical equilibrium, e.g. in a depletion region, absorption is larger than emission. This effect is also known as 'negative luminescence' [335] and plays a role particularly at elevated temperatures and in the infrared spectral region.

10.2.5 Recombination Dynamics

The carrier densities n and p, are decomposed into the densities n_0 and p_0 in thermodynamic equilibrium and the excess-carrier densities δn and δp, respectively

$$n = n_0 + \delta n \tag{10.13a}$$

$$p = p_0 + \delta p \,. \tag{10.13b}$$

Table 10.1. Bimolecular recombination coefficient B at room temperature for a number of semiconductors. Data for GaN from [336], Si from [337], SiC from [338], other values from [339]

Material	B (cm^3/s)
GaN	1.1×10^{-8}
GaAs	1.0×10^{-10}
AlAs	7.5×10^{-11}
InP	6.0×10^{-11}
InAs	2.1×10^{-11}
4H-SiC	1.5×10^{-12}
Si	1.1×10^{-14}
GaP	3.0×10^{-15}

Here, only neutral excitations are considered, i.e. $\delta n = \delta p$. Obviously the time derivative fulfills $\frac{\partial n}{\partial t} = \frac{\partial\,\delta n}{\partial t}$, and correspondingly for the hole density. The equation for the dynamics

$$\dot{n} = \dot{p} = -Bnp + G_{\mathrm{th}} = -B(np - n_0 p_0) = -B(np - n_{\mathrm{i}}^2) \tag{10.14}$$

can be written as

$$\frac{\partial\,\delta p}{\partial t} = -B\left(n_0\,\delta p + p_0\,\delta n + \delta n\,\delta p\right)\,. \tag{10.15}$$

The general solution of (10.15) is given by

$$\delta p(t) = \frac{(n_0 + p_0)\delta p(0)}{[n_0 + p_0 + \delta p(0)]\exp\left(Bt(n_0 + p_0)\right) - \delta p(0)}\,. \tag{10.16}$$

In the following, we discuss some approximate solutions of (10.15). First, we treat the case of a small (neutral) excitation, i.e. $\delta n = \delta p \ll n_0, p_0$. The dynamic equation is in this case

$$\frac{\partial\,\delta p}{\partial t} = -B\left(n_0 + p_0\right)\delta p\,. \tag{10.17}$$

Then the decay of the excess-carrier density is exponential with a time constant (lifetime) τ given by

$$\tau = \frac{1}{B\left(n_0 + p_0\right)}\,. \tag{10.18}$$

In an n-type semiconductor additionally $n_0 \gg p_0$, and thus the minority carrier lifetime τ_p is

$$\tau_p = \frac{1}{Bn_0} . \quad (10.19)$$

If the nonequilibrium carrier densities are large, i.e. $n \approx p \gg n_0, p_0$, e.g. for strong injection, the kinetics obeys

$$\frac{\partial \delta p}{\partial t} = -B(\delta p)^2 , \quad (10.20)$$

and the transient has the form

$$\delta p(t) = \frac{\delta p(0)}{1 + Bt\delta p(0)} , \quad (10.21)$$

where $\delta p(0)$ is the excess hole density at time $t = 0$. Such a decay is called *hyperbolic* and the recombination is bimolecular. The exponential decay time is formally $\tau^{-1} = B\delta p(t)$ and is thus time and density dependent.

10.2.6 Lasing

The net rate for stimulated emission and absorption is

$$r_{\mathrm{st}}(E) - r_{\mathrm{abs}}(E) = \left[1 - \exp\left(\frac{E - (F_{\mathrm{n}} - F_{\mathrm{p}})}{kT}\right)\right] \quad (10.22)$$

$$\times \int_{E_{\mathrm{C}}}^{E+E_{\mathrm{V}}} \mathrm{d}E_{\mathrm{e}} B D_{\mathrm{e}}(E_{\mathrm{e}}) f_{\mathrm{e}}(E_{\mathrm{e}}) D_{\mathrm{h}}(E_{\mathrm{e}} - E) f_{\mathrm{h}}(E_{\mathrm{e}} - E) N_{\mathrm{ph}}(E) .$$

The net rate at photon energy $E = \hbar\omega$ is only larger than zero (i.e. dominating stimulated emission) when

$$F_{\mathrm{n}} - F_{\mathrm{p}} > E \geq E_{\mathrm{g}} . \quad (10.23)$$

When the difference of the quasi-Fermi levels is larger than the bandgap, the carrier population is inverted, i.e. close to the band edges the conduction-band states are more strongly populated with electrons than the valence-band states, as shown in Fig. 10.3. An incoming optical wave of energy E will then be net amplified by stimulated emission. Equation (10.23) is also called the thermodynamic laser condition. We note that lasing requires further conditions as discussed in Sect. 20.4.

10.3 Free-Exciton Recombination

The observation of free-excitons is limited for semiconductors with a small exciton binding energies (such as in GaAs) to low temperatures. However, for large exciton binding energy, recombination from free-excitons is observed even at room temperature, as shown in Fig. 10.4 for ZnO.

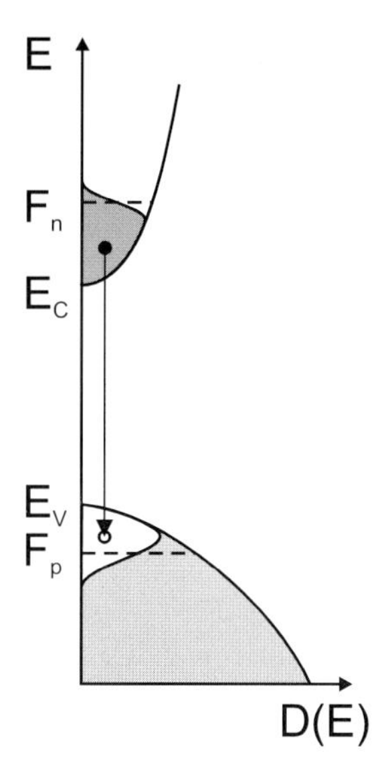

Fig. 10.3. Charge-carrier distribution during inversion, necessary for lasing. *Shaded areas* are populated with electrons. A stimulated transition between an electron and a hole is indicated

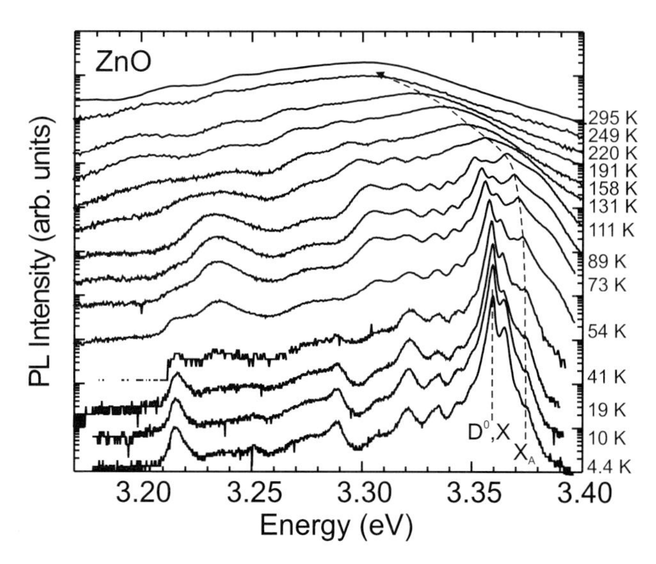

Fig. 10.4. Temperature-dependent luminescence spectra of a ZnO thin film (on sapphire). At low temperatures, the spectra are dominated by donor-bound exciton transitions (Al^0,X)). The *vertical dashed line* indicates the low-temperature position of the donor-bound exciton transition (D^0,X). The *curved dashed line* visualizes the energy position of the free-exciton transition (X_A) that becomes dominant at room temperature

10.4 Bound-Exciton Recombination

Excitons can localize at impurities or other potential fluctuations and subsequently recombine. Excitons can be bound to neutral or ionized impurities (donors and acceptors). The transition energy $\hbar\omega$ of an exciton bound to a neutral impurity is

$$\hbar\omega = E_{\mathrm{g}} - E_{\mathrm{X}}^{\mathrm{b}} - Q \,, \tag{10.24}$$

where Q is the binding energy of the exciton to the impurity. A transition involving an exciton bound to a neutral donor is denoted (D^0,X); correspondingly (D^+,X), also denoted as (h,D^0), and (A^0,X). In Si, the binding energy to the impurity is about one tenth of the binding energy of the impurity (Haynes's rule [340]), i.e. $Q/E_{\mathrm{D}}^{\mathrm{b}}$ and $Q/E_{\mathrm{A}}^{\mathrm{b}} \approx 0.1$ (Fig. 10.5a). In GaP the approximate relations $Q = 0.26E_{\mathrm{D}}^{\mathrm{b}} - 7\,\mathrm{meV}$ and $Q = 0.056E_{\mathrm{A}}^{\mathrm{b}} + 3\,\mathrm{meV}$ have been found [341]. For donors in ZnO $Q = 0.40E_{\mathrm{D}}^{\mathrm{b}} - 5.34\,\mathrm{meV}$ holds (Fig. 10.5b) [342].

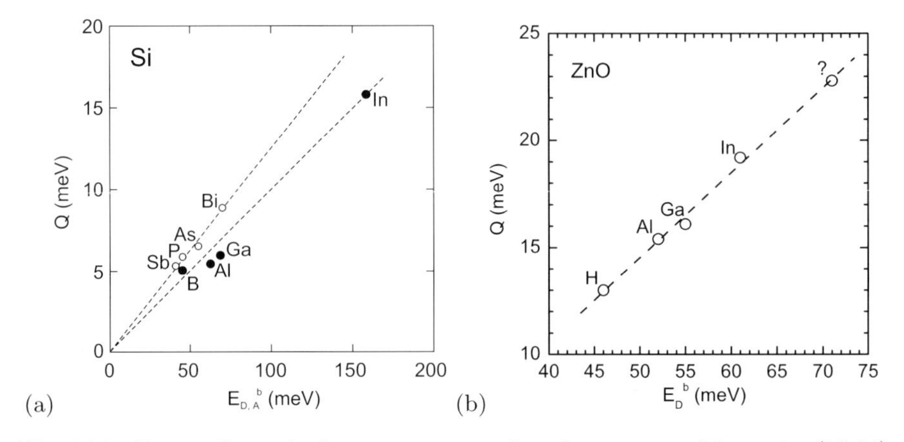

Fig. 10.5. Energy Q required to remove an exciton from a neutral impurity (10.24) as a function of the ionization energy $E_{\mathrm{D}}^{\mathrm{b}}$ (*open circles*) or $E_{\mathrm{A}}^{\mathrm{b}}$ (*solid circles*) of the involved impurity in (**a**) silicon (experimental data from [340]) and (**b**) ZnO (experimental data from [342])

In Fig. 10.6, the recombination spectrum of GaAs:C is shown that exhibits recombination from excitons bound to the acceptor (carbon) and shallow donors. The exciton is more strongly bound to an ionized donor (D^+) than to a neutral donor.

The peak labeled $(\mathrm{D}^0,\mathrm{X})_{2\mathrm{s}}$ in Fig. 10.6 is called a two-electron satellite (TES) [344]. High-resolution spectra of the TES in GaAs [186,345] are shown in Fig. 10.7a. The TES recombination is a (D^0,X) recombination that leaves the donor in an excited state as schematically shown in Fig. 10.7b. Therefore a hydrogen-like series with $n = 2, 3, \ldots$ is observed with energies

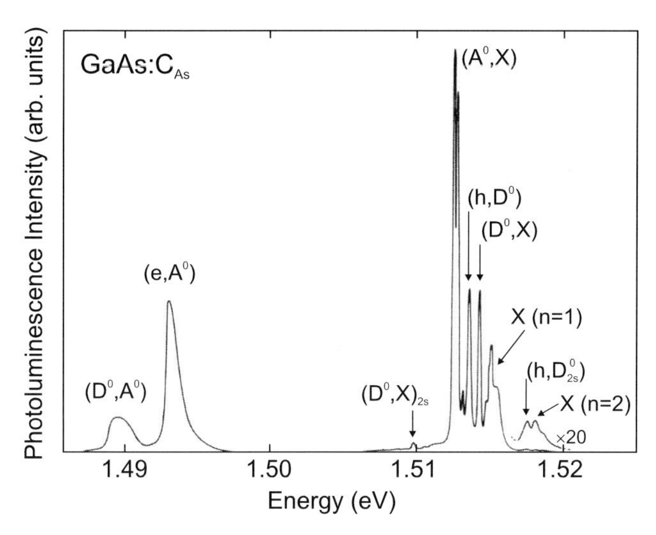

Fig. 10.6. Photoluminescence spectrum ($T = 2\,\mathrm{K}$, $D = 10\,\mathrm{mW\,cm^{-2}}$) of GaAs:C$_{\mathrm{As}}$ ($N_{\mathrm{A}} = 10^{14}\,\mathrm{cm^{-3}}$) with donor- and acceptor-related bound-exciton recombination around 1.512 eV, (e,A^0), (h,D^0) and (D^0,A^0) pair and free-exciton recombination. Adapted from [343]

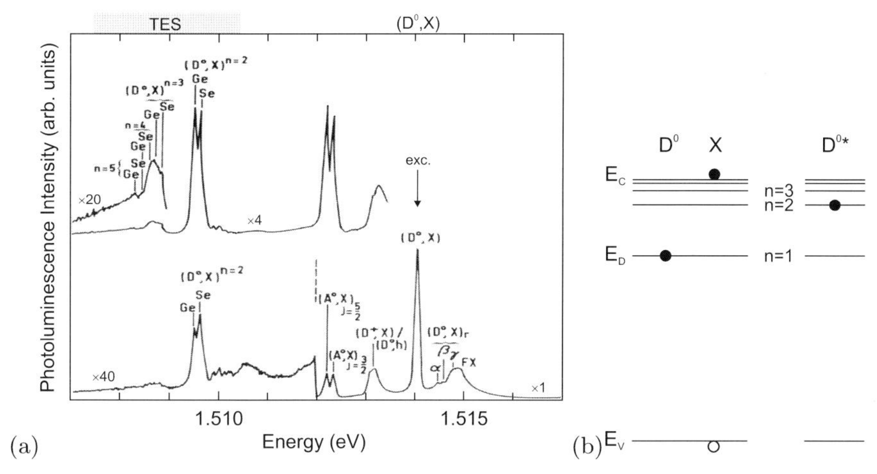

Fig. 10.7. (**a**) Photoluminescence spectrum ($T = 1.5\,\mathrm{K}$, $D = 50\,\mathrm{mW\,cm^{-2}}$) of high-purity GaAs with two donors (Ge and Se/Sn). The lower spectrum has been excited 6 meV above the bandgap, the upper spectrum has been resonantly excited with the laser set to the (D^0,X) transition and exhibits $n = 2, 3, 4$, and 5 TES transitions. α,β,γ denote excited (hole rotational) states of the (D^0,X) complex. Adapted from [345]. (**b**) Schematic representation of the $n = 2$ TES process, *left*: initial, *right*: final state

$$E_{\rm TES}^{n} = E_{({\rm D}^0,{\rm X})} - E_{\rm D}^{\rm b}\left(1 - \frac{1}{n^2}\right) . \tag{10.25}$$

The effect of isotope disorder on the sharpness and splitting of impurity states has been investigated in [346,347]. The recombination of excitons bound to Al, Ga and In in natural silicon (92.23% ^{28}Si, 4.67% ^{29}Si, 3.10% ^{30}Si) is split into three lines due to the valley-orbit splitting [348] of electron states at the band minimum. Each of these (A^0,X) lines is split by 0.01 cm^{-1} for Si:Al due to a symmetry reduction of the 4-fold degenerate A^0 ground state, as observed in the presence of applied axial strain or an electric field. The comparison to spectra from enriched ^{28}Si shows that the observed splitting without external perturbation is due to isotope disorder that causes random strains and splits the A^0 ground state into two doublets [347]. Similarly, the (unsplit) phosphorus-induced (D^0,X) transition in enriched Si is found to be much sharper ($<40\,\mu$eV) than in natural Si (330 µeV) [346].

10.5 Alloy Broadening

The bound-exciton recombination peak in a binary compound is spectrally fairly sharp (Sect. 10.4), even in the presence of isotope disorder. In an alloy (see Sect. 3.7), the random distribution of atoms (with different atomic order number Z) causes a significant broadening effect of the luminescence (and absorption) line, the so-called *alloy broadening* [349]. As an example, we treat $Al_xGa_{1-x}As$. The exciton samples, at different positions of the lattice, different coordinations of Ga and Al atoms. If the experiment averages over these configurations, an inhomogeneously broadened line is observed.

The cation concentration $c_{\rm c}$ for the zincblende lattice is given as

$$c_{\rm c} = 4\,a_0^{-3} , \tag{10.26}$$

where a_0 is the lattice constant. $c_{\rm c} = 2.21 \times 10^{22}\,{\rm cm}^{-3}$ for $Al_xGa_{1-x}As$ in the entire composition range $0 \le x \le 1$ since the lattice constant does not vary significantly. In a random alloy, the probability $p(N)$ to find exactly N Ga atoms in a given volume V (with a total of $c_{\rm c}V$ cations) is given by the binomial distribution

$$p(N) = \binom{c_{\rm c}V}{N} x^N (1-x)^{c_{\rm c}V-N} . \tag{10.27}$$

The sampling volume for a luminescence event is the exciton volume (cf. (9.28)) that is given for the free-exciton as

$$V_{\rm ex} = \frac{4\pi}{3} a_{\rm X}^3 = \frac{4\pi}{3}\left(\frac{m_0}{m_{\rm r}^*}\epsilon_{\rm s} a_{\rm B}\right)^3 . \tag{10.28}$$

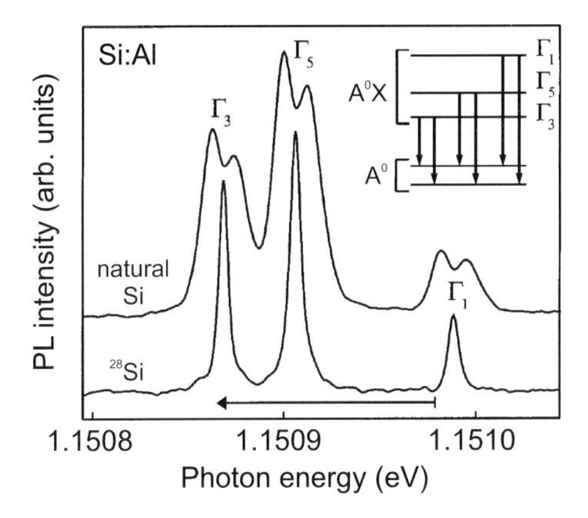

Fig. 10.8. High-resolution photoluminescence (PL) spectra of (A^0,X) recombination in natural and ^{28}Si-enriched silicon doped with aluminum (T = 1.8 K). The ^{28}Si PL spectrum is shifted up in energy by 0.114 meV, as indicated by the arrow, to compensate for the shift in bandgap. The *inset* shows a level scheme for the recombination in natural silicon. Adapted from [347], reprinted with permission, ©2002 APS

In GaAs there are about 1.6×10^5 cations in the exciton volume. In $\mathrm{Al}_x\mathrm{Ga}_{1-x}\mathrm{As}$, there are on average $x c_{\mathrm{c}} V_{\mathrm{ex}}$ Al atoms in the exciton volume. The fluctuation is given by the standard deviation of the binomial distribution

$$\sigma_{\mathrm{x}}^2 = \frac{x(1-x)}{c_{\mathrm{c}} V_{\mathrm{ex}}} \ . \tag{10.29}$$

The corresponding energetic broadening (full width at half-maximum) of the spectral line Δ_{E} is

$$\Delta_{\mathrm{E}} = 2.36 \frac{\partial E_{\mathrm{g}}}{\partial x} \sigma_{\mathrm{x}} = 2.36 \frac{\partial E_{\mathrm{g}}}{\partial x} \left[\frac{x(1-x)}{c_{\mathrm{c}} V_{\mathrm{ex}}} \right]^{1/2} \ . \tag{10.30}$$

This theoretical dependence is shown for $\mathrm{Al}_x\mathrm{Ga}_{1-x}\mathrm{As}$ in Fig. 10.9 together with experimental data.

10.6 Phonon Replica

The momentum selection rule for free-exciton recombination allows only excitons with $\mathbf{K} \approx 0$ (for $\mathbf{K}$ cf. (9.26)) to recombine. The fine structure of this recombination is connected to polariton effects (cf. Sect. 9.4.8). Excitons with large $\mathbf{K}$ can recombine if a phonon or several phonons are involved that provide the necessary momentum $\mathbf{q} = \mathbf{K}_1 - \mathbf{K}_2$, with $\mathbf{K}_1$ ($\mathbf{K}_2$) being the wavevector of the initial (intermediate) exciton state (Fig. 10.10). A so-called

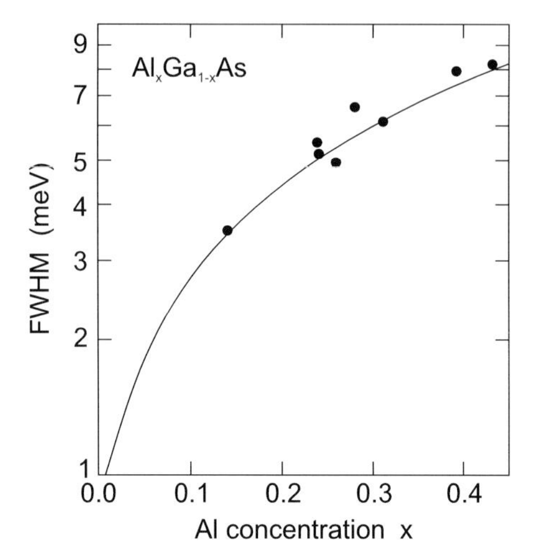

Fig. 10.9. Spectral width of the (A^0,X) recombination photoluminescence line of excitons bound to carbon acceptors in $Al_xGa_{1-x}As$ with various Al content within the direct-bandgap regime. *Full circles* are experimental data at $T = 2$ K and low excitation density from various sources, *solid line* is theory according to (10.30). Experimental data from [349]

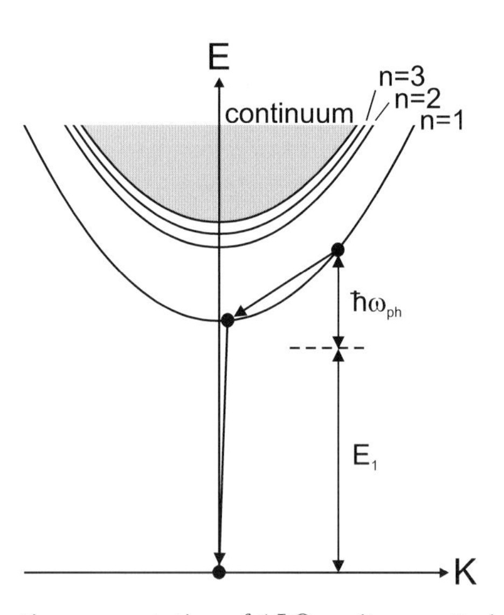

Fig. 10.10. Schematic representation of 1 LO exciton scattering of an exciton at $\mathbf{K} \neq 0$ to an intermediate state with $\mathbf{K} \approx 0$ and subsequent radiative decay. $\hbar\omega_{ph}$ represents the phonon energy and E_1 the energy of the emitted photon

zero-phonon line at energy E_0 is then accompanied by *phonon replica* below E_0 at integer multiples (at low temperature) of the (LO) phonon energy $\hbar\omega_{\mathrm{ph}}$

$$E_n = E_0 - n\hbar\omega_{\mathrm{ph}} \,. \tag{10.31}$$

Phonon replica have been observed in many polar semiconductors such as CdS [350] and ZnSe [351]. A sequence of such phonon replica, as observed in GaN [352], is depicted in Fig. 10.11a.

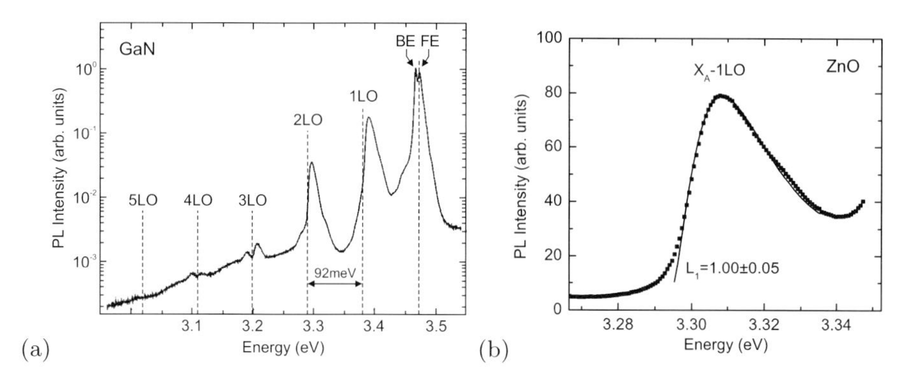

Fig. 10.11. (**a**) Photoluminescence spectrum of GaN (grown on SiC substrate) at $T = 50\,\mathrm{K}$. In addition to emission from free (FE) and bound (BE) excitons several phonon replica (labeled as 1LO–5LO) are observed. *Vertical dashed lines* indicate energy positions of multiple LO-phonon energies ($\hbar\omega_{\mathrm{LO}} = 92\,\mathrm{meV}$) below the FE peak. Adapted from [352]. (**b**) Photoluminescence spectrum of 1LO phonon-assisted recombination peak at $T = 103\,\mathrm{K}$ (from the data of Fig. 10.4). Data points (*dots*) and lineshape fit (*solid line*) according to (10.32) with the parameters $L_1 = 1.00 \pm 0.05$ and $E_1 = 3.2947\,\mathrm{eV}$ (and background)

The lineshape of the n-th phonon-assisted line is proportional to the exciton population at a given excess energy, which is proportional to the density of states and the Boltzmann distribution function [353]

$$I_n(E_{\mathrm{ex}}) \propto E_{\mathrm{ex}}^{1/2} \exp\left(-\frac{E_{\mathrm{ex}}}{kT}\right) w_n(E_{\mathrm{ex}}) \,. \tag{10.32}$$

Here, E_{ex} represents the exciton kinetic energy. The factor $w_n(E_{\mathrm{ex}})$ accounts for the $\mathbf{q}$-dependence of the matrix element. It is typically expressed as

$$w_n(E_{\mathrm{ex}}) \propto E_{\mathrm{ex}}^{L_n} \,. \tag{10.33}$$

Accordingly, the energy separation ΔE_n of the energy of the peak maximum of phonon replica from E_0 is given by

$$\Delta E_n = E_n - E_0 = n\hbar\omega_{\mathrm{ph}} + \left(L_n + \frac{1}{2}\right) kT \,. \tag{10.34}$$

It is found theoretically that $L_1 = 1$ and $L_2 = 0$. These relations are experimentally verified and, e.g., fulfilled for the peaks of Fig. 10.11a. A lineshape fit for the 1 LO phonon-assisted transition in ZnO is shown in Fig. 10.11b.

In Fig. 10.12a the 'green band' emission of ZnO is shown as presented in [354]. This band is mostly attributed to a Cu impurity; recently, evidence has grown from isotope decay and annealing studies that it is related to the zinc vacancy [355] (Fig. 10.12b). The zero phonon line is followed by many replica with a maximum at about 6 LO phonons. The intensity I_N of the N-th replica is given by [356, 357]

$$I_N \propto \exp(-S) \frac{S^N}{N!} , \tag{10.35}$$

where S is the so-called Huang–Rhys parameter. In [355], a coupling parameter of $S = 6.9$ has been determined.

Equation (10.35) is obtained from the consideration of transitions in the configuration diagram [356, 358] (Fig. 10.13), using the Born–Oppenheimer approximation, i.e. the electronic wavefunctions are separated from the vibrational wavefunctions, the Frank–Condon principle, i.e. that optical transitions occur with the positions of the nuclei fixed and thus vertical in the configuration diagram Fig. 10.13 and assuming low temperatures, i.e. only the lowest state is (partially) occupied. The Huang–Rhys parameter, the average number of phonons involved in the transition, is related to the displacement $\delta q = q_1 - q_0$ of the two configurations

$$S = \frac{C \delta q^2}{2\hbar\omega_{\mathrm{ph}}} , \tag{10.36}$$

where C is the curvature of the parabola, $C = \mathrm{d}^2E/\mathrm{d}q^2$.

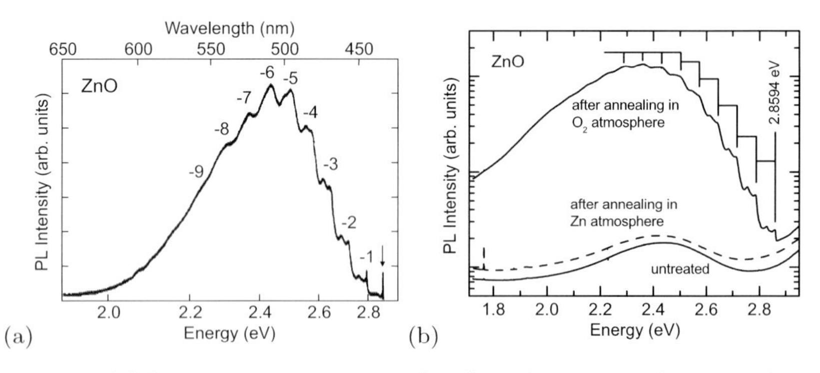

Fig. 10.12. (**a**) Luminescence spectrum of ZnO in the visible. The *arrow* denotes the zero-phonon line at 2.8590 eV. The numbers of the phonon replica are labeled. Adapted from [354]. (**b**) Luminescence spectra (*solid lines*) of a ZnO bulk crystal before ('untreated') and after annealing in O_2 atmosphere at $T = 1073$ K. After annealing in Zn atmosphere at the same temperature, the green band disappears again (*dashed line*). From [355]

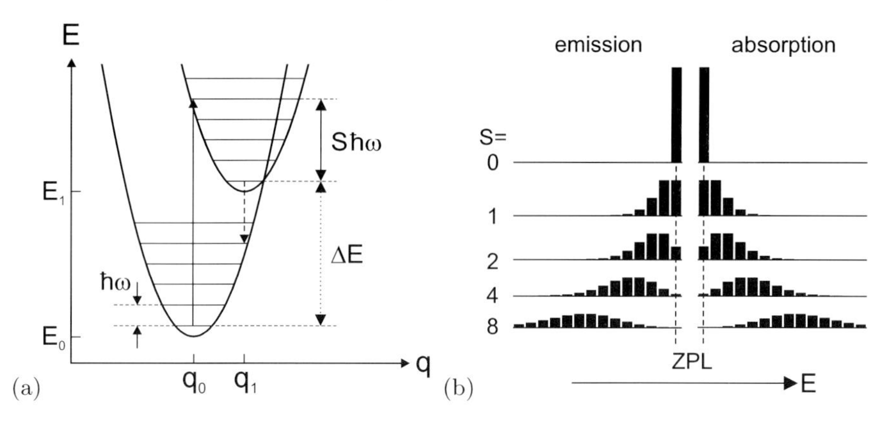

Fig. 10.13. (**a**) Configuration diagram of two states that differ in their configuration coordinate by $\delta q = q_1 - q_0$. Both are coupled to phonons of energy $\hbar\omega$. The absorption maximum (*solid vertical line*) and emission maximum (*dashed vertical line*) are shifted with respect to the zero-phonon line (*dotted vertical line*) with energy $E_1 - E_0$. The Huang–Rhys parameter is $S \sim 4$. (**b**) Intensity of zero-phonon line ('ZPL') and phonon replica (10.35) for emission and absorption processes with different values of the Huang–Rhys parameter S as labeled

For small $S \ll 1$, we are in the weak coupling regime and the zero-phonon line is the strongest. In the strong coupling regime, $S > 1$, the maximum is (red-) shifted from the zero-phonon line. We note that in absorption, phonon replica occur on the high-energy side of the zero-phonon absorption. For large S the peak intensities are close to a Gaussian.

10.7 Donor–Acceptor Pair Transitions

Optical transitions can occur between neutral donors and acceptors. The (spatially indirect) donor–acceptor pair (DAP) recombination is present in (partially) compensated semiconductors and follows the scheme $D^0A^0 \rightarrow D^+A^-eh \rightarrow D^+A^- + \gamma$, where γ is a photon with the energy $\hbar\omega$. The energy of the emitted photon is given by

$$\hbar\omega = E_{\mathrm{g}} - E_{\mathrm{D}}^{\mathrm{b}} - E_{\mathrm{A}}^{\mathrm{b}} - \frac{1}{4\pi\epsilon_0}\frac{e^2}{\epsilon_{\mathrm{r}}R}\,, \tag{10.37}$$

where R is the distance between the donor and the acceptor for a specific pair. Since R is discrete, the DAP recombination spectrum consists of several discrete lines. If the donor and acceptor occupy the same sublattice, e.g. O and C both substituting P sites in GaP, the spatial distance of the donor and acceptor is $R(n) = a_0\sqrt{n/2}$, where a_0 is the lattice constant and n is an integer. However, for certain 'magic' numbers $n = 14, 30, 46, \ldots$ no lattice points exist and therefore the corresponding lines are missing (labeled 'G'

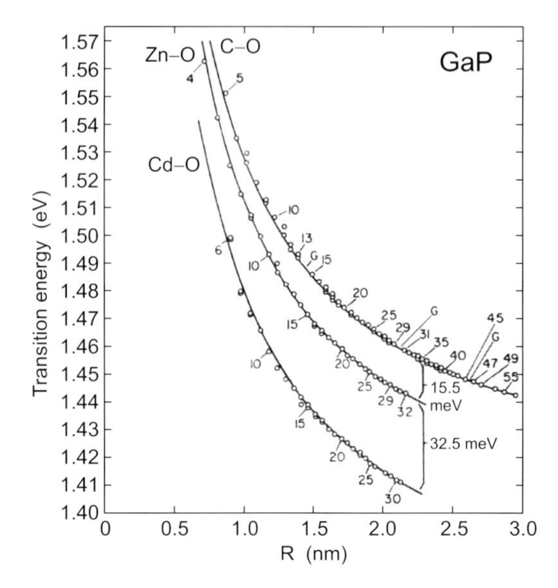

Fig. 10.14. Transition energies in GaP ($T = 1.6$ K) of the donor–acceptor recombination involving the deep oxygen donor and C, Zn, and Cd acceptors, respectively. The lines follow (10.37) for $E_{\mathrm{g}}^{\mathrm{GaP}} = 2.339$ eV, $\epsilon_{\mathrm{r}} = 11.1$ and $(E_{\mathrm{D}}^{\mathrm{b}})_{\mathrm{O}} = 893$ meV, $(E_{\mathrm{A}}^{\mathrm{b}})_{\mathrm{C}} = 48.5$ meV, $(E_{\mathrm{A}}^{\mathrm{b}})_{\mathrm{Zn}} = 64$ meV, and $(E_{\mathrm{A}}^{\mathrm{b}})_{\mathrm{Cd}} = 96.5$ meV. Predicted missing modes for GaP:C,O are labeled with 'G'. Adapted from [359]

in Fig. 10.14). No such gaps exist in DA spectra where donors and acceptors occupy different sublattices, e.g. GaP:O,Zn (see also Fig. 10.14). In this case, the spatial separation is given by $R(n) = a_0\sqrt{n/2 - 5/16}$. If significant broadening is present, the lines are washed out and a donor–acceptor pair band forms.

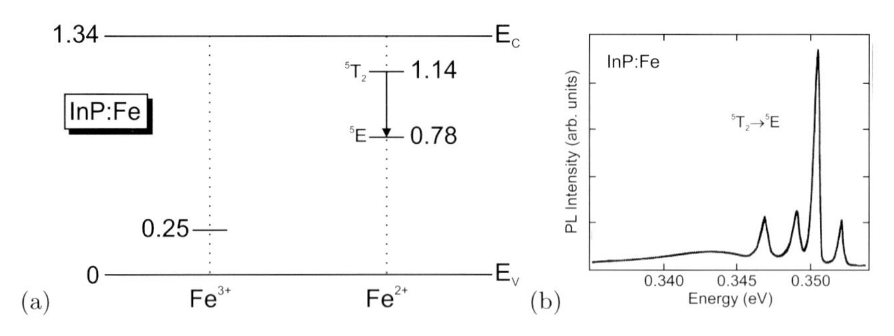

Fig. 10.15. (**a**) Schematic band diagram of InP with levels of Fe impurities in the 3+ and 2+ charge states at low temperature. All energies are given in eV. The *arrow* denotes the optical transition from an excited Fe^{2+} state to the Fe^{2+} ground state. (**b**) Photoluminescence spectrum (at $T = 4.2$ K) of InP:Fe sample with [Fe]=5×10^{16} cm^{-3}. Part (**b**) adapted from [361].

10.8 Inner-Impurity Recombination

The transitions of electrons between different states of an impurity level can be nonradiative or radiative. As an example, the radiative transition of electrons in the Fe^{2+} state in InP $^5T_2 \rightarrow ^5E$ (Fig. 10.15) and its fine structure were observed first in [360] at around 0.35 eV.

10.9 Auger Recombination

In competition with the radiative, bimolecular recombination is the Auger recombination (Fig. 10.16). In the Auger process, the energy that is released during the recombination of an electron and hole is not emitted with a photon but, instead, transferred to a third particle. This can be an electron (eeh, Fig. 10.16a) or a hole (hhe, Fig. 10.16b). The energy is eventually transferred nonradiatively from the hot third carrier via phonon emission to the lattice. The probability for such process is $\propto n^2p$ if two electrons are involved and $\propto np^2$ if two holes are involved. The Auger process is a three-particle process and becomes likely for high carrier density, either through doping, in the presence of many excess carriers, or in semiconductors with small bandgap. Auger recombination is the inverse of the impact ionization (see Sect. 8.5.3).

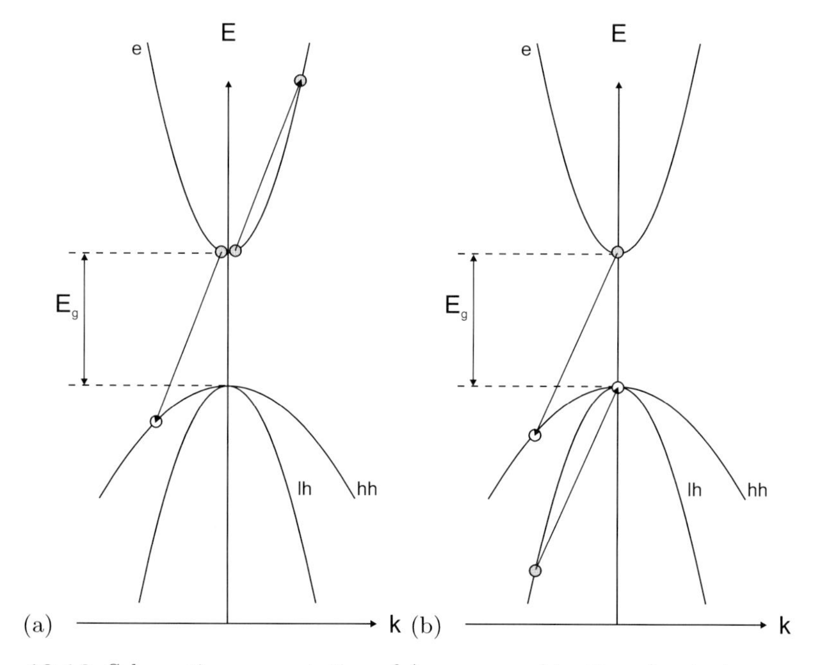

Fig. 10.16. Schematic representation of Auger recombination. An electron recombines with a hole and transfers the energy to (**a**) another electron in the conduction band, (**b**) another electron in the valence band

In thermodynamic equilibrium the rates for Auger recombination and thermal Auger generation must be equal, thus

$$G_{\mathrm{th}} = C_{\mathrm{n}} n_0^2 p_0 + C_{\mathrm{p}} n_0 p_0^2 \,, \tag{10.38}$$

where C_{n} and C_{p} denote the Auger recombination coefficients. The equation for the dynamics in the presence of excess carriers (if solely Auger recombination is present) is given as

$$\frac{\partial\,\delta n}{\partial t} = G_{\mathrm{th}} - R = -C_{\mathrm{n}}(n^2 p - n_0^2 p_0) - C_{\mathrm{p}}(n p^2 - n_0 p_0^2) \,. \tag{10.39}$$

The Auger recombination rate typically used in SRH kinetics is

$$r_{\mathrm{Auger}} = (C_{\mathrm{n}} n + C_{\mathrm{p}} p)(np - n_0 p_0) \,. \tag{10.40}$$

Typical values for the Auger recombination coefficients are given in Table 10.2.

Table 10.2. Auger recombination coefficients for some semiconductors. Data for InSb from [362], SiC from [338], others from [339]

Material	C_{n} (cm^6/s)	C_{p} (cm^6/s)
4H-SiC	5×10^{-31}	2×10^{-31}
Si, Ge	2.8×10^{-31}	9.9×10^{-32}
GaAs, InP	5.0×10^{-30}	3.0×10^{-30}
InSb	1.2×10^{-26}	

10.10 Band–Impurity Recombination

Another recombination process is the capture of carriers by impurities. This process is in competition with all other recombination processes, e.g. the radiative recombination and the Auger mechanism. The band–impurity recombination is the inverse process to the carrier release from impurities. It is particularly important at low carrier densities, for high dopant concentration and in indirect semiconductors since for these the bimolecular recombination is slow. The theory of capture on and recombination involving impurities is called Shockley–Read–Hall (SRH) kinetics [333,334]. An example of radiative band–impurity recombination (of the type shown in Fig. 10.17a) is shown in Fig. 10.6 for the (e,A^0) recombination at the carbon acceptor in GaAs.

We consider electron traps (see Fig. 10.17) with a concentration N_{t} with an energy level E_{t}. In thermodynamic equilibrium they have an electron population

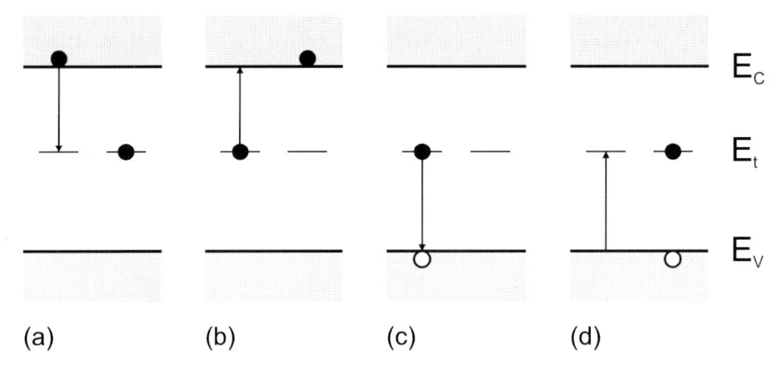

Fig. 10.17. Band-to-impurity processes at an impurity with one level (*left*: initial, *right*: final state in each part): (**a**) electron capture (from conduction band), (**b**) electron emission (into conduction band), (**c**) hole capture (from valence band), (**d**) hole emission (into valence band). The *arrows* indicate the transition of the electron

$$f_{\mathrm{t}}^{0} = \frac{1}{\exp\left(\frac{E_{\mathrm{t}}-E_{\mathrm{F}}}{kT}\right)+1}\,, \tag{10.41}$$

where f_{t} is the nonequilibrium population of the trap. Then the capture rate r_{c} is proportional to the density of unoccupied traps and the electron concentration, $r_{\mathrm{c}} \propto nN_{\mathrm{t}}(1-f_{\mathrm{t}})$. The proportionality factor has the form $v_{\mathrm{th}}\sigma_{\mathrm{n}}$, where v_{th} is the thermal velocity $v_{\mathrm{th}} = (3kT/m^*)^{1/2} \approx 10^7$ cm/s at room temperature and σ_{n} is the capture cross section that is of the order of the lattice constant a_0^2, typically $\sim 10^{-15}\mathrm{cm}^2$. In order to make the following calculation more transparent, we put the effective-mass ratio $(m_0/m^*)^{1/2}$ into σ in the following and thus have the same thermal velocity $v_{\mathrm{th}} = (3kT/m_0)^{1/2}$ for electrons and holes. The capture rate of electrons is

$$r_{\mathrm{c}} = v_{\mathrm{th}}\sigma_{\mathrm{n}} nN_{\mathrm{t}}(1-f_{\mathrm{t}})\,. \tag{10.42}$$

The emission rate from filled traps is

$$g_{\mathrm{c}} = e_{\mathrm{n}} N_{\mathrm{t}} f_{\mathrm{t}}\,, \tag{10.43}$$

where e_{n} denotes the emission probability. In a similar way, the emission and capture rates for holes can be written:

$$r_{\mathrm{v}} = v_{\mathrm{th}}\sigma_{\mathrm{p}} pN_{\mathrm{t}} f_{\mathrm{t}} \tag{10.44}$$

$$g_{\mathrm{v}} = e_{\mathrm{p}} N_{\mathrm{t}}(1-f_{\mathrm{t}})\,. \tag{10.45}$$

In thermodynamical equilibrium, capture and generation rates are equal, i.e. $r_{\mathrm{c}} = g_{\mathrm{c}}$ and $r_{\mathrm{v}} = g_{\mathrm{v}}$. Thus, the emission probability is

$$e_{\mathrm{n}} = v_{\mathrm{th}}\sigma_{\mathrm{n}} n_0 \frac{1-f_{\mathrm{t}}^0}{f_{\mathrm{t}}^0}\,. \tag{10.46}$$

Using $\frac{1-f_t^0}{f_t^0} = \exp\left(\frac{E_t - E_F}{kT}\right)$, (7.15) and (7.16) the emission probabilities can be written as

$$e_n = v_{th}\sigma_n n_t \tag{10.47}$$

$$e_v = v_{th}\sigma_p p_t \ , \tag{10.48}$$

with $n_t = N_C \exp\left(\frac{E_t - E_C}{kT}\right)$ and $p_t = N_V \exp\left(-\frac{E_t - E_V}{kT}\right)$. We note that (cf. (7.20))

$$n_t p_t = N_C N_V \exp\left(-\frac{E_C - E_V}{kT}\right) = n_0 p_0 \ . \tag{10.49}$$

The temperature dependence of the thermal velocity is $\propto T^{1/2}$, the temperature dependence of the band-edge density of states is $\propto T^{3/2}$ (7.10) and (7.13). Thus, the temperature dependence of the emission rate is (apart from the exponential term) $\propto T^2$ if σ is temperature independent.

Charge conservation requires $r_c - r_v = g_c - g_v$, also in non-equilibrium. From this we obtain the population function in nonequilibrium:

$$f_t = \frac{\sigma_n n + \sigma_p p}{\sigma_n(n + n_t) + \sigma_p(p + p_t)} \ . \tag{10.50}$$

The recombination rate r_{b-i} of the band–impurity recombination is then

$$r_{b-i} = -\frac{\partial \delta n}{\partial t} = r_c - g_c \tag{10.51}$$

$$= \frac{\sigma_n \sigma_p v_{th} N_t}{\sigma_n(n + n_t) + \sigma_p(p + p_t)} (np - n_0 p_0) \ .$$

Using the 'lifetimes' $\tau_{p_0} = (\sigma_n v_{th} N_t)^{-1}$ and $\tau_{n_0} = (\sigma_p v_{th} N_t)^{-1}$, this is typically written as

$$r_{b-i} = \frac{1}{\tau_{p_0}(n + n_t) + \tau_{n_0}(p + p_t)} (np - n_0 p_0) \ . \tag{10.52}$$

For an n-type semiconductor the Fermi level is above E_t and the traps are mostly full. Thus hole capture is the dominating process. The equation for the dynamics simplifies to

$$\frac{\partial \delta p}{\partial t} = -\frac{p - p_0}{\tau_{p_0}} \ . \tag{10.53}$$

Thus, an exponential decay with minority-carrier lifetime τ_{p_0} (or τ_{n_0} for p-type material) occurs.

A recombination center is most effective when it is close to the middle of the bandgap (midgap level). The condition $\partial r_{b-i}/\partial E_t = 0$ leads to the trap energy E_t^{max} with the maximum recombination rate being located at

$$E_{\mathrm{t}}^{\max} = \frac{E_{\mathrm{C}} + E_{\mathrm{V}}}{2} - kT \ln \left(\frac{\sigma_{\mathrm{n}} N_{\mathrm{C}}}{\sigma_{\mathrm{p}} N_{\mathrm{V}}} \right) . \tag{10.54}$$

The curvature $\partial^2 r_{\mathrm{b-i}}/\partial E_{\mathrm{t}}^2$ at $E_{\mathrm{t}}^{\max}$ is proportional to $-(np - n_0 p_0)$ and thus indeed is negative in the presence of excess carriers. The maximum recombination rate (for $E_{\mathrm{t}} = E_{\mathrm{t}}^{\max}$) is

$$\begin{aligned} r_{\mathrm{b-i}}^{\max} &= \frac{\sigma_{\mathrm{n}} \sigma_{\mathrm{p}} v_{\mathrm{th}}}{2 n_{\mathrm{i}} (\sigma_{\mathrm{n}} \sigma_{\mathrm{p}})^{1/2} + n \sigma_{\mathrm{n}} + p \sigma_{\mathrm{p}}} N_{\mathrm{t}} \, (np - n_{\mathrm{i}}^2) \\ &= (\sigma_{\mathrm{n}} \sigma_{\mathrm{p}})^{1/2} v_{\mathrm{th}} \, N_{\mathrm{t}} \, \gamma \left(\frac{np}{n_{\mathrm{i}}} - n_{\mathrm{i}} \right) \approx \sigma v_{\mathrm{th}} E_{\mathrm{t}} \, \frac{np - n_{\mathrm{i}}^2}{n + p + 2 n_{\mathrm{i}}} \, , \end{aligned} \tag{10.55}$$

with

$$\gamma^{-1} = 2 + \frac{n \sigma_{\mathrm{n}} + p \sigma_{\mathrm{p}}}{n_{\mathrm{i}} (\sigma_{\mathrm{n}} \sigma_{\mathrm{p}})^{1/2}} \approx 2 + \frac{n + p}{n_{\mathrm{i}}} \, , \tag{10.56}$$

where the approximations are for $\sigma = \sigma_{\mathrm{n}} = \sigma_{\mathrm{p}}$.

The energetic width (FWHM) ΔE_{t} for which the recombination rate has dropped by a factor of 2 from this maximum value is

$$\Delta E_{\mathrm{t}} = 2kT \ln \left(\frac{1 + 2\gamma + (1 + 4\gamma)^{1/2}}{2\gamma} \right) . \tag{10.57}$$

For large excess carrier concentration $n + p \gg n_{\mathrm{i}}$ and $\gamma \approx n_{\mathrm{i}}/(n + p) \ll 1$, the width is

$$\Delta E_{\mathrm{t}} \approx 2kT \ln \gamma^{-1} \approx 2kT \ln \left(\frac{n + p}{n_{\mathrm{i}}} \right) \tag{10.58}$$

which can be significant for sufficiently high temperature. In the case of a constant density of impurity states (per energy) D_{t} around midgap the total recombination rate is (for $\Delta E_{\mathrm{t}} \ll E_{\mathrm{g}}$)

$$\begin{aligned} r_{\mathrm{b-i}} &= \int r_{\mathrm{b-i}}^{\max} \frac{D_{\mathrm{t}}}{E_{\mathrm{t}}} \frac{1}{1 + \gamma \, [2 \sinh(\hat{E}/(2kT))]^2} \, \mathrm{d}\hat{E} \\ &= 4kT \, r_{\mathrm{b-i}}^{\max} \frac{D_{\mathrm{t}}}{E_{\mathrm{t}}} \frac{\operatorname{arctanh}(\sqrt{1 - 4\gamma})}{\sqrt{1 - 4\gamma}} \, , \end{aligned} \tag{10.59}$$

with $\hat{E} = E - E_{\mathrm{t}}$. For large excess carrier concentration $\gamma \ll 1$:

$$r_{\mathrm{b-i}} \approx 2kT r_{\mathrm{b-i}}^{\max} \frac{D_{\mathrm{t}}}{E_{\mathrm{t}}} \ln \gamma^{-1} \, . \tag{10.60}$$

With $\sigma = \sigma_{\mathrm{n}} = \sigma_{\mathrm{p}}$ we find

$$r_{\mathrm{b-i}} \approx 2kT \sigma v_{\mathrm{th}} D_{\mathrm{t}} \frac{np}{n + p} \ln \left(\frac{n + p}{n_{\mathrm{i}}} \right) . \tag{10.61}$$

For a neutral excitation $n = p \gg n_{\rm i}$ the rate is

$$r_{\rm b-i} \approx kT\sigma v_{\rm th} D_{\rm t}\, n \,\ln\left(\frac{2n}{n_{\rm i}}\right) , \tag{10.62}$$

dominated by n. In the case of an unipolar excitation ($np \gg n_{\rm i}$ und $n \gg p, n_{\rm i}$), the minority carrier density p is the dominating factor,

$$r_{\rm b-i} \approx 2kT\sigma v_{\rm th} D_{\rm t}\, p \,\ln\left(\frac{n}{n_{\rm i}}\right) . \tag{10.63}$$

In the case of small carrier density $n + p \ll n_{\rm i}$ (10.59) becomes [364]

$$r_{\rm b-i} \approx \pi kT r_{\rm b-i}^{\rm max} \frac{D_{\rm t}}{E_{\rm t}} \approx -\frac{\pi}{2} kT\sigma v_{\rm th} D_{\rm t} n_{\rm i} . \tag{10.64}$$

A typical example is gold in silicon. The minority carrier lifetime decreases from 2×10^{-7} s to 2×10^{-10} s upon increase of the Au concentration from 10^{14} to 10^{17} cm^{-3}. The incorporation of recombination centers is an important measure for the design of high-frequency devices [363]. A reduction in minority-carrier lifetime can also be achieved by irradiation with high-energy particles and the subsequent generation of point defects with energy levels at midgap.

10.11 Field Effect

The emission of electrons from a trap is thermally activated with an ionization energy $E_{\rm i} = E_{\rm C} - E_{\rm t}$. If the trap is in a strong electric field $\mathcal{E}$, the emission probability can change. An acceptor-like trap after removal of the electron is neutral and its potential is short range. A donor has a long-range Coulomb potential after ionization. In an electric field, these potentials are modified as visualized in Fig. 10.18. Various additional processes can now occur.

10.11.1 Thermally Activated Emission

For the δ-like potential the ionization energy remains unchanged. For the Coulomb potential the barrier in the field direction is lowered by

$$\Delta E_{\rm i} = e \left(\frac{e\mathcal{E}}{\pi\epsilon_r}\right)^{1/2} . \tag{10.65}$$

The emission probability is increased in the field by $\exp(\Delta E_{\rm i}/kT)$. This effect is called the Poole–Frenkel effect and can be quite important. For silicon and $\mathcal{E} = 2 \times 10^5$ V/cm and $\Delta E_{\rm i} = 100$ meV a 50-fold increase of emission rate results at room temperature.

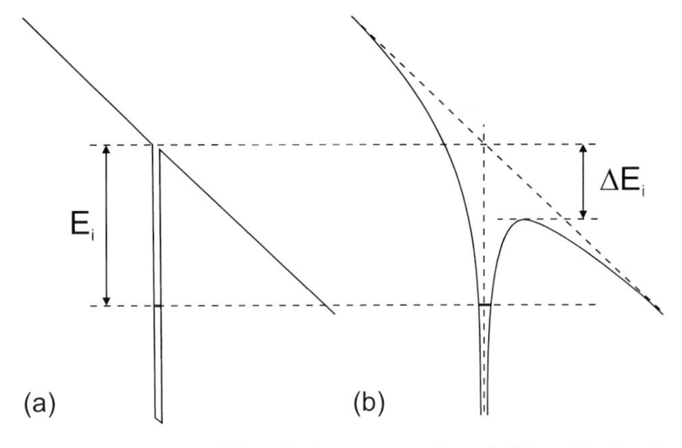

Fig. 10.18. Field effect at (**a**) a δ-like potential and (**b**) a Coulomb potential

10.11.2 Direct Tunneling

Carriers can tunnel from the trap level through the barrier in the field direction into the conduction band. This process is temperature independent. The transmission factor of a barrier is (in WKB approximation) proportional to $\exp\left(-\frac{2}{\hbar}\int\sqrt{2m[V(x)-E]}\mathrm{d}E\right)$. The emission probability for a triangular barrier is then

$$e_{\mathrm{n}} = \frac{e\mathcal{E}}{4(2m^* E_{\mathrm{i}})^{1/2}} \exp\left(-\frac{4(2m^*)^{1/2} E_{\mathrm{i}}^{3/2}}{3e\hbar\mathcal{E}}\right) . \tag{10.66}$$

In the case of a Coulomb-like potential the argument of the exponent in (10.66) needs to be multiplied by a factor $1 - (\Delta E_{\mathrm{i}}/E_{\mathrm{i}})^{5/3}$ with ΔE_{i} from (10.65).

10.11.3 Assisted Tunneling

In a thermally assisted tunneling process the electron on the trap level is first excited to a virtual level $E_{\mathrm{t}} + E_{\mathrm{ph}}$ by phonon absorption and then tunnels out of the trap (photon-assisted tunneling). From the energetically higher level the tunneling rate is higher. The probability is proportional to $\exp(E_{\mathrm{ph}}/kT)$. The additional energy can also be supplied by a photon (photon-assisted tunneling).

10.12 Multilevel Traps

Traps with multiple levels in the bandgap have generally similar but more complicated dynamics as compared to single-level traps. Lifetimes are an average over negatively and positively charged states of the trap.

10.13 Surface Recombination

A surface is typically a source of recombination, e.g. by midgap levels induced by the break of crystal symmetry. The recombination at surfaces [364] is modeled as a recombination current

$$j_{\mathrm{s}} = -eS(n_{\mathrm{s}} - n_0) \,, \tag{10.67}$$

where n_{s} is the carrier density at the surface and S is the so-called surface recombination velocity.

The surface recombination velocity for GaAs is shown in Fig. 10.19. For InP, if the surface Fermi level is pinned close to midgap, the surface recombination velocity increases from $\sim 5 \times 10^3$ cm/s for a doping level of $n \sim 3 \times 10^{15}$ cm^{-3} to $\sim 10^6$ cm/s for a doping level of $n \sim 3 \times 10^{18}$ cm^{-3} [365]. For Si, the surface recombination rate depends on the treatment of the surface and lies in the range between 10–10^4 cm/s [366, 367]. The Si-SiO_2 interface can exhibit $S \leq 0.5$ cm/s. Time-resolved measurements and detailed modeling for Si have been reported in [368].

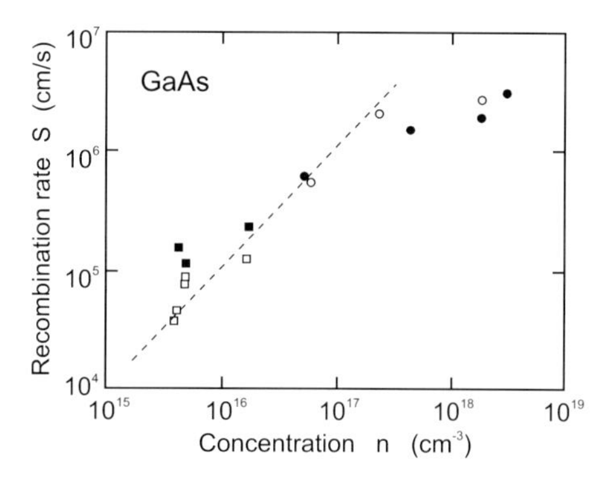

Fig. 10.19. Surface recombination velocity for GaAs as a function of n-type doping concentration. Different experimental points correspond to different surface treatment methods. *Dashed line* is a guide to the eye. Experimental data from [369]

10.14 Excess-Carrier Profiles

In this section, some typical excess-carrier profiles (in one-dimensional geometry) are discussed that arise from certain excitation conditions. The excess-carrier density Δp (here holes in an n-type semiconductor, i.e. $\Delta p = p_{\mathrm{n}} - p_{\mathrm{n_0}}$) is determined by the diffusion equation (cf. (8.51a))

$$D_{\mathrm{p}} \frac{\partial^2 \Delta p}{\partial x^2} + \frac{\Delta p}{\tau_{\mathrm{p}}} = -G(x) \,. \tag{10.68}$$

The generation of excess carriers in a semi-infinite piece of semiconductor occurs only at the surface at $x = 0$. The generation is zero everywhere and the excitation is incorporated via the boundary condition $\Delta p(x = 0) = \Delta p_0$. The general solution for the homogeneous equation (10.68), i.e. $G = 0$, is

$$\Delta p(x) = C_1 \exp\left(-\frac{x}{L_\mathrm{p}}\right) + C_2 \exp\left(\frac{x}{L_\mathrm{p}}\right) , \tag{10.69}$$

with the diffusion length $L_\mathrm{p} = \sqrt{D_\mathrm{p}\tau_\mathrm{p}}$. Taking the boundary condition $\Delta p_\mathrm{n}(x \to \infty) = 0$ the solution is

$$\Delta p(x) = \Delta p_0 \exp\left(-\frac{x}{L_\mathrm{p}}\right) . \tag{10.70}$$

In order to connect Δp_0 with the total generation rate per unit area G_tot, we calculate

$$G_\mathrm{tot} = \int_0^\infty \frac{\Delta p(x)}{\tau_\mathrm{p}} \mathrm{d}x = \frac{\Delta p_0 L_\mathrm{p}}{\tau_\mathrm{p}} = \Delta p_0 \sqrt{\frac{D_\mathrm{p}}{\tau_\mathrm{p}}} . \tag{10.71}$$

Now, the generation is taken as

$$G(x) = G_0 \exp(-\alpha x) , \tag{10.72}$$

i.e. due to light absorption with the (wavelength-dependent) absorption coefficient α. The total generation rate is

$$G_\mathrm{tot} = \int_0^\infty G(x)\mathrm{d}x = \frac{G_0}{\alpha} . \tag{10.73}$$

The solution of (10.68) is the sum of the homogeneous solution (10.69) and a particular solution that is given by

$$\Delta p(x) = C \exp(-\alpha x) . \tag{10.74}$$

The constant C is determined to be

$$C = \frac{G_0 \tau_\mathrm{p}}{1 - \alpha^2 L_\mathrm{p}^2} . \tag{10.75}$$

Therefore, the solution is

$$\Delta p(x) = C_1 \exp\left(-\frac{x}{L_\mathrm{p}}\right) + C_2 \exp\left(\frac{x}{L_\mathrm{p}}\right) + \frac{G_0 \tau_\mathrm{p}}{1 - \alpha^2 L_\mathrm{p}^2} \exp(-\alpha x) . \tag{10.76}$$

Using again $\Delta p_\mathrm{n}(x \to \infty) = 0$ (leading to $C_2 = 0$) and a recombination velocity S at the front surface, i.e.

$$-eS\Delta p_0 = -eD_\mathrm{p}\left[\frac{\mathrm{d}\Delta p}{\mathrm{d}x}\right]_{x=0} . \tag{10.77}$$

The solution is given as

$$\Delta p(x) = \frac{G_0\tau_\mathrm{p}}{1-\alpha^2L_\mathrm{p}^2}\left[\exp\left(-\alpha x\right) - \frac{S+\alpha D_\mathrm{p}}{S+D_\mathrm{p}/L_\mathrm{p}}\exp\left(-\frac{x}{L_\mathrm{p}}\right)\right] . \tag{10.78}$$

For $S=0$ the solution is

$$\Delta p(x) = \frac{G_0\tau_\mathrm{p}}{1-\alpha^2L_\mathrm{p}^2}\left[\exp\left(-\alpha x\right) - \alpha L_\mathrm{p}\exp\left(-\frac{x}{L_\mathrm{p}}\right)\right] , \tag{10.79}$$

and then for $\alpha L_\mathrm{p} \gg 1$, (10.70) is recovered. This dependence is the excess-carrier profile if the absorption is strong, which is a tendency for short wavelengths.

11 Heterostructures

11.1 Introduction

Heterostructures consist of (at least two) different materials. The geometry of the interfaces between the two materials can be complicated. The simplest case is a planar interface, i.e. a layered system. A metal–semiconductor junction is generally a heterostructure. However, we will use the term mostly for structures of various semiconductors. Most of the heterostructures discussed here are epitaxial, i.e. fabricated by the successive epitaxy of the various layers on a substrate. Another method to fabricate heterostructures of different (and dissimilar) materials is wafer bonding that is briefly discussed in Sect. 11.8.

Many modern semiconductor devices rely on heterostructures, such as the heterobipolar transistor (HBT), the high electron mobility transistor (HEMT), lasers and nowadays also light-emitting diodes. Shockley had already considered heterostructures in his 1951 patent for pn-junctions. For the development and the realization of heterostructures H. Kroemer and Zh.I. Alferov were awarded the 2000 Physics Nobel Prize. The properties of charge carriers in layers that are part of heterostructures can be quite different from those in bulk material, e.g. extremely high mobility, high radiative recombination efficiency or novel states of matter, as revealed in the quantum Hall effects.

11.2 Growth Methods

Since the thickness of layers in the active part of heterostructures has to be controlled to monolayer precision and the thickness of layers can go down to the single monolayer range, special growth methods have been developed. Among these molecular beam epitaxy (MBE), chemical vapor deposition (CVD) and metal-organic vapor phase epitaxy (MOVPE) are the most common for Si, Ge, III–V and II–VI semiconductors. In particular, oxide semiconductors are also fabricated with pulsed laser deposition (PLD). Liquid phase epitaxy (LPE) used to be very important for the fabrication of LEDs but has lost its role largely to MOVPE.

MBE is performed in an ultrahigh vacuum (UHV) chamber, pumped by getter pumps and cryoshrouds. The source materials are evaporated from effusion cells and directed towards the heated substrate. If the source materials are supplied as a gas stream, the method is called gas-source MBE (GSMBE). If metal-organic compounds are used as precursors, the method is denoted as MOMBE. The atoms impinge on the substrate with thermal energy and are first physisorbed. After diffusion on the surface they either desorb or they are chemisorbed, i.e. incorporated into the crystal. In order to obtain high spatial homogeneity of material properties such as composition, thickness and doping, the substrate is rotated during deposition.

During CVD and MOVPE the heated substrate is in a gaseous environment. The transport gas is typically H_2, N_2 or O_2. Precursor materials are hydrides such as silane, germane, arsine or phosphine (SiH_4, GeH_4, AsH_3, PH_3) and (for MOVPE) metal-organic compounds, such as, e.g. trimethylgallium (TMG). Due to the toxicity of the hydrides, alternative, i.e. less-toxic and less-volatile compounds are used, such as TBAs ($(CH_3)_3CAsH_2$). The crystal growth occurs after pyrolysis and catalysis of the compounds close to or on the substrate surface. All remaining C and H atoms (and whatever else that is not incorporated into the crystal) leave the reactor and are neutralized and stopped in a scrubber.

Thin-film epitaxy is mostly performed on wafers, i.e. thin circular slices of substrate material. The most common substrate materials are Si (currently up to 300 mm diameter), GaAs (up to 6 inch), InP (up to 4 inch) and sapphire (up to 6 inch). Typical wafer thickness is 300–500 μm. Also, very thin, flexible Si wafers (8–10 μm) have been developed. A wafer is cut from a large single cylindrical crystal that is fabricated with suitable growth techniques such as Czochralski (CZ) growth (1918) modified by Teal and Little [370]. In CZ growth the crystal is pulled from a seed crystal out of a melt of previously polycrystalline, pure or doped material. All dislocations stop in the narrow neck between the seed and the main body of the cylinder. The diameter of the crystal is controlled by the pulling rate and the heating power. For the growth of III–V compound semiconductors liquid encapsulated CZ (LEC) growth has been developed to counteract the high volatility of the growth-V component. During LEC growth the melt is completely covered with molten boric oxide (B_2O_3). The keys to optimization of the crystal growth process are numerical modeling and computer control. In Fig. 11.1 a large CZ silicon crystal and a smaller LEC GaAs crystal (boule) are shown. For details on other important fabrication methods for bulk crystals, including float-zone (FZ) or vertical gradient freeze (VGF), we refer to the literature [371]. Significant expertise is necessary for cutting, grinding and polishing (lapping) wafers for epitaxy.

For semiconductors, the wafer is marked with *flats* to indicate orientation and doping. In Fig. 11.2 the standard flats are shown for (001)- and (111)-oriented material. The primary flat is longer than the secondary flat, e.g. 32

Fig. 11.1. (**a**) Silicon single crystal for 300-mm diameter wafers after opening of the crucible. From [372] with permission. (**b**) GaAs single crystal (boule) for 4-inch wafers and some cut and polished wafers

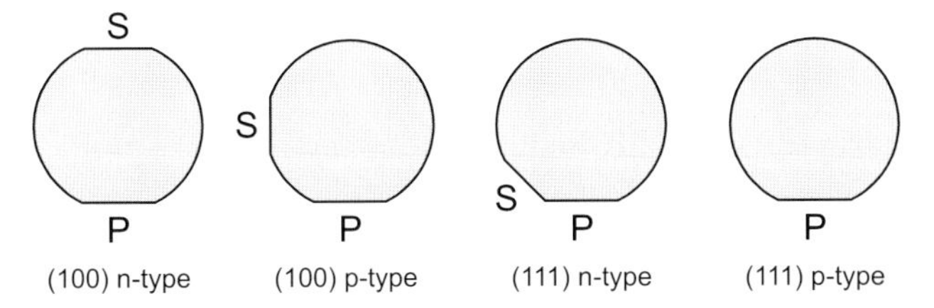

Fig. 11.2. Schematic semiconductor wafer geometry for various orientations and doping with the primary (P) and secondary (S) flats indicated.

and 12 mm, respectively, for a 4-inch (100-mm) diameter wafer. The front surface, on which the epitaxy is performed, typically undergoes an elaborate cleaning and polishing process.

One prerequisite for making high-quality heterostructures with thin layers is a flat surface. Even if the polished substrate is not perfect, flat interfaces can be achieved with the growth of appropriate superlattice buffer layers (Fig. 11.3). Roughness can exist on all length scales and is typically investigated using atomic force microscopy scans. However, it is possible to create surfaces that exhibit large, essentially monoatomically flat terraces between individual surface steps.

In-situ monitoring is of importance to obtain information about the growth process while it is underway. Using the information in a feedback loop it is possible to achieve in-situ control of the process, e.g. for precise determination of growth rates or layer thickness. Techniques are RHEED (reflection high-energy electron diffraction) [373] (only for UHV systems) and RAS (reflection anisotropy spectroscopy) [374, 375].

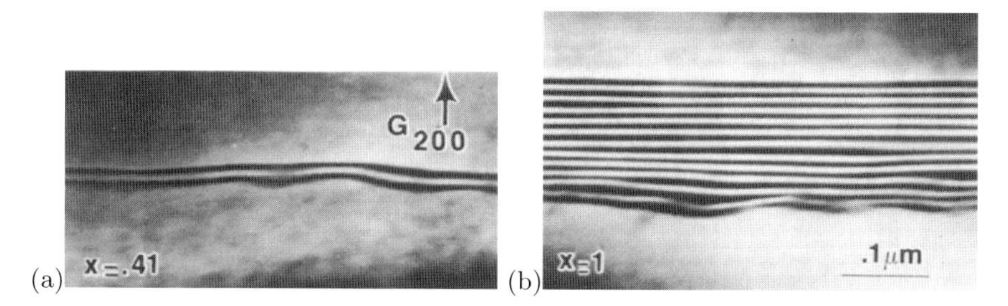

Fig. 11.3. Cross-sectional TEM of MBE-grown $Al_xGa_{1-x}As/GaAs$ heterostructures for (**a**) $x = 0.41$ and (**b**) $x = 1.0$. Using an AlAs/GaAs superlattice an excellent flattening of substrate roughness is achieved. From [376]

11.3 Material Combinations

11.3.1 Pseudomorphic Structures

Heterostructures can generally be made from any sequence of materials. However a mismatch in lattice constant (or a different crystal structure) leads to strains and stresses that are of the order of 10^3 atmos for strains of 1% ($\sigma \sim C\epsilon$, $C \approx 5 \times 10^{10}$ Pa) as discussed in Sect. 5.3.2. The total strain energy is $\propto C\epsilon^2$. Above a critical thickness $d_c \propto \epsilon^{-1}$ (see Sect. 5.3.6) defects, e.g. misfit dislocations (relaxing strain with their edge components), are generated. The cross-hatch pattern due to misfit dislocations is shown in Fig. 11.4. There are a number of semiconductor combinations that are lattice matched and thus can be grown with arbitrary thickness. $Al_xGa_{1-x}As$ is closely lattice matched to GaAs for all Al concentrations. See Fig. 6.14 for lattice-matched pairs, e.g. $In_{0.53}Ga_{0.47}As/InP$. Often, thin layers of lattice-mismatched materials with thickness smaller than the critical thickness are used.

11.3.2 Heterosubstrates

If homosubstrates are not available or very expensive, semiconductors are also grown on dissimilar substrates, e.g. GaN and ZnO on sapphire (Al_2O_3) or SiC. For ZnO, homosubstrates have recently been produced with 3-inch diameter [377]. In many cases, the integration of III–V- or II–VI-based semiconductors for optoelectronic applications on silicon for electronic devices

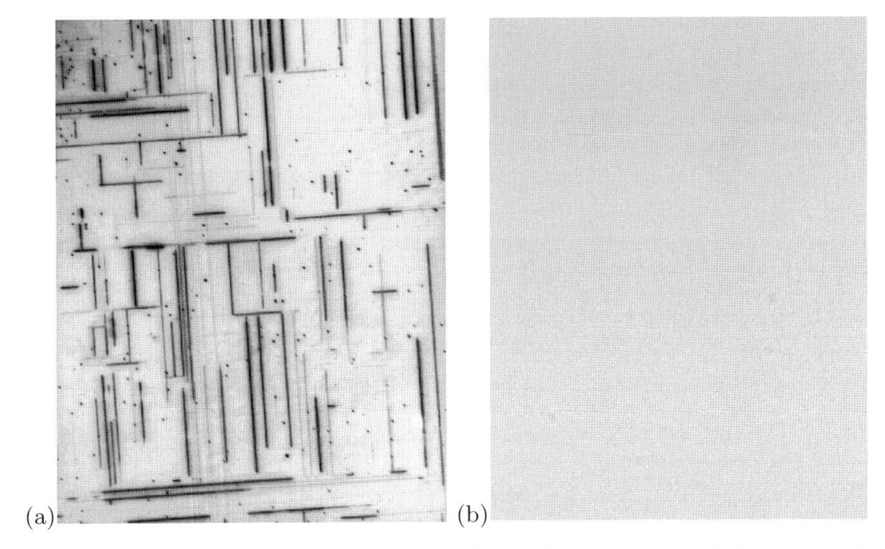

Fig. 11.4. Surface (**a**) of a supercritical, plastically relaxed and (**b**) of a pseudomorphic $In_xGa_{1-x}As$ film on GaAs. In (**a**) the cross-hatch pattern is due to misfit dislocations along [110] and $[1\bar{1}0]$

is desirable, such as GaAs/Si, InP/Si, GaN/Si or ZnO/Si. For such combinations the epitaxial relationship, i.e. the alignment of the crystallographic directions of both materials, which can have different space groups, has to be considered [378]. Some examples are given in Table 11.1. In Fig. 11.5, X-ray diffraction data are shown from a ZnO layer on *c*-oriented sapphire. The hexagonal ZnO lattice is rotated by 30° with respect to the trigonal sapphire lattice. In the case of growth of ZnO on Si(111) an amorphous SiO_x layer forms at the interface such that the crystallographic information of the substrate is lost. The ZnO grains exhibit random in-plane orientation (Fig. 11.6).

Table 11.1. Epitaxial relationship for various film/substrate combinations, ZnO (or GaN) on *c*-, *a*- and *r*-sapphire and Si(111)

ZnO	Al_2O_3 [00.1]	ZnO	Al_2O_3 [11.0]	ZnO	Al_2O_3 $[\bar{1}0.2]$	ZnO / GaN	Si [111]
[00.1] [11.0]	[00.1] [01.0]	[00.1] [11.0]	[11.0] [00.1]	[11.0] [00.1]	$[\bar{1}0.2]$ $[0\bar{1}.1]$	[00.1] — / $[2\bar{1}.0]$	[111] $[\bar{1}10]$

The initial growth steps can determine the orientation in polar materials. GaN directly grown on *c*-Al_2O_3 grows with N-face orientation (see Fig. 3.18). The high surface mobility of Ga allows nitrogen to take its preferred position in the first atomic layer. Even under Ga-rich conditions the N atoms can kick-

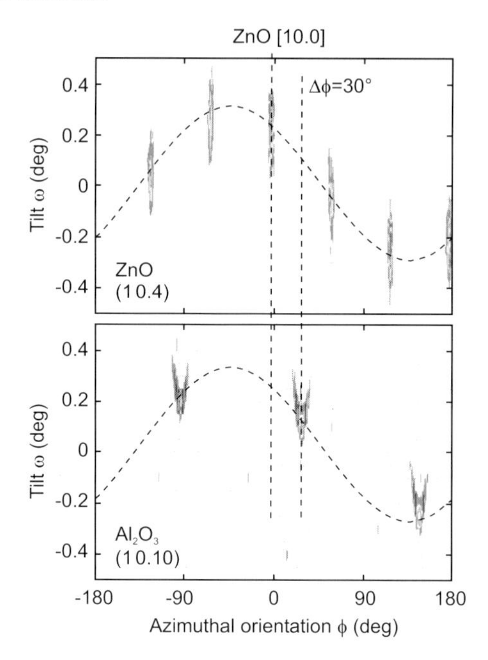

Fig. 11.5. X-ray diffraction intensity from the asymmetric ZnO (10.4) (*upper panel*) and the sapphire (10.10) (*lower panel*) reflections as a function of the azimuthal sample orientation (rotation angle ϕ around the [00.1] axis). The peaks appear at different tilt angles ω due to an overall tilt of the mounted sample (*dashed sinusoidal lines*). The ZnO [00.1] axis is not tilted with respect to the sapphire [00.1] direction

off the Ga from its favorite site on the surface. If an AlN buffer is used the strong bond between Al and oxygen leads to an Al atomic layer at the interface and subsequent GaN grows with a Ga-face [379]. On a sapphire substrate with lateral AlN patterns, laterally orientation-modulated GaN can be grown. At the juncture of the phases an inversion domain boundary forms [380].

Now, the nucleation and the initial film growth are important and determine the film quality. Several techniques have been developed to overcome common problems. A typical strategy is the growth of a low-temperature nucleation layer. The defect density can be reduced in parts of the structure using epitaxial lateral overgrowth (ELO) [382]. The defects thread only from the limited contact area of the layer with the substrate and the part of the layer away from the mask ('wing') is free of defects (Fig. 11.8).

The cracking of thick films during cooling due to the mismatch of the thermal expansion coefficients (Fig. 11.9) can be avoided by the introduction of suitable stress-relaxing layers [383, 384].

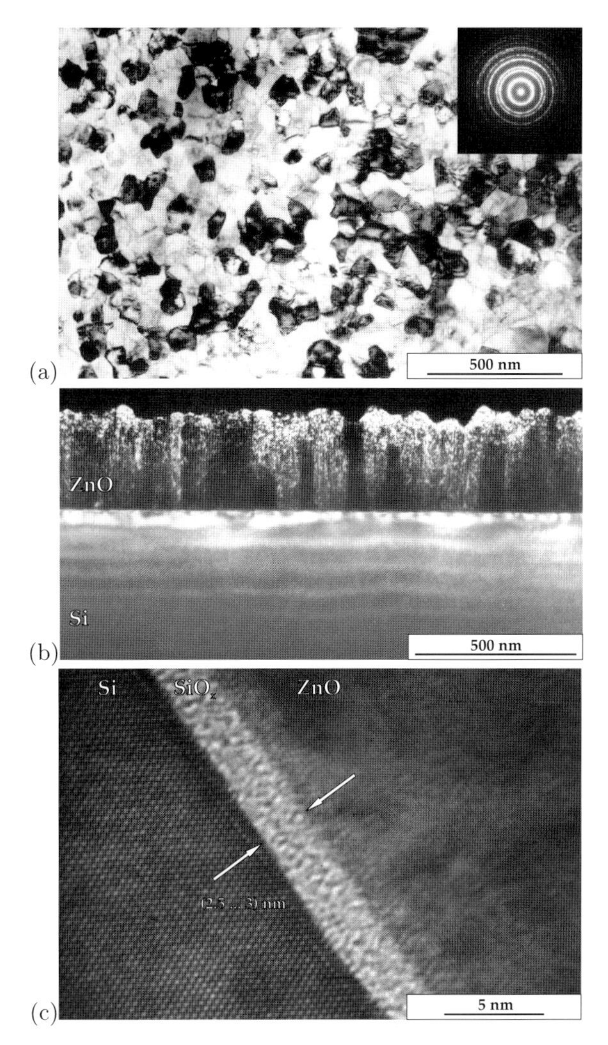

Fig. 11.6. (**a**) Plan-view TEM image (*inset*: electron diffraction diagram averaged over several grains) of ZnO on Si(111). (**b**) Cross-sectional TEM of the same sample. (**c**) High-resolution cross-sectional image of the ZnO/SiO$_x$/Si interface region

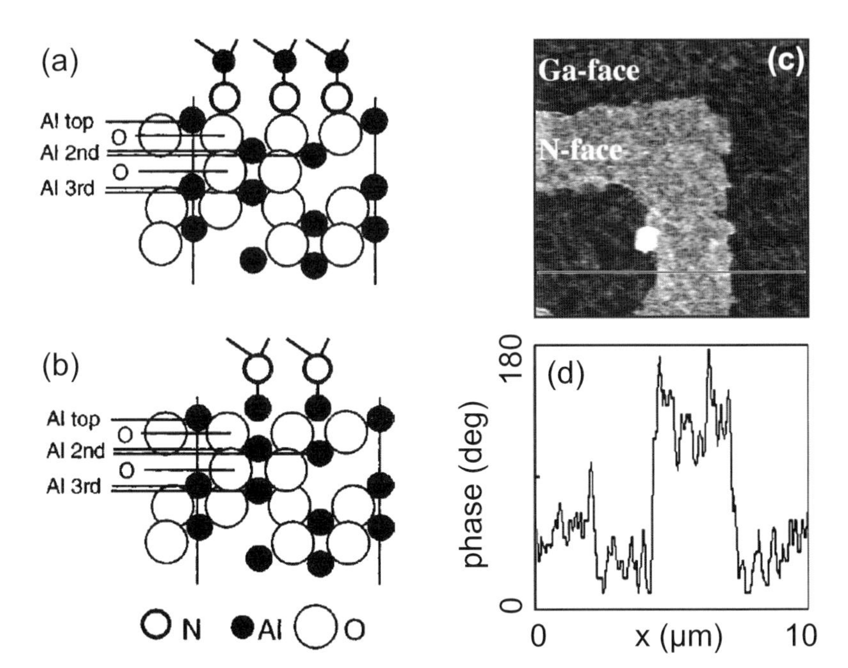

Fig. 11.7. Side view of the heterointerfaces between AlN and *c*-oriented sapphire with nitrogen (**a**) and Al (**b**) being the first layer. Adapted from [379]. (**c**) Phase image of piezoresponse force microscopy (PFM) of lateral polarity GaN heterostructure. (**d**) Linescan of phase signal along *white line* in part (**c**), adapted from [381]. Part (**c**) reprinted with permission from [381], ©2002 AIP

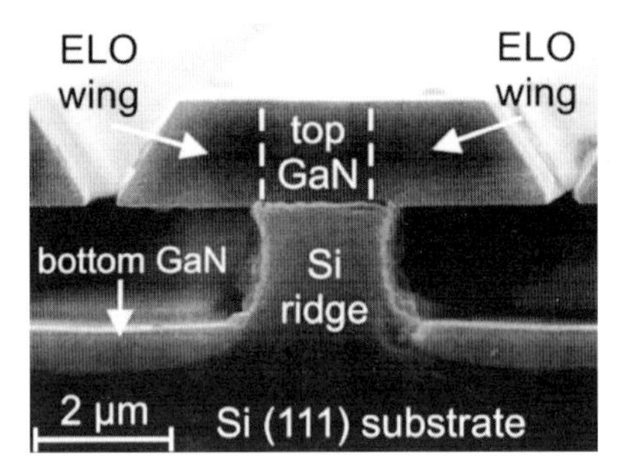

Fig. 11.8. SEM cross-sectional image of GaN grown on a structured Si(111) substrate. The laterally grown wings extend about 2.5 µm over the grooves. The thickness of the GaN layer is 0.5 µm on the bottom of the grooves, while it is 1.4 µm on top of the ridges. Reprinted with permission from [382], ©2001 AIP

Fig. 11.9. 1×1 mm^2 top view with a differential interference contrast microscope of a 1.3-µm thick GaN layer grown on Si(111). Reproduced with permission from [383], ©2000 IPAP

11.4 Band Lineup in Heterostructures

In heterostructures, semiconductors with different bandgaps are combined. The relative position of conduction and valence band (band alignment) is determined by the electron affinities as shown in Fig. 11.10. It can lead to different types of heterostructures. In Fig. 11.11, the band alignment for type-I and type-II heterostructures are shown. In the type-I structure (straddeled band lineup) the lower conduction-band edge and the higher valence-band edge are both in the material with smaller bandgap. Thus, electrons and holes will localize there. In the type-II structure a staggered lineup is present and electrons and holes will localize in different materials. The technologically most relevant are type-I structures.

In a type-I heterostructure, the conduction- and valence-band discontinuities are given, respectively, by

$$\Delta E_{\mathrm{C}} = \chi_1 - \chi_2 \tag{11.1a}$$

$$\Delta E_{\mathrm{V}} = (\chi_1 + E_{\mathrm{g}_1}) - (\chi_2 + E_{\mathrm{g}_2}) \,. \tag{11.1b}$$

Depending on the layer sequence of high- and low-bandgap materials various configurations, as shown in Fig. 11.12 have obtained special names, such as single heterointerface, quantum well (QW), multiple quantum well (MQW), superlattice (SL). In the extreme case the layer is only one monolayer thick (Fig. 11.13) and the concept of layer and interface blurs. Such atomically precise layer sequences are mastered nowadays for a variety of material systems such as AlGaAs/GaAs/InAs, InP/InGaAs, Si/SiGe and also $BaTiO_3/SrTiO_3$.

The design of heterostructures to fulfill a certain device functionality or to have certain physical properties is called 'bandgap engineering'.

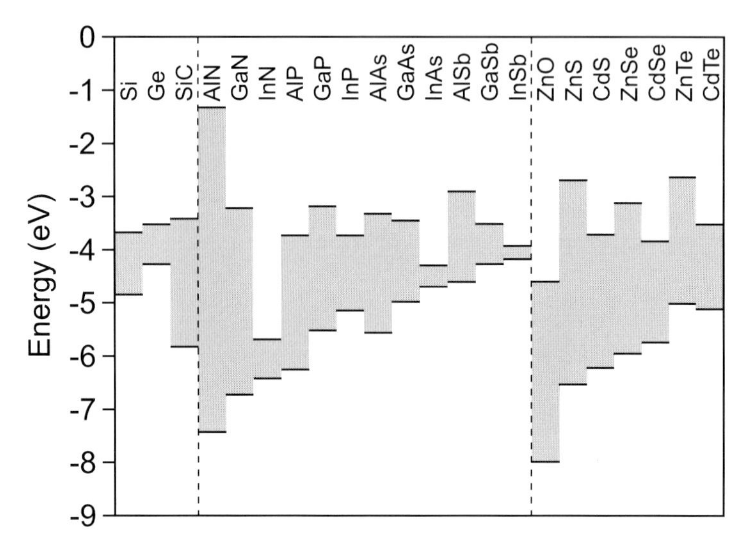

Fig. 11.10. Position of conduction and valence-band edges for a variety of semiconductors. Based on values from [385]

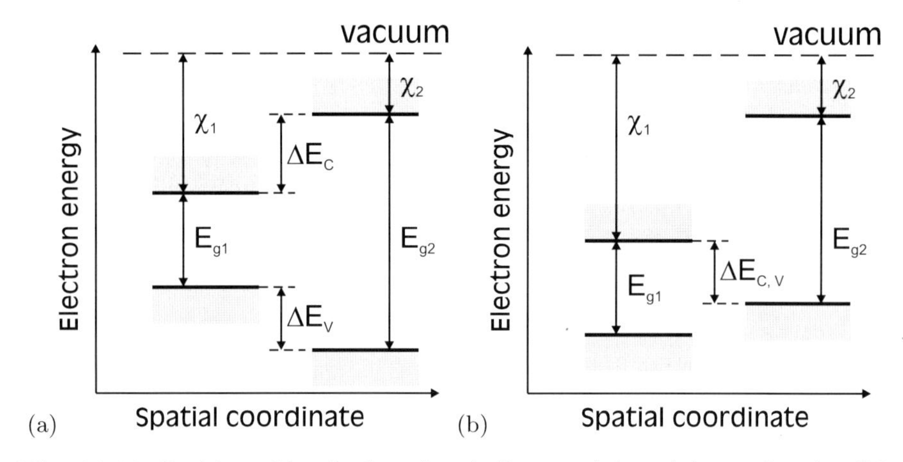

Fig. 11.11. Position of band edges (band alignment) in a (**a**) type-I and a (**b**) type-II heterostructure.

11.5 Energy Levels in Heterostructures

11.5.1 Quantum Well

The energy in a single quantum well of thickness L_z (along the growth direction z) can be calculated with the quantum-mechanical particle-in-a-box model. In the envelope function approximation (Appendix F) the wavefunction is written as a product of the Bloch function and the envelope function $\chi(z)$.

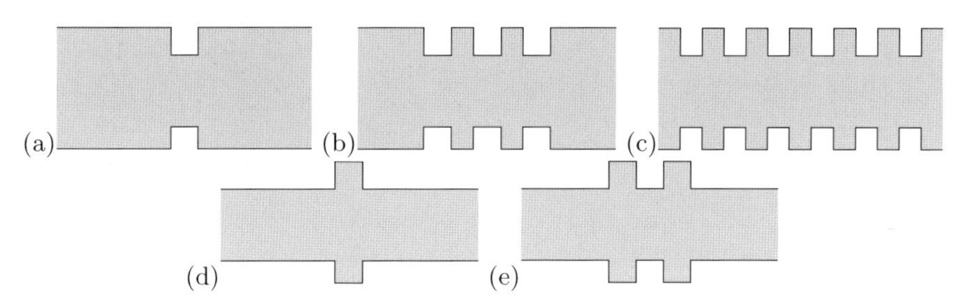

Fig. 11.12. Heterostructures with different layer sequences (bandgap engineering). (**a**) quantum well (QW), (**b**) multiple quantum well (MQW), (**c**) superlattice (SL), (**d**) single-barrier tunneling structure, (**e**) double-barrier tunneling structure

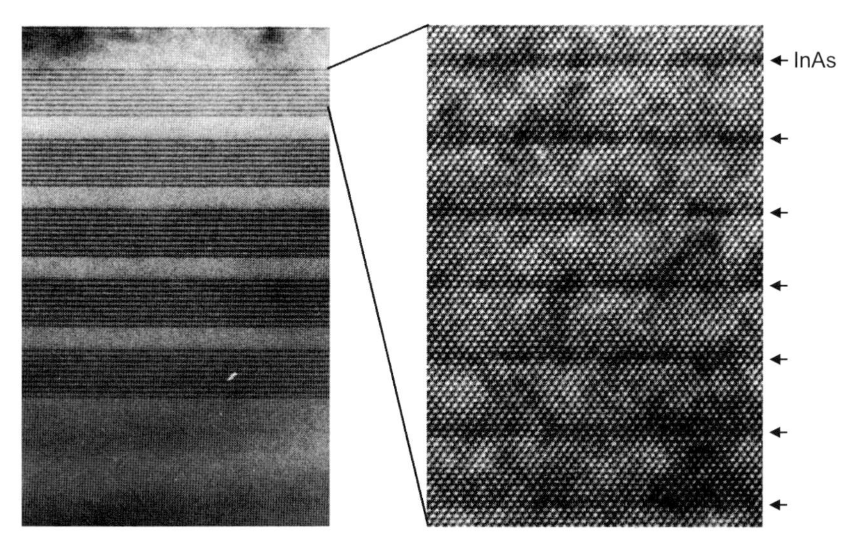

Fig. 11.13. Cross-sectional TEM of a MOVPE-grown short-period superlattice (SPS) of InAs layers in $GaAs_{1-x}N_x$. In high resolution (*right image*), the individual rows of atoms can be seen. The InAs layers are indicated by *arrows*. From [386]

$$\Psi^{\mathrm{A,B}}(\mathbf{r}) = \exp\left(\mathrm{i}\mathbf{k}_{\perp}\mathbf{r}\right) u_{n\mathbf{k}}(\mathbf{r})\chi_n(z) \,, \tag{11.2}$$

where 'A' and 'B' denote the two different materials. The envelope function χ fulfills, approximately, the one-dimensional Schrödinger-type equation,

$$\left(-\frac{\hbar^2}{2m^*}\frac{\partial^2}{\partial z^2} + V_{\mathrm{c}}(z)\right)\chi_n(z) = E_n\chi_n(z) \,, \tag{11.3}$$

where m^* denotes the effective mass. V_{c} is the confinement potential determined by the band discontinuities. Typically, $V_{\mathrm{c}} = 0$ in the well and $V_0 > 0$ outside in the barrier. E_n are the resulting energy values of the quantized levels. In the case of infinite barriers ($V_0 \to \infty$, Fig. 11.14a) the boundary conditions $\chi(0) = \chi(L_z) = 0$ yield

$$E_n = \frac{\hbar^2}{2m^*}\left(\frac{n\pi}{L_z}\right)^2 \tag{11.4}$$

$$\chi_n(z) = A_n \sin\left(\frac{n\pi z}{L_z}\right) , \tag{11.5}$$

where E_n is called the confinement energy. For finite barrier height (Fig. 11.14b) the calculation leads to a transcendental equation

$$\tan\left[\left(\frac{m_\mathrm{w} E L_z^2}{2\hbar^2}\right)^{1/2}\right] = \left(\frac{m_\mathrm{b}}{m_\mathrm{w}}\frac{V_0 - E}{E}\right)^{1/2} , \tag{11.6}$$

where m_w (m_b) is the effective masses in the well (barrier). The wavefunction tunnels into the barrier. While for infinite barrier height the lowest level diverges for $L_z \to 0$, for finite barrier height $E_1 \to V_0$.

In the plane the carriers are still free and have a two-dimensional dispersion. Thus, each quantized level contributes $m^*/(\pi\hbar^2)$ (6.72) to the density of states at each subband edge E_n.

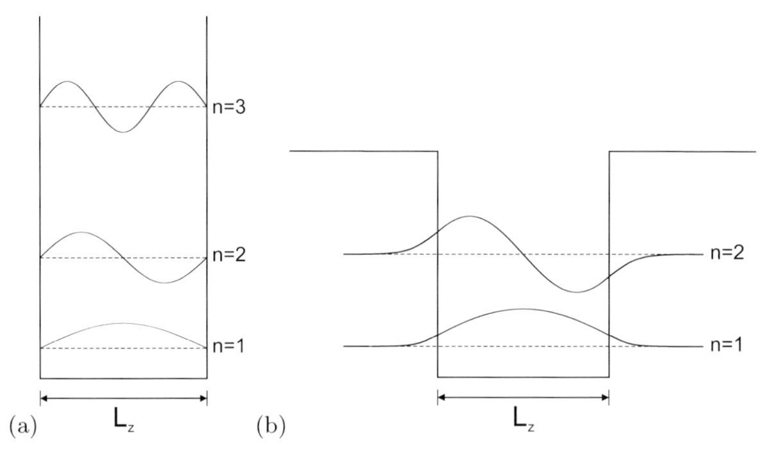

Fig. 11.14. Schematic energy levels and wavefunctions in (**a**) a potential well with infinite barriers, and (**b**) for finite barrier height

For holes, the situation is a little more complicated than that for electrons (Fig. 11.15). First, the degeneracy of heavy and light holes is lifted since their mass enters the confinement energy. The effective hole masses along the z direction, i.e. those that enter (11.3), are

$$\frac{1}{m_\mathrm{hh}^z} = \gamma_1 - 2\gamma_2 \tag{11.7a}$$

$$\frac{1}{m_\mathrm{lh}^z} = \gamma_1 + 2\gamma_2 . \tag{11.7b}$$

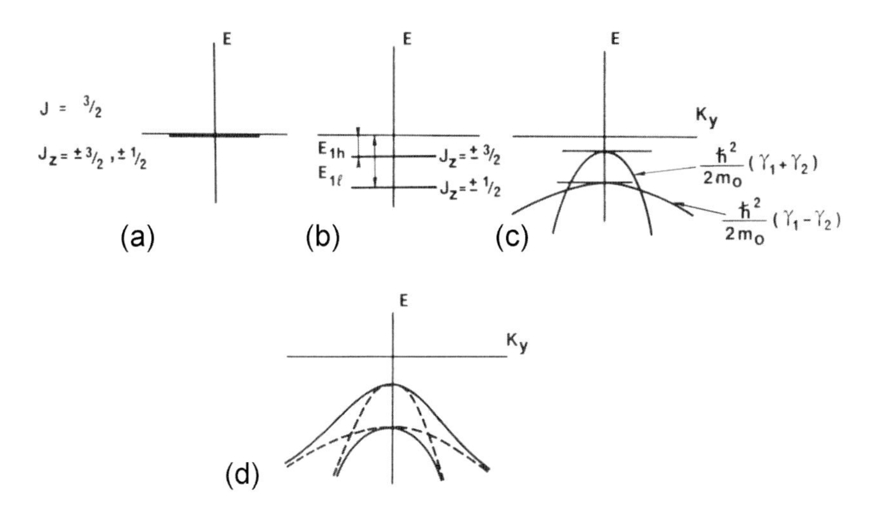

Fig. 11.15. Schematic representation of the development of hole levels in a quantum well: (**a**) bulk, (**b**) splitting at the subband edge, (**c**) dispersion (mass reversal), (**d**) anticrossing behavior. From [376]

The light holes have the higher quantization energy. The angular momentum is quantized along the z direction. The transverse masses for the dispersion in the interface plane are

$$\frac{1}{m_{\mathrm{hh}}^{xy}} = \gamma_1 + \gamma_2 \tag{11.8a}$$

$$\frac{1}{m_{\mathrm{lh}}^{xy}} = \gamma_1 - \gamma_2 \, . \tag{11.8b}$$

Now the heavy hole, i.e. the $J_z = \pm\frac{3}{2}$ state, has the smaller mass and the light hole ($J_z = \pm\frac{1}{2}$) the larger (Fig. 11.15c). However, this consideration is only an approximation since the lifting of degeneracy and the dispersion have to be treated on the same level. Higher terms of the perturbation calculation lead to band mixing and remove the band crossing that seems to originate from the situation at the Γ point. In reality, the bands show anticrossing behavior and are strongly deformed. The hole dispersion in a superlattice and the anticrossing behavior is shown in Fig. 11.16.

Experimentally observed transition energies in quantum wells of varying thickness are shown in Fig. 11.17 and are in good agreement with the theoretical calculation. We note that for infinite barriers optical transitions are only allowed between confined electron and hole states with the same quantum number n. For finite barriers this selection rule becomes relaxed, and other transitions become partially allowed, e.g. e_1–hh_3. The optical matrix element from the Bloch part of the wavefunction, which was isotropic for (cubic) bulk

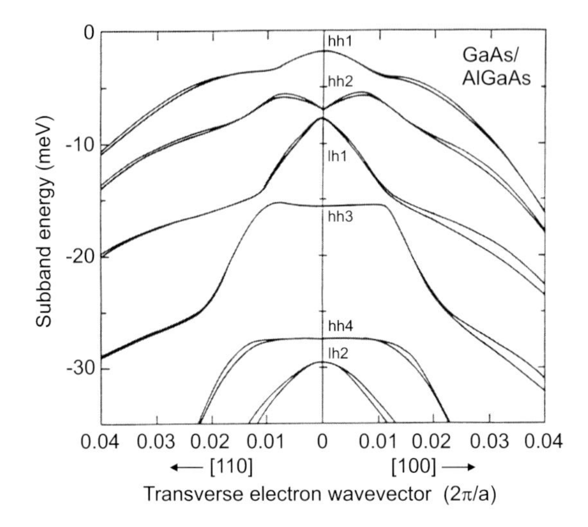

Fig. 11.16. Hole dispersion in a 68-ML GaAs/71 ML $Al_{0.25}Ga_{0.75}As$ superlattice (numerical calculation). The double curves originate from a lifting of time-reversal symmetry at $\mathbf{k} \neq 0$. Reprinted with permission from [387], ©1985 APS

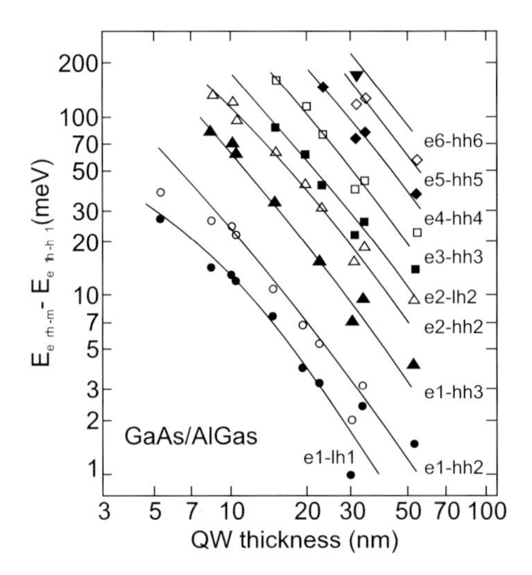

Fig. 11.17. Observed electron–hole transitions (energy difference to the first e–h transition from excitation spectroscopy) in GaAs/AlGaAs quantum wells of varying thickness. *Lines* are theoretical model. Data from [389]

material (9.16), is anisotropic for quantum wells. TE (TM) polarization is defined with the electromagnetic field in (perpendicular to) the plane of the quantum well (Fig. 11.18a). At the subband edge, i.e. for in-plane wavevector $k_{||} = 0$ the matrix elements for the various polarizations and propagation directions are given in Table 11.2. The matrix elements averaged over all in-

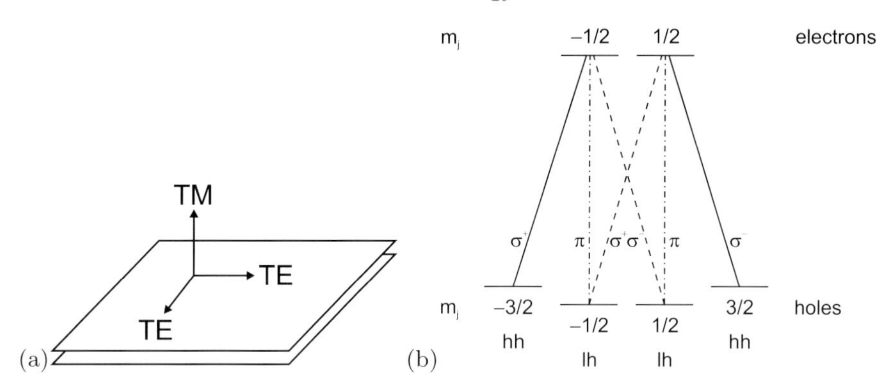

Fig. 11.18. (**a**) Directions of electric-field vector relative to the quantum-well plane for TE and TM polarization. (**b**) Optical selection rules for band–band transitions in a quantum well. If the (in-plane averaged) relative strength of the e–hh transitions (*solid lines*) is 1, the relative strength of the TE-polarized e–lh transitions (*dashed lines*) is 1/3 and that of the TM-polarized e–lh transitions (*dash-dotted lines*) is 4/3

Table 11.2. Squared momentum matrix elements $|\langle c|\hat{\mathbf{e}}\cdot\mathbf{p}|v\rangle|^2$ in a quantum well for various propagation directions in units of M_b^2. The quantum-well normal is along **z**

	propagation	$\hat{\mathbf{e}}_x$ (TE)	$\hat{\mathbf{e}}_y$ (TE)	$\hat{\mathbf{e}}_z$ (TM)
	x	–	1/2	0
e–hh	**y**	1/2	–	0
	z	1/2	1/2	–
	x	–	1/6	2/3
e–lh	**y**	1/6	–	2/3
	z	1/6	1/6	–
	x	–	1/3	1/3
e–so	**y**	1/3	–	1/3
	z	1/3	1/3	–

plane directions for TE-polarization are $3/2M_b^2$ ($1/2M_b^2$) for the electron to heavy (light) hole transition. For TM polarization the values are 0 and $2M_b^2$, respectively [388]. The optical selection rules are shown in Fig. 11.18 (cf. Fig. 9.8 for bulk material). For propagation along the quantum-well plane, the ratio between the strength of the TE polarized e–hh and e–lh transitions is 3:1.

The confinement potential squeezes charge carriers bound to impurities closer to the ion. Therefore, the binding energy increases as shown in Fig. 11.19. This behavior can be modeled theoretically with good precision.

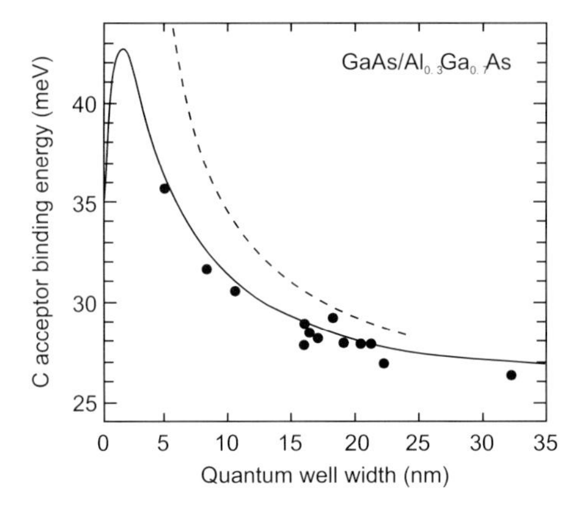

Fig. 11.19. Experimental values for the acceptor binding energy in $GaAs/Al_{0.3}Ga_{0.7}As$ quantum wells (*solid circles*) from [390] as a function of well width. *Solid line* is theory (variational calculation) for the well-center acceptor including top four valence bands and finite barriers, *dashed line* is hydrogen-like model with infinite barrier height. Adapted from [391]

It makes a difference whether the impurity is located at the center or the interface of the quantum well.

The confinement potential also squeezes electrons and holes in the exciton closer together and thus increases their Coulomb interaction. The binding energy of the quantum-well exciton is thus larger than in bulk material and depends on the well width (Fig. 11.20). In the simple hydrogen-like model with infinite barriers the exciton binding energy is 4 times the bulk binding

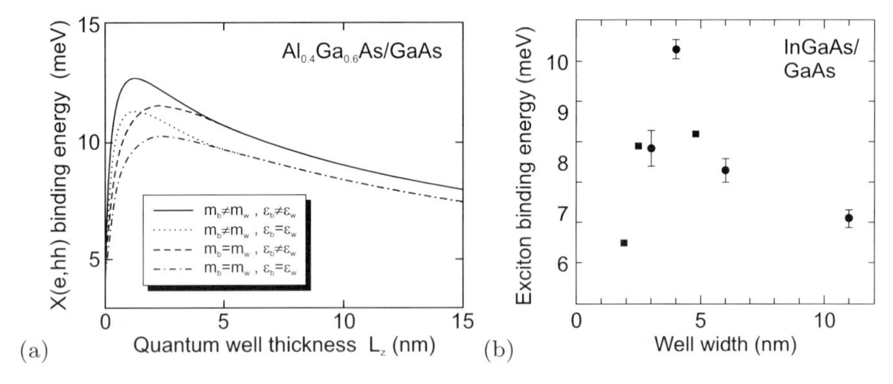

Fig. 11.20. (**a**) Theoretical (variational) calculation (*solid line*) of the heavy-hole exciton binding energy vs. QW thickness in a $GaAs/Al_{0.4}Ga_{0.6}As$ quantum well (using also different approximations, *other lines*). Adapted from [392]. (**b**) Experimental exciton binding energy in $In_xGa_{1-x}As/GaAs$ quantum wells of different thickness. *Circles*: data and error bars from [393], x unspecified, *squares*: data from [394], $x = 0.18$

energy in the limit $L_z \to 0$. In a realistic calculation the effect of different dielectric constants in the well and barrier (image charge effect) need to be considered.

11.5.2 Superlattices

In a superlattice, the barrier thickness is so small that carriers can tunnel in neighboring wells or, in other terms, that there exists a significant wavefunction overlap between adjacent wells. This leads to a band structure (Fig. 11.21), similar to the Kronig–Penney model (Sect. 6.4). For the superlattice the bands are called minibands, the gaps are called minigaps. The density of states does not make a step at the subband edge but follows an arccos function. The modification of the density of states, as seen in the absorption spectrum, are shown in Fig. 11.22 for 1, 2, 3 and 10 coupled wells.

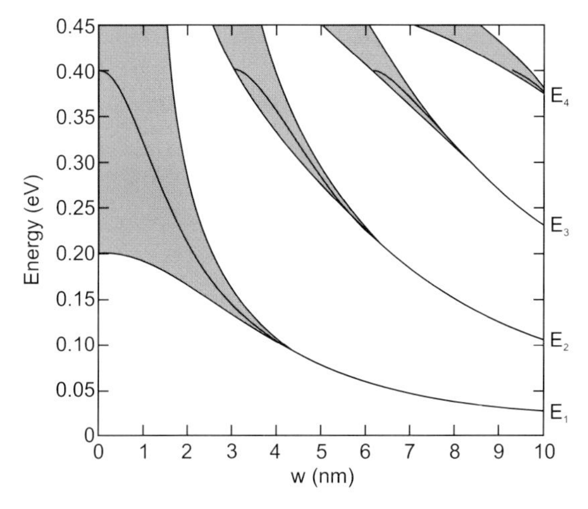

Fig. 11.21. Band structure of a superlattice with a potential depth of 0.4 eV and well and barrier width w ($L_{\mathrm{QW}} = L_{\mathrm{barr}}$). Adapted from [395]

11.5.3 Single Heterointerface Between Doped Materials

We consider a single heterointerface between n-doped materials. As an example we take n-AlGaAs/n-GaAs (Fig. 11.23). First, we consider the materials without contact, forming a type-I structure. In thermodynamic equilibrium the system must have a constant Fermi level. Thus, charge is transferred from the region close to the interface from AlGaAs to GaAs. This results in the formation of a triangular potential well in the GaAs close to the interface. A two-dimensional electron gas (2DEG) forms in this potential well (Fig. 11.24). The charge transfer in thermodynamic equilibrium adjusts the

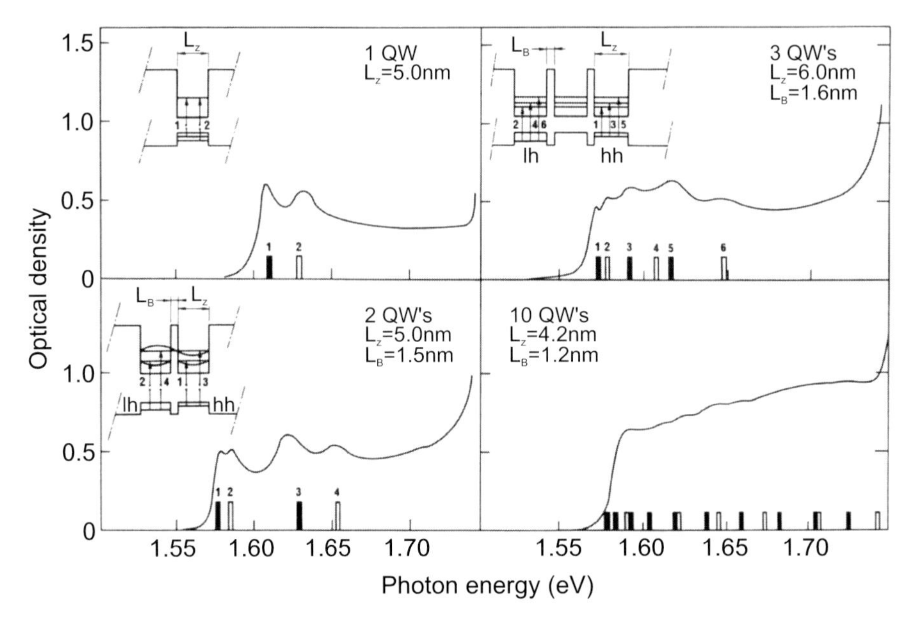

Fig. 11.22. Absorption spectra of a single, double, triple and ten coupled quantum wells. Theoretically predicted transitions with heavy (light) holes are labeled with *filled* (*empty*) *bars* at their respective transition energies. Adapted from [396]

band bending and the charge density (quantized levels in the well) in such a way that they are self-consistent. The Poisson equation and the Schrödinger equation are simultaneously fulfilled. Numerically, both equations are iteratively solved and the solution is altered until it is self-consistent, i.e. it fulfills both equations.

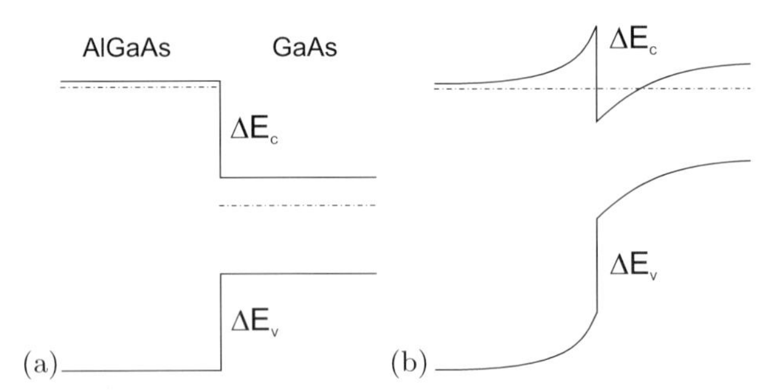

Fig. 11.23. Schematic formation of a triangular potential well in a n-AlGaAs/n-GaAs heterostructure, (**a**) before and (**b**) after equilibration of Fermi levels

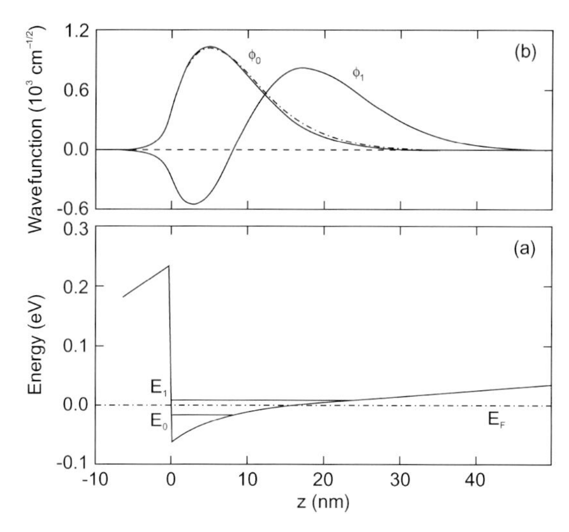

Fig. 11.24. (**a**) Conduction-band edge at a $GaAs/Al_{0.3}Ga_{0.7}As$ heterointerface ($T = 0\,K$) with two confined states at E_0 and E_1 marked with *solid horizontal lines*. In the GaAs channel there are $5 \times 10^{11}\,cm^{-2}$ electrons. The barrier height is 300 meV, $N_D^{GaAs} = 3 \times 10^{14}\,cm^{-3}$. The position of the Fermi level E_F is at $E = 0$ and indicated with a *dash-dotted line*. (**b**) Envelope wavefunctions ϕ_0 and ϕ_1 of the two confined states, *dash-dotted line*: calculation without exchange and correlation for state at E_0. Adapted from [398]

If the region of the 2DEG is not doped, the electron gas exists without any dopant atoms and ionized impurity scattering no longer exists. This concept is called *modulation doping*. Mobilities up to $3.1\times10^7\,cm^2/V\,s$ have been realized (Fig. 11.25). The theoretical limits of mobility in a 2DEG at modulation-doped AlGaAs/GaAs heterointerfaces are discussed in detail in [397].

11.6 Recombination in Quantum Wells

The energy of exciton recombination in quantum wells is blue-shifted with respect to that in bulk material due to the quantum-confinement energies of electrons and holes (Fig. 11.26). The electron–hole recombination lineshape in quantum wells is given by the product of the joint density of states and the Boltzmann function (when Boltzmann statistics apply). The JDOS is given by a step function (Heavyside function $H(E)$).

$$I(E) \propto H(E - E_{11}) \exp\left(-\frac{E}{kT}\right) , \qquad (11.9)$$

where $E_{11} = E_g + E_{e1} + E_{h1}$ represents the energy separation of the E1–H1 subband edges as shown in Fig. 11.26. An experimental spectrum

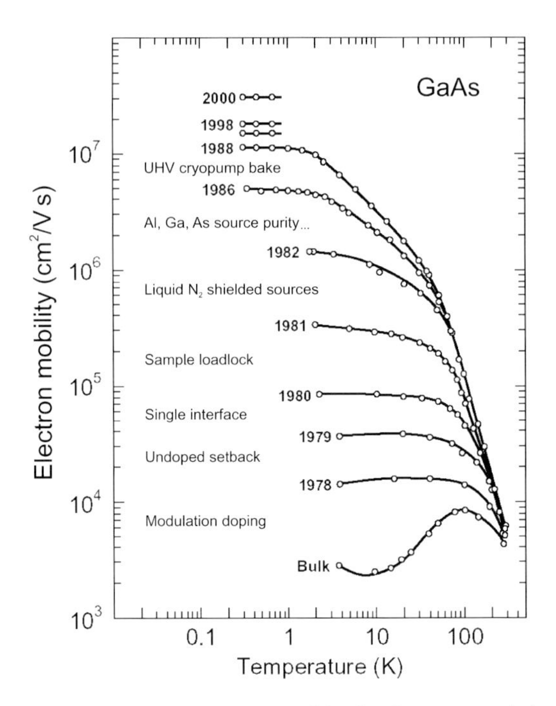

Fig. 11.25. Progress in the achievement of high electron mobility in GaAs, annotated with the technical innovation responsible for the improvement. Reprinted from [399], ©2003, with permission from Elsevier

(Fig. 11.27a) shows that excitonic effects influence the recombination lineshape even at room temperature [400].

At high carrier densities when the electron (quasi-) Fermi level is above the electron subband edge, the spectrum broadens and reflects the Fermi–Dirac distribution (Fig. 11.27b). At low temperatures a many-body effect, multiple electron–hole scattering with electrons at the Fermi edge, leads to an additional peak, termed *Fermi-edge singularity* that is discussed in [401].

Inhomogeneous broadening effects the recombination lineshape. Since the interfaces of the QW are not ideally flat, the exciton averages over different quantum-well thicknesses within its volume. Also, e.g. for the GaAs/AlGaAs system, the wavefunction in the (binary) quantum well tunnels into the barrier, the amount depending on the QW width, and there 'sees' the alloy broadening (see Sect. 10.5). The problem of exciton dynamics in a potential with random fluctuations has been treated in detail [402, 403].

A simplified picture is as follows: At low temperatures the excitons populate preferentially the potential minima. A simple lineshape[1] of the QW absorption or joint density of states is given by a step function (cf. Table 9.2)

[1]neglecting excitonic enhancement

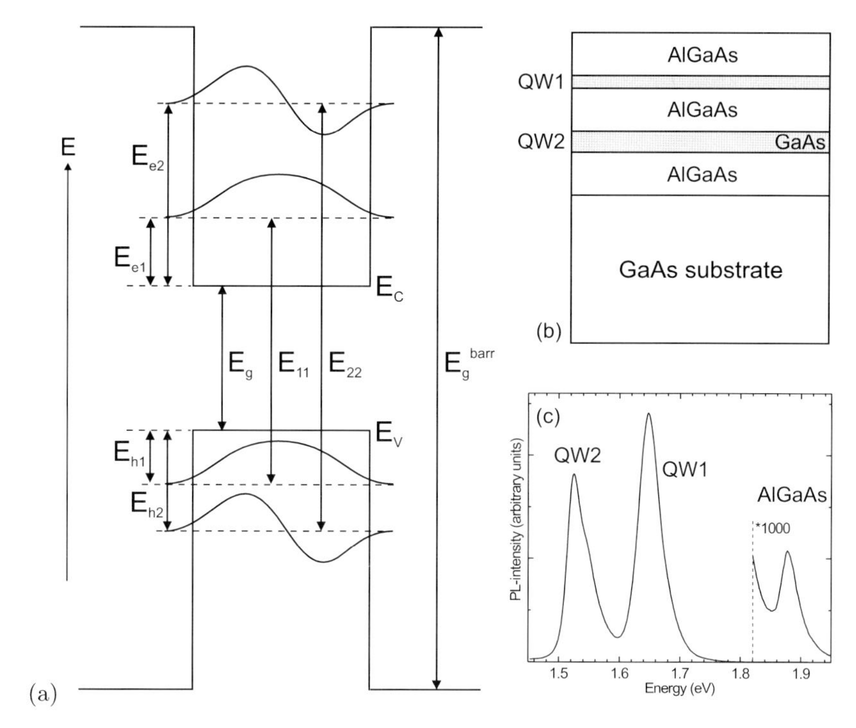

Fig. 11.26. (**a**) Schematic energy diagram of a quantum well with confined electron (e1, e2) and hole (h1, h2) states and recombination between them at energies E_{11} and E_{22}. (**b**) Schematic sample structure with two GaAs/$Al_xGa_{1-x}As$ quantum wells with thicknesses 3 nm and 6 nm. (**c**) Photoluminescence spectrum ($T = 300$ K) of the structure from part (**b**). A small amount of barrier luminescence appears at 1.88 eV, according to $x = 0.37$ (cf. Fig. 6.13c)

at the QW band edge E_0. The inhomogeneous broadening has a Gaussian probability distribution $p(\delta E) \propto \exp[-(\delta E)^2/2\sigma^2)]$ with δE being the deviation from the QW band edge $\delta E = E - E_0$. The resulting lineshape is given by the convolution of the Gaussian with the unperturbed absorption spectrum yielding an error-function-like spectrum[2] as shown in Fig. 11.28a.

For complete thermalization the level population is given by the Boltzmann function. The recombination spectrum is given by the product of the absorption spectrum (or JDOS) and the Boltzmann function. It is (red-) shifted with respect to E_0 by about[3]

$$\Delta E(T) = -\sigma^2/kT \,. \tag{11.10}$$

This shift between emission and absorption is also called the Stoke's shift.

[2]The error function is defined as $\mathrm{erf}(x) = \frac{2}{\sqrt{\pi}} \int_0^z \exp -t^2 \mathrm{d}t$.

[3]Formula (11.10) is exact for the product of a Gaussian and the Boltzmann function.

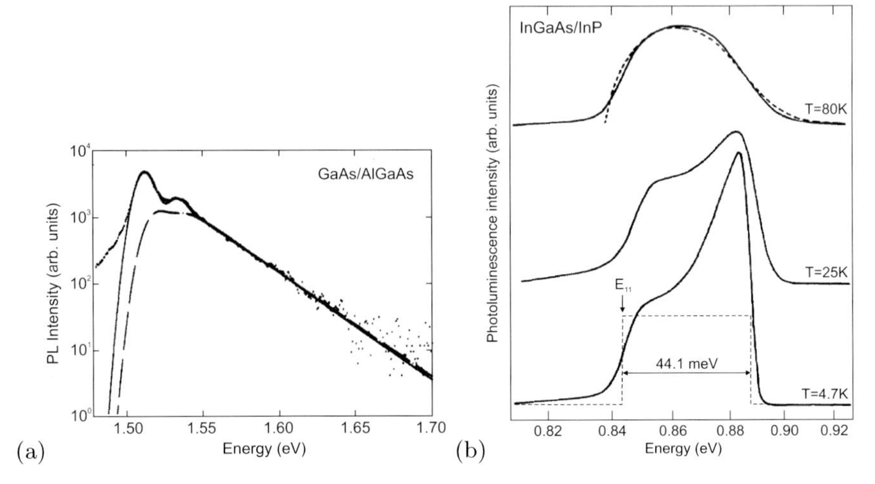

Fig. 11.27. (**a**) Photoluminescence spectrum of a 5-nm GaAs/AlGaAs quantum well at $T = 300\,\mathrm{K}$. The *solid* (*dashed*) line is fit with (without) excitonic effects. The two peaks are due to transitions involving heavy and light holes. Adapted from [400]. (**b**) Photoluminescence spectra at three different temperatures as labeled of a 10-nm modulation-doped InGaAs/InP quantum well with an electron sheet density $n_\mathrm{s} = 9.1 \times 10^{11}\,\mathrm{cm}^{-2}$. The electron quasi-Fermi level is $F_\mathrm{n} - (E_\mathrm{C} + E_\mathrm{e1}) = 44.1\,\mathrm{meV}$ from the subband edge. The *dashed line* in the $T = 80\,\mathrm{K}$ spectrum is the lineshape from JDOS and a Fermi–Dirac distribution without enhancement at the Fermi edge. Adapted from [401]

Within their lifetime, limited at least by radiative recombination, the excitons are typically unable to reach the energy position required by the Boltzmann function, but only a local minimum. Thus, their thermalization may be incomplete due to insufficient lateral diffusion. This effect is particularly important at low temperatures when thermal emission into adjacent deeper potential minima is suppressed. In this case the red-shift is smaller than expected from (11.10). A numerical simulation [402] yields such behavior of the energy position of the recombination line as shown in Fig. 11.28b. Simultaneously, the width of the recombination spectrum also exhibits a minimum (Fig. 11.28c). These findings are in agreement with experiments [404, 405].

A potential fluctuation can localize an exciton laterally at low temperatures [400] and behave like a quantum dot (see Sect. 13.3). Localized and delocalized excitons are separated by a boundary called the mobility edge [406]. The transition between the two regimes is a Mott transition [407].

Under certain growth conditions, quantum wells with piecewise very flat interfaces can be fabricated. The thickness difference between such regions (with lateral extension in the µm range) is an integer monolayer. Accordingly, the recombination spectrum yields several, typically two or three, discrete lines.

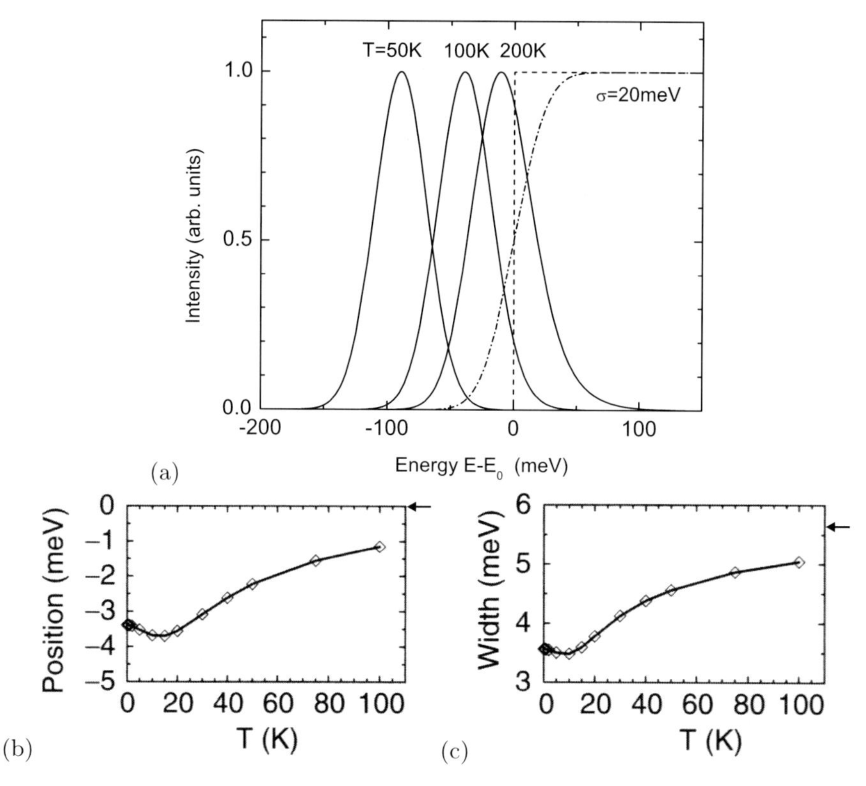

Fig. 11.28. (**a**) Recombination spectra (*solid lines*, scaled to same height) of a model quantum well for different temperatures as labeled and complete thermalization, *dashed* (*dash-dotted*) *line* is unperturbed (inhomogeneously broadened by $\sigma = 20\,\mathrm{meV}$) shape of the QW absorption edge. The energy scale is relative to the energy position of the unperturbed QW absorption edge at E_0. (**b**) Theoretical energy position and (**c**) linewidth of exciton recombination from a model disordered quantum well. The high-temperature limits are marked by arrows. Parts (**b**,**c**) adapted from [402]

11.7 Isotope Superlattices

A special type of heterostructure is the modulation of the isotope content. The first kind of heterostructures made like this were $^{70}\mathrm{Ge}_n/^{74}\mathrm{Ge}_n$ symmetric superlattices [408]. Figure 11.29 shows phonon energies determined from Raman spectroscopy for various layer numbers n. The modes are classified by $^{70}\mathrm{LO}_m$ and $^{74}\mathrm{LO}_m$ denoting the material in which the amplitude is maximal and m being the number of maxima in that medium.[4] Such modes are visualized in Fig. 11.30a for a $(^{69}\mathrm{GaP})_{16}/(^{71}\mathrm{GaP})_{16}$ superlattice. Theoretical mode energies as a function of the superlattice period are shown in Fig. 11.30b.

[4] Only modes with odd m are Raman-active.

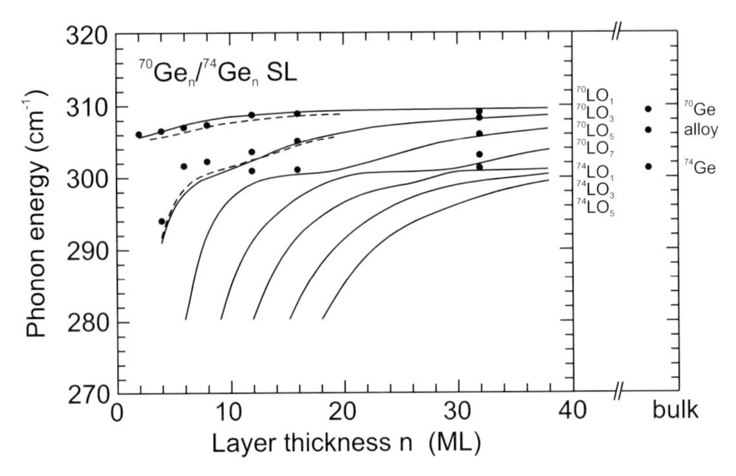

Fig. 11.29. Measured (*full circles*) and theoretical (*solid lines*) confined LO phonon energies in $^{70}\mathrm{Ge}_n/^{74}\mathrm{Ge}_n$ superlattices vs. the layer thickness (number of monolayers) n. The *dashed lines* represent a calculation that considers intermixing at the interfaces. On the right, the energies of bulk modes for isotopically pure ^{70}Ge and ^{74}Ge are shown together with that of an $^{70}\mathrm{Ge}_{0.5}{}^{74}\mathrm{Ge}_{0.5}$ alloy. Adapted from [408]

11.8 Wafer Bonding

Wafer bonding is a fairly recently developed method to join different and dissimilar materials. Two wafers of the respective materials are put together face to face and are adequately fused. The idea is to not only 'glue' the wafers together with a sticky (and compliant) organic material, but to form strong atomic bonds between the two materials with possibly a perfect interface. In some cases, the interface needs to allow charge-carrier transport through it. Less stringent conditions need to be met for photon transport.

Mechanical deficiencies such as surface roughness, dust particles and the like must be avoided in the wafer-bonding process since they result in voids. Several methods have been developed for bonding various materials [96, 97]. Such processes are successful for large substrate sizes. With proper processing, ideal interfaces can be created, as shown in Fig. 11.31. Such structures, if made between a p-doped and a n-doped semiconductor, show diode characteristics.

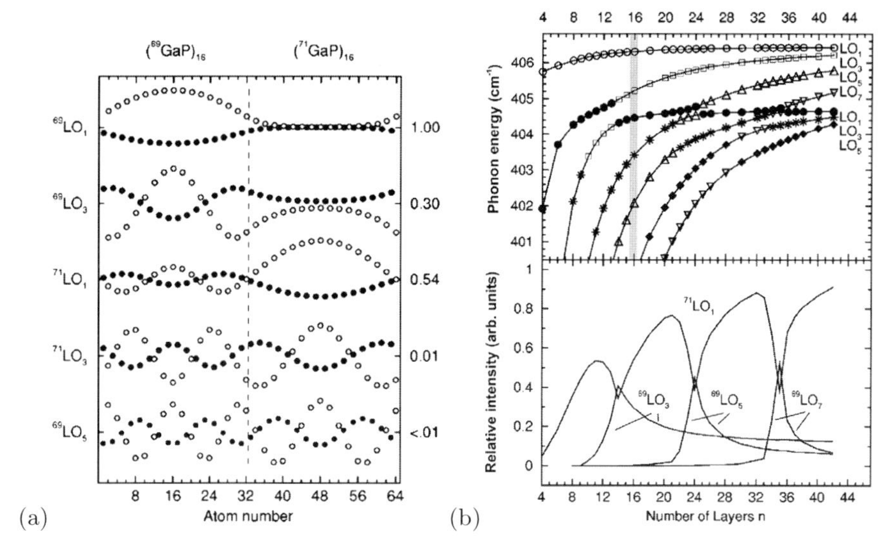

Fig. 11.30. (**a**) Atomic displacements [Ga (*filled dots*) and P (*open circles*)] of odd-index LO modes in a $(^{69}GaP)_{16}/(^{71}GaP)_{16}$ superlattice unit cell. These modes have even parity with respect to midlayer planes, which are at atom numbers 16 and 48 in this example. The labels on the left identify the predominant character of the mode, those on the right give the relative Raman intensities with respect to that of the $^{69}LO_1$ mode. The tick marks on the vertical axis indicate zero displacement of the respective mode. (**b**) *Upper panel*: Energies and characters of odd-index LO phonon modes in GaP isotope SLs as calculated within the planar bond charge model for the case of ideal interfaces. $^{69}LO_m$ modes are shown as *open symbols*; $^{71}LO_m$ modes as *full symbols*. The *shaded area* marks $n = 16$ for which the atomic displacements of the modes are shown in part (**a**). *Lower panel*: Calculated intensities of the modes relative to that of the $^{69}LO_1$ phonon mode. Adapted from [95], reprinted with permission, ©1999 APS

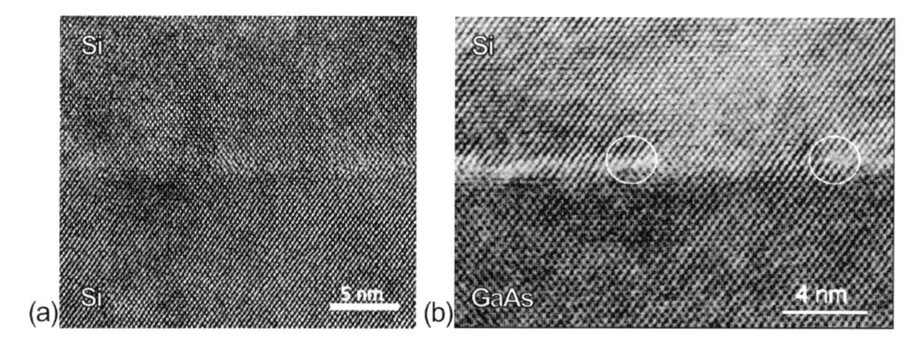

Fig. 11.31. High-resolution TEM images of wafer-bonded (**a**) Si-Si and (**b**) GaAs-Si interfaces. *White circles* indicate the position of misfit dislocations. Part (**a**) reprinted from [98], ©2003, with permission from Elsevier. Part (**b**) reprinted with permission from [99], ©1998 AIP

12 External Fields

The energy levels of the solid and its optical and electronic properties depend on external electric and magnetic fields. In high magnetic fields and at low temperatures the quantum Hall effects give evidence for new states of matter in many-body systems.

12.1 Electric Fields

12.1.1 Bulk Material

The center-of-mass motion of the exciton is not influenced by an electric field. The Hamilton operator for the relative motion of an electron–hole pair of reduced mass μ along z in the presence of an electric field $\mathcal{E}$ along the z direction is

$$\hat{H} = -\frac{\hbar^2}{2\mu}\Delta - e\mathcal{E}z \; . \tag{12.1}$$

Here, the Coulomb interaction, leading to the formation of bound exciton states, is neglected. In the plane perpendicular to the field (here the z direction) the solutions for the relative motion are just plane waves.

In the electric field the bands are tilted (Fig. 12.1), i.e. there is no longer an overall bandgap. Accordingly, the wavefunctions are modified and have exponential tails in the energy gap.

After separation of the motion in the (x,y) plane the Schrödinger equation for the motion in the z direction is

$$\left(-\frac{\hbar^2}{2\mu}\frac{\mathrm{d}^2}{\mathrm{d}z^2} - e\mathcal{E}z - E_z\right)\phi(z) = 0 \; , \tag{12.2}$$

which is of the type

$$\frac{\mathrm{d}^2 f(\xi)}{\mathrm{d}\xi^2} - \xi f(\xi) = 0 \; , \tag{12.3}$$

with $\xi = \frac{E_z}{\Theta} - z\left(\frac{2\mu e}{\hbar^2}\mathcal{E}\right)^{1/3}$ and the optoelectronic energy $\Theta = \left(\frac{e^2\mathcal{E}^2\hbar^2}{2\mu}\right)^{1/3}$. The solution of (12.3) is given by the Airy function Ai (see Fig. 12.2):

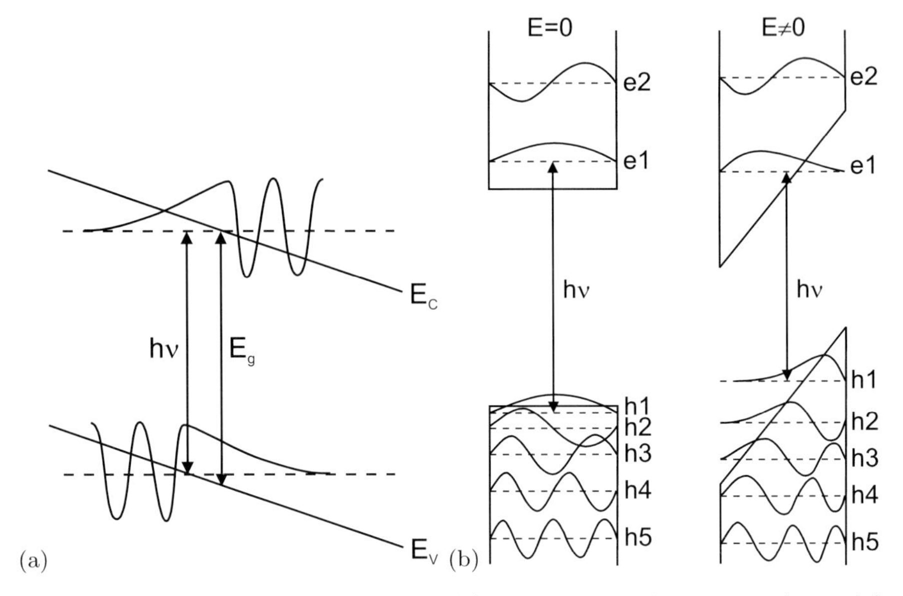

Fig. 12.1. Impact of an electric field on (**a**) bulk material (tilt of bands) and (**b**) a quantum well (quantum confined Stark effect, QCSE)

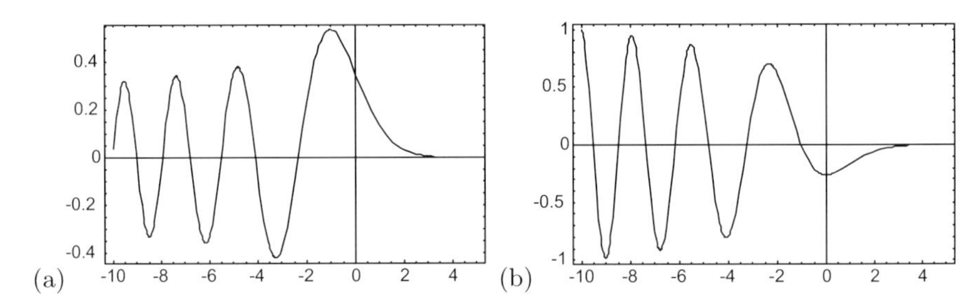

Fig. 12.2. (**a**) Airy function Ai(x), (**b**) Ai'(x)

$$\phi_{E_z}(\xi) = \frac{(e\mathcal{E})^{1/2}}{\Theta} Ai(\xi) \; . \tag{12.4}$$

The prefactor guaranties the orthonormality (with regard to the E_z). The absorption spectrum is then given by

$$\alpha(\omega, E) \propto \frac{1}{\omega} \Theta^{1/2} \pi \left[Ai^{'2}(\eta) - \eta Ai^2(\eta) \right] \; , \tag{12.5}$$

with $\eta = (E_g - E)/\Theta$ and $Ai'(x) = \mathrm{d}Ai(x)/\mathrm{d}x$.

Optical transitions below the bandgap become possible that are photon-assisted tunneling processes. The below-bandgap transitions have the form of an exponential tail. Additionally, oscillations develop above the bandgap, the so-called Franz–Keldysh oscillations (FKO) (Fig. 12.3a).

The absorption spectrum scales with the optoelectronic energy Θ. The energy position of the FKO peaks E_n is periodic with ($\nu \sim 0.5$)

$$(E_n - E_{\mathrm{g}})^{3/2} \propto (n - \nu)\mathcal{E}\mu^{-1/2} \,. \tag{12.6}$$

A nonperiodicity can indicate a nonparabolicity of the mass. For a given mass the electric field strength can be determined. Well-pronounced oscillations are only present for homogeneous fields.

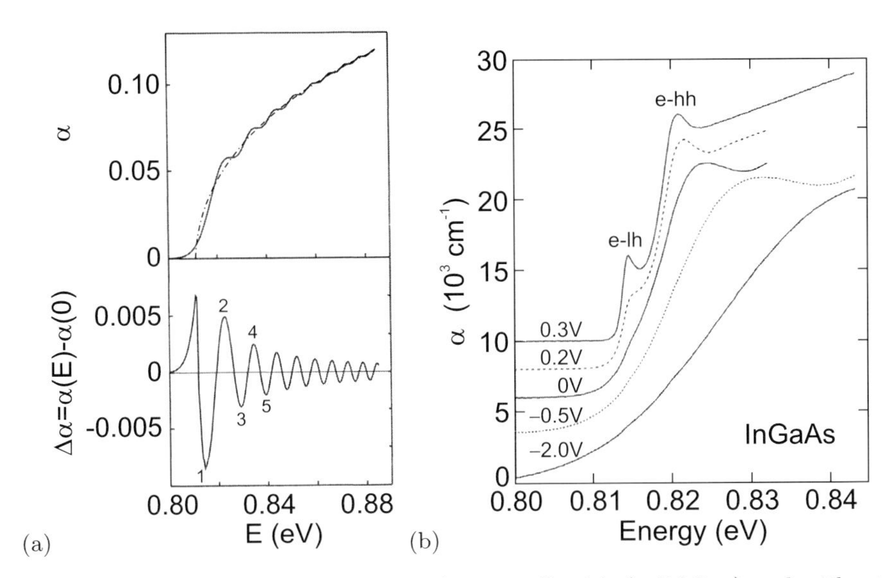

Fig. 12.3. (**a**) Theoretical absorption (*top panel*) with (*solid line*) and without (*dash-dotted line*) electric field for a volume semiconductor (without Coulomb interaction) and theoretical change of absorption (*bottom panel*). (**b**) Experimental absorption spectra of InGaAs on InP at $T = 15\,\mathrm{K}$ for various applied voltages as labeled. Adapted from [410]

Experimental spectra show additionally the peaks due to excitonic correlation (Fig. 12.3b) at low field strength. At higher fields the FKO evolve and the amplitude of the excitonic peaks decreases because the excitons are ionized in the field.

12.1.2 Quantum Wells

In a quantum well an electric field along the confinement direction (z direction) causes electrons and holes to shift their mean position to opposite interfaces (Fig. 12.1b). However, excitons are not ionized due to the electric field. With increasing field (for both field directions) the energy position of the absorption edge and the recombination energy is reduced. This is the quantum confined Stark effect (QCSE). Corresponding experimental data are shown in Fig. 12.4a(i–v). The shift depends quadratically on the electric field since the exciton has no permanent dipole moment (mirror symmetry of the quantum well). Thus, only the second-order Stark effect is present in which the field first induces a dipole $\mathbf{p} = \alpha\mathcal{E}$. This dipole interacts with the field with an energy $E = -\mathbf{p} \cdot \mathcal{E} = -\alpha\mathcal{E}^2$.

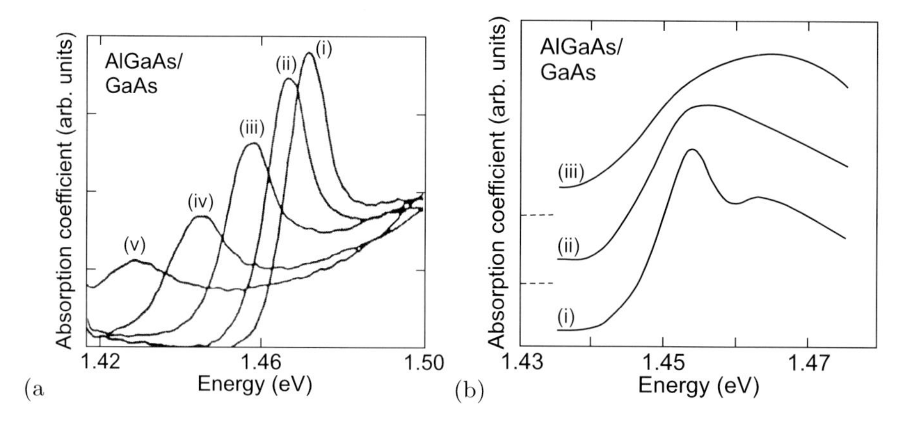

Fig. 12.4. Impact of electric fields on the absorption spectrum of a 9.4-nm AlGaAs/GaAs quantum well. (**a**): Electric field along the [001] growth direction, (**i**)–(**v**): $\mathcal{E} = 0$, 0.6, 1.1, 1.5, and 2×10^5 V/cm. (**b**): Electric field within the interface plane, (**i,ii,iii**): $\mathcal{E} = 0$, 1.1, and $2{\times}10^5$ V/cm

If the field is within the quantum-well interface plane, the field leads to the ionization of excitons. The loss of the excitonic peak is visualized in the spectra in Fig. 12.4b(i–iii).

12.2 Magnetic Fields

In magnetic fields, electrons (or holes) perform a cyclotron motion with frequency $\omega_c = eB/m^*$, i.e. a motion perpendicular to the magnetic field on a line of constant energy in **k**-space. This line is the intersection of a plane perpendicular to the magnetic field and the respective isoenergy surface in **k**-space. For semiconductors with anisotropic mass, such as Si and Ge, the quantum theory of cyclotron resonance has been given in [409]. The 'free' cyclotron motion can only occur between two scattering events. Thus, a significant cyclotron motion and the connected magnetotransport properties are only possible when

- $\omega_c\tau \gg 1$, i.e. when the average scattering time τ is sufficiently large. This requires high mobility.

- the magnetic field is sufficiently strong and the temperature sufficiently low, i.e. $\hbar\omega_c \gg kT$, such that thermal excitations do not scatter electrons between different Landau levels.
- the cyclotron path is free of geometric obstructions.

An external magnetic field also produces a splitting of the spin states. For the electron, the energy splitting ΔE is given by

$$\Delta E = g_e^* \mu_B B \; , \tag{12.7}$$

where B is the magnetic-field amplitude and g_e^* the (effective) electron g-factor. This value differs from the free-electron value in vacuum of $g_e = 2.0023$ due to the presence of spin-orbit interaction. Values for g_e^* at low carrier density and low temperatures are 2 for Si, 1.2 for InP and ZnSe, −1.65 for CdTe, −0.44 for GaAs, and −50 for InSb. In [411] the temperature dependence of g_e^* in GaAs, InP and CdTe is also measured and discussed. The electron g-factor increases in thin GaAs/AlGaAs quantum wells [412].

12.2.1 Free-Carrier Absorption

The absorption of free carriers was treated in Sect. 9.6 without the presence of a static magnetic field. Solving (8.19) for a static magnetic field $\mathbf{B} = \mu_0 \mathbf{H}$ with $\mathbf{H} = H\,(h_x, h_y, h_z)$ and a harmonic electric field $\mathbf{E} \propto \exp(-i\omega t)$ yields for the dielectric tensor (cf. (9.50))

$$\epsilon = \frac{i}{\epsilon_0 \omega} \sigma \; , \tag{12.8}$$

by comparing $\mathbf{j} = \sigma \mathbf{E} = qN\mathbf{v}$

$$\epsilon(\omega) = -\omega_p^{*2} \left[(\omega^2 + i\omega\gamma)\mathbf{1} - i\omega_c \begin{pmatrix} 0 & -h_z & h_y \\ h_z & 0 & -h_z \\ -h_y & h_z & 0 \end{pmatrix} \right]^{-1} \; , \tag{12.9}$$

where $\mathbf{1}$ denotes the (3 by 3) unity matrix and $\gamma = 1/\tau = m^*\mu/q$ is the damping parameter with μ representing the optical carrier mobility. The (unscreened) plasma frequency is given by

$$\omega_p^{*2} = n \frac{e^2}{\epsilon_0 m^*} \; . \tag{12.10}$$

The free-carrier cyclotron frequency is

$$\omega_c = e \frac{\mu_0 H}{m^*} \; . \tag{12.11}$$

If the effective mass is treated as a tensor, $1/m^*$ is replaced by $\mathbf{m}^{*-1}$ in (12.10) and (12.11). For zero magnetic field the classical Drude theory for one carrier species is recovered (cf. (9.52a))

$$\epsilon(\omega) = -\frac{\omega_{\mathrm{p}}^{*2}}{\omega(\omega + \mathrm{i}\gamma)} . \tag{12.12}$$

With the magnetic field perpendicular to the sample surface, i.e. $\mathbf{B} = \mu_0(0, 0, H)$ the magneto-optic dielectric tensor simplifies to (cf. (8.20d))

$$\epsilon(\omega) = \frac{-\omega_{\mathrm{p}}^{*2}}{\omega^2} \begin{pmatrix} \tilde{\epsilon}_{xx} & \mathrm{i}\tilde{\epsilon}_{xy} & 0 \\ -\mathrm{i}\tilde{\epsilon}_{xy} & \tilde{\epsilon}_{xx} & 0 \\ 0 & 0 & \tilde{\epsilon}_{zz} \end{pmatrix} \tag{12.13a}$$

$$\tilde{\epsilon}_{xx} = \frac{1 + \mathrm{i}\gamma/\omega}{(1 + \mathrm{i}\gamma/\omega)^2 - (\omega_{\mathrm{c}}/\omega)^2} \tag{12.13b}$$

$$\tilde{\epsilon}_{zz} = \frac{1}{(1 + \mathrm{i}\gamma/\omega)} \tag{12.13c}$$

$$\tilde{\epsilon}_{xy} = \frac{\omega_{\mathrm{c}}/\omega}{(1 + \mathrm{i}\gamma/\omega)^2 - (\omega_{\mathrm{c}}/\omega)^2} . \tag{12.13d}$$

The in-plane component ϵ_{xx} provides information about ω_{p}^* and γ, i.e. n and μ, the effective mass, are known. Additionally, the antisymmetric tensor component ϵ_{xy} is linear in the cyclotron frequency and provides q/m. This subtle but finite birefringence depends on the strength (and orientation) of the magnetic field and can be experimentally determined in the infrared using magneto-ellipsometry [413, 414]. Such 'optical Hall effect' experiment allows the determination of the carrier density n, the mobility μ, the carrier mass[1] m^* and the sign of the carrier charge $\mathrm{sgn}(q)$ with optical means. The electrical Hall effect (Sect. 8.4) can reveal n, μ and $\mathrm{sgn}(q)$ but cannot reveal the carrier mass.

12.2.2 Energy Levels in Bulk Crystals

In a 3D electron gas (the magnetic field is along z, i.e. $\mathbf{B} = [0, 0, B]$) the motion in the (x, y) plane is described by Landau levels. Quantum mechanically they correspond to levels of a harmonic oscillator. The magnetic field has no impact on the motion of electrons along z, such that in this direction a free dispersion relation $\propto k_z^2$ is present. The energy levels are given as

$$E_{nk_z} = \left(n + \frac{1}{2}\right) \hbar\omega_{\mathrm{c}} + \frac{\hbar^2}{2m} k_z^2 . \tag{12.14}$$

Thus, the states are on concentric cylinders in **k**-space (Fig. 12.5a). The populated states of the 3D electron gas (at 0 K) lie within the Fermi vector of length k_{F}. For the 3D system the density of states at the Fermi energy

[1] We note that mobility and effective mass defined and measured in this way may be referred to as 'optical'. Other definitions and approaches to the mobility or effective mass may give different results.

is a square root function of the Fermi energy (6.70). In the presence of a magnetic field the density of states diverges every time that a new cylinder (with a one-dimensional density of states, (6.74)) touches the Fermi surface at E_{F}. In real systems, the divergence will be smoothed, however, a pronounced peak or the periodic nature of the density of states is often preserved.

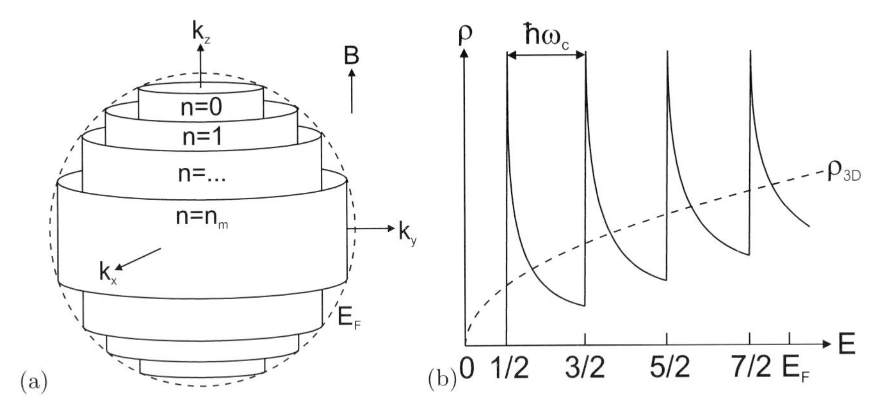

Fig. 12.5. 3D electron gas in an external magnetic field. (**a**) Allowed states in **k**-space for magnetic field along the z direction. (**b**) Density of states (DOS) ρ vs. energy (in units of $\hbar\omega_c$). *Dashed line* is three-dimensional DOS without magnetic field. Based on [376]

The period is given by the number n_{m} of cyclotron orbits (Landau levels) within the Fermi surface.

$$\left(n_{\mathrm{m}} + \frac{1}{2}\right) \hbar\omega_{\mathrm{c}} = E_{\mathrm{F}} \, . \tag{12.15}$$

If the number of carriers is constant, the density of states at the Fermi energy at varying magnetic field varies periodically with $1/B$. From the conditions $\left(n_{\mathrm{m}} + \frac{1}{2}\right) \hbar \frac{eB_1}{m} = E_{\mathrm{F}}$ and $\left(n_{\mathrm{m}} + 1 + \frac{1}{2}\right) \hbar \frac{eB_2}{m} = E_{\mathrm{F}}$ with $\frac{1}{B_2} = \frac{1}{B_1} + \frac{1}{\Delta B}$ we find

$$\frac{1}{\Delta B} = \frac{e\hbar}{m^* E_{\mathrm{F}}} \, . \tag{12.16}$$

This periodicity is used to determine experimentally, e.g., the properties of the Fermi surface in metals using the Shubnikov–de Haas oscillations (of the magnetoresistance) or the de Haas–van Alphén effect (oscillation of the magnetic susceptibility).

12.2.3 Energy Levels in a 2DEG

In a 2D electron gas (2DEG), e.g. in a quantum well or a potential well at a modulation-doped heterointerface, a free motion in z is not possible and

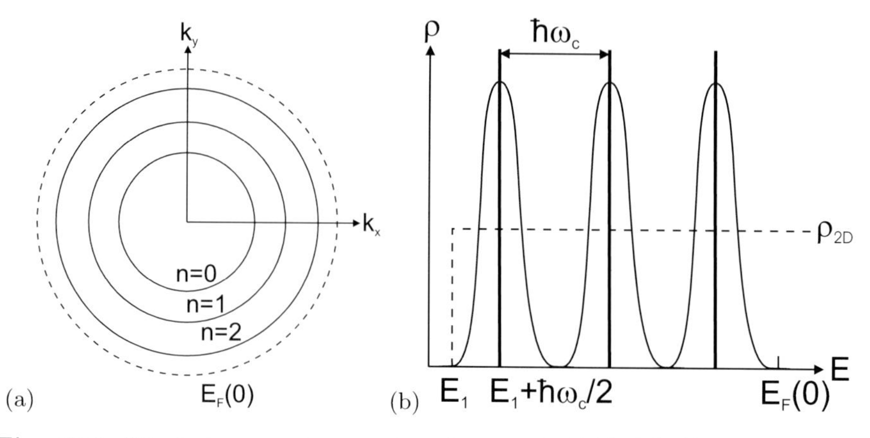

Fig. 12.6. 2D electron gas in an external magnetic field. (**a**) Allowed states in **k**-space. (**b**) Density of states (DOS) ρ vs. energy. *Dashed line* is two-dimensional DOS without magnetic field. *Thick vertical lines*: δ-like DOS without broadening, *curves*: broadened DOS. Based on [376]

k_z is quantized. The energy levels (for each 2D subband) are only given by the cyclotron energy (Fig. 12.6a). The density of states is a sequence of δ-like peaks (Fig. 12.6b). Each peak contributes (degeneracy $\hat{g}$ of a Landau level) a total number of

$$\hat{g} = \frac{eB}{h} \tag{12.17}$$

states (per unit area without spin degeneracy and without the degeneracy of the band extremum). In reality, disorder effects lead to an inhomogeneous broadening of these peaks. The states in the tails of the peaks correspond to states that are localized in real space.

Also, in a 2D system several physical properties exhibit an oscillatory behavior as a function of Fermi level, i.e. with varying electron number, and as a function of the magnetic field (periodic with $1/B$) at fixed Fermi energy, i.e. at fixed electron number (Fig. 12.7).

12.2.4 Shubnikov–de Haas Oscillations

From the 2D density of states (per unit area including spin degeneracy) $D^{2\mathrm{D}}(E) = m^*/\pi\hbar^2$ (6.72) the sheet density of electrons n_s can be expressed as a function of the Fermi level (at $T = 0\,\mathrm{K}$ without spin degeneracy)

$$n_\mathrm{s} = \frac{m^*}{2\pi\hbar^2}(E_\mathrm{F} - E_\mathrm{C}) \; . \tag{12.18}$$

Using (12.16) we thus find (without spin degeneracy, without valley degeneracy), that the period of $1/B$ is $\propto n_\mathrm{s}$:

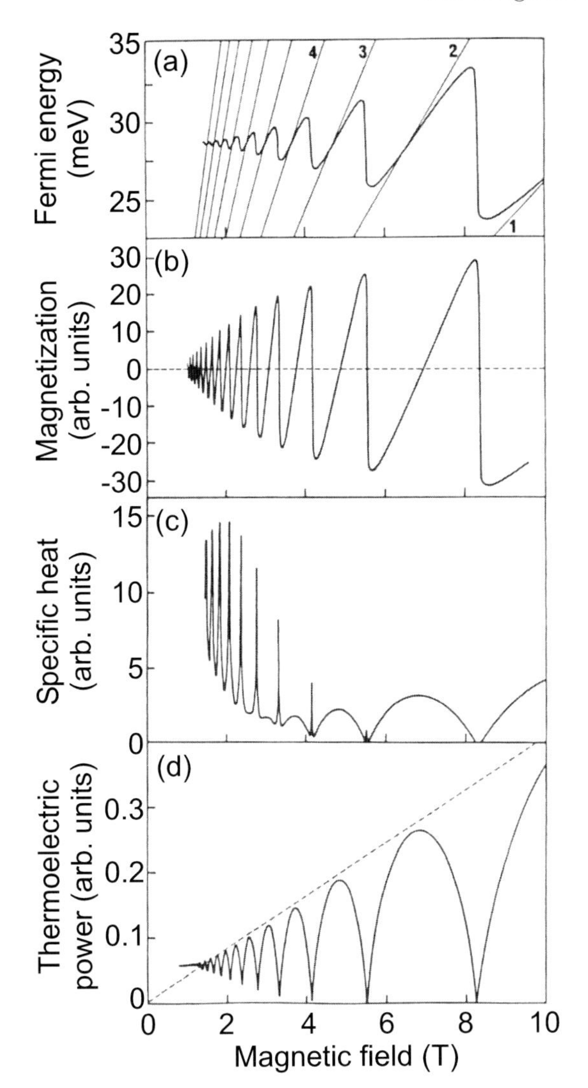

Fig. 12.7. Oscillatory (theory, $T = 6\,\mathrm{K}$) behavior of a 2DEG (GaAs/AlGaAs) in a magnetic field: (**a**) Fermi level, (**b**) magnetization, (**c**) specific heat, (**d**) thermoelectric power. A Gaussian broadening of 0.5 meV was assumed. Adapted from [376,415]

$$\frac{1}{\Delta B} = \frac{e}{h} n_\mathrm{s} \; . \tag{12.19}$$

The carrier density of a 2DEG can therefore be determined from the oscillations of magnetoresistance, and is proportional to the density of states at the Fermi level (Shubnikov–de Haas effect). A corresponding measurement with varying field and fixed electron density is shown in Fig. 12.8. The periodicity with $1/B$ is obvious. Since only the component of the magnetic field

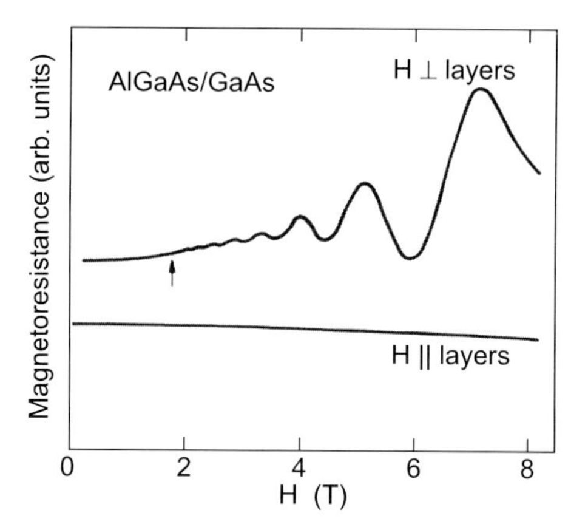

Fig. 12.8. Shubnikov–de Haas oscillations at a modulation-doped AlGaAs/GaAs heterostructure with a 2DEG, $n = 1.7 \times 10^{17}\,\mathrm{cm}^{-2}$ and $\mu = 11\,400\,\mathrm{cm}^2/\mathrm{V\,s}$. Data from [416]

perpendicular to the layer affects the (x, y) motion of the carriers, no effect is observed for the magnetic field parallel to the layer.

In another experiment the carrier density was varied at constant field (Fig. 12.9). The electron density in an inversion layer in p-type silicon is (linearly) varied with the gate voltage of a MOS (metal–oxide–semiconductor) structure (inset in Fig. 12.9, for MOS diodes see Sect. 21.5). In this experiment, the Fermi level was shifted through the Landau levels. The equidistant peaks show that indeed each Landau level contributes the same number of states.

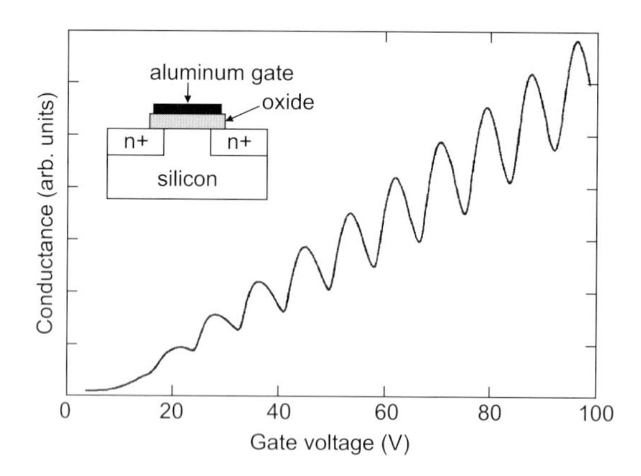

Fig. 12.9. Shubnikov–de Haas oscillations of a 2DEG at the (100) surface of p-type silicon ($100\,\Omega\,\mathrm{cm}$) at a magnetic field of 33 kOe and $T = 1.34\,\mathrm{K}$. The *inset* shows schematically the contact geometry. Data from [417]

12.3 Quantum Hall Effect

In high magnetic fields, at low temperatures and for high-mobility, 2D electron gases exhibit a deviation from the classical behavior. We recall that the classical Hall effect (i.e. considering the Lorentz force, classical Drude theory), the generation of a field $\mathcal{E}_y$ perpendicular to a current flow j_x (see Sect. 8.4), was described with the conductivity tensor σ (here, for the (x, y)-plane only)

$$\sigma = \frac{\sigma_0}{1+\omega_c^2\tau^2}\begin{pmatrix} 1 & \omega_c\tau \\ -\omega_c\tau & 1 \end{pmatrix} \tag{12.20a}$$

$$\sigma_{xx} = \sigma_0 \frac{1}{1+\omega_c^2\tau^2} \rightarrow 0 \tag{12.20b}$$

$$\sigma_{xy} = \sigma_0 \frac{\omega_c\tau}{1+\omega_c^2\tau^2} \rightarrow \frac{ne}{B}\,, \tag{12.20c}$$

where σ_0 is the zero-field conductivity $\sigma_0 = ne^2\tau/m^*$ (8.5). The arrows denote the limit for $\omega_c\tau \rightarrow \infty$, i.e. large fields. The resistivity tensor $\rho = \sigma^{-1}$ is given by

$$\rho = \begin{pmatrix} \rho_{xx} & \rho_{xy} \\ -\rho_{xy} & \rho_{xx} \end{pmatrix} \tag{12.21a}$$

$$\rho_{xx} = \frac{\sigma_{xx}}{\sigma_{xx}^2+\sigma_{xy}^2} \rightarrow 0 \tag{12.21b}$$

$$\rho_{xy} = \frac{-\sigma_{xy}}{\sigma_{xx}^2+\sigma_{xy}^2} \rightarrow -\frac{B}{ne}\,. \tag{12.21c}$$

12.3.1 Integral QHE

Experiments yield strong deviations from the linear behavior of the transverse resistivity $\rho_{xy} = \mathcal{E}_y/j_x = BR_\mathrm{H}$ with the Hall coefficient $R_\mathrm{H} = -1/(ne)$ with increasing magnetic field is observed at low temperatures for samples with high carrier mobility, i.e. $\omega_c\tau \gg 1$ (Fig. 12.10). In Fig. 12.11a,b, Hall bars are shown for 2DEGs in silicon metal–oxide–semiconductor field-effect transistor (Si-MOSFET) electron inversion layers and at GaAs/AlGaAs heterostructures, respectively.

The Hall resistivity exhibits extended Hall plateaus with resistivity values that are given by

$$\rho_{xy} = \frac{1}{i}\frac{h}{e^2}\,, \tag{12.22}$$

i.e. integer fractions of the quantized resistance $\rho_0 = h/e^2 = 25812.807\,\Omega$, which is also called the von-Klitzing constant. In Fig. 12.10, a spin splitting is seen for the $n = 1$ Landau level (and a small one for the $n = 2$). We note

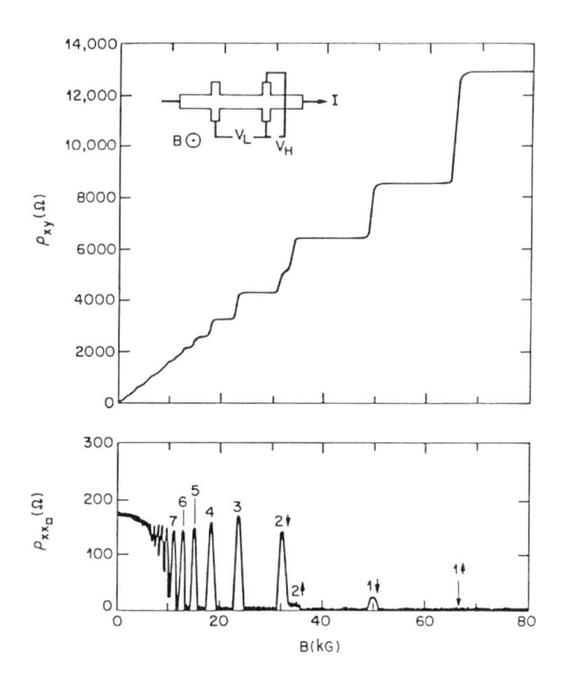

Fig. 12.10. Hall resistivity ρ_{xy} and longitudinal resistivity ρ_{xx} for a modulation-doped GaAs/AlGaAs heterostructure ($n = 4 \times 10^{11}\,\mathrm{cm}^{-2}$, $\mu = 8.6 \times 10^4\,\mathrm{cm}^2/\mathrm{V\,s}$) at 50 mK as a function of magnetic field (10 kG=1 T). The numbers refer to the quantum number and spin polarization of the Landau level involved. The inset shows schematically the Hall bar geometry, V_L (V_H) denotes the longitudinal (Hall) voltage drop. Reprinted with permission from [421], ©1982 APS

that the topmost Hall plateau is due to the completely filled $n = 0$ Landau level; the resistance is $\rho_0/2$ due to the spin degeneracy of 2.

The integral quantum Hall effect, first reported in [418,419], and the value for ρ_0 are found for a wide variety of samples and conditions regarding sample temperature, electron density or mobility of the 2DEG and the materials of the heterostructure.

Within the plateau the resistivity is well defined within 10^{-7} or better up to 4×10^{-9}. A precise determination allows for a new normal for the unit Ohm [422], being two orders of magnitude more precise than the realization in the SI system, and an independent value for the fine-structure constant $\alpha = \frac{e^2}{\hbar c} \frac{1}{4\pi\epsilon_0}$. At the same time, the longitudinal resistivity, starting from the classical value for small magnetic fields, exhibits oscillations and eventually it is zero for the plateaus in ρ_{xy}. For ρ_{xx} values of $10^{-10}\,\Omega/\Box$ have been measured, which corresponds to $10^{-16}\,\Omega/\mathrm{cm}$ for bulk material, a value three orders of magnitude smaller than for any nonsuperconductor.

The simplest explanation is that the conductivity is zero when a Landau level is completely filled and the next is completely empty, i.e. the Fermi level

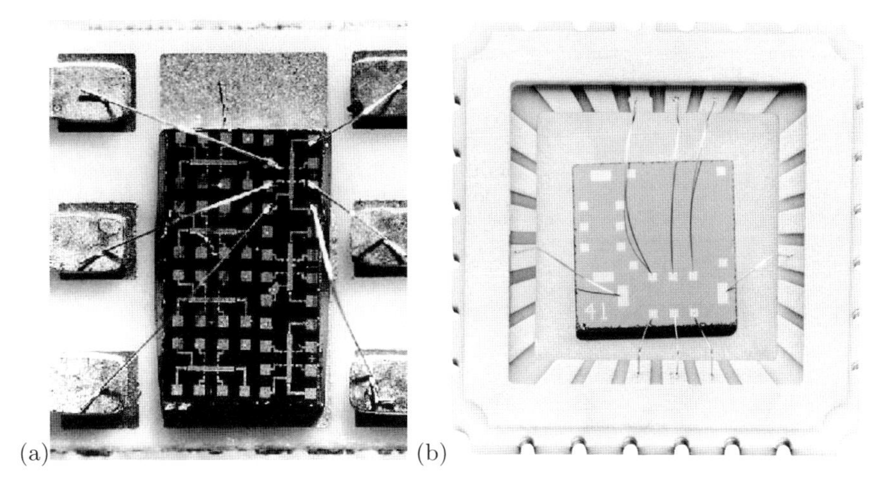

Fig. 12.11. (**a**) Silicon MOS (metal–oxide–semiconductor) structure of K. v. Klitzing's et al. original experiments. (**b**) GaAs/AlGaAs heterostructure sample grown with molecular beam epitaxy for QHE measurements, chip carrier and bond wires. Reprinted with permission from [423]

lies between them. The temperature is small, i.e. $kT \ll \omega_c \tau$, such that no scattering between Landau levels can occur. Thus no current, similar to a completely filled valence band, can flow. The sheet carrier density n_s is given by counting the i filled Landau levels (degeneracy according to (12.17)) as

$$n_s = i\frac{eB}{h} . \tag{12.23}$$

In the transverse direction energy dissipation takes place and the Hall resistivity $\rho_{xy} = B/(n_s e)$ takes the (scattering-free) values given in (12.22).

However, this argument is too simple as it will not explain the extension of the plateaus. As soon as the system has one electron more or less, the Fermi energy will (for a system with δ-like density of states) be located in the upper or lower Landau level, respectively. Then, the longitudinal conductivity should no longer be zero and the Hall resistivity deviates from the integer fraction of ρ_0.

Therefore (in the localization model) an inhomogeneous (Gaussian) broadening of the density of states is assumed. Additionally, the states in the tails of the distribution are considered nonconducting, i.e. localized, while those around the peak are considered conductive. This mobility edge is schematically shown in Fig. 12.12a–c.

When the Fermi level crosses the density of states of the broadened Landau level (upon increase of the magnetic field), it first populates localized states and ρ_{xx} remains at zero. When the Fermi level crosses the delocalized, conducting states, the longitudinal conductivity shows a peak and the Hall conductivity increases from the plateau value by e^2/h to the next plateau.

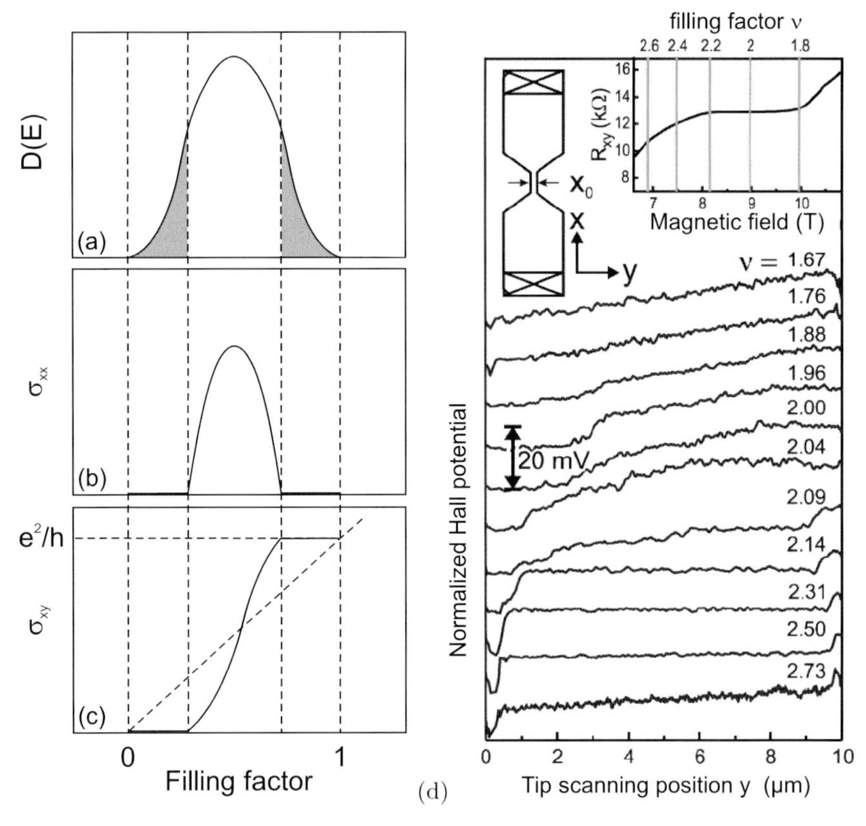

Fig. 12.12. (**a**) Density of states $D(E)$, (**b**) longitudinal conductivity σ_{xx} and (**c**) Hall conductivity σ_{xy} for a Landau level as a function of the filling factor $\mu = n/d$, where n is the electron density and d the degeneracy of the level. In (**a**) the filling factor relates via the position of the Fermi level to an energy scale. The *grey areas* in (**a**) denote localized states. The *slanted dashed line* in (**c**) has a slope of ne/B. Adapted from [420]. (**d**) Normalized Hall potential profile for different magnetic fields around filling factor $\nu = 2$. The overall voltage drop corresponds to 20 mV. The *insets* show the sample geometry and transport data. The 2DEG is from a GaAs/$Al_{0.33}Ga_{0.67}As$ modulation-doped heterostructure, $n_s = 4.3 \times 10^{11}\,cm^{-2}$, $\mu = 5 \times 10^5\,cm^2/Vs$, $T = 1.4\,K$. Adapted from [427]

When the Fermi level lies within the upper nonconducting states, the Hall resistivity remains constant in the next plateau. For the sample in Fig. 12.10 it has been estimated that 95% of the Landau level states are localized [421].

In the localization model it is still astonishing that the samples exhibit the step of e^2/h in the conductivity as if *all* electrons on the Landau level contribute to conduction, cf. (12.23). According to the calculations in [425] the Hall current lost by the localized states is compensated by an extra current by the extended states. The behavior of the electrons in the quantum

Hall regime ('quantum Hall liquid') can be considered similar to that of an incompressible fluid where obstructions lead the fluid to move with increased velocity. Nevertheless, the single particle picture seems to be insufficient to model the IQHE.

Another model for the explanation of the QHE, supported now with plenty of experimental evidence, is the edge state model where quantized one-dimensional conductivity of edge channels, i.e. the presence of conductive channels along the sample boundaries, is evoked [426]. Due to depletion at the boundary of the sample, the density of the 2DEG varies at the edge of the sample and 'incompressible' stripes develop for which $\partial\mu/\partial n_{\mathrm{s}} \to \infty$. When the filling factor is far from an integer, the Hall voltage is found to vary linearly across the conductive channel and the current is thus homogeneous over the sample (Fig. 12.12d). In the Hall plateau, the Hall voltage is flat in the center of the channel and exhibits drops at the edges, indicating that the current flows along the boundary of the sample (edge current) [427] in agreement with predictions from [428]. Although the current pattern changes with varying magnetic field, the Hall resistivity remains at its quantized value.

The most fundamental arguments for the explanation of the IQHE come from gauge invariance and the presence of a macroscopic quantum state of electrons and magnetic flux quanta [429]. This model holds as long as there are any extended states at all.

12.3.2 Fractional QHE

For very low temperatures and in the extreme quantum limit, novel effects are observed when the kinetic energy of the electrons is smaller than their Coulomb interaction. New quantum Hall plateaus are observed at various fractional filling factors $\nu = p/q$. We note that the effects of the fractional quantum Hall effect (FQHE) in Fig. 12.13 mostly arise for magnetic fields beyond the $n = 1$ IQHE plateau. The filling factor $\nu = n/(eB/h)$ is now interpreted as the number of electrons per magnetic flux quantum $\phi_0 = h/e$.

The effects of the FQHE cannot be explained by single-electron physics. The plateaus at fractional fillings ν occur when the Fermi energy lies within a highly degenerate Landau (or spin) level and imply the presence of energy gaps due to many-particle interaction and the result of correlated 2D electron motion in the magnetic field.

A decisive role is played by the magnetic flux quanta. The presence of the magnetic field requires the many-electron wavefunction to assume as many zeros per unit area as there are flux quanta penetrating it. The decay of the wavefunction has a length scale of the magnetic length $l_0 = \sqrt{\hbar/(eB)}$. Since the magnetic field implies a 2π phase shift around the zero, such an object is also termed a *vortex*, being the embodiment of the magnetic flux quanta in the electron system. Such a vortex represents a charge deficit (compared to a homogeneous charge distribution) and thus electrons and vortices attract each other. If a vortex and an electron are placed onto each other, considerable

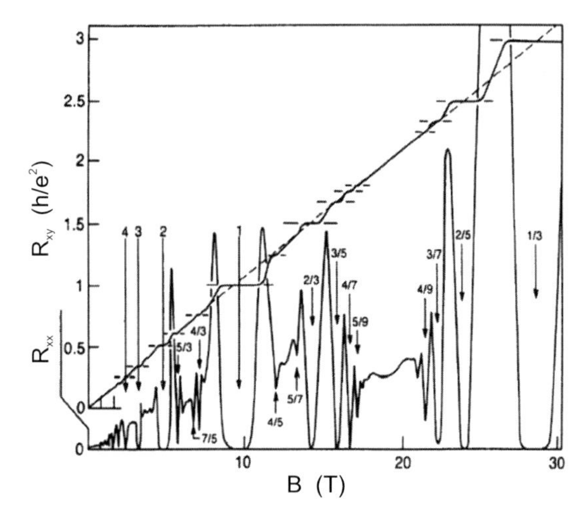

Fig. 12.13. Hall resistance R_{xy} and magnetoresistance R_{xx} of a two-dimensional electron system (GaAs/AlGaAs heterostructure) of density $n = 2.33 \times 10^{11}\,\mathrm{cm}^{-2}$ at a temperature of 85 mK, vs. magnetic field B. Numbers identify the filling factor ν, which indicates the degree to which the sequence of Landau levels is filled with electrons. Plateaus are due to the integral ($\nu = i$) quantum Hall effect (IQHE) and fractional ($\nu = p/q$) quantum Hall effect (FQHE). Adapted from [424], reprinted with permission, ©1990 AAAS

Coulomb energy is gained. At $\nu = 1/3$, there are three times more vortices than there are electrons, each vortex representing a charge deficit of $1/3\,e$. Such a system is described with many-particle wavefunctions, such as the Laughlin theory for $\nu = 1/q$ [429] and novel quasi-particles called *composite fermions* [430, 431] for other fractional fillings. For further reading we refer readers to [432] and references therein.

12.3.3 Weiss Oscillations

In Fig. 12.14, measurements are shown for a Hall bar in which an array of antidots (in which no conduction is possible) has been introduced by dry etching. The antidot size is 50 nm (plus depletion layer) and the period is 300 nm. These obstructions for the cyclotron motion lead to a modification of the magnetotransport properties.

Before etching of the antidot array the 2DEG has a mean free path length of 5–10 µm at 4 K for the mobility of $\approx 10^6\,\mathrm{cm}^2/\mathrm{V\,s}$. At low magnetic fields there is a strong deviation of the Hall resistivity from the straight line to which the QHE levels converge. Similarly, ρ_{xx} shows a strong effect as well.

These effects are related to commensurability effects between the antidot lattice and the cyclotron resonance path. When the cyclotron orbit is equal to the lattice period, electrons can fulfill a circular motion around one antidot (pinned orbit, Fig. 12.14b) that leads to a reduction of conductivity. At

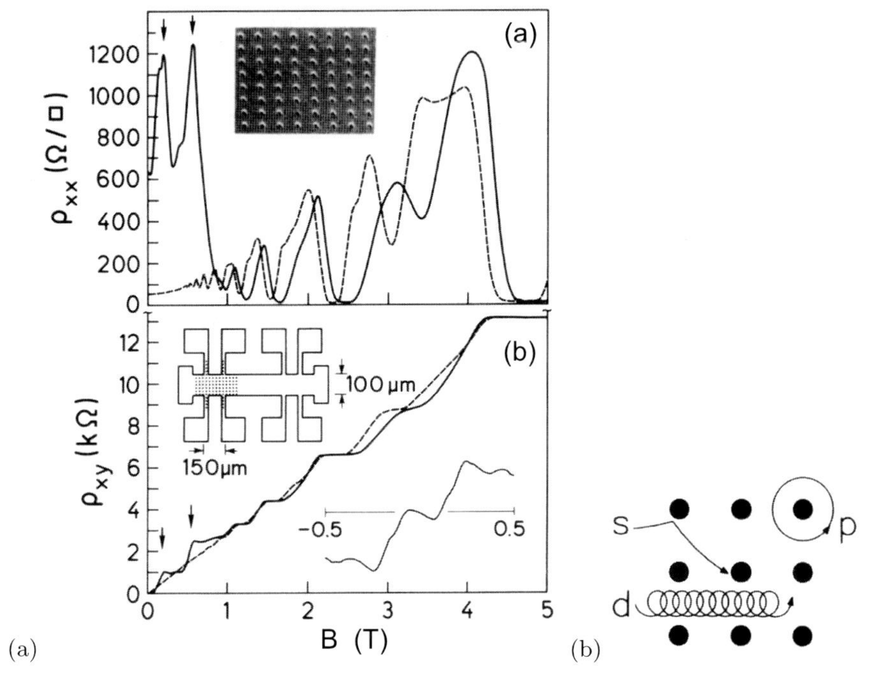

Fig. 12.14. Weiss oscillations: (**a**) magnetoresistance and (**b**) Hall resistance of an antidot lattice (*inset* in (**a**)) with pattern (*solid lines*) and without pattern (*dashed lines*) at $T = 1.5\,\mathrm{K}$. (**b**) Schematic of the different orbits: 'p': pinned, 'd': drifting, 's': scattered. Reprinted with permission from [433], ©1991 APS

high fields, drifting orbits for which the cyclotron orbit is much smaller than the lattice period occur. At small fields, scattering orbits also contribute for which the cyclotron radius is large and the electron has antidots from time to time. Resonances in the Hall resistivity have been found due to pinned orbits enclosing 1, 2, 4, 9 or 21 antidots.

13 Nanostructures

The principles of physics,
as far as I can see,
do not speak against the possibility
of maneuvering things atom by atom.
R.P. Feynman, 1959 [434]

13.1 Introduction

When the structural size of functional elements enters the size range of the de Broglie matter wavelength, the electronic and optical properties are dominated by quantum-mechanical effects. The most drastic impact can be seen from the density of states (Fig. 13.1).

The quantization in a potential is ruled by the Schrödinger equation with appropriate boundary conditions. These are simplest if an infinite potential is assumed. For finite potentials, the wavefunction leaks out into the barrier. Besides making the calculation more complicated (and more realistic), this allows electronic coupling of nanostructures. Via the Coulomb interaction, a coupling is even given if there is no wavefunction overlap. In the following, we will discuss some of the fabrication techniques and properties of quantum wires (QWR) and quantum dots (QD). In particular for the latter, several textbooks can also be consulted [435, 436].

13.2 Quantum Wires

13.2.1 Preparation Methods

Many methods have been proposed and realized to fabricate quantum wires. In the following, we will restrict ourselves to the growth of QWRs on corrugated substrates, fabrication using cleaved-edge overgrowth (CEO) and nanowhiskers.

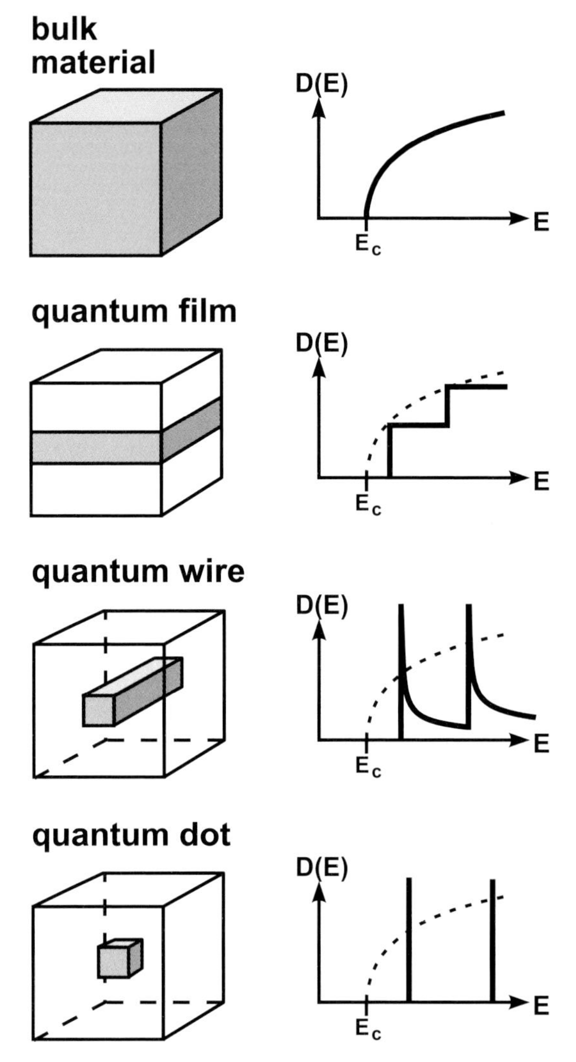

Fig. 13.1. Schematic geometry and density of states for 3D, 2D, 1D and 0D electronic systems

V-groove Quantum Wires

Quantum wires with high optical quality, i.e. high recombination efficiency and well-defined spectra, can be obtained by employing epitaxial growth on corrugated substrates. The technique is shown schematically in Fig. 13.2. A V-groove is etched, using, e.g., an anisotropic wet chemical etch, into a GaAs substrate. The groove direction is along $[1\bar{1}0]$. Even when the etched pattern is not very sharp on the bottom, subsequent growth of AlGaAs sharpens the apex to a self-limited radius ρ_l of the order of 10 nm. The side facets of the groove are {111}A. Subsequent deposition of GaAs leads to a larger upper radius $\rho_u > \rho_l$ of the heterostructure. The GaAs QWR formed in the bottom of the groove is thus crescent-shaped as shown in Fig. 13.3. A thin GaAs layer also forms on the side facets (sidewall quantum well) and on the top of the ridges. Subsequent growth of AlGaAs leads to a resharpening of the V-groove to the initial, self-limited value ρ_l. The complete resharpening after a sufficiently thick AlGaAs layer allows vertical stacking of crescent-shaped QWRs of virtually identical size and shape, as shown in Fig. 13.4. In this sense, the self-limiting reduction of the radius of curvature and its recovery during barrier-layer growth leads to self-ordering of QWR arrays whose structural parameters are determined solely by growth parameters. The lateral pitch of such wires can be down to 240 nm.

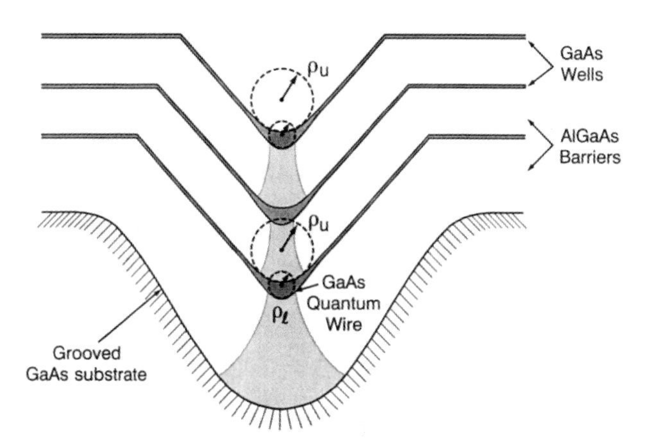

Fig. 13.2. Schematic cross section of a GaAs/AlGaAs heterostructure grown on a channeled substrate, illustrating the concept of self-ordered quantum-wire fabrication. From [437]

To directly visualize the lateral modulation of the bandgap, a lateral cathodoluminescence (CL) linescan perpendicular across the wire is displayed in Fig. 13.5. In Fig. 13.5a, the secondary electron (SE) image of the sample from Fig. 13.3 is shown in plan view. The top ridge is visible in the upper and lower parts of the figure, while in the middle the sidewalls with the QWR in the center are apparent. In Fig. 13.5b, the CL spectrum along a linescan perpendicular to the wire (as indicated by the white line in Fig. 13.5a) is displayed. The x-axis is now the emission wavelength, while the y-axis is the lateral position along the linescan. The CL intensity is given on a logarithmic scale to display the full dynamic range. The top QW shows almost no variation in bandgap energy ($\lambda = 725$ nm); only directly at the edge close to the sidewall does a second peak at lower energy ($\lambda = 745$ nm) appear, indicating a thicker region there. The sidewall QW exhibits a recombination

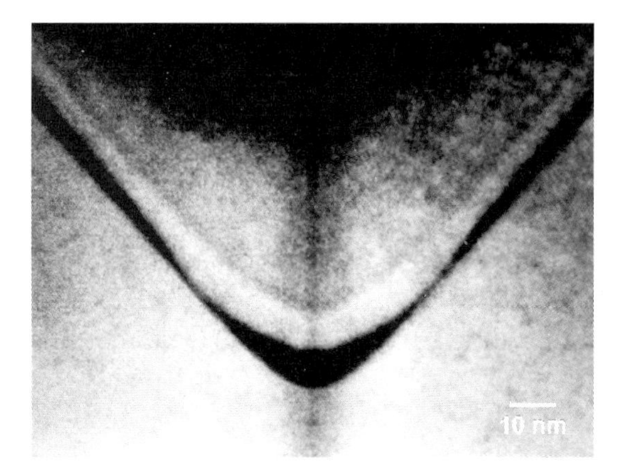

Fig. 13.3. Transmission electron microscopy cross-sectional image of a crescent-shaped single GaAs/AlGaAs quantum wire. From [438]

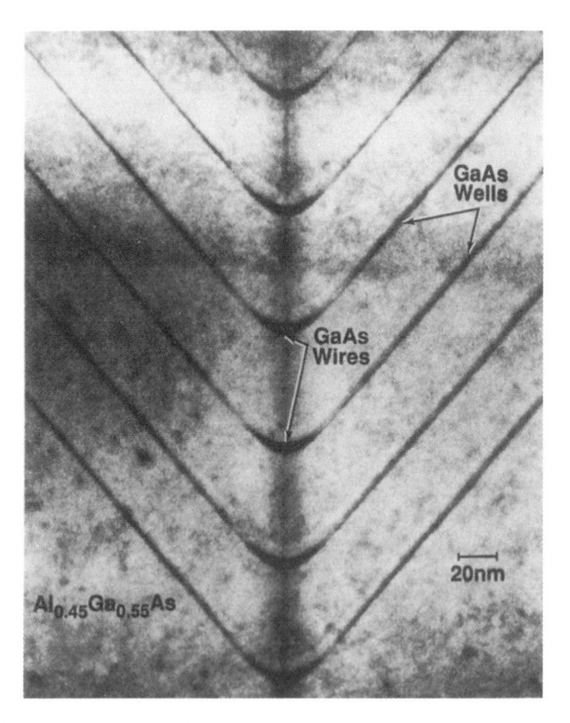

Fig. 13.4. TEM cross-sectional image of a vertical stack of identical GaAs/AlGaAs crescent-shaped QWRs. From [437]

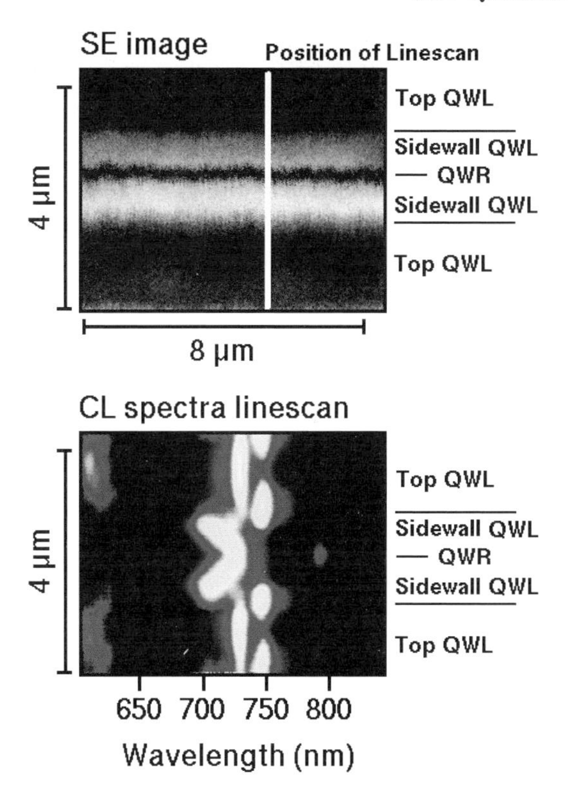

Fig. 13.5. (**a**) Plan-view SE image of single QWR (sample A), showing top and sidewall with QWR in the center. The *white line* indicates the position of the linescan on which the CL spectra linescan (**b**) has been taken at $T = 5\,\mathrm{K}$. The CL intensity is given on a logarithmic grey scale to display the full dynamic range as a function of wavelength and position along the *white line* in (**a**). From [438]

wavelength of 700 nm at the edge to the top QW, which gradually increases to about 730 nm at the center of the V-groove. This directly visualizes a linear tapering of the sidewall QW from about 2.1 nm thickness at the edge to 3 nm in the center. The QWR luminescence itself appears at about 800 nm.

After fast capture from the barrier into the QWs and, to a much smaller extent corresponding to its smaller volume, into the QWR, excess carriers will diffuse into the QWR via the adjacent sidewall QW and the vertical QW. The tapering of the sidewall QW induces an additional drift current.

Cleaved-Edge Overgrowth Quantum Wires

Another method to create quantum wires of high structural perfection is cleaved-edge overgrowth (CEO) [439], shown schematically in Fig. 13.6. First, a layered structure is grown (single or multiple quantum wells or superlattice).

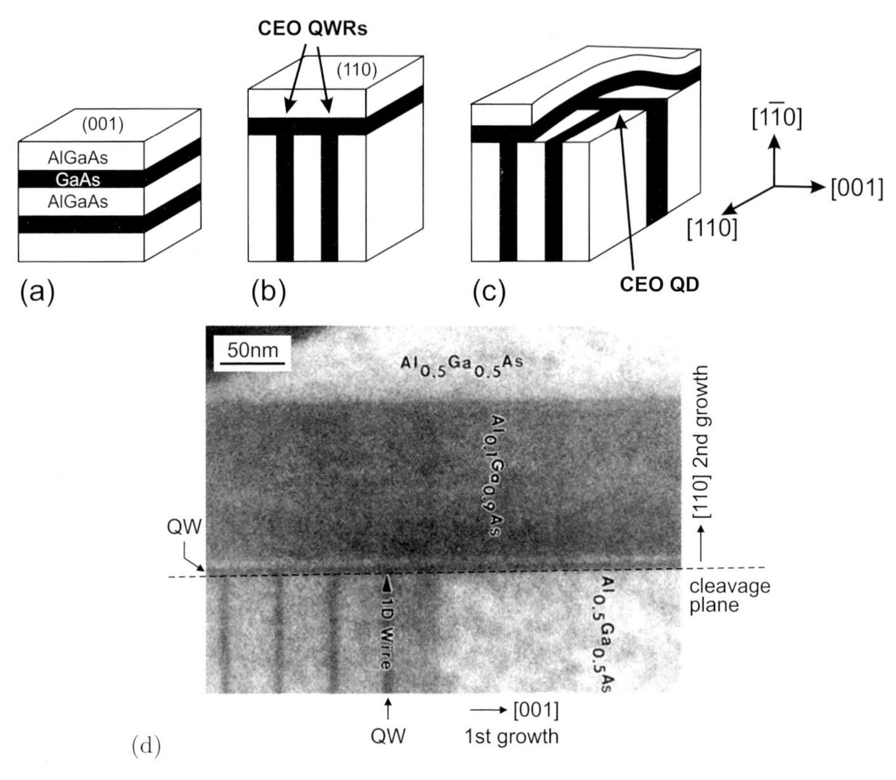

Fig. 13.6. Principle of CEO quantum wires and 2-fold CEO quantum dots. Part (**a**) depicts a layered structure (quantum wells or superlattice), (**b**) describes the growth on the cleaved facet used for fabrication of quantum wires. In (**c**) a second cleave and growth on top of the plane allows the fabrication of quantum dots. From [440]. (**d**) Cross-sectional TEM image of CEO GaAs/AlGaAs quantum wires. Two quantum wells (QW) and the QWR at their junction are labeled. The first epitaxy was from left to right. The second epitaxy step was on top of the cleavage plane (*dashed line*) in the upward direction. Adapted from [441], reprinted with permission, ©1997 APS

Then, a {110} facet is fabricated by cleaving (in vacuum) and epitaxy is continued on the cleaved facet. At the junctures of the {110} layer and the original quantum wells QWRs form. Due to their cross-sectional form they are also called *T-shaped* QWRs. A second cleave and another growth step allow fabrication of CEO quantum dots [440, 441] (Fig. 13.6c).

Nanowhiskers

Whiskers are primarily known as thin metal spikes and have been investigated in detail [442]. Semiconductor whiskers can be considered as (fairly short) quantum wires. They have been reported for a number of materials, such as

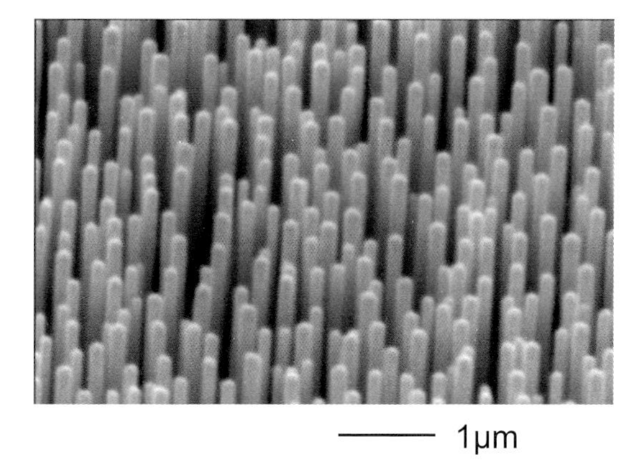

Fig. 13.7. Array of ZnO nanowhiskers on sapphire, fabricated using thermal evaporation

Si, GaAs, InP and ZnO. A field of ZnO whiskers is shown in Fig. 13.7. If heterostructures are incorporated along the whisker axis [443], quantum dots or tunneling barriers can be created (Fig. 13.8). The nanocrystal can also act as a nanolaser [444, 445].

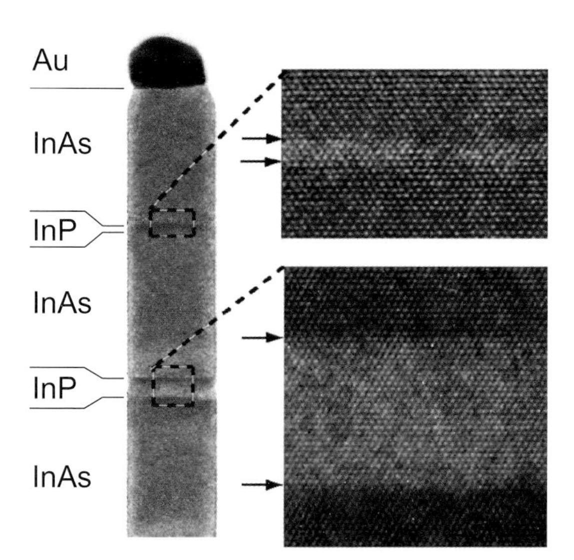

Fig. 13.8. TEM image of a part of an InAs whisker 40 nm in diameter that contains InP barriers. The zooms show sharp interfaces. On top of the whisker is a gold droplet from the so-called vapor–liquid–solid growth mechanism. The whisker axis is [001], the viewing direction is [110]. Adapted from [443], reprinted with permission, ©2002 AIP

13.2.2 Quantization in Two-Dimensional Potential Wells

The motion of carriers along the quantum wire is free. In the cross-sectional plane the wavefunction is confined in two dimensions. The simplest case is for constant cross section along the wire. However, generally the cross section along the wire may change and therefore induce a potential variation along the wire. Such potential variation will impact the carrier motion along the longitudinal direction. Also, a twist of the wire along its axis is possible.

In Fig. 13.9, the electron wavefunctions in a V-groove GaAs/AlGaAs QWR are shown. Further properties of V-groove QWRs have been reviewed in [446]. In Fig. 13.10, the excitonic electron and hole wavefunctions are shown for a (strained) T-shaped QWR.

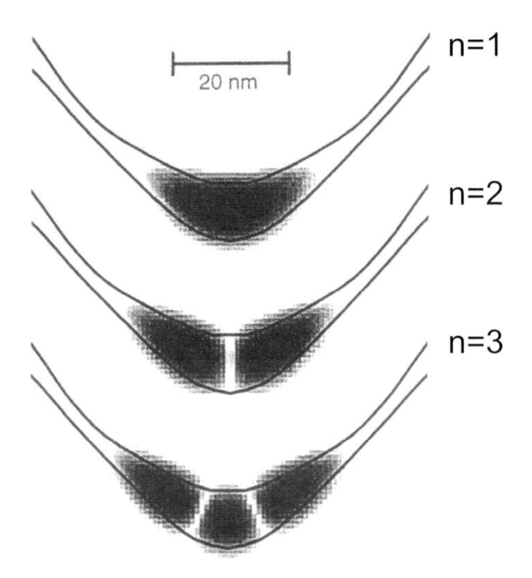

Fig. 13.9. Electron wavefunctions ($|\Psi|^2$ on logarithmic grey scale) for the first three confined levels for the QWR of Fig. 13.3. From [438]

13.3 Quantum Dots

13.3.1 Quantization in Three-Dimensional Potential Wells

The solutions for the d-dimensional ($d = 1$, 2, or 3) harmonic oscillator, i.e. the eigenenergies for the Hamiltonian

$$\hat{H} = \frac{\mathbf{p}^2}{2m} + \sum_{i=1}^{d} \frac{1}{2} m \omega_0^2 x_i^2 \tag{13.1}$$

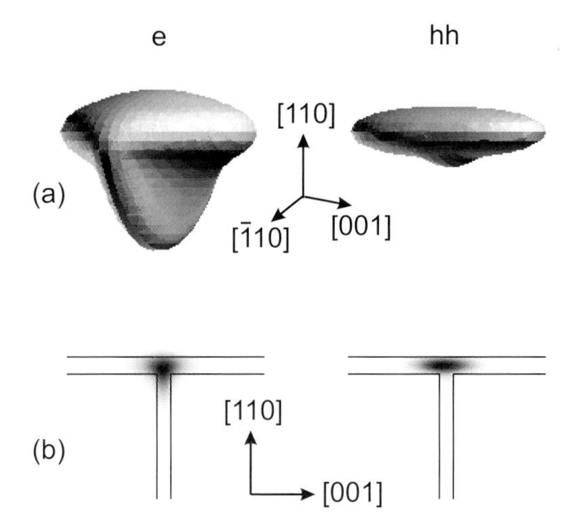

Fig. 13.10. (**a**) Three-dimensional view of the electron and (heavy) hole part of the excitonic wavefunction in a 4 nm × 5 nm T-shaped $In_{0.2}Ga_{0.8}As/GaAs$ QWR. The orbitals correspond to 70% probability inside. (**b**) Cross section through the electron and hole orbitals in their center along the wire direction. Reprinted with permission from [447], ©1998 APS

are given by

$$E_n = \left(n + \frac{d}{2}\right) \hbar\omega_0 \,, \tag{13.2}$$

with $n = 0, 1, 2, \ldots$. More detailed treatments can be found in quantum-mechanics textbooks.

Next, we discuss the problem of a particle in a centrosymmetric finite potential well with different masses m_1 in the dot and m_2 in the barrier. The Hamiltonian and the potential are given by

$$\hat{H} = \nabla \frac{\hbar^2}{2m} \nabla + V(r) \tag{13.3}$$

$$V(r) = \begin{cases} -V_0 & , r \leq R_0 \\ 0 & , r > R_0 \end{cases} \,. \tag{13.4}$$

The wavefunction can be separated into radial and angular components $\Psi(\mathbf{r}) = R_{nlm}(r)\, Y_{lm}(\theta, \phi)$, where Y_{lm} are the spherical harmonic functions. For the ground state ($n = 1$) the angular momentum l is zero and the solution for the wavefunction (being regular at $r = 0$) is given by

$$R(r) = \begin{cases} \frac{\sin(kr)}{kr} & , r \le R_0 \\ \frac{\sin(kR_0)}{kR_0} \exp\left(-\kappa(r - R_0)\right) & , r > R_0 \end{cases} \tag{13.5a}$$

$$k^2 = \frac{2m_1(V_0 + E)}{\hbar^2} \tag{13.5b}$$

$$\kappa^2 = -\frac{2m_2 E}{\hbar^2} \; . \tag{13.5c}$$

From the boundary conditions that both $R(r)$ and $\frac{1}{m}\frac{\partial R(r)}{\partial r}$ are continuous across the interface at $r = R_0$, the transcendental equation

$$kR_0 \cot(kR_0) = 1 - \frac{m_1}{m_2}(1 + \kappa R_0) \tag{13.6}$$

is obtained. From this formula the energy of the single particle ground state in a spherical quantum dot can be determined. For a given radius, the potential needs a certain strength $V_{0,\min}$ to confine at least one bound state; this condition can be written as

$$V_{0,\min} < \frac{\pi^2 \hbar^2}{8m^* R_0^2} \tag{13.7}$$

for $m_1 = m_2 = m^*$. For a general angular momentum l, the wavefunctions are given by spherical Bessel functions j_l in the dot and spherical Hankel functions h_l in the barrier. Also, the transcendental equation for the energy of the first excited level can be given:

$$kR_0 \cot(kR_0) = 1 + \frac{k^2 R_0^2}{\frac{m_1}{m_2}\frac{2+2\kappa R_0+\kappa^2 R_0^2}{1+\kappa R_0} - 2} \; . \tag{13.8}$$

In the case of infinite barriers ($V_0 \to \infty$), the wavefunction vanishes outside the dot and is given by (normalized)

$$R_{nml}(r) = \sqrt{\frac{2}{R_0^3}} \frac{j_l(k_{nl} r)}{j_{l+1}(k_{nl} R_0)} \; , \tag{13.9}$$

where k_{nl} is the n-th zero of the Bessel function j_l, e.g. $k_{n0} = n\pi$. With two-digit precision the lowest levels are determined by

k_{nl}	$l = 0$	$l = 1$	$l = 2$	$l = 3$	$l = 4$	$l = 5$
$n = 0$	3.14	4.49	5.76	6.99	8.18	9.36
$n = 1$	6.28	7.73	9.10	10.42		
$n = 2$	9.42					

The $(2l{+}1)$ degenerate energy levels E_{nl} are ($V_0 = \infty$, $m = m_1$):

$$E_{nl} = \frac{\hbar^2}{2m} \frac{k_{nl}^2}{R_0^2} \,. \tag{13.10}$$

The 1s, 1p, and 1d states have smaller eigenenergies than the 2s state.

A particularly simple solution is given for a cubic quantum dot of side length a_0 and infinite potential barriers. One finds the levels $E_{n_x n_y n_z}$:

$$E_{n_x n_y n_z} = \frac{\hbar^2}{2m} \pi^2 \frac{n_x^2 + n_y^2 + n_z^2}{a_0^2} \,, \tag{13.11}$$

with n_x, n_y, $n_z = 1, 2, \ldots$. For a sphere, the separation between the ground and first excited state is $E_1 - E_0 \approx E_0$, for a cube and a two-dimensional harmonic oscillator it is exactly E_0. For a three-dimensional harmonic oscillator this quantity is $E_1 - E_0 = 2E_0/3$.

For realistic quantum dots a full three-dimensional simulation of strain, piezoelectric fields and the quantum-mechanical confinement must be performed [448, 450]. In Fig. 13.11, the lowest four electron and hole wavefunctions in a pyramidal InAs/GaAs quantum dot (for the strain distribution cf. Fig. 5.19 and for the piezoelectric fields cf. Fig. 14.15) are shown. The figure shows that the lowest hole states have dominantly heavy-hole character and contain admixtures of the other hole bands.

13.3.2 Electrical and Transport Properties

The classical electrostatic energy of a quantum dot with capacity C that is capacitively coupled to a gate (Fig. 13.12) at a bias voltage V_{g} is given by

$$E = \frac{Q^2}{2C} - Q\alpha V_{\mathrm{g}} \,, \tag{13.12}$$

where α is a dimensionless factor relating the gate voltage to the potential of the island and Q is the charge of the island.

Mathematically, minimum energy is reached for a charge $Q_{\mathrm{min}} = \alpha C V_{\mathrm{g}}$. However, the charge has to be an integer multiple of e, i.e. $Q = Ne$. If V_{g} has a value, such that $Q_{\mathrm{min}}/e = N_{\mathrm{min}}$ is an integer, the charge cannot fluctuate as long as the temperature is low enough, i.e.

$$kT \ll \frac{e^2}{2C} \,. \tag{13.13}$$

Tunneling into or out of the dot is suppressed by the Coulomb barrier $e^2/2C$, and the conductance is very low. Analogously, the differential capacitance is small. This effect is called *Coulomb blockade*. Peaks in the tunneling current (Fig. 13.13b), conductivity (Fig. 13.13a) and the capacitance occur,

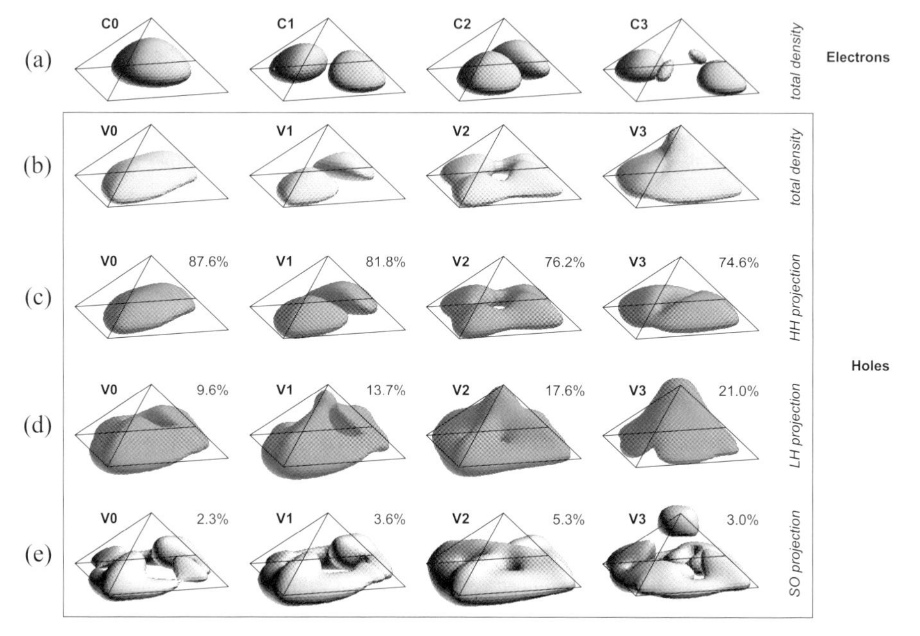

Fig. 13.11. Isosurface plots (25% of maximum value) of the total probability densities (**a**,**b**) and valence-band projections (**c**)–(**e**) of bound electron (**a**) and hole (**b**)–(**e**) states in a model pyramidal InAs/GaAs quantum dot with base length $b = 11.3$ nm. The percentages are the integrals of the projections to the bulk heavy, light and split-off hole bands, respectively, and the isosurfaces show the corresponding projection shapes. For each valence-band state the difference from 100% is the integral $\int_{-\infty}^{\infty} |\psi_{s\uparrow}|^2 + |\psi_{s\downarrow}|^2 \mathrm{d}^3\boldsymbol{r}$ of the s-type (conduction band) Bloch function projection (not shown). Reprinted with permission from [449]

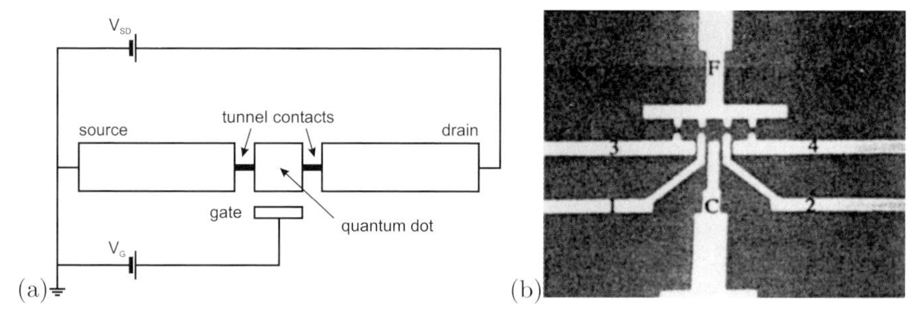

Fig. 13.12. (**a**) Schematic drawing of a quantum dot with tunnel contacts and gate electrode. (**b**) Realization with an in-plane gate structure. The distance between 'F' and 'C' (gate electrode) is 1 µm. Electron transport occurs from a 2DEG between 3/F to 4/F through the quantum points contacts 1/3 and 2/4. Part (**b**) reprinted with permission from [451]

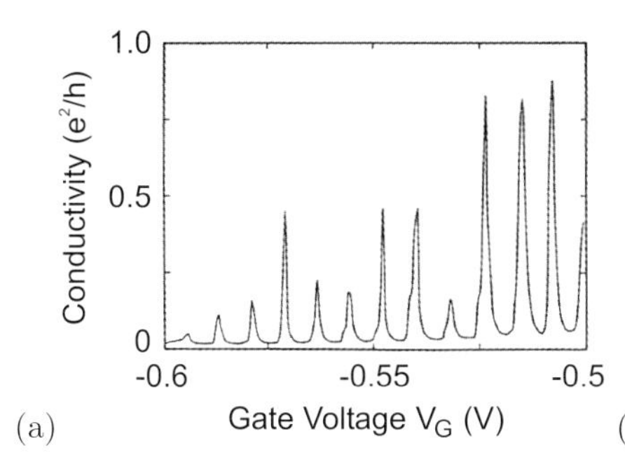

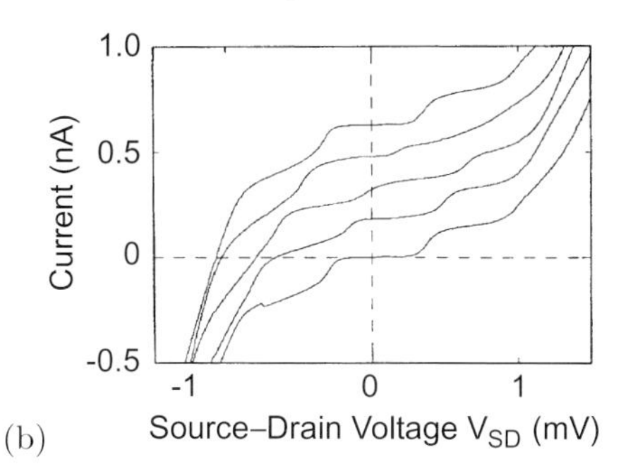

Fig. 13.13. (**a**) Conductivity (Coulomb oscillations) and (**b**) current–voltage diagram at different gate voltages (Coulomb staircase, shifted vertically for better readability) of a tunnel junction with a quantum dot as in Fig. 13.12. Adapted from [451]

when the gate voltage is such that the energies for N and $N+1$ electrons are degenerate, i.e. $N_{\mathrm{min}} = N + \frac{1}{2}$. The expected level spacing is

$$e\alpha \Delta V_{\mathrm{g}} = \frac{e^2}{C} + \Delta\epsilon_N \;, \tag{13.14}$$

where $\Delta\epsilon_N$ denotes the change in lateral (kinetic) quantization energy for the added electron. e^2/C will be called the charging energy in the following.

A variation of the source–drain voltage (for a given gate voltage) leads to a so-called Coulomb staircase since more and more channels of conductivity contribute to the current through the device (Fig. 13.14).

A lot of research so far has been done on lithographically defined systems where the lateral quantization energies are small and smaller than the Coulomb charging energy. In this case periodic oscillations are observed, especially for large N. A deviation from periodic oscillations for small N and a characteristic shell structure (at $N = 2, 6, 12$) consistent with a harmonic oscillator model ($\hbar\omega_0 \approx 3\,\mathrm{meV}$) has been reported for $\approx$ 500-nm diameter mesas (Fig. 13.15). In this structure, a small mesa has been etched and contacted (top contact, substrate back contact and side gate). The quantum dot consists of a 12-nm $\mathrm{In_{0.05}Ga_{0.95}As}$ quantum well that is laterally constricted by the 500-nm mesa and vertically confined due to 9- and 7.5-nm thick $\mathrm{Al_{0.22}Ga_{0.68}As}$ barriers. By tuning the gate voltage the number of electrons can be varied within 0 and 40. Measurements are typically carried out at a sample temperature of 50 mK.

The levels split as a function of the magnetic field due to radial magnetic quantum number l. Thus, e.g. the 7th and 8th electrons (dashed line in Fig. 13.16a) undergo a transition from (n,l)=(0,2) to (0,$-$1) at 1.3 T and to (0,3) at 3 T.

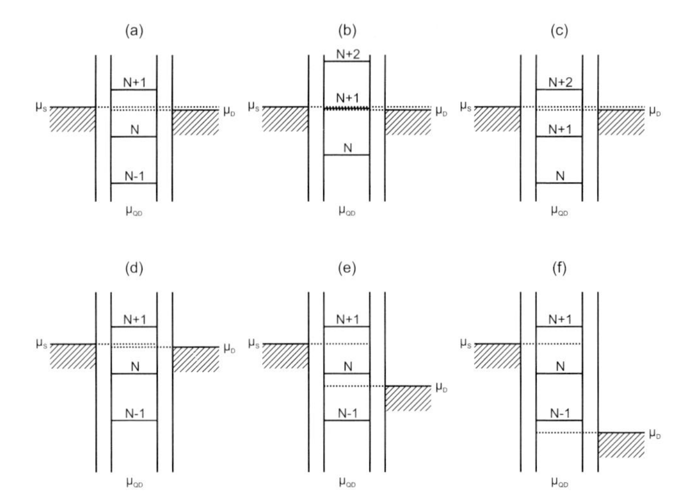

Fig. 13.14. Chemical potentials of source and drain and of a quantum dot in between them. (**a**), (**b**), and (**c**) show the sequence for a variation of the gate voltage and visualize the origin of the Coulomb oscillations (cf. Fig. 13.13a). (**d**), (**e**) and (**f**) visualize a variation of the source–drain voltage and the origin of the Coulomb staircase (cf. Fig. 13.13b)

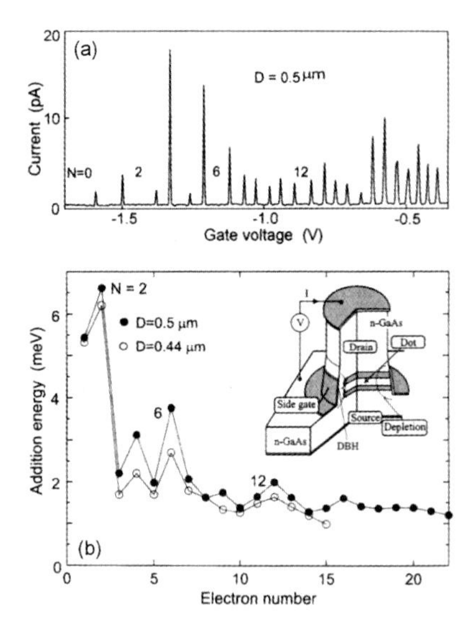

Fig. 13.15. (**a**) Coulomb oscillations in the current vs. gate voltage at $B = 0\,\mathrm{T}$ observed for a $D = 0.5\,\mu\mathrm{m}$ dot. (**b**) Addition energy vs. electron number for two different dots with $D = 0.5$ and $0.44\,\mu\mathrm{m}$. The *inset* shows a schematic diagram of the device. The dot is located between the two heterostructure barriers. Reprinted with permission from [452], ©1996 APS

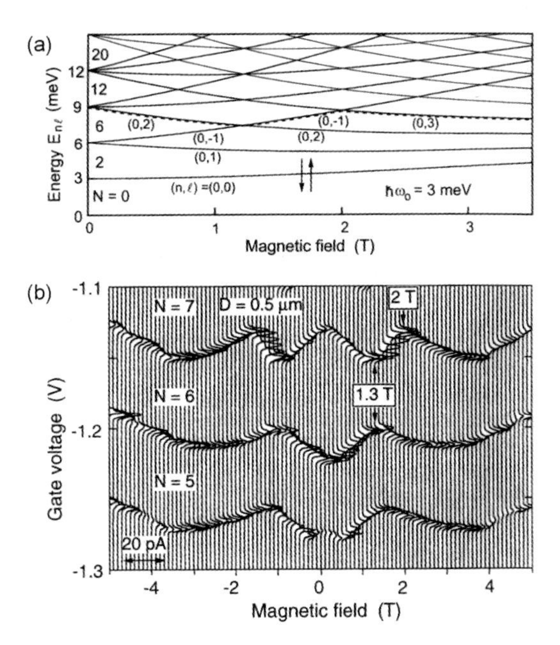

Fig. 13.16. (**a**) Calculated single-particle energy vs. magnetic field for a parabolic potential with $\hbar\omega_0 = 3\,\mathrm{meV}$. Each state is 2-fold spin degenerate. (**b**) Evolution of the fifth, sixth, and seventh current peaks with B field from 25 to 5 T observed for the $D = 0.5\,\mu\mathrm{m}$ dot. The original data consists of current vs. gate voltage traces for different magnetic fields, which are offset and rotated by 90°. Reprinted with permission from [452], ©1996 APS

In the sample shown in Fig. 13.17, self-assembled QDs are positioned in the channel under a split-gate structure. In a suitable structure, tunneling through a single QD is resolved.

In small self-assembled quantum dots single-particle level separations can be larger than or similar to the Coulomb charging energy. Classically, the capacity for a metal sphere of radius R_0 is given as

$$C_0 = 4\pi\epsilon_0\epsilon_{\mathrm{r}} R_0 \,, \tag{13.15}$$

e.g. $C_0 \approx 6\,\mathrm{aF}$ for $R_0 = 4\,\mathrm{nm}$ in GaAs, resulting in a charging energy of 26 meV. Quantum mechanically, the charging energy is given in first-order perturbation theory by

$$E_{21} = \langle 00|W_{\mathrm{ee}}|00\rangle = \int\!\!\int \Psi_0^2(\mathbf{r}_{\mathrm{e}}^1) W_{\mathrm{ee}}(\mathbf{r}_{\mathrm{e}}^1, \mathbf{r}_{\mathrm{e}}^2) \Psi_0^2(\mathbf{r}_{\mathrm{e}}^2) \mathrm{d}^3\mathbf{r}_{\mathrm{e}}^1\, \mathrm{d}^3\mathbf{r}_{\mathrm{e}}^2 \,, \tag{13.16}$$

where W_{ee} denotes the Coulomb interaction of the two electrons and Ψ_0 the ground state (single particle) electron wavefunction. The matrix element gives an upper bound for the charging energy since the wavefunctions will

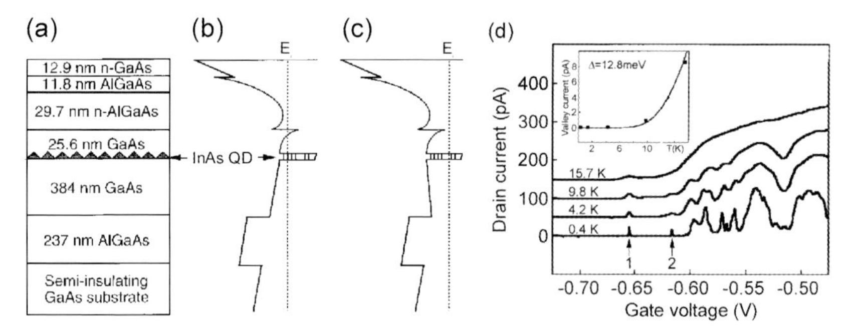

Fig. 13.17. (**a**) Schematic layer sequence of epitaxial structure comprising a n-AlGaAs/GaAs heterointerface with a two-dimensional electron gas and a layer of InAs/GaAs quantum dots. (**b**) and (**c**) are corresponding band diagrams with no gate bias and gate voltage below the critical value, respectively. (**d**) Experimental dependence of drain current on gate voltage in a split-gate structure at a drain source voltage of 10 μV. *Inset*: Dependence of valley current on temperature (*squares*) with theoretical fit. Reprinted with permission from [453], ©1997 AIP

rearrange to lower their overlap and the repulsive Coulomb interaction. For lens-shaped InAs/GaAs quantum dots with radius 25 nm a charging energy of about 30 meV has been predicted.

13.3.3 Self-Assembled Preparation

The preparation methods for QDs split into bottom-down (lithography and etching) and bottom-up (self-assembly) methods. The latter achieve typically smaller sizes and require less effort (at least concerning the machinery).

Using artificial patterning, based on lithography and etching (Fig. 13.18), quantum dots of arbitrary shape can be made (Fig. 13.19). Due to defects introduced by high-energy ions during reactive ion etching the quantum efficiency of such structures is very low when they are very small. Using wet-chemical etching techniques the damage can be significantly lowered but not completely avoided. Since the QDs have to compete with other structures that can be made structurally perfect, this is not acceptable.

Template growth is another technique for the formation of nanostructures. Here, a mesoscopic structure is fabricated by conventional means. The nanostructure is created using size-reduction mechanisms, e.g. faceting, (Fig. 13.20). This method can potentially suffer from low template density, irregularities of the template, and problems of reproducibility.

Another successful route to nanocrystals is the doping of glasses with subsequent annealing (color filters). When nanocrystals are prepared in a sol-gel process, the nanoparticles are present as a colloid in wet solution (Fig. 13.21). With the help of suitable stabilizing agents they are prevented from sticking to each other and can be handled in ensembles and also individually.

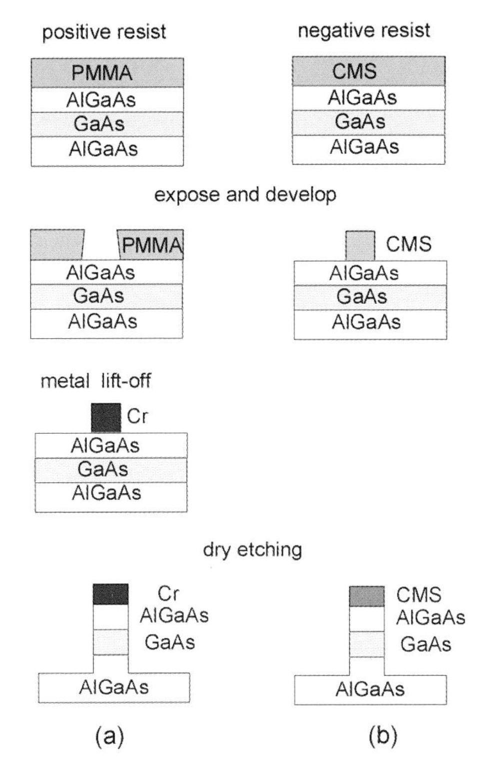

Fig. 13.18. Lithography and etching techniques for the fabrication of semiconductor structures

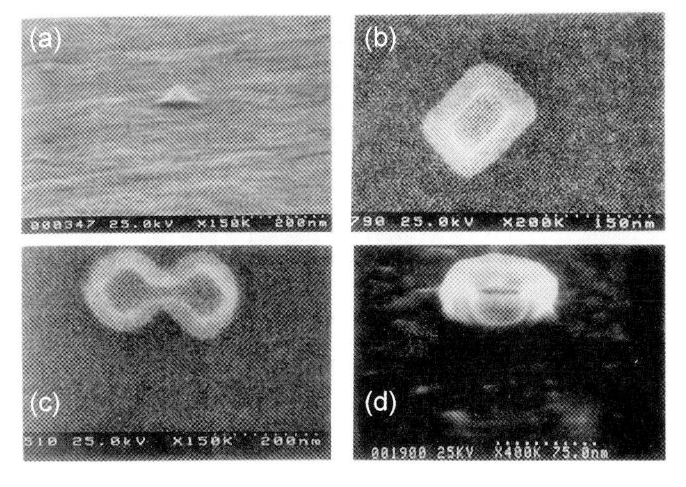

Fig. 13.19. Quantum dots of various shapes created by lithography and etching techniques. From [454]

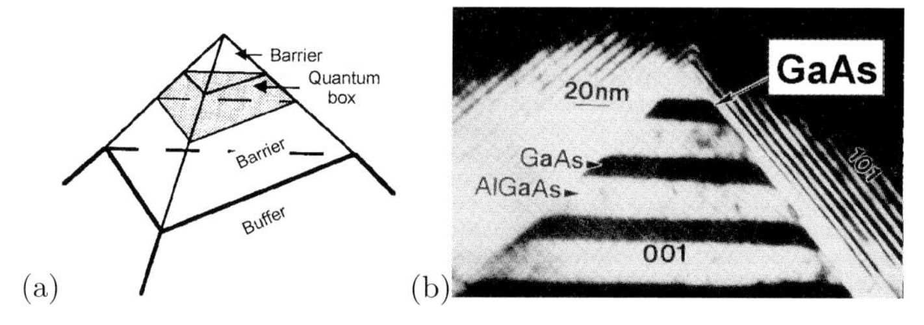

Fig. 13.20. (**a**) Schematic representation of growth on top of a predefined template, (**b**) cross-sectional TEM of quantum dot formation at the apex. Reprinted with permission from [455], ©1992 MRS

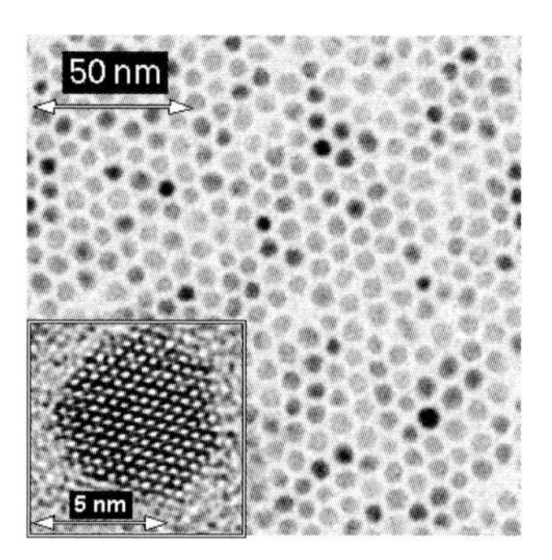

Fig. 13.21. CdSe colloidal nanoparticles. From [458]

The self-assembly (or self-organization) relies on strained heterostructures that achieve energy minimization by island growth. Additional ordering mechanisms lead to ensembles that are homogeneous in shape and size [459, 460] (Fig. 13.22).

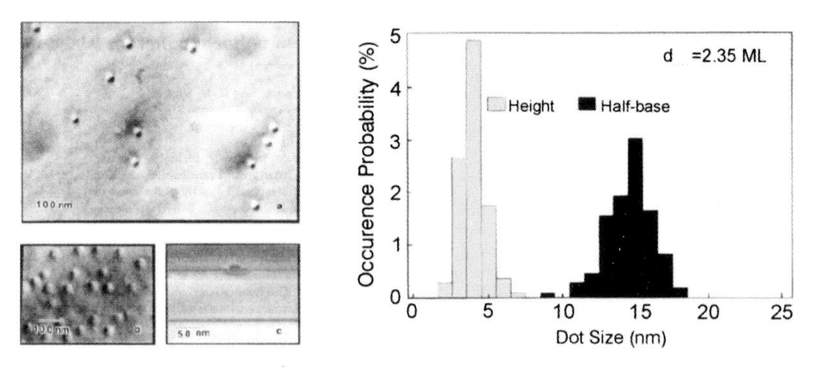

Fig. 13.22. Self-organized formation of InGaAs/GaAs quantum dots during epitaxy. *Left*: Plan-view and cross-sectional transmission electron micrographs. *Right*: Histogram of vertical and lateral size of the quantum dots. Reprinted with permission from [461], ©1993 AIP

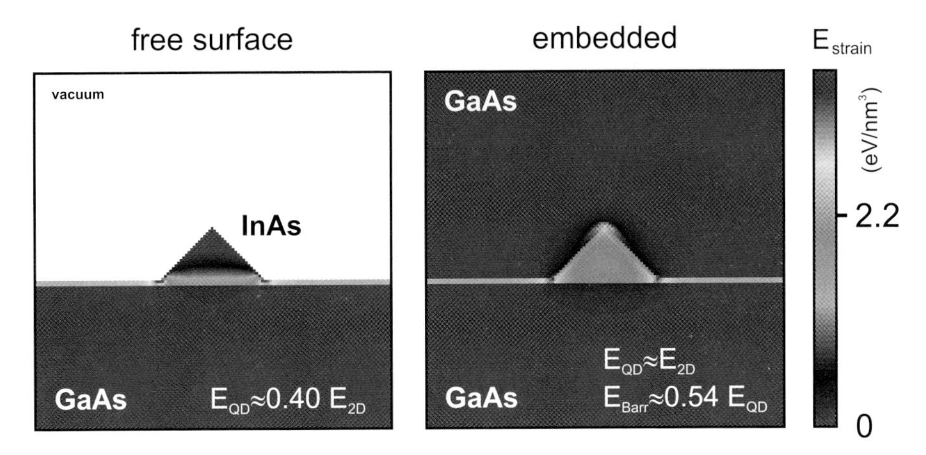

Fig. 13.23. Distribution of strain energy for (*left*) uncapped island and (*right*) island embedded in host matrix. Numerical values are for InAs/GaAs

When a flat layer of a semiconductor is grown on top of a flat substrate the layer suffers a tetragonal distortion. Strain can only relax along the growth direction (Fig. 13.23). If the strain energy is too large (highly strained layer or large thickness), plastic relaxation via dislocation formation occurs. If there is island geometry, strain can relax in all three directions and about 50% more strain energy can relax, thus making this type of relaxation energetically favorable. When the island is embedded in the host matrix, the strain energy is similar to the 2D case and the matrix becomes strained (metastable state).

When such QD layers are vertically stacked, the individual quantum dots grow on top of each other (Fig. 13.24) if the separation is not too large (Fig. 13.26). This effect is due to the effect of the underlying QD. In the case of InAs/GaAs (compressive strain), the buried QD stretches the surface

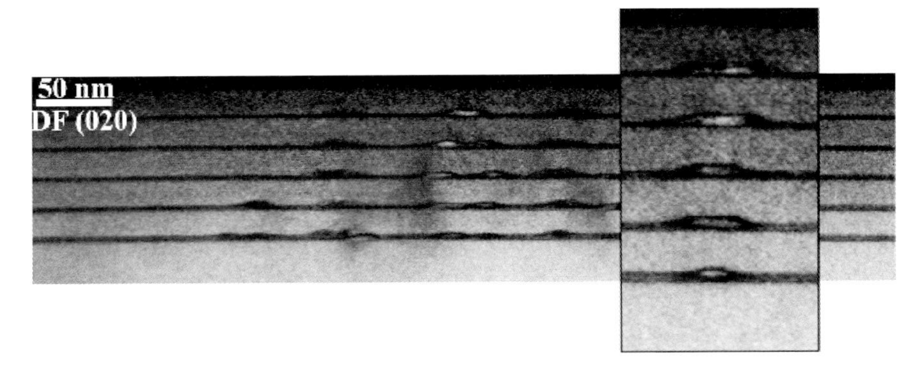

Fig. 13.24. Cross-sectional TEM image of a stack of five layers of quantum dots. Due to strain effects, vertical arrangement is achieved

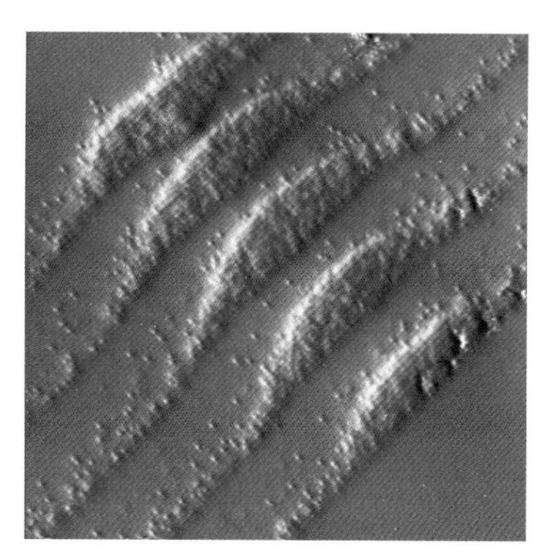

Fig. 13.25. Cross-sectional STM image of a stack of five InAs quantum dots in a GaAs matrix. Individual In atoms can be observed in-between the wetting layers and the quantum dots. Each quantum dot layer was formed by growing 2.4 ML of InAs. The intended distance between the quantum dot layers was 10 nm. Image size is $55 \times 55\,\mathrm{nm}^2$. Reprinted with permission from [462], ©2003 AIP

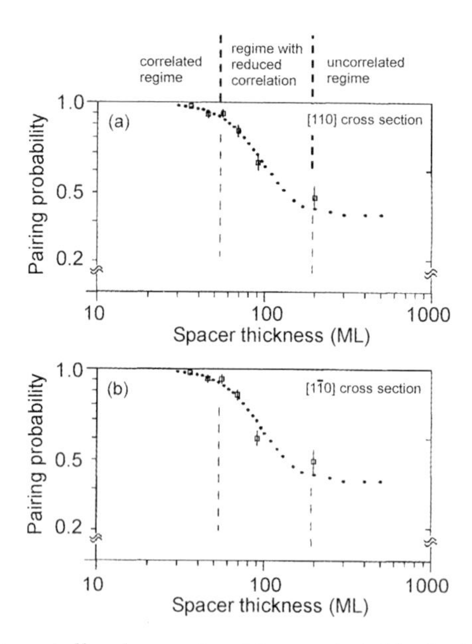

Fig. 13.26. Experimentally observed pairing probability in MBE-grown stacks of InAs/GaAs quantum dots as a function of the spacer-layer thickness. Data are taken from (**a**) (110) and (**b**) (1–10) cross-sectional TEM images. The *filled circles* are fit to data from theory of correlated island formation under strain fields. Reprinted with permission from [463], ©1995 APS

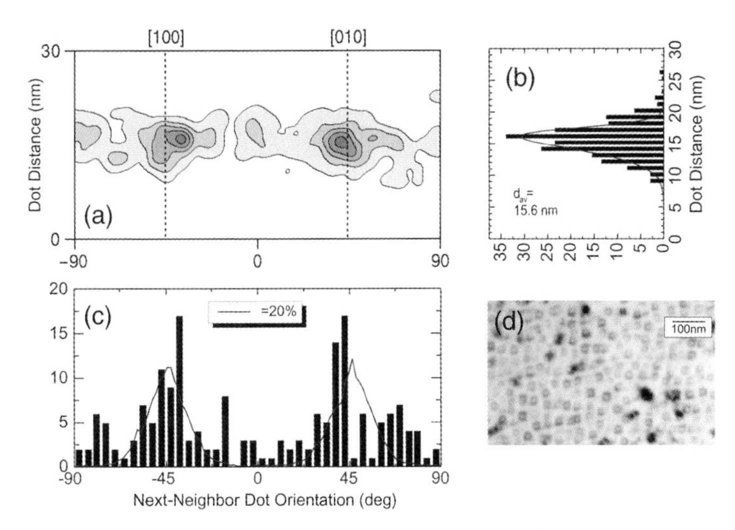

Fig. 13.27. Lateral ordering of QD array. (**d**) Plan-view TEM of QD array on which the statistical evaluation is based. (**a**) Two-dimensional histogram of QDs as a function of the nearest-neighbor distance and direction, (**b**,**c**) projections of part (**a**). *Solid lines* in (**b**) and (**c**) are theory for square array with $\sigma = 20\%$ deviation from ideal position

above it (tensile surface strain). Thus, atoms impinging in the next QD layer find a smaller strain right on top of the buried QDs. In STM images of the cross section through (XSTM) such a stack (Fig. 13.25) individual indium atoms are visible and the shape can be analyzed in detail [462].

The vertical arrangement can lead to further ordering since a homogenization in lateral position takes place. If two QDs in the first layers are very close, their strain fields overlap and the second layer 'sees' only one QD.

The lateral (in-plane) ordering of the QDs with respect to each other occurs in square or hexagonal patterns and is mediated via strain interaction through the substrate. The interaction energy is fairly small, leading only to short-range in-plane order [459] as shown in Fig. 13.27. The in-plane ordering can be improved up to the point that regular one- or two-dimensional arrays form or individual quantum dots are placed on designated positions using directed self-assembly [435]. Among others, dislocation networks buried under the growth surface of the nanostructure, surface patterning and modification have been used to direct the QD positioning.

13.3.4 Optical Properties

The optical properties of QDs are related to their electronic density of states. In particular, optical transitions are allowed only at discrete energies due to the zero-dimensional density of states.

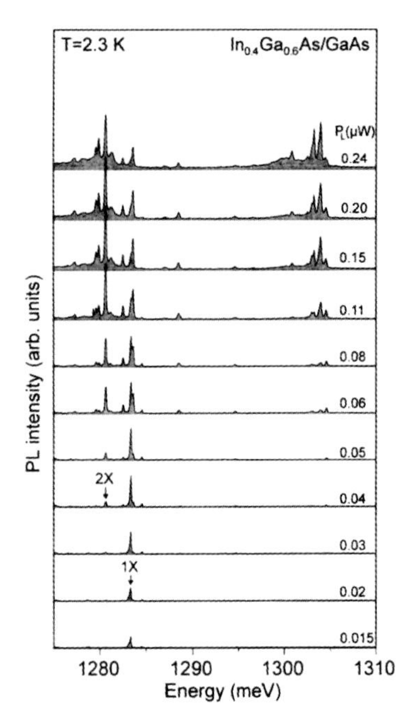

Fig. 13.28. Optical emission spectra of a single InGaAs/GaAs QD at different laser excitation levels. From [464]

Photoluminescence from a single QD is shown in Fig. 13.28. The δ-like sharp transition is strictly true only in the limit of small carrier numbers ($\ll 1$ exciton per dot on average) since otherwise many-body effects come into play that can encompass recombination from charged excitons or multi-excitons. At very low excitation density the recombination spectrum consists only of the one-exciton (X) line. With increasing excitation density small satellites on either side of the X-line develop that are attributed to charged excitons (trions) X^+ and X^-. On the low-energy side, the biexciton (XX) appears. Eventually, the excited states are populated and a multitude of states contribute with rich fine structure. The charging state of the exciton can be controlled in a field-effect structure. The recombination energy is modified due to Coulomb and exchange effects with the additional carriers.

In charge-tunable quantum dots [456] and rings [457] exciton emission has been observed in dependence of the number of additional electrons. The electron population can be controlled in a Schottky-diode-like structure through the manipulation of the Fermi level with the bias voltage. At high negative bias all charge carriers tunnel out of the ring and no exciton emission is observed. A variation of the bias then leads to an average population with $N = 1, 2, 3, \ldots$ electrons. The recombination of additional laser-excited excitons depends (due to the Coulomb interaction) on the number of the electrons present. The singly negatively charged exciton X^- is also called a trion.

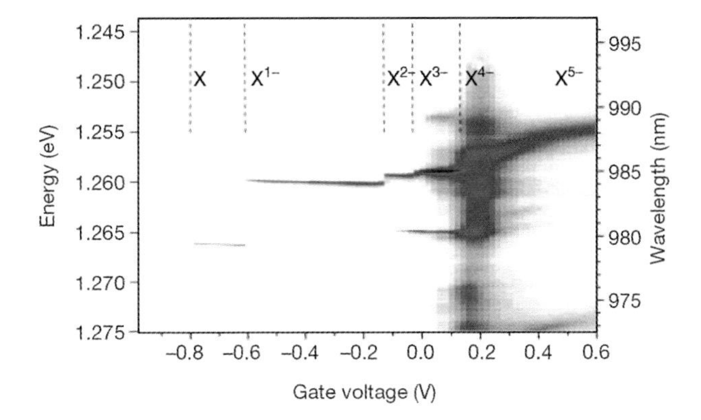

Fig. 13.29. Luminescence of charged excitons from a single quantum ring at $T = 4.2\,\mathrm{K}$ vs. the bias voltage with which the number of electrons in the quantum dot N is tuned from zero to $N > 3$. Adapted from [457], reprinted with permission from Nature, ©2000 Macmillan Magazines Limited

In a QD ensemble, optical transitions are inhomogeneously broadened due to fluctuations in the QD size and the size dependence of the confinement energies (Fig. 13.30). Interband transitions involving electrons and holes suffer from the variation of the electron and hole energies:

$$\sigma_E \propto \left(\left| \frac{\partial E_{\mathrm{e}}}{\partial L} \right| + \left| \frac{\partial E_{\mathrm{h}}}{\partial L} \right| \right) \delta L \, . \tag{13.17}$$

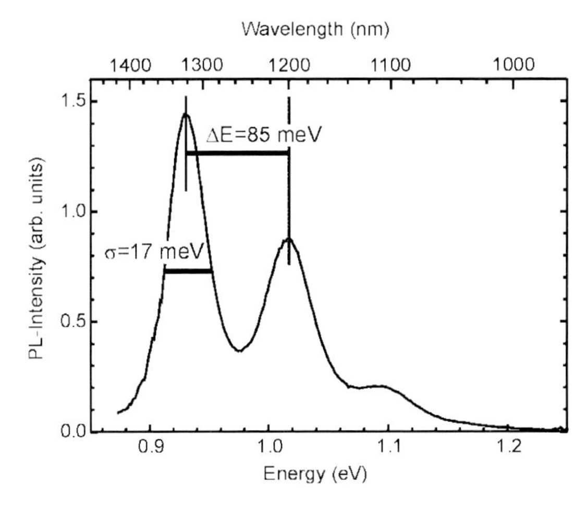

Fig. 13.30. Ensemble photoluminescence spectrum ($T = 293\,\mathrm{K}$, excitation density $500\,\mathrm{W/cm^2}$) of InAs/GaAs QDs

A typical relative size inhomogeneity of σ_L/L of 7% leads to several tens of meV broadening. Additional to broadening due to different sizes fluctuations of the quantum dot shape can also play a role. The confinement effect leads to an increase of the recombination energy with decreasing quantum-dot size. This effect is nicely demonstrated with colloidal quantum dots of different size as shown in Fig. 13.31.

Fig. 13.31. Luminescence (under UV excitation) from flasks of colloidal CdTe quantum dots with increasing size (from left to right). From [458]

14 Polarized Semiconductors

14.1 Introduction

Semiconductors can have an electric polarization. Such polarization can be induced by an external electric field (Fig. 14.1a). This phenomenon, i.e. that the semiconductor is dielectric, has been discussed already in Chap. 9. In this chapter, we discuss pyroelectricity, i.e. a spontaneous polarization without an external field (Fig. 14.1b), ferroelectricity, i.e. pyroelectricity with a hysteresis (Fig. 14.1c) and piezoelectricity, i.e. a polarization due to external stress.

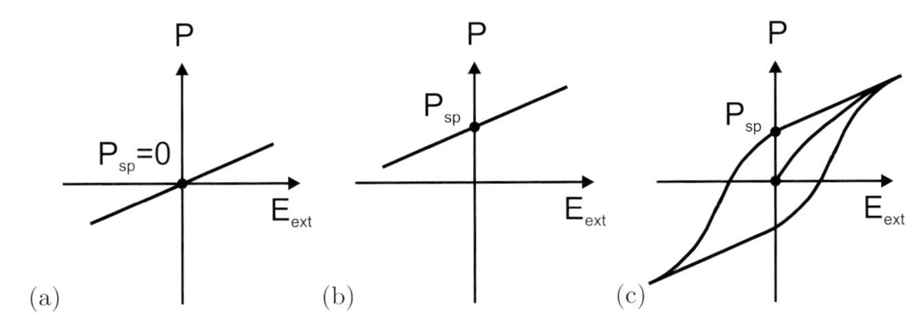

Fig. 14.1. Schematic representation of the polarization vs. external electric field dependence for (**a**) dielectric, (**b**) pyroelectric and (**c**) ferroelectric semiconductors

14.2 Spontaneous Polarization

The reason for a spontaneous polarization P_{sp} (without external electric field) is the static, relative shift of positive and negative charges in the unit cell. For a slab of semiconductor material (thus ignoring depolarization effects present in other geometries), the polarization causes polarization charges located at the upper and lower surfaces (Fig. 14.2a). The polarization vector $\mathbf{P}$ points from the negative to the positive charge. The electric field due to the polarization charges has the opposite direction. In the absence of free charges, the Maxwell equation $\nabla \cdot \mathbf{D} = 0$ yields for piecewise constant fields at a planar interface (Fig. 14.2b) $(\mathbf{D}_2 - \mathbf{D}_1) \cdot \mathbf{n}_{12} = 0$ where $\mathbf{n}_{12}$ is the surface normal pointing from medium 1 to medium 2. Therefore, the polarization charge $\sigma_\mathrm{P} = \epsilon_0 \nabla \cdot \mathbf{E}$ is given by

$$\sigma_\mathrm{P} = -(\mathbf{P}_2 - \mathbf{P}_1) \cdot \mathbf{n}_{12} \, . \tag{14.1}$$

Polarization charges develop at interfaces where the polarization is discontinuous, e.g. an interface between two semiconductors with different spontaneous polarization. Vacuum (at a surface) represents a special case with $\mathbf{P} = 0$.

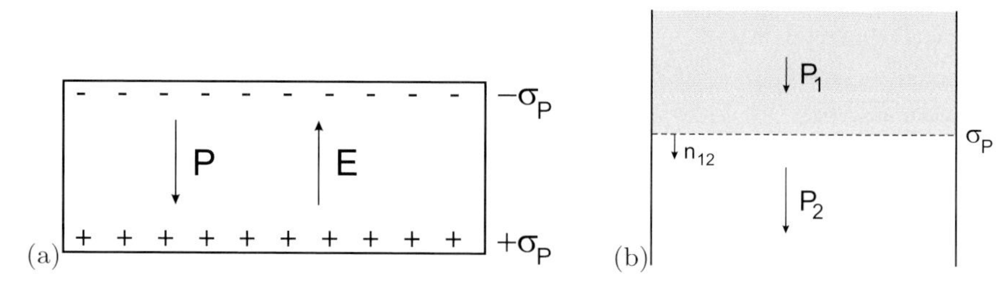

Fig. 14.2. (**a**) Surface polarization charges σ_P on a slab of semiconductor material with polarization. The electric field is given by $\mathbf{E} = -\mathbf{P}/\epsilon_0$. (**b**) Polarization charge σ_P at an interface between two semiconductors with different polarization. In the depicted situation σ is negative

For cubic zincblende structure semiconductors, P_sp is typically fairly small. The anisotropy of the wurtzite structure allows for sizeable effects. The main cause is the nonideality of the cell-internal parameter u (see Sect. 3.4.5).

14.3 Ferroelectricity

Ferroelectric semiconductors exhibit a spontaneous polarization in the ferroelectric phase and zero spontaneous polarization in the paraelectric phase. As a function of temperature, the ferroelectric material undergoes a phase transition from the high-temperature paraelectric phase into the ferroelectric phase. There can be further phase transitions between different ferroelectric phases that differ in the direction of the polarization. The literature until 1980 is summarized in [504]. A more recent treatment can be found in [505].

$PbTiO_3$ has perovskite structure (cf. Sect. 3.4.8). It exhibits a phase transition at $T_\mathrm{C} = 490\,°\mathrm{C}$ from the cubic into the (ferroelectric) tetragonal phase as shown in Fig. 14.4a. Mostly the cell symmetry changes, while the cell volume remains almost constant. A more complicated situation arises for $BaTiO_3$. At $120\,°\mathrm{C}$ the transition into the ferroelectric phase occurs (Fig. 14.4b) that is tetragonal with the polarization in the [100] direction. At $-5\,°\mathrm{C}$ and $-90\,°\mathrm{C}$ transitions occur into an orthorhombic and a rhombohedral (trigonal) phase, respectively. The largest polarization is caused by

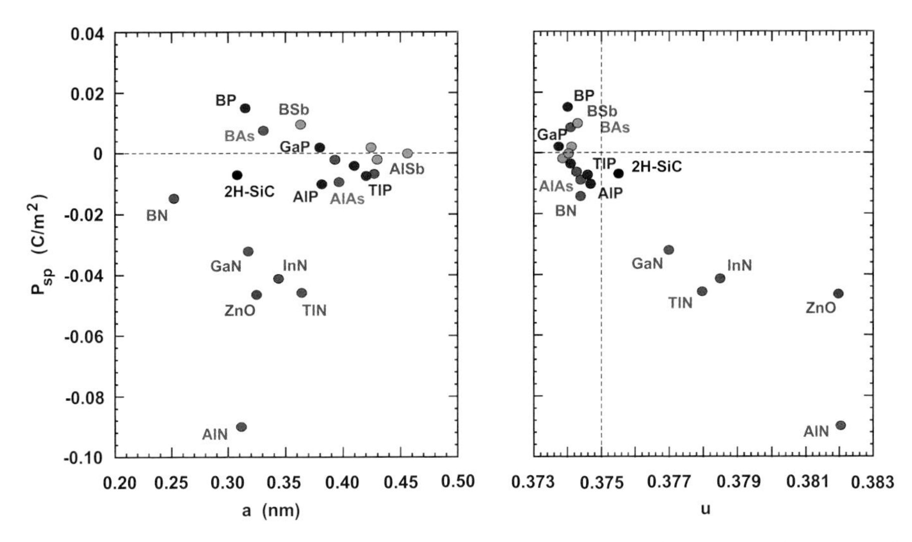

Fig. 14.3. Spontaneous polarization P_{sp} for various semiconductors as a function of the lattice constant a (left) and the cell-internal parameter u (right). Based on [503]

a displacement of the negatively (O) and positively (Ba, Ti) charged ions of the unit cell by $\delta \approx 0.02\,\mathrm{nm}$ (Fig. 14.5). Such an origin of the spontaneous polarization is called a *displacement* transition. However, other mechanisms (not discussed here) can also be responsible for ferroelectricity.

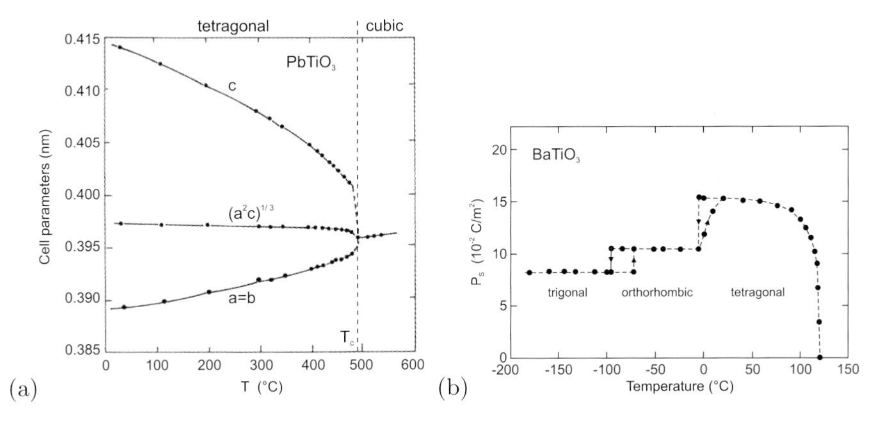

Fig. 14.4. (**a**) Cell parameters of $PbTiO_3$ as a function of temperature. Adapted from [506]. (**b**) Phase transitions of $BaTiO_3$ as a function of temperature. The spontaneous polarization P_{S} points along $\langle 100 \rangle$, $\langle 110 \rangle$ and $\langle 111 \rangle$ in the tetragonal (C_{4v}), orthorhombic (C_{2v}) and trigonal (C_{3v}, rhombohedral) phase, respectively. Adapted from [507]

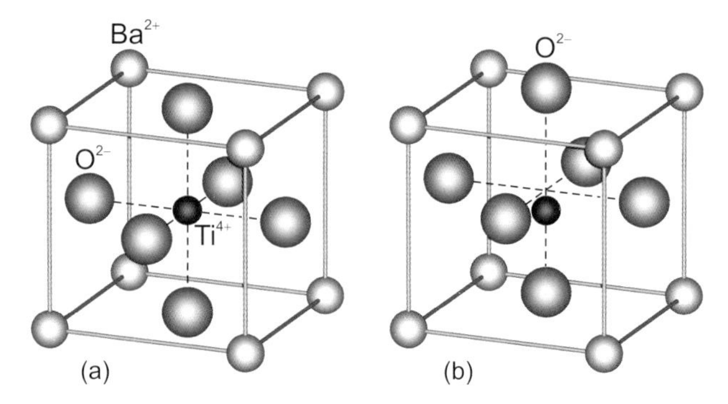

Fig. 14.5. (**a**) Crystal structure of $BaTiO_3$ (cf. Fig. 3.23). (**b**) Schematic tetragonal deformation below the Curie temperature, generating a dipole moment

14.3.1 Materials

A large class of ferroelectric semiconductors are of the type ABO_3, where A stands for a cation with larger ionic radius and B for an anion with smaller ionic radius. Many ferroelectrics have perovskite ($CaTiO_3$) structure. They are $A^{2+}B^{4+}O_3^{2-}$, e.g. (Ba,Ca,Sr) (Zi,Zr)O_3 or $A^{1+}B^{5+}O_3^{2-}$, e.g. (Li,Na,K) (Nb,Ta)O_3. Ferroelectrics can also be alloyed. Alloying in the B component yields, e.g. $PbTi_xZr_{1-x}O_3$ also called PZT. PZT is widely used for piezoelectric actuators. Also, alloying in the A component is possible, e.g. $Ba_xSr_{1-x}TiO_3$.

Another class of ferroelectrics are $A^{V}B^{VI}C^{VII}$ compounds, such as SbSI, SbSBr, SbSeI, BiSBr. These materials have a width of the forbidden band in the $\sim 2\,eV$ range. A further class of ferroelectric semiconductors are $A_2^{V}B_3^{VI}$ compounds, such as Sb_2S_3.

14.3.2 Soft Phonon Mode

The finite displacement of the sublattices in the ferroelectric means that the related lattice vibration has no restoring force. The displacement is, however, finite due to higher-order terms (anharmonicity). Thus, for $T \to T_C$ $\omega_{TO} \to 0$. Such a mode is called a *soft phonon mode*. The decrease of the phonon frequency is shown in Fig. 14.6a for SbSI.

From the LST relation (5.42), it follows that the static dielectric function must increase strongly. The increase is $\propto (T - T_C)^{-1}$ (Fig. 14.6b).

14.3.3 Phase Transition

In the case of ferroelectrics, the order parameter for the Ginzburg–Landau theory of phase transitions is the spontaneous polarization P. The free energy F of the ferroelectric crystal is written in terms of the free energy of the paraelectric phase F_0 and is expanded in powers of P (here up to P^6) as

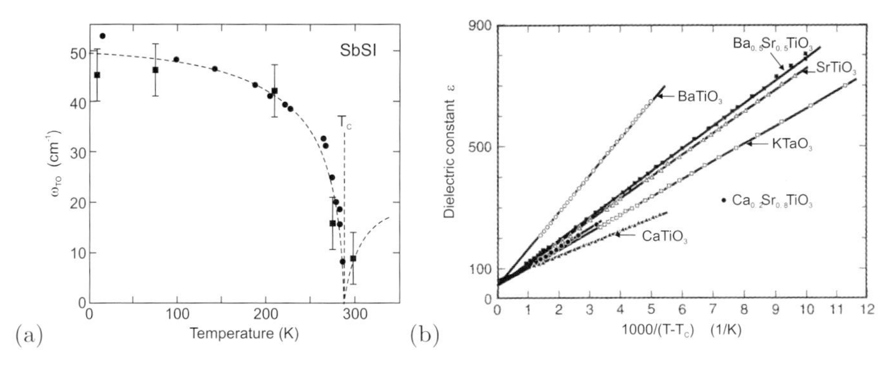

Fig. 14.6. (**a**) Decrease of the transverse phonon mode of SbSI close to the Curie temperature of $T_C = 288\,\mathrm{K}$. The *dashed curved line* represents a $|T - T_C|^{1/2}$ dependence. Adapted from [508]. (**b**) Dielectric constant of various perovskites vs. $1/(T - T_C)$ in the paraelectric phase ($T > T_C$). Adapted from [509]

$$F = F_0 + \frac{1}{2}\alpha P^2 + \frac{1}{4}\beta P^4 + \frac{1}{6}\gamma P^6 \; . \tag{14.2}$$

In this equation, we have neglected effects due to charge carriers, an external electric field or external stresses and we assume homogeneous polarization. In order to obtain a phase transition, it has to be assumed that α has a zero at a certain temperature T_C and we assume (expanding only to the linear term)

$$\alpha = \alpha_0 (T - T_C) \; . \tag{14.3}$$

Second-Order Phase Transition

For modeling a second-order phase transition, we set $\gamma = 0$. Thus, the free energy has the form (Fig. 14.7a)

$$F = F_0 + \frac{1}{2}\alpha P^2 + \frac{1}{4}\beta P^4 \; . \tag{14.4}$$

The equilibrium condition with regard to the free energy yields a minimum for

$$\frac{\partial F}{\partial P} = \alpha P + \beta P^3 = 0 \tag{14.5a}$$

$$\frac{\partial^2 F}{\partial P^2} = \alpha + 3\beta P^2 > 0 \; . \tag{14.5b}$$

Equation (14.5a) yields two solutions. $P = 0$ corresponds to the paraelectric phase. $P^2 = -\alpha/\beta$ is the spontaneous polarization in the ferroelectric phase. The condition from (14.5b) yields that $\alpha > 0$ in the paraelectric phase,

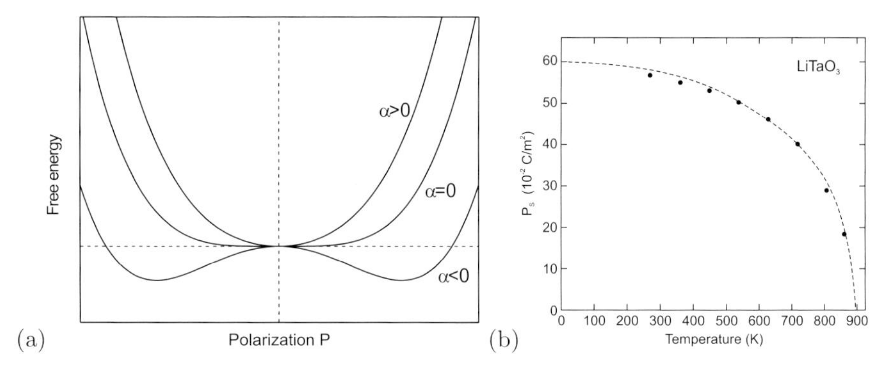

Fig. 14.7. (**a**) Schematic plot of the free energy vs. spontaneous polarization for a second-order phase transition. $\alpha > 0$ ($\alpha < 0$) corresponds to the paraelectric (ferroelectric) phase. (**b**) Spontaneous polarization of $LiTaO_3$ as a function of temperature exhibiting a second-order phase transition. The *dashed line* is theory with suitable parameters. Adapted from [510]

while $\alpha < 0$ in the ferroelectric phase. Also, $\beta > 0$ below the Curie temperature (β is assumed to be temperature independent in the following). Using (14.3), the polarization is given as (Fig. 14.7b)

$$P^2 = \frac{\alpha_0}{\beta}(T - T_C)\,. \tag{14.6}$$

Therefore, the entropy $S = -\frac{\partial F}{\partial T}$ and the discontinuity ΔC_p of the heat capacity $C_p = T\left(\frac{\partial S}{\partial T}\right)_p$ at the Curie point T_C are given by

$$S = S_0 + \frac{\alpha_0^2}{\beta}(T - T_C) \tag{14.7a}$$

$$\Delta C_p = \frac{\alpha_0^2}{\beta} T_C\,, \tag{14.7b}$$

with $S_0 = -\frac{\partial F_0}{\partial T}$ being the entropy of the paraelectric phase. This behavior is in accordance with a second-order phase transition with vanishing latent heat (continuous entropy) and a discontinuity of the heat capacity. The dielectric function in the paraelectric phase is $\propto 1/\alpha$ and in the ferroelectric phase $\propto -1/\alpha$. The latter relation is usually written as the Curie–Weiss law

$$\epsilon = \frac{C}{T - T_C}\,. \tag{14.8}$$

First-Order Phase Transition

When the P^6 term is included in (14.2) ($\gamma \neq 0$), a first-order phase transition is modeled. However, in order to obtain something new, compared to the previous consideration, now $\beta < 0$ (and $\gamma > 0$) is necessary. The dependence of the free energy on P is schematically shown in Fig. 14.8a for various values of α. The condition $\frac{\partial F}{\partial P} = 0$ yields

$$\alpha P + \beta P^3 + \gamma P^5 = 0\,, \tag{14.9}$$

with the solutions $P = 0$ and

$$P^2 = -\frac{\beta}{2\gamma}\left(1 + \sqrt{1 - \frac{4\alpha\gamma}{\beta^2}}\right)\,. \tag{14.10}$$

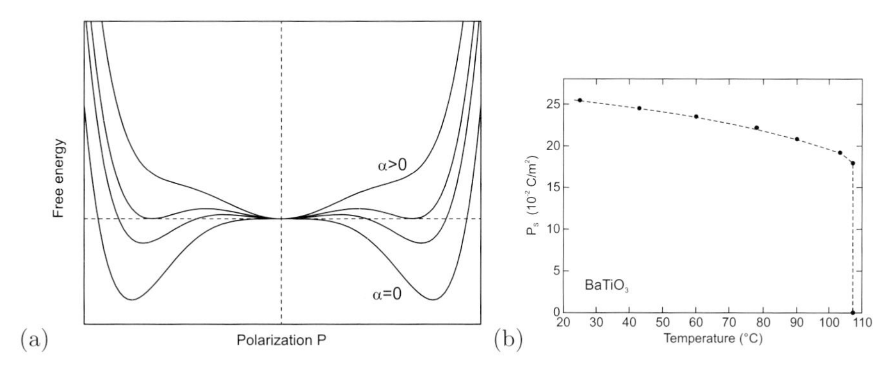

Fig. 14.8. (**a**) Schematic plot of the free energy vs. spontaneous polarization for a first-order phase transition. The lowest curve is for $\alpha = 0$, the others are for $\alpha > 0$. (**b**) Spontaneous polarization in $BaTiO_3$ as a function of temperature exhibiting a first-order phase transition. The *dashed line* is a guide to the eye. Adapted from [511]

For a certain value of α, i.e. at a certain temperature $T = T_1$, the free energy is zero for $P = 0$ *and* also for another value $P = P_0$ (second curve from the top in Fig. 14.8a). From the condition

$$\frac{1}{2}\alpha(T_1)P_0^2 + \frac{1}{4}\beta P_0^4 + \frac{1}{6}\gamma P_0^6 = 0\,, \tag{14.11}$$

the values for P_0 and α at the transition temperature $T = T_1$ are given by

$$P_0^2 = -\frac{3}{4}\frac{\beta}{\gamma} \tag{14.12a}$$

$$\alpha(T_1) = \frac{3}{16}\frac{\beta^2}{\gamma} > 0\,. \tag{14.12b}$$

The schematic dependence of P at the phase transition temperature T_1 is depicted in Fig. 14.8b.

For $T \leq T_1$ the absolute minimum of the free energy is reached for finite polarization $P > P_0$. However, between $F(P = 0)$ and the minimum of the free energy an energy barrier (second lowest curve in Fig. 14.8) is present for T close to T_1. The energy barrier disappears at the Curie–Weiss temperature T_0. At the phase transition temperature, the entropy has a discontinuity

$$\Delta S = \alpha_0 P_0^2 , \tag{14.13}$$

that corresponds to a latent heat $\Delta Q = T\Delta S$. Another property of the first-order phase transition is the occurrence of hysteresis in the temperature interval between T_1 and T_0

$$\Delta T \approx T_1 - T_0 = \frac{1}{4\alpha_0} \frac{\beta^2}{\gamma} , \tag{14.14}$$

in which an energy barrier is present to hinder the phase transition. For decreasing temperature, the system tends to remain in the paraelectric phase. For increasing temperature, the system tends to remain in the ferroelectric phase. Such behavior is observed for $BaTiO_3$, as shown in Fig. 14.4b.

14.3.4 Domains

Similar to ferromagnets, ferroelectrics form domains with different polarization directions in order to minimize the total energy by minimizing the field energy outside the crystal. The polarization can have different orientations, 6 directions for P along $\langle 100 \rangle$ (tetragonal phase), 12 directions for P along $\langle 110 \rangle$ (orthorhombic phase) and 8 directions for P along $\langle 111 \rangle$ (rhombohedral phase). In Fig. 14.9, such domains are visualized for $BaTiO_3$. Due to the restricted geometry, domain formation in thin films is different from that in bulk material.

Domains can also be artificially created by so-called poling. The ferroelectric semiconductor is heated to the paraelectric phase. With electrodes,

Fig. 14.9. Ferroelectric domains in a $BaTiO_3$ single crystal visualized by birefringence contrast. Reprinted with permission from [512], ©1949 APS

appropriate electric fields are applied and the material is cooled. The polarization is then frozen in the ferroelectric phase. The domains of a periodically poled structure in $LiNbO_3$ (PPLN) are shown in Fig. 14.10b. The nonlinear optical properties in such structures can be used for efficient second harmonic generation (SHG).

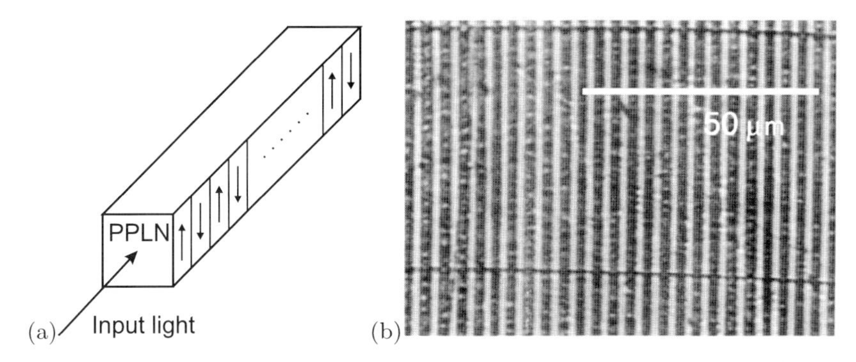

Fig. 14.10. (**a**) Scheme of PPLN (perodically poled lithium niobate), *arrows* denote the direction of spontaneous polarization. (**b**) Polarization microscopy image (vertical stripes are domains, horizontal dark lines are scratches)

14.3.5 Optical Properties

The first-order phase transition of $BaTiO_3$ manifests itself also in a discontinuity of the bandgap (Fig. 14.11). The coefficient $\partial E_g/\partial T$ for the temperature dependence of the bandgap is also different in the para- and ferroelectric phases.

14.4 Piezoelectricity

External stress causes atoms in the unit cell to shift with respect to each other. In certain directions, such a shift can lead to a polarization. Generally, all ferroelectric materials are piezoelectric. However, there are piezoelectric materials that are not ferroelectric, e.g. quartz, GaAs and GaN. Piezoelectricity can occur only when no center of inversion is present. Thus, e.g., GaAs is piezoelectric along $\langle 111 \rangle$, but Si is not. Also, the cubic perovskite structure (in the paraelectric phase) is not piezoelectric. Generally, the piezoelectric polarization is related to the strains via the tensor e_{ijk} of the piezoelectric modules

$$P_i = e_{ijk} \epsilon_{jk} \; . \tag{14.15}$$

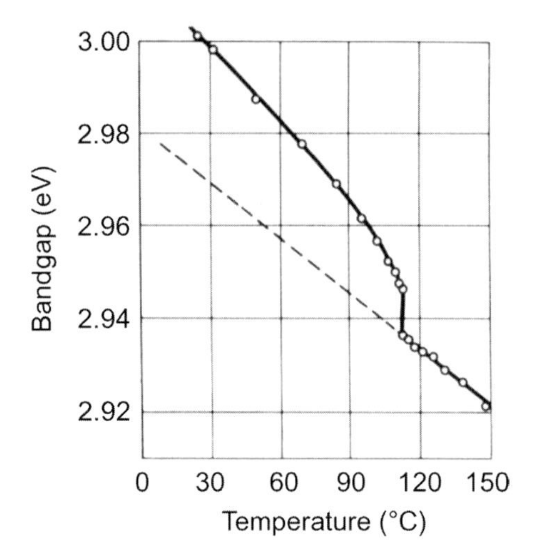

Fig. 14.11. Temperature dependence of the bandgap in $BaTiO_3$ (for polarized light with $\mathbf{E} \perp c$). Experimental data from [513]

In zincblende semiconductors, the polarization (with respect to $x = [100]$, $y = [010]$, $z = [001]$) is due to shear strains only and is given as

$$\mathbf{P}_{\rm pe} = 2e_{13} \begin{pmatrix} \epsilon_{yz} \\ \epsilon_{xz} \\ \epsilon_{xy} \end{pmatrix} . \tag{14.16}$$

The strain in pseudomorphic heterostructures (see Sect. 5.3.2) can cause piezoelectric polarization in a piezoelectric semiconductor. In zincblende, the main effect is expected when the growth direction is along [111] and the strain has a purely shear character. In this case, the polarization is in the [111] direction, i.e. perpendicular to the interface ($P_\perp$). For the [001] growth direction, no piezoelectric polarization is expected. For the [110] growth direction, the polarization is found to be parallel to the interface ($P_\|$). The situation is shown for various orientations of the growth direction in Fig. 14.12.

In wurtzite crystals, the piezoelectric polarization (with respect to $x = [2\text{–}1.0]$, $y = [01.0]$, $z = [00.1]$) is given by

$$\mathbf{P}_{\rm pe} = \begin{pmatrix} 2e_{15}\epsilon_{yz} \\ 2e_{15}\epsilon_{xz} \\ e_{31}(\epsilon_{xx} + \epsilon_{yy}) + e_{33}\epsilon_{zz} \end{pmatrix} . \tag{14.17}$$

Therefore, the polarization (along c) for biaxial strain in heteroepitaxy (5.62) on the [00.1] surface is

$$P_{\rm PE} = 2\epsilon_\| \left(e_{31} - \frac{C_{13}}{C_{33}} e_{33} \right) , \tag{14.18}$$

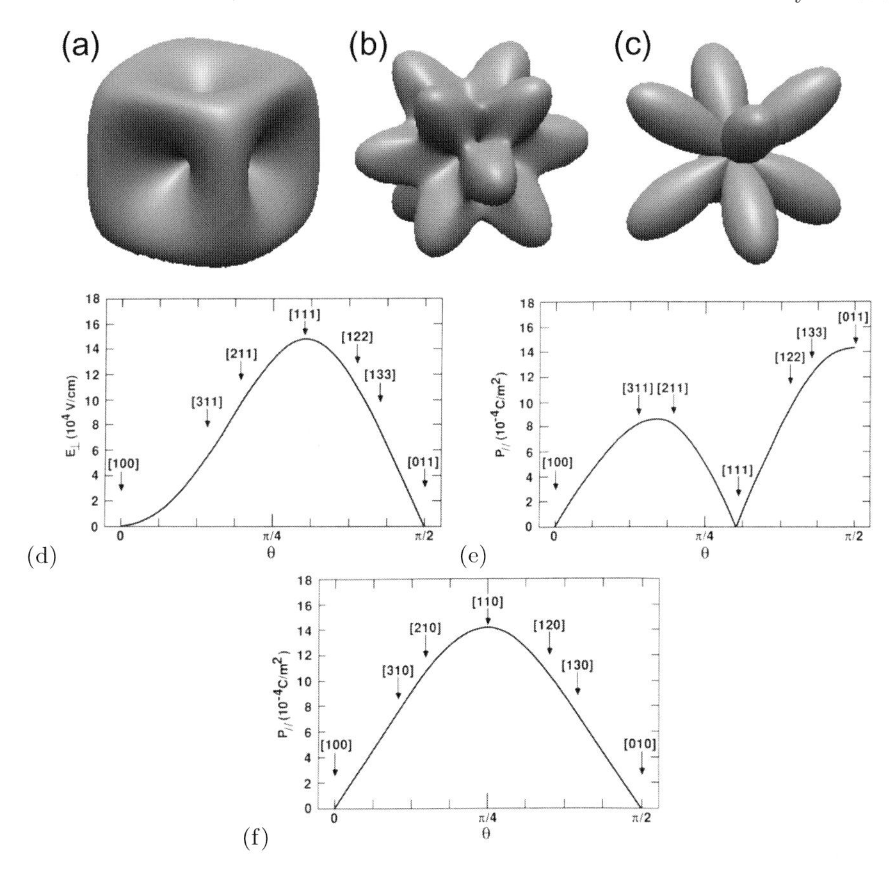

Fig. 14.12. Three-dimensional view of the (**a**) total, (**b**) longitudinal and (**c**) transverse polarization in uniaxially compressed GaAs. Reprinted with permission from [516], ©1994 APS. (**d**) Longitudinal electric field $E_\perp$ (perpendicular to the interface) and (**e**) transverse polarization $P_\parallel$ (parallel to the interface) in the InGaAs layer of a GaAs/In$_{0.2}$Ga$_{0.8}$As superlattice with joint in-plane lattice constant (obtained from energy minimization, 1.4% lattice mismatch, the InGaAs is under compressive and the GaAs under tensile strain). The layer thicknesses of the GaAs and InGaAs layers are identical. The quantities are shown for various orientations of the growth direction. The vector of the growth direction varies in the (01–1) plane ($\phi = \pi/4$) with polar angle θ reaching from [100] (0°) over [111] to [011] (90°). Image (**f**) depicts the transverse polarization $P_\parallel$ ($P_\perp = 0$ in this geometry) for growth directions in the (001) plane ($\phi = 0$). Parts (**d,e,f**) reprinted with permission from [514], ©1988 AIP

where $\epsilon_\parallel = (a - a_0)/a_0$ is the in-plane strain. The dependence of the magnitude for GaN on the in-plane strain is shown in Fig. 14.13 together with the polarization for uniaxial stress along [00.1] and hydrostatic strain. In the latter two cases, the polarization is smaller.

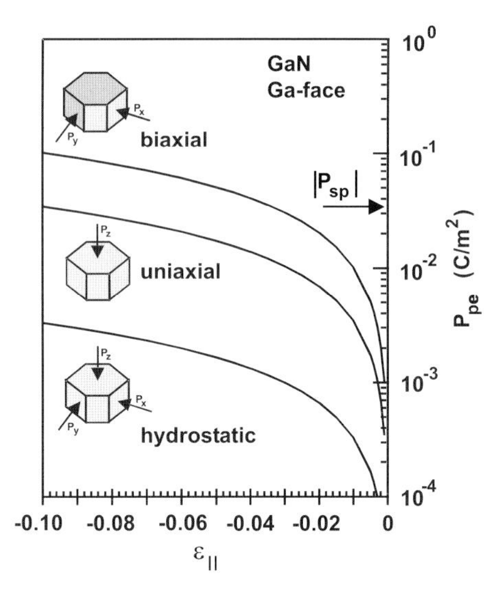

Fig. 14.13. Piezoelectric polarization P_{pe} in GaN (Ga-face) vs. in-plane strain $\epsilon_{\|} = (a - a_0)/a_0$ for biaxial, uniaxial and hydrostatic strain. The value of the spontaneous polarization P_{sp} is indicated by an *arrow*. From [515]

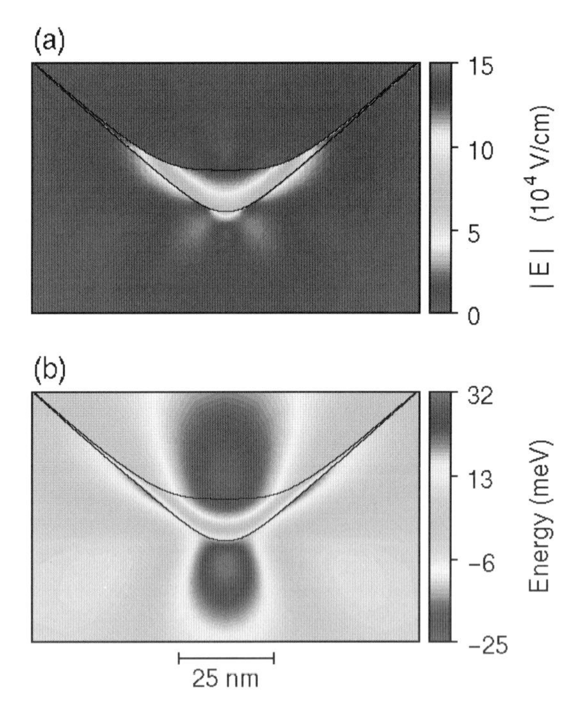

Fig. 14.14. (**a**) Electric field and (**b**) additional confinement potential for electrons due to piezoelectric charges for a strained $In_{0.2}Ga_{0.8}As/GaAs$ quantum wire. Reprinted with permission from [516], ©1994 APS

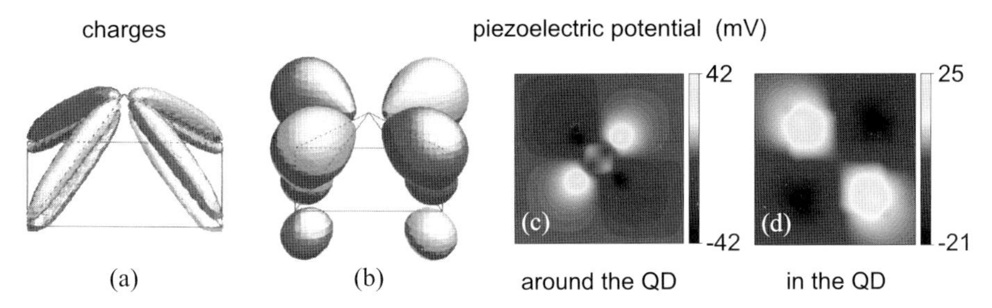

Fig. 14.15. (**a**) Piezoelectric charges and (**b**)–(**d**) resulting Coulomb potential for InAs/GaAs quantum dot with base length $b = 12\,\mathrm{nm}$. (**a**) Isosurfaces corresponding to volume charges $\pm 0.3\,\mathrm{e\,nm}^{-3}$. (**b**) Isosurfaces for the Coulomb potential at $\pm 30\,\mathrm{meV}$. (**c**,**d**) Cross section through the Coulomb potential somewhat above the wetting layer in two different magnifications, (**d**) is a zoom into (**c**). The InAs/GaAs interface is visible in (**d**) due to the image charge effect. Parts (**a**) and (**b**) reprinted with permission from [117], ©1995 APS

The strain distribution around zincblende strained quantum wires [516] and quantum dots [117] contains shear components and thus generates piezoelectric fields. In Fig. 14.14, the electric field and potential due to the piezoelectric charges are shown for a strained $In_{0.2}Ga_{0.8}As$/GaAs quantum wire. In Fig. 14.15, the piezoelectric charges and potential are shown for the quantum dot from Fig. 5.19. The piezoelectric potential has quadrupole character and thus reduces the symmetry of the QD (to C_{2v}) [117].[1] Piezoelectric effects are particularly important in wurtzite nanostructures [517].

[1]The strain distribution has C_{2v} symmetry for a square-based pyramid for zincblende materials. The energy levels and wavefunctions are more strongly impacted by the piezoelectric effects than by the strain asymmetry [448, 450]

15 Magnetic Semiconductors

15.1 Introduction

Magnetic semiconductors exhibit spontaneous magnetic order. Even ferromagnetism, important for spin polarization, as needed in spinelectronics (also called spintronics), can occur below the Curie temperature that is characteristic of the material. Magnetic semiconductors can be binary compounds such as EuTe (antiferromagnetic) or EuS (ferromagnetic). Another class of magnetic semiconductors contains paramagnetic ions in doping concentration (typically $< 10^{21}\,\mathrm{cm}^{-1}$) or alloy concentration x (typically $x \geq 0.1\%$). Such materials are termed *diluted magnetic semiconductors* (DMS). The incorporation of the magnetic atoms leads first to conventional alloy effects, such as the modification of the lattice constant, the carrier concentration or the bandgap. The status of the field up to the mid-1980s can be found in [518], mostly focused on II–VI DMS. A review of work on III–V based materials for spintronics, mostly GaAs:Mn, can be found in [519]. A 2003 review of wide-bandgap ferromagnetic semiconductors is given in [520].

15.2 Magnetic Semiconductors

In a magnetic semiconductor, one sublattice is populated with paramagnetic ions. The first two ferromagnetic semiconductors discovered were $CrBr_3$ [521] in 1960 and EuO [522] one year later. Europium monoxide has an ionic $Eu^{2+}O^{2-}$ character, such that the electronic configuration of europium is $[Xe]4f^7 5d^0 6s^0$ and that of oxygen is $1s^2 2s^2 2p^6$. Some properties of europium chalcogenides are summarized in Table 15.1.

EuO can be modeled as a Heisenberg ferromagnet with dominant nearest- and next-nearest Eu–Eu interactions [523]. The Heisenberg exchange parameters J_1 and J_2 for these four compounds are shown in Fig. 15.1. In the nearest-neighbor interaction a 4f electron is excited to the 5d band, experiences an exchange interaction with the 4f spin on a nearest neighbor and returns to the initial state. This mechanism generally leads to ferromagnetic exchange. The next-nearest-neighbor interaction is weakly ferromagnetic (EuO) or antiferromagnetic (EuS, EuSe, EuTe). In the superexchange process, electrons are transferred from the anionic p states to the 5d states of the Eu^{2+} cations, resulting in an antiferromagnetic coupling.

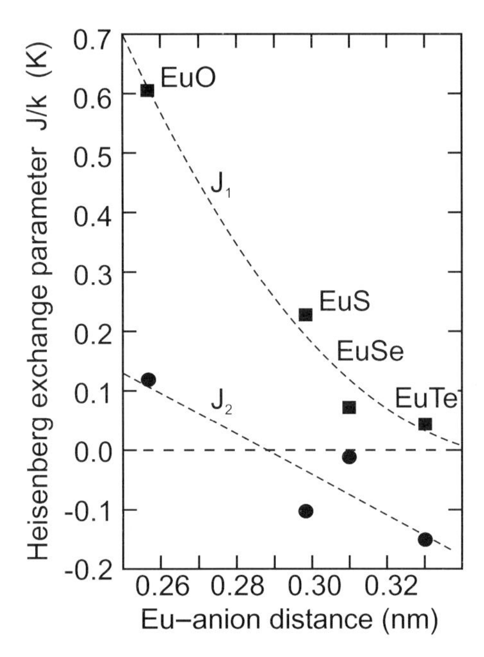

Fig. 15.1. Heisenberg nearest (J_1, *squares*) and next-nearest (J_2, *circles*) exchange parameters (in units of $J_{1,2}/k_B$) for the Eu chalcogenides vs. the Eu–anion distance. *Dashed lines* are guides to the eye. Experimental data from [524]

Table 15.1. Material properties of Eu chalcogenides. 'F' ('AF') denotes ferromagnetic (antiferromagnetic) order. T_N (T_C) denotes the Néel (Curie) temperature. Data collected in [524]

material	E_g (eV)	magnetic order	T_N, T_C (K)
EuO	1.12	F	69.3
EuS	1.65	F	16.6
EuSe	1.8	AF	4.6
		F	2.8
EuTe	2.00	AF	9.6

15.3 Diluted Magnetic Semiconductors

In Table 15.2, the transition metals and their electron configurations are summarized. The 3d transition metals are typically used for magnetic impurities in DMS due to their partially filled 3d shell. Due to Hund's rule, the spins on the 3d shell are filled in parallel for the first five electrons up to half filling (in order to allow the electrons to get out of their way in real space). Thus, the atoms have a sizeable spin and a magnetic moment. The spin of Mn is $S = 5/2$. Most transition metals have a $4s^2$ configuration that makes them isovalent in II–VI compounds. We note that Zn has a complete 3d shell and thus no net spin. In Fig. 15.2, an overview of the crystallographic properties is given for Mn-alloyed II–(Se,S,Te,O) based DMS [525] (DMS with Se, S, and Te have been discussed in [526]).

As an example, the properties of $Hg_{1-x}Mn_xTe$ are discussed. This alloy is semiconducting (positive bandgap ϵ_0) for $x > 0.075$ and a zero-gap material (negative interaction gap ϵ_0) for smaller Mn concentration (cf. Fig. 6.29). The transitions between the Γ_6 and Γ_8 bands can be determined with magnetoabsorption spectra in the infrared [527]. In Fig. 15.3a, the magnetic field dependence of transition energies between different Landau levels is shown that can be extrapolated to yield the interaction gap. The interaction gap is shown in Fig. 15.3b as a function of the Mn concentration.

For small Mn concentrations, the DMS behaves like a paramagnetic material. For larger concentrations, the Mn atoms have increasing probability to be directly neighbored by another Mn atom and suffer superexchange interaction (cf. Eq. (3.19b)). At a certain critical concentration x_c, the cluster size becomes comparable with the size of the sample. If interaction up to the first, second or third neighbor are taken into account for a fcc lattice, the critical concentrations are given by x_c =0.195, 0.136, and 0.061, respectively [528]. The nearest-neighbor interaction between Mn atoms in such DMS

Table 15.2. 3d, 4d and 5d transition metals and their electron configurations. Note that Hf^{72} has an incomplete 4f-shell with $4f^{14}$

Sc^{21}	Ti^{22}	V^{23}	Cr^{24}	Mn^{25}	Fe^{26}	Co^{27}	Ni^{28}	Cu^{29}	Zn^{30}
3d	$3d^2$	$3d^3$	$3d^5$	$3d^5$	$3d^6$	$3d^7$	$3d^8$	$3d^{10}$	$3d^{10}$
$4s^2$	$4s^2$	$4s^2$	4s	$4s^2$	$4s^2$	$4s^2$	$4s^2$	4s	$4s^2$
Y^{39}	Zr^{40}	Nb^{41}	Mo^{42}	Tc^{43}	Ru^{44}	Rd^{45}	Pd^{46}	Ag^{47}	Cd^{48}
4d	$4d^2$	$4d^4$	$4d^5$	$4d^6$	$4d^7$	$4d^8$	$4d^{10}$	$4d^{10}$	$4d^{10}$
$5s^2$	$5s^2$	5s	5s	5s	5s	5s	–	5s	$5s^2$
La^{57}	Hf^{72}	Ta^{73}	W^{74}	Re^{75}	Os^{76}	Ir^{77}	Pt^{78}	Au^{79}	Hg^{80}
5d	$5d^2$	$5d^3$	$5d^4$	$5d^5$	$5d^6$	$5d^9$	$5d^9$	$5d^{10}$	$5d^{10}$
$6s^2$	$6s^2$	$6s^2$	$6s^2$	$6s^2$	$6s^2$	–	6s	6s	$6s^2$

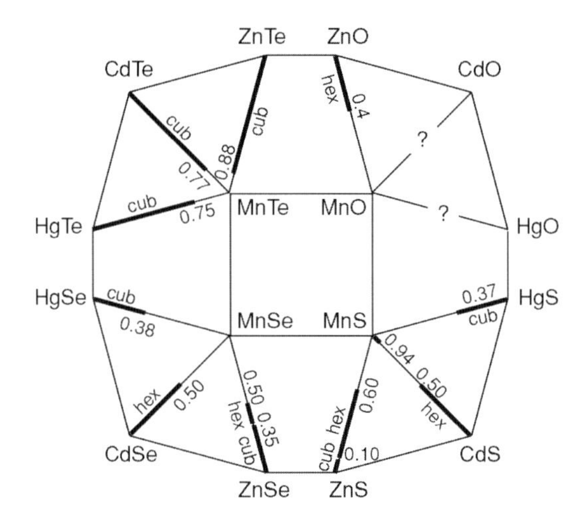

Fig. 15.2. Diagrammatic overview of $A^{II}_{1-x}Mn_xB^{VI}$ alloys and their crystal structures. The *bold lines* indicate ranges of the molar fraction x for which homogeneous crystal phases form. 'Hex' and 'Cub' indicate wurtzite and zincblende, respectively. From [525]

as (Zn,Cd,Hg)Mn(S,Se,Te) was found to be antiferromagnetic,[1] i.e. neighboring spins are aligned antiparallel. Due to frustration of antiferromagnetic long-range order on a fcc lattice, an antiferromagnetic spin glass forms. The transition temperature T_C between the paramagnetic and spin-glass phases of $Hg_{1-x}Mn_xTe$ is shown in Fig. 15.4.

In III–V compounds, the 3d transition metals represent an acceptor if incorporated on the site of the group-III element as, e.g., in the much investigated compound $Ga_{1-x}Mn_xAs$. This material will be used in the following to discuss some properties of magnetic semiconductors. It seems currently well understood and has a fairly high Curie temperature of $T_C \approx 160$ K. Ferromagnetism in a diluted magnetic semiconductor is believed to be caused by indirect exchange through itinerant charge carriers. The ferromagnetic coupling can be invoked by the Ruderman–Kittel–Kasuya–Yoshida (RKKY) interaction, i.e. the spins of the paramagnetic ions are aligned via interaction with the free carriers in the semiconductor. A related concept is the double exchange[2] [529–531] in which carriers move in a narrow Mn-derived d-band (for d-wave character cf. Fig. 7.13c). Such a mechanism was first invoked for PbSnMnTe [532]. Later, ferromagnetism was discovered in InMnAs [533] and GaMnAs [534]. In (In,Ga)MnAs a Mn ion (spin up) spin polarizes the surrounding hole gas (spin down), which has been supplied from the Mn

[1] Such superexchange leads to antiferromagnetic interaction if the bond angle is 'close' to 180°.

[2] This model is also called the Zener model.

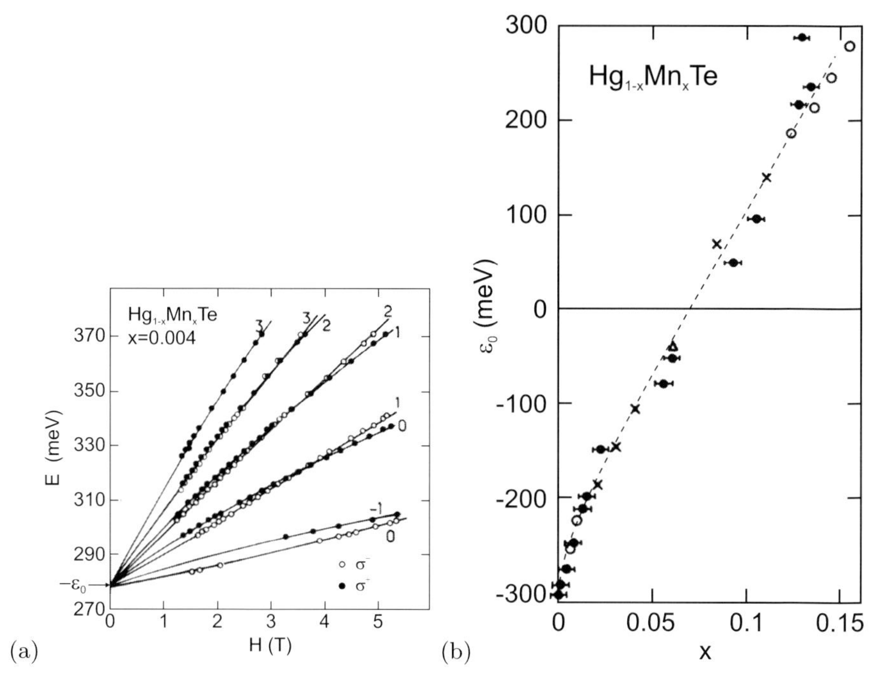

Fig. 15.3. (**a**) Energies of $\Gamma_6 \rightarrow \Gamma_8$ transitions vs. magnetic field for $Hg_{0.996}Mn_{0.004}Te$ at $T = 2\,K$. *Symbols* are experimental values for two polarization directions as indicated. Numbers denote quantum numbers of transitions. *Solid lines* are theoretical fits. (**b**) Interaction gap vs. Mn concentration for $Hg_{1-x}Mn_xTe$ at $T = 4.2\,K$. Various *symbols* represent data from different authors and methods. *Dashed line* is a guide to the eye. Adapted from [527]

acceptors. This mechanism lowers the energy of the coupled system. The interaction

$$H = -\beta N_0 S s \tag{15.1}$$

between the Mn d-shell electrons ($S = 5/2$) and the p-like free holes ($s = 1/2$) is facilitated by p–d hybridization of the Mn states. N_0 denotes the concentration of cation sites. The coupling via electrons is much weaker (coupling coefficient α). The holes interact with the next Mn ion and polarize it (spin up), thus leading to ferromagnetic order. The ferromagnetic properties are evident from the hysteresis shown in Fig. 15.6a. Without the carrier gas such interaction is not present and the material is only paramagnetic. Theoretical results for the Curie temperature of various p-type semiconductors are shown in Fig. 15.5. Generally, the quest for higher Curie temperatures (well above room temperature) is underway and wide-bandgap materials such as GaN or ZnO doped with transition metals have shown some encouraging results. Mn-substituted chalcopyrite semiconductors are analyzed theoretically

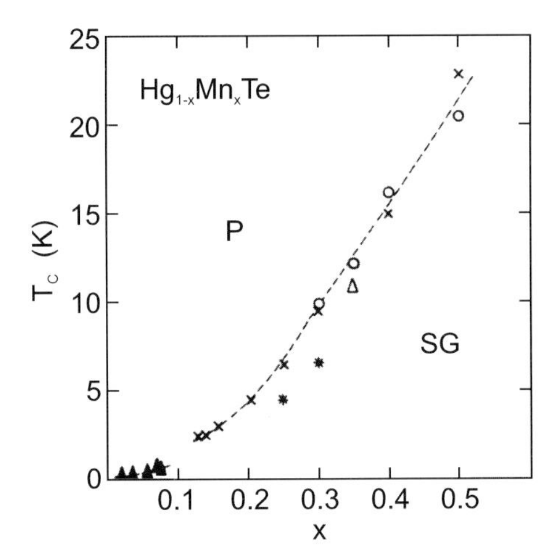

Fig. 15.4. Magnetic phase diagram of $Hg_{1-x}Mn_xTe$, 'P' ('SG') denotes the paramagnetic (spin glass) phase. Various *symbols* represent data from different authors and methods. *Dashed line* is a guide to the eye. Adapted from [527]

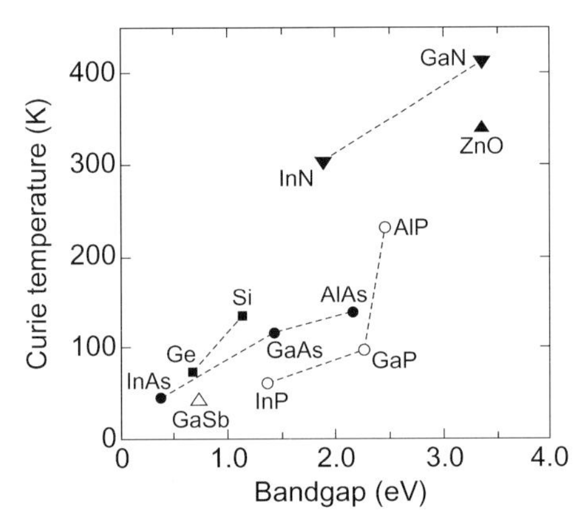

Fig. 15.5. Computed values of the Curie temperature T_C for various p-type semiconductors plotted vs. the bandgap (*dashed lines* are guides to the eye). All materials contain 5% Mn on the cation sublattice and a hole concentration of $p = 3.5 \times 10^{20}\,\mathrm{cm}^{-3}$. Values for T_C taken from [531]

in [535] and are predicted to exhibit less-stable ferromagnetism than III–V semiconductors of comparable bandgap.

The carrier density and thus magnetic properties in a DMS can be controlled in a space-charge region (cf. Sect. 18.2.2) as demonstrated in [536]. In Fig. 15.6, results are shown for hydrogen- (deuterium-) passivated GaMnAs that exhibits ferromagnetism as 'as-grown' thin film. The deuterium is incorporated in similar concentration as the Mn, assumes a back-bond position (forming a H–As–Mn complex) and compensates the hole gas from the Mn (cf. Sect. 7.8). The low-temperature conductivity drops nine orders of magnitude [537]. Such material displays only paramagnetic behavior. An optimal Mn concentration for ferromagnetic $Ga_{1-x}Mn_xAs$ is around $x = 0.05$. For smaller Mn concentrations, the hole density is too small and the Curie temperature drops; for larger Mn concentrations, the structural properties of the alloy degrade (phase separation into GaAs and MnAs.[3])

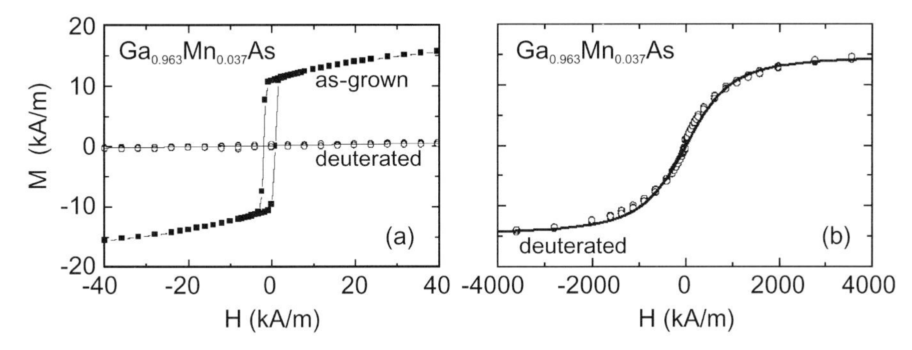

Fig. 15.6. Magnetization M vs. magnetic field H for $Ga_{0.963}Mn_{0.037}As$ at low temperature. (**a**) Comparison of as-grown (*full squares*) and deuterated (*open circles*) thin film with magnetic field in the layer plane at $T = 20\,K$. (**b**) Magnetization of the deuterated sample at $T = 2\,K$ for larger magnetic fields. *Solid line* is Brillouin function for $g = 2$ and $S = 5/2$. Adapted from [537]

Magnetic hysteresis has been found in n-conducting Mn-doped ZnO [538, 539] (Fig. 15.7). Such material is interesting due to its small spin-orbit coupling. The exchange mechanism is under debate.

15.4 Spintronics

Spintronics (as opposed to electronics) is an emerging field that uses the electron *spin* rather than its *charge* for transport, processing and storage of information. Prototype devices are the spin transistor and the spin LED. It remains to be seen whether spintronics can be developed to its theoretically envisioned potential and will play a commercially important role in the course of microelectronics. The spin degree of freedom also promises potential for quantum information processing due to its weak coupling to charge and phonons and the resulting long dephasing time.

---

[3]MnAs is a ferromagnetic metal. MnAs clusters can be a problem since they create ferromagnetic properties but not in the way the DMS is supposed to work.

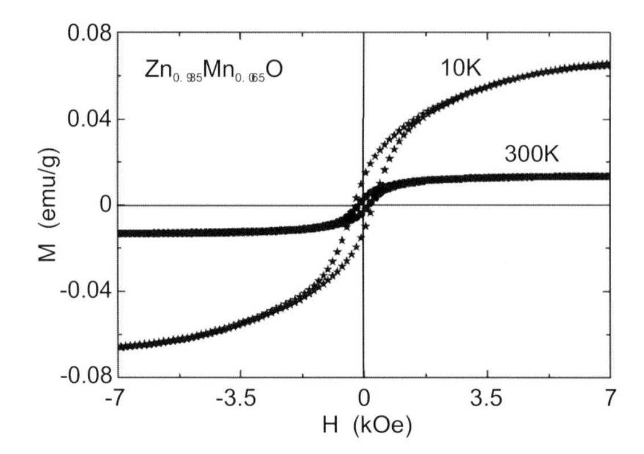

Fig. 15.7. Magnetization M vs. magnetic field H for $Zn_{0.935}Mn_{0.065}O$ thin film at $T = 10$ and 300 K. A hysteresis is obvious for both temperatures

15.4.1 Spin Transistor

In this device (for regular transistors cf. Chap. 21), spin-polarized electrons are injected from contact 1, transported through a channel and detected in contact 2. During the transport, the spin rotates (optimally by π) such that the electrons cannot enter contact 2 that has the same magnetization as contact 1. The spin rotation is caused by spin-orbit interaction due to the electric field under the gate contact. This effect is called the Rashba effect and is purely relativistic [540]. As channel material, a semiconductor with strong spin-orbit coupling such as InAs or (In,Ga)Sb is preferable. However, the use of narrow-gap semiconductors and the increase of spin scattering at elevated temperatures [541] make the realization of such a transistor at room temperature difficult.

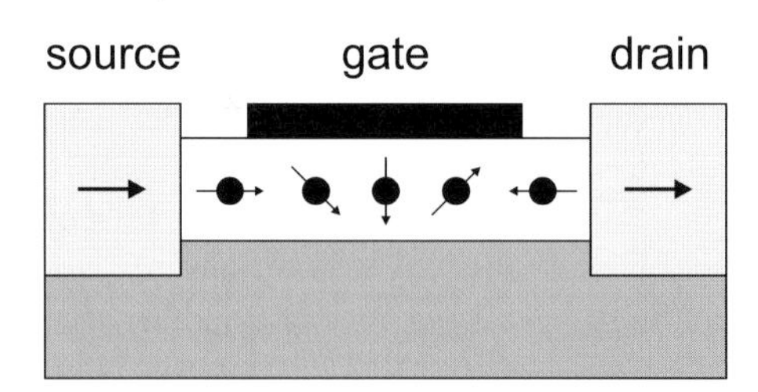

Fig. 15.8. Scheme of spin transistor after the proposal of [542]. Source and drain are ferromagnets with their magnetization shown schematically as arrows. The channel under the gate transports electrons whose spin rotates in the electric field under the gate

15.4.2 Spin LED

In a spin LED (for LEDs see Sect. 20.3), the injection of spin-polarized carriers into the active layer leads to circularly polarized luminescence. The spin alignment can be achieved with semimagnetic semiconductors grown on top of the active layer or via spin injection from a ferromagnetic metal into the semiconductor (for metal–semiconductor junctions see Sect. 18.2). In Fig. 15.9a, a Fe/AlGaAs interface is shown.

Ideally, the spin-polarized electrons from the ferromagnetic metal tunnel into the semiconductor and transfer to the recombination region. Subsequently, the emission is circularly polarized (Fig. 11.18b). The degree of circular polarization is

$$P_\sigma = \frac{I_{\sigma+} - I_{\sigma-}}{I_{\sigma+} + I_{\sigma-}} , \tag{15.2}$$

with $I_{\sigma\pm}$ being the intensity of the respective polarization. The degree of polarization depends on the magnetization of the metal. For the saturation magnetization of Fe, the maximum polarization is about 30% at $T = 4.5\,\mathrm{K}$ (Fig. 15.9b) [543]. The interface and its structural nonideality of the interface presumably prevent the spin injection from being 100% efficient [544].

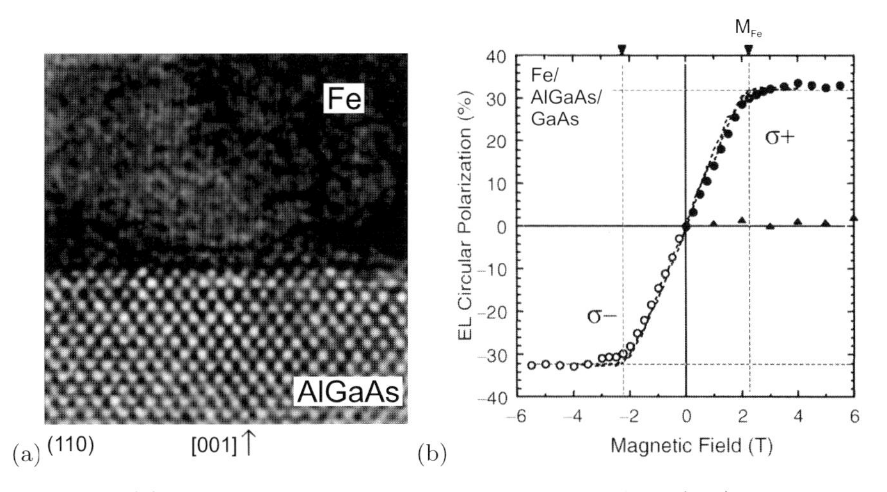

Fig. 15.9. (**a**) Transmission electron microscopy image of the (110) cross section of the Fe/AlGaAs interface of a spin LED. The *vertical lines* in Fe are the (110) planes with 0.203 nm distance. (**b**) Magnetic-field dependence of the circular polarization ratio P_σ at $T = 4.5\,\mathrm{K}$ (15.2) (*filled* and *empty* circles) and the out-of-plane component of the Fe-film magnetization (*dashed line*, scaled to the maximum of P_σ). Reproduced from [544] by permission of the MRS Bulletin

16 Organic Semiconductors

Organic semiconductors are based on carbon compounds. The main difference from inorganic semiconductors is the sp^2 bond (see Sect. 2.2.3) as present in graphite. Diamond, although consisting of 100% carbon, is not considered an organic semiconductor. We note that carbon can form further interesting structures based on sp^2 bonds, such as carbon nanotubes or fullerenes. Carbon nanotubes are graphene sheets rolled up to form cylinders. Their walls can consist of a single graphene sheet (SWNT, single-wall carbon nanotube) or of several sheets (MWNT, multiwall carbon nanotube). Fullerenes are soccer-ball-like molecules such as C_{60}.

16.1 Materials

The prototype organic molecule is the benzene molecule with its ring-like structure (Fig. 2.8). The p_z orbitals are partially filled and there is a separation between HOMO and LUMO (Fig. 2.11). A similar consideration is valid for polymers. The coupling of orbitals along the polymer chain leads to broadening of the π and π^* states into a (filled) valence and an (empty) conduction band, respectively (Fig. 16.1).

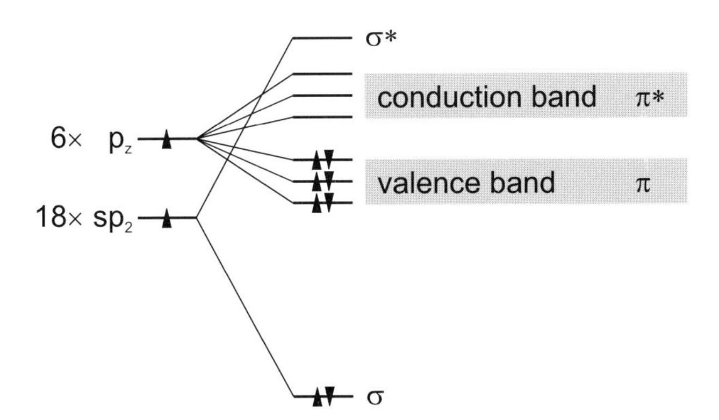

Fig. 16.1. Schematic band structure of a polymer originating from the states of the benzene molecule (cf. Fig. 2.11)

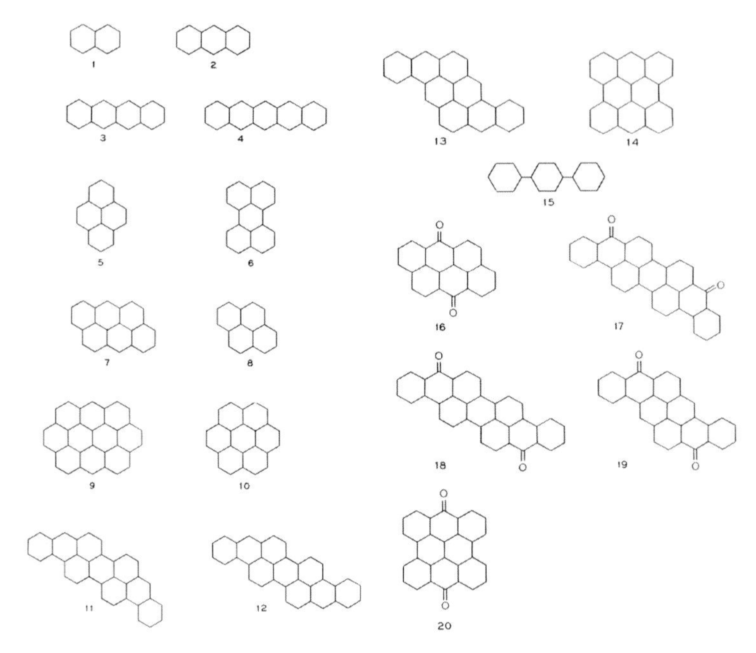

Fig. 16.2. Various organic compounds: 1: naphtalene, 2: anthracene, 3: tetracene, 4: pentacene, 5: pyrene, 6: perylene, 7: anthanthrene, 8: chrysene, 9: ovalene, 10: coronene, 11: violanthrene, 12: isoviolanthrene, 13: pyranthrene, 14: m-dinaphthanthrene, 15: p-terphenyl, 16: anthanthrone, 17: violanthrone, 18: isoviolanthrone, 19: pyranthrone, 20. m-dinaphthanthrone

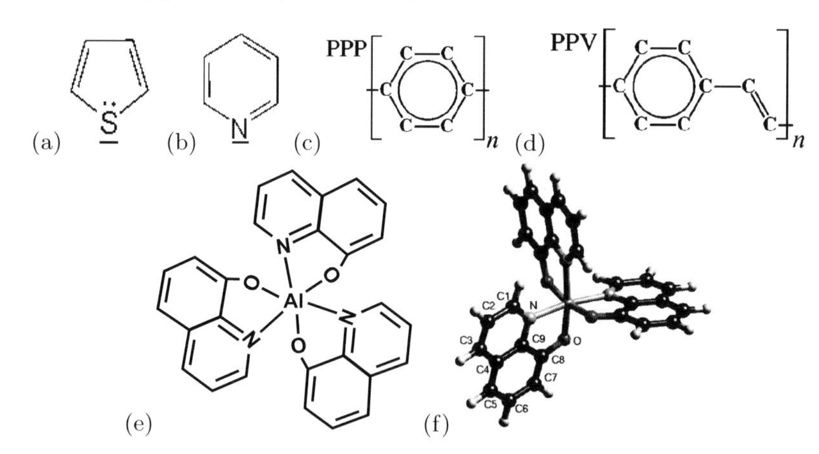

Fig. 16.3. Organic compounds: (**a**) thiophene, (**b**) pyridine, (**c**) poly-(p-phenyl), (**d**) poly-(p-phenylvinyl), (**e**) Alq3 (tris-(8-hydroxyquinolate)-aluminum) and (**f**) a three-dimensional view of the Alq3 molecule. Part (**f**) reprinted with permission from [466], ©1998 AIP

There is a large number of organic, semiconducting molecules that differs by the number of benzene rings (Fig. 16.2), the substitution of carbon atoms by nitrogen or sulfur (Fig. 16.3a,b), the polymerization (Fig. 16.3c) or the substitution of hydrogen atoms by side groups (Fig. 16.3d). Since PPV is insoluble, typically derivatives such as MEH-PPV[1] [465] that are soluble in organic solvents are used. Compared to benzene, the substitution of one carbon atom by nitrogen (pyridine) represents doping with one electron. In Fig. 16.4, the most important building blocks of organic molecules are shown.

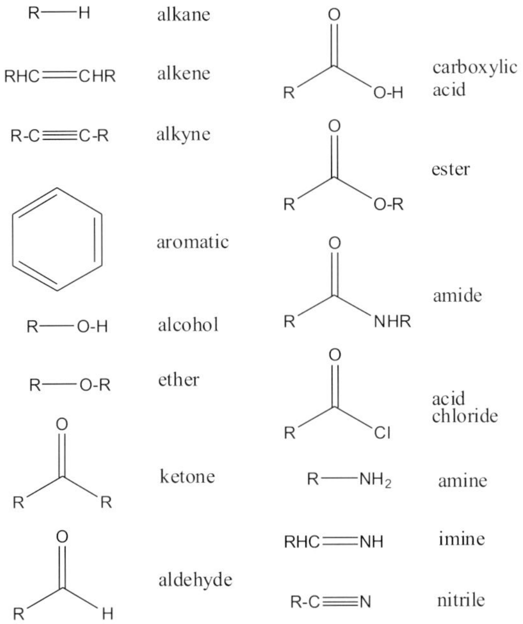

Fig. 16.4. Building blocks of organic molecules, 'R'=alkyl group, i.e. CH_3 (methyl-), CH_3CH_2 (buthyl-), ...

Small organic molecules can crystallize into solids due to van-der-Waals interaction. In Fig. 16.5, the monoclinic unit cell of an anthracene crystal is shown.

16.2 Properties

For organic semiconductors, the conductivity within a molecule, e.g. a long polymer chain, and the conductivity between different molecules have to be distinguished. A magnetic field induces a circular current within the benzene

[1] 2-ethoxy,5-(2'-ethyl-hexyloxy)-1,4-phenylene vinylene

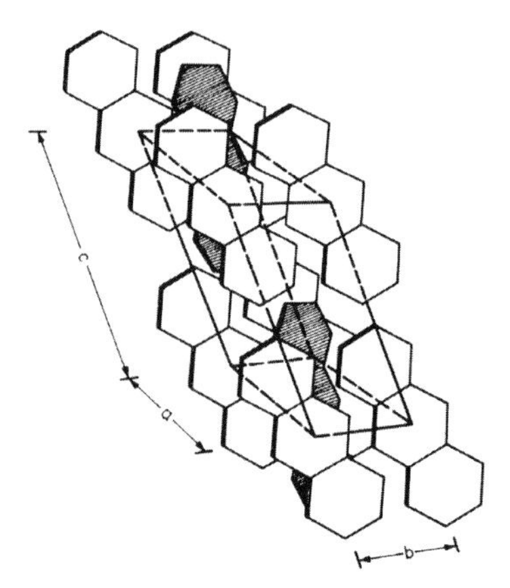

Fig. 16.5. Monoclinic unit cell of anthracene crystal

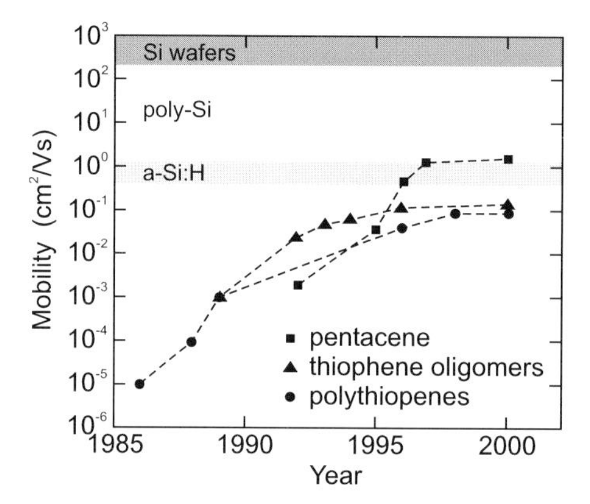

Fig. 16.6. Historic development of the experimentally achieved mobility of organic semiconductors

molecule. This current will flow unhindered (diamagnetic effect) as long as the HOMO/LUMO gap is large enough to prevent thermal excitations. The conduction between different molecules occurs via *hopping*. Typically, the conductivity is thermally activated according to

$$\sigma = \sigma_0 \exp\left(-\frac{E}{kT}\right) , \tag{16.1}$$

where E is an energy of the order of 1 eV. The photoconductivity is related to the absorption spectrum (Fig. 16.7). The mobility is mostly much smaller than $1\,\mathrm{cm}^2/\mathrm{V\,s}$ (Fig. 16.6) and thus much smaller than that of crystalline silicon and rather comparable to that of amorphous silicon.

The density of excited (empty) states of the Alq3 molecule is shown in Fig. 16.8 together with the orbitals associated with the four prominent states.

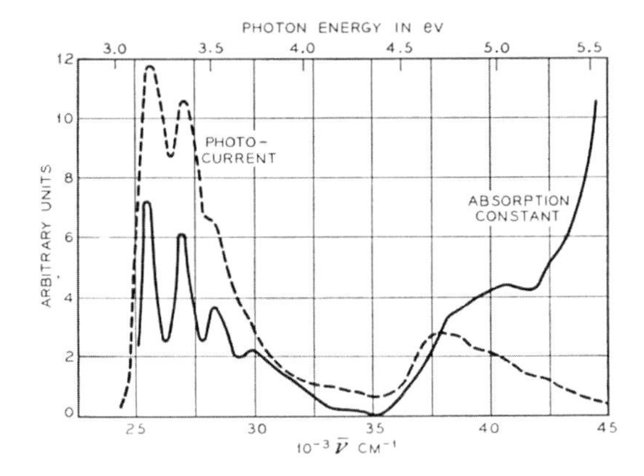

Fig. 16.7. Photoconductivity and absorption spectrum of anthracene

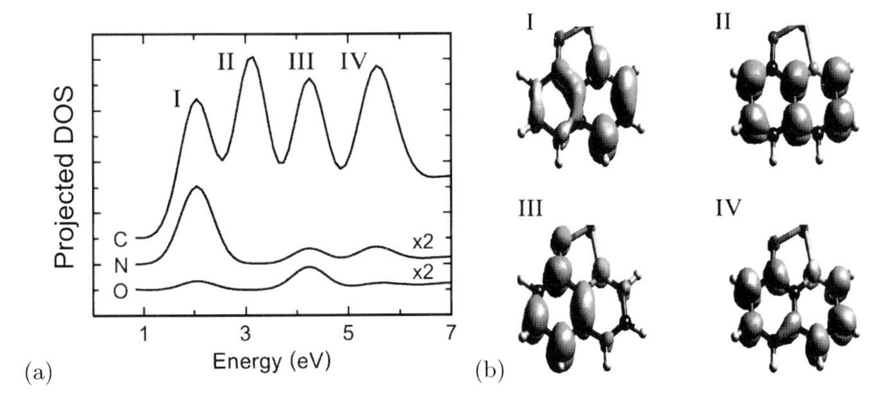

Fig. 16.8. (**a**) Projected density of states (for C, N, and O) of excited states in an Alq3 molecule. The origin of the energy axis is the HOMO level. (**b**) Orbitals for the four states labeled I–IV in (**a**). Reprinted with permission from [466], ©1998 AIP

The lowest orbital is the LUMO and leads to the visible luminescence of the Alq3 in the red.

Organic molecules can emit light efficiently and are thus useful for light emitters. In Fig. 16.9, the photoluminescence (PL) and the PL excitation (PLE) spectra of poly-thiophene are shown. The recombination is below the bandgap of 2.1–2.3 eV on an excitonic level at 1.95 eV. There are several phonon replica whose separation of 180 meV corresponds to the C–C stretching mode. The PLE demonstrates that the PL at 1.83 eV can be excited via the exciton level.

The theoretical band structure of poly-thiophene is shown in Fig. 16.10a. The Brillouin zone is one-dimensional. The situation I corresponds to a single molecular chain, the situation II pertains to the chain embedded in a medium

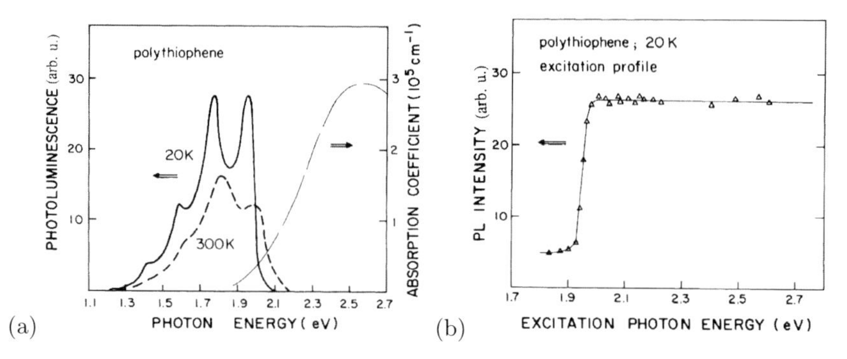

Fig. 16.9. (**a**) Photoluminescence (PL) and absorption spectra, (**b**) PL excitation (PLE) spectrum ($E_{\mathrm{det}} = 1.83\,\mathrm{eV}$) of poly-thiophene. Reprinted with permission from [467], ©1987 APS

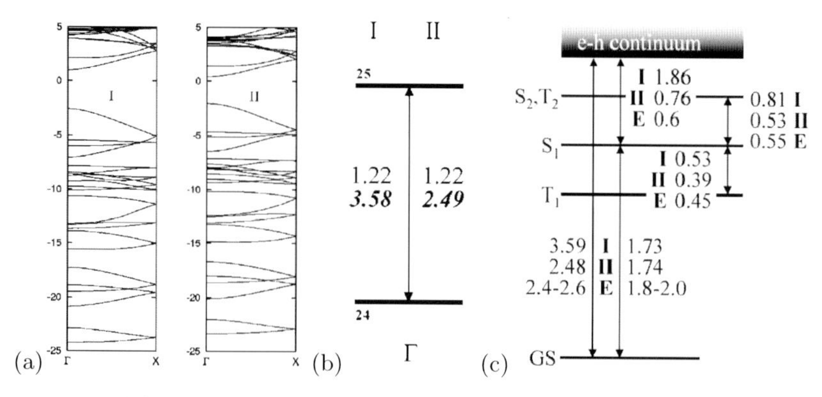

Fig. 16.10. (**a**) Band structure of poly-thiophene ('I': naked chain, 'II': chain in a dielectric medium ($\epsilon = 3$)), (**b**) single-particle energies and bandgap, (**c**) exciton levels ('E': experimental values). Reprinted with permission from [468], ©2002 APS

with a dielectric constant $\epsilon = 3$. The predicted bandgaps are 3.6 and 2.5 eV, respectively. The exciton binding energy is about 0.5 eV. The exciton is a Frenkel exciton that has a small extension and is localized. The high binding energy is favorable for radiative recombination since the exciton is stable at room temperature. For photovoltaic applications, it is unfavorable since it has to be overcome in order to separate electrons and holes (after absorption). Generally, intrachain excitons (as here) and interchain excitons, where electron and the hole are localized on different chains, are distinguished.

17 Dielectric Structures

17.1 Photonic-Bandgap Materials

17.1.1 Introduction

A structure with a so-called *photonic* bandgap (PBG) exhibits an energy range (color range) in which photons cannot propagate in any direction. In the photonic bandgap, there are no optical modes, no spontaneous emission and no vacuum (zero-field) fluctuations. We recollect that spontaneous emission is not a necessary occurrence: Looking at Fermi's golden rule (9.9) for the transition probability integrated over all final states

$$w(E) = \frac{2\pi}{\hbar} |M|^2 \rho_{\mathrm{f}}(E) \; , \tag{17.1}$$

we see that the decay rate depends on the density ρ_{f} of final states at energy E. In the case of spontaneous emission, this is the (vacuum) density D_{em} of electromagnetic modes (per energy per volume) that varies $\propto \omega^2$:

$$D_{\mathrm{em}}(E) = \frac{8\pi}{(hc)^3} E^2 \; . \tag{17.2}$$

In a homogeneous optical medium c must be replaced with c/n.

If the bandgap of a PBG is tuned to the electronic gap of a semiconductor, the spontaneous emission, and also induced emission, can be suppressed. Thus, one mode has to be left by 'doping' the structure. In this mode all emission will disappear and an efficient single-mode (monochromatic) LED or 'zero-threshold' laser could be built. A schematic comparison of the band structure (dispersion) of electrons and photons is given in Fig. 17.1.

17.1.2 General 1D Scattering Theory

The formation of a photonic bandgap in a one-dimensional dielectric can be calculated to a large extent analytically and thus with direct insight. Let $n(x)$ be the spatially varying index of refraction (no losses or nonlinear optical effects). The one-dimensional wave equation (Helmholtz equation) reads for the electric field $\mathcal{E}$

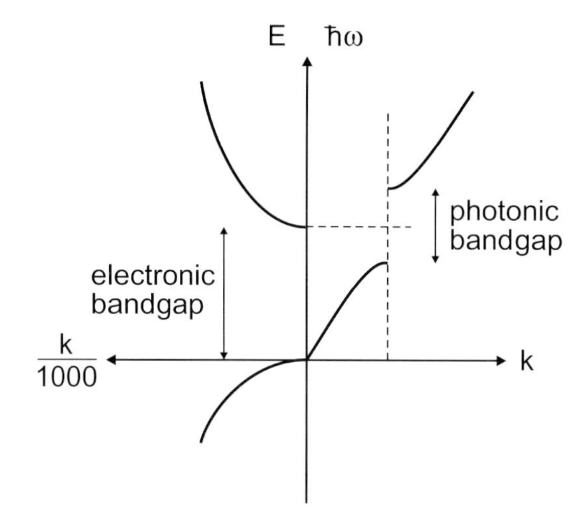

Fig. 17.1. *Right*: electromagnetic dispersion with a forbidden gap at the wavevector of the periodicity. *Left*: Electron-wave dispersion typical of a direct-gap semiconductor, the *dots* representing electrons and holes. When the photonic bandgap straddles the electronic bandgap, electron–hole recombination into photons is inhibited since the photons have no place to go (zero final density of states)

$$\frac{\partial^2 \mathcal{E}(x)}{\partial x^2} + n^2(x)\frac{\omega^2}{c^2}\mathcal{E}(x) = 0 \,. \tag{17.3}$$

A comparison with a one-dimensional Schrödinger equation

$$\frac{\partial^2 \Psi(x)}{\partial x^2} - \frac{2m}{\hbar^2}\left[V(x) - E\right]\Psi(x) = 0 \tag{17.4}$$

shows that the Helmholtz equation corresponds to the quantum-mechanical wave equation of zero external potential V and a spatially modulated mass, i.e. a case that is usually not considered in quantum mechanics.

Let us consider now the amplitude a_k of the k eigenvector. The eigenvalue is ω_k. The one-dimensional mode density $\rho(\omega)$ (per energy and per unit length) is

$$\rho(\omega) = \frac{\mathrm{d}k}{\mathrm{d}\omega} \,, \tag{17.5}$$

which is the inverse of the group velocity.

We follow one-dimensional scattering theory as presented in [469]. At this point we do not rely on any specific form of $n(x)$ (Fig. 17.2a). The (complex) transmission coefficient t for any index structure is

$$t = x + \mathrm{i}y = \sqrt{T}\exp(\mathrm{i}\phi) \,, \tag{17.6}$$

where $\tan\phi = y/x$. ϕ is the total phase accumulated during propagation through the structure. It can be written as the product of the physical thickness of the structure d and the effective wave number k. Hence we have the

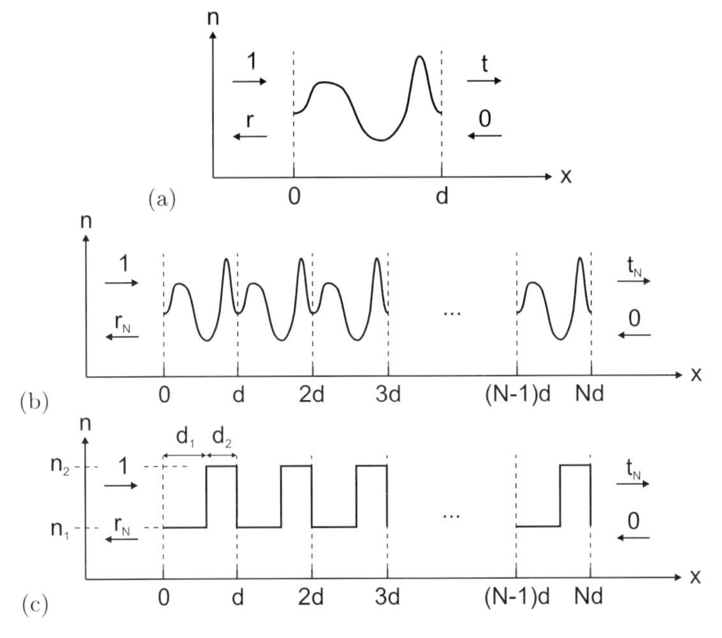

Fig. 17.2. 1D scattering problem: (**a**) General scattering of an index of refraction distribution, (**b**) N-period stack, (**c**) two-layer (quarter-wave) stack.

dispersion relation

$$\frac{\mathrm{d}}{\mathrm{d}\omega} \tan(kd) = \frac{\mathrm{d}}{\mathrm{d}\omega} \left(\frac{y}{x} \right) . \tag{17.7}$$

Evaluating the derivative we find

$$\frac{1}{\cos^2(kd)} d \frac{\mathrm{d}k}{\mathrm{d}\omega} = \frac{y'x - x'y}{x^2} , \tag{17.8}$$

where the prime denotes derivation with respect to ω. Using the relation $\cos^2 \theta = (1 + \tan^2 \theta)^{-1}$, we obtain the general expression

$$\rho(\omega) = \frac{\mathrm{d}k}{\mathrm{d}\omega} = \frac{1}{d} \frac{y'x - x'y}{x^2 + y^2} . \tag{17.9}$$

17.1.3 Transmission of an N-Period Potential

Now, the behavior of N periods of a given index distribution $n(x)$ within a thickness d of one period (Fig. 17.2b) is investigated. The scattering matrix $\mathbf{M}$ connects the intensity at $x = 0$ with that at $x = d$. We use the column vector $\mathbf{u} = \begin{pmatrix} u^+ \\ u^- \end{pmatrix}$ containing the right- and left-going waves (labeled '+' and '−', respectively), $u^\pm = f^\pm \exp(\pm \mathrm{i}kx)$,

$$\mathbf{u}(0) = \mathbf{M}\,\mathbf{u}(d) \,. \tag{17.10}$$

Using the boundary conditions $\mathbf{u}(0) = (1, r)$ and $\mathbf{u}(d) = (t, 0)$, we find that $\mathbf{M}$ has the structure

$$\mathbf{M} = \begin{pmatrix} 1/t & r^*/t^* \\ r/t & 1/t^* \end{pmatrix} \,. \tag{17.11}$$

The conservation of energy requires that $\det \mathbf{M} = (1 - R)/T = 1$. The eigenvalue equation for $\mathbf{M}$ is

$$\mu^2 - 2\mu \mathrm{Re}(1/t) + 1 = 0 \,. \tag{17.12}$$

The two eigenvalues $\mu^\pm$ are related by $\mu^+ \mu^- = \det \mathbf{M} = 1$. If we consider an infinite, periodic structure, we know from Bloch's theorem that the eigenvector varies between unit cells only via a phase factor, i.e. $|\mu| = 1$. Therefore, the eigenvalues can be written as

$$\mu^\pm = \exp(\pm \mathrm{i}\beta) \,, \tag{17.13}$$

where β corresponds to the Bloch phase of a hypothetical infinite periodic structure. This phase β should not be confused with ϕ defined earlier, which is associated with the unit cell transmission. We find the condition

$$\mathrm{Re}(1/t) = \cos\beta \tag{17.14}$$

for the Bloch phase. Since every matrix obeys its own eigenvalue equation, we have also ($\mathbf{1}$ being the unity matrix) from (17.12)

$$\mathbf{M}^2 - 2\mathbf{M}\cos\beta + \mathbf{1} = 0 \,. \tag{17.15}$$

By induction one can show that the N-period case has the scattering matrix

$$\mathbf{M}^N = \mathbf{M}\frac{\sin(N\beta)}{\sin\beta} - \mathbf{1}\frac{\sin((N-1)\beta)}{\sin\beta} \,. \tag{17.16}$$

The solution for the finite period case can be written in terms of the Bloch phase of the infinite potential. The transmission and reflection of the N-period system are given by

$$\frac{1}{t_n} = \frac{1}{t}\frac{\sin(N\beta)}{\sin\beta} - \frac{\sin((N-1)\beta)}{\sin\beta} \tag{17.17a}$$

$$\frac{r_n}{t_n} = \frac{r}{t}\frac{\sin(N\beta)}{\sin\beta} \,. \tag{17.17b}$$

The transmission of intensity can be written as ($T = t^* t$)

$$\frac{1}{T_N} = 1 + \frac{\sin^2(N\beta)}{\sin^2\beta}\left[\frac{1}{T} - 1\right] \,. \tag{17.18}$$

Again, up to this point no specific distribution of the index of refraction within the unit cell has been specified.

From (17.17a), a general formula for the mode density $\rho_N(\omega)$ (17.9) of the N-stack can be obtained as [469]

$$\rho_N = \frac{1}{Nd} \frac{\frac{\sin(2N\beta)}{2\sin\beta}\left[\eta' + \frac{\eta\xi\xi'}{1-\xi^2}\right] - \frac{N\eta\xi'}{1-\xi^2}}{\cos^2(N\beta) + \eta^2\left[\frac{\sin(N\beta)}{\sin\beta}\right]^2} , \tag{17.19}$$

where $\xi = x/T = \cos\beta$ and $\eta = y/T$.

17.1.4 The Quarter-Wave Stack

A quarter-wave stack, also known as a Bragg mirror, exhibits a one-dimensional photonic bandgap. One period consists of two regions with thickness and index of refraction (d_1, n_1) and (d_2, n_2), respectively (Fig. 17.2c). In the quarter-wave stack each region has an optical thickness of $\lambda/4$ (the wave accumulates in each region a phase of $\pi/2$) for a particular wavelength λ_0 or (midgap) frequency ω_0. Thus, the condition reads

$$n_1 d_1 = n_2 d_2 = \frac{\lambda_0}{4} = \frac{\pi}{2}\frac{c}{\omega_0} . \tag{17.20}$$

The transmission of an arbitrary two-layer cell is

$$t = \frac{T_{12}\exp(\mathrm{i}(p+q))}{1 + R_{12}\exp(2\mathrm{i}q)} , \tag{17.21}$$

where $p = n_1 d_1 \omega/c$ and $q = n_2 d_2 \omega/c$ are the phases accumulated in the two layers, respectively. The values of T_{12} and R_{12} are given from the Fresnel formula (9.7) as

$$T_{12} = \frac{4 n_1 n_2}{(n_1+n_2)^2} \tag{17.22a}$$

$$R_{12} = \frac{(n_1-n_2)^2}{(n_1+n_2)^2} . \tag{17.22b}$$

For the quarter-wave stack ($p = q = \pi/2$), we obtain for (17.21)

$$t = \frac{T_{12}\exp(\mathrm{i}\pi\tilde{\omega})}{1 + R_{12}\exp(\mathrm{i}\pi\tilde{\omega})} , \tag{17.23}$$

where $\tilde{\omega} = \omega/\omega_0$ is the frequency scaled to the midgap value.

The transmission of a single two-layer cell is

$$T = \frac{T_{12}^2}{1 - 2R_{12}\cos(\pi\tilde{\omega}) + R_{12}^2} , \tag{17.24}$$

and the Bloch phase is given by

$$\cos\beta = \xi = \frac{\cos(\pi\tilde{\omega}) - R_{12}}{T_{12}} \tag{17.25a}$$

$$\eta = \frac{\sin(\pi\tilde{\omega})}{T_{12}} \ . \tag{17.25b}$$

For the N-period quarter-wave stack the transmission is given by

$$T_N = \frac{1+\cos\beta}{1+\cos\beta + 2(R_{12}/T_{12})\sin^2(N\beta)} \ . \tag{17.26}$$

A bandgap forms. Within the bandgap, the density of modes is lowered, at the edges it is enhanced (Figs. 17.3 and 17.4). The transmission at midgap decreases exponentially as $\propto (n_i/n_j)^{2N}$, where $n_i < n_j$.

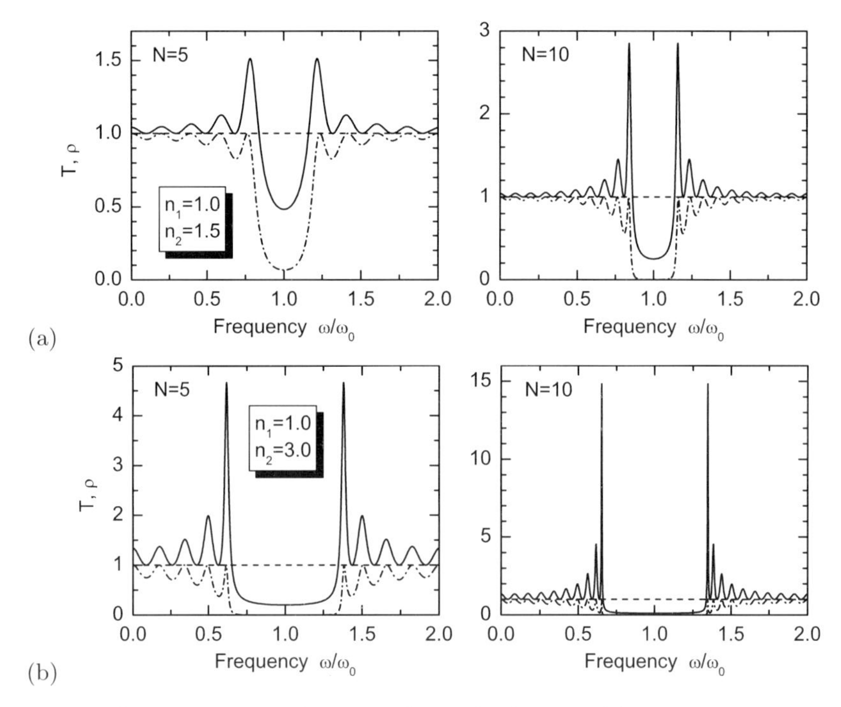

Fig. 17.3. Quarter-wave stack with indices of refraction (**a**) $n_1, n_2 = 1.0, 1.5$ and (**b**) 1.0, 3.0. *Solid lines*: dimensionless density of modes ρ_N (17.19), *dashed lines*: transmission T_N (17.26) for two different numbers of pairs $N = 5$ (*left panels*) and 10 (*right panels*) vs. the dimensionless frequency $\tilde{\omega}$

In the limit of large N the complete width $\Delta\tilde{\omega}$ of the bandgap is

$$\cos\left(\frac{\pi}{2}\Delta\tilde{\omega}\right) = 1 - 2\left(\frac{n_1 - n_2}{n_1 + n_2}\right)^2 \ . \tag{17.27}$$

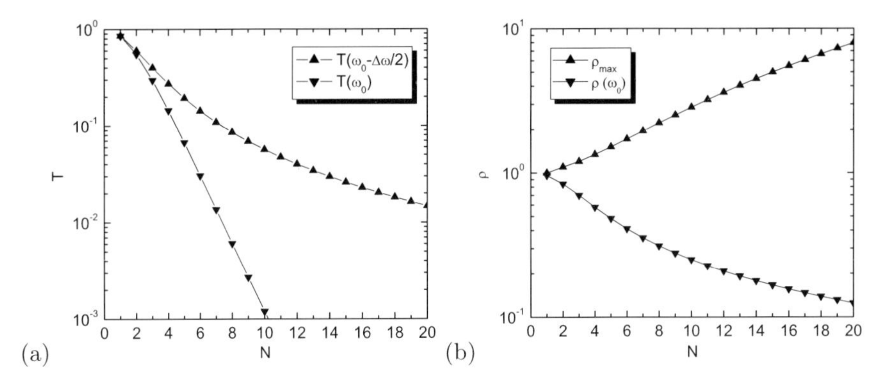

Fig. 17.4. Quarter-wave stack with indices of refraction n_1, n_2 = 1.0, 1.5: (**a**) Transmission T_N at midgap ($\tilde{\omega} = 1$, *down triangles*) and at the band edge ($\tilde{\omega} = 1 - \Delta\tilde{\omega}/2$, *up triangles*) vs. number of pairs N. (**b**) Dimensionless density of modes ρ_N at maximum near the band edge and at midgap vs. number of pairs N

If $|n_1 - n_2| \ll n_1 + n_2$, we find

$$\Delta\tilde{\omega} \approx \frac{4}{\pi} \frac{|n_1 - n_2|}{n_1 + n_2} \, . \tag{17.28}$$

The principle of the quarter-wave stack is scalable to frequencies other than visible light. As an example, a Mo/Si Bragg mirror with a period of 6.7 nm is shown in Fig. 17.5. Such a mirror works in the extreme UV and is used for soft X-ray optics, possibly in advanced lithography systems.

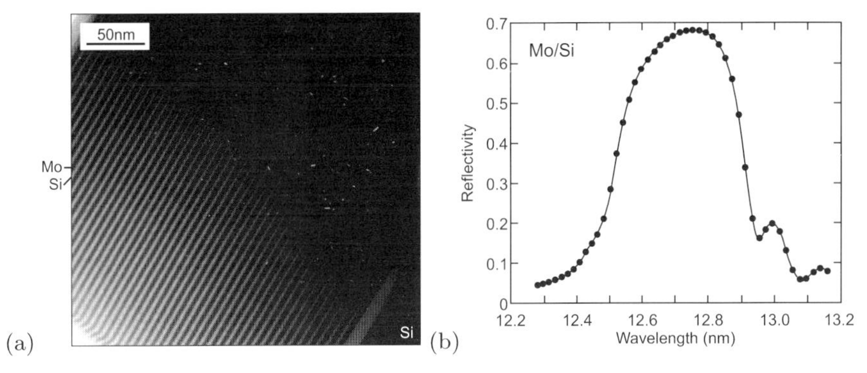

Fig. 17.5. (**a**) Cross-sectional TEM of Mo/Si superlattice with 2.7 nm Mo and 4.0 nm Si (period: 6.7 nm) on Si(001) substrate. From [470]. (**b**) Reflection spectrum for a SL with period of 6.5 nm and 88.5° angle of incidence. Data points are shown as *circles*, the *solid line* is a fit with a period of 6.45 nm. Adapted from [471]

17.1.5 Formation of a 3D Band Structure

For applications, 3D (or at least 2D) photonic bandgap structures are needed. Further details can be found in [472, 473]. In [474] planar, cylindrical and spherical Bragg mirrors are discussed.

Since we want a photonic bandgap that is present for *all* directions of propagation, a Brillouin zone with a shape close to a sphere is preferable. Then, the main directions are at similar k-values (Fig. 17.6). One of the best suited is the fcc lattice. Since the L point is centered at ≈14% lower frequency than the X point, the forbidden gaps for different directions must be, however, sufficiently wide to create a forbidden frequency band overlapping at all points along the surface of the Brillouin zone. For example, the bcc lattice has a Brillouin zone that is more asymmetric than that of the fcc lattice (cf. Fig. 3.31) and thus is less suited for the creation of a photonic bandgap.

Maxwell's equations (zero charge density) for monochromatic waves $\propto \exp(\mathrm{i}\omega t)$

$$\nabla \cdot \mathbf{D} = 0 \tag{17.29a}$$

$$\nabla \times \mathbf{E} = \mathrm{i}\frac{\mu\omega}{c}\mathbf{H} \tag{17.29b}$$

$$\nabla \times \mathbf{H} = \mathrm{i}\frac{\omega}{c}\mathbf{D} \tag{17.29c}$$

$$\nabla(\mu\mathbf{H}) = 0\,, \tag{17.29d}$$

together with $\mathbf{D}(\mathbf{r}) = \epsilon(\mathbf{r})\mathbf{E}(\mathbf{r})$ and $\mu = 1$ are combined into the wave equation

$$\nabla \times \left[\epsilon^{-1}(\mathbf{r})\nabla \times \mathbf{H}(\mathbf{r})\right] + \frac{\omega^2}{c^2}\mathbf{H}(\mathbf{r}) = 0\,. \tag{17.30}$$

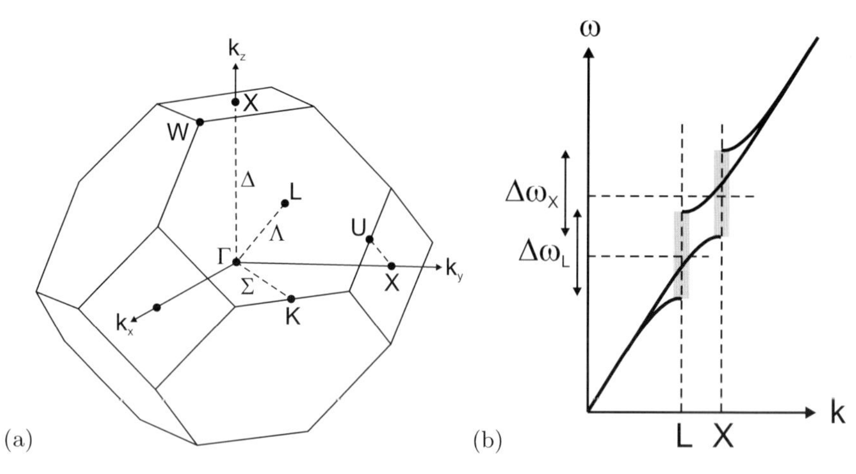

Fig. 17.6. The face-centered cubic Brillouin zone (**a**) in reciprocal space. Schematic forbidden gaps (**b**) at the L and X points

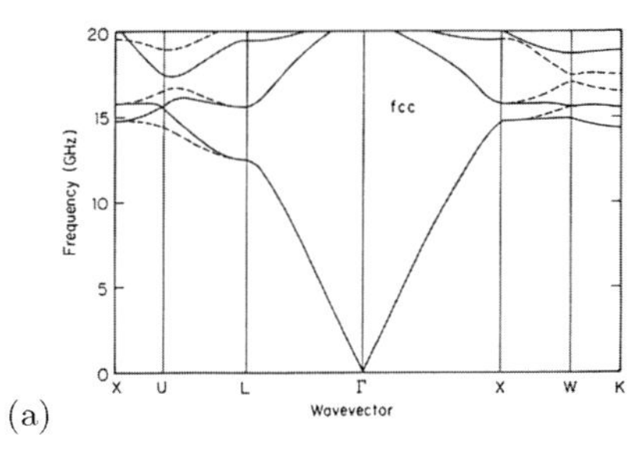

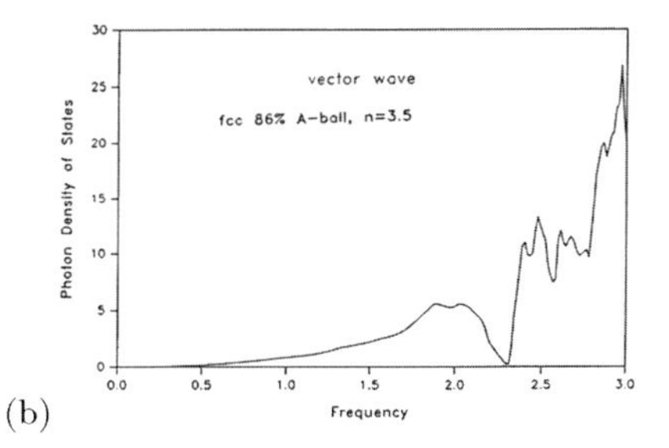

Fig. 17.7. (**a**) Calculated photonic bandgap structure along symmetry lines in the fcc Brillouin zone of a dielectric structure composed of air spheres in a dielectric background of refractive index $n = 3.5$. The filling ratio is 86% air and 14% dielectric material. *Dotted* and *solid lines* represent coupling to s- and p-polarized light, respectively. Reprinted with permission from [475], ©1990 APS. (**b**) Density of states for the band structure of part (**a**)

This equation is numerically solved for planar waves with wavevector $\mathbf{k}$. In the following, results are shown for various structures. In a fcc lattice of air spheres in a dielectric medium with $n = 3.6$ (a typical semiconductor), no bandgap can be achieved (Fig. 17.7a), only a pseudogap (Fig. 17.7b) appears.

In a diamond lattice (two fcc lattices shifted by $1/4\ \langle 111 \rangle$), a complete photonic bandgap is possible [475] (Fig. 17.8). Recently, a periodic array of spirals (Fig. 17.9) has been predicted to exhibit a large photonic bandgap [476]. Glancing-angle deposition [477] (GLAD) is a way to realize such structures. Another method to fabricate structures with arbitrary geom-

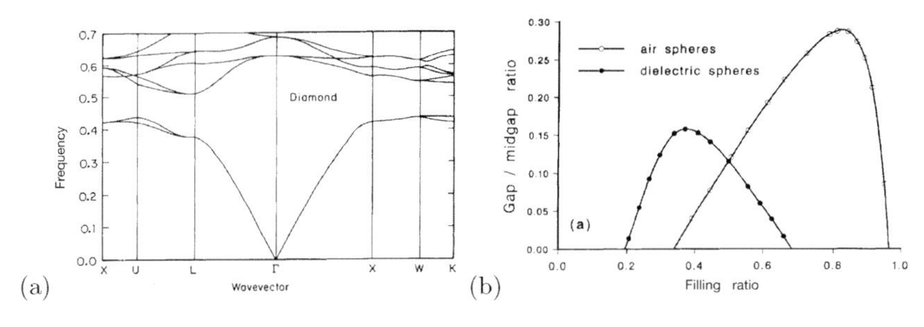

Fig. 17.8. (**a**) Calculated photonic band structure for a diamond dielectric structure consisting of overlapping air spheres in a dielectric material with $n = 3.6$. Filling ratio of air is 81%. The frequency is given in units of c/a, a being the cubic lattice constant of the diamond lattice and c being the velocity of light. (**b**) Gap-to-midgap frequency ratio for the diamond structure as a function of filling ratio for dielectric spheres $n = 3.6$ in air (*solid circles*) and air spheres in dielectric n. Optimal case: air spheres with 82% filling ratio. Reprinted with permission from [475], ©1990 APS

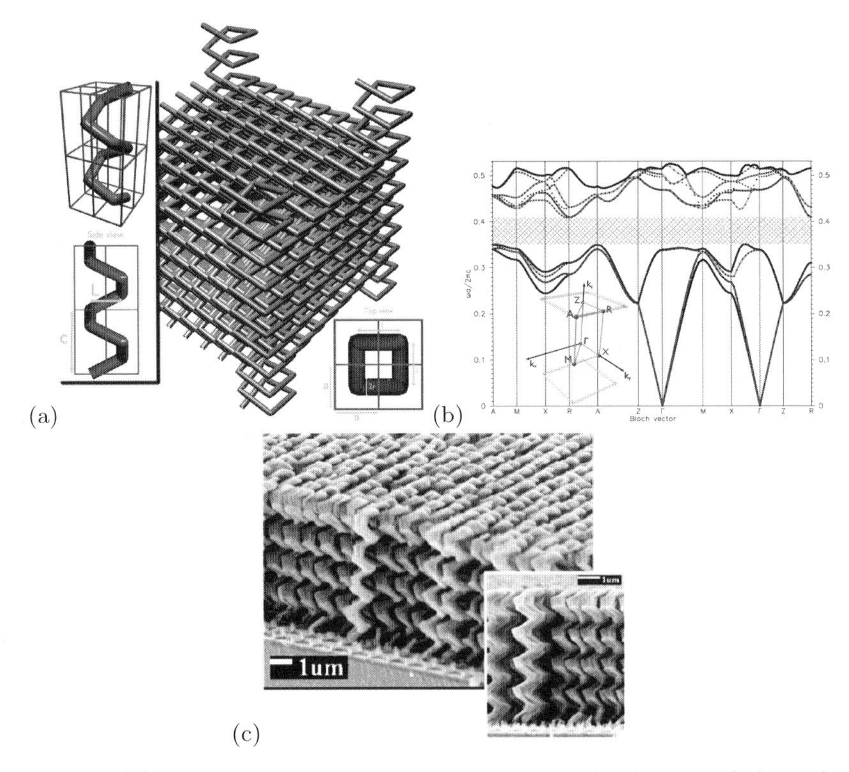

Fig. 17.9. (**a**) Tetragonal square-spiral photonic crystal. The crystal shown here has a solid filling fraction of 30%. For clarity, spirals at the corners of the crystal are highlighted with a different shade and height. The tetragonal lattice is characterized by lattice constants a and b. The geometry of the square spiral is illustrated in the insets and is characterized by its width, L, cylinder radius, r, and pitch, c. The *top left inset* shows a single spiral coiling around four unit cells. (**b**) Band structure for the direct structure crystal characterized by $[L,C,r]=[1.6,1.2,0.14]$ and a spiral filling factor $f_{\mathrm{spiral}} = 30\%$. The lengths are given in units of a, the lattice constant. The width of the PBG is 15.2% relative to the center frequency for background dielectric constant $\epsilon_{\mathrm{b}} = 1$ and spiral dielectric constant $\epsilon_{\mathrm{s}} = 11.9$. The positions of high-symmetry points are illustrated in the *inset*. Reprinted with permission from [476], ©2001 AAAS. (**c**) Oblique and edge views of a tetragonal square spiral structure grown using the GLAD (glancing-angle deposition) process. Reprinted with permission from [477], ©2002 ACS

etry within a material is two-photon lithography or two-photon holography. Another path to PBG structures are so-called inverted opals. First, a close-packed structure of spheres, e.g. monodisperse silica spheres, is fabricated by sedimentation or self-assembly. The gaps are filled with a high-index medium and the template is subsequently removed, e.g. by etching or dissolving. The resulting structure is shown in Fig. 17.10a. Such a structure has a photonic bandgap (Fig. 17.10b) if the refractive index is sufficiently high (> 2.85) [478].

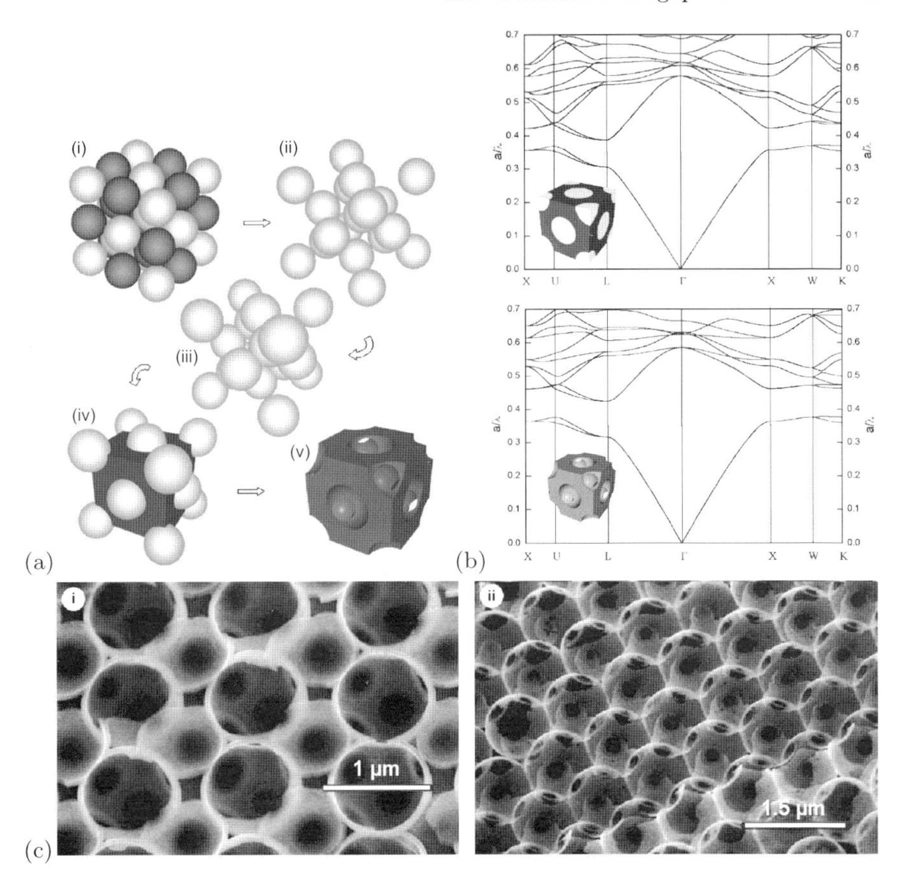

Fig. 17.10. (**a**) Cartoon showing, in five steps, the fabrication of an inverse diamond structure with a full photonic bandgap. First, (i) a mixed body-centered cubic lattice is assembled (ii) after which the latex sublattice is removed; (iii) then the structure is sintered to a filling fraction of ~50% after that (iv) silicon or germanium infiltration takes place and finally (v) silica elimination. (**b**) Photonic band diagrams of (upper panel) a silicon/silica composite diamond opal and (lower panel) made of air spheres in silicon resulting from the removal of the silica spheres from the former. The filling fraction for silicon is 50%. The inset shows the corresponding real space structures. Reprinted with permission from [479], ©2001 AIP. (**c**) SEM images of internal facets of silicon inverse opal: (i) (110) facet, (ii) (111) facet. Adapted from [480], reprinted with permission from Nature, ©2000 Macmillan Magazines Limited

17.1.6 Defect Modes

Similar to a perfect periodic atomic arrangement leading to the formation of the electronic band structure, a perfectly periodic dielectric structure leads to the photonic band structure. As we know from semiconductor physics, much of the interesting physics and numerous applications lie in defect modes, i.e.

localized electronic states due to doping and recombination at such centers. The equivalent in PBG structures are point defects (one unit missing) or line defects (a line of units, straight, bend or with sharp angles, missing). Such defects create localized states, i.e. regions for light localization. In the case of line defects we deal with waveguides that can be conveniently designed and could help to reduce the size of photonic and optoelectronic integrated circuits.

1D Model

We revisit our 1D scattering theory and create now a 'defect'. A simple defect is the change of the width of the center n_2-region in a quarter-wave stack. For the numerical example, we choose $N = 11$, $n_1 = 1$, $n_2 = 2$.

In Fig. 17.11, the transmission curves are shown for the undisturbed quarter-wave stack and the microcavity with $n_2 d_2^{\mathrm{center}} = 2\lambda_0/4 = \lambda_0/2$. A highly transmissive mode at $\omega = \omega_0$ arises that is quite sharp with $\Delta\omega = 3\times10^{-4}$. Thus, the quality factor Q, also called the Q-factor or finesse,

$$Q = \frac{\omega_0}{\Delta\omega} , \tag{17.31}$$

with ω_0 being the resonance frequency and $\Delta\omega$ being the linewidth, is 3.3×10^3 in this case.

If the thickness is varied (Fig. 17.12), the mode shifts away from the center. A similar scenario arises for higher-order $n_l/2$-cavities, e.g. $n_2 d_2^{\mathrm{center}} = 4\lambda_0/4 = \lambda_0$ (Fig. 17.13).

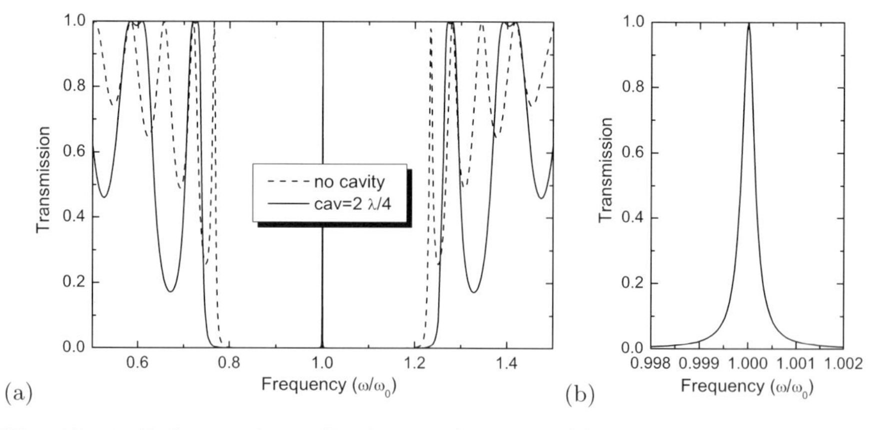

Fig. 17.11. Defect mode in 1D photonic bandgap: (**a**) Transmission of $N = 11$ quarter-wave stack exhibiting a photonic bandgap ($n_1 = 1$, $n_2 = 2$) (*dashed line*) and of microcavity (*solid line*) with center n_2-region of width $\lambda_0/2$ (instead of $\lambda_0/4$). (**b**) Relative width of mode is about 3×10^{-4}

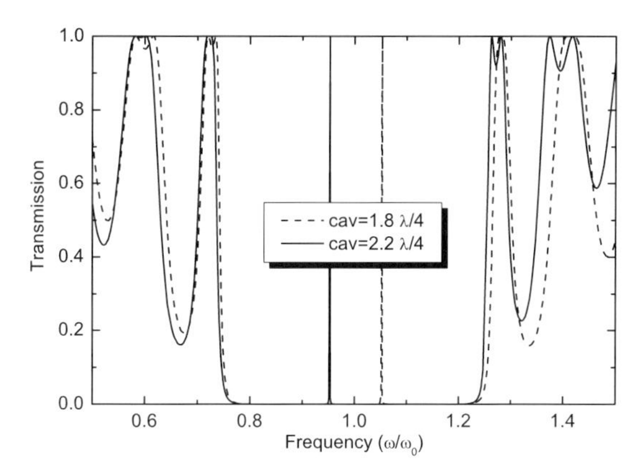

Fig. 17.12. Transmission of $N = 11$ quarter-wave stack ($n_1 = 1$, $n_2 = 2$) with center n_2-region of widths $1.8\,\lambda_0/4$ (*dashed line*) and $2.2\,\lambda_0/4$ (*solid line*)

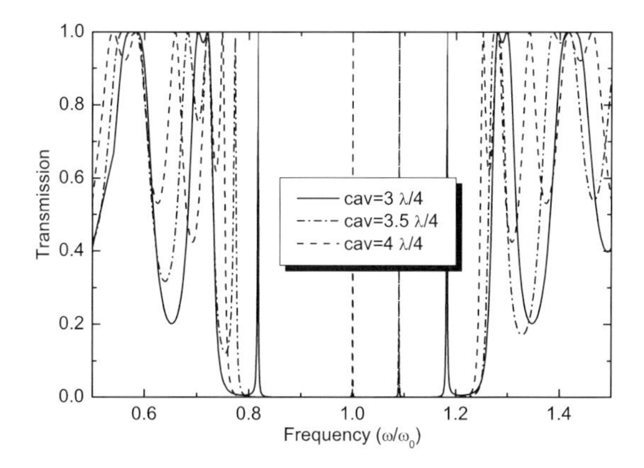

Fig. 17.13. Transmission of $N = 11$ quarter-wave stack ($n_1 = 1$, $n_2 = 2$) with center n_2-regions of widths $3\,\lambda_0/4$ (*solid line*), $3.5\,\lambda_0/4$ (*dash-dotted line*) and $4\,\lambda_0/4$ (*dashed line*)

2D or 3D Defect Modes

An example of 2D waveguiding is shown in Fig. 17.14. Point defects can be used for high-finesse wavelength filtering. Emitters surrounded by a photonic bandgap material with a defect mode can emit into the defect mode only, leading to spectrally filtered, highly directional emission.

17.1.7 Coupling to an Electronic Resonance

In a vertical-cavity surface-emitting lasers (see Sect. 20.4.14), an optical defect mode in a 1D dielectric structure is coupled to an electronic excitation, such as an exciton in a quantum well or dot. In the simplest picture, the

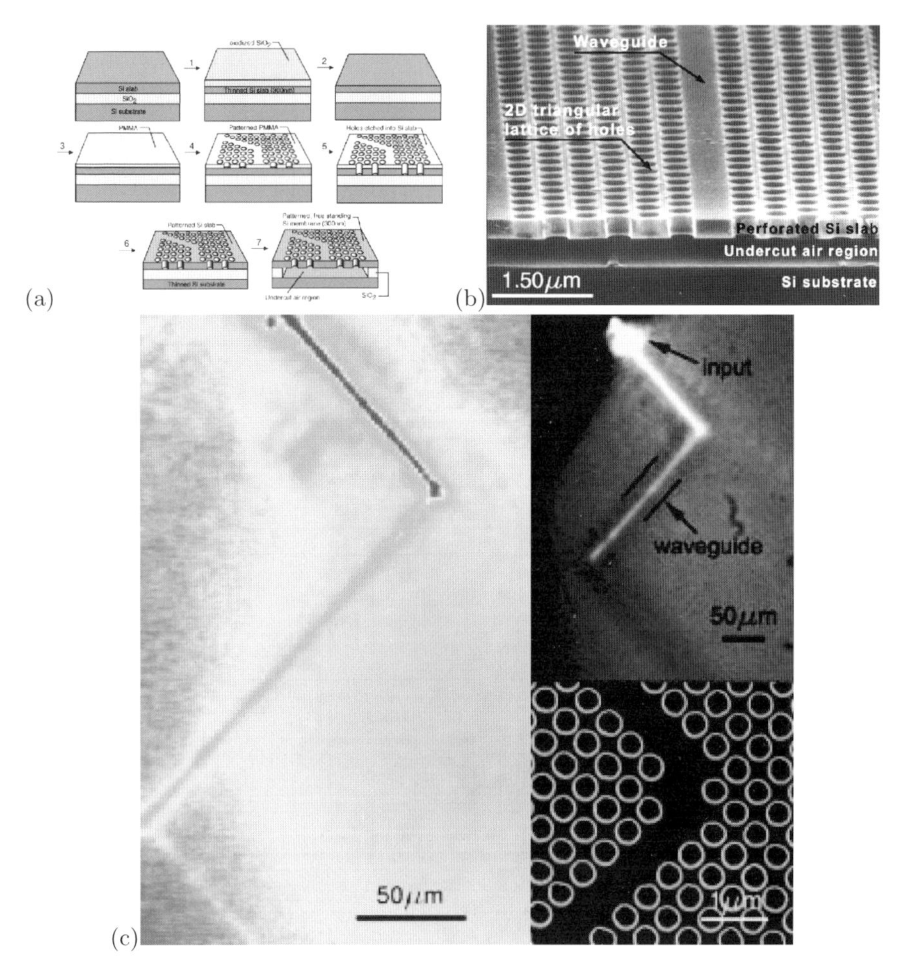

Fig. 17.14. 2D photonic bandgap waveguide structure. (**a**) Fabrication principle, (**b**) SEM image of the structure, (**c**) light guiding at a 90° bend. Reprinted with permission from [481], ©2000 AIP

oscillator must emit its radiation into the cavity mode since other modes do not exist in the Bragg band. Thus, the emission energy is given and fixed by the cavity mode. However, the photon mode (field oscillator) and the electronic oscillator form a coupled system that generally must be described using quantum electrodynamics. Energy is periodically exchanged between the two oscillators with the Rabi frequency. An analogous phenomenon is investigated in the field of atom–cavity interactions. A necessary condition for the observation of such an oscillation is that the radiation energy remains long enough in the cavity that can be expressed as [482, 483] (cf. (17.37))

$$\alpha d \gg 1 - R \approx \pi / Q \, , \tag{17.32}$$

where α is the absorption coefficient of the electronic transition, d is the length of the absorbing medium, R is the reflectivity of the cavity mirror and Q is the finesse of the cavity given in (17.31). This situation is called the *strong coupling* regime since it leads to anticrossing behavior of the cavity mode and electronic resonance. In the *weak coupling* regime for small absorption, the resonances cross (within their linewidth). For resonance, the emission intensity of the oscillator into the cavity mode is enhanced and its lifetime is reduced (Purcell effect), which is discussed in Sect. 17.2.2.

The transmission T of a Fabry–Perot cavity with two (equal and lossless) mirrors of transmission $T_\mathrm{m} = 1 - R_\mathrm{m}$ is given by

$$T(\omega) = \frac{T_\mathrm{m}^2 \exp\left(-2L\alpha(\omega)\right)}{\left|1 - R_\mathrm{m} \exp\left(\mathrm{i}2n^* L\omega/c\right)\right|^2} , \tag{17.33}$$

with the complex index of refraction $n^* = n_\mathrm{r} + \mathrm{i}\kappa = \sqrt{\epsilon}$ and $\alpha = 2\omega\kappa/c$ (cf. (9.53)). For an empty cavity, i.e. a (small) background absorption α_B and a background index of refraction $n_\mathrm{r} = n_\mathrm{B}$, the resonances occur when the phase shift $2n_\mathrm{B}L\omega/c$ is an integer multiple of 2π, i.e. for

$$\omega_\mathrm{m} = m\frac{\pi c}{n_\mathrm{B} L} , \tag{17.34}$$

with $m \geq 1$ being a natural number. In the vicinity of the resonance, i.e. for $\omega = \omega_\mathrm{m} + \delta\omega$, we can expand $\exp\left(2n_\mathrm{B}L\omega/c\right) \approx 1 + \mathrm{i}2n_\mathrm{B}L\delta\omega/c$ and obtain from (17.33) a Lorentzian for the transmission

$$T(\omega) \approx \frac{T_\mathrm{m}^2 \exp\left(-2L\alpha(\omega)\right)}{\left|1 - R_\mathrm{m}(1 + \mathrm{i}2n_\mathrm{B}L\,\delta\omega/c)\right|^2} = \frac{(T_\mathrm{m}/R_\mathrm{m})^2 \exp\left(2L\alpha(\omega)\right)}{(\delta\omega)^2 + \gamma_\mathrm{c}^2} . \tag{17.35}$$

The frequency width (HWHM) γ_c of the empty-cavity resonance is given by

$$\gamma_\mathrm{c} = \frac{1 - R'}{R'} \frac{c}{2n_\mathrm{B} L} , \tag{17.36}$$

where $R' = R_\mathrm{m} \exp\left(-2L\alpha\right)$. Thus, the decay rate (photon loss from the cavity) is proportional to $T_\mathrm{m} + \alpha_\mathrm{B} L$ if both terms are small. The quality factor of the cavity resonance m is given by

$$Q = \frac{\omega_\mathrm{m}}{2\gamma_\mathrm{c}} \approx \frac{m\pi}{1 - R} . \tag{17.37}$$

Now, the electronic resonance is put into the cavity leading to a change in the dielectric function to (cf. (C.11))

$$\epsilon = n_\mathrm{B}^2 \left[1 + \frac{f}{1 - \frac{\omega^2 + \mathrm{i}\omega\Gamma}{\omega_0^2}}\right] , \tag{17.38}$$

where the index of refraction due to the electronic resonance is given by $n(\omega) = \sqrt{\epsilon}$ and (C.13a,b). For resonance of the cavity mode and the electronic oscillator, i.e. $\omega_{\mathrm{m}} = \omega_0$, the solution for the cavity resonance condition $2n_{\mathrm{r}}\omega L/c = m2\pi$ is obtained, using (17.34), from

$$n_{\mathrm{r}}(\omega) = m\frac{\pi c}{\omega L} = n_{\mathrm{B}}\frac{\omega_{\mathrm{m}}}{\omega} . \tag{17.39}$$

A graphical solution (Fig. 17.15a) yields three intersections of the left and right hands of (17.39). The very high absorption at the central solution ($\omega = \omega_0$) results in very low transmission. The other two solutions[1] yield the frequencies of the coupled normal mode peaks. For $f \ll 1$, we use (C.13a) in (17.39) and find for the splitting $\pm\Omega_0/2$ of the two modes

$$\Omega_0^2 = f\omega_0^2 - \Gamma^2 . \tag{17.40}$$

This frequency is called the Rabi frequency. If the dielectric function of the oscillator is put into (17.33), the splitting is found to be

$$\Omega_0^2 = f\omega_0^2 - (\Gamma - \gamma_{\mathrm{c}})^2 . \tag{17.41}$$

A splitting will only be observable if $\Omega_0 \gg \Gamma, \gamma_{\mathrm{c}}$. If the two resonances ω_{c} and ω_0 are detuned by $\Delta = \omega_{\mathrm{c}} - \omega_0$, the splitting Ω of the transmission peaks shows the typical anticrossing behavior of two coupled oscillators

$$\Omega^2 = \Omega_0^2 + \Delta^2 . \tag{17.42}$$

In the experiment, typically the electronic resonance remains fixed at ω_0 and the cavity resonance is detuned by variation of the cavity length across the wafer (Fig. 17.15b). A detailed theory of cavity polaritons is given in [484]. The nonlinear optics of normal mode coupling in semiconductor microcavities is reviewed in [485].

17.2 Microscopic Resonators

17.2.1 Microdiscs

A microdisc is a cylindrical resonator with a thickness d that is small compared to the radius R. It can be fabricated from semiconductors and semiconductor heterostructures using patterning and material-selective etching. With underetching a mostly free-standing disc can be made that resides on a post (Fig. 17.16).

---

[1]These solutions only occur for sufficient oscillator strength $f > (\Gamma/\omega_0)^2$, i.e. in the strong coupling regime where $\Omega_0^2 > 0$. The absorption coefficient at ω_0 must be larger than $\Gamma n_\infty/c$.

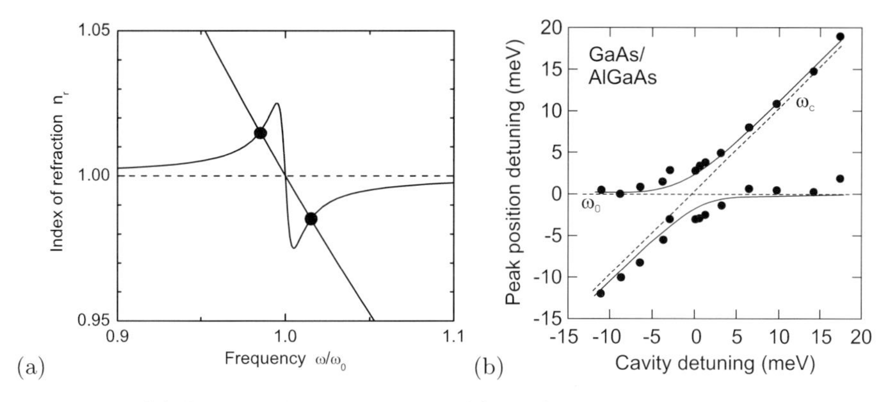

Fig. 17.15. (**a**) Graphical representation of (17.39) with the two solutions marked with *circles* for $n_\infty = 1$ (*dashed line*), $f = 10^{-3}$, $\Gamma/\omega_0 = 10^{-2}$ and $\omega_0 = \omega_m$. (**b**) Reflectivity peak positions (experimental data (*circles*) at $T = 5\,\mathrm{K}$) vs. cavity detuning $\omega_c - \omega_0$ for a cavity with two GaAs/AlGaAs Bragg mirrors (24/33 pairs for the front/bottom mirror) and five embedded quantum wells whose resonances are closely matched. *Solid lines* are a theoretical fit according to (17.42) with $\Omega_0 = 4.3\,\mathrm{meV}$. The *dashed lines* show the electronic resonance ω_0 and the cavity resonance ω_c. Part (**b**) adapted from [483]

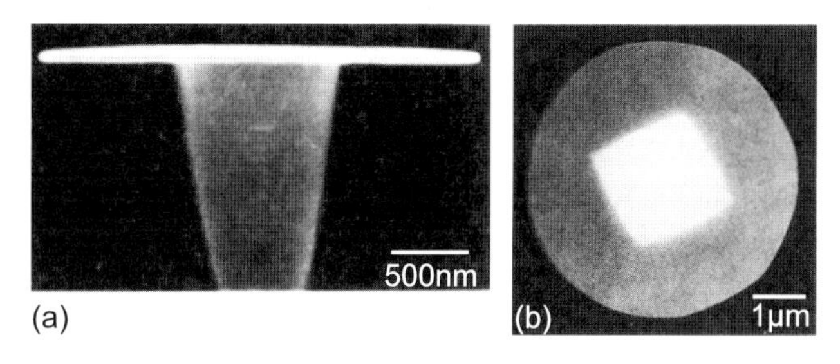

Fig. 17.16. (**a**) Side view of a 3-µm diameter disc containing one 10-nm InGaAs quantum well between 20-nm InGaAsP barriers standing on an InP pillar that has been selectively underetched using HCl. (**b**) Top view SEM image of a 5-µm diameter InGaAsP microdisc. The pedestal shape is a rhombus due to anisotropic etching of the HCl. Adapted from [486], reprinted with permission, ©1992 AIP

The coordinate system is (ρ, ϕ, z) with the z direction being perpendicular to the disc area. Typically, the disc is so thin that there is only one node along z. Solving the wave equation in this geometry [487], the modes are characterized by two numbers (m, l). m describes the number of zeros along the azimuthal direction ϕ with the field amplitude being proportional to $\exp(\pm im\phi)$. Thus, except for $m = 0$, the modes are simply degenerate. Modes with $\mathcal{E}_z = 0$ are called TE modes. This is the preferred polarization of emission. The number l denotes the number of zeros in the radial direction. Only for modes with $|m| = 1$, is the intensity nonzero on the axis, i.e. for $\rho = 0$. All other modes have vanishing intensity in the disc center.

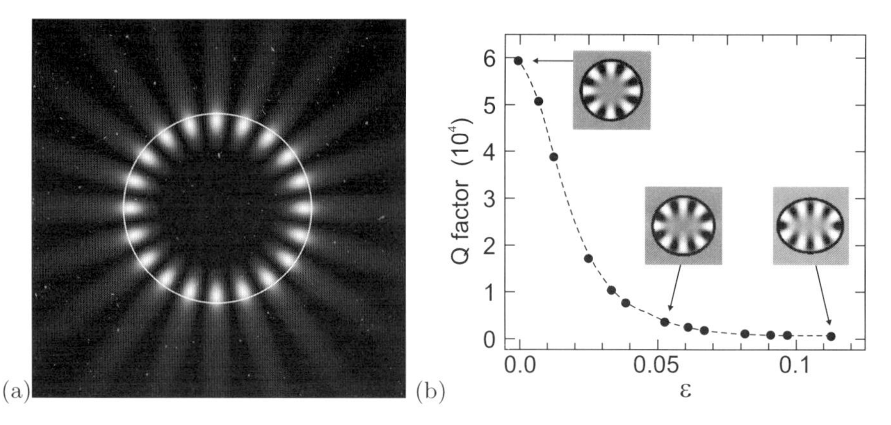

Fig. 17.17. (**a**) Field intensity for whispering gallery mode $(10, 0)$ (TM-polarized) for a circle with 1 µm radius (shown as *white line*) and $n = 1.5$. The image size is $4 \times 4\,\mu\mathrm{m}^2$. (**b**) Theoretical quality factor of a 2-µm InP microdisc as a function of the deformation parameter (17.44). The *insets* show (8,0) whispering gallery modes at a wavelength of 1.55 µm for $n = 3.4$. Part (**b**) adapted from [489]

The light intensity in whispering gallery modes is preferentially concentrated along the circumference of the disc as shown in Fig. 17.17a. Since the light can only escape via evanescent waves, the light is well 'captured' in such a mode. The Q-factor (17.31) is extremely high and takes values of several 10^4. In order to couple light out of such a disc, deformed resonators, e.g. with a defect in the form of protrusions [488], were devised. Deformed resonators are discussed in more detail in the next section.

17.2.2 Purcell Effect

According to Fermi's golden rule (17.1), the probability of an optical transition depends on the density of available optical modes (final states). If the density of modes is enhanced compared to its vacuum value (17.2) at a resonance of an optical cavity, the lifetime of the electronic state decreases by the Purcell factor [490]

$$F_{\mathrm{P}} = \frac{3}{4\pi^2} Q \frac{(\lambda/n)^3}{V} , \tag{17.43}$$

where n is the refractive index of the medium, Q is the quality factor of the cavity resonance and V is the effective mode volume.[2] Experiments on the emission of quantum dots (that generally provide small absorption and thus allow for the weak coupling regime) in etched micropillars containing a microcavity (Fig. 17.18a) have shown that indeed the luminescence decay

[2] V is given by the spatial integral of the vacuum field intensity for the cavity mode, divided by its maximum value.

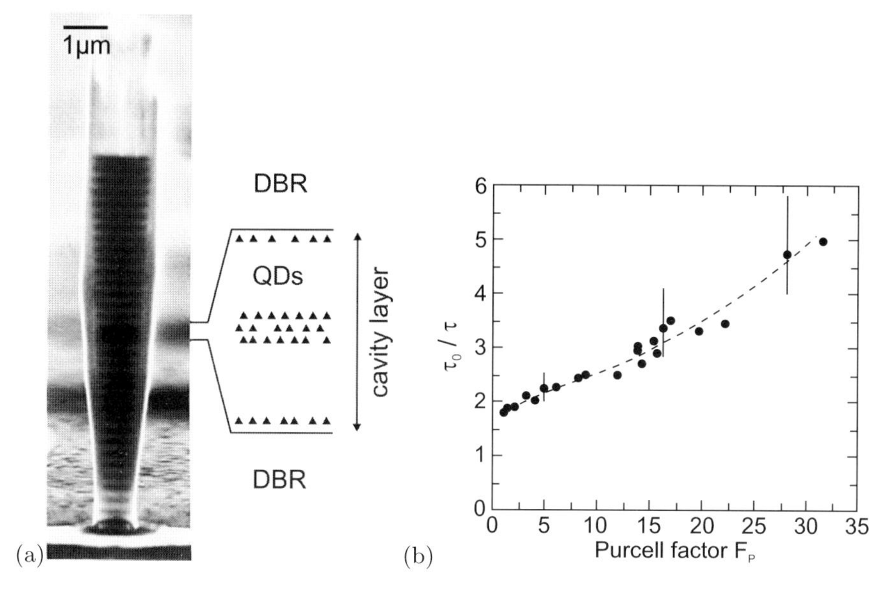

Fig. 17.18. (**a**) Micropillar with MBE-grown GaAs/AlAs DBRs and a cavity containing five layers of InAs quantum dots as indicated. The pillar has been prepared by reactive ion etching. Reprinted with permission from [491], ©1998 APS. (**b**) Experimental decay time τ of on-resonance quantum dot luminescence scaled by off-resonance lifetime $\tau_0 = 1.1\,$ns (close to lifetime in a QD in bulk) for a variety of micropillars with different Purcell factors F_{P}. The *error bars* correspond to the measurement accuracy of the decay time ($\pm 70\,$ps), the *dashed line* is a guide to the eye. Adapted from [491]

is faster for cavities with large Purcell factor (Fig. 17.18b) [491]. The resonance of cavity mode and emitter leads to an enhanced emission intensity as shown in Fig. 17.19 for the exciton emission of a single quantum dot in a microdisc [492].

17.2.3 Deformed Resonators

The whispering gallery modes in circular (or spherical) cavities are long-lived and emission goes into all angles. Light escape is based only on the exponentially slow process of evanescent leakage (neglecting disorder effects such as surface roughness). In order to overcome the isotropic light emission, the resonator needs to be deformed. This can be accomplished with an ellipsoidal shape, i.e.

$$r(\phi) = R\,[1 + \epsilon \cos \phi] \;, \tag{17.44}$$

where $1 + 2\epsilon$ is the aspect ratio of the ellipse. The increased radiation leads to a decrease of the Q-factor as shown in Fig. 17.17b. Also, a new decay

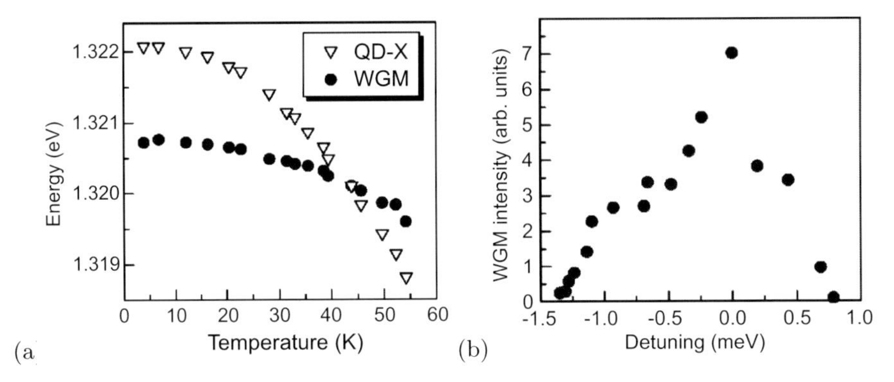

Fig. 17.19. (**a**) Temperature dependence of the energy positions of the whispering gallery mode (WGM) of a 5-µm diameter AlGaAs/GaAs microdisc ($Q = 6500$) and the single-exciton resonance of a single InAs quantum dot contained within the disc. (**b**) Intensity of WGM mode as a function of the detuning $E_{\mathrm{WGM}} - E_{\mathrm{QD-X}}$ from the QD single exciton resonance. The excitation density was 15 Wcm^{-2} for all data. Adapted from [492]

process, refractive escape, becomes possible. A ray that is initially in a whispering gallery trajectory diffuses in phase space until finally an angle smaller than the critical angle for total reflection (9.3) is reached. The ray dynamics becomes partially chaotic [493].

One other possible deformation of the circular disc geometry is a 'flattened quadrupole' as shown in Fig. 17.20a. This shape can be parameterized by a deformation parameter ϵ and the angle-dependent radius $r(\phi)$ given by

$$r(\phi) = R\left[1 + 2\epsilon \cos^2(2\phi)\right]^{1/2} . \tag{17.45}$$

For small deformation, the whispering gallery modes become chaotic and exhibit preferred emission along the long axis of the resonator (Fig. 17.20b). For larger deformations ($\epsilon \geq 0.14$), a stronger and qualitatively different directionality occurs in the shape of a bow-tie [494] as shown in Fig. 17.20c. The optical laser power extracted from deformed resonators was found to increase exponentially with ϵ; for $\epsilon = 0.2$ it was 50 times larger than for the circular resonator.

Another modification that can be applied to the microdisc in order to increase outcoupling of light, is the spiral resonator [495] as shown in Fig. 17.21a. The radius is parameterized by

$$r(\phi) = R\left[1 + \frac{\epsilon}{2\pi}\phi\right] . \tag{17.46}$$

The experimental emission pattern is displayed in Fig. 17.21b. It exhibits a maximum along the direction of the tangent at the radius step. The simulated near-field intensity of such an emission mode is shown in Fig. 17.21c. In a spiral laser, ray dynamics is also chaotic [497].

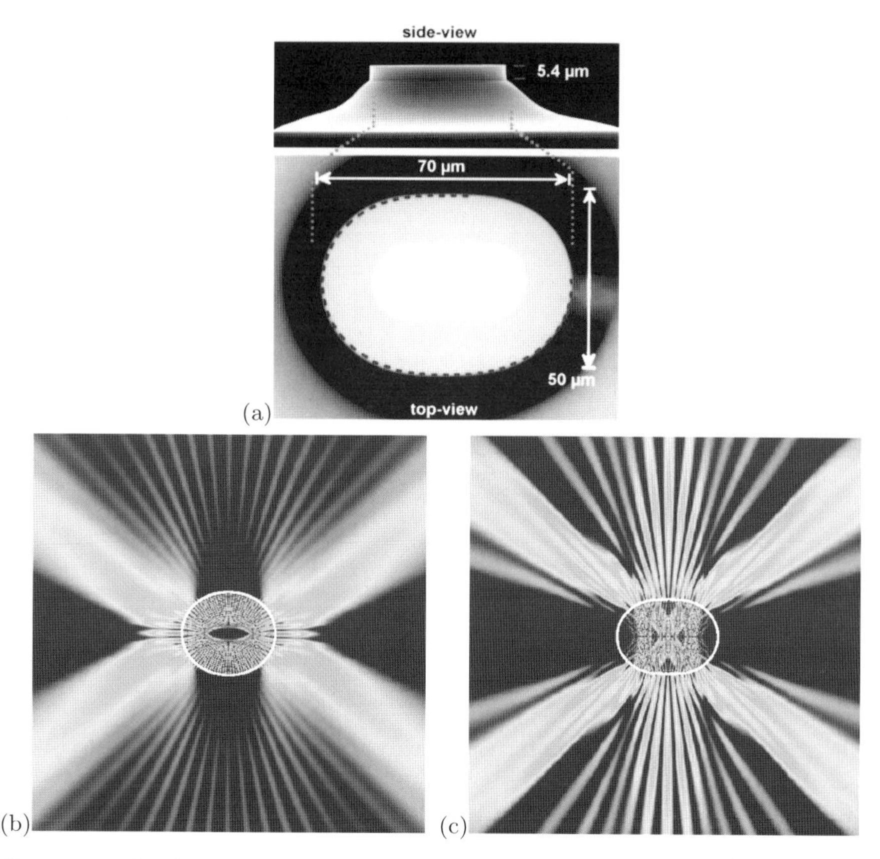

Fig. 17.20. (**a**) SEM image of a quadrupolar cylinder laser with deformation parameter $\epsilon \approx 0.16$ on a sloped InP pedestal. The *light grey area* in the top view is the electrical contact. (**b**) Simulated near-field intensity pattern of a chaotic whispering gallery mode for $\epsilon = 0.06$ and $n = 3.3$. (**c**) Simulated near-field intensity pattern of a bow-tie mode for $\epsilon = 0.15$. The length of the minor axis for (**b**) and (**c**) is 50 µm. Reprinted with permission from [494], ©1998 AAAS

17.2.4 Hexagonal Cavities

Hexagonal cavities develop, e.g., in microcrystals of wurtzite semiconductors (with the c-axis along the longitudinal axis of the pillar). In Fig. 17.22a, a ZnO tapered hexagonal resonator (needle) is shown. Whispering gallery modes modulate the intensity of the green ZnO luminescence [499].[3] In a simple plane-wave model, the resonance condition is given by

[3]We note that besides the green luminescence as in Fig. 10.12, an unstructured green band also occurs that is observed here. Its origin may be linked to the oxygen vacancy [498].

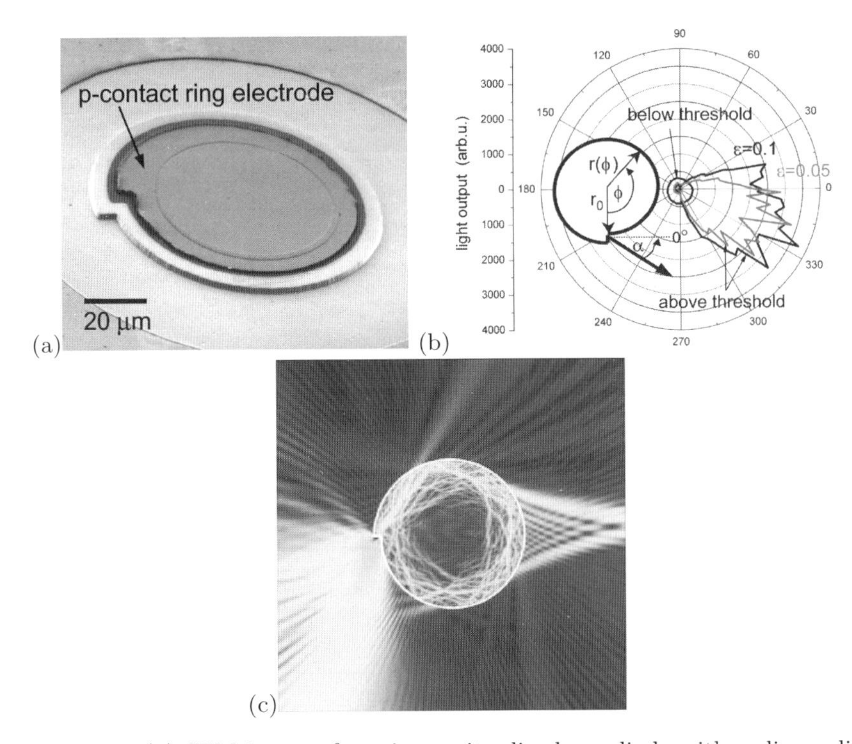

Fig. 17.21. (**a**) SEM image of a microcavity disc laser diode with a disc radius of 50 µm. The p-contact ring electrode defines the areas through which carriers are injected into the microdisc and where stimulated emission can take place. (**b**) Radial distribution of the light output from the spiral-shaped microdisc laser diode measured below and above threshold. The radius of the spiral microdisc was $r_0 = 250\,\mu\mathrm{m}$ and the deformation parameters were $\epsilon = 0.05$ (*grey*) and $\epsilon = 0.10$ (*black*). An emission beam at an angle of $\alpha = 0°$ corresponds to a direction normal to the notch surface as shown in the *inset*. Below the laser threshold, the emission pattern is essentially isotropic and independent of the deformation parameter. Above the threshold, directional emission is clearly observed with the emission direction at a tilt angle $\alpha \approx 25°$. The measured divergence angle of the far-field pattern is $\sim 75°$ for $\epsilon = 0.10$ and $\sim 60°$ for $\epsilon = 0.05$. Reprinted with permission from [496], ©2004 AIP. (**c**) Simulated near-field intensity pattern of an emission mode with $nkR \approx 200$ for deformation $\epsilon = 0.10$. Reprinted with permission from [495], ©2003 AIP

$$6R_{\mathrm{i}} = \frac{hc}{nE}\left[N + \frac{6}{\pi}\arctan\left(\beta\sqrt{3n^2-4}\right)\right] , \tag{17.47}$$

where R_{i} is the radius of the inner circle (Fig. 17.22d), n is the index of refraction, N is the mode number and β is given by $\beta_{\mathrm{TM}} = 1/n$ ($\beta_{\mathrm{TE}} = n$) for TM (TE) polarization, respectively. Due to birefringence, $n_{||}$ ($n_{\perp}$) has to be used as the index of refraction for TM (TE) polarization. A detailed numerical

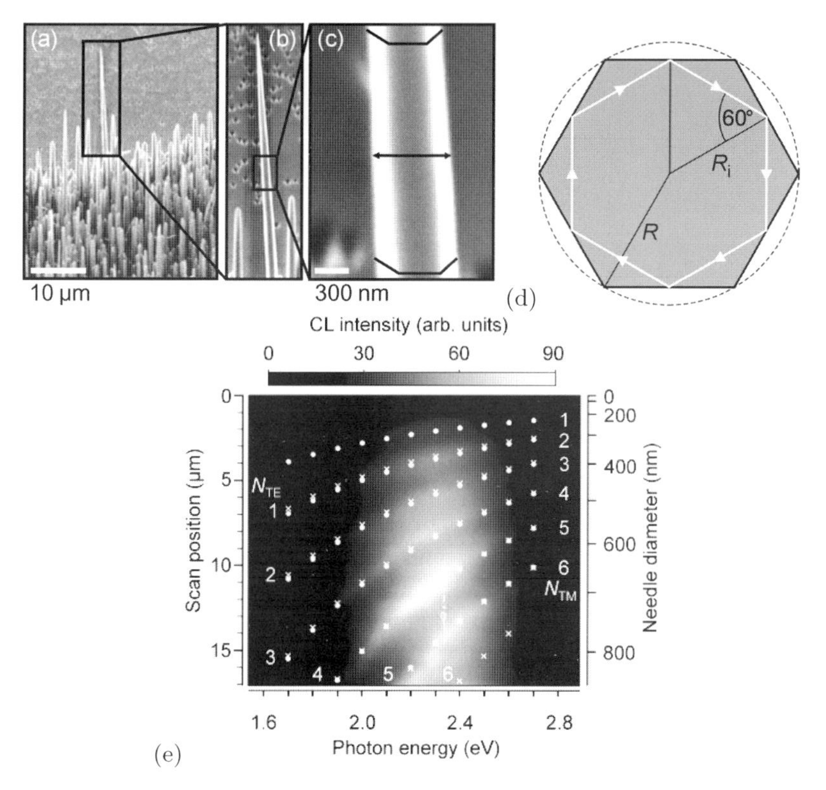

Fig. 17.22. (**a**)–(**c**) SEM images of ZnO nanoneedle fabricated by pulsed laser deposition. (**d**) Schematic geometry of cross-sectional plane. R_i (R) is the radius of the incircle (circumscribing circle). The circumference of the inscribed *white hexagon*, representing the path of a whispering gallery mode, has a length of $6R_i$. (**e**) Two-dimensional plot of spectra recorded along a linescan along the needle's longitudinal axis. The left vertical axis shows the linescan position x, the right one refers to the respective needle diameter D. The spectral maxima, i.e. the measured WGM energies, appear as bright belts going from the bottom left corner to the right upper one. With decreasing diameter, all resonances shift systematically to higher energies. The *white dots* give theoretical TM-resonance energy positions obtained from (17.47), *white crosses* give the same for TE-polarization. Reprinted with permission from [499], ©2004 APS

treatment can be found in [500]. From the energy positions of the TE and TM resonances, the birefringence of the resonator can be determined [501].

A $N = 26$ whispering gallery mode of a hexagonal resonator is shown in Fig. 17.23c,d. The 6-fold symmetric emission stems from the edges of the hexagon. While whispering gallery resonators have typically mode numbers $N \gg 1$, in such hexagonal resonators the whispering gallery modes could be followed down to $N = 1$ [499] as shown in Fig. 17.22a,b,e.

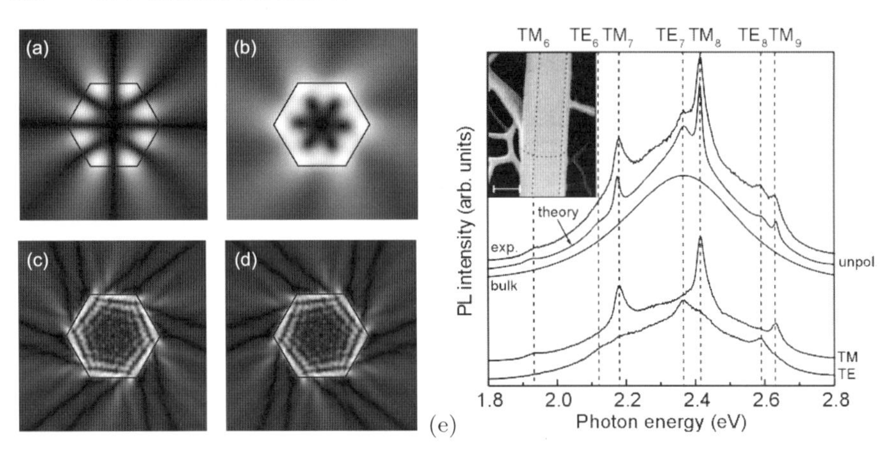

Fig. 17.23. Simulated near-field intensity pattern of modes in a cavity with hexagonal cross section (absolute value of electric field in linear grey scale): Modes ($N = 4$) with (**a**) symmetry $-a$ and (**b**) mode 4+ (nomenclature from [502]) for $n = 2.1$ and $kR = 3.1553 - \mathrm{i}0.0748$. Modes (**c**) 26− and (**d**) 26+ for $n = 1.466$ and $kR = 22.8725 - \mathrm{i}0.1064$. The displayed modes have a chiral pattern. Emission originates mostly from the corners. (**e**) Micro-photoluminescence spectra of a single ZnO nanopillar. The three topmost curves are unpolarized. The curve labeled 'bulk' shows the unmodulated luminescence of the green luminescence in bulk. The line labeled 'exp.' shows the experimental μ-PL spectrum of the investigated nanopillar. The experimental spectra recorded for TM- and TE-polarization, respectively, are shown in the lowest two curves. The curve labeled 'theory' displays the theoretical luminescence spectra. *Dashed vertical lines* are guides to the eye referring to the spectral position of the dominating WGMs. The *inset* shows a SEM image of the investigated pillar, the scale bar has a length of 500 nm. The *dotted lines* show the position of the edges of the hexagonal resonator obtained from topography contrast

Part II

Applications

18 Diodes

18.1 Introduction

One of the simplest[1] semiconductor devices is the diode. It is a so-called two-terminal device, i.e. a device with two leads. The most prominent property of a diode is the rectifying current–voltage (I–V) characteristic. This function was initially realized with vacuum tubes (Fig. 18.1); a heated filament emits electrons that are transferred through vacuum to the anode if it is on a positive potential. The semiconductor diode technology led to a tremendous miniaturization, integration with other devices (in planar technology) and cost reduction.

We distinguish[2] *unipolar* diodes, for which the *majority* carriers cause the effects (e.g. metal–semiconductor diodes), and *bipolar* diodes in which *minority* carriers play the decisive role, e.g. in the pn-junction diode.

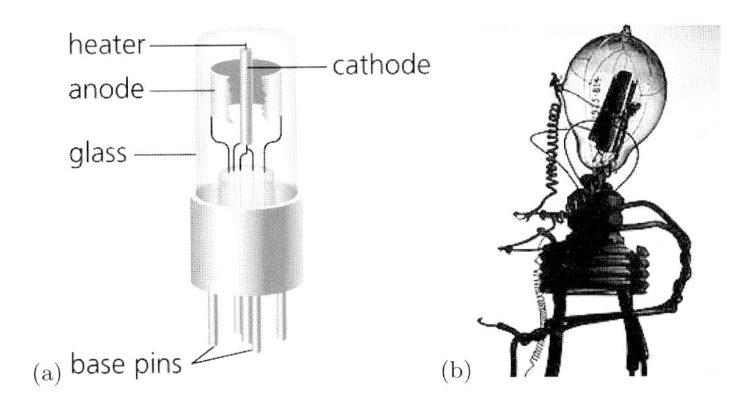

Fig. 18.1. (**a**) Schematic image of a vacuum diode. The electron current flows from the heated cathode to the anode when the latter is at a positive potential. (**b**) John A. Fleming's first diode 'valve', 1904

[1]The simplest device is a resistor made from a homogeneous piece of semiconductor, used, e.g., as a part of an integrated circuit or as a photoresistor as discussed in Sect. 19.2.

[2]This distinction is not only made for diodes but also many other semiconductor devices such as transistors, see Chap. 21.

18.2 Metal–Semiconductor Contacts

The metal–semiconductor contact was investigated in 1874 by F. Braun (see Sect. 1.1). For metal sulfides, e.g. $CuFeS_2$, he found nonohmic behavior. We remark here that we treat first metal–semiconductor contacts with rectifying properties. Later it becomes understandable that metal–semiconductor contacts can also be used as ohmic contacts, i.e. contacts with a very small contact resistance. Rectifying metal–semiconductor contacts are also called Schottky diodes. A very important variation are metal–insulator–semiconductor diodes for which an insulator, mostly an oxide, is sandwiched between the metal and the semiconductor. Such diodes are treated in Sect. 18.3. Reviews on Schottky diodes can be found in [545–548].

18.2.1 Band Diagram in Equilibrium

The metal and the semiconductor have generally different positions of the Fermi levels relative to the vacuum level. When the metal is in contact with the semiconductor, charges will flow in such a way that in thermodynamic equilibrium the Fermi level is constant throughout the structure.[3] In the following we treat two limiting cases: The contact of a metal with a semiconductor without any surface states (Schottky–Mott model) and a contact where the semiconductor has a very high density of surface states (Bardeen model).

The position of the Fermi level in the metal is given by the work function $W_{\mathrm{m}} = -e\phi_{\mathrm{m}}$ that is shown in Fig. 18.2 for various metals. The work function reflects the atomic shell structure; minima of the work function exist for group-I elements. The work function is the energy difference between the vacuum level (an electron is at rest in an infinite distance from the metal surface) and the metal Fermi level. Since the electron density in the metal conduction band is very high, the position of the metal Fermi level does not change considerably when charge is exchanged between the metal and the semiconductor.

Ideal Band Structure

When the metal and the semiconductor are not in contact (Fig. 18.3a), the metal is characterized by its work function $W_{\mathrm{m}} = E_{\mathrm{vac}} - E_{\mathrm{F}} = -e\phi_{\mathrm{m}}$ and the semiconductor by its electron affinity χ_{sc}. $E_{\mathrm{vac}} - E_{\mathrm{C}} = -e\chi_{\mathrm{sc}}$ is defined as the energy difference between the vacuum level and the conduction-band edge.[4] First, we assume that $|\phi_{\mathrm{m}}| > |\chi_{\mathrm{sc}}|$. For an n-type semiconductor, the

[3] This situation is similar to the heterostructure interface (Sect. 11.5.3), with the metal, however, having a very short screening length.

[4] The use of the Fermi level for the semiconductor here is not suitable since it depends strongly on the doping and temperature.

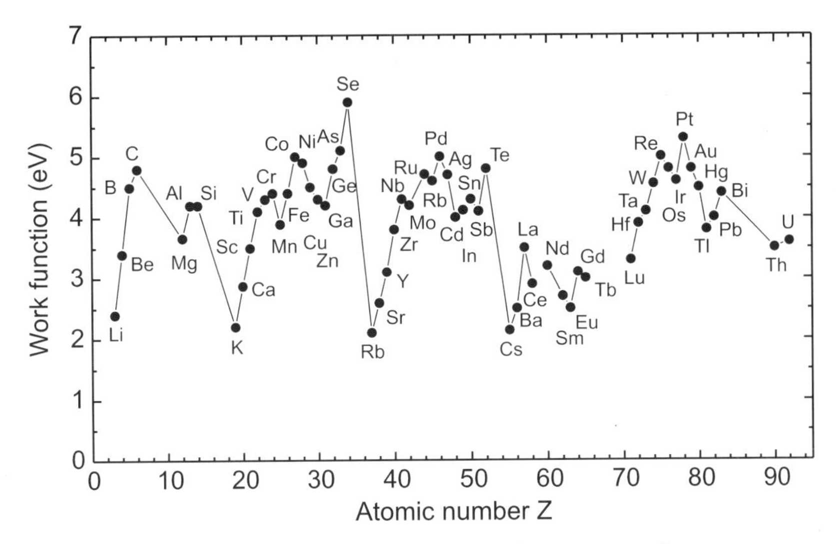

Fig. 18.2. Work function W_m of various metals

Table 18.1. Values of the work function W_m of various metals

Z	element	W_m (eV)	Z	element	W_m (eV)	Z	element	W_m (eV)
3	Li	2.4	37	Rb	2.1	64	Gd	3.1
4	Be	3.4	38	Sr	2.59	65	Tb	3.0
5	B	4.5	39	Y	3.1	66	Dy	-
6	C	4.8	40	Zr	3.8	67	Ho	-
12	Mg	3.66	41	Nb	4.3	68	Er	-
13	Al	4.2	42	Mo	4.2	69	Tm	-
14	Si	4.2	44	Ru	4.71	70	Yb	-
19	K	2.2	45	Rh	4.6	71	Lu	3.3
20	Ca	2.87	46	Pd	5.0	72	Hf	3.9
21	Sc	3.5	47	Ag	4.7	73	Ta	4.1
22	Ti	4.1	48	Cd	4.0	74	W	4.55
23	V	4.3	49	In	4.12	75	Re	5.0
24	Cr	4.4	50	Sn	4.3	76	Os	4.8
25	Mn	3.89	51	Sb	4.1	77	Ir	4.6
26	Fe	4.4	52	Te	4.8	78	Pt	5.3
27	Co	5.0	55	Cs	2.14	79	Au	4.8
28	Ni	4.9	56	Ba	2.5	80	Hg	4.49
29	Cu	4.5	57	La	3.5	81	Tl	3.8
30	Zn	4.3	58	Ce	2.9	82	Pb	4.0
31	Ga	4.2	59	Pr	-	83	Bi	4.4
32	Ge	4.8	60	Nd	3.2	90	Th	3.5
33	As	5.1	62	Sm	2.7	92	U	3.6
34	Se	5.9	63	Eu	2.5			

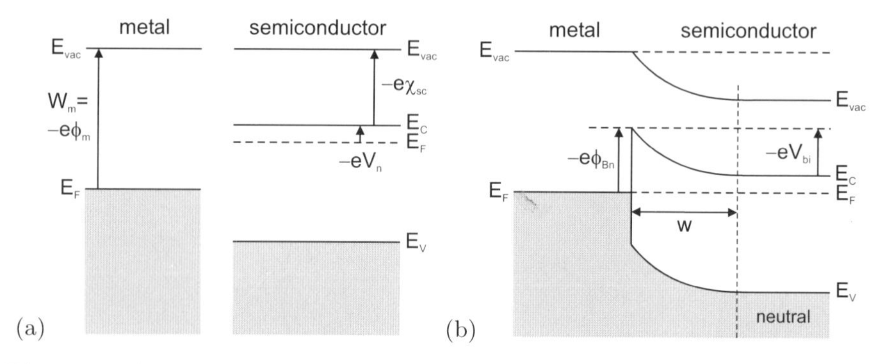

Fig. 18.3. Schematic band structure of a metal–semiconductor junction that is dominated by bulk properties of the semiconductor. (**a**) no contact, (**b**) metal and semiconductor in contact. w denotes the width of the depletion layer. Outside the depletion layer the semiconductor is neutral. $\phi_{\mathrm{B,n}}$ denotes the Schottky barrier height, V_{bi} denotes the built-in voltage

energy difference between the Fermi level and the conduction band is denoted as $-eV_{\mathrm{n}} = E_{\mathrm{C}} - E_{\mathrm{F}}$ ($V_{\mathrm{n}} < 0$ for nondegenerate semiconductors). Thus, the position of the semiconductor Fermi level is given as $E_{\mathrm{F}} = E_{\mathrm{vac}} + e(\chi_{\mathrm{sc}} + V_{\mathrm{n}})$.

The potential difference $\phi_{\mathrm{m}} - (\chi_{\mathrm{sc}} + V_{\mathrm{n}})$ is called the contact potential. If the metal and semiconductor are connected to each other via a conductive link,[5] the Fermi levels will equilibrate. For the case of Fig. 18.3 electrons will flow from the semiconductor to the metal. The negative surface charge of the metal is compensated by a positive charge (due to D^+) in the semiconductor in the vicinity of the surface. If the metal to semiconductor distance is reduced to zero (and surface effects such as nonmatching bonds, surface states, etc., play no role), a (Schottky) barrier of height ϕ_{Bn}

$$\phi_{\mathrm{Bn}} = \phi_{\mathrm{m}} - \chi_{\mathrm{sc}} \tag{18.1}$$

forms at the interface. The subscript 'n' stands for the contact on an n-type semiconductor. In the semiconductor there exists a positively charged region that is called the *depletion layer* or space-charge layer. Its extension (w in Fig. 18.3d) and properties will be discussed in Sect. 18.2.2.

For a contact on a p-type semiconductor the barrier ϕ_{Bp} (to the valence band) is (see Fig. 18.4a)

$$\phi_{\mathrm{Bp}} = -\frac{E_{\mathrm{g}}}{e} + \phi_{\mathrm{m}} - \chi_{\mathrm{sc}} \,. \tag{18.2}$$

Between the surface of the metal and the bulk part of the semiconductor there is a potential drop

[5] This consideration is a gedanken experiment and not the metal–semiconductor contact itself.

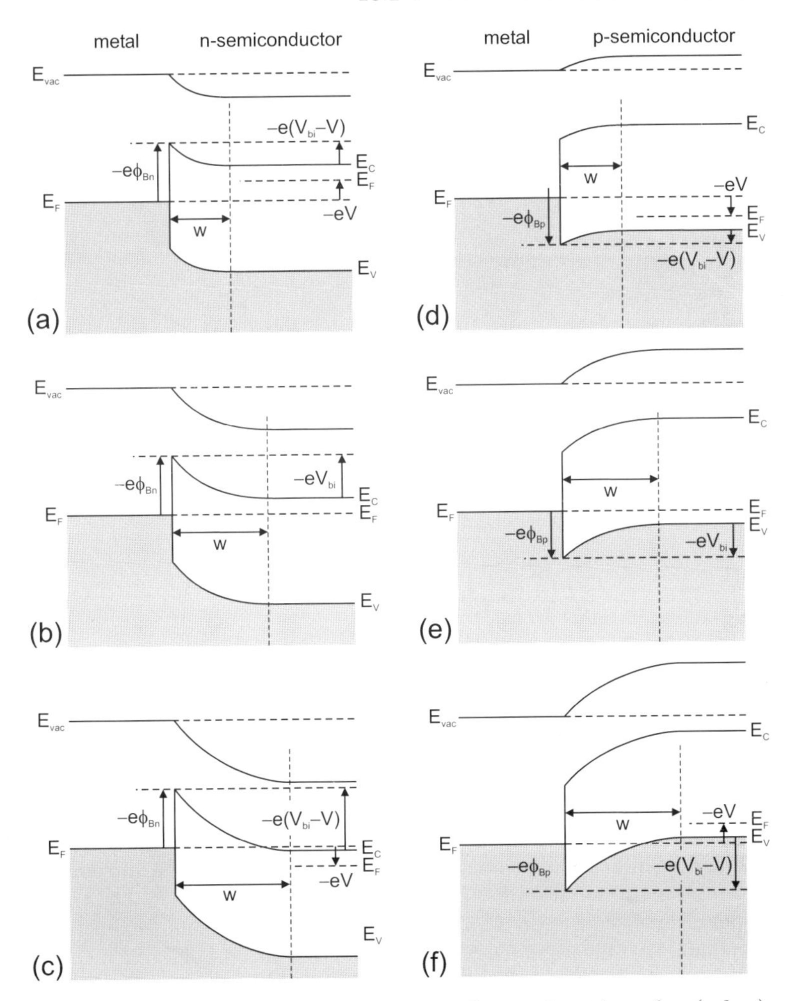

Fig. 18.4. Band structures of metal–semiconductor junctions for (**a**,**b**,**c**) an n-type semiconductor and (**d**,**e**,**f**) a p-type semiconductor. (**b**,**e**) in thermodynamic equilibrium, (**a**,**d**) with forward bias ($V > 0$), (**c**,**f**) with reverse bias ($V < 0$)

$$V_{\mathrm{bi}} = -(\phi_{\mathrm{Bn}} - V_{\mathrm{n}}) = -\phi_{\mathrm{m}} + \chi_{\mathrm{sc}} + V_{\mathrm{n}} \ , \tag{18.3}$$

which is termed the built-in potential (or diffusion voltage). The exact form of the voltage drop, the so-called band bending, will be discussed in Sect. 18.2.2.

The surface index is defined as

$$S = \partial \phi_{\mathrm{Bn}} / \partial \phi_{\mathrm{m}} \ . \tag{18.4}$$

From the present consideration, the same semiconductor with metals of varying work function should result in $S = 1$.

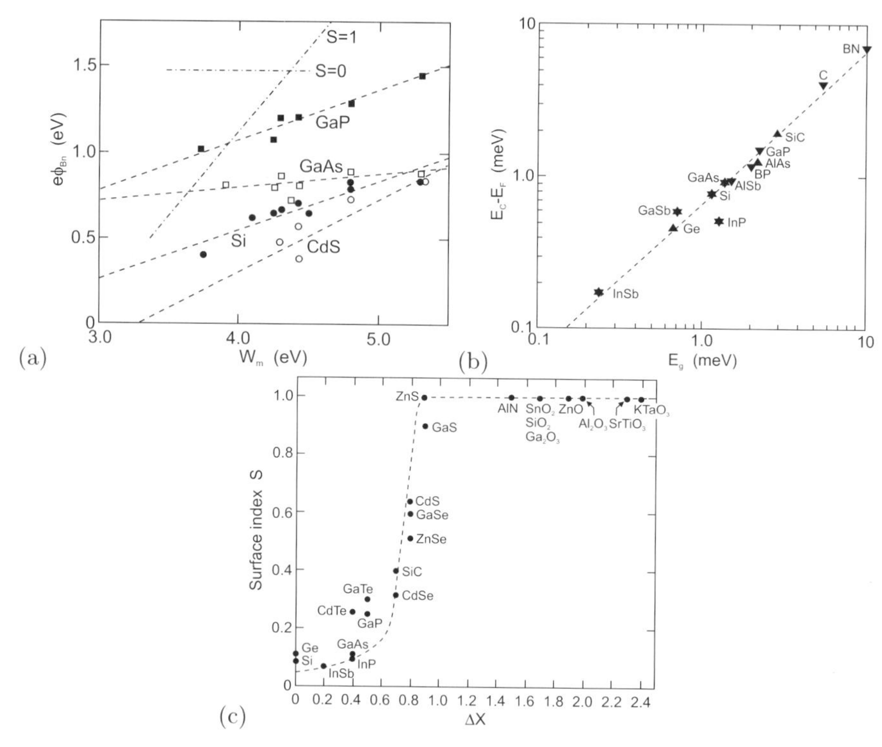

Fig. 18.5. (**a**) Experimental Schottky barrier heights $\phi_{\rm Bn}$ vs. metal work function $\phi_{\rm m}$ for various metal–semiconductor junctions, *filled squares*: GaP, *empty squares*: GaAs, *filled circles*: Si, *empty circles*: CdS. *Dashed lines* are guides to the eye, *dash-dotted lines* indicate dependencies for $S = 1$ and $S = 0$. Data from [549]. (**b**) $E_{\rm C} - E_{\rm F}$ at the metal–semiconductor interface vs. the bandgap $E_{\rm g}$ for Au Schottky contacts on various semiconductors. The *dashed line* represents $E_{\rm C} - E_{\rm F} = 2E_{\rm g}/3$. Data from [550]. (**c**) Surface index S vs. the electronegativity difference ΔX between the species of compound semiconductors. *Dashed line* is a guide to the eye. Adapted from [551]

Band Structure in the Presence of Surface States

Experimental data shown in Fig. 18.5a, however, show a different behavior with smaller slope. For GaAs, e.g., the barrier height is almost independent of the metal work function. Thus, a different model is needed for realistic Schottky diodes. A rule of thumb for the dominantly covalent semiconductors is that for n-type material the barrier height is 2/3 of the bandgap and for p-type material 1/3 of the bandgap, such that $E_{\rm C} - E_{\rm F} \approx 2E_{\rm g}/3$ (Fig. 18.5b). Only for ionic semiconductors does $S \approx 1$ (Fig. 18.5c).

If the semiconductor has a large density of states at its surface, there is a space-charge region already without the metal. Surface traps are filled up to the Fermi level (Fig. 18.6a). The size of the band bending in the semicon-

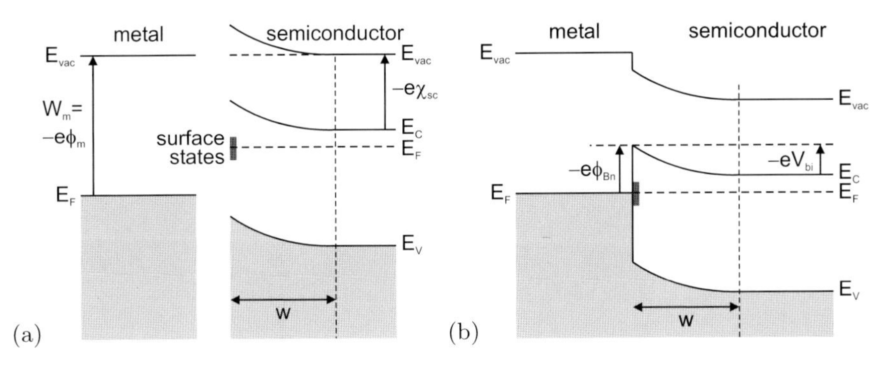

Fig. 18.6. Schematic band structure of a metal–semiconductor junction that is dominated by surface properties of the semiconductor. (**a**) no contact; due to pinning of the Fermi level at surface states of the semiconductor, a depletion-layer of width w is already present. (**b**) Metal and semiconductor in contact

ductor will be denoted as ϕ_{Bn} since it will turn out below as the Schottky barrier height. If the density of surface states is very high, the charge carriers moving from the semiconductor into the metal upon contact formation are accommodated in the surface states and the position of the Fermi level at the semiconductor surface changes only very little. Thus, the space-charge region is not modified and it is identical to the surface depletion region. The Schottky barrier height is then given by the band bending at the (bare) semiconductor surface ϕ_{Bn} (Fig. 18.6d) and does not depend on the metal work function at all. For this case we find for the surface index $S = \partial\phi_{\mathrm{Bn}}/\partial\phi_{\mathrm{m}} = 0$.

For actual metal–semiconductor contacts the surface index S is between 0 and 1 and a theory involving the semiconductor band structure and midgap (surface) states (MIGS) is needed [236]. For Si, the experimental result is $S = 0.27$. The corresponding density of surface states is $D_{\mathrm{s}} = 4 \times 10^{13}\,\mathrm{cm}^{-2}\,\mathrm{eV}^{-1}$.

18.2.2 Space-Charge Region

The width w of the space-charge region is calculated next. First, we make the so-called *abrupt approximation*. In this approximation (Schottky–Mott model), the charge density ρ in the space-charge region ($0 \leq x \leq w$) is given by the doping, i.e. $\rho = eN_{\mathrm{D}}$. Outside the space-charge region the semiconductor is neutral, i.e. $\rho = 0$ and the electrical field is zero, i.e. $\mathrm{d}V/\mathrm{d}x = 0$. As further boundary conditions the potential at the interface is $V(0) = -V_{\mathrm{bi}} < 0$. The potential drop in the space-charge region is determined by the one-dimensional Poisson equation

$$\frac{\mathrm{d}^2 V}{\mathrm{d}^2 x} = -\frac{\rho}{\epsilon_s}\,, \tag{18.5}$$

where ϵ_s is the static dielectric constant of the semiconductor. Using the ansatz $V = V_0 + V_1 x + V_2 x^2$ we find

$$V(x) = -V_{\mathrm{bi}} + \frac{eN_{\mathrm{D}}}{\epsilon_s}\left(wx - \frac{1}{2}x^2\right) . \tag{18.6}$$

The electric field strength is

$$\mathcal{E}(x) = -\frac{eN_{\mathrm{D}}}{\epsilon_s}(w - x) = \mathcal{E}_{\mathrm{m}} + \frac{eN_{\mathrm{D}}}{\epsilon_s}x , \tag{18.7}$$

with the maximum field strength $\mathcal{E}_{\mathrm{m}} = -eN_{\mathrm{D}}w/\epsilon_s$ at $x = 0$. From the condition $V(w_0) = 0$ we obtain w as

$$w_0 = \left[\frac{2\epsilon_s}{eN_{\mathrm{D}}}V_{\mathrm{bi}}\right]^{1/2} . \tag{18.8}$$

The thermal distribution of the majority carriers must be treated with a little more care. The dependence of the charge-carrier density $\rho = N_{\mathrm{D}}^{+} - n$ on the potential V is ($\beta = e/kT$)

$$\rho = eN_{\mathrm{D}}\left[1 - \exp\left(\beta V\right)\right] . \tag{18.9}$$

We note that for the depletion layer $V \leq 0$ and $n \leq N_{\mathrm{D}}$. The charge difference $\Delta\rho$ (due to the tail of the thermal distribution of the majority charge carriers in the depletion layer) between the real distribution (18.9) to the abrupt approximation model with constant charge density ($\rho = eN_{\mathrm{D}}$) in the depletion layer is

$$\Delta\rho(x) = -eN_{\mathrm{D}}\exp\left(\beta V(x)\right) . \tag{18.10}$$

The integration of $\Delta\rho$ over the depletion layer yields that the voltage drop V_{bi} across the depletion layer needs to be corrected by ΔV

$$\Delta V = \int_0^w \left[\int_0^x \frac{-\Delta\rho(x')}{\epsilon_{\mathrm{s}}}\,\mathrm{d}x'\right]\mathrm{d}x = \frac{1}{\beta}\left[1 - \exp(-\beta V_{\mathrm{bi}})\right] \approx \frac{kT}{e} . \tag{18.11}$$

The approximation is valid for $\beta V_{\mathrm{bi}} \gg 1$. Therefore, (18.8) is corrected to

$$w_0 = \left[\frac{2\epsilon_s}{eN_{\mathrm{D}}}\left(V_{\mathrm{bi}} - \frac{kT}{e}\right)\right]^{1/2} . \tag{18.12}$$

When a potential V_{ext} is applied externally to the diode, (18.12) is modified by the change in the interface boundary condition, $V(0) = -V_{\mathrm{bi}} + V_{\mathrm{ext}}$. The band structure is shown schematically in Fig. 18.4b for a forward bias and in Fig. 18.4c for a reverse bias. Therefore, we finally obtain for the depletion-layer width

$$w = \left[\frac{2\epsilon_s}{eN_{\mathrm{D}}}\left(V_{\mathrm{bi}} - V_{\mathrm{ext}} - \frac{kT}{e}\right)\right]^{1/2} . \tag{18.13}$$

Now we can also give explicitly the value of the maximum electrical field (at $x = 0$)

$$\begin{aligned}\mathcal{E}_\mathrm{m} &= -\left[\frac{2eN_\mathrm{D}}{\epsilon_s}\left(V_\mathrm{bi} - V_\mathrm{ext} - \frac{kT}{e}\right)\right]^{1/2} \\ &= -\frac{2}{w}\left(V_\mathrm{bi} - V_\mathrm{ext} - \frac{kT}{e}\right) .\end{aligned} \tag{18.14}$$

We note that so far the barrier height is independent of the applied bias voltage. In the next section, it is shown that this is actually not the case.

18.2.3 Schottky Effect

The barrier height is reduced by the image-charge effect that has been neglected so far. An electron (charge $q = -e$) at position x in the semiconductor is facing a metal surface. The metal surface is at zero position (Fig. 18.7a). The potential distribution of the free charge is modified since the metal surface is an equipotential surface. The potential distribution outside the metal is identical to that if an *image* charge $-q$ were located at $-x$. This image charge exerts a force (image force F_im) on the electron

$$F_\mathrm{im} = -\frac{q^2}{16\,\pi\epsilon_0\epsilon_\mathrm{s}x^2} , \tag{18.15}$$

where ϵ_s is the relative dielectric constant of the semiconductor. In order to bring an electron to x from infinity the work E_im

$$E_\mathrm{im} = \int_\infty^x F_\mathrm{im}\,\mathrm{d}x = -\frac{q^2}{16\,\pi\epsilon_0\epsilon_\mathrm{s}x} \tag{18.16}$$

is needed. This image potential energy is shown in Fig. 18.7a. The total energy E_tot (solid line in Fig. 18.7a) of the electron in the presence of an electric field $\mathcal{E}$ is given by

$$E_\mathrm{tot} = q\mathcal{E}x - \frac{q^2}{16\,\pi\epsilon_0\epsilon_\mathrm{s}x} . \tag{18.17}$$

The maximum of this function ($\mathrm{d}E_\mathrm{tot}/\mathrm{d}x = 0$) is at x_m

$$x_\mathrm{m} = \left[\frac{e}{16\,\pi\epsilon_0\epsilon_\mathrm{s}\mathcal{E}}\right]^{1/2} . \tag{18.18}$$

The barrier (now labeled ϕ_{B_0}) is reduced by $\Delta\phi$ given by

$$\Delta\phi = \left[\frac{e\mathcal{E}}{4\,\pi\epsilon_0\epsilon_\mathrm{s}}\right]^{1/2} = 2\mathcal{E}x_\mathrm{m} . \tag{18.19}$$

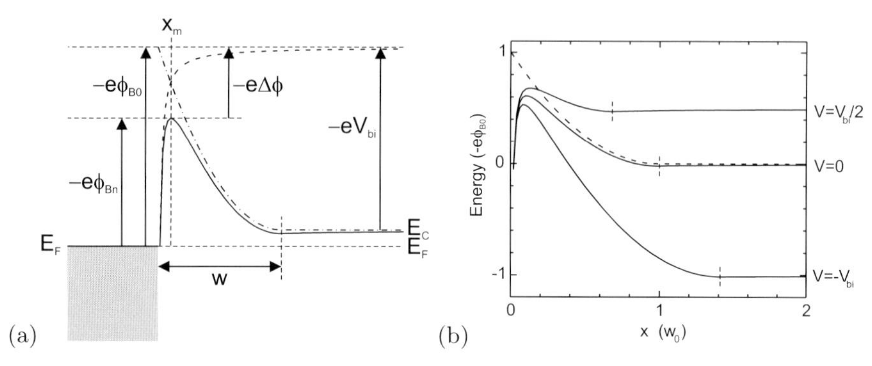

Fig. 18.7. (**a**) Energy of a particle with respect to the metal surface (*dashed line*), conduction band in semiconductor depletion layer (*dash-dotted line*) and combined effect (*solid line*). The image charge energy lowers the potential barrier $-e\phi_{B_0}$ by the amount $-e\Delta\phi$ to $-e\phi_{Bn}$. (**b**) Conduction band on the semiconductor side of a metal–semiconductor junction at various bias voltages ($V = 0$, $V = +V_{bi}/2$, and $V = -V_{bi}$ as labeled) taking into account the Schottky effect. The width of the depletion layer is indicated with a *short vertical dashed line*. The barrier height without Schottky effect is ϕ_{B0}. The *dashed line* is the situation without Schottky effect for zero bias

With the field in the vicinity of the interface given by $\mathcal{E}_m$ from (18.15), the barrier reduction is[6]

$$\Delta\phi = \left[\frac{e^3 N_D}{8\,\pi^2 \epsilon_0^3 \epsilon_s^3}\left(V_{bi} - V_{ext} - \frac{kT}{e}\right)\right]^{1/4} . \tag{18.20}$$

For $\epsilon_s = 1$ (vacuum) and a field strength of 10^5 V/cm the maximum position is at $x_m = 6$ nm and the barrier reduction is $\Delta\phi = 0.12$ V. For 10^7 V/cm, $x_m = 1$ nm and $\Delta\phi = 1.2$ V. For semiconductors with $\epsilon_s \sim 10$ the effect is smaller (Fig. 18.8). The Schottky effect depends on the bias voltage as visualized in Fig. 18.7b and therefore the barrier height depends on the applied bias voltage.

18.2.4 Capacitance

The total space charge Q (per unit area) in the semiconductor is

$$Q = eN_D w = \left[2eN_D\epsilon_s\left(V_{bi} - V_{ext} - \frac{kT}{e}\right)\right]^{1/2} . \tag{18.21}$$

---

[6] The term ϵ_s^3 is technically $\epsilon_s\epsilon_d^2$ where ϵ_d is the image-force dielectric constant. ϵ_d is equal to ϵ_s if the transit time of an electron from the metal to the maximum of the potential energy is sufficiently long to build up the dielectric polarization of the semiconductor [545].

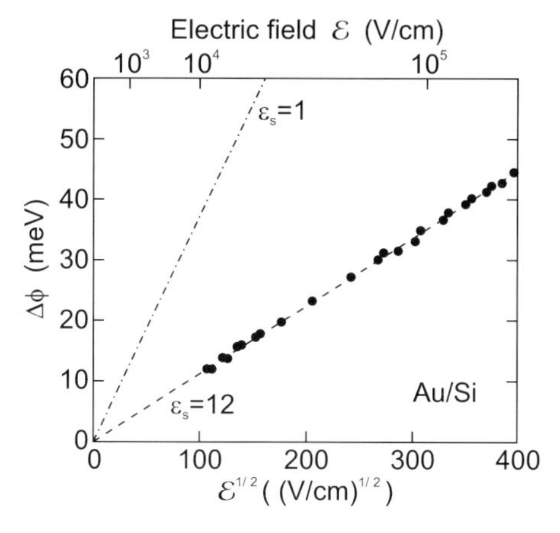

Fig. 18.8. Electric-field dependence of the image charge lowering of the Schottky barrier. *Dash-dotted* line is for vacuum dielectric constant, *dashed line* is (18.19) for ϵ_s =12. Adapted from [552]

Therefore, the capacity $C = |\mathrm{d}Q/\mathrm{d}V|$ (per unit area) of the space charge region[7] is given by

$$C = \left[\frac{e N_\mathrm{D} \epsilon_s}{2 \left(V_\mathrm{bi} - V_\mathrm{ext} - \frac{kT}{e} \right)} \right]^{1/2} = \frac{\epsilon_s}{w} \,. \tag{18.22}$$

Equation (18.22) can also be written as

$$\frac{1}{C^2} = \frac{2 \left(V_\mathrm{bi} - V_\mathrm{ext} - \frac{kT}{e} \right)}{e N_\mathrm{D} \epsilon_s} \,. \tag{18.23}$$

If $1/C^2$ is measured as a function of the bias voltage (C–V spectroscopy), it should be linearly dependent on the bias voltage if the doping concentration is homogeneous (Fig. 18.9a). The doping concentration can be determined from the slope via

$$N_\mathrm{D} = -\frac{2}{e \epsilon_s} \left[\frac{\mathrm{d}}{\mathrm{d}V_\mathrm{ext}} \left(\frac{1}{C^2} \right) \right]^{-1} , \tag{18.24}$$

(see Fig. 18.9b) and the built-in voltage V_bi from the extrapolation to $1/C^2 = 0$. The Schottky barrier height can be determined from this [553] via

[7] A dc bias voltage is set and the capacitance is probed with an ac voltage with small amplitude. A typical frequency of the ac voltage is 1 MHz. However, this might correlate to a characteristic time constant of the system. In this case, another ac frequency is preferred. Probing the capacitance as a function of the ac frequency is called admittance spectroscopy.

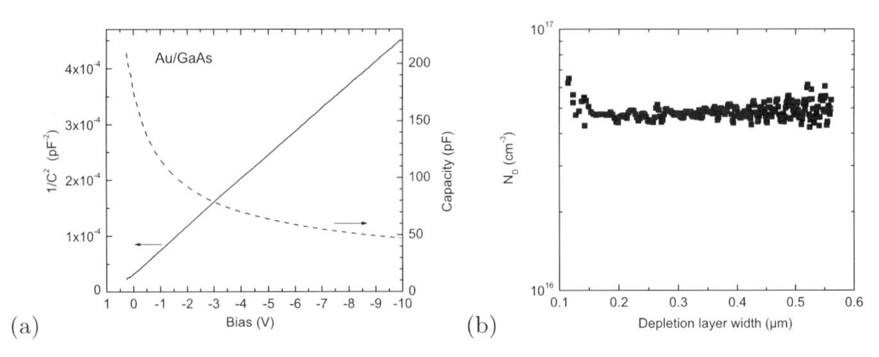

Fig. 18.9. (**a**) Capacity C (*dashed line*) and $1/C^2$ (*solid line*) vs. bias voltage dependence for an Au/GaAs Schottky diode (2-μm MOVPE-grown GaAs:Si on a n-GaAs substrate) at room temperature. From the extrapolation to $1/C^2 = 0$ and (18.23) we obtain $V_{\mathrm{bi}} = 804 \pm 3\,\mathrm{meV}$. (**b**) Donor concentration ($N_{\mathrm{D}} = 4.8 \times 10^{16}\,\mathrm{cm}^{-3}$) determined via (18.24) from the $1/C^2$ plot vs. the depletion layer width (calculated using (18.22))

$$\phi_{\mathrm{B_n}} = V_{\mathrm{bi}} + V_{\mathrm{n}} + \frac{kT}{e} - \Delta\phi\,, \tag{18.25}$$

where $\Delta\phi$ is the barrier lowering due to the image force effect between the flat-band and the zero-bias cases.

We note that for inhomogeneous doping the depth profile of the doping can be determined by C–V spectroscopy. The $1/C^2$ vs. bias curve is then no longer a straight line and exhibits a varying slope. At a given bias voltage, the charge (ionized donors or acceptors) at the boundary of the space-charge region is tested by the capacitance measurement. However, this principle works only if the depth of the space-charge region actually changes with the bias voltage. The method can therefore not be directly applied to such systems like δ-doped layers or quantum wells.

18.2.5 Current–Voltage Characteristic

The current transport through a metal–semiconductor junction is dominated by the majority charge carriers, i.e. electrons (holes) in the case of an n-type (p-type) semiconductor, respectively.

In Fig. 18.10, the possible transport mechanisms are visualized for an n-type semiconductor. Thermionic emission 'above' the barrier involves the hot electrons from the thermal distribution and will be important at least at high temperatures. Tunneling 'through' the barrier will be important for thin barriers, i.e. at high doping ($w \propto N_{\mathrm{D}}^{-1/2}$, cf. (18.13)). 'Pure' tunneling for electrons close to the (quasi-) Fermi level, also called field emission, and thermionic field emission, i.e. tunneling of electrons with higher energies, are distinguished. Also, recombination in the depletion layer and hole injection from the metal are possible.

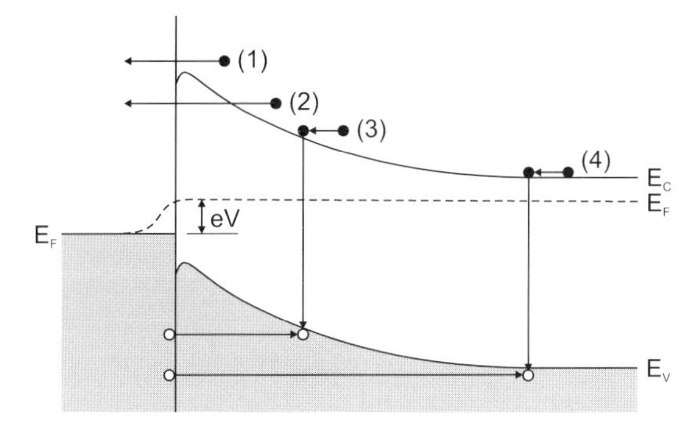

Fig. 18.10. Transport mechanisms at metal–semiconductor junctions. (1) Thermionic emission ('above' the barrier) (2) tunneling ('through' the barrier), (3) recombination in the depletion layer, (4) hole injection from metal

The transport of electrons above the barrier can be described with diffusion theory [554, 555] or thermionic-emission theory [556]. In both cases the barrier height is large compared to kT. For thermionic emission (typically relevant for semiconductors with high mobility) the current is limited by the emission process and an equilibrium (constant electron quasi-Fermi level) is established throughout the depletion layer and ballistic transport is considered. In diffusion theory (for low mobility) a thermal equilibrium between metal and semiconductor electrons is established in the interface plane and the current is limited by diffusion and drift in the depletion region.

Thermionic Emission

The current density per unit area $j_{\mathrm{s}\to\mathrm{m}}$ of electrons that flow from the semiconductor into the metal is due to the hot electrons from the thermal distribution function

$$j_{\mathrm{s}\to\mathrm{m}} = \int_{E_\mathrm{F}-e\phi_{\mathrm{Bn}}}^{\infty} -ev_{\mathrm{x}}\mathrm{d}n \, . \tag{18.26}$$

The integral starts at the lowest possible energy, the top of the Schottky barrier (no tunneling allowed in this model!). The electron density $\mathrm{d}n$ in a small energy interval $\mathrm{d}E$ is

$$\mathrm{d}n = D(E)f(E)\mathrm{d}E \, . \tag{18.27}$$

The carrier velocity is obtained from

$$E = E_\mathrm{C} + \frac{1}{2}m^* v^2 \, . \tag{18.28}$$

For a bulk semiconductor and the Boltzmann distribution ($-e\phi_{\mathrm{B}} \gg kT$)

$$\mathrm{d}n = 2\left(\frac{m^*}{h}\right)\exp\left(\frac{eV}{kT}\right)\exp\left(-\frac{m^*v^2}{2kT}\right)4\pi v^2\mathrm{d}v\,. \tag{18.29}$$

Using $4\pi v^2 \mathrm{d}v = \mathrm{d}v_{\mathrm{x}}\,\mathrm{d}v_{\mathrm{y}}\,\mathrm{d}v_{\mathrm{z}}$ and integrating over all velocities in y and z directions and v_{x} from the minimum velocity necessary to pass the barrier $v_{\mathrm{min,x}} = (2e(V_{\mathrm{bi}} - V)/m^*)^{1/2}$ to ∞, the current density is found to be

$$j_{\mathrm{s}\to\mathrm{m}} = A^*T^2\exp\left(\frac{e\phi_{\mathrm{Bn}}}{kT}\right)\exp\left(\frac{eV}{kT}\right)\,, \tag{18.30}$$

with A^* being the Richardson constant given by

$$A^* = \frac{4\pi e m^* k_{\mathrm{B}}^2}{h^3} = \frac{eN_{\mathrm{C}}\bar{v}}{4T^2}\,, \tag{18.31}$$

where $\bar{v}$ is the average thermal velocity in the semiconductor. A^* for electrons in vacuum is $120\,\mathrm{A\,cm^{-2}\,K^{-2}}$. A similar result is obtained for the thermionic emission of electrons from a metal into vacuum. Since ϕ_{Bn} is negative, the saturation current increases with increasing temperature.

If the bias is changed, the current from the semiconductor to the metal increases in the forward direction because the energy difference between the quasi-Fermi level and the top of the barrier is reduced. The current is reduced for reverse bias. The barrier from the metal into the semiconductor remains constant (except for the Schottky effect whose impact on the current–voltage characteristic is discussed next). Therefore the current from the metal into the semiconductor is constant and can be obtained from the condition $j = 0$ for zero bias. Therefore the current–voltage characteristic in the thermionic-emission model is

$$\begin{aligned} j &= A^*T^2\exp\left(\frac{e\phi_{\mathrm{Bn}}}{kT}\right)\left[\exp\left(\frac{eV}{kT}\right) - 1\right] \\ &= j_{\mathrm{s}}\left[\exp\left(\frac{eV}{kT}\right) - 1\right]\,. \end{aligned} \tag{18.32}$$

The prefactor

$$j_{\mathrm{s}} = A^*T^2\exp\left(\frac{e\phi_{\mathrm{Bn}}}{kT}\right) \tag{18.33}$$

is called the saturation current density. This dependence is the ideal diode characteristic and is shown in Fig. 18.11. The temperature dependence of the saturation current j_{s} can be written as

$$\ln\left(\frac{j_{\mathrm{s}}}{T^2}\right) = \ln A^* + \frac{e\phi_{\mathrm{Bn}}}{kT} \tag{18.34}$$

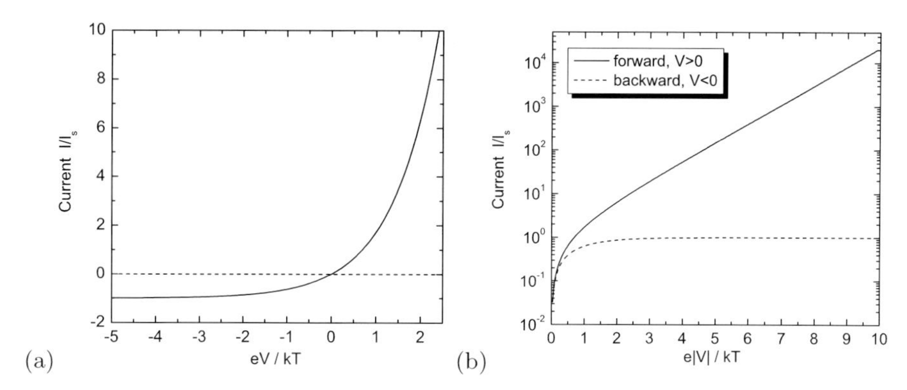

Fig. 18.11. Ideal diode I–V characteristics $I = I_s\,(\exp(eV/kT) - 1)$ (**a**) in linear plot and (**b**) semilogarithmic plot

by transforming (18.33). The plot of $\ln\left(j_s/T^2\right)$ to vs. $1/T$ is called a Richardson plot and allows the barrier height and the Richardson constant to be determined from a linear fit.

If the Schottky effect, i.e. the change of barrier height with bias voltage, is considered, the semilogarithmic slope of the forward I–V characteristic is no longer $V_0^{-1} = e/kT$ but can be expressed as $V_0^{-1} = e/nkT$, n being a dimensionless parameter termed the ideality factor.[8] n is given by

$$n = 1 + \frac{\partial \phi_B}{\partial V}\ . \tag{18.35}$$

Values for n obtained this way (using (18.20)) are smaller than 1.03. For GaAs and $N_D = 10^{17}\,\mathrm{cm}^{-3}$ $n = 1.02$. With regard to V_0 and its temperature dependence we refer also to Fig. 18.13 and the related discussion.

In [557] the effect of a spatial inhomogeneity of the diffusion voltage and thus the barrier height is investigated. The barrier height $\phi_{Bn}(x,y)$ across the contact area is assumed to have a Gaussian probability distribution $p(\phi_{Bn})$ with a mean value $\bar{\phi}_{Bn}$ and a standard deviation σ_ϕ. It turns out that the barrier height ϕ_{Bn}^C responsible for the capacity, and thus the diffusion voltage determined by C–V spectroscopy, is given by the spatial average, i.e. $\phi_{Bn}^C = \bar{\phi}_{Bn}$. The barrier height ϕ_{Bn}^j determining the current–voltage characteristics (cf. (18.32)) via

$$\begin{aligned} j &= A^* T^2 \left[\exp\left(\frac{eV}{kT}\right) - 1\right] \int \exp\left(\frac{e\phi_{Bn}}{kT}\right) p(\phi_{Bn})\mathrm{d}\phi_{Bn} \\ &= A^* T^2 \exp\left(\frac{e\phi_{Bn}^j}{kT}\right)\left[\exp\left(\frac{eV}{kT}\right) - 1\right] \end{aligned} \tag{18.36}$$

is given by

[8]Obviously $n = 1$ for the ideal characteristic. Otherwise $n \geq 1$.

$$\phi_{\mathrm{Bn}}^{\mathrm{j}} = \bar{\phi}_{\mathrm{Bn}} - \frac{e\sigma_{\phi}^{2}}{2kT} \,. \tag{18.37}$$

Thus, the barrier height determined from the current–voltage characteristic underestimates the spatial average of the barrier height.[9] The Richardson plot (18.34) is now modified (and is nonlinear in $1/T$) to

$$\ln\left(\frac{j_{\mathrm{s}}}{T^2}\right) = \ln A^* + \frac{e\phi_{\mathrm{Bn}}}{kT} - \frac{e^2\sigma_{\phi}^2}{2k^2T^2} \,. \tag{18.38}$$

Temperature Dependence

Figure 18.12a shows the temperature-dependent I–V characteristics of a Pd/ZnO Schottky diode. A straightforward evaluation according to (18.32) results in a barrier height of about 700 meV and a Richardson constant that is orders of magnitude smaller than the theoretical value of $32\,\mathrm{A\,K^{-2}\,cm^{-2}}$ (for $m_{\mathrm{e}}^* = 0.27$). A fit of the temperature-dependent data with (18.37), as shown in Fig. 18.12b, results in $\bar{\phi}_{\mathrm{Bn}} = 1.1\,\mathrm{eV}$, in agreement with the (temperature-independent) value obtained from CV spectroscopy, and $\sigma_{\phi} = 0.13\,\mathrm{V}$ [558].

The temperature dependence of the ideality factor n is given by [559]

$$n = \frac{1}{1 - \rho_2 + e\rho_3/(2kT)} \,, \tag{18.39}$$

where ρ_2 (ρ_3) is the (temperature-independent) proportionality coefficient of the bias dependence of the mean barrier height (standard deviation), i.e. $\rho_2 = \partial\bar{\phi}_{\mathrm{Bn}}/\partial V$ ($\rho_3 = \partial\sigma_{\phi}^2/\partial V$). The fit of $1/n - 1$ vs. $1/T$ (Fig. 18.12c) yields $\rho_2 = -0.025$ and $\rho_3 = -0.028\,\mathrm{V}$.

The forward I–V characteristic of an Au/GaAs Schottky diode reported in [560, 561] is shown in Fig. 18.13a at various temperatures. The current amplitude decreases with decreasing temperature due to the temperature dependence of the saturation current (18.32). Also, the slope V_0^{-1} of the characteristic $j = j_{\mathrm{s}} \exp(V/V_0)$ varies with temperature. Looking at the temperature dependence of V_0, it is described as $V_0 = k(T + T_0)/e$ rather than with an ideality factor n in the form of $V_0 = nkT/e$. In other words, the ideality factor follows a temperature dependence $n = 1 + T_0/T$. In view of (18.39), such behavior means for small T_0 that $n \approx 1/(1 - T_0/T)$ and thus $\rho_2 = 0$ and $\rho_3 = 2kT_0/e$. For $T_0 = 45\,\mathrm{K}$, ρ_3 is $0.008\,\mathrm{V}$, which is a fairly small value. Thus, the temperature behavior of the diode is due to the narrowing of the Gaussian distribution of barrier height with bias voltage [557].

Diffusion Theory

In diffusion theory the current density is considered in the presence of a carrier-density and electric-field gradient. In the Boltzmann approximation

[9] This phenomenon is similar to the red-shift of luminescence lines (Stoke's shift) due to thermalization in the presence of disorder, cf. Sect. 11.6.

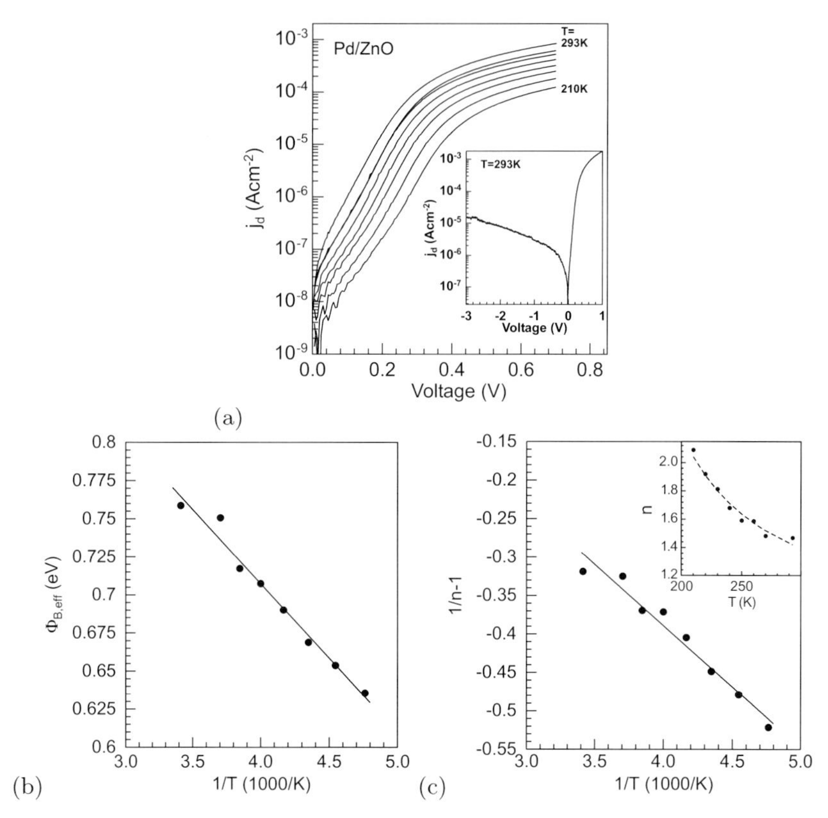

Fig. 18.12. (**a**) Forward I–V characteristic of Pd/ZnO Schottky diodes for various temperatures. Diode temperatures are 210 K, 220 K, 230 K, 240 K, 250 K, 260 K, 270 K, and 293 K. The *inset* shows the current density vs. voltage for 293 K on a semilogarithmic scale. (**b**) Effective barrier height $\phi^{\mathrm{j}}_{\mathrm{Bn}}$ vs. the inverse temperature. The *solid line* is a linear fit according to (18.37) yielding the standard deviation $\sigma_\phi = 0.13\,\mathrm{V}$ and the mean barrier height $\bar{\phi}_{\mathrm{Bn}} = 1.1\,\mathrm{eV}$. (**c**) Plot of $1/n - 1$ vs. the inverse temperature. The *solid line* is a linear fit of the data yielding the voltage deformation coefficients $\rho_2 = -0.025$ and $\rho_3 = -0.028\,\mathrm{V}$. The *inset* shows the experimentally determined n factors and the n factors calculated from (18.39) using the voltage-deformation coefficients obtained from the linear fit (*dashed line*)

the electron current is given by (8.46a). In stationary equilibrium the current density is constant, i.e. independent of x. Assuming that the carrier density has its equilibrium values at $x = 0$ and $x = w$, we find after integration and using (18.6)

$$\begin{aligned} j &= -e\mu_n N_{\mathrm{C}} \mathcal{E}_{\mathrm{m}} \exp\left(\frac{e\phi_{\mathrm{Bn}}}{kT}\right) \left[\exp\left(\frac{eV}{kT}\right) - 1\right] \\ &= j_{\mathrm{s}} \left[\exp\left(\frac{eV}{kT}\right) - 1\right] . \end{aligned} \tag{18.40}$$

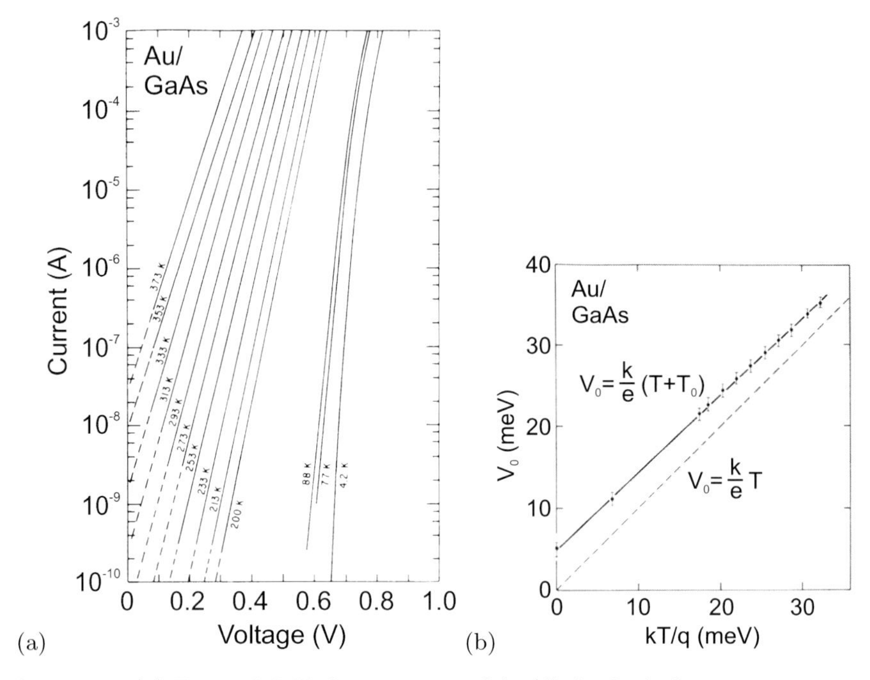

Fig. 18.13. (**a**) Forward I–V characteristic of Au/GaAs diode for various temperatures. (**b**) Temperature dependence of the voltage V_0. The experimental data are fitted with $T_0 = 45 \pm 8\,\mathrm{K}$. Adapted from [560]

Therefore, also in this case the ideal diode characteristic is obtained, however, with a different saturation current. The ideality factor in diffusion theory is $n = 1.06$ (for $-e\phi_{\mathrm{Bn}} \gtrsim 15\,kT$) [547].

Combined Theory

A combination of both theories [563] considers both mechanisms to be in series. The current can then be expressed as

$$j = \frac{eN_{\mathrm{C}}v_{\mathrm{r}}}{1 + v_{\mathrm{r}}/v_{\mathrm{D}}} \exp\left(\frac{e\phi_{\mathrm{Bn}}}{kT}\right)\left[\exp\left(\frac{eV}{kT}\right) - 1\right] \tag{18.41a}$$

$$= A^{**}T^2\left[\exp\left(\frac{eV}{kT}\right) - 1\right] \tag{18.41b}$$

$$= j_{\mathrm{s}}\left[\exp\left(\frac{eV}{kT}\right) - 1\right] .$$

Here $v_{\mathrm{r}} = \bar{v}/4$ is a 'recombination velocity' [562] at the top of the barrier according to $j = v_{\mathrm{r}}(n - n_0)$, n_0 being the equilibrium electron density at the top of the barrier and $\bar{v}$ is the average thermal velocity in the semiconductor.

v_D is an effective diffusion velocity describing the transport of electrons from the edge of the depletion layer ($x = w$) to the top of the barrier ($x = x_\mathrm{m}$). It is defined as

$$v_\mathrm{D} = \left[\int_{x_\mathrm{m}}^{w} \frac{-e}{\mu_n kT} \exp\left(-\frac{-e\phi_\mathrm{Bn} - E_\mathrm{C}(x)}{kT} \right) \mathrm{d}x \right]^{-1} . \tag{18.42}$$

In [563] μ_n has been assumed to be independent of the electric field. This assumption is potentially not realistic. If $v_\mathrm{D} \gg v_\mathrm{r}$, thermionic theory applies and we obtain (18.32). The case $v_\mathrm{r} \gg v_\mathrm{D} \sim \mu_n \mathcal{E}_\mathrm{m}$ relates to diffusion theory and we recover (18.40).

The constant A^{**} in (18.41b) is called the effective Richardson constant. Its calculated dependence on the electric field is shown in Fig. 18.14 for Si. At room temperature for most Ge, Si and GaAs Schottky diodes the thermionic emission of majority carriers is the dominating process.

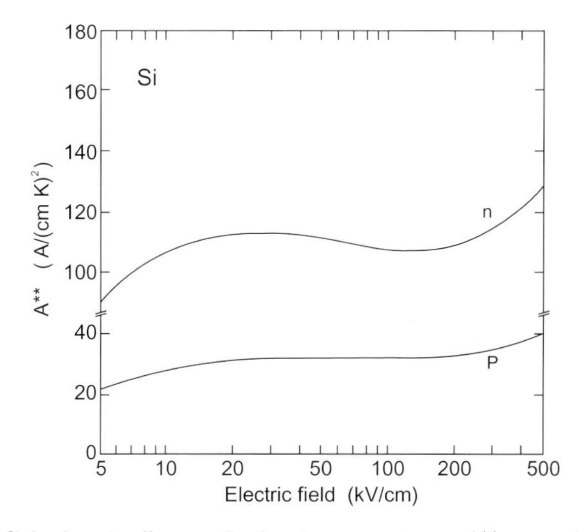

Fig. 18.14. Calculated effective Richardson constant A^{**} as a function of the electric field for a metal–Si diode at $T = 300\,\mathrm{K}$ for a (n-type, upper curve or p-type, lower curve) doping of $10^{16}\,\mathrm{cm}^{-3}$. Adapted from [564]

Tunneling Current

At high doping the width of the depletion layer becomes small and tunneling processes become more probable. Also at low temperatures, when thermionic emission is very small, tunneling processes can dominate the transport between metal and semiconductor. One process is tunneling of electrons close to the Fermi level of the semiconductor. This process is called field emission (F) and is at least important for degenerate semiconductors at very low temperatures. If the temperature is raised, electrons are excited to higher energies where they encounter a thinner barrier. The tradeoff between thermal energy and barrier width selects an electron energy E_{m} above the conduction-band edge for which the current is largest. This process is known as thermionic field emission (TF). For very high temperatures enough carriers can overcome the barrier completely and we are back in the thermionic emission regime. The validity of the two regimes is shown in Fig. 18.15 for Au/GaAs Schottky diodes as a function of doping concentration (n-type) and temperature.

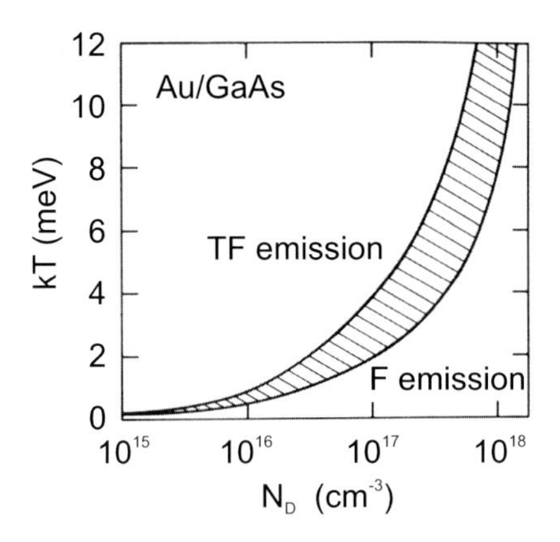

Fig. 18.15. Calculated conditions for thermionic field ('TF') and field ('F') emission in a Au/GaAs Schottky diode as a function of temperature and doping concentration. Adapted from [547]

In the field-emission regime the forward current is given by

$$j = j_{\mathrm{s}} \exp\left(\frac{eV}{E_{00}}\right) , \tag{18.43}$$

with the characteristic energy parameter E_{00} given by

$$E_{00} = \frac{e\hbar}{2}\left(\frac{N_{\mathrm{D}}}{m^* \epsilon_{\mathrm{s}}}\right)^{1/2} . \tag{18.44}$$

The saturation current is $j_{\mathrm{s}} \propto \exp\left(e\phi_{\mathrm{Bn}}/E_{00}\right)$. In Fig. 18.16, the forward characteristic of a highly doped Au/Si is shown. The experimental value of $E_{00} = 29\,\mathrm{meV}$ agrees well with the theoretical expectation of $E_{00} = 29.5\,\mathrm{meV}$.

In the reverse direction the I–V characteristic under field emission is given by

$$j = \frac{4e\pi m^*}{h^3} E_{00}^2 \frac{V_{\mathrm{bi}} - V}{\phi_{\mathrm{Bn}}} \exp\left(-\frac{2(-\phi_{\mathrm{Bn}})^{3/2}}{3E_{00}\, e(V_{\mathrm{bi}} - V)^{1/2}}\right) . \tag{18.45}$$

From Fig. 18.16b, a barrier height of 0.79 eV is deduced.

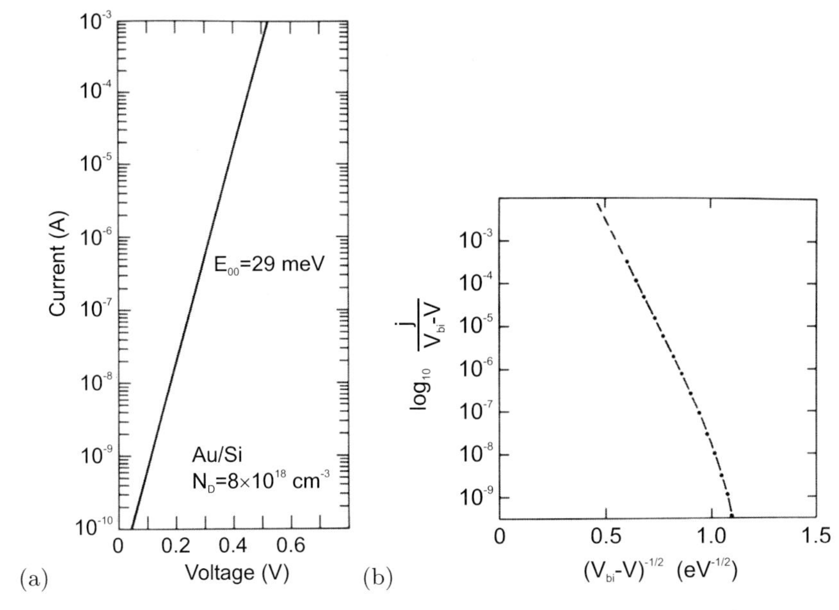

Fig. 18.16. (**a**) Forward and (**b**) reverse I–V characteristic of a Au/Si Schottky diode at 77 K. The doping concentration of the Si was $N_D = 8 \times 10^{18}\,\mathrm{cm}^{-3}$. Adapted from [547]

In the TF-emission regime the current–voltage characteristic is given by

$$j = j_s \exp\left(\frac{eV}{E_0}\right) , \tag{18.46}$$

with

$$E_0 = E_{00} \coth(E_{00}/kT) , \tag{18.47}$$

where E_{00} is given by (18.44). The energy for maximum TF emission E_m is given by $E_m = e(V_{bi} - V)/\cosh^2(E_{00}/kT)$. The coth dependence of E_0 is shown in Fig. 18.17 for an Au/GaAs diode.

A Schottky diode can suffer from nonideality such as series and parallel ohmic resistance [559]. These effects are discussed in some detail below for pn-diodes in Sect. 18.4.4 and apply similarly to Schottky diodes.

18.2.6 Ohmic Contacts

Although an ohmic contact does not have a diode characteristic, it can be understood from the previous remarks. An ohmic contact will have a small contact resistance for both current directions. The voltage drop across the

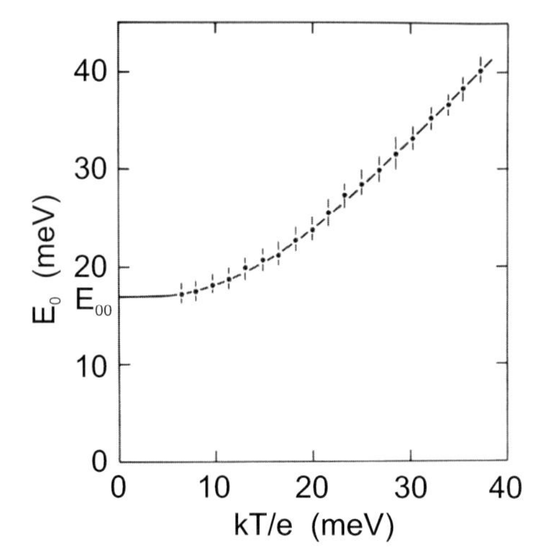

Fig. 18.17. Temperature dependence of E_0 of an Au/GaAs diode with $N_D = 5 \times 10^{17}\,\mathrm{cm}^{-3}$. The *solid line* is the theoretical dependence for thermionic emission according to (18.47) with $N_D = 6.5 \times 10^{17}\,\mathrm{cm}^{-3}$ and $m^* = 0.07$. Adapted from [547]

contact will be small compared to the voltage drop in the active layer (somewhere else). The contact resistance R_c is defined as the differential resistance at $V = 0$

$$R_c = \left(\frac{\partial I}{\partial V} \right)^{-1}_{V=0} . \tag{18.48}$$

At low doping, the transport is dominated by thermionic emission (18.32). In this case R_c is given by

$$R_c = \frac{k}{eA^*T} \exp\left(-\frac{e\phi_{Bn}}{kT} \right) . \tag{18.49}$$

A small barrier height will lead to small contact resistance.

For high doping R_c is determined by the tunneling current and is proportional to

$$R_c \propto \exp\left(-\frac{e\phi_{Bn}}{E_{00}} \right) . \tag{18.50}$$

The contact resistance decreases exponentially with the doping. A theoretical calculation and experimental data are compared in Fig. 18.18 for contacts on Si. The two mechanisms, low barrier height and high doping, for the formation of Ohmic contacts are summarized schematically in Fig. 18.19. Ohmic contacts on wide-bandgap semiconductors are difficult, since metals with sufficiently small work function are mostly not available.

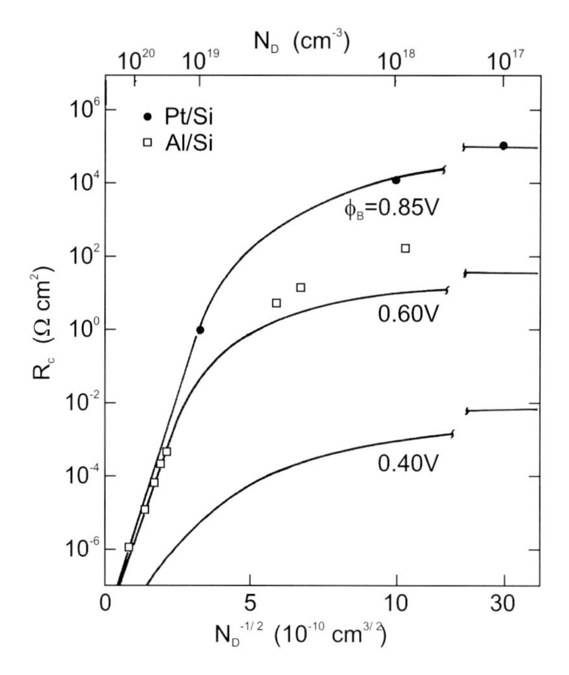

Fig. 18.18. Theoretical and experimental values of specific contact resistances at $T = 300\,\mathrm{K}$ for Al/n-Si [565] and PtSi/n-Si [566] contacts as a function of donor concentration. *Solid lines* are theoretical dependencies for different values of the barrier height as labeled. Adapted from [183]

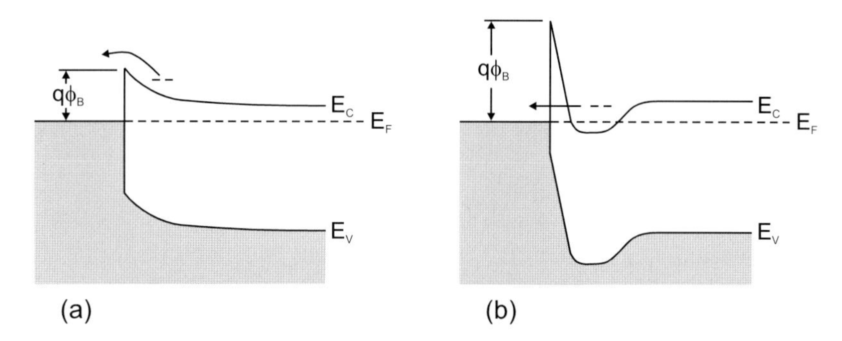

Fig. 18.19. Schematic conditions for the formation of an Ohmic contact, (**a**) low barrier height and (**b**) high doping (thin depletion layer)

Ohmic contacts are typically prepared by evaporating a contact metal containing the doping material for the semiconductor, e.g. Au/Zn for a contact on p-type GaAs [567] and Au/Ge for a contact [568] on n-type GaAs. The contact is alloyed around 400–500 °C (see Fig. 18.20) above the eutectic temperature of $T_{\mathrm{eu}} = 360\,°\mathrm{C}$ (for Au/Ge) to form a eutectic liquid in which the dopant can quickly diffuse. When the eutectic liquid cools it forms a solid, a highly doped semiconductor layer underneath the metal. The liquid-phase reactions can lead to inhomogeneous contacts. On n-type GaAs Pd/Ge/Au contacts have been reported to have superior structural quality [569]. Ohmic contacts for a number of different semiconductors are reviewed in [570, 571].

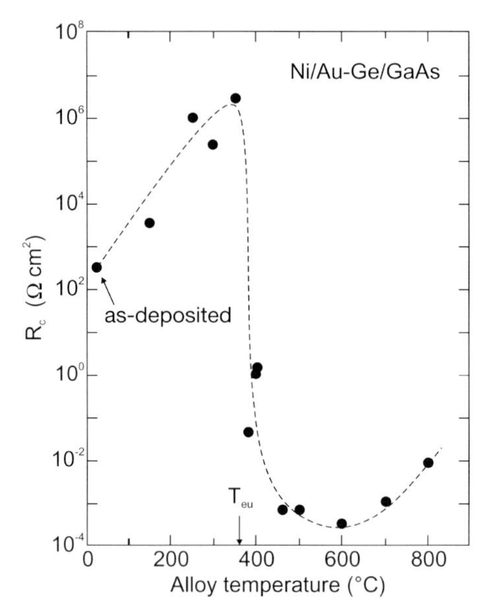

Fig. 18.20. Specific contact resistance for Ni/Au-Ge on n-type epitaxial GaAs for varying alloying temperatures (2 min). *Arrow* at T_{eu} denotes the eutectic temperature of Au-Ge. Adapted from [572]

18.2.7 Metal Contacts to Organic Semiconductors

Also, for organic semiconductors the metal contact plays a vital role, either for carrier injection or for manipulation of the space-charge region. The position of the Fermi level has been determined for various organic semiconductors as shown in Fig. 18.21. These data have been obtained from measurements on metal–semiconductor–metal structures (MSM, see also Sect. 19.3.5) as shown in Fig. 18.22a. The thin (50 nm) organic layer is fully depleted, thus the built-in field inside the semiconductor is constant. The built-in field is measured by applying an external dc bias and finding the external potential at which the electroabsorption signal vanishes. Figure 18.22b shows the measured electroabsorption signal $\Delta T/T$ (relative change of transmission T) and the optical density of an MEH-PPV film as a function of photon energy for an Al/MEH-PPV/Al structure. The exciton absorption peak is found at 2.25 eV. The bias at which the built-in field vanishes can then be determined for various other metals in metal/MEH-PPV/Al structures. Figure 18.21 summarizes such results for various metals and three organic semiconductors. The plot of the Fermi level position vs. the metal work function (Figs. 18.22d,e) shows that the metals investigated do not introduce interface states in the single-particle gap that pin the Schottky barrier (cf. Fig. 18.5 for inorganic semiconductors). An electron trap, such as C_{60} in MEH-PPV, can pin the Fermi level of the n-contact metal and leads to a change of the built-in potential [575].

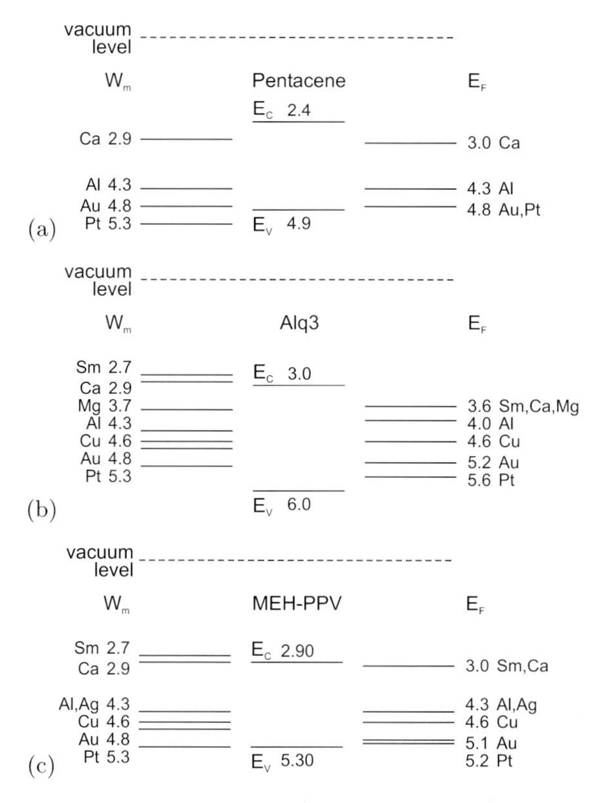

Fig. 18.21. Measured Fermi energies E_F (*labeled data* in eV) and the work functions W_m of various metals contacting (**a**) pentacene, (**b**) Alq3 and (**c**) MEH-PPV. E_C (E_V) denotes the energy position of the electron (hole) polaron. Measured data for E_F for MEH-PPV from [573], other from [574]. Data for W_m from Table 18.1

18.3 Metal–Insulator–Semiconductor Diodes

In a metal–insulator–semiconductor (MIS) diode an insulator is sandwiched between the metal and the semiconductor. Subsequently, a MIS contact has

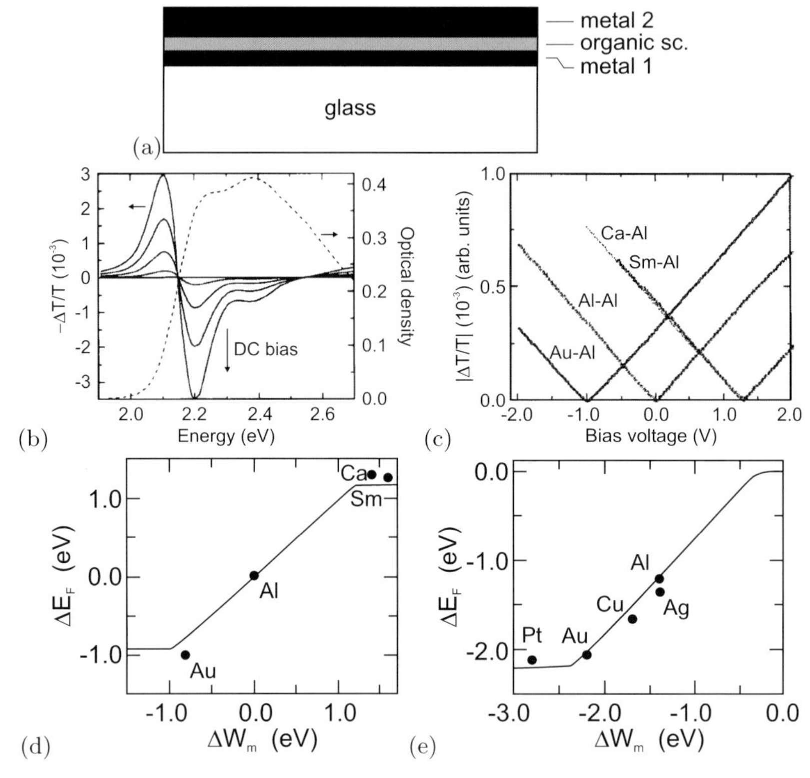

Fig. 18.22. (**a**) Schematic MSM structure with organic semiconductor (sc.) on transparent glass substrate. Metal 1 is thin and semitransparent. Thickness of organic semiconductor (polymer or small molecules) is about 50 nm. (**b**) Electroabsorption spectra of Al/MEH-PPV/Al structure at four dc bias voltages (*solid lines*) and optical density spectrum (*dashed line*). (**c**) Magnitude of the electroabsorption response at 2.1 eV as a function of bias for metal/MEH-PPV/Al structures. (**d,e**) Calculated (*solid lines*) and experimental (*points*) potential difference across (**d**) metal/MEH-PPV/Al structures and (**e**) metal/MEH-PPV/Ca structures as a function of the work-function difference of the contacts. Parts (**b**)–(**e**) adapted from [573]

zero dc conductance. The semiconductor typically has an ohmic back contact. As insulator, often the oxide of the respective semiconductor, is used. In particular SiO_2 on Si has been technologically advanced (Fig. 18.23). In the latter case, the diode is called a MOS (metal–oxide–semiconductor) diode. This structure has great importance for the investigation of semiconductor surfaces and overwhelming importance for semiconductor technology (planar integration of electronic circuits, CMOS technology). Also, CCDs (Sect. 19.3.8) are based on MIS diodes.

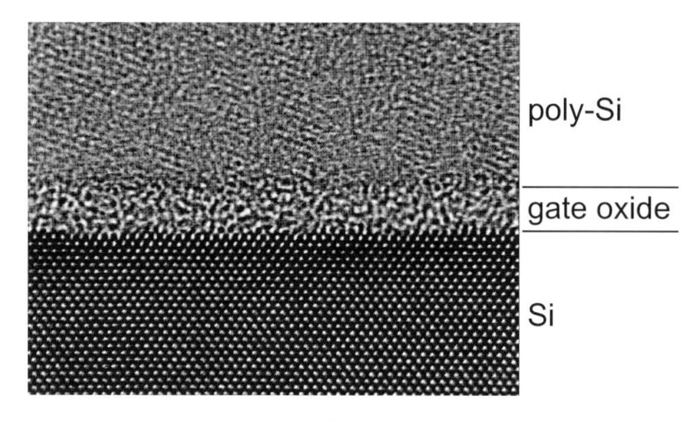

Fig. 18.23. High-resolution transmission electron microscopy image of a 1.6-nm thick gate oxide between poly-Si (see Sect. 21.5.4) and crystalline Si. From [576]

18.3.1 Band Diagram for Ideal MIS Diode

An ideal MIS diode has to fulfill the following three conditions:

(i) (as shown in Fig. 18.24) without external bias the energy difference ϕ_{ms} between the work function of the metal and the semiconductor

$$\phi_{\mathrm{ms}} = \phi_{\mathrm{m}} - \left(\chi_{\mathrm{s}} - \frac{E_{\mathrm{g}}}{2e} \pm \Psi_{\mathrm{B}}\right) \tag{18.51}$$

is zero ($\phi_{\mathrm{ms}} = 0$). The '+' ('−') sign in (18.51) applies to a p-type, Fig. 18.24b (n-type, Fig. 18.24a) semiconductor. Ψ_{B} is the difference between the intrinsic and actual Fermi level, $\Psi_{\mathrm{B}} = |E_{\mathrm{i}} - E_{\mathrm{F}}|/e > 0$.

(ii) The only charges present are those in the semiconductor and the opposite charge is on the metal surface close to the insulator.

(iii) There is no dc current between the metal and the semiconductor, i.e. the conductivity of the insulator is zero.

When an ideal MIS diode is biased, three general cases – accumulation, depletion and inversion – can occur (Fig. 18.25). We discuss these first for the p-type semiconductor.

Figure 18.25a shows the accumulation case for a negative voltage at the metal.[10] Part of the voltage drops across the insulator, the rest across the semiconductor. The valence band is bent upwards towards the Fermi level. The quasi-Fermi level in the semiconductor, however, is constant since no dc current flows.[11]

[10]This poling is a forward bias of the respective Schottky diode since the positive pole is at the p-type semiconductor.

[11]We note that in order to reach the situations shown in Fig. 18.25 from the zero bias case of Fig. 18.24, a current must have flowed since charge carriers are

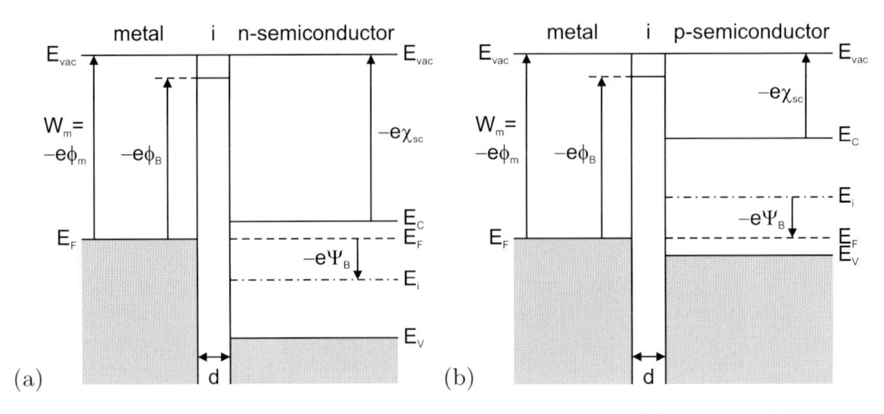

(a) (b)

Fig. 18.24. Band diagram of an ideal MIS diode with (**a**) n- and (**b**) p-type semiconductor at external bias $V = 0$. The insulator ('i') thickness is d as labeled. The *dash-dotted line* represents the intrinsic Fermi level E_i

Since the charge-carrier (hole) density depends exponentially on the energy separation $E_\mathrm{F} - E_\mathrm{V}$, a charge accumulation (of holes) occurs in the semiconductor in the vicinity of the interface to the insulator.

In Fig. 18.25b the depletion case is shown. Now a moderate reverse voltage, i.e. a positive bias to the metal, is applied. A depletion of majority charge carriers occurs in the semiconductor close to the insulator. The quasi-Fermi level in the semiconductor remains beneath the intrinsic level ($E_\mathrm{i} \approx E_\mathrm{C} + E_\mathrm{g}/2$), i.e. the semiconductor remains p-type everywhere. If the voltage is increased further to large values, the quasi-Fermi level intersects the intrinsic level and lies *above* E_i close to the insulator. In this region, the electron concentration becomes larger than the hole concentration and we have the inversion case. The inversion is called 'weak' if the Fermi level is still close to E_i. The inversion is called 'strong' when the Fermi level lies close to the conduction-band edge.

The corresponding phenomena occur for n-type semiconductors for the opposite signs of the voltage with electron accumulation and depletion. In the inversion case, $p > n$ close to the insulator.

18.3.2 Space-Charge Region

Now we calculate the charge and electric field distribution in an ideal MIS diode, following the treatment in [577]. We introduce the potential Ψ that measures the separation of the intrinsic bulk Fermi level and the actual intrinsic level E_i, i.e. $-e\Psi(x) = E_\mathrm{i}(x) - E_\mathrm{i}(x \to \infty)$ (see Fig. 18.26). Its value

---

redistributed. Figure 18.25 depicts the stationary equilibrium after transient voltage switch-on effects have subsided. The time, however, that is needed in order to reach the stationary equilibrium from zero bias (thermal equilibrium) may be very long (days, see Sect. 19.3.8).

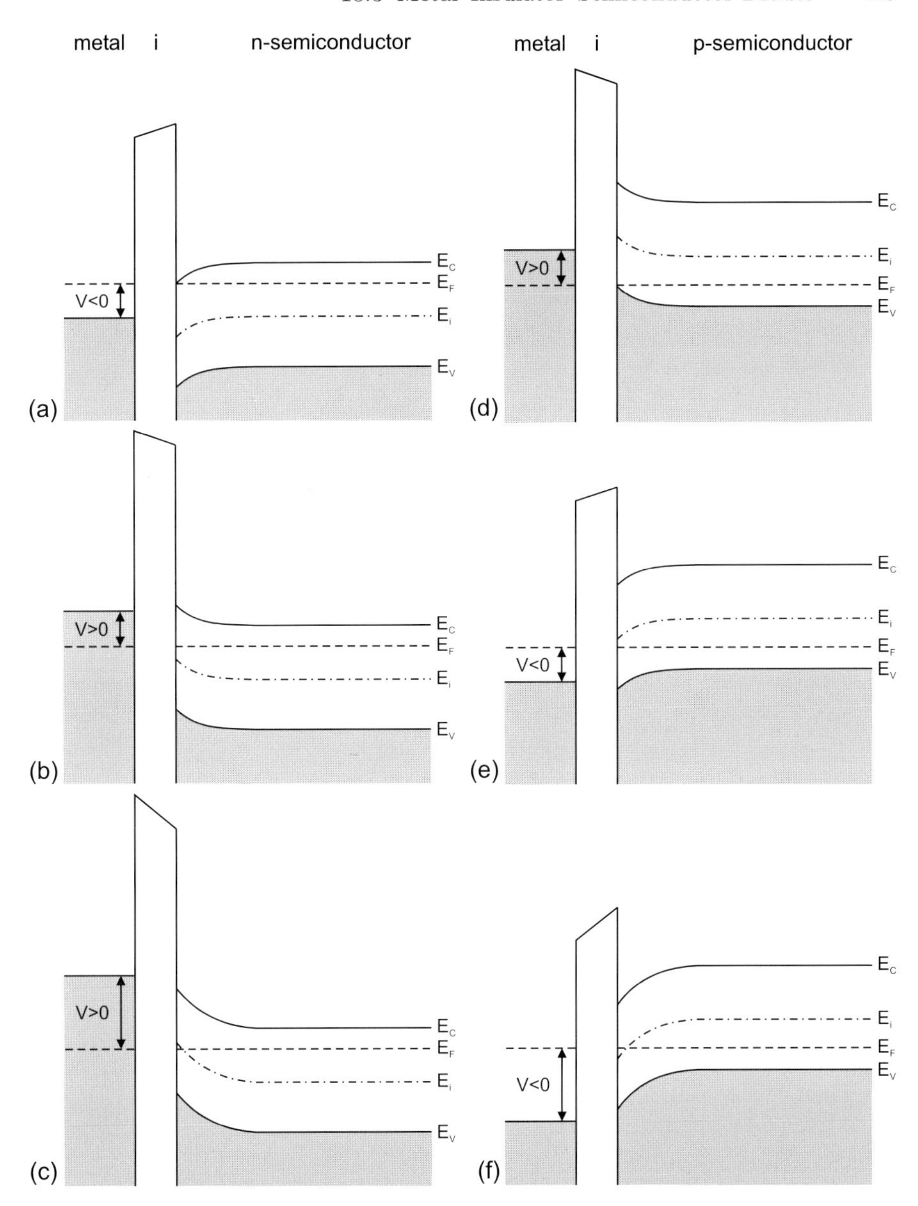

Fig. 18.25. Band diagram of ideal MIS diodes with (**a**,**b**,**c**) n-type and (**d**,**e**,**f**) p-type semiconductors for $V \neq 0$ in stationary equilibrium for the cases (**a**,**d**) accumulation, (**b**,**e**) depletion and (**c**,**f**) inversion

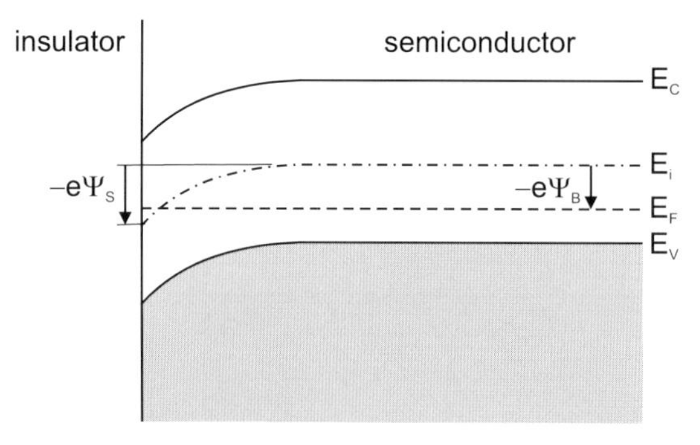

Fig. 18.26. Band diagram at the surface of a p-type semiconductor of a MIS diode. Accumulation occurs for $\Psi_{\mathrm{s}} < 0$, depletion for $\Psi_{\mathrm{s}} > 0$ and inversion (as shown here) for $\Psi_{\mathrm{s}} > \Psi_{\mathrm{B}} > 0$

at the surface is termed Ψ_{s}, the surface potential. The value is positive, i.e. $\Psi_{\mathrm{s}} > 0$, if the intrinsic Fermi level at the surface is below the bulk Fermi level.

The electron and hole concentrations are given (for a p-type semiconductor) as

$$n_{\mathrm{p}} = n_{\mathrm{p0}} \exp(\beta\Psi) \tag{18.52a}$$

$$p_{\mathrm{p}} = p_{\mathrm{p0}} \exp(-\beta\Psi) \ , \tag{18.52b}$$

where n_{p0} (p_{p0}) are the bulk electron (hole) concentrations, respectively, and $\beta = e/kT > 0$.

Therefore, the net free charge is given by

$$n_{\mathrm{p}} - p_{\mathrm{p}} = n_{\mathrm{p0}} \exp(\beta\Psi) - p_{\mathrm{p0}} \exp(-\beta\Psi) \ . \tag{18.53}$$

The electron and hole concentrations at the surface are denoted with an index 's' and are given by[12]

$$n_{\mathrm{s}} = n_{\mathrm{p0}} \exp(\beta\Psi_{\mathrm{s}}) \tag{18.54a}$$

$$p_{\mathrm{s}} = p_{\mathrm{p0}} \exp(-\beta\Psi_{\mathrm{s}}) \ . \tag{18.54b}$$

We use the Poisson equation $\frac{\mathrm{d}^2\Psi}{\mathrm{d}x^2} = -\frac{\rho}{\epsilon_{\mathrm{s}}}$ with the charge given by

$$\rho(x) = e\left[p(x) - n(x) + N_{\mathrm{D}}^{+}(x) - N_{\mathrm{A}}^{-}(x)\right] \ . \tag{18.55}$$

As boundary condition we employ that far away from the surface (for $x \to \infty$) there is charge neutrality (cf. (7.47)), i.e.

[12] Ψ_{s} represents the voltage drop across the semiconductor that will be discussed in more detail in Sect. 18.3.3. In this sense, Ψ_{s} for the MIS diode is related to $V_{\mathrm{bi}} - V$ for the Schottky contact.

$$n_{\rm po} - p_{\rm po} = N_{\rm D}^{+} - N_{\rm A}^{-} \,, \tag{18.56}$$

and that $\Psi = 0$. We note that $N_{\rm D}^{+} - N_{\rm A}^{-}$ must be constant throughout the completely ionized, homogeneous semiconductor. Therefore (18.56) (but not charge neutrality) holds everywhere in the semiconductor. Using (18.53) the Poisson equation reads

$$\frac{\partial^2 \Psi}{\partial x^2} = -\frac{e}{\epsilon_{\rm s}} \{p_{\rm po}\left[\exp(-\beta\Psi) - 1\right] - n_{\rm po}\left[\exp(\beta\Psi) - 1\right]\} \,. \tag{18.57}$$

The Poisson equation is integrated and with the notations

$$\mathcal{F}(\Psi) = \left[(\exp(-\beta\Psi) + \beta\Psi - 1) + \frac{n_{\rm po}}{p_{\rm po}}(\exp(\beta\Psi) - \beta\Psi - 1)\right]^{1/2} \tag{18.58a}$$

$$L_{\rm D} = \left[\frac{\epsilon_{\rm s} kT}{e^2 p_{\rm po}}\right]^{1/2} = \left[\frac{\epsilon_{\rm s}}{e\beta p_{\rm po}}\right]^{1/2} \,, \tag{18.58b}$$

with $L_{\rm D}$ being the Debye length for holes, the electric field can be written as

$$\mathcal{E} = -\frac{\partial \Psi}{\partial x} = \pm \frac{\sqrt{2}kT}{eL_{\rm D}} \mathcal{F}(\Psi) \,. \tag{18.59}$$

The positive (negative) sign is for $\Psi > 0$ ($\Psi < 0$), respectively. At the surface, $\Psi_{\rm s}$ will be taken as the value for Ψ. The total charge $Q_{\rm s}$ per unit area creating the surface field

$$\mathcal{E}_{\rm s} = -\frac{\partial \Psi}{\partial x} = \pm \frac{\sqrt{2}kT}{eL_{\rm D}} \mathcal{F}(\Psi_{\rm s}) \tag{18.60}$$

is given by Gauss's law as $Q_{\rm s} = -\epsilon_{\rm s}\mathcal{E}_{\rm s}$.

The dependence of the space-charge density from the surface potential[13] is depicted in Fig. 18.27. If $\Psi_{\rm s}$ is negative, $\mathcal{F}$ is dominated by the first term in (18.58a) and the space charge is positive (accumulation) and proportional to $Q_{\rm s} \propto \exp(-\beta|\Psi|/2)$. For $\Psi_{\rm s} = 0$ the (ideal) MIS diode has the flat-band condition and the space charge is zero. For $0 \leq \Psi_{\rm s} \leq \Psi_{\rm B}$ the space charge is negative (depletion) and $\mathcal{F}$ is dominated by the second term in (18.58a), i.e. $\Psi_{\rm s} \propto \sqrt{\Psi_{\rm s}}$. For $\Psi_{\rm s} \gg \Psi_{\rm B}$ we are in the (weak) inversion regime and the space charge is dominated by the fourth term in (18.58a), $Q_{\rm s} \propto -\exp(-\beta\Psi/2)$. Strong inversion starts at about $\Psi_{\rm s}^{\rm inv} \approx 2\Psi_{\rm B} = 2kT\ln(N_{\rm A}/n_{\rm i})/e$.

For the case of strong inversion the band diagram is shown in Fig. 18.28 together with the charge, field and potential. The total voltage drop V across the MIS diode is

$$V = V_{\rm i} + \Psi_{\rm s} \,, \tag{18.61}$$

[13]We note that we discuss the space-charge region now only with regard to $\Psi_{\rm s}$, the voltage drop across the semiconductor, and the dependence of $\Psi_{\rm s}$ on the bias of the diode will be discussed in the next section.

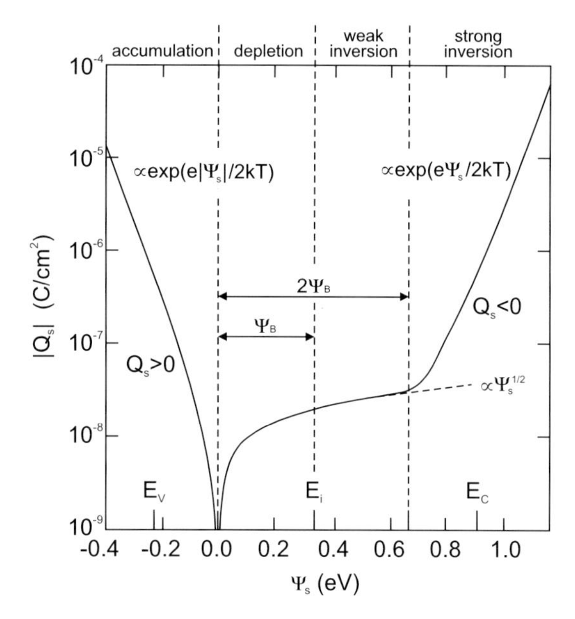

Fig. 18.27. Dependence of the space charge on the surface potential Ψ_s for p-type silicon with $N_\mathrm{A} = 4 \times 10^{15}\,\mathrm{cm}^{-3}$ at $T = 300\,\mathrm{K}$. The flat-band condition is present for $\Psi_\mathrm{s} = 0$

with V_i being the voltage drop across the insulator. In the case of inversion, the charge (per unit area) in the space-charge region

$$Q_\mathrm{s} = Q_\mathrm{d} + Q_\mathrm{n} \tag{18.62}$$

is composed of the depletion charge (ionized acceptors)

$$Q_\mathrm{d} = -ewN_\mathrm{A}\ , \tag{18.63}$$

with w being the width of the depletion region, and the inversion charge Q_n, which is present only close to the interface.

The metal surface carries the opposite charge

$$Q_\mathrm{m} = -Q_\mathrm{s} \tag{18.64}$$

due to global charge neutrality. The insulator itself does not contribute charges in the case of an ideal MIS diode.

18.3.3 Capacity

The insulator represents a capacitor with the dielectric constant ϵ_i and a thickness d. Therefore, the capacity is

$$C_\mathrm{i} = \frac{\epsilon_\mathrm{i}}{d}\ . \tag{18.65}$$

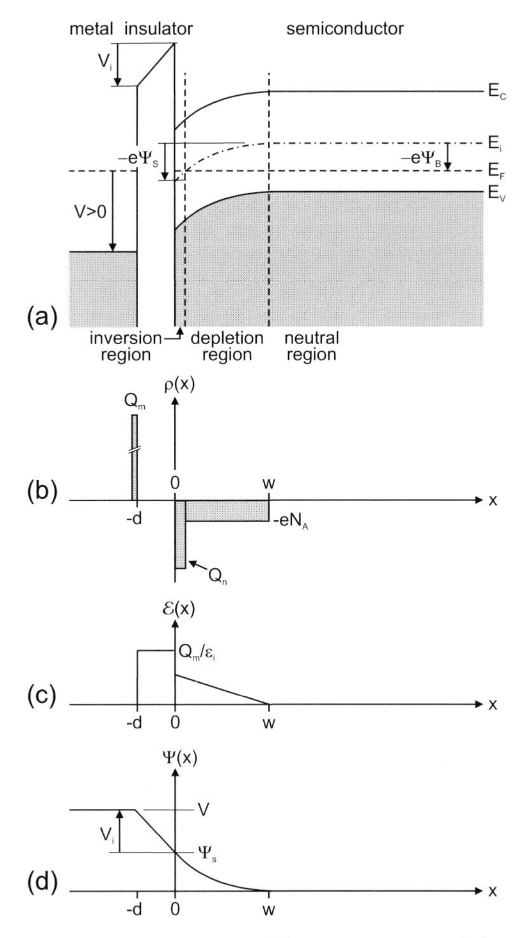

Fig. 18.28. Ideal MIS diode at inversion: (**a**) band diagram, (**b**) charge distribution, (**c**) electric field and (**d**) potential

Between the charges $-Q_\mathrm{s}$ and Q_s the field strength $\mathcal{E}_\mathrm{i}$ in the insulator is

$$\mathcal{E}_\mathrm{i} = \frac{|Q_\mathrm{s}|}{\epsilon_\mathrm{i}} . \tag{18.66}$$

The voltage drop V_i across the insulator is given by

$$V_\mathrm{i} = \mathcal{E}_\mathrm{i}\, d = \frac{|Q_\mathrm{s}|}{C_\mathrm{i}} . \tag{18.67}$$

The total capacity C of the MIS diode is given by the insulator capacitance in series with the capacitance C_d of the depletion layer

$$C = \frac{C_\mathrm{i}\, C_\mathrm{d}}{C_\mathrm{i} + C_\mathrm{d}} . \tag{18.68}$$

The capacity of the space-charge region varies with the applied bias (Fig. 18.29). For forward bias (accumulation), the capacity of the space-charge region is high. Therefore, the total capacity of the MIS diode is given by the insulator capacitance $C \approx C_{\mathrm{i}}$. When the voltage is reduced, the capacity of the space-charge region drops to $C_{\mathrm{d}} = \epsilon_{\mathrm{s}}/L_{\mathrm{D}}$ for the flat-band case ($\Psi_{\mathrm{s}} = 0$). For a high reverse voltage, the semiconductor is inverted at the surface and the space-charge region capacity is high again. In this case, the total capacity is given by $C \approx C_{\mathrm{i}}$ again.

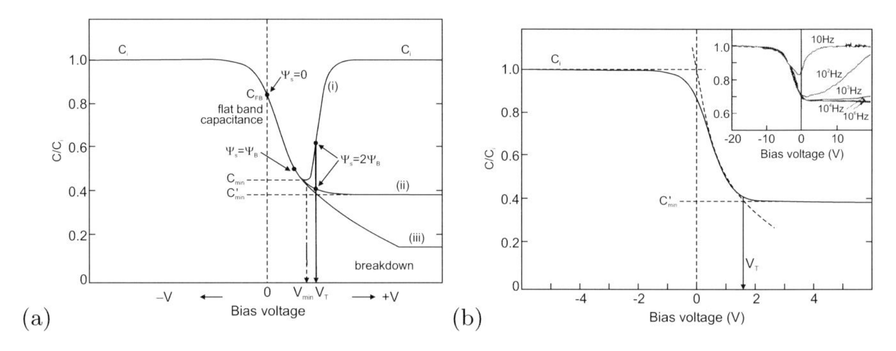

Fig. 18.29. (**a**) Schematic dependence of the capacity of a MIS diode on the bias for (**i**) low frequencies, (**ii**) high frequencies and (**iii**) deep depletion. (**b**) High-frequency capacity of a Si/SiO$_2$ diode ($N_{\mathrm{A}} = 1.45 \times 10^{16}\,\mathrm{cm}^{-3}$, $d = 200\,\nu$). The *inset* shows the frequency dependence for various frequencies as labeled. Adapted from [578]

The previous consideration assumes that the charge density in the semiconductor can follow changes of the bias sufficiently fast.[14] The inversion charge must disappear via recombination that is limited by the recombination time constant τ. For frequencies around τ^{-1} or faster, the charge in the inversion layer cannot follow and the capacitance of the semiconductor is given by the value $C_{\mathrm{d}} \cong \epsilon_{\mathrm{s}}/w_{\mathrm{m}}$. w_{m} (Fig. 18.30) is the maximum depletion-layer width present at the beginning of inversion (cf. (18.8))

$$w_{\mathrm{m}} \cong \left(\frac{2\epsilon_{\mathrm{s}}}{eN_{\mathrm{A}}} \Psi_{\mathrm{s}}^{\mathrm{inv}} \right)^{1/2} = \left(\frac{4\epsilon_{\mathrm{s}} kT \ln(N_{\mathrm{A}}/n_{\mathrm{i}})}{e^2 N_{\mathrm{A}}} \right)^{1/2} . \tag{18.69}$$

For further increased voltage (into the inversion regime), the electric field is screened by the inversion charge and the width of the depletion layer remains constant. Therefore, the total capacity in the inversion regime is given by

$$C \cong \frac{\epsilon_{\mathrm{i}}}{d + \frac{\epsilon_{\mathrm{i}}}{\epsilon_{\mathrm{s}}} w_{\mathrm{m}}} . \tag{18.70}$$

[14]Typically, a dc bias voltage V is set and the capacity is sampled with a small ac voltage of amplitude ΔV, with $\delta V \ll V$

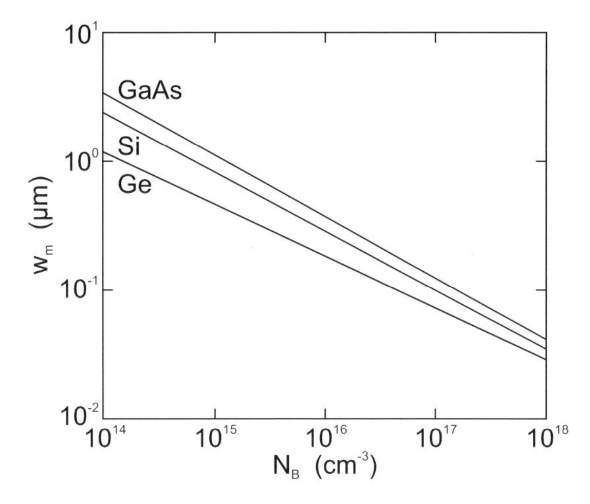

Fig. 18.30. Maximum width of the depletion layer $w_{\rm m}$ (18.69) at room temperature for deep depletion for GaAs, Si and Ge diodes as a function of bulk doping level

18.3.4 Nonideal MIS Diode

In a real, i.e. nonideal, MIS diode, the difference $\phi_{\rm ms}$ in the work functions of the metal and semiconductor (cf. (18.51)) is no longer zero. Therefore, the capacitance vs. voltage relation is shifted with respect to the ideal MIS diode characteristic by the *flat-band voltage shift* $V_{\rm FB}$

$$V_{\rm FB} = \phi_{\rm ms} - \frac{Q_{\rm ox}}{C_{\rm i}} \,. \tag{18.71}$$

Additionally, the flat-band voltage can be shifted by charges $Q_{\rm ox}$ in the oxide that have been neglected so far. Such charges can be trapped, i.e. fixed with regard to their spatial position, or mobile, e.g. ionic charges such as sodium.

For Al as metal ($\phi_{\rm m} = 4.1\,{\rm eV}$) and n-type Si ($\phi_{\rm s} = 4.35\,{\rm eV}$), the flat-band voltage shift is $\phi_{\rm ms} = -0.25\,{\rm V}$, as shown schematically in Fig. 18.31a for zero bias. $V_{\rm FB}$ is split into 0.2 eV and 0.05 eV for the oxide and the silicon, respectively. In Fig. 18.31b, the dependence of $\phi_{\rm ms}$ on the doping, conductivity type and metal is shown for various SiO_2–Si MIS diodes. An Au-SiO_2–Si diode with p-type Si and $N_{\rm A} \approx 10^{15}\,{\rm cm}^{-3}$ fulfills the condition of an ideal MIS diode with regard to $\phi_{\rm ms} = 0$.

18.4 Bipolar Diodes

A large class of diodes is based on pn-junctions. In a homo pn-junction an n-doped region is next to a p-doped region of the same semiconductor. Such

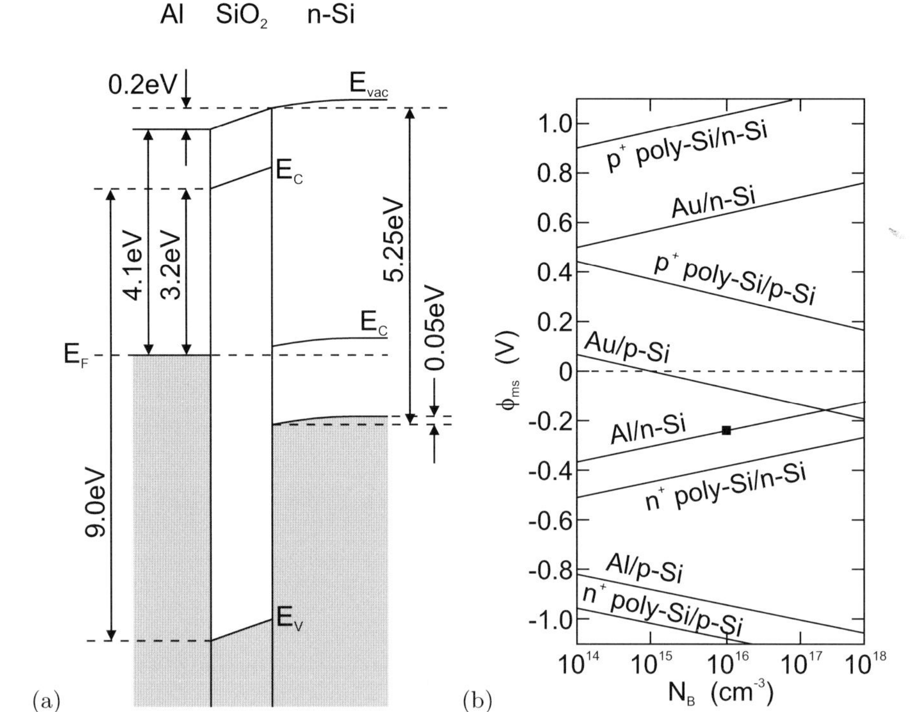

Fig. 18.31. (**a**) Schematic band diagram of an Al–SiO_2–Si (n-type) diode with 50 nm oxide thickness and $N_D = 10^{16}\,\mathrm{cm}^{-3}$ for zero bias. Based on data from [579]. (**b**) Difference of work functions ϕ_{ms} for SiO_2–Si MIS diodes and various doping levels and electrode materials (Al, Au and polycrystalline Si). The *square* represents the situation of part (**a**). Based on data from [580]

a device is called *bipolar*. At the junction a depletion region forms. The transport properties are determined by the minority carriers. An important variation is the pin-diode in which an intrinsic (or lowly doped) region is between the doped region (Sect. 18.5.8). If the differently doped regions belong to different semiconductor materials, the diode is a heterostructure pn-diode (Sect. 18.5.11).

18.4.1 Band Diagram

If the doping profile is arbitrarily sharp, the junction is called *abrupt*. This geometry is the case for epitaxial pn-junctions where the differently doped layers are grown on top of each other.[15] For junctions that are fabricated by diffusion, the abrupt approximation is suitable for alloyed, ion-implanted

[15]The choices of dopant and the growth conditions, in particular the temperature, need to be made such that no interdiffusion of the dopants takes place.

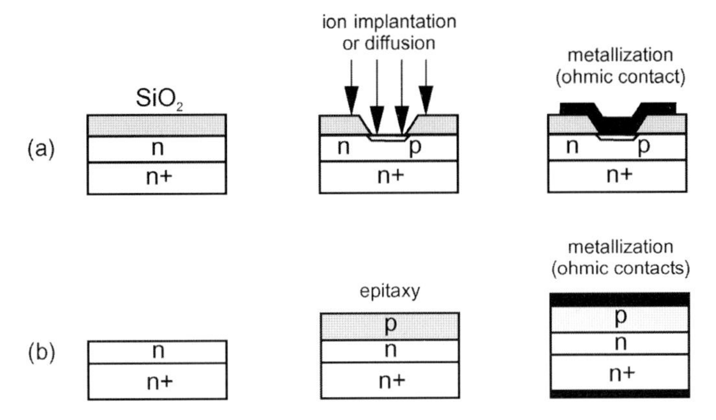

Fig. 18.32. Schematic fabrication technologies for bipolar diodes: (**a**) Planar junction with local impurity incorporation (diffusion from gas phase or ion implantation) through mask and contact metallization, (**b**) epitaxial junction

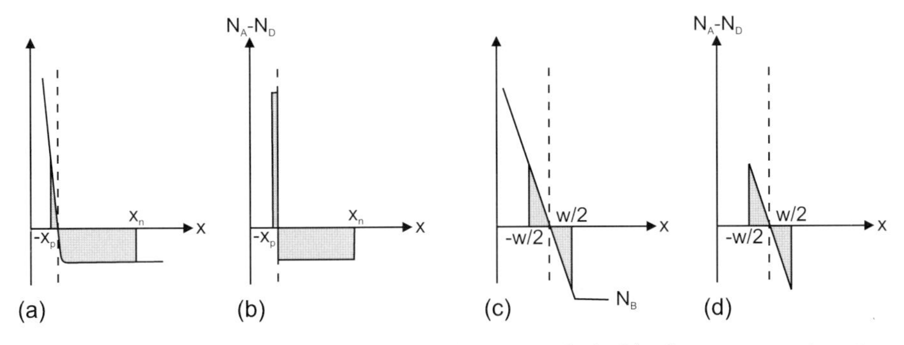

Fig. 18.33. Description of doping distribution with (**a**,**b**) abrupt approximation and with (**c**,**d**) linearly graded junction. (**a**,**c**) show real impurity concentration, (**b**,**d**) idealized doping profile

and shallow-diffused junctions. For deep-diffused junctions a linearly graded approximation is better, which is treated in more detail in [183]. If one doping level is much higher that the other, the junction is termed a *one-sided* (abrupt) junction. If $n \gg p$ ($p \gg n$), the junction is denoted as an n^+p-diode (p^+n-diode).

18.4.2 Space-Charge Region

We first consider the pn-diode in thermodynamical equilibrium, i.e. without external bias. No net current flows in this case. The electron current is given by (8.42a)

$$\mathbf{j}_{\mathrm{n}} = -e\mu_{\mathrm{n}} n \mathcal{E} - \mu_{\mathrm{n}} n \nabla E_{\mathrm{F}} \; , \tag{18.72}$$

and therefore in the absence of electrical fields and zero net electron and hole currents the Fermi energy is constant:

$$\nabla E_{\mathrm{F}} = 0 \,. \tag{18.73}$$

The built-in voltage V_{bi} is given by (see Fig. 18.34d)

$$eV_{\mathrm{bi}} = E_{\mathrm{g}} + eV_{\mathrm{n}} + eV_{\mathrm{p}} \,, \tag{18.74}$$

where V_{n} is the difference between conduction band and Fermi level on the n-side, $-eV_{\mathrm{n}} = E_{\mathrm{C}} - E_{\mathrm{F}}$. V_{p} is the difference between valence band and Fermi level on the p-side, $-eV_{\mathrm{p}} = E_{\mathrm{F}} - E_{\mathrm{V}}$. For the nondegenerate semiconductor $V_{\mathrm{n}}, V_{\mathrm{p}} < 0$ and (using (7.17), (7.15) and (7.16))

$$\begin{aligned} eV_{\mathrm{bi}} &= kT \ln\left(\frac{N_{\mathrm{C}} N_{\mathrm{V}}}{n_{\mathrm{i}}^2}\right) - \left[kT \ln\left(\frac{N_{\mathrm{C}}}{n_{\mathrm{n_0}}}\right) + kT \ln\left(\frac{N_{\mathrm{V}}}{p_{\mathrm{p_0}}}\right)\right] \\ &= kT \ln\left(\frac{p_{\mathrm{p_0}} n_{\mathrm{n_0}}}{n_{\mathrm{i}}^2}\right) = kT \ln\left(\frac{p_{\mathrm{p_0}}}{p_{\mathrm{n_0}}}\right) = kT \ln\left(\frac{n_{\mathrm{n_0}}}{n_{\mathrm{p_0}}}\right) \qquad (18.75\mathrm{a}) \\ &\cong kT \ln\left(\frac{N_{\mathrm{A}} N_{\mathrm{D}}}{n_{\mathrm{i}}^2}\right) \,. \qquad (18.75\mathrm{b}) \end{aligned}$$

The electron and hole densities on either side of the junction ($n_{\mathrm{p_0}}$ and $p_{\mathrm{p_0}}$ at $x = -x_{\mathrm{p}}$ and $n_{\mathrm{n_0}}$ and $p_{\mathrm{n_0}}$ at $x = x_{\mathrm{n}}$) are related to each other by (from rewriting (18.75a))

$$n_{\mathrm{p_0}} = n_{\mathrm{n_0}} \exp\left(-\frac{eV_{\mathrm{bi}}}{kT}\right) \tag{18.76a}$$

$$p_{\mathrm{n_0}} = p_{\mathrm{p_0}} \exp\left(-\frac{eV_{\mathrm{bi}}}{kT}\right) \,. \tag{18.76b}$$

Microscopically, the equilibration of the Fermi levels on the n- and p-side occurs via the diffusion of electrons and holes to the p- and n-side, respectively. The electrons and holes recombine in the depletion layer. Therefore, on the n-side the ionized donors and on the p-side the ionized acceptors remain (Fig. 18.34a). These charges build up an electrical field that works against the diffusion current (Fig. 18.34d). At thermal equilibrium the diffusion and drift currents cancel and the Fermi level is constant.

Values for the built-in potential are depicted in Fig. 18.35 for Si and GaAs diodes. The spatial dependence of the potential in the depletion layer is determined by the Poisson equation.

We assume here the complete ionization of the donors and acceptors. Also, we neglect at first majority carriers in the depletion layers on the n- and p-sides.[16] With these approximations, the Poisson equation in the depletion layers on the n- and p-side reads

[16] An abrupt decrease of the majority carrier density at the border of the space-charge region corresponds to zero temperature.

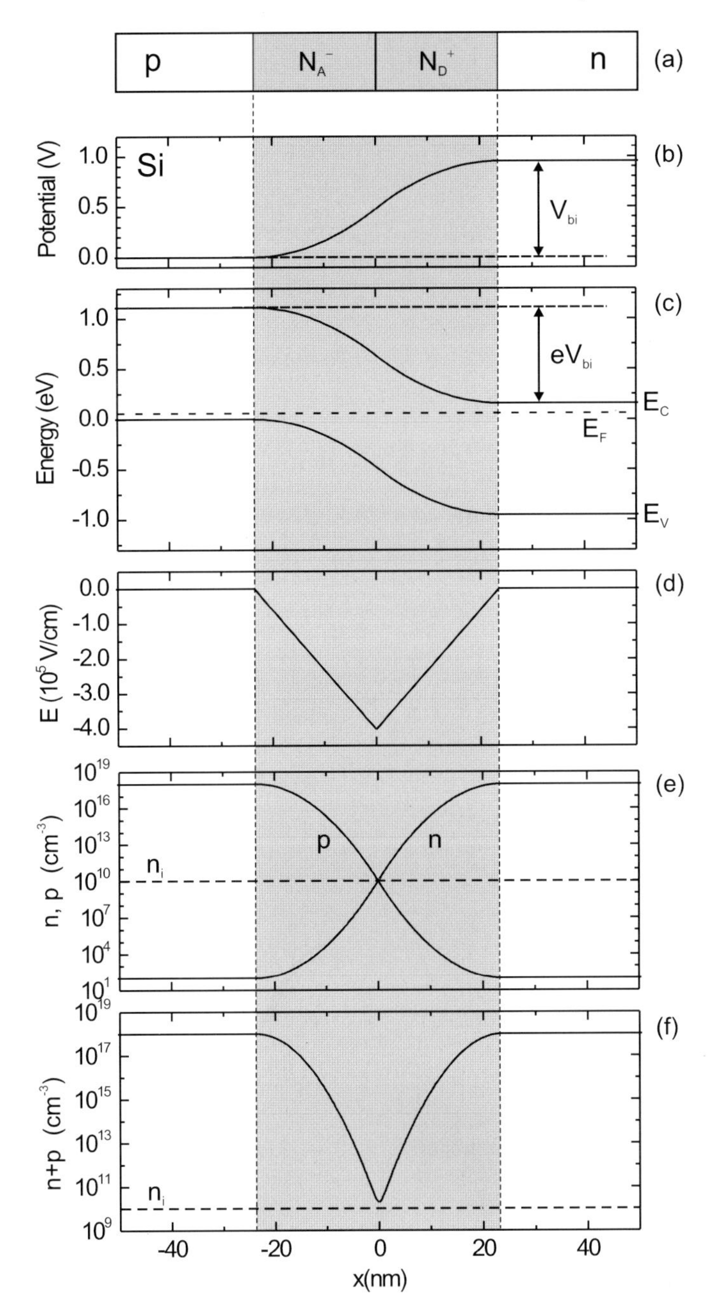

Fig. 18.34. (Symmetric) pn-junction (abrupt approximation) in thermal equilibrium (zero bias) for Si at room temperature and $N_{\mathrm{A}} = N_{\mathrm{D}} = 10^{18}\,\mathrm{cm}^{-3}$. (**a**) Schematic representation of p-doped and n-doped region with depletion layer (*grey area*) and fixed space charges, (**b**) diffusion potential, (**c**) band diagram with Fermi level (*dashed line*), (**d**) electric field, (**e**) free-carrier concentrations n and p and (**f**) total free-carrier density $n + p$

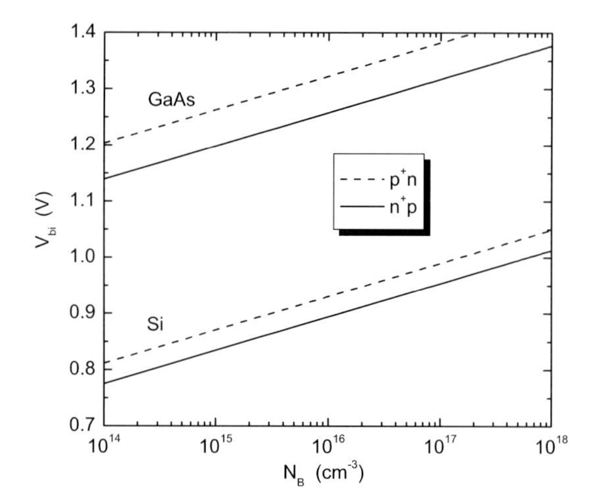

Fig. 18.35. Built-in voltage as a function of doping for one-sided Si and GaAs pn-diodes

$$\frac{\partial^2 V}{\partial x^2} = \frac{e N_{\mathrm{D}}}{\epsilon_{\mathrm{s}}}, \; 0 \leq x \leq x_{\mathrm{n}} \tag{18.77a}$$

$$\frac{\partial^2 V}{\partial x^2} = -\frac{e N_{\mathrm{A}}}{\epsilon_{\mathrm{s}}}, \; -x_{\mathrm{p}} \leq x \leq 0 \,. \tag{18.77b}$$

One integration yields (together with the boundary conditions that the field is zero at the boundaries of the depletion layer) the electrical field in the two regions

$$\mathcal{E}(x) = \frac{e}{\epsilon_{\mathrm{s}}} N_{\mathrm{D}}(x - x_{\mathrm{n}}), \; 0 \leq x \leq x_{\mathrm{n}} \tag{18.78a}$$

$$\mathcal{E}(x) = -\frac{e}{\epsilon_{\mathrm{s}}} N_{\mathrm{A}}(x + x_{\mathrm{p}}), \; -x_{\mathrm{p}} \leq x \leq 0 \,. \tag{18.78b}$$

The maximum field strength $\mathcal{E}_{\mathrm{m}}$ is present at $x = 0$ and is given by

$$\mathcal{E}_{\mathrm{m}} = -\frac{e N_{\mathrm{D}} x_{\mathrm{n}}}{\epsilon_{\mathrm{s}}} = -\frac{e N_{\mathrm{A}} x_{\mathrm{p}}}{\epsilon_{\mathrm{s}}} \,. \tag{18.79}$$

The continuity of the field at $x = 0$ is equivalent to the overall charge neutrality

$$N_{\mathrm{D}} x_{\mathrm{n}} = N_{\mathrm{A}} x_{\mathrm{p}} \,. \tag{18.80}$$

Another integration yields the potential (setting $V(x = 0) = 0$)

$$V(x) = -\mathcal{E}_{\mathrm{m}} \left(x - \frac{x^2}{2 x_{\mathrm{n}}} \right), \; 0 \leq x \leq x_{\mathrm{n}} \tag{18.81a}$$

$$V(x) = -\mathcal{E}_{\mathrm{m}} \left(x + \frac{x^2}{2 x_{\mathrm{p}}} \right), \; -x_{\mathrm{p}} \leq x \leq 0 \,. \tag{18.81b}$$

The built-in potential $V_{\mathrm{bi}} = V(x_{\mathrm{n}}) - V(-x_{\mathrm{p}}) > 0$ is related to the maximum field via

$$V_{\mathrm{bi}} = -\frac{1}{2}\mathcal{E}_{\mathrm{m}} w \ , \tag{18.82}$$

where $w = x_{\mathrm{n}} + x_{\mathrm{p}}$ is the total width of the depletion layer. The elimination of $\mathcal{E}_{\mathrm{m}}$ from (18.79) and (18.82) yields

$$w = \left[\frac{2\epsilon_{\mathrm{s}}}{e}\left(\frac{N_{\mathrm{A}}+N_{\mathrm{D}}}{N_{\mathrm{A}}N_{\mathrm{D}}}\right) V_{\mathrm{bi}}\right]^{1/2} . \tag{18.83}$$

For $\mathrm{p^+n}$ and $\mathrm{n^+p}$ junctions, the width of the depletion layer is determined by the lowly doped side of the junction

$$w = \left[\frac{2\epsilon_{\mathrm{s}}}{eN_{\mathrm{B}}} V_{\mathrm{bi}}\right]^{1/2} , \tag{18.84}$$

where N_{B} denotes the doping of the lowly doped side, i.e. N_{A} for a $\mathrm{n^+p}$ diode and N_{D} for a $\mathrm{p^+n}$ diode.

If the spatial variation of the majority carrier density is considered in more detail (and for finite temperature, cf. (18.11)), an additional term $-2kT/e = -2/\beta$ is added [577] to V_{bi}

$$w = \left[\frac{2\epsilon_{\mathrm{s}}}{e}\left(\frac{N_{\mathrm{A}}+N_{\mathrm{D}}}{N_{\mathrm{A}}N_{\mathrm{D}}}\right)\left(V_{\mathrm{bi}} - V - \frac{2kT}{e}\right)\right]^{1/2} . \tag{18.85}$$

Also, the external bias V has been included in the formula. If w_0 denotes the depletion layer width at zero bias, the depletion layer width for a given voltage V can be written as

$$w(V) = w_0 \left[1 - \frac{V}{V_{\mathrm{bi}} - 2/\beta}\right]^{1/2} \approx w_0 \left[1 - \frac{V}{V_{\mathrm{bi}}}\right]^{1/2} . \tag{18.86}$$

Using the Debye length (cf. (18.58b))

$$L_{\mathrm{D}} = \left(\frac{\epsilon_{\mathrm{s}} kT}{e^2 N_{\mathrm{B}}}\right)^{1/2} , \tag{18.87}$$

the depletion layer width for a one-sided diode can be written as (with $\beta = e/kT$)

$$w = L_{\mathrm{D}}\sqrt{2(\beta V_{\mathrm{bi}} - \beta V - 2)} \ . \tag{18.88}$$

The Debye length is a function of the doping level and is shown for Si in Fig. 18.36. For a doping level of $10^{16}\,\mathrm{cm^{-3}}$ the Debye length in Si is 40 nm at room temperature. For one-sided junctions the depletion layer width is about $6\,L_{\mathrm{D}}$ for Ge, $8\,L_{\mathrm{D}}$ for Si and $10\,L_{\mathrm{D}}$ for GaAs.

The external bias is counted positive if the '+' ('−') pole is at the p-side (n-side). The reverse voltage has opposite polarity. If a reverse bias is applied, the depletion layer width is increased (Fig. 18.37).

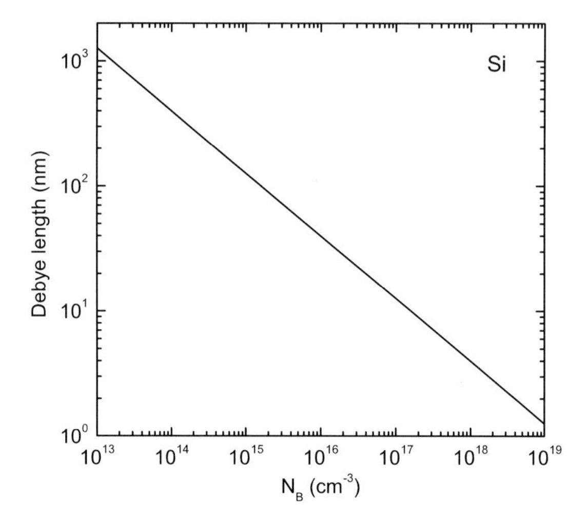

Fig. 18.36. Debye length in Si at room temperature as a function of the doping level N_{B} according to (18.87)

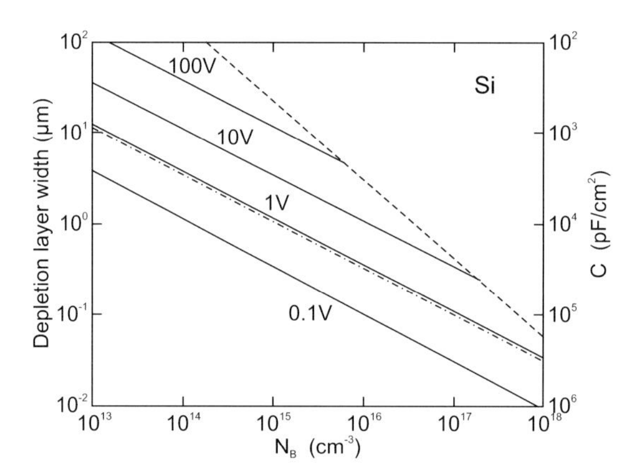

Fig. 18.37. Width of the depletion layer and capacitance per area for one-sided, abrupt Si junctions for various values of $V_{\mathrm{bi}} - V - 2kT/e$ as labeled. The *dash-dotted line* is for zero bias, the *dashed line* is the limit due to avalanche breakdown. Adapted from [183]

18.4.3 Capacitance

The capacity of the depletion layer is the charge change upon a change of the external bias. It is given as

$$C = \left| \frac{\mathrm{d}Q}{\mathrm{d}V} \right| = \frac{\mathrm{d}(eN_{\mathrm{B}}w)}{\mathrm{d}(w^2 eN_{\mathrm{B}}/2\epsilon_{\mathrm{s}})} = \frac{\epsilon_{\mathrm{s}}}{w} = \frac{\epsilon_{\mathrm{s}}}{\sqrt{2}L_{\mathrm{D}}} \left(\beta V_{\mathrm{bi}} - \beta V - 2\right)^{-1/2} . \quad (18.89)$$

Therefore, the capacity of the depletion layer is inversely proportional to the depletion-layer width (see the two scales in Fig. 18.37). A detailed treatment has been given in [581]. $1/C^2$ is proportional to the external bias

$$\frac{1}{C^2} = \frac{2L_\mathrm{D}^2}{\epsilon_\mathrm{s}^2} \left(\beta V_\mathrm{bi} - \beta V - 2\right) . \tag{18.90}$$

From C–V spectroscopy the doping level can be obtained from the slope

$$\frac{\mathrm{d}(1/C^2)}{\mathrm{d}V} = \frac{2\beta L_\mathrm{D}^2}{\epsilon_\mathrm{s}^2} = \frac{2}{e\epsilon_\mathrm{s} N_\mathrm{B}} . \tag{18.91}$$

From the extrapolation to the voltage for which $1/C^2 = 0$ the built-in voltage can be obtained.

18.4.4 Current–Voltage Characteristics

Ideal Current–Voltage Characteristics

Now, the currents in thermodynamical equilibrium ($V = 0$) and under bias are discussed. A diode characteristic will be obtained. We work at first with the following assumptions: abrupt junction, Boltzmann approximation, low injection, i.e. the injected minority-carrier density is small compared to the majority-carrier density, and zero generation current in the depletion layer, i.e. the electron and hole currents are constant throughout the depletion layer. In the presence of a bias, electrons and holes have quasi-Fermi levels and the carrier densities are given by (cf. (7.57a and b))

$$n = N_\mathrm{C} \exp\left(\frac{F_\mathrm{n} - E_\mathrm{C}}{kT}\right) \tag{18.92a}$$

$$p = N_\mathrm{V} \exp\left(-\frac{F_\mathrm{p} - E_\mathrm{V}}{kT}\right) . \tag{18.92b}$$

Using the intrinsic carrier concentration, we can write

$$n = n_\mathrm{i} \exp\left(\frac{F_\mathrm{n} - E_\mathrm{i}}{kT}\right) = n_\mathrm{i} \exp\left[\frac{e(\psi - \phi_\mathrm{n})}{kT}\right] \tag{18.93a}$$

$$p = n_\mathrm{i} \exp\left(-\frac{F_\mathrm{p} - E_\mathrm{i}}{kT}\right) = n_\mathrm{i} \exp\left[\frac{e(\phi_\mathrm{p} - \psi)}{kT}\right] , \tag{18.93b}$$

where ϕ and ψ are the potentials related to the (quasi-) Fermi level and the intrinsic Fermi levels, $-e\phi_\mathrm{n,p} = F_\mathrm{n,p}$ and $-e\psi = E_\mathrm{i}$. The potentials ϕ_n and ϕ_p can also be written as

$$\phi_\mathrm{n} = \psi - \frac{kT}{e} \ln\left(\frac{n}{n_\mathrm{i}}\right) \tag{18.94a}$$

$$\phi_\mathrm{p} = \psi + \frac{kT}{e} \ln\left(\frac{p}{p_\mathrm{i}}\right) . \tag{18.94b}$$

The product np is given by

$$np = n_{\mathrm{i}}^2 \exp\left[\frac{e(\phi_{\mathrm{p}} - \phi_{\mathrm{n}})}{kT}\right] . \tag{18.95}$$

Of course, at thermodynamical equilibrium (zero bias) $\phi_{\mathrm{p}} = \phi_{\mathrm{n}}$ and $np = n_{\mathrm{i}}^2$. For forward bias $\phi_{\mathrm{p}} - \phi_{\mathrm{n}} > 0$ (Fig. 18.38a) and $np > n_{\mathrm{i}}^2$. For reverse bias $\phi_{\mathrm{p}} - \phi_{\mathrm{n}} < 0$ (Fig. 18.38b) and $np < n_{\mathrm{i}}^2$.

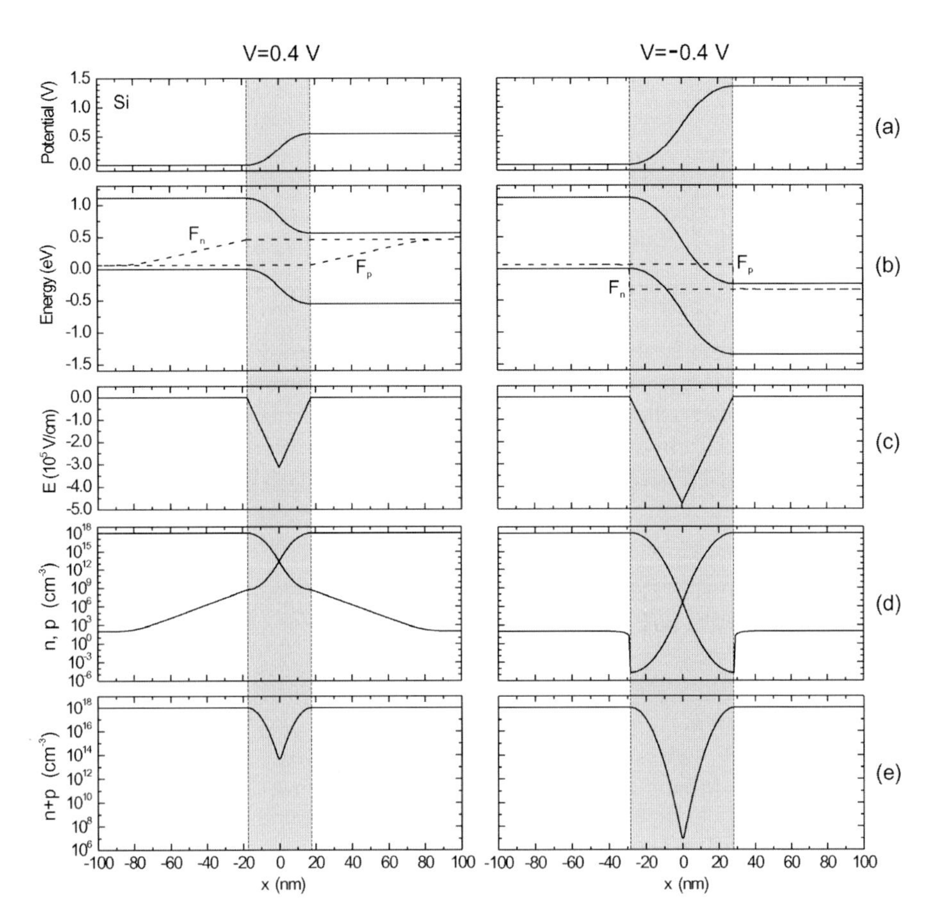

Fig. 18.38. (**a**) Diffusion potential, (**b**) band diagram, (**c**) electric field, (**d**) electron and hole concentrations and (**e**) $n + p$ under forward bias +0.4 V (*left panel*) and reverse bias (*right panel*) −0.4 V for a silicon pn-diode at room temperature with $N_{\mathrm{A}} = N_{\mathrm{D}} = 10^{18}\,\mathrm{cm}^{-3}$ (same as in Fig. 18.34). The *dashed lines* in (b) are the electron and hole quasi-Fermi levels F_{n} and F_{p}. The depletion layer is shown as the *grey area*. The diffusion length in the n- and p-type material is taken as 4 nm. This value is much smaller than the actual diffusion length (µm-range) and is chosen here only to show the carrier concentration in the depletion layer and the neutral region in a single graph

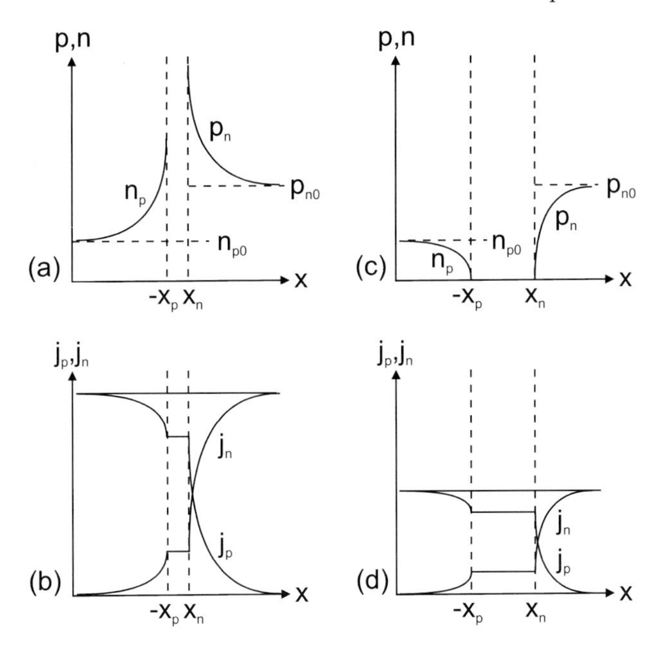

Fig. 18.39. Carrier densities (**a**,**c**) and current densities (**b**,**d**) (linear scales) in a pn-diode under (**a**,**b**) forward bias and (**c**,**d**) reverse bias

The electron current density (per unit area) is given by (8.46a) that reads here with $\mathcal{E} = \nabla\psi$ and n given by (18.93a) as[17]

$$\mathbf{j}_\mathrm{n} = -e\mu_\mathrm{n}\left(n\mathcal{E} + \frac{kT}{e}\nabla n\right) = en\mu_\mathrm{n}\nabla\phi_\mathrm{n} \, . \tag{18.96}$$

Similarly, we obtain for the hole current density (using (8.46b) and (18.93b))

$$\mathbf{j}_\mathrm{p} = e\mu_\mathrm{p}\left(p\mathcal{E} - \frac{kT}{e}\nabla p\right) = -ep\mu_\mathrm{p}\nabla\phi_\mathrm{p} \, . \tag{18.97}$$

The current through the depletion layer is constant (since no recombination/generation was assumed). The gradient of the quasi-Fermi levels in the depletion layer is very small and the quasi-Fermi levels $\phi_\mathrm{n,p}$ are practically constant. The change of carrier density is mostly due to the variation of ψ (or E_i).

Therefore, the voltage drop across the depletion layer is $V = \phi_\mathrm{p} - \phi_\mathrm{n}$ and (18.95) reads

$$np = n_\mathrm{i}^2 \exp\left(\frac{eV}{kT}\right) \, . \tag{18.98}$$

[17]We remind the reader that μ_n was defined as a negative number.

The electron density at the boundary of the depletion layer on the p-side (at $x = -x_{\mathrm{p}}$) is (using (18.98))

$$n_{\mathrm{p}} = \frac{n_{\mathrm{i}}^2}{p_{\mathrm{p}}} \exp\left(\frac{eV}{kT}\right) = n_{\mathrm{p}0} \exp\left(\frac{eV}{kT}\right) . \tag{18.99}$$

Similarly, the hole density on the n-side at $x = x_{\mathrm{n}}$ is given by

$$p_{\mathrm{n}} = p_{\mathrm{n}0} \exp\left(\frac{eV}{kT}\right) . \tag{18.100}$$

From the continuity equation and the boundary condition that far away from the depletion layer the hole density is $p_{\mathrm{n}0}$, the hole density on the n-side is given by

$$p_{\mathrm{n}}(x) - p_{\mathrm{n}0} = p_{\mathrm{n}0} \left[\exp\left(\frac{eV}{kT}\right) - 1\right] \exp\left(-\frac{x - x_{\mathrm{n}}}{L_{\mathrm{p}}}\right) , \tag{18.101}$$

where $L_{\mathrm{p}} = \sqrt{D_{\mathrm{p}}\tau_{\mathrm{p}}}$ is the hole (minority-carrier) diffusion length.

The hole current density at the boundary of the depletion layer on the n-side is

$$j_{\mathrm{p}}(x_{\mathrm{n}}) = -eD_{\mathrm{p}} \left.\frac{\partial p_{\mathrm{n}}}{\partial x}\right|_{x_{\mathrm{n}}} = \frac{eD_{\mathrm{p}}p_{\mathrm{n}0}}{L_{\mathrm{p}}} \left[\exp\left(\frac{eV}{kT}\right) - 1\right] . \tag{18.102}$$

Similarly, the electron current in the depletion layer is

$$j_{\mathrm{n}}(-x_{\mathrm{p}}) = \frac{eD_{\mathrm{n}}n_{\mathrm{p}0}}{L_{\mathrm{n}}} \left[\exp\left(\frac{eV}{kT}\right) - 1\right] . \tag{18.103}$$

The total current due to diffusion is

$$j_{\mathrm{d}} = j_{\mathrm{p}}(x_{\mathrm{n}}) + j_{\mathrm{n}}(-x_{\mathrm{p}}) = j_{\mathrm{s}} \left[\exp\left(\frac{eV}{kT}\right) - 1\right] , \tag{18.104}$$

with the saturation current given by

$$j_{\mathrm{s}}^{\mathrm{d}} = \frac{eD_{\mathrm{p}}p_{\mathrm{n}0}}{L_{\mathrm{p}}} + \frac{eD_{\mathrm{n}}n_{\mathrm{p}0}}{L_{\mathrm{n}}} . \tag{18.105}$$

This dependence is an ideal diode characteristic and the famous result from Shockley. For a one-sided ($\mathrm{p}^+\mathrm{n}$-) diode, the saturation current is

$$j_{\mathrm{s}}^{\mathrm{d}} \cong \frac{eD_{\mathrm{p}}p_{\mathrm{n}0}}{L_{\mathrm{p}}} \cong e\left(\frac{D_{\mathrm{p}}}{\tau_{\mathrm{p}}}\right)^{1/2} \frac{n_{\mathrm{i}}^2}{N_{\mathrm{B}}} . \tag{18.106}$$

The saturation depends via $D_{\mathrm{p}}/\tau_{\mathrm{p}}$ weakly on the temperature. The term n_{i}^2 depends on T, proportional to $T^3 \exp(-E_{\mathrm{g}}/kT)$, which is dominated by the exponential function.

If the minority carrier lifetime is given by the radiative recombination (10.19), the hole diffusion length is

$$L_{\mathrm{p}} = \left(\frac{D_{\mathrm{p}}}{B n_{\mathrm{n}0}} \right)^{1/2} . \tag{18.107}$$

For GaAs (Tables 8.1 and 10.1) with $N_{\mathrm{D}} = 10^{18}\,\mathrm{cm}^{-3}$, we find $\tau_{\mathrm{p}} = 10\,\mathrm{ns}$ and $L_{\mathrm{p}} \approx 3\,\mu\mathrm{m}$. For L_{n} we find $14\,\mu\mathrm{m}$, however, the lifetime at room temperature can be significantly shorter due to nonradiative recombination and subsequently also the diffusion length will be shorter (by about a factor of 10). For $L \sim 1\,\mu\mathrm{m}$, the diffusion saturation current is $j_{\mathrm{s}}^{\mathrm{d}} \sim 4 \times 10^{-20}\,\mathrm{A/cm^2}$.

The radiative recombination rate (band–band recombination, b–b) in the neutral n-region (as relevant for LEDs, see Sect. 20.3) is $B(np - n_{\mathrm{i}}^2) \approx B n_{\mathrm{n}0}(p_{\mathrm{n}}(x) - p_{\mathrm{n}0})$. Therefore, the recombination current $j_{\mathrm{d,n}}^{\mathrm{b-b}}$ in the neutral n-region is (using (18.101))

$$\begin{aligned} j_{\mathrm{d,n}}^{\mathrm{b-b}} &= e \int_{x_{\mathrm{n}}}^{\infty} B n_{\mathrm{i}}^2 \left[\exp\left(\frac{eV}{kT} \right) - 1 \right] \exp\left(-\frac{x - x_{\mathrm{n}}}{L_{\mathrm{p}}} \right) \mathrm{d}x \\ &= e B L_{\mathrm{p}} n_{\mathrm{i}}^2 \left[\exp\left(\frac{eV}{kT} \right) - 1 \right] . \end{aligned} \tag{18.108}$$

For the neutral region on the p-side a similar expression results. The total radiative recombination current from the neutral regions is

$$j_{\mathrm{d}}^{\mathrm{b-b}} = eB(L_{\mathrm{n}} + L_{\mathrm{p}}) n_{\mathrm{i}}^2 \left[\exp\left(\frac{eV}{kT} \right) - 1 \right] . \tag{18.109}$$

For GaAs, the saturation current for the radiative recombination in the neutral region

$$j_{\mathrm{s}}^{\mathrm{r,b-b}} = eB(L_{\mathrm{n}} + L_{\mathrm{p}}) n_{\mathrm{i}}^2 \tag{18.110}$$

is (Tables 7.1 and 10.1) for a diffusion length of $1\,\mu\mathrm{m}$ of $j_{\mathrm{s}}^{\mathrm{r,b-b}} \sim 4 \times 10^{-21}\,\mathrm{A/cm^2}$.

Since the (radiative) minority-carrier lifetime is inversely proportional to the majority-carrier density, the relevant diffusion length is that of the side with the lower doping level and is given by

$$L = \frac{1}{n_{\mathrm{i}}} \left(\frac{D_{\mathrm{B}} N_{\mathrm{B}}}{B} \right)^{1/2} , \tag{18.111}$$

where D_{B} is the minority-carrier diffusion coefficient on the lowly doped side. The radiative recombination current from the neutral region can be written as

$$j_{\mathrm{d}}^{\mathrm{b-b}} = e \left(B D_{\mathrm{B}} N_{\mathrm{B}} \right)^{1/2} n_{\mathrm{i}} \left[\exp\left(\frac{eV}{kT} \right) - 1 \right] . \tag{18.112}$$

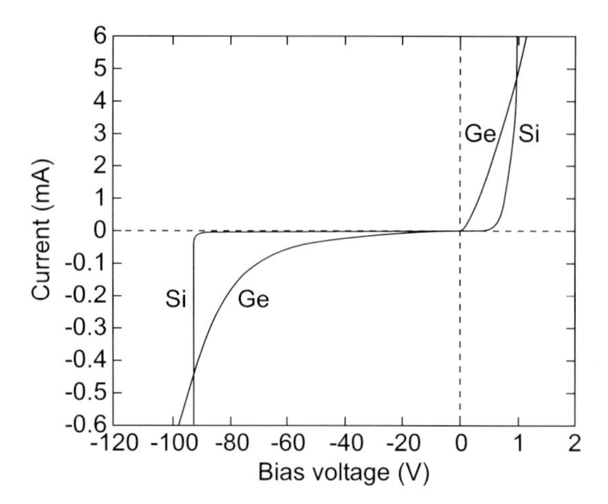

Fig. 18.40. Comparison of the characteristics of Ge and Si diodes at room temperature. Note the different scales in the forward and reverse regime

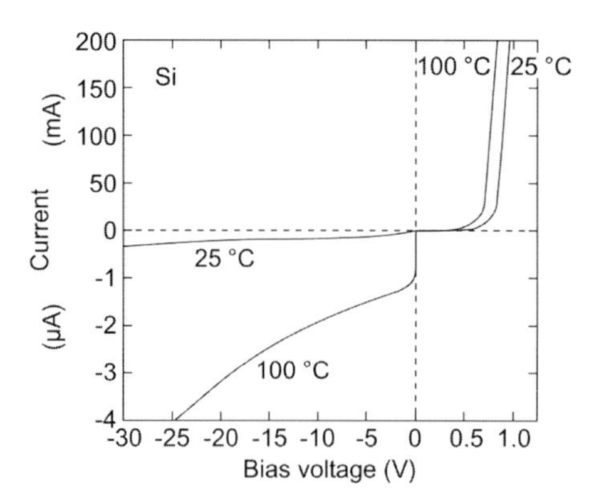

Fig. 18.41. Characteristics of a Si power diode at two temperatures, 25 °C and 100 °C

The I–V characteristic for pn-diodes from two semiconductors with different bandgap are shown in Fig. 18.40 (for Ge and Si). Both forward characteristics are exponential functions. The Si diode seems to have the steeper slope, however, the difference is only the smaller saturation current. The saturation current increases at higher temperature (Fig. 18.41).

Real I–V Characteristics

The ideal I–V characteristics are not observed for real diodes due to several reasons:

– besides the diffusion current, also a generation–recombination (G–R) current is present

- already for fairly small forward voltages, high injection conditions are present, i.e. $p_n \ll n_n$ is no longer valid
- the series resistance R_s of the diode is finite (ideally $R_s = 0$)
- the diode has a finite parallel (shunt) resistance R_p (ideally $R_p = \infty$)
- at high reverse voltage the junction breaks down; this phenomenon is treated in Sect. 18.4.5

First, we consider the generation–recombination current due to band–impurity (b–i) processes (see Sect. 10.10). Such recombination is nonradiative or at least does not produce photons with an energy close to the bandgap. The net rate is given by (10.52). For reverse voltage, the generation dominates the G–R current. For $n < n_i$ and $p < n_i$, the net recombination rate r is

$$r \cong \frac{\sigma_n \sigma_p v_{th} N_t}{\sigma_n \exp\left(\frac{E_t - E_i}{kT}\right) + \sigma_p \exp\left(\frac{E_i - E_t}{kT}\right)} n_i \equiv \frac{n_i}{\tau_e} , \tag{18.113}$$

where τ_e is the effective electron lifetime. The generation current density is given by

$$j_g = \frac{e n_i w}{\tau_e} . \tag{18.114}$$

Since the width of the depletion layer varies with the applied reverse bias V, we expect a dependence

$$j_g \propto (V_{bi} + |V|)^{1/2} . \tag{18.115}$$

The saturation current is given by the sum of the diffusion and generation parts

$$j_s = e \left(\frac{D_p}{\tau_p}\right)^{1/2} \frac{n_i^2}{N_D} + \frac{e n_i w}{\tau_e} . \tag{18.116}$$

In semiconductors with large n_i (narrow bandgap, e.g. Ge) the diffusion current will dominate; in Si (larger bandgap) the generation current can dominate.

The maximum of the recombination rate is present for $E_t \approx E_i$ (10.54). Then $n_t = p_t = n_i$ in (10.52). Assuming $\sigma = \sigma_n = \sigma_p$, the recombination rate is

$$r_{b-i} = \sigma v_{th} N_t \frac{np - n_i^2}{n + p + 2n_i} . \tag{18.117}$$

Using (18.95) we can write

$$r_{b-i} = \sigma v_{th} N_t n_i \frac{n_i}{n + p + 2n_i} \left[\exp\left(\frac{eV}{kT}\right) - 1\right] . \tag{18.118}$$

The term $\zeta = \frac{n_\mathrm{i}}{n+p+2n_\mathrm{i}}$ is maximal for $n = p$, which is given (from (18.98)) by

$$n_\mathrm{mr} = p_\mathrm{mr} = n_\mathrm{i} \exp\left(\frac{eV}{2kT}\right) . \tag{18.119}$$

Since ζ cannot be analytically integrated, typically as an approximation the maximum rate

$$\zeta_\mathrm{mr} = \frac{n_\mathrm{i}}{n_\mathrm{mr} + p_\mathrm{mr} + 2n_\mathrm{i}} = \frac{1}{2}\frac{1}{1+\exp\left(\frac{eV}{2kT}\right)} \tag{18.120}$$

is integrated over the depletion layer [183]. This approximation yields a recombination current

$$j_\mathrm{mr} = \frac{e\sigma v_\mathrm{th} N_\mathrm{t} w n_\mathrm{i}}{2} \frac{\exp\left(\frac{eV}{kT}\right) - 1}{\exp\left(\frac{eV}{2kT}\right) + 1} \cong j_\mathrm{s}^\mathrm{mr} \exp\left(\frac{eV}{2kT}\right) , \tag{18.121}$$

with $j_\mathrm{s}^\mathrm{mr} = e\sigma v_\mathrm{th} N_\mathrm{t} w n_\mathrm{i}/2$ and the approximation being valid for $eV/kT \gg 1$. Thus the nonradiative band–impurity recombination is often said to cause an ideality factor of $n = 2$.

In order to evaluate the integral of ζ over the depletion layer

$$\chi = \int_{-x_\mathrm{p}}^{x_\mathrm{n}} \zeta \,\mathrm{d}x \tag{18.122}$$

in a good approximation, the dependence of the potential can be approximated as linear (constant-field approximation), i.e. using the local field $\mathcal{E}_\mathrm{mr}$ at the position where $n = p$ [582]. For a symmetric diode with $n_\mathrm{n_0} = p_\mathrm{p_0}$, this position is at $x = 0$; for a one-sided junction on the lower-doped side. $\mathcal{E}_\mathrm{mr}$ is given for $p_\mathrm{p_0} \leq n_\mathrm{n_0}$ by

$$\mathcal{E}_\mathrm{mr} = -\left(V_\mathrm{bi} - V\right)\left(1 + \frac{1}{\beta\left(V_\mathrm{bi} - V\right)} \ln\frac{p_\mathrm{p_0}}{n_\mathrm{n_0}}\right)^{1/2}\left(1 + \frac{p_\mathrm{p_0}}{n_\mathrm{n_0}}\right)^{1/2} \frac{\sqrt{2}}{w} . \tag{18.123}$$

For a symmetric diode (18.124a) holds, for a one-sided diode the approximation in (18.124b) holds

$$\mathcal{E}_\mathrm{mr} = -\frac{2}{w}\left(V_\mathrm{bi} - V\right) \propto \left(V_\mathrm{bi} - V\right)^{1/2} \tag{18.124a}$$

$$\mathcal{E}_\mathrm{mr} \cong -\frac{\sqrt{2}}{w}\left(V_\mathrm{bi} - V\right) \propto \left(V_\mathrm{bi} - V\right)^{1/2} . \tag{18.124b}$$

We note that for $V = 0$ (18.82) is recovered from (18.124a). Using the above approximation ζ is given by

$$\zeta = \frac{1}{2}\frac{1}{1+\exp\left(\beta V/2\right)\cosh(\beta\mathcal{E}_\mathrm{mr} x)} , \tag{18.125}$$

with $\beta = e/kT$. Since ζ decreases sufficiently fast within the depletion layer, the integration over the depletion layer can be extended to $\pm\infty$ and we obtain

$$\chi = \frac{2}{\beta \mathcal{E}_{\mathrm{mr}}} \left(\frac{1}{\exp(\beta V) - 1} \right)^{1/2} \arctan \left[\left(\frac{\exp(\beta V/2) - 1}{\exp(\beta V/2) + 1} \right)^{1/2} \right] . \tag{18.126}$$

We note that for $V = 0$, the integral takes the value $\chi = (\beta \mathcal{E}_{\mathrm{mr}})^{-1}$. The recombination current is now given by [582]

$$j_{\mathrm{r,b-i}} = \frac{2\sigma v_{\mathrm{th}} N_{\mathrm{t}} n_{\mathrm{i}} kT}{\mathcal{E}_{\mathrm{mr}}} \arctan \left[\left(\frac{\exp(\beta V/2) - 1}{\exp(\beta V/2) + 1} \right)^{1/2} \right] \left[\exp\left(\frac{eV}{kT} \right) - 1 \right]^{1/2} . \tag{18.127}$$

For large voltage the arctan term becomes $\pi/4$. For $eV/kT \gg 1$ the nonradiative recombination current can be written as

$$j_{\mathrm{r,b-i}} = j_{\mathrm{s}}^{\mathrm{r,b-i}} \exp\left(\frac{eV}{nkT} \right) , \tag{18.128}$$

with $j_{\mathrm{s}}^{\mathrm{r,b-i}} = e\sigma v_{\mathrm{th}} N_{\mathrm{t}} n_{\mathrm{i}} kT\pi/(2\mathcal{E}_{\mathrm{mr}})$. The voltage-dependent ideality factor n (semilogarithmic slope $n = \beta j_{\mathrm{r}}(V)/j_{\mathrm{r}}'(V)$) is close but not identical to 2 and is shown in Fig. 18.42 for various values of V_{bi}. The built-in voltage influences the logarithmic slope via the factor $1/\mathcal{E}_{\mathrm{mr}}$ in (18.127).

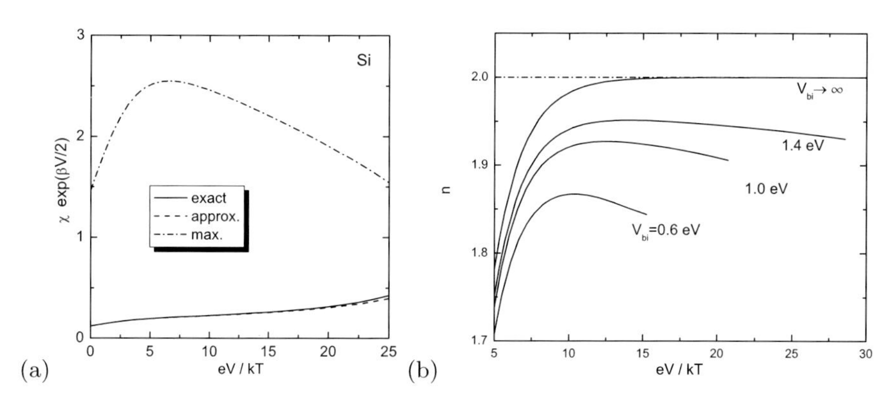

Fig. 18.42. (**a**) Integral χ (18.122) multiplied by $\exp(\beta V/2)$ in order to extract the differences on a linear scale. *Solid line*: Exact numerical calculation, *dash-dotted line*: standard approximation with constant maximum rate, *dashed line*: this work (approximation with constant field). As material parameters we have used room temperature and $n_{\mathrm{i}} = 10^{10}\,\mathrm{cm}^{-3}$ (Si), $n_{\mathrm{n_0}} = 10^{18}\,\mathrm{cm}^{-3}$ and $p_{\mathrm{p_0}} = 10^{17}\,\mathrm{cm}^{-3}$. (**b**) Logarithmic slope of band–impurity recombination current in the forward bias regime for various values of the built-in voltage $V_{\mathrm{bi}} = 0.6$, 1.0, and 1.4 eV and in the limit $V_{\mathrm{bi}} \to \infty$. Adapted from [582]

In the case of radiative band–band (b–b) recombination, the recombination rate is given by (10.14). Together with (18.98) and integrated over the depletion layer, the recombination current in the depletion layer is given by (cf. (18.109))

$$j_{\mathrm{r,b-b}} = eBwn_{\mathrm{i}}^2 \left[\exp\left(\frac{eV}{kT}\right) - 1\right] , \tag{18.129}$$

and exhibits an ideality factor of $n = 1$. Comparing (18.109) and (18.129), the dominating radiative-recombination current is determined by the ratio of w and $L_{\mathrm{n}}+L_{\mathrm{p}}$. Since in the forward direction, the depletion-layer width tends towards zero (for flat-band conditions), the radiative-recombination current is dominated by the recombination in the neutral region(s).

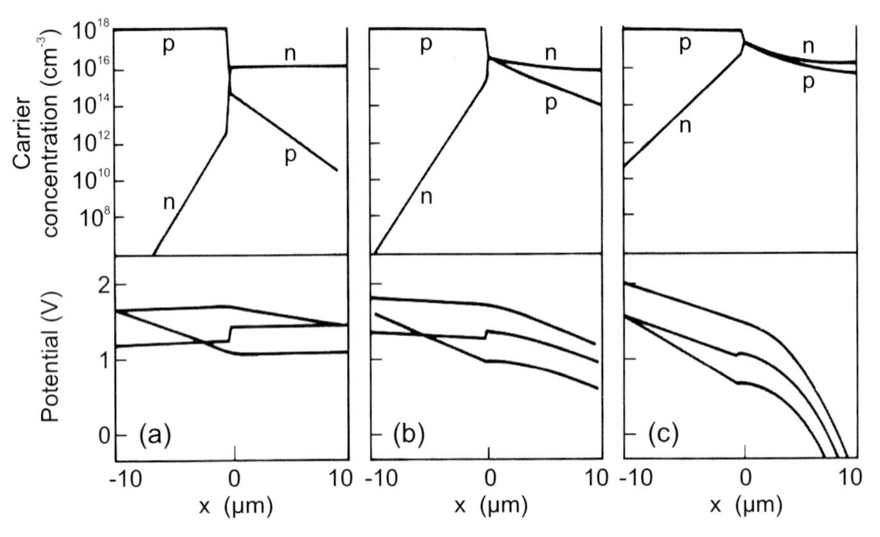

Fig. 18.43. Theoretical modeling of charge-carrier concentration, intrinsic Fermi level (potential) ψ and quasi-Fermi levels (with arbitrary offset) for a Si p$^+$n diode for various current densities: (**a**) $10\,\mathrm{A\,cm^{-2}}$, (**b**) $10^3\,\mathrm{A\,cm^{-2}}$ and (**c**) $10^4\,\mathrm{A\,cm^{-2}}$. $N_{\mathrm{A}} = 10^{18}\,\mathrm{cm^{-3}}$, $N_{\mathrm{D}} = 10^{16}\,\mathrm{cm^{-3}}$, $\tau_{\mathrm{n}} = 0.3\,\mathrm{ns}$, $\tau_{\mathrm{p}} = 0.84\,\mathrm{ns}$. Adapted from [583]

For high injection current (under forward bias), the injected minority-carrier density can become comparable with the majority-carrier density. In this case, diffusion *and* drift need to be considered. At large current density, the voltage drop across the junction is small compared to the ohmic voltage drop across the current path. In the simulation (Fig. 18.43), the high-injection effects start on the n-doped side because it has been modeled with the lower doping ($N_{\mathrm{D}} < N_{\mathrm{A}}$).

The series resistance R_{s} (typically a few Ohms) also effects the characteristic at low injection. The voltage drop across the junction is reduced by

$R_s\,I$. Thus, the I–V characteristic taking into account the effect of the series resistance is

$$I = I_s \left[\exp\left(\frac{e(V - R_s I)}{nkT} \right) - 1 \right] . \tag{18.130}$$

This equation is implicit with regard to I and can only be solved numerically. At high current, the resistance of the junction becomes very small (Fig. 18.44a) and the I–V characteristic is dominated by the series resistance and becomes linear.

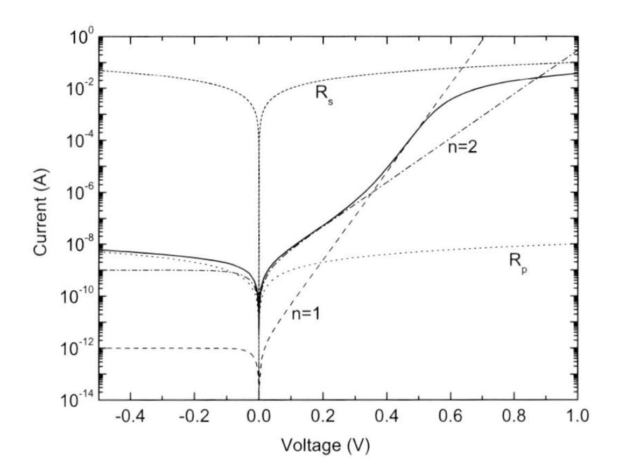

Fig. 18.44. Theoretical I–V characteristic of a diode at room temperature with saturation currents for the $n = 1$ and $n = 2$ processes of $I_s^{n=1} = 10^{-12}$ A and $I_s^{n=2} = 10^{-9}$ A and resistances $R_s = 10\,\Omega$, $R_p = 100\,\mathrm{M}\Omega$. *Dashed line*: Ideal diode with $n = 1$ characteristic only, *dash-dotted line*: only $n = 2$ process, *dotted line*: only parallel ohmic resistance, *short dashed line*: only series resistance, *solid line*: all effects combined as in (18.132)

The diode can also exhibit a parallel (shunt) resistance R_p, e.g. due to surface conduction between the contacts. Including the shunt resistance, the diode characteristic is

$$I = I_s \left[\exp\left(\frac{e(V - R_s I)}{nkT} \right) - 1 \right] + \frac{V - R_s I}{R_p} . \tag{18.131}$$

The shunt resistance can be evaluated best from the differential conductance in the reverse-voltage regime [559]. Due to a high surface-state density, the passivation of GaAs diodes can be difficult. Si can be very well passivated with low leakage current and high reliability.

Often, a clear distinction between the $n = 1$ and $n = 2$ regimes cannot be made. In this case, an intermediate ideality factor $1 \leq n \leq 2$ is fitted to the I–V characteristic as in (18.131). If the current can be separated into a $n = 1$ (diffusion) and a $n = 2$ (recombination–generation) process, the characteristic is given by (see Fig. 18.44) (for $V \gg kT/e$)

$$I = I_s^{n=1} \exp\left[\frac{e(V - R_s I)}{kT}\right] + I_s^{n=2} \exp\left[\frac{e(V - R_s I)}{2kT}\right] + \frac{V - R_s I}{R_p} . \tag{18.132}$$

In summary, the pn-diode has the equivalent circuit given in Fig. 18.45; the photocurrent source I_{ph} is discussed below in Sect. 19.3.

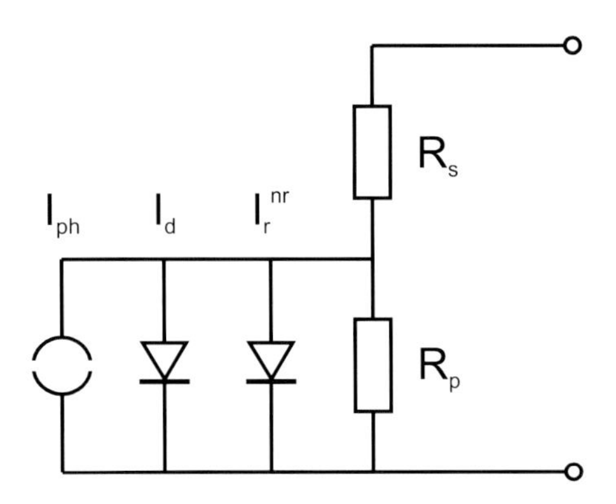

Fig. 18.45. Equivalent circuit of a pn-diode. Serial (R_s) and parallel (R_p) resistance and diode currents: I_d (due to diffusion, $n \approx 1$), I_r^{nr} (nonradiative recombination due to band–impurity recombination, $n \approx 2$) and an ideal current source due to photogeneration (as discussed in Sect. 19.3)

18.4.5 Breakdown

If a large voltage is applied in the reverse direction, the pn-junction breaks down. At breakdown, a small voltage increase leads to a dramatic increase of the current. There are three mechanisms that lead to breakdown: thermal instability, tunneling, and avalanche multiplication [584, 585].

Thermal Instability

The reverse current at large applied voltage leads to a power dissipation and heating of the junction. This temperature increase leads to a further increase of the saturation current (18.106). If the heat sink, e.g. the mounting of the chip, is not able to transport the heat away from the device, the current increases indefinitely. If not limited by a resistor, such a current can destroy the device. The thermal instability is particularly important for diodes with high saturation current, e.g. Ge at room temperature.

Tunneling

At large reverse bias, charge carriers can tunnel between conduction and valence band through the junction. A more detailed discussion will be given below in Sect. 18.5.9 about the tunneling diode. Since for the tunneling effect a thin barrier is necessary, breakdown due to tunneling is important for diodes where both sides are highly doped. For Si and Ge diodes, tunneling dominates the breakdown if the breakdown voltage $V_{\rm br}$ is $V_{\rm br} < 4E_{\rm g}/e$. For $V_{\rm br} > 6E_{\rm g}/e$ avalanche multiplication dominates. The intermediate regime is a mixed case.

With increasing temperature, the tunneling current can be achieved already with a smaller field (since the bandgap decreases with increasing temperature), thus the breakdown voltage *decreases* (negative temperature coefficient).

Avalanche Multiplication

Avalanche multiplication due to impact ionization is the most important mechanism for the breakdown of pn-diodes. It limits the maximum reverse voltage for most diodes and also the collector voltage in a bipolar transistor or the drain voltage in a field-effect transistor. Avalanche multiplication can be used for the generation of microwave radiation or for photon counting (see Sect. 19.3.6).

Impact ionization was discussed in Sect. 8.5.3. The most important parameters are the electron and hole ionization coefficients $\alpha_{\rm n}$ and $\alpha_{\rm p}$. For discussion of the diode breakdown, we assume that at $x = 0$ a hole current $I_{\rm po}$ enters the depletion layer. This current is amplified by the field in the depletion layer and impact ionization. At the end of the depletion layer ($x = w$), it is $M_{\rm p} I_{\rm po}$, i.e. $M_{\rm p} = I_{\rm p}(w)/I_{\rm p}(0)$. Similarly, an electron current is increased on its way from w to $x = 0$. The incremental change of the hole current due to electron–hole pairs generated along a line element $\mathrm{d}x$ is

$$\mathrm{d}I_{\rm p} = (I_{\rm p}\alpha_{\rm p} + I_{\rm n}\alpha_{\rm n})\mathrm{d}x \; . \tag{18.133}$$

The total current in the depletion layer is $I = I_{\rm p} + I_{\rm n}$ and is constant in stationary equilibrium. Therefore,

$$\frac{\mathrm{d}I_{\rm p}}{\mathrm{d}x} - (\alpha_{\rm p} - \alpha_{\rm n})\, I_{\rm p} = \alpha_{\rm n} I \; . \tag{18.134}$$

The solution is

$$I_{\rm p}(x) = I \frac{\frac{1}{M_{\rm p}} + \int_0^x \alpha_{\rm n} \exp\left[-\int_0^x (\alpha_{\rm p} - \alpha_{\rm n})\mathrm{d}x'\right] \mathrm{d}x}{\exp\left[-\int_0^x (\alpha_{\rm p} - \alpha_{\rm n})\mathrm{d}x'\right]} \; . \tag{18.135}$$

For $x = w$ we find for the multiplication factor

$$1 - \frac{1}{M_{\rm p}} = \int_0^w \alpha_{\rm n} \exp\left[-\int_0^x (\alpha_{\rm p} - \alpha_{\rm n})\mathrm{d}x'\right] \mathrm{d}x \; . \tag{18.136}$$

Avalanche breakdown is reached for $M_\mathrm{p} \to \infty$, i.e. when

$$\int_0^w \alpha_\mathrm{n} \exp\left[-\int_0^x (\alpha_\mathrm{p} - \alpha_\mathrm{n})\mathrm{d}x'\right] \mathrm{d}x = 1 \, . \tag{18.137}$$

A corresponding and equivalent equation is obtained when the consideration is started with the electron current. If $\alpha_\mathrm{p} = \alpha_\mathrm{n} = \alpha$, (18.137) simplifies to

$$\int_0^w \alpha \mathrm{d}x = 1 \, . \tag{18.138}$$

This means that per transit of one carrier through the depletion layer, on average another carrier is created such that the process just starts to diverge. The breakdown voltage for various semiconductor materials is shown in Fig. 18.46a as a function of the doping level. The depletion-layer width w at breakdown and the maximum electric field $\mathcal{E}_\mathrm{m}$ are depicted in Fig. 18.46b.

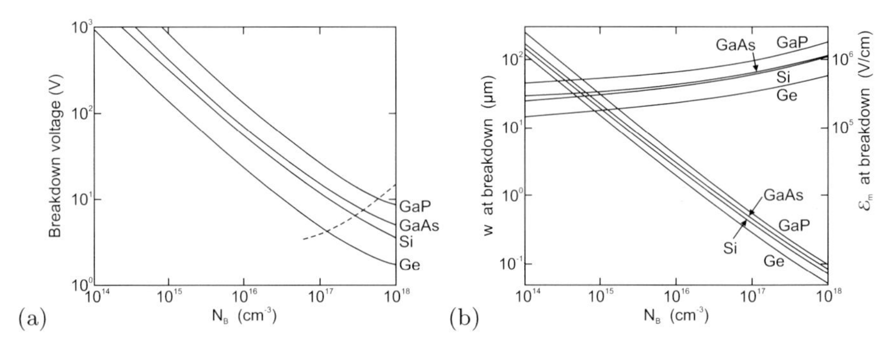

Fig. 18.46. (**a**) Avalanche breakdown voltage for one-sided abrupt junctions in Ge, Si, (100)-GaAs and GaP at $T = 300\,\mathrm{K}$. The *dashed line* indicates the limit of avalanche breakdown at high doping due to tunneling breakdown. (**b**) Depletion-layer width w at breakdown and maximum electric field $\mathcal{E}_\mathrm{m}$ for the same junctions. Adapted from [586]

In GaAs, the impact-ionization coefficients and therefore the breakdown voltage are direction dependent. At a doping of $N_\mathrm{B} = 10^{16}\,\mathrm{cm}^{-3}$, the breakdown voltage is the same for (001) and (111) orientation; for smaller doping the breakdown voltage of (001)-oriented GaAs is smaller, for larger doping that of GaAs (111) [587].

At higher temperatures, the charge carriers release their excess energy faster to the lattice.[18] Thus, less energy is available for impact ionization

[18]The scattering rate becomes higher with increasing temperature and, e.g., the mobility decreases, see Sect. 8.3.7, and the drift saturation velocity decreases, see Sect. 8.5.1.

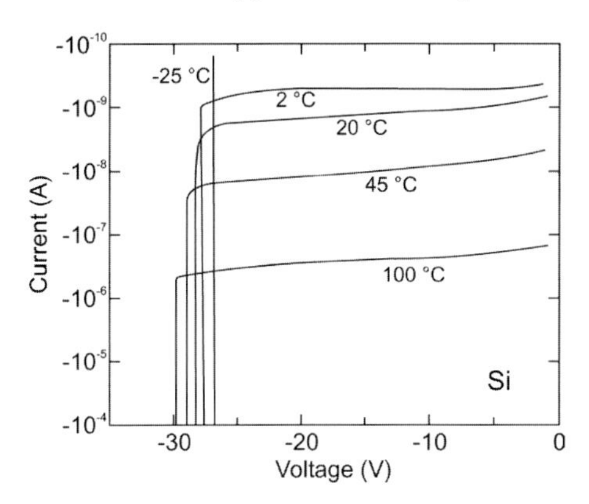

Fig. 18.47. Temperature dependence of a n^+p Si-diode with $N_B = 2.5 \times 10^{16}\,\mathrm{cm}^{-3}$ and a guard-ring structure (cf. Fig. 18.48d). The temperature coefficient $\partial V_{br}/\partial T$ is 0.024 V/K. Adapted from [588]

and the required electric field is higher. Therefore, the breakdown voltage *increases* with the temperature (Fig. 18.47). This behavior is opposite to tunneling diodes and the two processes can be distinguished in this way.

In planar structures (Fig. 18.48a), high electric fields are present at the parts with large curvature. At these sites breakdown will occur first and at much lower voltages than expected for a perfectly planar (infinite) structure. For devices that require high breakdown voltage, design changes have to be made. These include deep junctions (Fig. 18.48b) with a smaller curvature, a field-ring structure (Fig. 18.48c) in which an additional depletion layer is used to smooth the field lines and the often used guard ring (Fig. 18.48d) for which a circular region of low doping (and thus high breakdown voltage) is incorporated.

18.5 Applications and Special Diode Devices

In the following, various electronic applications of diodes are discussed. The most important special diode types are introduced. Optoelectronic applications (involving absorption and emission of photons) are treated below (Chap. 19).

18.5.1 Rectification

In a rectifier, the diode has to supply a high resistivity for one polarity of the bias and a low one to the other polarity. In Fig. 18.49a, a single-path rectification method is shown. Only the positive half-wave can pass the load resistor R_L (Fig. 18.49b). In Fig. 18.49c, the characteristic of a Si diode is shown. Of course, the voltage drop across the diode can only range in the

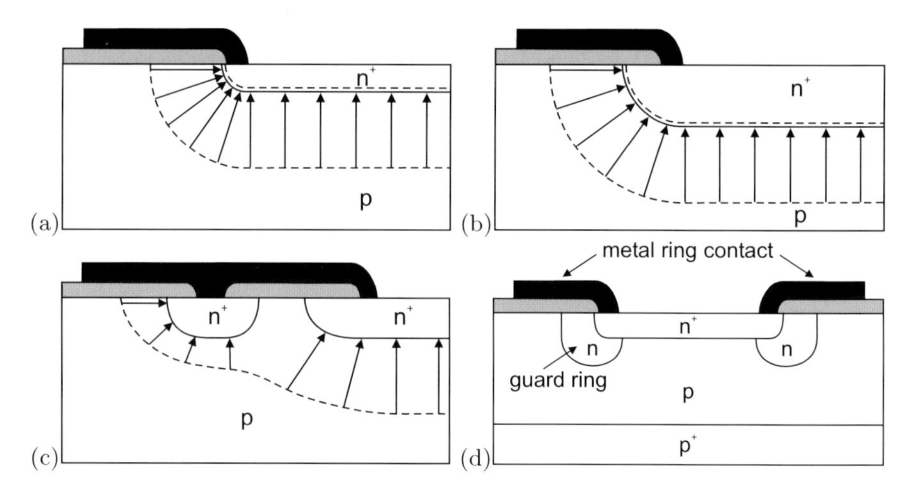

Fig. 18.48. (**a**) Large electric fields at large curvatures in a shallow junction. Avoidance of regions with large electric fields by (**b**) deep junction and (**c**) field-ring structure. (**d**) shows a guard-ring structure with circular, low-doped n region. *Grey area* denotes insulating material, arrows indicate field lines and the *dashed lines* indicate the extension of the depletion layers

1-V regime. In order to make the setup work, the load resistor has to be considered. The total current is given by $I = I_s [\exp(eU_d/nkT) - 1]$. The total voltage U is split between the voltage drop across the diode U_d and that over the load resistance $U_L = R_L I$. The current is therefore given by

$$I = \frac{U - U_d}{R_L} . \tag{18.139}$$

For sizeable currents the voltage drop across the diode U_d is between 0.7 and 1 V. The characteristic is linear between about 1 and 220 V (Fig. 18.49d). Typically, the voltage U_L is low-pass filtered with a capacitor parallel to the load resistor. The effective voltage is the peak voltage divided by 2.

The drawback of the single diode rectifier is that only the positive half-wave contributes to a dc signal. The setup in Fig. 18.49e (bridge rectifier) allows both half-waves to contribute to the dc signal. The effective voltage in this case is the peak voltage divided by $\sqrt{2}$.

The forward resistance in the static (R_f) and dynamic (r_f) case are (for $\beta V_f > 3$)

$$R_f = \frac{V_f}{I_f} \cong \frac{V_f}{I_s} \exp\left(-\frac{eV_f}{nkT}\right) \tag{18.140a}$$

$$r_f = \frac{\partial V_f}{\partial I_f} = \frac{nkT}{eI_s} \exp\left(\frac{eV_f}{nkT}\right) \cong \frac{nkT}{eI_f} . \tag{18.140b}$$

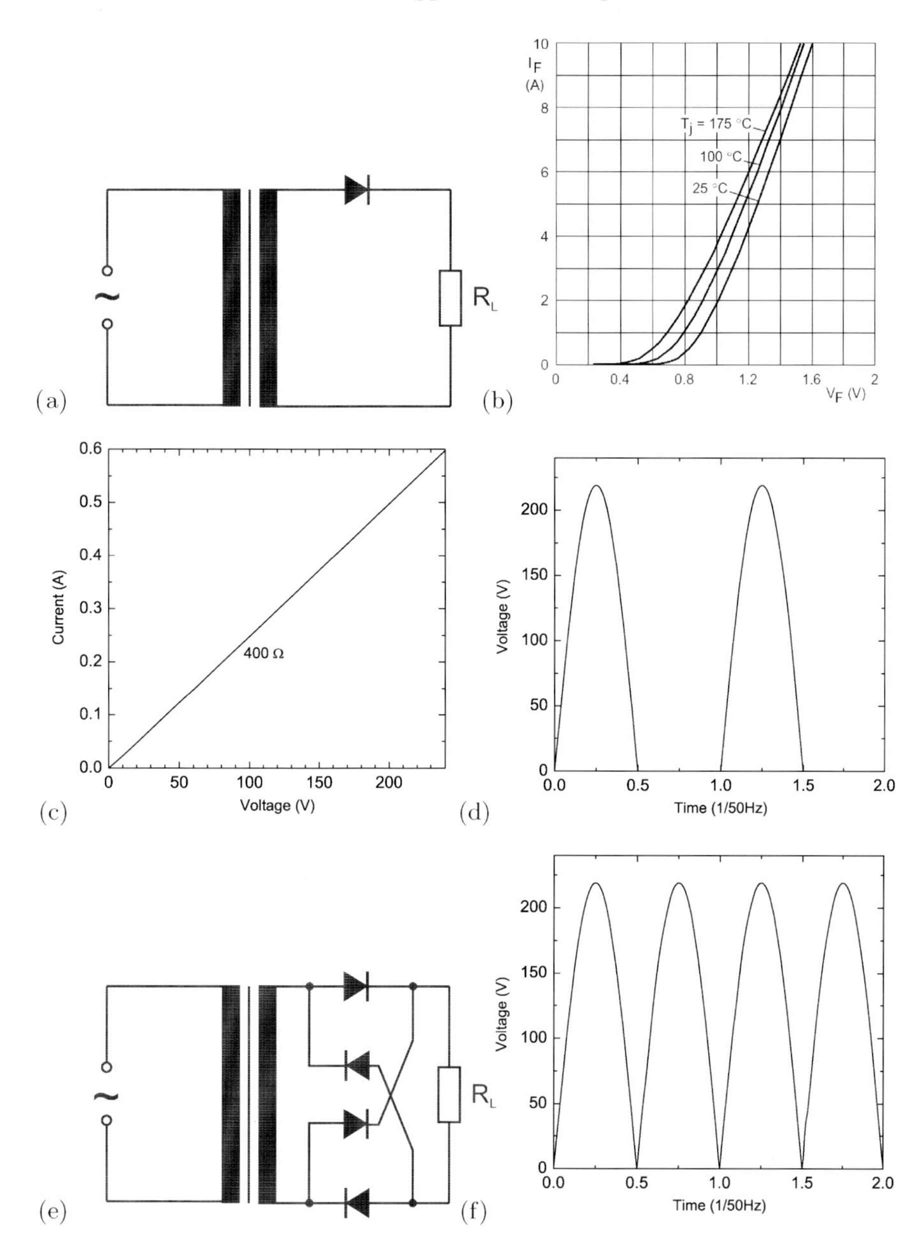

Fig. 18.49. (**a**) Single-path mains rectifier, (**b**) characteristics of Si diode (BYD127, Philips), (**c**) load characteristics of the mains rectifier ($R_\mathrm{L} = 400\,\Omega$), (**d**) voltage output of single-path mains rectifier. (**e**) depicts the schematic circuit of a bridge rectifier that works for both half-waves, (**f**) the resulting voltage output

For reverse bias we have ($\beta|V_\mathrm{r}| > 3$)

$$R_\mathrm{r} = \frac{V_\mathrm{r}}{I_\mathrm{r}} \cong \frac{V_\mathrm{r}}{I_\mathrm{s}} \tag{18.141a}$$

$$r_\mathrm{r} = \frac{\partial V_\mathrm{r}}{\partial I_\mathrm{r}} = \frac{nkT}{eI_\mathrm{s}} \exp\left(\frac{e|V_\mathrm{r}|}{nkT}\right) . \tag{18.141b}$$

Thus, the dc and ac rectifications ratios are given by

$$\frac{R_\mathrm{r}}{R_\mathrm{f}} = \exp\left(\frac{eV_\mathrm{f}}{nkT}\right) \tag{18.142a}$$

$$\frac{r_\mathrm{r}}{r_\mathrm{f}} = \frac{I_\mathrm{f}}{I_\mathrm{s}\exp\left(\frac{e|V_\mathrm{r}|}{nkT}\right)} . \tag{18.142b}$$

Rectifiers generally have slow switching speeds. A significant time delay arises from the necessary charge-carrier recombination when the diode switches from low (forward) to high (reverse) impedance. This poses typically no problem for line-frequency (50–60 Hz) applications. For fast applications, however, the minority-carrier lifetime needs to be reduced, e.g. by doping Si with Au (see Sect. 10.10).

18.5.2 Frequency Mixing

The nonlinear characteristic of the diode allows the mixing of frequencies, e.g. for second- (or higher-) harmonic generation, upconversion or demodulating of radio-frequency (RF) signals. A single balanced mixer is shown in Figs. 18.50a and b. The RF signal consists of a RF carrier frequency f_0 modulated with an intermediate frequency (IF) signal $f_\mathrm{IF}(t)$. It is mixed with a local oscillator (LO) that has a constant frequency f_LO outside the RF modulation bandwidth $f_0 \pm f_\mathrm{IF}$. The IF signal can be detected from the setup in Fig. 18.50a if filtered through a low-pass filter to avoid loss of power to the IF amplifier. The temperature dependence of the diode characteristic (via j_s and β) on mixing efficiency is typically less than 0.5 dB for a change in temperature of 100 K.

Problems of single-diode mixers are the radiation of local-oscillator power from the RF input port,[19] loss of sensitivity by absorption of input power in the local oscillator circuit, loss of input power in the intermediate frequency amplifier, and the generation of spurious output frequencies by harmonic mixing. Some of these problems can be solved by circuit techniques, but these circuits often introduce new problems. Most mixers therefore use multiple-diode techniques to provide a better solution of these problems. In Fig. 18.50c, the circuit diagram of a double balance mixer is shown. Even-order harmonics of both the LO and the signal frequency are rejected. This mixer does not

[19]that in military applications could make the mixer detectable by the enemy.

require a low-pass filter to isolate the IF circuit. The three ports are isolated from each other by the symmetry of the circuit. These mixers usually cover a broader frequency band than the others. Ratios as high as 1000:1 are available. Microwave equivalents (working at $f \gg 1\,\mathrm{GHz}$) of such mixer circuits are available. Bandwidth ratios as high as 40:1 are available at microwave frequencies in MMICs (millimeter-wave integrated circuits).

The common drawback of MMIC diodes is that they are obtained from the Schottky barriers used in field-effect transistors, that have inferior performance compared to discrete diodes. The use of pHEMT technology[20] for millimeter-wave applications provides diodes that differently from regular Schottky diodes, since they consist of a Schottky barrier in series with a heterojunction. In Fig. 18.50d, a MMIC 45 GHz mixer is shown using fast GaAs-based pHEMTs.

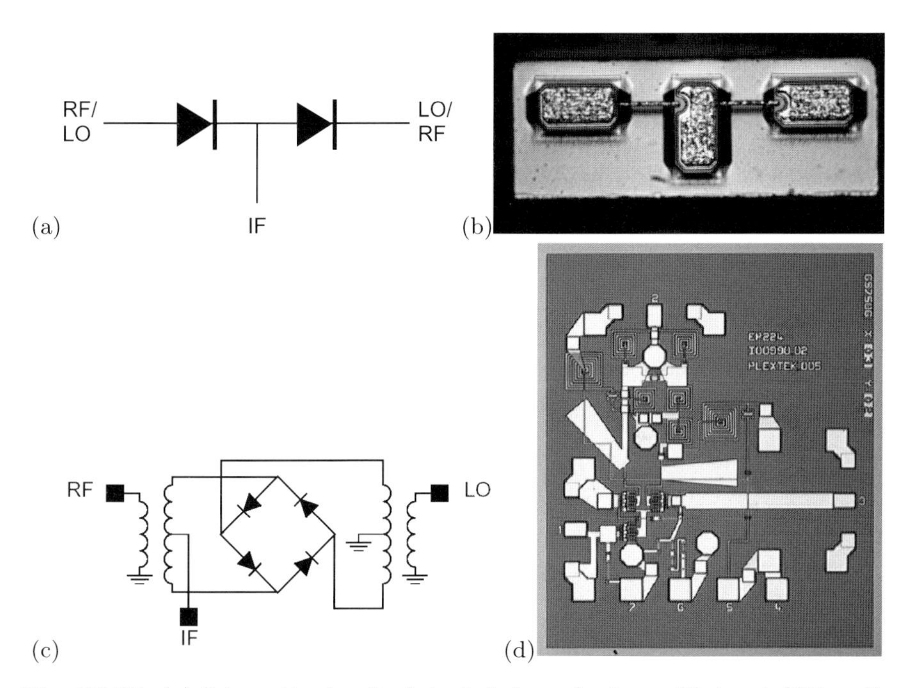

Fig. 18.50. (**a**) Schematic circuit of single balanced mixer with input (RF: radio frequency, LO: local oscillator) and output (IF: intermediate frequency). (**b**) Optical plan-view image ($300 \times 125\,\mu\mathrm{m}^2$) of a high-speed single balanced mixer with two GaAs Schottky diodes with opposite poling. The device properties are $R_\mathrm{s} = 5\,\Omega$, for $I = 1\,\mu\mathrm{A}$ a forward and reverse voltage of 0.7 and 6 V, respectively; the capacitance of each diode is 8 fF. Reprinted with permission from [589]. (**c**) Schematic circuit of a double balanced mixer. (**d**) Optical image ($1.65\,\mathrm{mm}^2$) of 40–45 GHz MMIC (Gilbert cell) mixer on GaAs basis using pHEMTs. Reprinted with permission from [590]

[20]pseudomorphic high electron mobility transistors, see Sect. 21.5.7.

18.5.3 Voltage Regulator

In a voltage regulator, the large variation of resistance with applied bias is used. This effect occurs in the forward direction and close to the breakdown voltage.

In Fig. 18.51a, a simple circuit is shown. When the input voltage V_{in} is increased, the current increases. The preresistor $R_1 = 5\,\mathrm{k}\Omega$ and the load resistor represent a voltage divider with $V_{\mathrm{in}} = IR_1 + V_{\mathrm{out}}$. The total current I is given by the two currents through the diode and the load resistor $I = I_{\mathrm{s}}\left[\exp(\beta V_{\mathrm{out}}/n) - 1\right] + V_{\mathrm{out}}/R_{\mathrm{L}}$. Therefore, the output voltage is implicitly given by

$$V_{\mathrm{out}}\left(1 + \frac{R_1}{R_{\mathrm{L}}}\right) = V_{\mathrm{in}} - R_1 I_{\mathrm{s}}\left[\exp\left(\frac{\beta V_{\mathrm{out}}}{n}\right) - 1\right] . \tag{18.143}$$

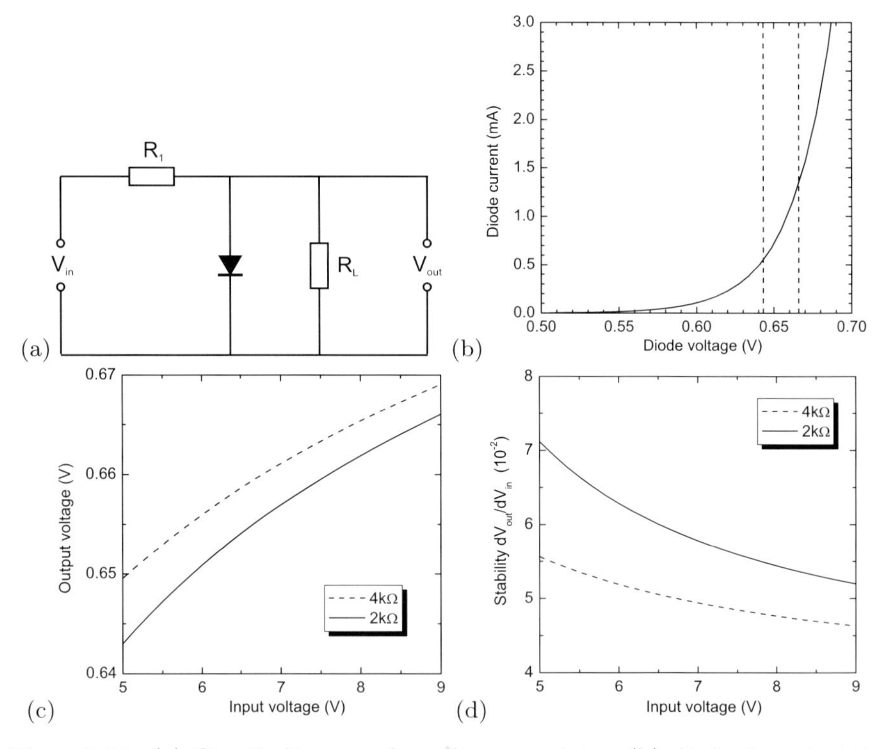

Fig. 18.51. (**a**) Circuit diagram of a voltage regulator, (**b**) diode characteristic ($n = 1$). The *vertical dashed lines* show the operation conditions for $R_{\mathrm{L}} = 2\,\mathrm{k}\Omega$ and $U_{\mathrm{E}} = 5$ and $9\,\mathrm{V}$ and thus the principle of voltage stabilization. (**c**) Output vs. input voltage and (**d**) stability (differential voltage ratio α, see text) for input voltage between 5 and $9\,\mathrm{V}$

A large current change is related to a fairly small change of the voltage across the diode, which at the same time is the output voltage. Therefore, a change in the input voltage causes only a small change in the output voltage.

We assume a diode with $n = 1$ and $I_S = 10^{-14}$ A with the characteristic shown in Fig. 18.51a. The numerical example in Fig. 18.51c is calculated for $R_L = 2\,\text{k}\Omega$ and $4\,\text{k}\Omega$, respectively. The output voltage varies by about 0.02 V if the input varies between 5 and 9 V. In Fig. 18.51d, the differential voltage change $\alpha = \frac{V_{in}}{V_{out}} \frac{\partial V_{out}}{\partial V_{in}}$ is shown.

In this way, voltage peaks can be filtered from the input voltage. If two antiparallel diodes are used, this principle works for both polarities. Instead of a diode in the forward direction, the very steep slope of the diode I–V characteristic at the breakdown can be used. Just before breakdown, the diode has a high resistance and the voltage drops at the load resistor. If the input voltage increases a little, the diode becomes conductive and shorts the additional voltage (the maximum allowed breakdown current needs to be observed!). Due to its small saturation current, typically Si diodes are used. The breakdown voltage can be designed via the diode parameters. Such diodes with defined breakdown voltage are called Z- or Zener diodes (see next section).

If the breakdown is due to tunneling (avalanche multiplication), the breakdown voltage decreases (increases) with temperature. If two diodes with positive and negative temperature coefficient are put in series, a very good temperature stability of the breakdown voltage of 0.002%/K can be achieved. Such diodes can be used to realize a reference voltage.

18.5.4 Zener Diodes

A Zener diode is designed to have a defined breakdown voltage. Zener diodes are available with a number of different standard breakdown voltages. Their characteristic is shown for reverse bias with the current shown positive. The characteristics of various Zener diodes for different breakdown voltages are shown in Fig. 18.52.

18.5.5 Varactors

A diode exhibits a voltage-dependent capacitance. This effect can be used to tune an oscillator using the diode bias (voltage-controlled oscillator, VCO). The equivalent circuit is shown in Fig. 18.53. The capacitance consists of a parasitic capacitance C_p due to mounting and bonding. This effect also causes a parasitic inductance. The series resistance due to mounting can typically be neglected. The variable junction capacitance C_j and the ohmic resistance R_s are bias dependent.

The $C(V)$ dependence has generally a power law with an exponent γ (which itself may depend on the bias voltage)

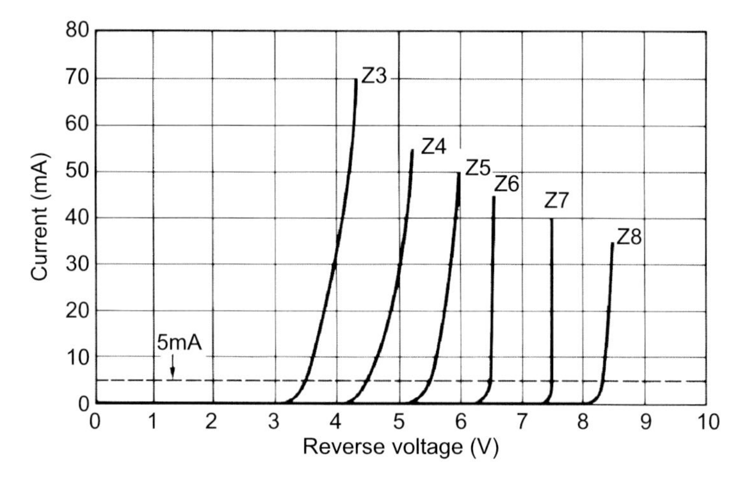

Fig. 18.52. Characteristics of a field of Zener diodes (at room temperature)

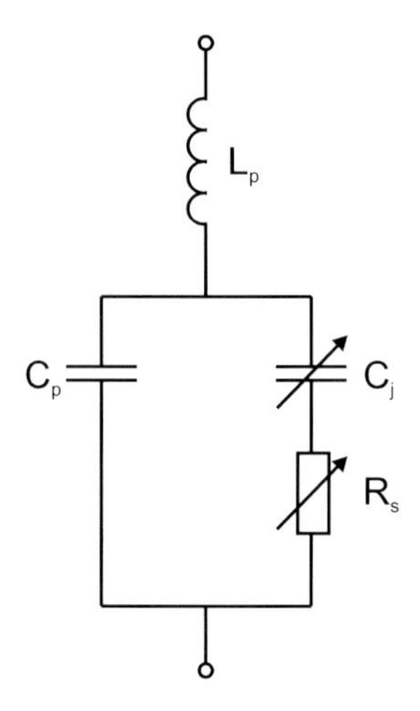

Fig. 18.53. Equivalent circuit of a varactor diode with parasitic capacitance C_p and inductance L_p and variable capacitance C_j and resistance R_s

$$C = \frac{C_0}{(1 + V/V_\mathrm{bi})^\gamma} , \tag{18.144}$$

where C_0 is the-zero bias capacitance. Since the frequency f of an LC oscillator circuit depends on $C^{-1/2}$ the frequency, f depends on the voltage as $f \propto V^{\gamma/2}$. Therefore, a $\gamma = 2$ dependence is most desirable.

For uniformly doped profiles, the capacitance depends with an inverse square root law on the applied voltage (18.90), i.e. $\gamma = 0.5$. Hyperabrupt junctions are typically made by ion implantation or epitaxy with graded impurity incorporation to create a special nonuniform doping profile (Fig. 18.54a). For

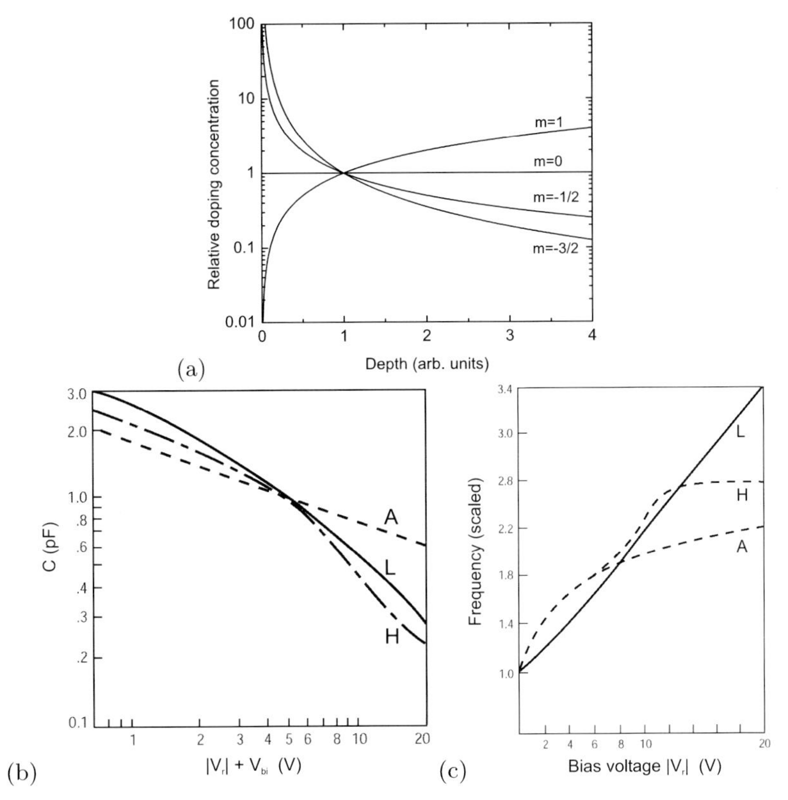

Fig. 18.54. (**a**) Donor-doping profile according to (18.145) in p^+n or Schottky diodes for $m = 0$ (abrupt junction), $m = 1$ (linearly graded junction) and two values with $m < 0$ (hyperabrupt junctions). (**b**) Bias dependence of capacitance for diodes with abrupt junction ('A', $\gamma = 0.5$), hyperabrupt junction ('H', $\gamma > 0.5$) and 'epilinear' junction ('L', $\gamma = 2$). (**c**) Frequency–voltage tuning characteristics (scaled to 1.0 for $V = 0$) for the three diode types. Parts (**b**) and (**c**) adapted from [591], reprinted with permission.

a doping profile

$$N_{\mathrm{B}}(z) = \hat{N}_{\mathrm{B}} \left(\frac{z}{z_0} \right)^m \tag{18.145}$$

the capacitance is given as

$$C = \left[\frac{e \hat{N}_{\mathrm{B}} \epsilon_{\mathrm{s}}^{m+1}}{(m+2) z_0^m (V_{\mathrm{bi}} - V)} \right]^{\frac{1}{m+2}} = \frac{C_0}{(1 + V/V_{\mathrm{bi}})^{\frac{1}{m+2}}} \, . \tag{18.146}$$

Ideally, $m = -3/2$ results is a linear frequency vs. voltage relation. The C–V characteristic of an implanted, hyperabrupt diode exhibits a part that has an exponent $\gamma = 2$ (Fig. 18.54b). A $\gamma = 2$ $C(V)$ dependence and therefore a linear $f(V)$ curve can be achieved over more than one octave using computer-controlled variable epitaxial-layer doping (Fig. 18.54c).

18.5.6 Fast-Recovery Diodes

Fast-recovery diodes are designed for high switching speeds. The switching speed from the forward to the reverse regime is given by the time $t_0 = t_1 + t_2$ with t_1 being the time to reduce the minority carrier density to the equilibrium value (e.g. $p_{\mathrm{n}} \rightarrow p_{\mathrm{n}_0}$) and t_2 being the time in which the current decreased exponentially. The time t_1 can be drastically reduced by incorporation of deep levels that act as recombination centers. A prominent example is Si:Au. However, this concept is limited since the reverse generation current, e.g. (18.127), depends on the trap density. For direct semiconductors, recombination times are short, e.g. 0.1 ns or less for GaAs. In silicon, they can be extremely long (up to ms) or at least 1–5 ns. Schottky diodes are suitable for high-speed applications since they are majority-carrier devices and minority-charge carrier storage can be neglected.

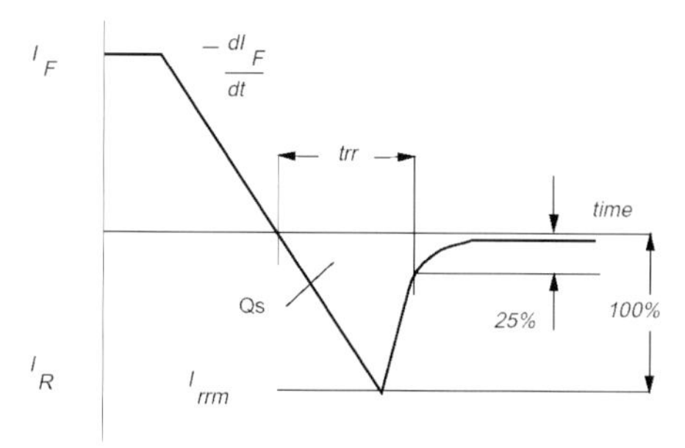

Fig. 18.55. Current vs. time trace for a (soft) fast-recovery diode. Reprinted with permission from [592]

18.5.7 Step-Recovery Diodes

This type of diode is designed to store charge in the forward direction. If polarity is reversed, the charge will allow conductance for a short while, ideally until a current peak is reached (Fig. 18.56a), and then cutoff the current very rapidly during the so-called snapback time T_{s} (Fig. 18.56b). The cutoff can be quite rapid, in the ps regime. These properties are used for pulse (comb) generation or as a gate in fast sampling oscilloscopes. In Si, only 0.5–5 µs are reached (fast-recovery diode, see previous section) while GaAs diodes can be used in the several tens of GHz regime.

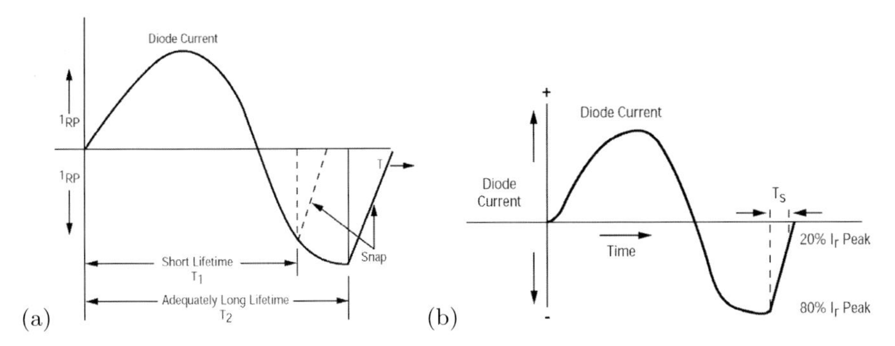

Fig. 18.56. (**a**) Current vs. time trace for a step-recovery diode and sinusoidal voltage input. The lifetime must be sufficiently large such that a current peak is reached. (**b**) Definition of the snapback time T_{s}. Reprinted with permission from [593]

Using a heterostructure GaAs/AlGaAs diode (see Sect. 18.5.11), as shown schematically in Fig. 18.57a, a steepening of a 15-V, 70-ps (10% to 90%) pulse to a fall time of 12 ps was observed (Fig. 18.57c). The forward current of the diode was 40 mA, supplied via a bias tee.

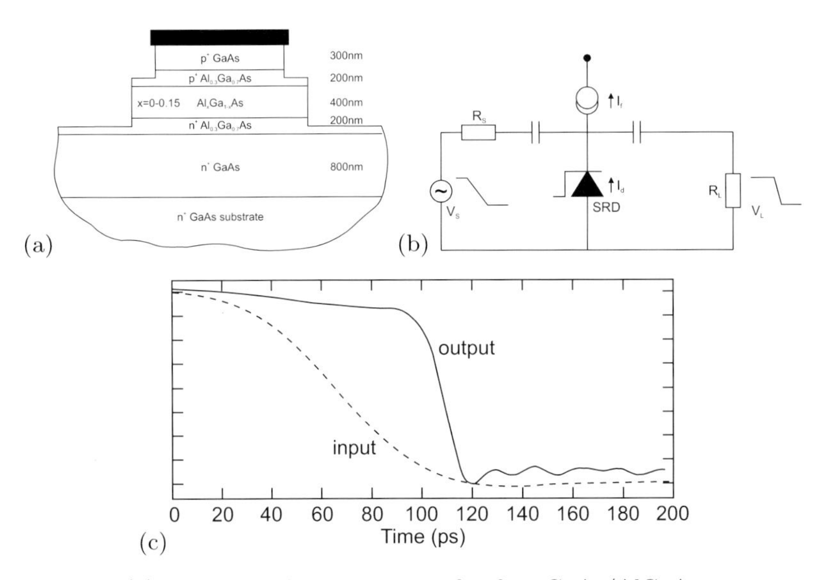

Fig. 18.57. (**a**) Schematic layer sequence for fast GaAs/AlGaAs step-recovery diode. (**b**) Circuit with input and output pulse. (**c**) Input (*dashed line*) and output (*solid line*) waveforms. Vertical division is 2 V. Adapted from [594]

18.5.8 pin-Diodes

In a pin-diode, an intrinsic (i), i.e. undoped region (with high resistivity) is located between the n- and the p-regions. Often, also a region with low n- or p-doping is used. In this case, the center region is denoted as a ν- or π-region, respectively. The fabrication of arbitrary doping profiles and an intrinsic region poses little problem for epitaxial diodes.

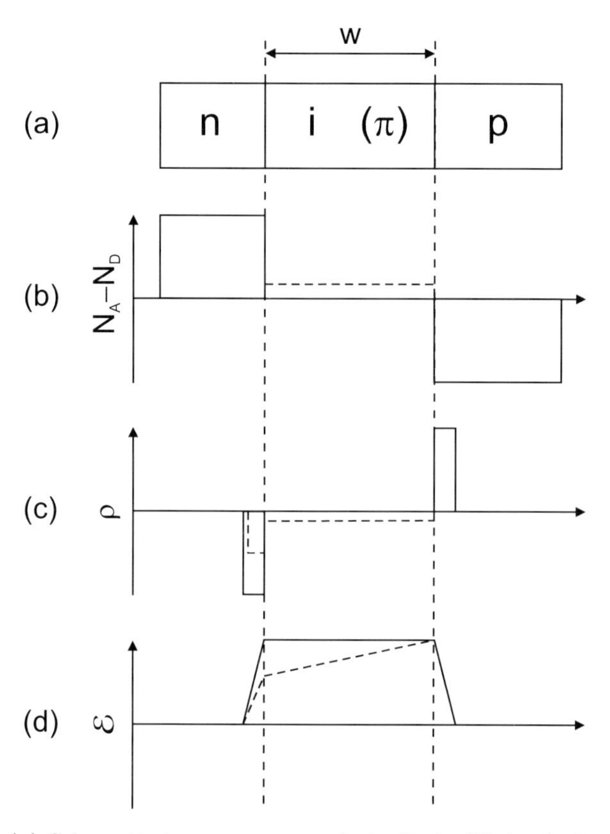

Fig. 18.58. (**a**) Schematic layer sequence of pin-diode ('i': intrinsic, 'π': lowly p-doped), (**b**) net impurity distribution $N_A - N_D$, (**c**) space charge and (**d**) electric field in a pin (*solid lines*) and a p–π–n (*dashed lines*) diode

Via the Poisson equation, the charge in the intrinsic layer is related to the electric field. If no dopants are present, there is a constant (maximum) field in the i-region at zero bias. If there is low doping, a field gradient exists.

The capacity for reverse bias is $\epsilon_s A/w$ and is constant starting at fairly small reverse bias (10 V). The series resistance is given by $R_s = R_i + R_c$. The contact resistance R_c dominates the series resistance for large forward bias.

18.5.9 Tunneling Diodes

For the invention of the tunneling diode and the explanation of its mechanism the 1973 Nobel Prize in Physics was awarded to L. Esaki. Eventually, the tunneling diode did not make the commercial breakthrough due to its high basis capacity. It is used for special microwave applications with low power consumption and for frequency stabilization.

First, the tunneling diode is a pn-diode. While the tunnel effect has already been discussed for Schottky diodes, it has not yet been considered by us for pn–diodes. We expect the tunnel effect to be important if the depletion-layer is thin, i.e. when the doping of both sides is high.

The doping is so high that the quasi-Fermi levels lie within the respective bands (Fig. 18.59). The degeneracy is typically several kT and the depletion layer width is in the 10 nm range.

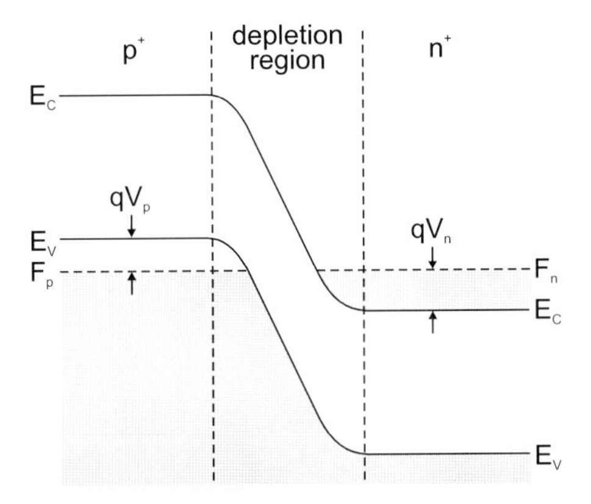

Fig. 18.59. Band diagram of a tunneling diode in thermodynamic equilibrium ($V = 0$). V_n and V_p characterize the degeneracies on the n- and p-side, respectively

In the forward direction, the I–V characteristic of the tunneling diode exhibits a maximum followed by a minimum and subsequently an exponential increase (Fig. 18.60a). As shown in Fig. 18.60b, the total current consists of three currents, the tunneling current, the excess current and the thermal (normal diode) current.

The $V = 0$ situation is again shown in Fig. 18.61b. Upon application of a small forward bias, electrons can tunnel from populated conduction-band states on the n-doped side into empty valance-band states (filled with holes) on the p-doped side (Fig. 18.61c). We note that this tunneling process is elastic. A similar situation, now with electrons tunneling from the valence band on the p-side into the conduction band on the n-side, is present for small reverse bias (Fig. 18.61a). Thus the rectifying behavior of the diode is lost. For larger forward bias, the bands are separated so far that the electrons coming from the n-doped side do not find final states on the p-doped side.

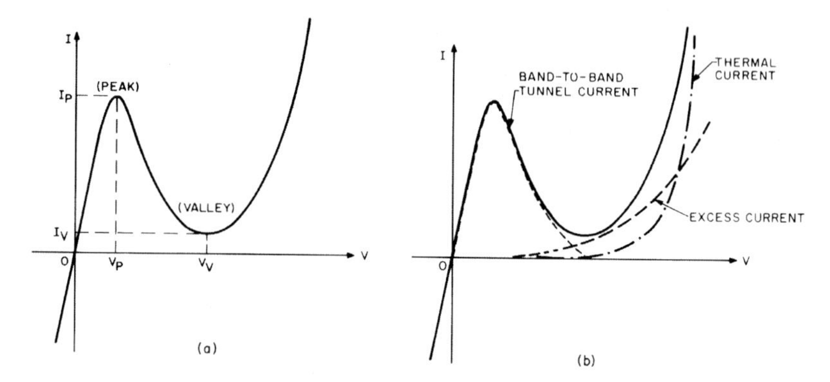

Fig. 18.60. (**a**) Static current–voltage characteristics of a typical tunneling diode. Peak and valley current and voltage are labeled. (**b**) The three components of the current are shown separately. Reprinted with permission from [183], ©1981 Wiley

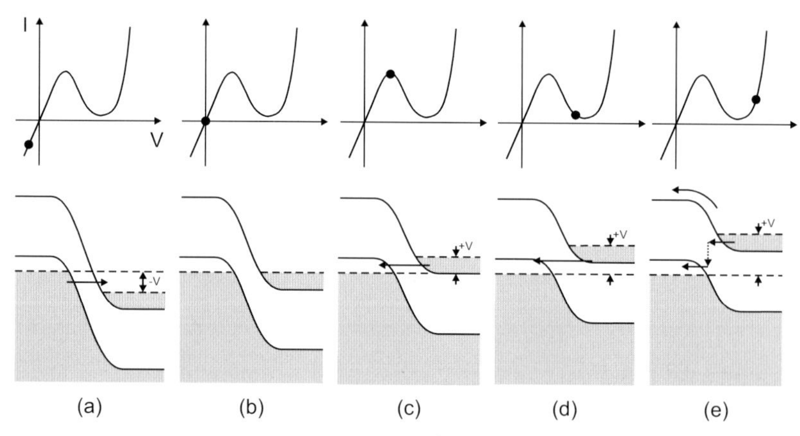

Fig. 18.61. I–V characteristics (*upper row*) and simplified band structure (*lower row*) of a tunneling diode at various bias voltages as indicated with a *dot* in the I–V plot. (**a**) Reverse bias, (**b**) in thermodynamic equilibrium ($V = 0$), (**c**) in the maximum of the tunneling current, (**d**) close to the valley and (**e**) forward bias with dominating thermal current. The tunneling current is indicated with *straight arrows*. In (**e**) the thermionic emission current (*curved arrow*) and the excess current with inelastic tunneling (*dotted arrow*) are shown.

Thus the tunneling current ceases (Fig. 18.61d). The current minimum is at a voltage $V = V_n + V_p > 0$. The thermal current is the normal diode diffusion current (Fig. 18.61e). Therefore, a minimum is present in the I–V characteristic. The excess current is due to inelastic tunneling processes through states in the bandgap and causes the minimum to not drop down to almost zero current.

The peak (V_p, I_p) and valley (V_v, I_v) structure of the characteristic leads to a region of negative differential resistance (NDR). I_p/I_v is termed the peak-to-valley ratio (Fig. 18.62). Peak-to-valley ratios of 8 (Ge), 12 (GaSb, GaAs), 4 (Si), 5 (InP) or 2 (InAs) have been reported (all at room temperature).

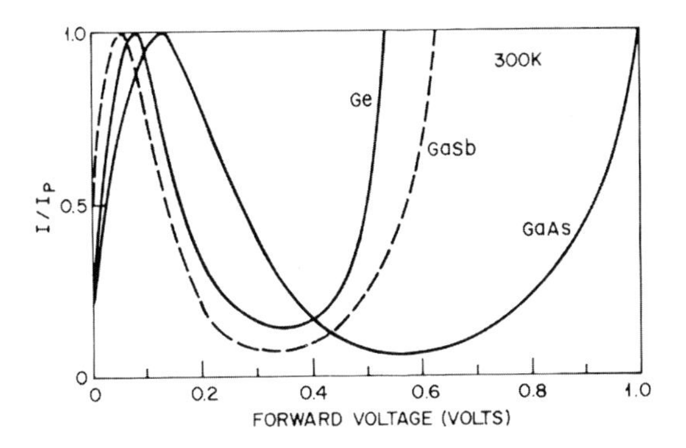

Fig. 18.62. Comparison of tunneling diodes based on Ge, GaSb and GaAs. Peak-to-valley ratios are 8 (Ge) and 12 (GaSb, GaAs). Reprinted with permission from [183], ©1981 Wiley

18.5.10 Backward Diodes

When the doping in a tunneling diode is nearly or not quite degenerate, the peak-to-valley ratio can be very small. Then the tunnel current flows mostly in the reverse direction (low resistance) and the forward direction has a higher resistance (with or without the NDR regime). Such diodes are called backward diodes. Since there is no minority-charge carrier storage, such diodes are useful for high-frequency applications.

18.5.11 Heterostructure Diodes

In a heterostructure diode, the n- and p-regions are made of different semiconductors. Such a diode is important in particular as an injection (emitter–base) diode in transistors. In Fig. 18.63, the band diagram is shown for a type-I heterostructure with the n (p) region having the smaller (larger) bandgap. Additionally to the built-in voltage, the barrier in the valence band is increased if the n-region is made from the semiconductor with larger bandgap Fig. 18.64. In this case, the (mostly undesired) hole current from the p- to the n-side is reduced. The peak in the conduction band poses potentially a greater barrier than the diffusion potential itself. The spike can be reduced by grading the materials across the heterojunction and creating a smooth transition of E_g between the materials. The effect of grading on the properties of the heterojunction is discussed in detail in [595].

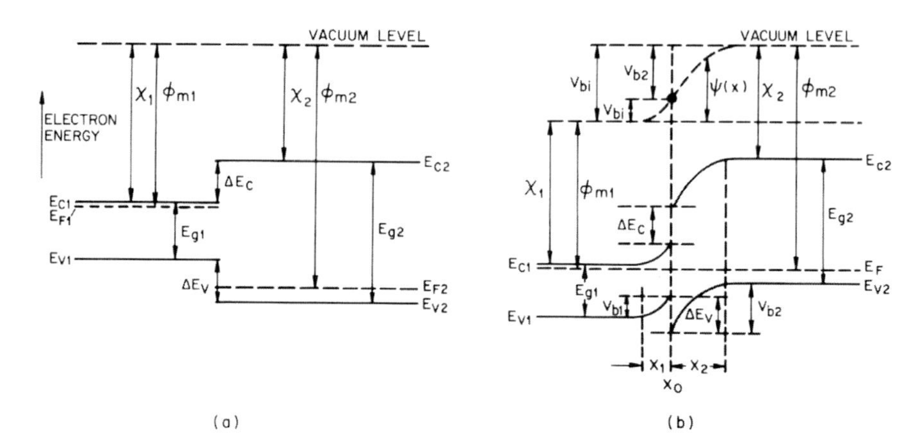

Fig. 18.63. Band diagram of a heterostructure diode (of the type n-GaAs/p-AlGaAs) (**a**) without contact of the n- and p-materials, (**b**) in thermodynamic equilibrium. Reprinted with permission from [183], ©1981 Wiley

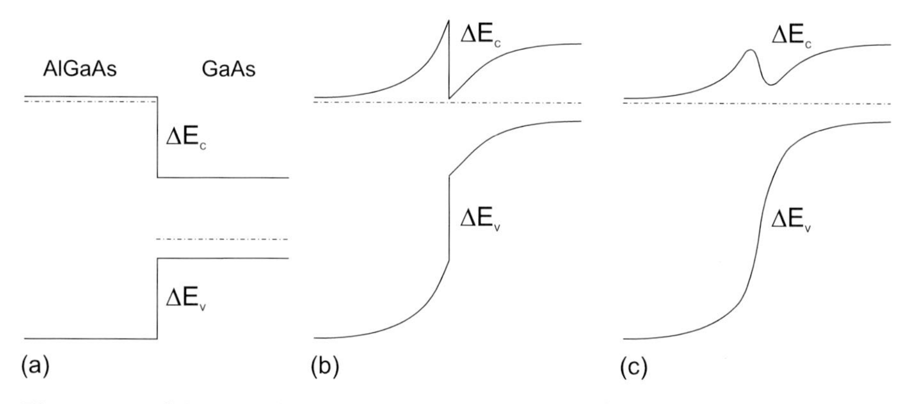

Fig. 18.64. Schematic band diagram of a n-AlGaAs/p-GaAs diode (**a**) without contact of the n- and p-materials, (**b**) in thermodynamic equilibrium, and (**c**) with graded Al composition at the heterointerface

19 Light-to-Electricity Conversion

19.1 Photocatalysis

The absorption of light in a semiconductor across the bandgap creates free electrons and holes. In particular, for small particle size in powders[1] these charge carriers can reach the surface of the semiconductor. At the surface they can react with chemicals. The hole can form $\bullet$OH radicals from OH^- attached to the bead. The electron can form $O_2\bullet^-$. These radicals can subsequently attack and detoxify, e.g., noxious organic pollutants in the solution surrounding the semiconductor. Such photocatalytic activity has been found, e.g., for TiO_2 and ZnO powders. A review of photocatalysis, in particular with TiO_2 particles and their surface modifications with metals and other semiconductors, is found in [596].

The efficiency of the photocatalytic activity depends on the efficiency of the charge separation (Fig. 19.1a). Any electron–hole pair that recombines within the bulk or the surface of the particle is lost for the catalytic activity. Thus, surfaces must exhibit a small density of recombination centers. Surface traps, however, can be beneficial for charge-carrier separation when they 'store' the charge-carrier rather than letting it recombine. Small particles are expected to exhibit more efficient charge-carrier separation than larger ones. Electrons at the surface can be donated and reduce an electron acceptor, typically oxygen, $A \rightarrow A^-$. A hole at the surface can oxidize a donor species, $D \rightarrow D^+$.

An example of increased photocatalytic activity are TiO_2 powders with deposited metal particles (such as Pt) for H_2 evolution and metal-oxide particles (such as RuO_2) for O_2 evolution (Fig. 19.1b). Such a system behaves as a short-circuited microscopic photoelectrochemical cell in which Pt is the cathode and RuO_2 is the anode [597]. Excitation with light energy above the bandgap in the TiO_2 particle (3.2 eV) injects electrons into the Pt particles and holes into the RuO_2 particles. Trapped electrons in Pt reduce water to hydrogen and trapped holes in RuO_2 oxidize water to oxygen.

The photocatalytic activity is also tied to the geometrical shape of the semiconductor. Generally, powders with nanosized grains have much higher

[1] 'Small' is here in relation to the diffusion length and does not need to be in the range where quantization effects (quantum dots) are present.

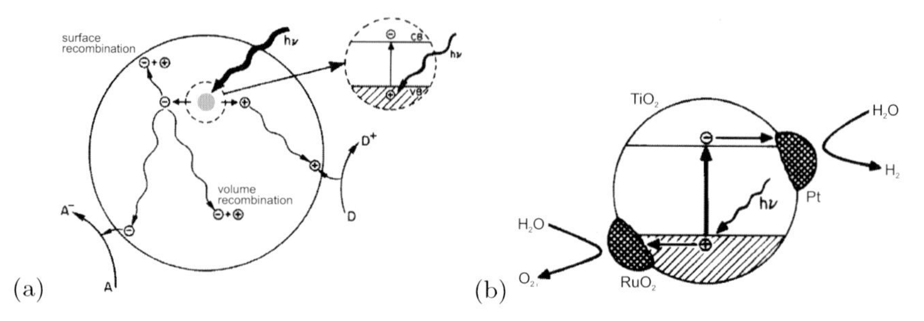

Fig. 19.1. (**a**) Principle of photocatalytic activity. Light absorption creates an electron–hole pair. The electron and hole diffuse and can recombine in the bulk or at the surface. Free carriers can react with species from the surrounding solution. (**b**) Schematic representation of photocatalytic activity of TiO_2 particle with Pt and RuO_2 depositions. Adapted from [596], reprinted with permission, ©1995 ACS

activity than those with microsized particles [598]. In Fig. 19.2 it is shown that nanosized object with high surface-to-volume ratio are more effective catalysts than rather compact surfaces.

In sun-protection cream only the UV absorption is wanted in UVA (330–420 nm) and UVB (260–330 nm) ranges. Subsequent photocatalysis on the skin and the presence of radicals are unwanted. Thus the semiconductor particles (∼10–200 nm diameter) are encapsulated in microbeads (∼1–10 µm diameter) of silica, PMMA or urethane, also improving ease of dispersion, aggregation, stability and skin feel.

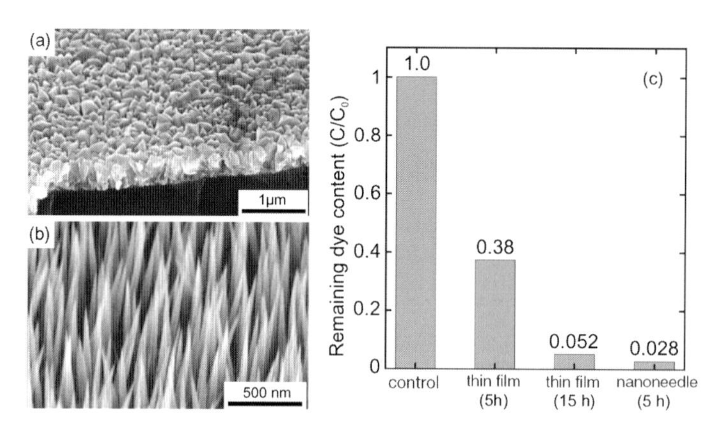

Fig. 19.2. SEM images of MOCVD-grown (**a**) ZnO thin film and (**b**) ZnO nanoneedle layer. (**c**) Comparison of the photocatalytic activity (decoloration of the dye Orange II in aqueous solution) of the ZnO thin film (irradiation with a Hg lamp for 5 h and 15 h) and the ZnO nanoneedles (irradiation 5 h). The sample labeled 'control' (scaled to 100%) is the start situation (absorption of the dye Orange II) without photocatalytic process. Adapted from [599], reprinted with permission

19.2 Photoconductors

19.2.1 Introduction

Charge carriers can be generated in the semiconductor through the absorption of light with a photon energy above or below the bandgap (Fig. 19.3). Absorption involving impurities occurs typically in the mid- and far-infrared spectral regimes (cf. Sect. 9.5). The additional charge carriers cause an increase in the conductivity (8.9).

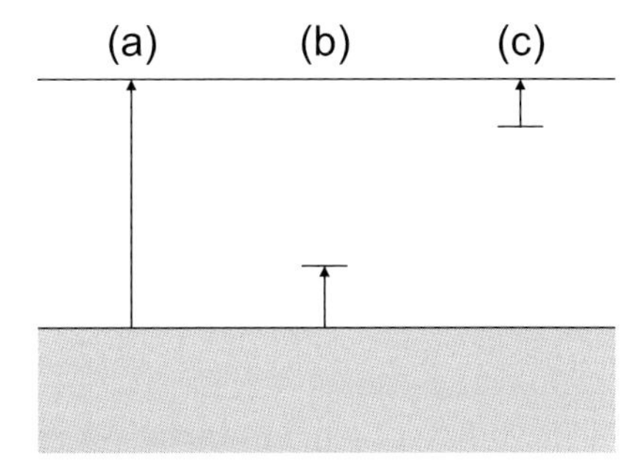

Fig. 19.3. Absorption and charge-carrier generation in a photoconductor: (**a**) band–band transition, (**b**) valence-band to acceptor and (**c**) donor to conduction-band transition

19.2.2 Photoconductivity Detectors

In stationary equilibrium for constant illumination of power $P_{\rm opt}$ and photon energy $E = h\nu$ the generation rate G is given by

$$G = \frac{n}{\tau} = \eta \, \frac{P_{\rm opt}/h\nu}{V} \,, \tag{19.1}$$

where V is the volume ($V = wdL$, cf. Fig. 19.4) and τ denotes the charge-carrier lifetime. η is the quantum efficiency, i.e. the average number of electron–hole pairs generated per incoming photon. The photocurrent between the electrodes is

$$I_{\rm ph} = \sigma \mathcal{E} wd \approx e\mu_{\rm n} n \mathcal{E} wd \,, \tag{19.2}$$

assuming that $\mu_{\rm n} \gg \mu_{\rm p}$ and with $\mathcal{E} = V/L$ denoting the electric field in the photoconductor, V being the voltage across the photoconductor. We can then also write

$$I_{\rm ph} = e \left(\eta \, \frac{P_{\rm opt}}{h\nu} \right) \left(\frac{\mu_{\rm n} \tau \mathcal{E}}{L} \right) . \tag{19.3}$$

With the primary photocurrent $I_\mathrm{p} = e\left(\eta \frac{P_\mathrm{opt}}{h\nu}\right)$ we deduce a gain

$$g = \frac{I_\mathrm{ph}}{I_\mathrm{p}} = \frac{\mu_\mathrm{n}\tau\mathcal{E}}{L} = \frac{\tau}{t_\mathrm{r}} , \tag{19.4}$$

where $t_\mathrm{r} = L/v_\mathrm{d}$ is the transit time through the photoconductor.

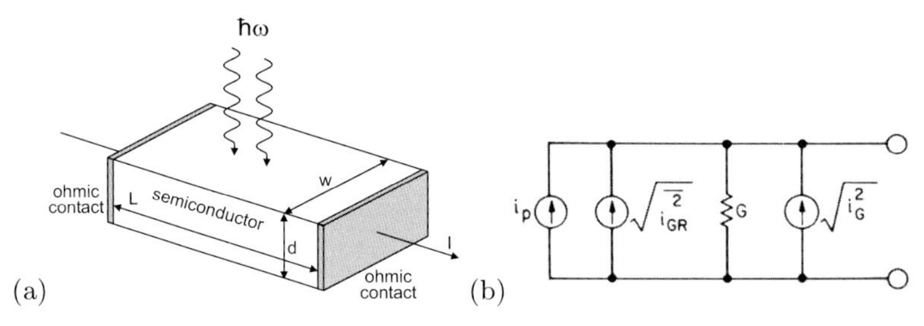

Fig. 19.4. (**a**) Scheme of photoconductor. (**b**) Equivalent circuit of photoconductor

Now we consider a modulated light intensity

$$P(\omega) = P_\mathrm{opt}\left[1 + m\exp(\mathrm{i}\omega t)\right] , \tag{19.5}$$

where m is between 0 and 1. For $m = 0$ it is a constant light power, for $m = 1$ the intensity is sinusoidally modulated between 0 and $P_\mathrm{max} = 2P_\mathrm{opt}$. The rms optical power[2] is given by $\sqrt{2}mP_\mathrm{opt}$. In the case of $m = 1$ this is equal to $P_\mathrm{max}/\sqrt{2}$.

The rms photocurrent is

$$i_\mathrm{ph} \approx \frac{e\eta m P_\mathrm{opt}}{\sqrt{2}h\nu}\frac{\tau}{t_\mathrm{r}}\frac{1}{(1+\omega^2\tau^2)^{1/2}} . \tag{19.6}$$

The thermal noise at a conductivity G is

$$i_\mathrm{G}^2 = 4kTGB , \tag{19.7}$$

with B being the bandwidth over which the noise is integrated. The generation–recombination noise (shot noise) is

$$i_\mathrm{GR}^2 = \frac{\tau}{t_\mathrm{r}}\frac{4qI_0}{1+\omega^2\tau^2}B \tag{19.8}$$

for the modulation frequency ω and I_0 being the current in stationary equilibrium. The equivalent circuit with the ideal photocurrent source and the noise currents is depicted in Fig. 19.4b.

[2]The rms value is the square root of the time average of the square of the power, $\sqrt{\langle P^2\rangle}$.

The signal-to-noise ratio of the power is then given by

$$S/N = \frac{i_{\mathrm{ph}}^2}{i_{\mathrm{G}}^2 + i_{\mathrm{GR}}^2} = \frac{\eta m^2 (P_{\mathrm{opt}}/h\nu)}{8B} \left[1 + \frac{kT}{e} \frac{t_{\mathrm{r}}}{\tau} (1 + \omega^2 \tau^2) \frac{G}{I_0} \right]^{-1} . \tag{19.9}$$

An important quantity is the noise equivalent power (NEP). This is the light power ($mP_{\mathrm{opt}}/\sqrt{2}$) for which the S/N ratio is equal to 1 (for $B = 1\,\mathrm{Hz}$). For infrared detectors the typical figure of merit is the detectivity D^* that is defined by

$$D^* = \frac{\sqrt{A\,B}}{\mathrm{NEP}} . \tag{19.10}$$

A denotes the area of the detector. The unit of D^* is $\mathrm{cm\,Hz^{1/2}\,W^{-1}}$. It should be given together with the modulation frequency. It can be given for monochromatic radiation of a particular wavelength λ or a blackbody spectrum of given temperature T_0.

19.2.3 Electrophotography

The principle of the Xerox copy machine is based on a photoconductive layer (Fig. 19.5). This layer is normally insulating such that both sides of the layer can be oppositely charged. If light hits the layer it becomes photoconductive and neutralizes locally. This requires a small lateral diffusion of charge carriers. Initially amorphous selenium ($E_{\mathrm{g}} = 1.8\,\mathrm{eV}$) was used. The conductivity in the dark of a-Se is $10^{16}\,\Omega/\mathrm{cm}$. Se was subsequently replaced by organic material. The highest performance is currently achieved with amorphous silicon.

On the charged areas of the photosensitive layer toner can be attached. The toner pattern is subsequently transferred to the copy sheet and fixated. A copy takes typically more than one rotation of the drum. The principle was invented in 1938 by Chester F. Carlson (1906–1968) with sulfur as the photoconductor.[3]

19.2.4 QWIPs

Quantum-well intersubband photodetectors (QWIPs) are based on the absorption of photons between two quantum well subbands (Fig. 19.6). A review can be found in [601]. Quantized electron or hole states can be used. Besides

[3] In 1947 the Haloid company bought the rights to this process, renamed itself XeroX and brought the first copy machine to the market in 1958 based on amorphous selenium. The word 'xerography' stems from the Greek word ξέρος (dry). The last 'X' in XeroX was added to mimic the name of the KodaK corporation.

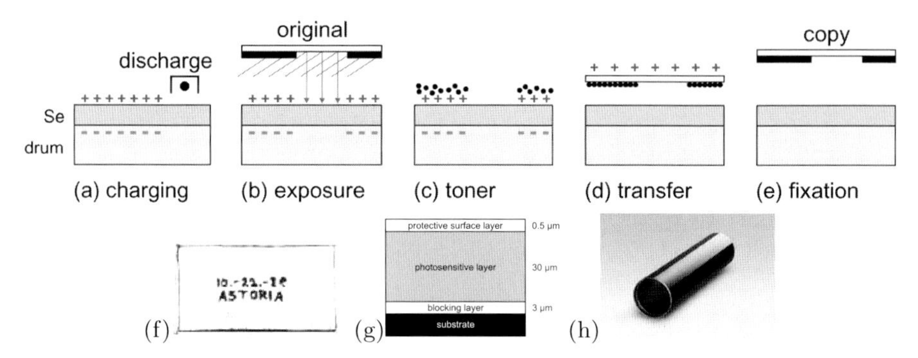

Fig. 19.5. Principle of xerography: (**a**) charging of the selenium-covered drum, (**b**) (reflection) exposure of the Se, exposed areas become uncharged, (**c**) toner addition, (**d**) toner transfer to paper for copy, and (**e**) fixation of the toner on the copy and preparation of drum for the next cycle. (**f**) First xerox copy (Oct. 22nd 1938). (**g**) Schematic cross section of coating of photosensitive drum. The indicated thicknesses are approximate. (**h**) Image of drum with photosensitive layer made from amorphous silicon. Part (**h**) from [600]

an oscillator strength for this transition, the lower level must be populated and the upper level must be empty in order to allow this process. The Fermi level is typically chosen by appropriate doping such that the lower subband is populated.

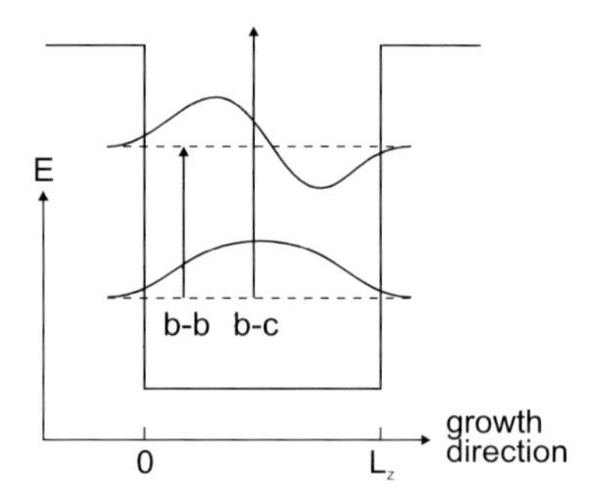

Fig. 19.6. Schematic level diagram of a quantum well. Optical intersubband transitions between the first and second quantized level (b–b) and the ground state and the continuum (b–c)

For infinite barrier height the energy separation between the first and second quantized levels (in the effective-mass theory) is (cf. (11.4))

$$E_2 - E_1 = 3\frac{\hbar^2}{2m^*}\frac{\pi^2}{L_z^2} \; . \tag{19.11}$$

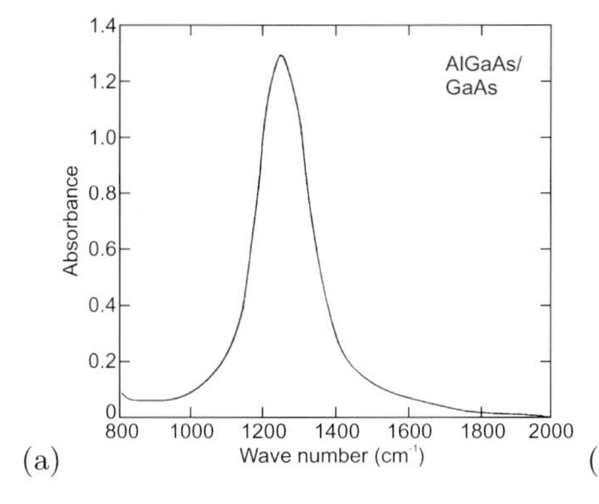

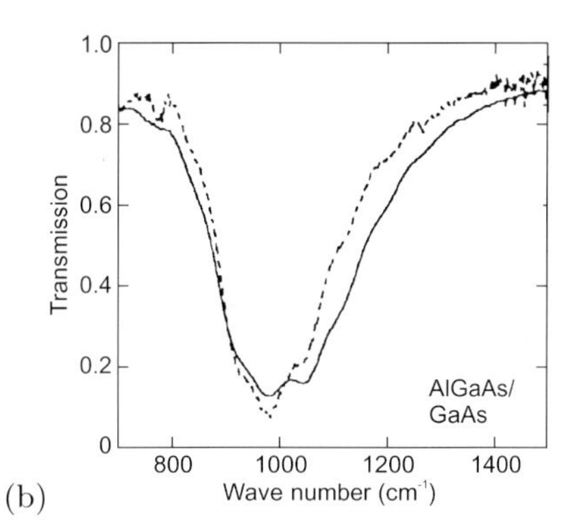

Fig. 19.7. (**a**) AlGaAs/GaAs QWIP absorption spectrum for multiple reflection geometry. Adapted from [601]. (**b**) Transmission of AlGaAs/GaAs QWIP (100 QWs) in double reflection geometry (45° angle of incidence). The well doping is $1.0 \times 10^{12}\,\mathrm{cm}^{-2}$ (*dashed line*) and $1.5 \times 10^{12}\,\mathrm{cm}^{-2}$ (*solid line*). Adapted from [602]

For real materials the barrier height determines the maximum transition energy. Typical absorption and transmission spectra of a QWIP structure are shown in Fig. 19.7. The spectral response is in the mid- or far-infrared.

The dipole matrix element $\langle z \rangle = \langle \Psi_2 | z | \Psi_1 \rangle$ can be easily calculated to be

$$\langle z \rangle = \frac{16}{9\pi^2}\, L_z \,. \tag{19.12}$$

The oscillator strength is about 0.96. The polarization selection rule causes the absorption to vary $\propto \cos^2\phi$, where ϕ is the angle between the electric-field vector and the z direction (Fig. 19.8). This means that for vertical incidence ($\phi = 90°$) the absorption vanishes. Thus schemes have been developed to allow for skew entry of the radiation (Fig. 19.9a). The strict selection rule can be relaxed by using asymmetric potential wells (breaking of mirror symmetry/parity), strained materials (band mixing) or quantum dots (lateral confinement). Also, a grating can be used to create a finite angle of incidence (Fig. 19.9b).

Besides a useful detectivity ($2 \times 10^{10}\,\mathrm{cm\,Hz^{1/2}/W}$ at 77 K) QWIPs have the advantage, e.g. against HgCdTe interband absorbers, that the highly developed GaAs planar technology is available for the fabrication of focal plane arrays (FPA) as shown in Fig. 19.10. A FPA is an image sensor (in the focal plane of an imaging infrared optics) and is used, e.g., for the detection of heat leaks in buildings or night surveillance. In particular, night vision support in cars may become a major market. A competing technology are bolometric arrays with thermally insulated pixels based on MEMS technology.

The carriers that have been optically excited into the upper state leave the QW by tunneling or thermionic emission. Also, a QWIP can be made based on the direct transfer from the (populated) subband into the continuum.

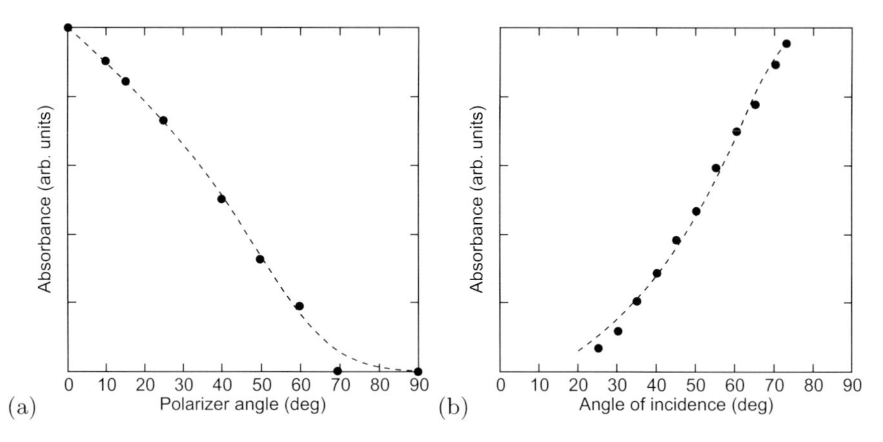

Fig. 19.8. Dependence of the QWIP response on (**a**) polarization and (**b**) angle of incidence. *Dashed lines* are guides to the eye. Adapted from [601]

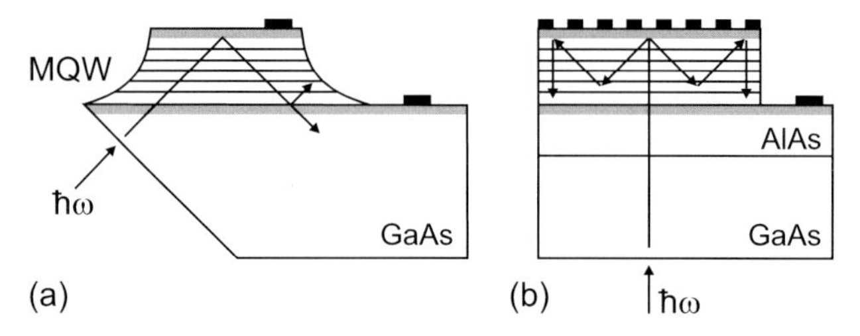

Fig. 19.9. QWIP geometries: (**a**) 45° edge coupled with multiple quantum-well (MQW) absorber and (**b**) grating coupled with GaAs substrate, AlAs reflector and metal grating on top. *Grey areas* are highly n-doped contact layers

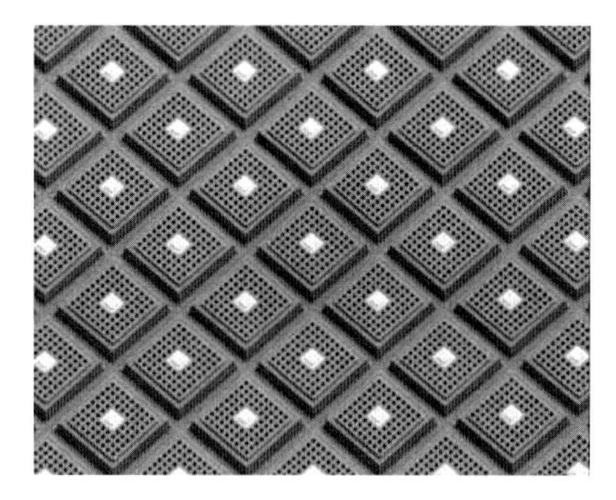

Fig. 19.10. Part of a 256 × 256 QWIP focal plane array (FPA) with grating coupler (area of one pixel: 37 μm^2). From [603]

The incoming infrared radiation creates a photocurrent density of

$$i_{\mathrm{ph}} = e\eta_{\mathrm{w}}\Phi\,, \tag{19.13}$$

where η_{w} is the quantum efficiency of a single quantum well (including the escape rate) and Φ is the photon flux per time and unit area. During the transport of the charge carriers through the barrier they can be (re-)captured by the QW with the probability p_{c}. The capture probability decreases exponentially with the applied bias. The total photocurrent density (including generation and recapture) is

$$j_{\mathrm{ph}} = (1 - p_{\mathrm{c}})j_{\mathrm{ph}} + i_{\mathrm{ph}} = \frac{i_{\mathrm{ph}}}{p_{\mathrm{c}}}\,. \tag{19.14}$$

If the quantum efficiency is small, the efficiency of N_{w} quantum wells $\eta \approx N_{\mathrm{w}}\,\eta_{\mathrm{w}}$. With this approximation the total photocurrent of N_{w} quantum wells is given by

$$j_{\mathrm{ph}} = e\eta\Phi g\,, \tag{19.15}$$

where g is termed the gain of the structure and is given by

$$g = \frac{1}{p_{\mathrm{c}}}\frac{\eta_{\mathrm{w}}}{\eta} \approx \frac{1}{N_{\mathrm{w}}p_{\mathrm{c}}}\,. \tag{19.16}$$

The dark current can be calculated from thermionic emission and agrees fairly well with experiment (Fig. 19.11a). When the voltage is increased further, avalanche multiplication can occur while the carriers are transported through the barrier(s). This mechanism provides further gain as shown in Fig. 19.11b.

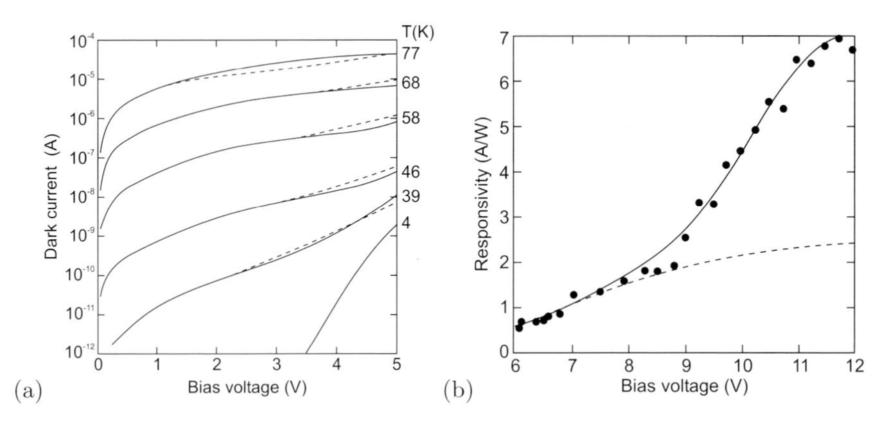

Fig. 19.11. (**a**) Dark current of a QWIP at 10.7 µm, experimental (*solid lines*) and theoretical (*dashed lines*) response. (**b**) QWIP responsivity as a function of the applied voltage. The *solid line* (*dashed line*) is the theoretical dependence with (without) the effect of avalanche multiplication. Adapted from [601]

19.2.5 Blocked Impurity-Band Detectors

Impurity absorption allows photoconductivity detectors in the mid- and far-infrared regions to be made. In particular, for THz spectroscopy in medicine and astronomy the extension to longer wavelengths is interesting. For conventional photoconductors the impurity concentration is well below the critical dopant concentration (cf. Sect 7.5.5). Long-wavelength response can be achieved by going to impurity/host systems with smaller ionization energy, such as Si:B (45 meV) → Ge:As (12.7 meV) → GaAs:Te (5.7 meV). By applying stress to Ge the energy separation between impurity and conduction bands can be lowered and subsequently the detector response is shifted towards longer wavelengths.

For high doping the impurity level broadens to an impurity band and thus allows smaller ionization energy and thus stronger long-wavelength detector response. However, conduction in the impurity band leads to dark current and makes such detectors unfeasible. In a blocked impurity band (BIB) detector [604, 605] an additional intrinsic blocking layer is sandwiched between the absorption layer and the contact (Fig. 19.12a). Such a structure is similar to a MIS diode, the insulator being the intrinsic semiconductor. We assume in the following an n-type semiconductor, such as Si:As or GaAs:Te, but also p-type BIBs can be made, e.g., from Ge:Ga.

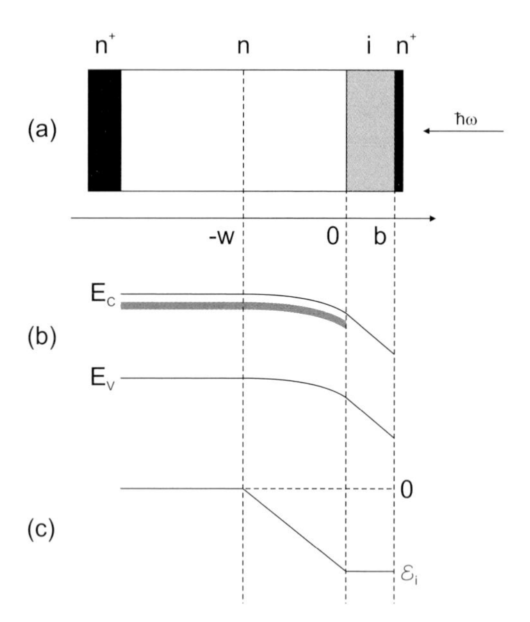

Fig. 19.12. (**a**) Structure of BIB photodetectors. Highly doped contact layers (*black*), doped semiconductor (*white*) and blocking (intrinsic) layer (*grey*). (**b**) Band diagram under small forward bias. *Shaded area* represents the donor impurity band. (**c**) Electric field in the structure

The semiconductor is highly doped (N_D) and partly compensated (N_A). Typically, the acceptor concentration must be small, about $10^{12}\,\mathrm{cm}^{-3}$, and controls the formation of the electric field as shown below. The doping is so high that the impurities form an impurity band. Some of the electrons recombine with the acceptors $N_\mathrm{A}^- = N_\mathrm{A}$ and leave some donors charged $N_\mathrm{D}^+ = N_\mathrm{A}$. For GaAs, e.g., the donor concentration in the doped semiconductor is $> 10^{16}\,\mathrm{cm}^{-3}$ and $\sim 10^{13}\,\mathrm{cm}^{-3}$ in the i-layer.

Under an external forward bias V, i.e. the positive pole is at the insulator, part of the applied voltage drops over the blocking layer of thickness b. If ideally no charges are present here, the electric field is constant. In the n-doped material electrons move in the impurity band towards the insulator, forming neutral donors in an electron accumulation layer of thickness w in the presence of the charged acceptors N_A^-. This layer is the absorption layer. The mechanism can also be considered as if positive charge (the charged donors, N_D^+) moves (via hopping conduction) towards the back contact. In the literature the layer close to the insulator is thus also termed a 'depletion layer'. The band diagram and the electric field are shown in Fig. 19.12b,c. Due to the blocking layer the carriers on the donors in the n-type material cannot spill via the impurity band into the contact but must be lifted (by infrared photoabsorption) into the conduction band.

From the Poisson equation the electric field is given by

$$\mathcal{E}(x) = -\frac{e}{\epsilon_\mathrm{s}} N_\mathrm{A}\,(w + x)\,, \quad -w \leq x \leq 0 \tag{19.17a}$$

$$\mathcal{E}(x) = -\frac{e}{\epsilon_\mathrm{s}} N_\mathrm{A} w = \mathcal{E}_\mathrm{i}, \; 0 \leq x \leq b\,. \tag{19.17b}$$

The voltage drops across the blocking layer V_b and the doped semiconductor V_s fulfill

$$V = V_\mathrm{b} + V_\mathrm{s}\,. \tag{19.18}$$

Integration of the fields yields

$$V_\mathrm{s} = \frac{e}{\epsilon_\mathrm{s}} N_\mathrm{A} \frac{w^2}{2} \tag{19.19a}$$

$$V_\mathrm{b} = \frac{e}{\epsilon_\mathrm{s}} N_\mathrm{A} w b\,. \tag{19.19b}$$

Substituting (19.19a,b) into (19.18) results in the width of the 'depletion layer'

$$w = \left(\frac{2\epsilon_\mathrm{s} V}{e N_\mathrm{A}} + b^2\right)^{1/2} - b\,. \tag{19.20}$$

The high dopant concentration allows for much thinner absorption layers than in a conventional photoconductivity detector, making it less susceptible to background high-energy cosmic radiation. The recombination in the depletion layer is negligible. Detector performance is modeled in [606].

19.3 Photodiodes

19.3.1 Introduction

The principle of the photodiode is the interband absorption of light in the depletion layer of a diode (or the i-zone of a pin-diode) and the subsequent separation of electrons and holes by the electric field. There are opposite requirements for fast detectors (thin depletion layer) and efficient detectors (complete light absorption, sufficiently thick depletion layer). For this reason generally semiconductors with high absorption coefficient are most suited (Fig. 19.13). In Fig. 19.14 the quantum efficiency and detectivity D^* of various semiconductor detectors are compared.

A diode can be operated without bias (photovoltaic mode) using the built-in field. An improvement in the speed of a pn-diode is achieved with a reverse bias since it increases the field strength in the depletion layer. However, the reverse bias is below the breakdown voltage. Operation near breakdown is exploited in the avalanche photodiode (APD). In the following we will discuss pn-, pin-, MS- (Schottky-), MSM- and heterostructure-diodes and APDs.

19.3.2 pn Photodiodes

The most important figures of merit are the quantum efficiency, responsivity, noise equivalent power (NEP) and the response speed.

If the depletion layer is hit by a photon flux with a generation rate G_0 (i.e. electron–hole pairs per unit volume per unit time) the photogenerated current is added to the diffusion current. The photocurrent density j_{p} (per unit area) is

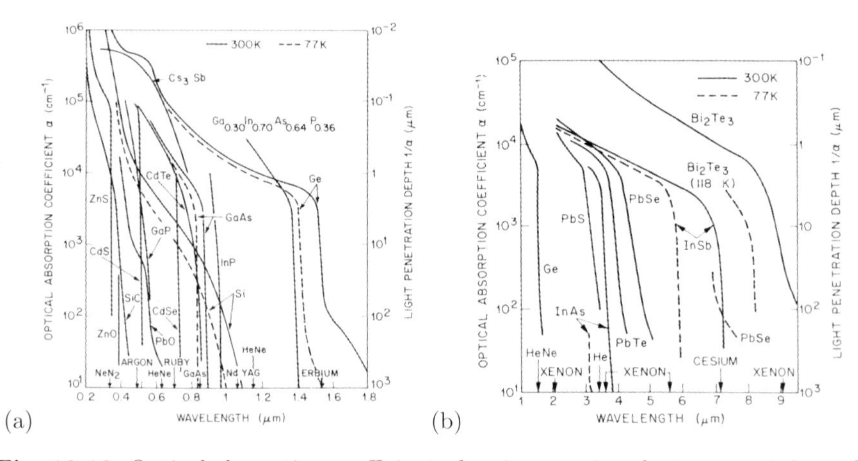

Fig. 19.13. Optical absorption coefficient of various semiconductor materials used for photodetectors in (**a**) the UV, visible and near-infrared range and (**b**) the mid-infrared spectral range. From [607]

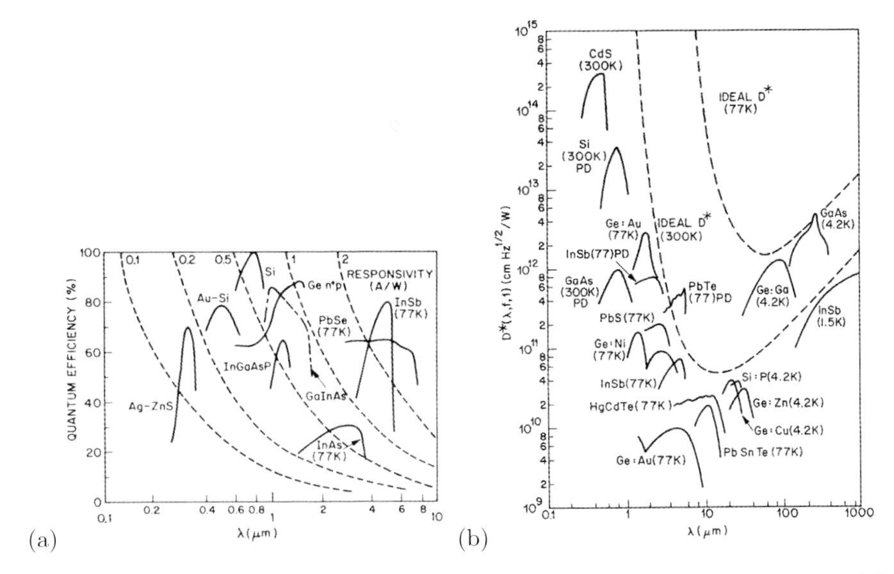

Fig. 19.14. (**a**) Quantum efficiency and sensitivity of various photodetectors. (**b**) Detectivity D^* of various photoconductors and photodiodes. Reprinted with permission from [183], ©1981 Wiley

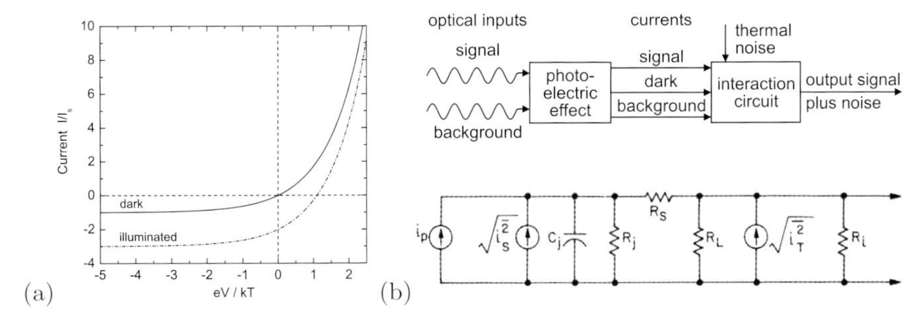

Fig. 19.15. (**a**) Schematic dark and illuminated I–V-characteristics of a photodiode (for the case $j_p = -2j_s$). (**b**) Schematic representation of currents in a photodiode and equivalent circuit. Part (**b**) adapted from [608]

$$j_{\mathrm{ph}} = -eG_0L_{\mathrm{p}} \tag{19.21}$$

for a p$^+$n-diode. In order to obtain this result the diffusion and continuity equations have to be solved for the depletion region[4]. Equation (19.21) means that the dark I–V characteristic is shifted by j_{ph} as shown in Fig. 19.15a. The number of electron–hole pairs that are generated per photon of energy $h\nu$ by the absorption of the (monochromatic) light power P_{opt} is

[4]This derivation is done in Sect. 19.3.3. Equation (19.21) is obtained from (19.31) for vanishing thickness w and $\alpha L_{\mathrm{p}} \ll 1$.

$$\eta = \frac{I_{\mathrm{ph}}/e}{P_{\mathrm{opt}}/(h\nu)} , \qquad (19.22)$$

where $I_{\mathrm{ph}} = Aj_{\mathrm{ph}}$ is the photogenerated current over the surface A. For a modulated light intensity P_{opt} must be replaced by $mP_{\mathrm{opt}}/\sqrt{2}$. The equivalent circuit including noise sources for a photodiode is shown in Fig. 19.15b.

Random processes lead to shot noise $\langle i_{\mathrm{S}}^2\rangle$. Besides the photocurrent I_{ph} itself, the background radiation (I_{B}, in particular for infrared detectors) and the thermal generation (dark current, I_{D}) of carriers contribute:

$$\langle i_{\mathrm{S}}^2\rangle = 2e\,(I_{\mathrm{ph}} + I_{\mathrm{B}} + I_{\mathrm{D}})\,B , \qquad (19.23)$$

with B being the bandwidth. Additionally, the parallel resistances cause thermal noise

$$\langle i_{\mathrm{T}}^2\rangle = 4kTB/R_{\mathrm{eq}} . \qquad (19.24)$$

The resistance R_{eq} is given by the resistance of the depletion layer (junction) R_{j}, the load R_{L} and the input of the amplifier R_{i} as $R_{\mathrm{eq}}^{-1} = R_{\mathrm{j}}^{-1} + R_{\mathrm{L}}^{-1} + R_{\mathrm{i}}^{-1}$. The series resistance R_{s} of the photodiode can be usually ignored in this context.

For a fully modulated signal ($m = 1$) the signal-to-noise ratio of the photodiode is given by ($i_{\mathrm{ph}} = e\eta mP_{\mathrm{opt}}/(\sqrt{2}h\nu)$)

$$S/N = \frac{i_{\mathrm{ph}}^2}{\langle i_{\mathrm{S}}^2\rangle^2 + \langle i_{\mathrm{T}}^2\rangle^2} = \frac{\left(e\eta P_{\mathrm{opt}}/h\nu\right)^2/2}{2e\,(I_{\mathrm{ph}} + I_{\mathrm{B}} + I_{\mathrm{D}})\,B + 4kTB/R_{\mathrm{eq}}} . \qquad (19.25)$$

Therefore the NEP is given by

$$\mathrm{NEP} = \frac{2h\nu B}{\eta}\left[1 + \left(1 + \frac{I_{\mathrm{eq}}}{eB}\right)^{1/2}\right] . \qquad (19.26)$$

The current I_{eq} is given by $I_{\mathrm{eq}} = I_{\mathrm{B}} + I_{\mathrm{D}} + 2kT/(eR_{\mathrm{eq}})$. If $I_{\mathrm{eq}}/eB \ll 1$, the NEP is determined by the shot noise of the signal itself. In the other limit $I_{\mathrm{eq}}/eB \gg 1$ the detection is limited by the background radiation or thermal noise. In this case, the NEP is (for $B = 1\,\mathrm{Hz}$, in $\mathrm{W\,cm^2\,Hz^{1/2}}$)

$$\mathrm{NEP} = \sqrt{2}\frac{h\nu}{\eta}\left(\frac{I_{\mathrm{eq}}}{e}\right)^{1/2} . \qquad (19.27)$$

In Fig. 19.16 the situation is shown for a silicon photodiode as a function of R_{eq}. The diode has a quantum efficiency of 75% at $\lambda = 0.77\,\mu\mathrm{m}$. A high value of $R_{\mathrm{eq}} \sim 1\,\mathrm{G\Omega}$ is necessary to ensure detection limited by dark current.

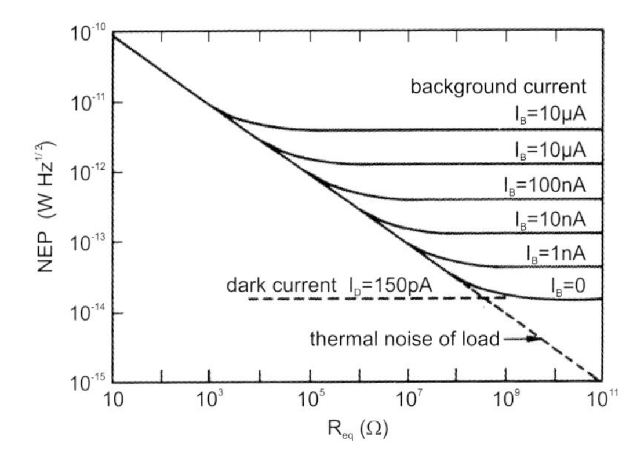

Fig. 19.16. NEP as a function of the resistance R_{eq} for a Si photodiode. Adapted from [608]

19.3.3 pin Photodiodes

The depletion layer in pn-diodes is relatively thin such that the incident light is not completely absorbed. An almost complete absorption of light can be achieved by using a thick intrinsic absorption layer. The field in the intrinsic region is constant or slowly varying linearly (Fig. 18.58). The generation rate per unit area decreases exponentially (Lambert–Beer's law) as shown in Fig. 19.17c:

$$G(x) = G_0 \exp(-\alpha x) \, . \tag{19.28}$$

The initial generation rate $G_0 = \Phi_0 \alpha$ is given by the incident photon flux per unit area Φ_0 and the reflectivity of the surface R as $\Phi_0 = P_{\mathrm{opt}}(1-R)/(Ah\nu)$.

The drift current in the i-region collects all those carriers (if recombination in the depletion layer is neglected). The electron drift current is given by

$$j_{\mathrm{dr}} = -e \int_0^w G(x)\mathrm{d}x = e\Phi_0 \left[1 - \exp(-\alpha w)\right] \, , \tag{19.29}$$

with w being the thickness of the depletion layer that is approximately the same as the thickness of the i-region. In the bulk (neutral) region ($x > w$) the minority-carrier density is determined by drift and diffusion (as in (10.79)). The diffusion current density at $x = w$ is given by

$$j_{\mathrm{diff}} = -e\Phi_0 \frac{\alpha L_{\mathrm{p}}}{1+\alpha L_{\mathrm{p}}} \exp(-\alpha w) + e p_{\mathrm{n}0} \frac{D_{\mathrm{p}}}{L_{\mathrm{p}}} \, . \tag{19.30}$$

The total current $j_{\mathrm{tot}} = j_{\mathrm{diff}} + j_{\mathrm{dr}}$ is given by

$$j_{\mathrm{tot}} = -e\Phi_0 \left[1 - \frac{\exp(-\alpha w)}{1+\alpha L_{\mathrm{p}}}\right] + e p_{\mathrm{n}0} \frac{D_{\mathrm{p}}}{L_{\mathrm{p}}} \, . \tag{19.31}$$

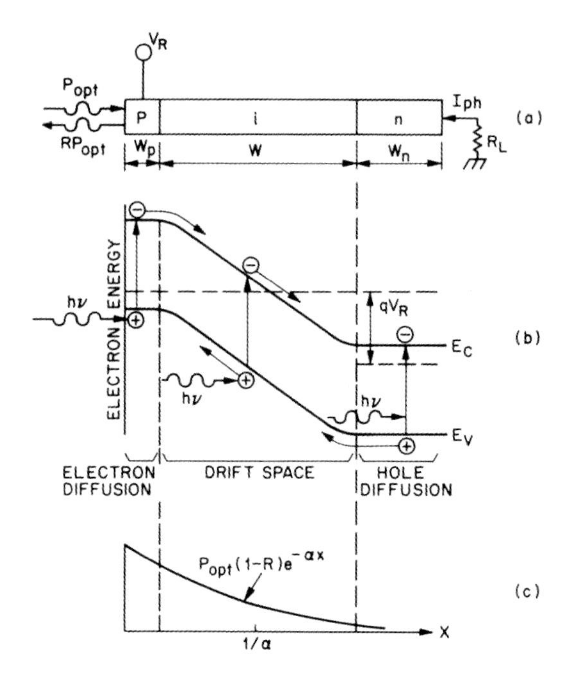

Fig. 19.17. (**a**) Schematic cross section of pin-diode, (**b**) schematic band structure under reverse bias and (**c**) profile of carrier generation due to light absorption. From [607]

The first part is due to the photocurrent, the second is due to the diffusion current known from the pn-diode. In normal operation, the second can be neglected compared to the first. The quantum efficiency is

$$\eta = \frac{j_{\mathrm{tot}}/e}{P_{\mathrm{opt}}/h\nu} = (1-R)\left[1 - \frac{\exp(-\alpha w)}{1+\alpha L_{\mathrm{p}}}\right] . \qquad (19.32)$$

For a high quantum efficiency of course a low reflection and a high absorption, i.e. $\alpha w \gg 1$, are necessary.

However, for $w \gg 1/\alpha$ the transit time through the depletion layer $t_{\mathrm{r}} \approx w/v_{\mathrm{s}}$ (at sufficiently high field, v_{s} being the drift-saturation velocity) increases too much. The 3 dB cutoff frequency $f_{3\,\mathrm{dB}}$ (Fig. 19.18) is

$$f_{3\,\mathrm{dB}} \cong \frac{2.4}{2\pi t_{\mathrm{r}}} \cong \frac{0.4\, v_{\mathrm{s}}}{w} . \qquad (19.33)$$

Therefore a tradeoff exists between the quantum efficiency and the response speed of the pin-photodiode (Fig. 19.18). Choosing $w \cong 1/\alpha$ is a good compromise.

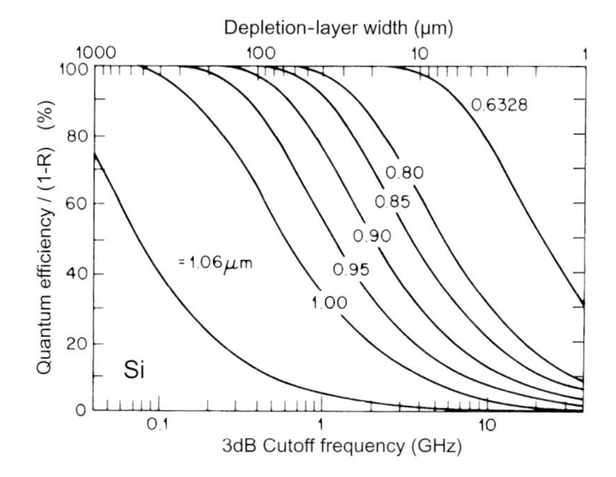

Fig. 19.18. Quantum efficiency and 3 dB cutoff frequency of a Si pin-diode at $T = 300\,\mathrm{K}$ for various wavelengths of input radiation. Adapted from [608]

19.3.4 Position-Sensing Detector

In a position-sensing detector (PSD) two electrodes are placed at opposite edges of a photodetector. The current output depends linearly on the beam position in between the electrodes, similar to a voltage divider. If two pairs of electrodes, one on the front and one on the back of the detector, are fabricated in orthogonal directions (Fig. 19.19a), the beam position can be measured in both x and y directions.

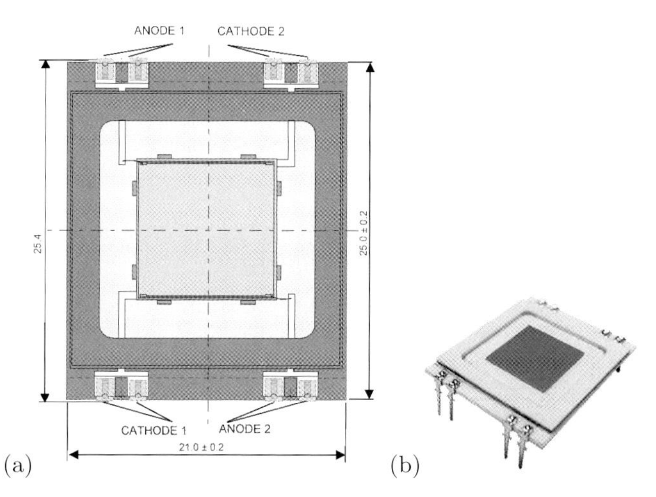

Fig. 19.19. (**a**) Scheme of two-dimensional position-sensing detector (PSD), (**b**) image of PSD. From [609]

19.3.5 MSM Photodiodes

A MSM photodiode consists of a piece of semiconductor between two Schottky contacts (MS contacts). These are typically arranged laterally (as shown in Fig. 19.24b) but will first be considered at the front and back of the semiconductor [610]. The band structure in thermodynamic equilibrium is shown in Fig. 19.20.

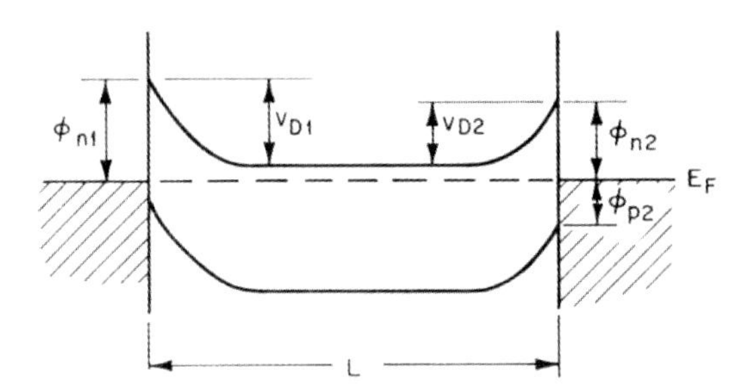

Fig. 19.20. Band diagram of a MSM structure with an n-type semiconductor in thermal equilibrium. From [610]

In the general case two different metals with two different barriers ϕ_{n1}, ϕ_{n2} and built-in voltage V_{D1}, V_{D2} are considered. If a voltage is applied across the MSM diode, one of the junctions is biased in the forward, the other in the reverse direction. We assume in Fig. 19.21 that the voltage biases the first contact in the reverse direction, i.e. the '+' pole is on the left contact. The applied voltage V is split between the two contacts, the larger voltage will drop at the reverse-biased contact (here: $V_1 > V_2$)

$$V = V_1 + V_2 \ . \tag{19.34}$$

The electron current $j_{\mathrm{n_2}}$ arises from thermionic emission at contact 2. Due to current continuity (without recombination since we inject majority charge carriers) this is also the current through contact 1, i.e.

$$j_{\mathrm{n1}} = j_{\mathrm{n2}} \ . \tag{19.35}$$

The reverse current at contact 1 is

$$j_{\mathrm{n1}} = A_{\mathrm{n}}^* T^2 \exp\left(-\beta\phi_{\mathrm{n1}}\right) \exp\left(\beta\Delta\phi_{\mathrm{n1}}\right) \left[1 - \exp\left(-\beta V_1\right)\right] \ , \tag{19.36}$$

where $\Delta\phi_{\mathrm{n1}}$ is the barrier reduction due to the Schottky effect (Sect. 18.2.3 and (18.20)). The forward current at contact 2 is

$$j_{\mathrm{n2}} = -A_{\mathrm{n}}^* T^2 \exp\left(-\beta\phi_{\mathrm{n2}}\right) \exp\left(\beta\Delta\phi_{\mathrm{n2}}\right) \left[1 - \exp\left(\beta V_2\right)\right] \ . \tag{19.37}$$

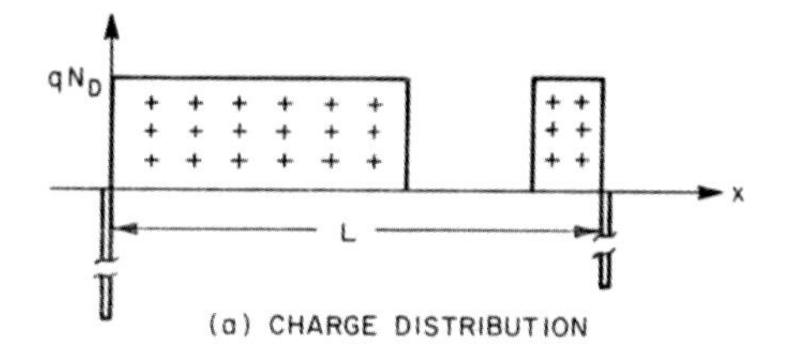

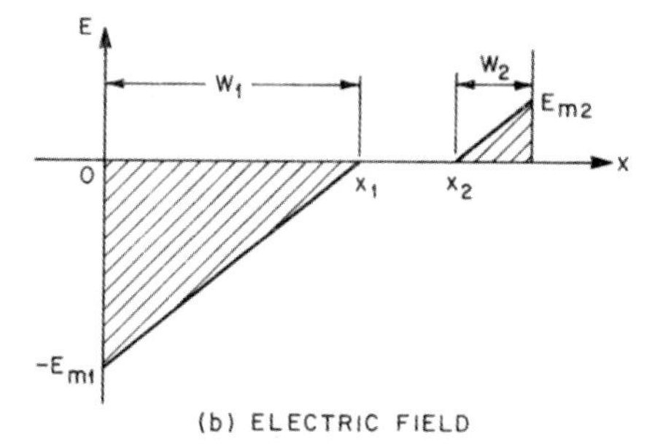

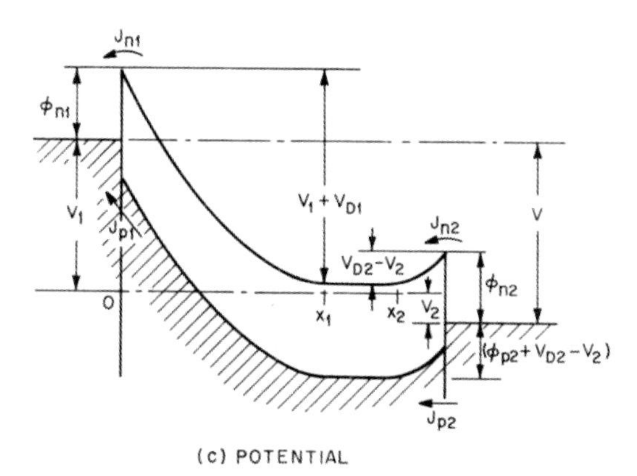

Fig. 19.21. (**a**) Charge-carrier distribution, (**b**) electric field and (**c**) potential for a MSM structure under bias ($V < V_{\mathrm{RT}}$). From [610]

For a symmetric structure, i.e. $\phi_{\mathrm{n1}} = \phi_{\mathrm{n2}}$ and $V_{\mathrm{D1}} = V_{\mathrm{D2}} = V_{\mathrm{D}}$, (19.35)–(19.37) yield together with (18.20)

$$\left(\frac{e^3 N_{\mathrm{D}}}{8\pi^2 \epsilon_{\mathrm{s}}^3}\right)^{1/4} \left[(V_{\mathrm{D}} + V_1)^{1/4} - (V_{\mathrm{D}} - V_2)^{1/4}\right] = \frac{1}{\beta} \ln\left[\frac{\exp(\beta V_2) - 1}{1 - \exp(-\beta V_1)}\right] . \tag{19.38}$$

Together with (19.34) a numerical or graphical solution can be found. Initially (for small voltages) the injected hole current (from contact 2) is much smaller than the electron current and diffusion occurs in the neutral region.

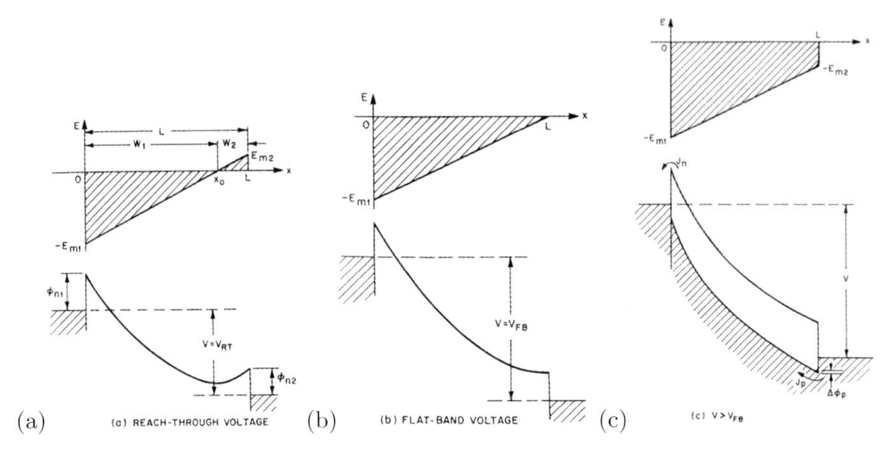

Fig. 19.22. Electric field (*upper parts*) and band diagram (*lower parts*) in a MSM diode for various bias conditions: (**a**) at reach-through voltage V_{RT}, (**b**) at flat-band voltage V_{FB} and (**c**) for $V > V_{\mathrm{FB}}$. From [610]

The reach-through voltage V_{RT} is reached when the width of the neutral region is reduced to zero (Fig. 19.22a). At the juncture of the two depletion regions inside the semiconductor material the electric field is zero and changes sign. For a larger voltage V_{FB} flat-band conditions are present at contact 2, i.e. the electric field is zero at contact 2 (Fig. 19.22b). At even larger voltage V_{B} breakdown occurs.

At $V = V_{\mathrm{RT}}$ we have

$$w_1 = \left[\frac{2\epsilon_{\mathrm{s}}}{eN_{\mathrm{D}}}(V_1 + V_{\mathrm{D1}})\right]^{1/2} \tag{19.39a}$$

$$w_2 = \left[\frac{2\epsilon_{\mathrm{s}}}{eN_{\mathrm{D}}}(V_{\mathrm{D2}} - V_2)\right]^{1/2} \tag{19.39b}$$

$$L = w_1 + w_2 , \tag{19.39c}$$

and therefore (with (19.34))

$$V_{\mathrm{RT}} = \frac{eN_{\mathrm{D}}}{2\epsilon_{\mathrm{s}}}L^2 - L\left[\frac{2eN_{\mathrm{D}}}{\epsilon_{\mathrm{s}}}(V_{\mathrm{D2}} - V_2)\right] - \Delta V_{\mathrm{D}} , \tag{19.40}$$

with $\Delta V_{\mathrm{D}} = (V_{\mathrm{D1}} - V_{\mathrm{D2}})$, vanishing for a symmetric MSM structure. At and after reach-through the electric field varies linearly from 0 to L within the semiconductor. The point of zero electric field shifts towards contact 2. At the flat-band voltage this point has reached the contact 2 and the width of the depletion layer at contact 2 is zero. This condition leads (as long as no breakdown occurred) to

$$V_{\mathrm{FB}} = \frac{eN_{\mathrm{D}}}{2\epsilon_{\mathrm{s}}}L^2 - \Delta V_{\mathrm{D}} . \tag{19.41}$$

The maximum electric field is at contact 1 and is given (for $V > V_{\rm FB}$) by

$$\mathcal{E}_{\rm m1} = \frac{V + V_{\rm FB} + 2\Delta V_{\rm D}}{L} \; . \tag{19.42}$$

If in a part of the structure the critical field $\mathcal{E}_{\rm B}$ for impact ionization is reached (this will be at contact 1, since the field is highest there), the diode breaks down. Therefore the breakdown voltage is given by

$$V_{\rm B} \approx \mathcal{E}_{\rm B} L - V_{\rm FB} - 2\Delta V_{\rm D} \; . \tag{19.43}$$

The current–voltage characteristic for a Si-MSM structure is shown in Fig. 19.23. At small voltages only small currents flow since one contact is in reverse bias. The hole current is much smaller than the electron current. Only those holes that diffuse through the neutral region contribute to the hole current. After reach-through the barrier $\phi_{\rm p2} + V_{\rm D2} - V_2$ for hole injection is strongly reduced that leads to strong hole injection. Beyond the flat-band voltage the hole current increases only weakly since a lowering of the barrier occurs only via the Schottky effect. For high fields ($V > V_{\rm FB}$, before breakdown) the hole current is

$$j_{\rm p1} = A_{\rm p}^* T^2 \exp\left(-\beta\phi_{\rm p2}\right) \exp\left(\beta\Delta\phi_{\rm p2}\right) = j_{\rm p,s} \exp\left(\beta\Delta\phi_{\rm p2}\right) \; , \tag{19.44}$$

and the total current is

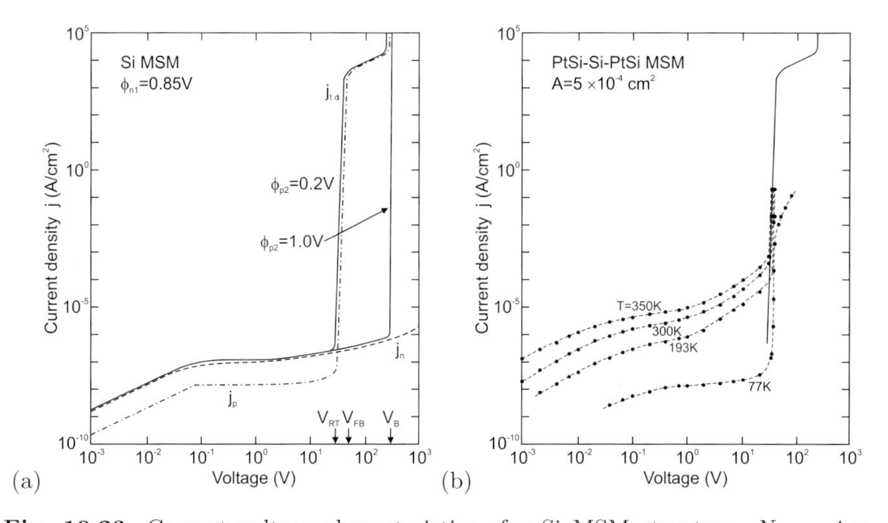

Fig. 19.23. Current voltage characteristics of a Si MSM structure, $N_{\rm D} = 4 \times 10^{14}\,{\rm cm}^{-3}$, $L = 12\,\mu{\rm m}$ (thin polished, $\langle 111 \rangle$-oriented wafer). (**a**) Theory for $\phi_{\rm n1} = 0.85\,{\rm V}$ and two different values of $\phi_{\rm p2}$ for T=300 K. Total current (*solid lines*), electron current (*dashed line*) and hole current (*dash-dotted line*). (**b**) Experiment on PtSi-Si-PtSi MSM structure (contact area: $5 \times 10^{-4}\,{\rm cm}^{-2}$) for four different temperatures as labelled (for $\phi_{\rm p2} \approx 0.2\,{\rm V}$). *Dashed lines* are guides to the eye, *solid line* is high current part of theory from part (**a**). Adapted from [610]

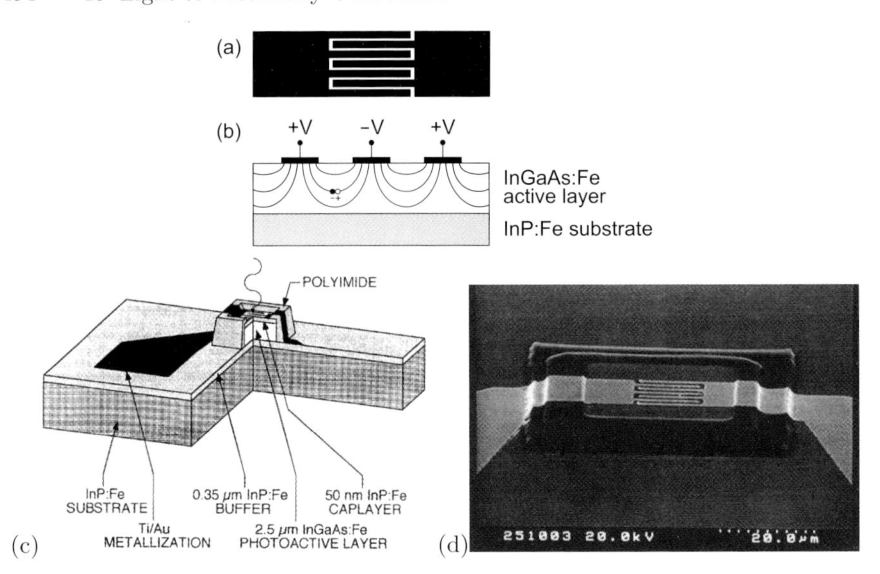

Fig. 19.24. Scheme of MSM photodetector with interdigital contacts in (**a**) plan view and (**b**) cross section. In part (**b**), the electrical field lines are shown schematically together with an electron–hole pair ready to be separated. (**c**) Scheme of a MSM mesa structure, (**d**) SEM image of an InGaAs/InP MSM mesa photodetector. Parts (**c**) and (**d**) from [611]

$$j = j_{\mathrm{n,s}} \exp\left(\beta \Delta\phi_{\mathrm{n1}}\right) + j_{\mathrm{p,s}} \exp\left(\beta \Delta\phi_{\mathrm{p2}}\right) , \tag{19.45}$$

with $j_{\mathrm{n,s}} = A_{\mathrm{n}}^* T^2 \exp\left(-\beta\phi_{\mathrm{n1}}\right)$ and $j_{\mathrm{p,s}} = A_{\mathrm{p}}^* T^2 \exp\left(-\beta\phi_{\mathrm{p2}}\right)$.

In a MSM photodetector the metal contacts are typically formed in an interdigitated structure on the semiconductor surface (Fig. 19.24). These contacts shield some of the active area from photons. An increase in quantum efficiency can be achieved with transparent contacts (e.g. ZnO or ITO) and an antireflection (AR) coating.

The dark current is given by (19.45) and is minimal when electron and hole saturation currents are equal. This conditions leads to the optimal barrier height

$$\phi_{\mathrm{n}} = E_{\mathrm{g}} - \phi_{\mathrm{ph}} = \frac{1}{2}\frac{kT}{e} \ln\left(\frac{m_{\mathrm{e}}}{m_{\mathrm{hh}}}\right) + \frac{1}{2} E_{\mathrm{g}} \tag{19.46}$$

close to middle of the bandgap. For InP and optimal barrier $\phi_{\mathrm{n}} = 0.645\,\mathrm{eV}$ a dark current density of $0.36\,\mathrm{pA/cm^2}$ is expected for a field of $10\,\mathrm{V/\mu m}$. For deviating barrier height the current increases exponentially. The current–voltage characteristic of an InGaAs:Fe MSM photodetector is shown in Fig. 19.25 for a dark environment and various illumination levels.

The time-dependent response of a MSM photodetector depends on the drift time of the carriers, i.e. the time that a created electron and hole need

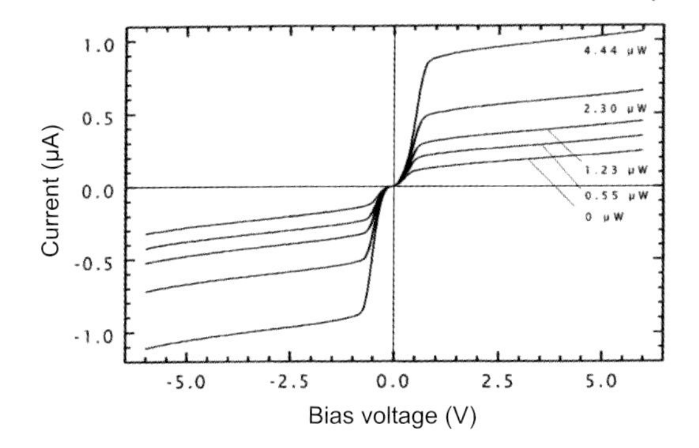

Fig. 19.25. dc I–V characteristic of an InGaAs/InP MSM photodetector (InP:Fe/InGaAs:Fe/InP:Fe, finger separation 1 µm, $\lambda = 1.3$ µm) under illumination for dark environment (0 µW) and various illumination levels as labeled. Adapted from [611]

to arrive at their respective contacts. In Fig. 19.26 a simulation is shown for a MSM detector. The current has two components, a fast one due to the electrons and a slow one due to the holes that have the lower mobility and smaller drift saturation velocity. A similar dependence is found in experiment (Fig. 19.27a). For longer wavelengths the detector is slower since they penetrate deeper into the material and thus the charge carriers have a longer path to the contacts (see scheme in Fig. 19.24b). An important role is played by the finger separation; smaller finger separation ensures a more rapid carrier collection (Fig. 19.27b). In [612] a bandwidth of 300 GHz was demonstrated for 100 nm/100 nm finger width and separation for LT-GaAs[5] and bulk GaAs, limited by the RC time constant. For 300 nm/300 nm fingers and a LT-GaAs a bandwidth of 510 GHz (pulsewidth of 0.87 ps) was reported, which is faster than the intrinsic transit time (1.1 ps) and not limited by the RC time constant (expected pulse width 0.52 ps), due to the recombination time (estimated to be 0.2 ps).

19.3.6 Avalanche Photodiodes

In an avalanche photodiode (APD) intrinsic amplification due to carrier multiplication (through impact ionization) in a region with high electric field is used to increase the photocurrent. The field is generated by a high reverse bias in the diode. In an ideal APD only one type of carrier is multiplied, resulting in the lowest noise. If electrons are injected into the field region at $x = 0$ (Fig. 19.28a), the multiplication factor for electrons is

$$M_{\mathrm{n}} = \exp(\alpha_{\mathrm{n}} w) , \tag{19.47}$$

[5]LT: grown at *low temperature*, i.e. containing many defects that reduce the carrier lifetime.

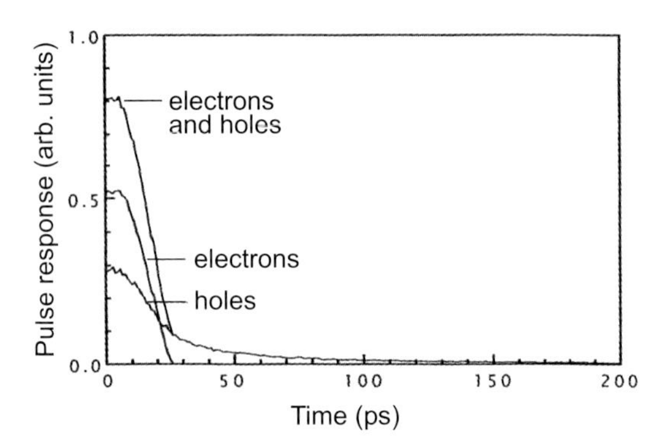

Fig. 19.26. Simulation of the time-dependent response of an InGaAs:Fe MSM photodetector to a short light pulse. Adapted from [611]

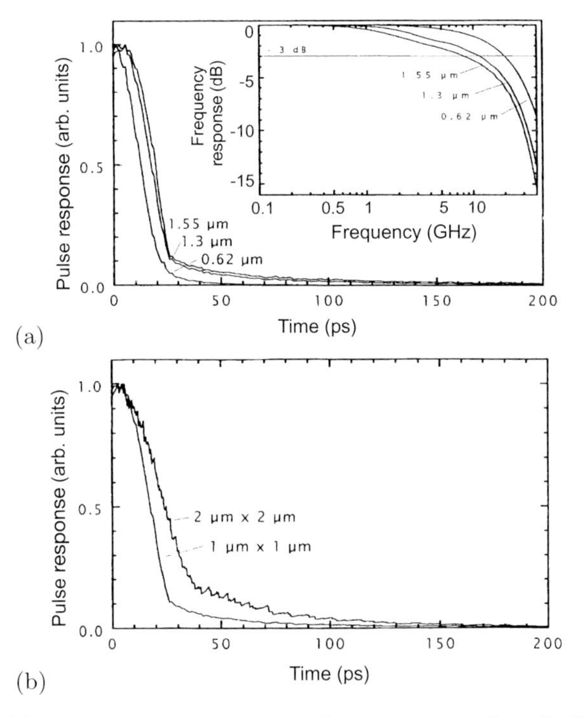

Fig. 19.27. (**a**) Experimental time-dependent response of an InGaAs:Fe MSM photodetector to a short light pulse for three different wavelengths, *inset* shows the frequency response from a Fourier transform. (**b**) Response of the MSM for two different finger widths and separations (both 1 or 2 μm, respectively), InGaAs layer thickness 2 μm, $\lambda = 1.3$ μm and bias voltage 10 V. Adapted from [611]

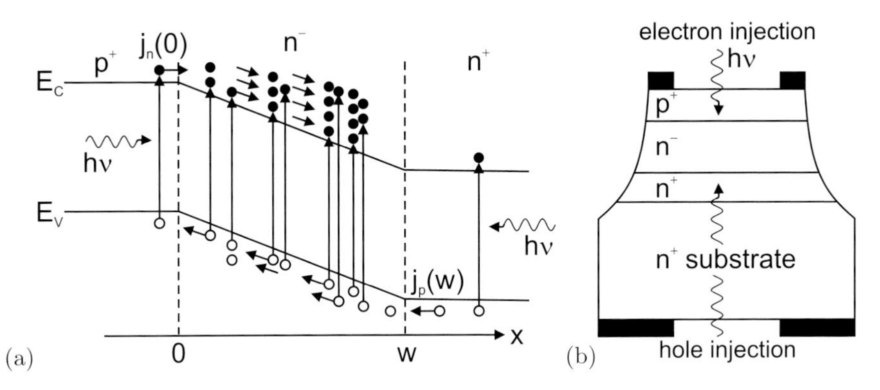

Fig. 19.28. Schematic band structure (**a**) and schematic device setup (**b**) of an avalanche photodiode (APD). Adapted from [388]

for $\alpha_{\rm p} = 0$. Typically, both carrier types suffer multiplication. If the electron and hole impact ionization coefficients are the same ($\alpha_{\rm n} = \alpha_{\rm p} = \alpha$), the multiplication factor for electrons and holes M is given by

$$M = \frac{1}{1 - \alpha w} \,. \tag{19.48}$$

The rms value of the current noise is the same as in the case of the pn-diode (19.23), only that now the gain M is added

$$\langle i_{\rm S}^2 \rangle = 2e \left(I_{\rm ph} + I_{\rm B} + I_{\rm D} \right) \langle M^2 \rangle B \,. \tag{19.49}$$

The term $\langle M^2 \rangle$ is written as $\langle M \rangle^2 \, F(M)$ with $F(M) = \langle M^2 \rangle / \langle M \rangle^2$ being the *excess noise factor* that describes the additional noise introduced by the random nature of the impact ionization. For multiplication started with electron injection, it is given by

$$F(M) = kM + (1-k) \left(2 - \frac{1}{M} \right) , \tag{19.50}$$

with $k = \alpha_{\rm p}/\alpha_{\rm n}$. For hole injection starting the multiplication (19.50) holds with k substituted by $k' = \alpha_{\rm n}/\alpha_{\rm p}$. In Fig. 19.29a the excess noise factor is shown vs. the average multiplication for various values of k and k'.

Experimental data are shown in Fig. 19.29b for a Si APD. For short wavelengths absorption is preferential at the surface (n-region) and we have the case of hole injection. The data for the excess noise factor are fairly well fit with $k' \approx 5$. For longer wavelengths, the data for electron injection are fit by $k \approx 0.2 = 1/k'$.

For a fully modulated signal the signal-to-noise ratio is given by

$$S/N = \frac{(e\eta P_{\rm opt}/h\nu)^2/2}{2e \left(I_{\rm ph} + I_{\rm B} + I_{\rm D} \right) F(M) B + 4kTB/(R_{\rm eq} M^2)} \,. \tag{19.51}$$

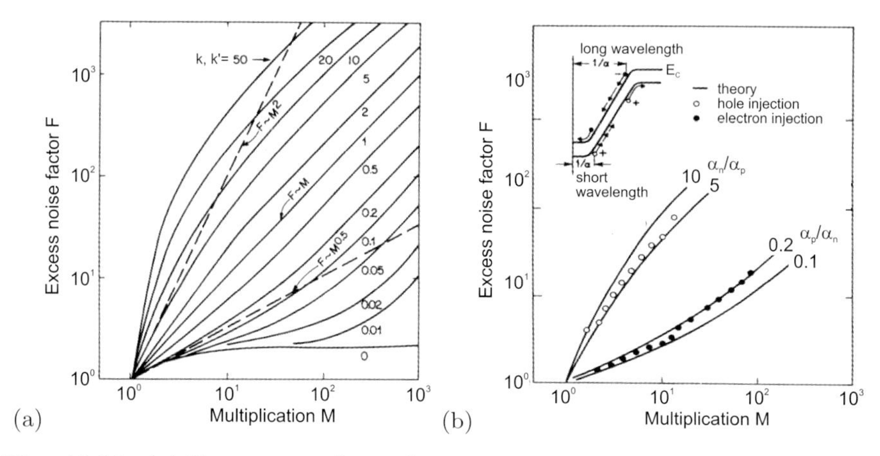

Fig. 19.29. (**a**) Excess noise factor for various values of the ratio of ionization coefficients k or k'. Adapted from [613]. (**b**) Experimental results for F for a Si APD with 0.1 µA primary current. The *empty* (*full*) symbols are for short (long) wavelengths [primary hole (electron) current]. The *inset* shows the schematic band diagram of the np-diode under reverse bias. Adapted from [614]

If S/N is limited by thermal noise, the APD concept leads to a drastic improvement of noise.

A particular APD structure is known as a *solid-state multiplier*. It has separate absorption and amplification regions (SAM structure). In the low-field region the light is absorbed. One type of carrier is transported with the drift field $\mathcal{E}_\mathrm{d}$ to the multiplication region in which a large field $\mathcal{E}_\mathrm{m}$ is present and multiplication occurs. In Fig. 19.30a a homo-APD with SAM structure is shown. Regions with different electric field are created by a special doping profile.[6] A π-p-π structure leads to regions with homogeneous low and high field strengths.

In the case of a heterostructure-APD with SAM structure (Fig. 19.30b) absorption (of light with sufficiently long wavelength with an energy smaller the the InP bandgap) takes place only in the InGaAs layer. Since no light is absorbed in the multiplication region, the device functions similarly for front and back illumination.

19.3.7 Traveling-Wave Photodetectors

In a standard photodetector there was a tradeoff between the thickness of the absorption layer and the speed of the detector. In a traveling-wave photodetector the light absorption occurs in a waveguide such that for sufficient length L all incident light is absorbed. Complete absorption is achieved ('long' waveguide) if $L \gg (\Gamma\alpha)^{-1}$, α being the absorption coefficient and $\Gamma \leq 1$ be-

[6] employing Poisson's equation $\partial(\epsilon_\mathrm{s}(x)\mathcal{E}(x))/\partial x = \rho(x)$.

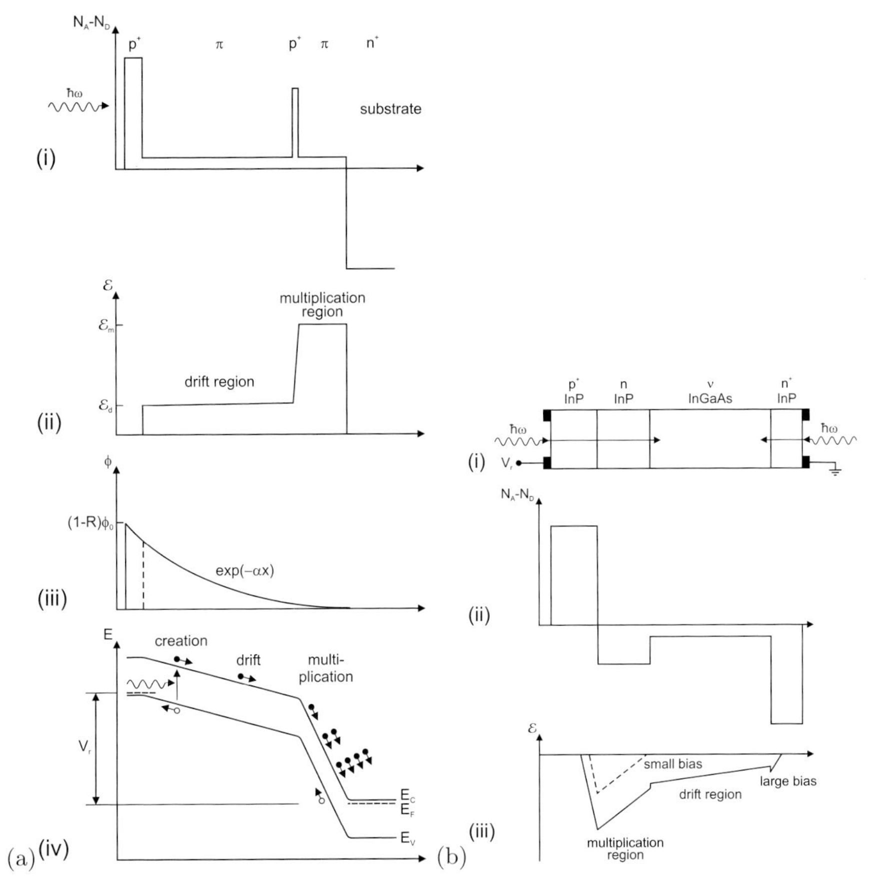

Fig. 19.30. (**a**) Homo-APD with SAM structure. (**i**) doping profile, (**ii**) electrical field, (**iii**) photon flux or electron–hole pair generation rate and (**iv**) schematic band diagram under reverse voltage V_{r} with charge-carrier transport. The multiplication is for $\alpha_{\mathrm{n}} \gg \alpha_{\mathrm{p}}$. Adapted from [183]. (**b**) (**i**) Scheme of an InP/InGaAs hetero-APD with SAM structure, (**ii**) doping profile and (**iii**) electric field for small (*dashed line*) and large (*solid line*) reverse bias V_{r}. Adapted from [388]

ing the optical confinement factor, the geometrical overlap of the optical mode with the cross section of the absorbing medium (see also Sect. 20.4.4).

The electrical connections are designed along this waveguide on the sides (coplanar layout, Fig. 19.31a) or above and below (Fig. 19.31b). The limitation due to a RC time constant is now replaced by the velocity match of the light wave $v_{\mathrm{opt}} = c/n$ and the traveling electric wave in the contact lines $v_{\mathrm{el}} \approx 1/\sqrt{LC}$. While the two waves travel along the waveguide, energy is transferred from the light wave to the electric wave. The 3 dB bandwidth due to velocity mismatch B_{vm} (for impedance- matched, long waveguides) is given by

$$B_{\mathrm{vm}} = \frac{\Gamma\alpha}{2\pi} \frac{v_{\mathrm{opt}}\, v_{\mathrm{el}}}{v_{\mathrm{opt}} - v_{\mathrm{el}}} \,. \tag{19.52}$$

For a MSM structure, whose electrode separation has been designed with a self-aligned process (without extensive effort in lateral patterning) by an etch depth of a few 100 nm (Fig. 19.32), 3 dB cutoff frequencies beyond 500 GHz have been achieved (Fig. 19.33). The quantum efficiency of this detector was still 8.1%.

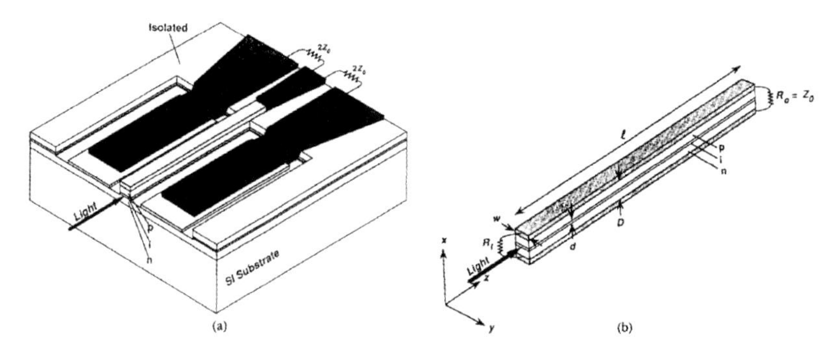

Fig. 19.31. Scheme of a traveling-wave photodetector (with pin structure). (**a**) coplanar contacts and (**b**) contacts as parallel plates. Reprinted with permission from [615], ©1992 IEEE

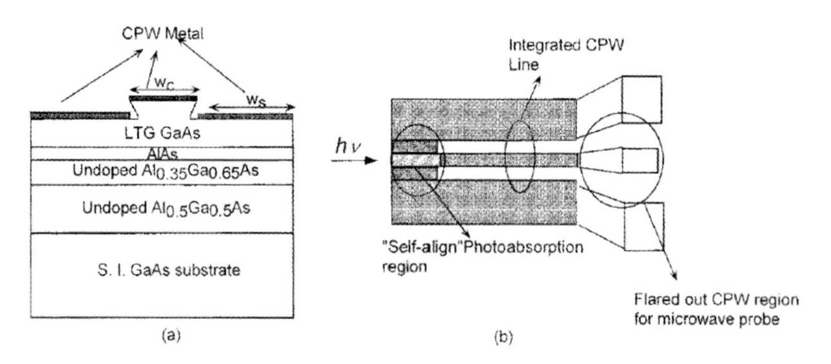

Fig. 19.32. Scheme of a MSM traveling-wave photodetector in (**a**) cross section and (**b**) plan view. Reprinted with permission from [616], ©2001 IEEE

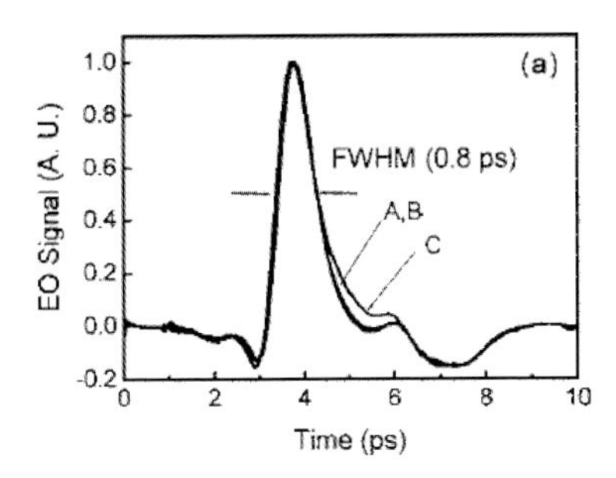

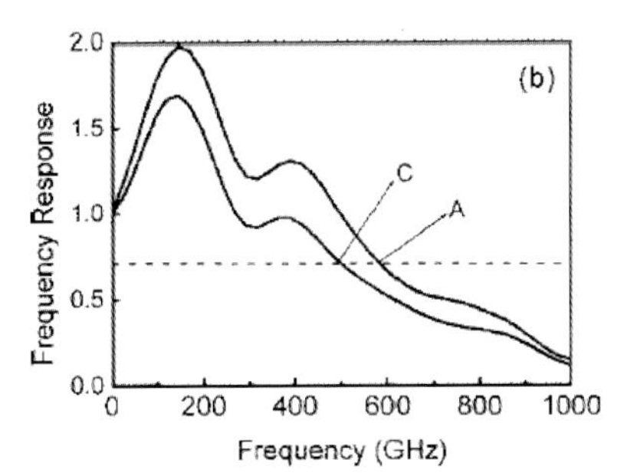

Fig. 19.33. (**a**) Pulse response and (**b**) frequency response (Fourier transform of time response) of a MSM traveling-wave photodetector (bias 5 V) for various illumination intensities, A: 1 mW, B: 1.8 mW and C: 2.2 mW. From [616]

19.3.8 Charge Coupled Devices

The concept of the charge coupled device (CCD), an array of connected photodetectors serving as an image sensor, was devised by W.S. Boyle and G.E. Smith [617] (Fig. 19.34). As textbook for further details [618] may serve.

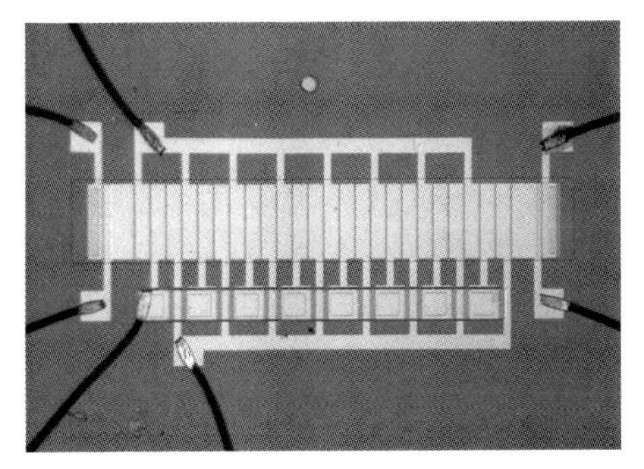

Fig. 19.34. First 8-bit charge coupled device (1970). The chip (size: $1.5 \times 2.5\,\mathrm{mm}^2$) consists of 24 closely packed MOS capacitors (narrow rectangles in the center grid). The thick rectangles at either end of the grid are input/output terminals.

A MIS diode (mostly a silicon-based MOS diode) can be designed as a light detector. The diode is operated in deep depletion. When a large reverse voltage is applied, initially a depletion layer is formed and the bands are strongly bent as shown in Fig. 19.35b. We note that in this situation the semiconductor is not in thermodynamic equilibrium (as it is in Fig. 19.35d) when the quasi-Fermi level is constant throughout the semiconductor. The inversion charge has yet to build up.

There are three mechanisms to generate the inversion charge. (a) generation–recombination, (b) diffusion from the depletion-layer boundary and (c)

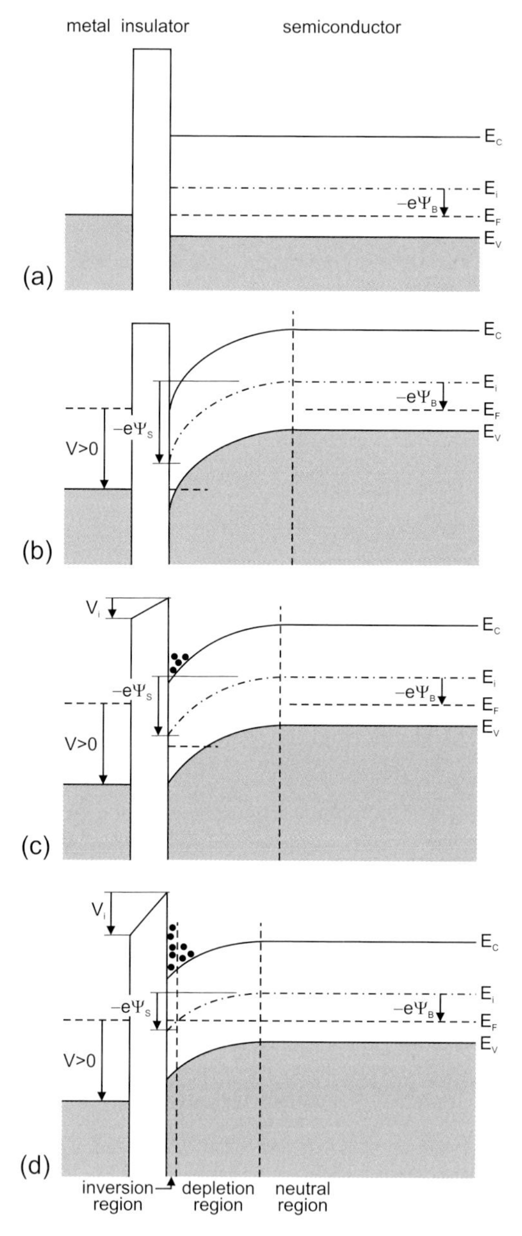

Fig. 19.35. Ideal MIS-diode (with p-type semiconductor) as photodetector (principle of a CCD pixel). (**a**) Without bias (cf. Fig. 18.24b). (**b**) Immediately after an external (reversely poled) voltage $V > 0$ has been applied, the surface potential is $\Psi_s = V$ and no charges have moved yet. (**c**) Strong depletion (still not in thermodynamic equilibrium) with signal charge and reduced surface potential $\Psi_s < V$. (**d**) The semiconductor in equilibrium (E_F is constant) with depletion and inversion layer (cf. Fig. 18.28). For all diagrams, $V = V_i + \Psi_s$

carrier generation by light absorption. Mechanisms (a) and (b) represent dark currents for the photodetector. The conductivity due to these two processes is shown in Fig. 19.36 and slowly builds up the inversion charge. Two temperature regimes are obvious; at low temperatures the generation dominates ($\propto n_\mathrm{i} \propto \exp(-E_\mathrm{g}/2kT)$), at high temperatures the diffusion ($\propto n_\mathrm{i}^2 \propto \exp(-E_\mathrm{g}/kT)$). The latter process can be strongly suppressed by cooling the device.

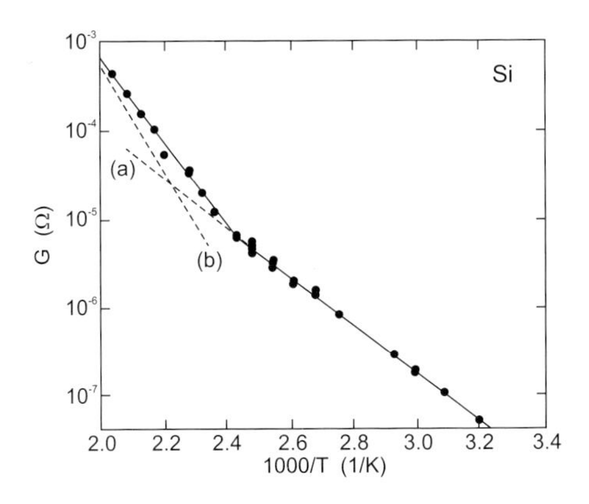

Fig. 19.36. Conductivity of a n-Si/SiO$_2$-diode as a function of temperature ($1/T$). The slope of the *dashed lines* is (a) $\sim 0.56\,\mathrm{eV}$ ($\approx E_\mathrm{g}/2$) and (b) $\sim 1.17\,\mathrm{eV}$ ($\approx E_\mathrm{g}$). Adapted from [619]

The gate voltage V_G and the surface potential Ψ_s are related to each other via

$$V_\mathrm{G} - V_\mathrm{FB} = V_\mathrm{i} + \Psi_\mathrm{s} = \frac{eN_\mathrm{A}w}{C_\mathrm{i}} + \frac{eN_\mathrm{A}w^2}{2\epsilon_\mathrm{s}} , \tag{19.53}$$

where w is the width of the depletion layer. w will be larger than w_m in thermodynamic equilibrium. The first term in the sum is $|Q_\mathrm{s}|/C_\mathrm{i}$ and the second is obtained by integrating the Poisson equation for the constant charge density $-eN_\mathrm{A}$ across the depletion layer. The elimination of w yields

$$V_\mathrm{G} - V_\mathrm{FB} = \Psi_\mathrm{s} + \frac{1}{C_\mathrm{i}}\sqrt{2e\epsilon_\mathrm{s}N_\mathrm{A}\Psi_\mathrm{s}} . \tag{19.54}$$

If light is absorbed in the depletion layer (process (c)), the hole (for p-Si) drifts towards the bulk material. The electron is stored as part of the signal charge Q_sig close to the oxide semiconductor interface (Fig. 19.35b).

$$V_\mathrm{G} - V_\mathrm{FB} = \frac{Q_\mathrm{sig}}{C_\mathrm{i}} + \frac{eN_\mathrm{A}w}{C_\mathrm{i}} + \Psi_\mathrm{s} . \tag{19.55}$$

As a consequence of the increase in signal charge the potential well becomes shallower (19.55). For each gate voltage there is a maximum charge

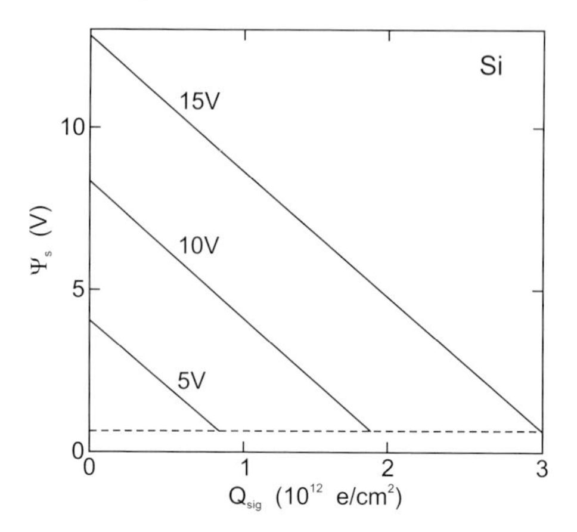

Fig. 19.37. Surface potential as a function of the signal charge Q_{sig} for various values of the bias $V_{\mathrm{G}} - V_{\mathrm{FB}}$ as labeled for a $\mathrm{SiO_2}$/p-Si diode with $N_{\mathrm{A}} = 10^{15}\,\mathrm{cm}^{-3}$ and an oxide thickness of 100 nm. The *dashed line* represents the limit for inversion given by $\Psi_{\mathrm{s}} = 2\Psi_{\mathrm{B}} \approx 0.6\,\mathrm{V}$. Adapted from [620]

(well capacity). The maximum signal charge is reached for $\Psi_{\mathrm{s}} \approx 2\Psi_{\mathrm{B}}$ (Fig. 19.37).

In a charge coupled device (CCD) many light-sensitive MIS diodes, as described above, are fabricated in matrix form to create an image sensor. Upon application of a gate voltage they accumulate charge depending on the local exposure to light. The read out of this charge occurs by shifting the charge through the array to a read-out circuit. Therefore charge is transferred from one pixel to the next. Several schemes have been developed for this task. The three-phase clocking is shown schematically in Fig. 19.38. Other clocking schemes involve four, two or only one electrode per pixel [621].

Since the CCD sensor has many pixels (e.g. up to 4096) along a line, the charge transfer must be highly efficient. The transfer of charge carriers occurs via thermal (regular) diffusion, self-induced drift and the effect of the fringing field (inset of Fig. 19.39). The time constant with which the charge carriers move due to diffusion (in a p-type semiconductor) is

$$\tau_{\mathrm{th}} = \frac{4L^2}{\pi^2 D_{\mathrm{n}}}\,, \tag{19.56}$$

where L is the length of the electrode. For a sufficiently large charge packet the self-induced drift due to Coulomb repulsion is important. The decay of charge is then given by

$$\frac{Q(t)}{Q(0)} = \frac{t_0}{t + t_0}\,, \tag{19.57}$$

with $t_0 = \pi L^3 W_{\mathrm{e}} C_{\mathrm{i}} / (2\mu_{\mathrm{n}} Q(0))$. W_{e} is the width of the electrode. This dependence is shown as a dashed line in Fig. 19.39a. The last electrons are efficiently

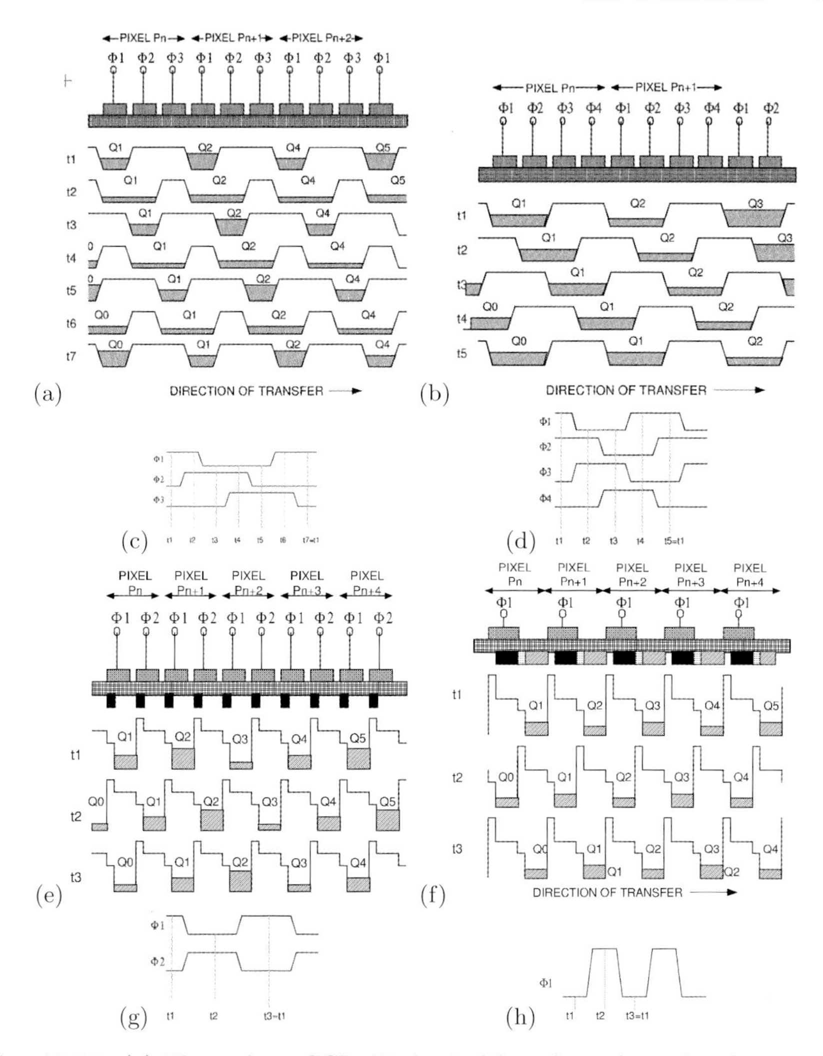

Fig. 19.38. (**a**) Three-phase CCD. Each pixel has three electrodes that can be switched independently (phases 1–3). (**b**,**e**,**f**) Schematic of CCDs with four, two or one phase, respectively. (**c**) (t_1) Charge accumulated after light exposure. A lateral potential well is formed along the row of pixels by the voltages at the three electrodes, e.g. $P_1 = P_3 = 5\,\mathrm{V}$, $P_2 = 10{,}\mathrm{V}$. (t_2–t_7) transfer of charge, (t_7) has the same voltages as (t_1), the charge has been moved one pixel to the right. (**d**,**g**,**h**) Timing schemes for 4-, 2- and 1-phase CCDs, respectively. From [621]

transferred by the drift induced by the fringing field of the electrodes (solid line in Fig. 19.39a). The origin of the fringing field is schematically shown in Fig. 19.39b; the minimum fringing field shown is $2 \times 10^3\,\mathrm{V/cm}$. In about

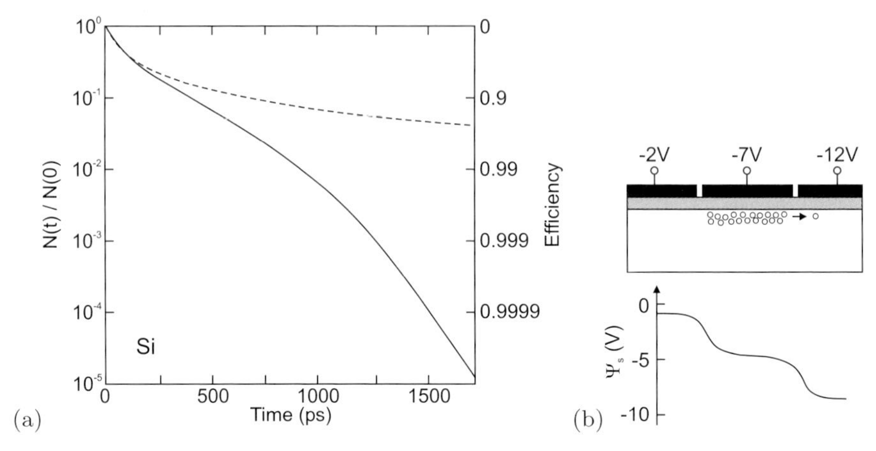

Fig. 19.39. (**a**) Efficiency of charge transfer with (*solid* line) and without (*dashed* line) the effect of the fringing field. (**b**) Schematic plot of the CCD electrodes and bias with the fringing field for a three-electrode CCD, oxide thickness 200 nm, doping $N_{\mathrm{D}} = 10^{15}\,\mathrm{cm}^{-3}$. The electrode pitch is 4 µm and the gap between electrodes is 200 nm wide. Adapted from [622]

1–2 ns practically all $(1 - 10^{-5})$ charges are transferred. This enables clock rates of several 10 MHz.

For the clocking of the CCD the lateral variation of potential depth with the applied gate voltage is used. In Fig. 19.40 it is shown how a lateral variation of doping or oxide thickness creates a lateral potential well. Such structures are used to confine the row of pixels against the neighboring rows (channel stops, Fig. 19.41). In order to avoid carrier loss at the interface between the oxide and the semiconductor a buried-channel structure is used (Fig. 19.42).

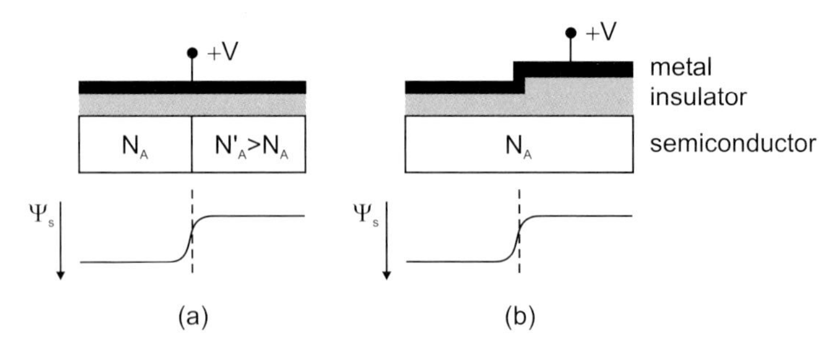

Fig. 19.40. Creation of a lateral potential well (barrier) in a MIS structure with (**a**) varying doping via diffusion or implantation and (**b**) varying (stepped) oxide thickness. *Upper row* shows schematic geometry, *lower row* depicts schematic lateral variation of the surface potential

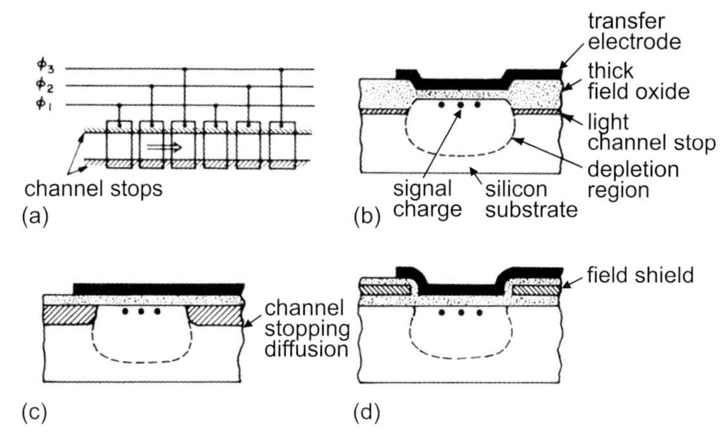

Fig. 19.41. (**a**) Schematic image of channel isolation. Cross section for channel isolation by (**b**) variation of oxide thickness, (**c**) highly doped region and (**c**) field effect. Adapted from [623]

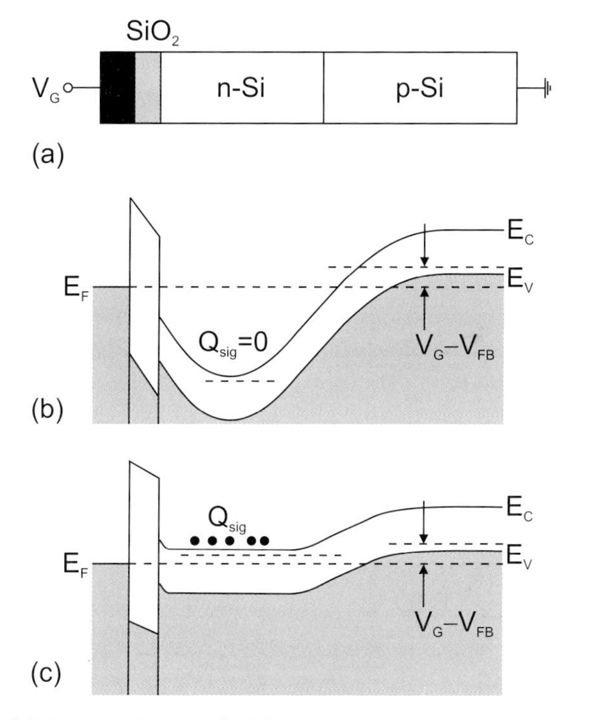

Fig. 19.42. (**a**) Schematic layers of MIS diode with buried-channel structure. Band diagram (**b**) after application of reverse voltage V_G and (**c**) with signal charge Q_{sig}. Adapted from [624]

For front illumination parts of the contact electrodes shield the active area of the device. Higher sensitivity (in particular in the UV) is achieved for back illumination. For this purpose the chip is thinned (polished). This process is expensive and makes the chip mechanically less stable. For red/infrared wavelengths typically interference fringes occur for such thinned chips due to the small thickness. An increase in efficiency for front illumination can be achieved with an onchip microlens (Fig. 19.43).

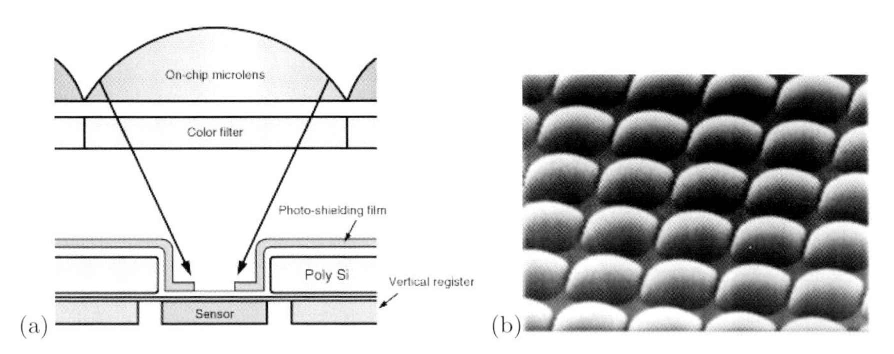

Fig. 19.43. (**a**) Scheme for enhancement of CCD efficiency for front illumination by application of an onchip microlens. (**b**) SEM image of an array of such microlenses. From [625]

For color imaging the CCD is covered with a three-color mask (Fig. 19.44a). On average there are one blue and one red pixel and two green pixels since green is the most prominent color in typical lighting situations. Thus each pixel delivers monochromatic information; RGB images are generated using suitable image software. Alternatives in high-end products are the use of a beam splitter, static color filters and three CCD chips, one for each color (Fig. 19.44b), or the time-sequential recording of three monochromatic images using one CCD chip and a rotating color-filter wheel (Fig. 19.44c).

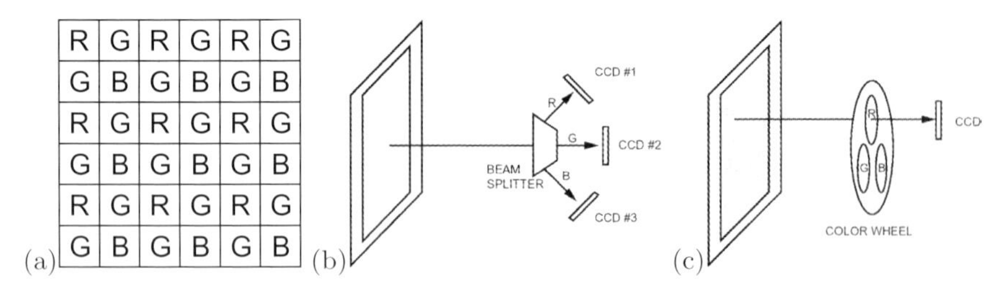

Fig. 19.44. (**a**) Arrangement of colors in a CCD color filter ('R': red, 'G': green, 'B': blue). Color splitting with (**b**) static color filters and (**c**) rotating color wheel. Parts (**b**) and (**c**) from [621]

19.3.9 Photodiode Arrays

An array of photodiodes is also suitable to create an image sensor. During illumination each diode charges a capacitor that is read out with suitable electronics. Based on CMOS technology (see Sect. 21.5.4) cheap image sensors can be made that show currently, however, inferior performance to CCDs. The built-in electronics allows simple outward connections (Fig. 19.45).

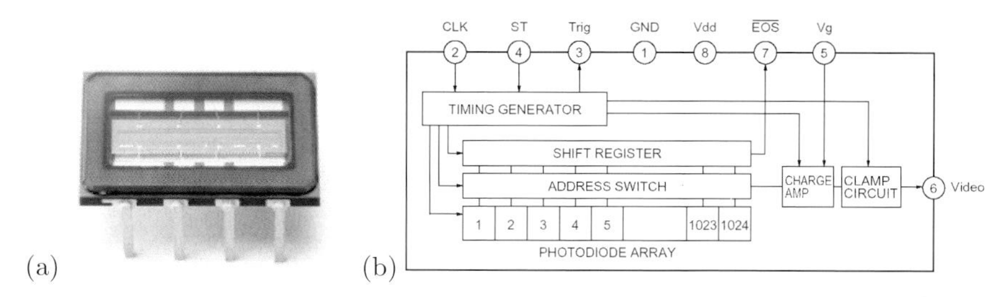

Fig. 19.45. (**a**) CMOS linear array sensor in a 8-pin package. (**b**) Block diagram, the built-in timing generator allows operation only with start and clock pulse inputs. Reprinted with permission from [626]

The three-color CCD image sensor did not offer RGB color information at each pixel. Therefore the spatial resolution of a color image is not directly given by the pixel distance. This is not a very dramatic drawback since human vision is more sensitive to intensity contrast than color contrast. However, RGB color information for each pixel would be desirable, giving higher resolution, in particular in professional photography. Such a sensor has been fabricated employing the wavelength dependence of the silicon absorption coefficient (Fig. 19.13). Blue light has the shortest and red light the largest penetration depth. By stacking three photodiodes on top of each other (Fig. 19.46) photocurrents at different penetration depth are recorded that can be used to generate a RGB value for each pixel.

In Fig. 19.47a a 16-channel array of silicon avalanche photodiodes is shown. It features a quantum efficiency of $> 80\%$ between 760 and 910 nm. The pixel size is $648 \times 208\,\mu\text{m}^2$ on a $320\,\mu$m pitch. The gain is 100 and the rise time 2 ns.

The InGaAs photodiode array in Fig. 19.47b is hybridized with CMOS read-out integrated circuits. It is useful for detection in the spectral range 0.8–1.7 μm. The asymmetric diode size of $25 \times 500\,\mu$m is designed for use in the focal plane of a monochromator.

Another special type of photodiode array is the four-quadrant detector. A light beam generates four photocurrents I_{a}, I_{b}, I_{c}, I_{d} of the respective parts (Fig. 19.48a). A beam deviation in the horizontal or vertical direction can be detected from the (signed) signals $(I_{\mathrm{a}}+I_{\mathrm{d}})-(I_{\mathrm{b}}+I_{\mathrm{c}})$ or $(I_{\mathrm{a}}+I_{\mathrm{b}})-(I_{\mathrm{c}}+I_{\mathrm{d}})$, respectively. We note that these signals can also be normalized to the total beam intensity $I_{\mathrm{a}}+I_{\mathrm{b}}+I_{\mathrm{c}}+I_{\mathrm{d}}$.

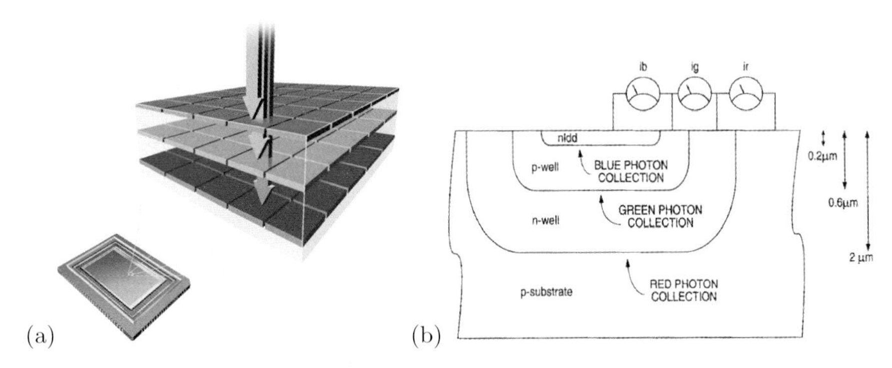

Fig. 19.46. (**a**) Scheme of image sensor with depth-dependent light detection. From [627]. (**b**) Schematic layer sequence for three-color pixel. i_{b}, i_{g} and i_{r} denote the photocurrents for blue, green and red light, respectively. Adapted from [628]

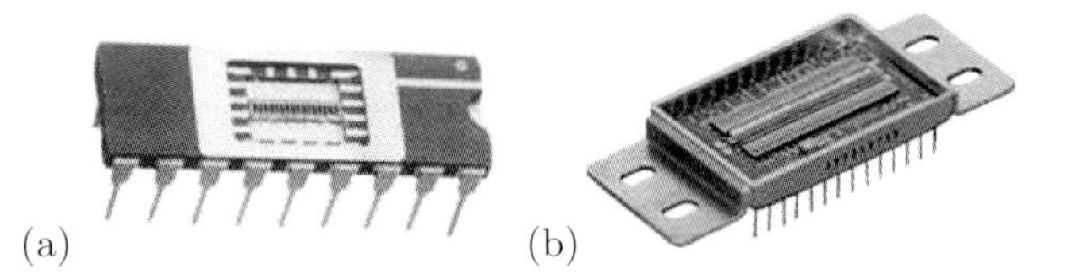

Fig. 19.47. (**a**) Array of 16 silicon APDs. From [629]. (**b**) 1024-pixel InGaAs photodiode array. From [630]

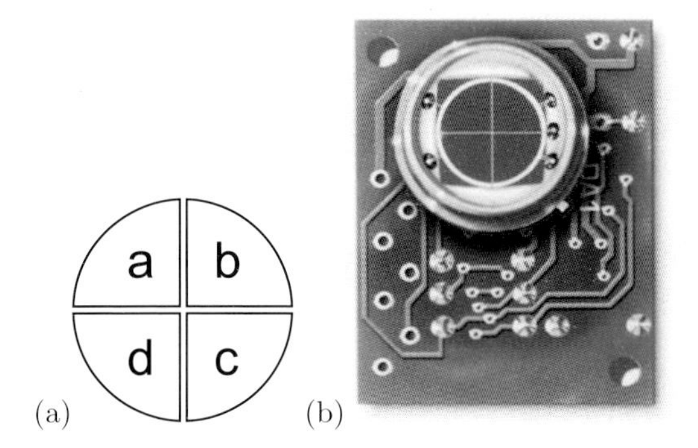

Fig. 19.48. (**a**) Scheme of four-quadrant photodetector with sections 'a', 'b', 'c' and 'd', (**b**) image of four-quadrant silicon photodetector with circuit board. Part (**b**) from [631].

19.4 Solar Cells

Solar cells are light detectors, mostly photodiodes, that are optimized for the (large-area) conversion of solar radiation (light) into electrical energy.

19.4.1 Solar Radiation

The sun has three major zones, the core with a temperature of 1.56×10^7 K and a density of 100 g/cm^3 in which 40% of the mass is concentrated and 90% of the energy is generated, the convective zone with a temperature of 1.3×10^5 K and a density of 0.07 g/cm^3, and the photosphere with a temperature of 5800 K and low density ($\sim 10^{-8}$ g/cm^3). The radius is 6.96×10^8 m and is about 100 times larger than that of the earth (6.38×10^6 m). The distance sun–earth is 1.496×10^{11} m. The angle under which the sun disk appears on earth is 0.54°. An energy density of 1367 ± 7 W/m^2 arrives at the earth in front of its atmosphere.

This value and the according spectrum of the sun's emission, which is similar to a blackbody with temperature 5800 K (Fig. 19.50), is termed air mass zero (AM0). The total energy that reaches the earth from the sun is 1.8×10^{17} W per year. This value is 10^4 times the world's primary energy need.

Air mass zero (AM0) is important for solar cells in satellites. When the solar spectrum passes the earth's atmosphere, it is changed with regard to its shape and the total energy density due to gas absorption (ozone, water, CO_2, ...). Depending on the meridian of the sun γ (Fig. 19.49), the spectrum on the surface of the earth is termed AMx with $x = 1/\sin\gamma$. In spring and fall (March 21st and September 21st), Leipzig (51°42′N latitude) has about AM1.61. At the summer (June 21st) and winter (December 21st) solstices the air mass in Leipzig is AM1.13 ($\gamma = 61.8°$) and AM3.91 ($\gamma = 14.8°$), respectively. Additionally, the duration of sunshine and thus the light power density is regionally different across the earth due to climate and weather (Fig. 19.51). For AM1.5, the incident power density is 844 W/m^2.

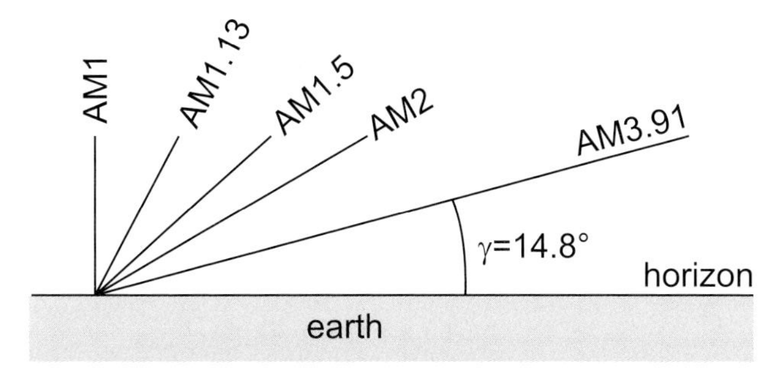

Fig. 19.49. Schematic path of sunlight through the atmosphere and definition of the air mass AMx

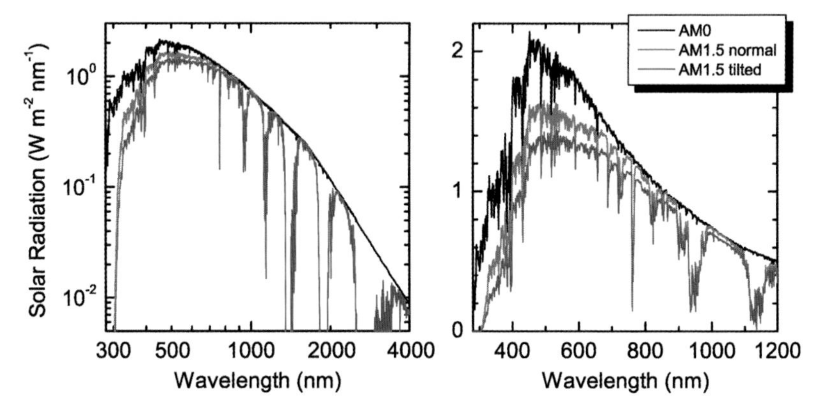

Fig. 19.50. Solar spectra (power per area and wavelength interval) for AM0 (*black line*, extraterrestrial irradiance) and AM1.5 (sun at 41.8° elevation above horizon) for direct normal irradiance (*blue line*) and global total irradiance (*red line*) on a sun facing surface (tilted 37° towards the equator). Left (right) graph in log-log (linear) scales

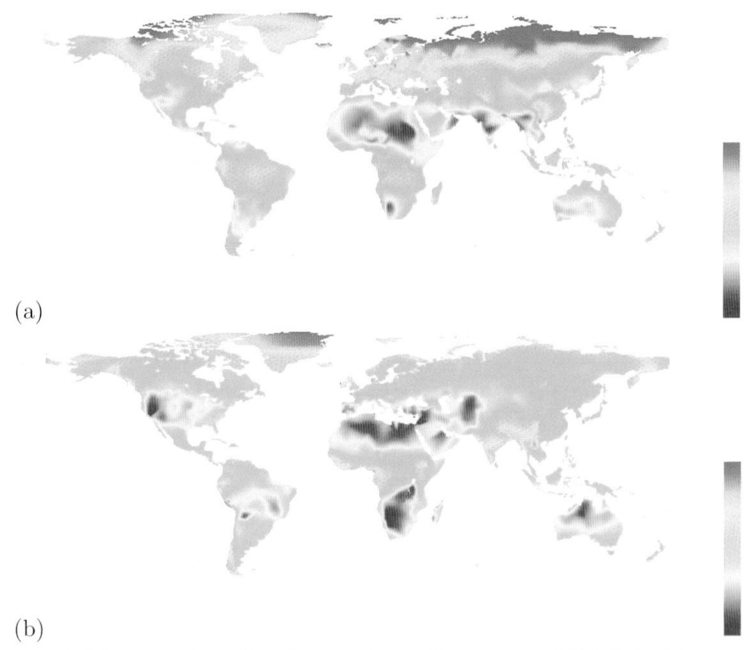

Fig. 19.51. Global sunshine distribution in (**a**) January and (**b**) July. The sunshine fraction is the actual number of bright sunshine hours over the potential number, and is thus expressed as a percentage figure. The color scale reaches from 0 to 100%. The sunshine data are in a 0.5 degree grid based on data from [632]

19.4.2 Ideal Solar Cells

When a solar cell made from a semiconductor with a bandgap E_g is irradiated by the sun, only photons with $h\nu \geq E_\mathrm{g}$ contribute to the photocurrent and the output power. The I–V characteristic under illumination (Fig. 19.52) is given by

$$I = I_\mathrm{s}\left[\exp\left(\frac{eV}{kT}\right) - 1\right] - I_\mathrm{L}\,, \tag{19.58}$$

with I_L being the current due to generation of excess carriers by the absorption of the sunlight. The saturation current density is given by (18.105) and (18.106)

$$j_\mathrm{s} = I_\mathrm{s}/A = eN_\mathrm{C}N_\mathrm{V}\left[\frac{1}{N_\mathrm{A}}\sqrt{\frac{D_\mathrm{n}}{\tau_\mathrm{n}}} + \frac{1}{N_\mathrm{D}}\sqrt{\frac{D_\mathrm{p}}{\tau_\mathrm{p}}}\right]\exp\left(-\frac{E_\mathrm{g}}{kT}\right)\,, \tag{19.59}$$

with A being the cell area.

The voltage at $I = 0$ is termed the *open-circuit* voltage V_oc, the current at $V = 0$ is termed the *short-circuit* current $I_\mathrm{sc} = I_\mathrm{L}$ (Fig. 19.52). Only a part of the rectangle $I_\mathrm{sc} \times V_\mathrm{oc}$ can be used for power conversion. By choice of the load resistance R_L, the work point is set. At I_m and V_m, the generated power $P_\mathrm{m} = I_\mathrm{m}V_\mathrm{m}$ is maximal. The filling factor F is defined as the ratio

$$F = \frac{I_\mathrm{m}V_\mathrm{m}}{I_\mathrm{sc}V_\mathrm{oc}}\,. \tag{19.60}$$

The open-circuit voltage is given by

$$V_\mathrm{oc} = \frac{kT}{e}\ln\left(\frac{I_\mathrm{L}}{I_\mathrm{s}} + 1\right) \cong \frac{kT}{e}\ln\left(\frac{I_\mathrm{L}}{I_\mathrm{s}}\right) \tag{19.61}$$

and increases with increasing light power and decreasing dark current. The output power is

$$P = IV = I_\mathrm{s}V\left[\exp\left(\frac{eV}{kT}\right) - 1\right] - I_\mathrm{L}V\,. \tag{19.62}$$

The condition $\mathrm{d}P/\mathrm{d}V = 0$ yields the optimal voltage at which the solar cell has to be operated and is given by the implicit equation ($\beta = e/kT$)

$$V_\mathrm{m} = \frac{1}{\beta}\ln\left(\frac{I_\mathrm{L}/I_\mathrm{s} + 1}{1 + \beta V_\mathrm{m}}\right) = V_\mathrm{oc} - \frac{1}{\beta}\ln\left(1 + \beta V_\mathrm{m}\right)\,. \tag{19.63}$$

The current at maximum power is

$$I_\mathrm{m} = I_\mathrm{L}\left(1 - \frac{1 - \beta V_\mathrm{m}I_\mathrm{s}/I_\mathrm{L}}{1 + \beta V_\mathrm{m}}\right) \cong I_\mathrm{L}\left(1 - \frac{1}{\beta V_\mathrm{m}}\right)\,. \tag{19.64}$$

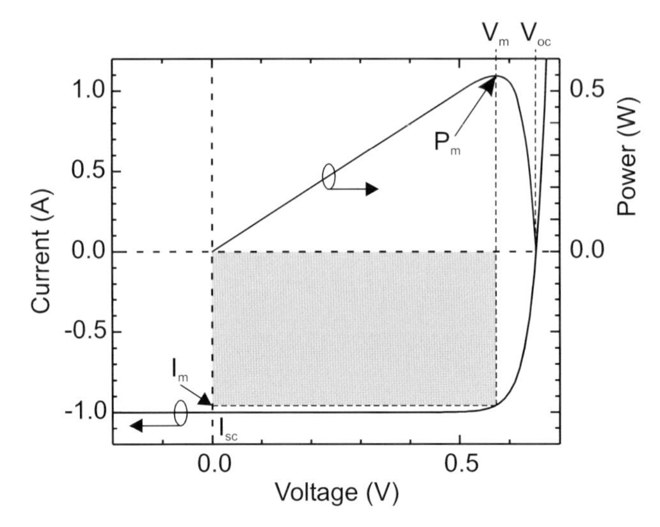

Fig. 19.52. Schematic I–V characteristics of a solar cell under illumination (*left scale*) and extracted power (*right scale*). The *grey area* is the maximum power rectangle with $P_{\mathrm{m}} = I_{\mathrm{m}} V_{\mathrm{m}}$.

E_{m} is the energy that is delivered per photon at the load resistor at the power maximum. The maximum power is $P_{\mathrm{m}} = I_{\mathrm{L}} E_{\mathrm{m}}/e$ and E_{m} is given by

$$E_{\mathrm{m}} \cong e \left[V_{\mathrm{oc}} - \frac{1}{\beta} \ln \left(1 + \beta V_{\mathrm{m}}\right) - \frac{1}{\beta} \right] . \tag{19.65}$$

The ideal solar cell has a (power) conversion efficiency $\eta = P_{\mathrm{m}}/P_{\mathrm{in}}$ that can be determined from Fig. 19.53a.

The right curve (1) in Fig. 19.53a shows the integral number n_{ph} of photons in the solar spectrum (per area and time) with an energy larger than a given one (E_{g}). For a given value of n_{ph}, the left curve (2) represents the value of E_{m}. The efficiency is the ratio of $E_{\mathrm{m}} n_{\mathrm{ph}}$ and the area under curve (1). The efficiency as a function of the bandgap is shown in Fig. 19.54a. It has a fairly broad maximum such that many semiconductors can be used for solar cells, in principle. The maximum theoretical efficiency for a single junction is 31% for nonconcentrated sunlight (AM1.5). This limit corresponds to the classic Shockley–Queisser limit [633], assuming radiative recombination as the only charge-carrier recombination mechanism. In [634], the limit for a single material is found to be 43% for an optimally tailored band structure that allows carrier multiplication by optically excited hot carriers. When the sunlight is concentrated, e.g. by a lens, the efficiency increases (Fig. 19.54b). The short-circuit current increases linearly. The effect is mostly due to the increase of the open-circuit voltage. For $C = 1000$, the maximum theoretical efficiency for a single-junction solar cell is 38%.

A further increase of efficiency can be achieved with multiple junctions using various materials for absorption. In a tandem cell (two junctions), the

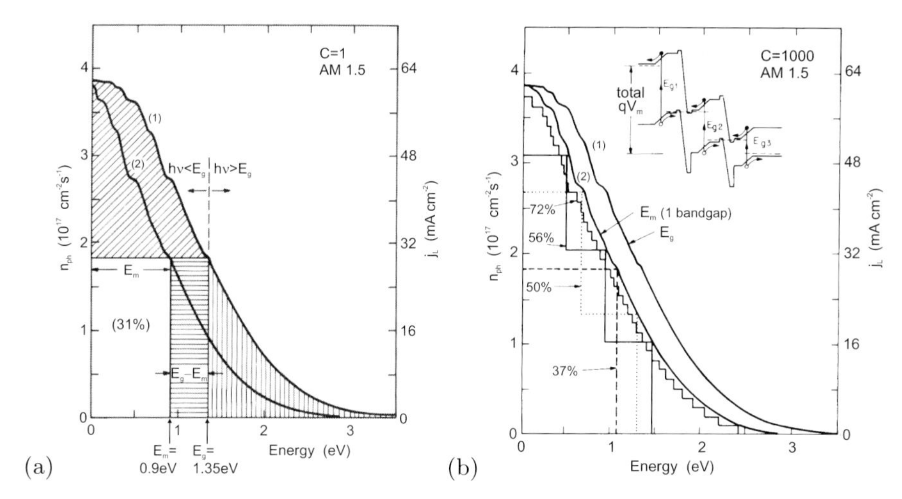

(a) (b)

Fig. 19.53. (**a**) Number of photons n_{ph} per area and time in the sun spectrum (AM1.5, $C = 1$ sun) with an energy larger than a cutoff energy (*curve* 1) and graphical method to determine the quantum efficiency (from *curve* 2). Adapted from [635]. (**b**) Number of photons in concentrated solar spectrum (AM1.5, $C = 1000$ sun) with an energy larger than a given energy and graphical method to determine the quantum efficiency of multijunction solar cells. Adapted from [183] after [635]

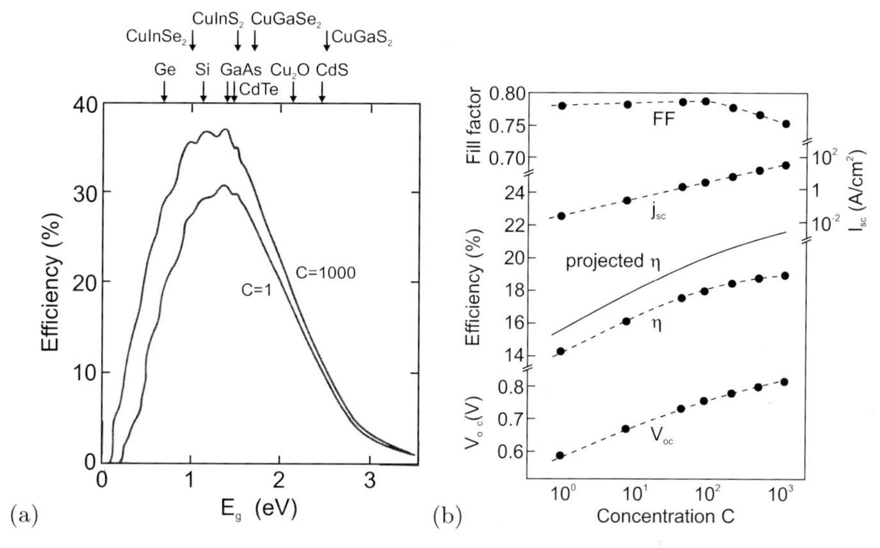

(a) (b)

Fig. 19.54. (**a**) Ideal quantum efficiency of solar cells (single junction) as a function of bandgap and light concentration C. The bandgaps of some important semiconductors are denoted by *arrows*. Adapted from [635]. (**b**) Properties of a silicon solar cell (water cooled) as a function of light concentration. *Solid line* is theoretically projected efficiency, *dashed lines* are guides to the eye. Adapted from [183].

upper layer absorbs the higher-energy photons in a wide-bandgap material. The lower-bandgap material makes use of the low-energy photons. Thus, the cell works with two different values of E_{m} (Fig. 19.53b). With bandgaps of 1.56 eV and 0.84 eV, an efficiency of 50% can be reached theoretically. With three materials 56%, and for a large number of materials 72% is the limit. Between the junctions, tunneling diodes (Sect. 18.5.9) must be used to allow carrier transport through the entire structure. It is a nontrivial task to fabricate multiple heterojunctions due to incompatibilities of the lattice constants. Besides heteroepitaxy, wafer bonding can also be used for fabrication. A lattice-matched InGaP/GaAs/InGaAsN cell seems a viable solution for high-efficiency solar cells.

19.4.3 Real Solar Cells

For a real solar cell, the effect of parallel resistance R_{sh} (shunt resistance due to leakage current, e.g. by local shorts of the solar cell) and serial resistance R_{s} (due to ohmic loss) must be considered. The I–V characteristic is then (cf. (18.131))

$$\ln\left(\frac{I+I_{\mathrm{L}}}{I_{\mathrm{s}}} - \frac{V-IR_{\mathrm{s}}}{I_{\mathrm{s}}R_{\mathrm{sh}}} + 1\right) = \frac{e}{kT}\left(V - IR_{\mathrm{s}}\right) . \qquad (19.66)$$

The serial resistance affects the filling factor more strongly than the shunt resistance (Fig. 19.55). Therefore, it is frequently enough to consider R_{s} only and use (cf. (18.130))

$$I = I_{\mathrm{s}}\left[\exp\left(\frac{e(V-IR_{\mathrm{s}})}{kT}\right) - 1\right] - I_{\mathrm{L}} . \qquad (19.67)$$

In the example of Fig. 19.55, a serial resistance of 5 Ω reduces the filling factor by a factor of about four.

19.4.4 Design Refinements

In order to collect electrons most efficiently, a back surface field is used (Fig. 19.56). A higher-doped region at the back contact creates a potential barrier and reflects electrons back to the front contact.

An important point for optimization is the management of the reflection at the solar cell surface. First, dielectric antireflection (AR) multilayers can be used. These layers should have a broad AR spectrum. Additionally, a textured surface, as schematically shown in Fig. 19.57a, e.g. created by anisotropic etching of Si (001) (Fig. 19.57c), reduces reflection (Fig. 19.57d), giving reflected photons a second chance for penetration (Fig. 19.57b). The reflectivity of bare Si, 35%, can be reduced to 2%. An AM0 efficiency of 15% was reached using textured cells.

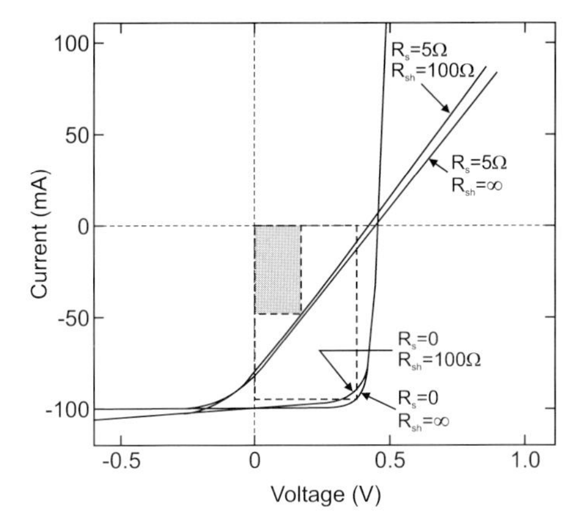

Fig. 19.55. *I–V* characteristics of a solar cell considering shunt and series resistances R_s and R_{sh}, respectively. The efficiency of the real cell (*shaded* power rectangle) is less than 30% of that of the ideal cell. Adapted from [636]

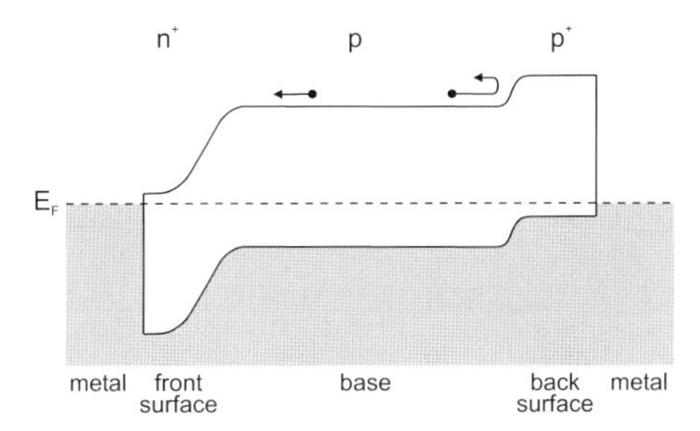

Fig. 19.56. Increase of the carrier collection efficiency by a back surface field. Adapted from [637]

During its course over the sky during the day, the sun changes its angle towards a fixed solar cell.[7] A tracking mechanism can optimize the angle of incidence during the day and increase the overall efficiency of the solar cell (Fig. 19.58).

19.4.5 Solar-Cell Types

Silicon is the most frequently used material for solar cells. Cells based on single-crystalline silicon (wafers) have the highest efficiency but are the most expensive (Fig. 19.59a). Polycrystalline silicon (Fig. 19.59b) is cheaper but offers less performance. Material design is oriented towards increasing the grain size. These solar cells are also called 'first-generation' photovoltaics. Thin sheets of crystalline silicon drawn from a melt between two seed crystals in a modified CZ growth (sheet silicon or ribbon silicon) allow cheaper

[7] We are of course aware that the earth rather rotates around the sun.

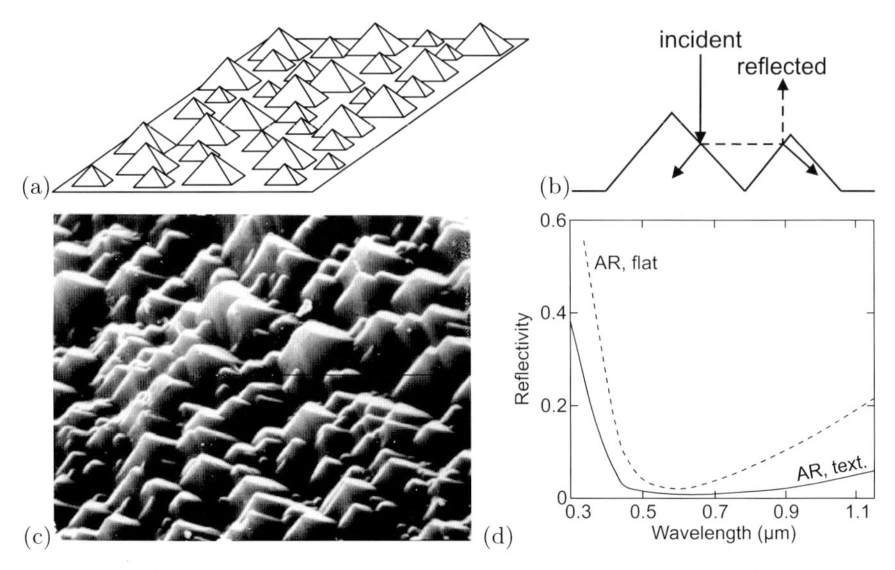

Fig. 19.57. (**a**) Schematic surface topology of a textured solar cell. (**b**) Exemplary light path. (**b**) SEM image of a textured surface (pyramid base: $\sim 5\,\mu m$). (**c**) Reflectivity of antireflection-coated flat (*dashed line*) and textured (*solid line*) surface). Parts (**a**) and (**d**) adapted from [638]

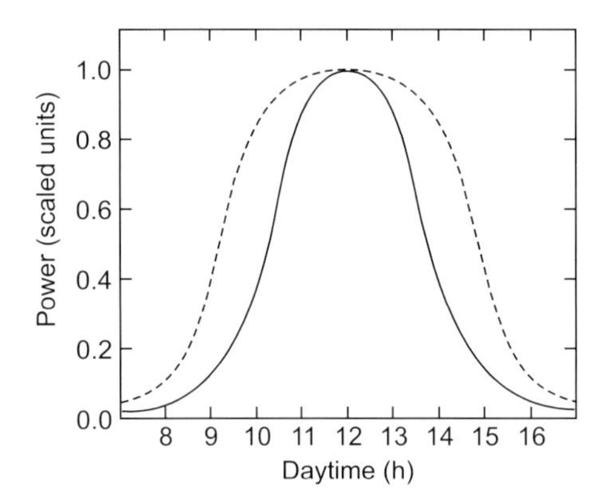

Fig. 19.58. Power generation of a solar cell vs. time (in daytime hours) for a stationary setup facing the sun at constant angle (*solid line*) and mounting with tracking (*dashed line*) to optimize the angle towards the sun. Adapted from [639]

production compared to cells based on 'traditional' polished wafers cut from a large silicon rod.

Even cheaper are solar cells from amorphous silicon (Fig. 19.59c). Since silicon is an indirect semiconductor, a fairly thick layer is needed for light absorption. If direct semiconductors are used, a thin layer ($d \approx 1\,\mu m$) is sufficient for complete light absorption. Such cells are called thin-film solar cells. A typical material class used in this type of cell are chalcopyrites, such as $CuInSe_2$ (CIS). The bandgap is around 1 eV, which is not optimal. An improvement can be achieved by adding Ga and/or S that increases the band gap, $Cu(In,Ga)(Se,S)_2$ (CIGS), to 1.2–1.6 eV. Using CIGS, an efficiency of 15% has been reported. Thin-film solar cells can be fabricated on glass substrate or on flexible polymer substrate such as Kapton[8] (Fig. 19.60a,b). Also here, optimization of the grain size is important (Fig. 19.60b). As the front contact, a transparent conductive oxide (TCO), such as ITO ($InSnO_2$) or ZnO:Al, is used. Thin-film and amorphous silicon solar cells are also termed 'second-generation' photovoltaics.

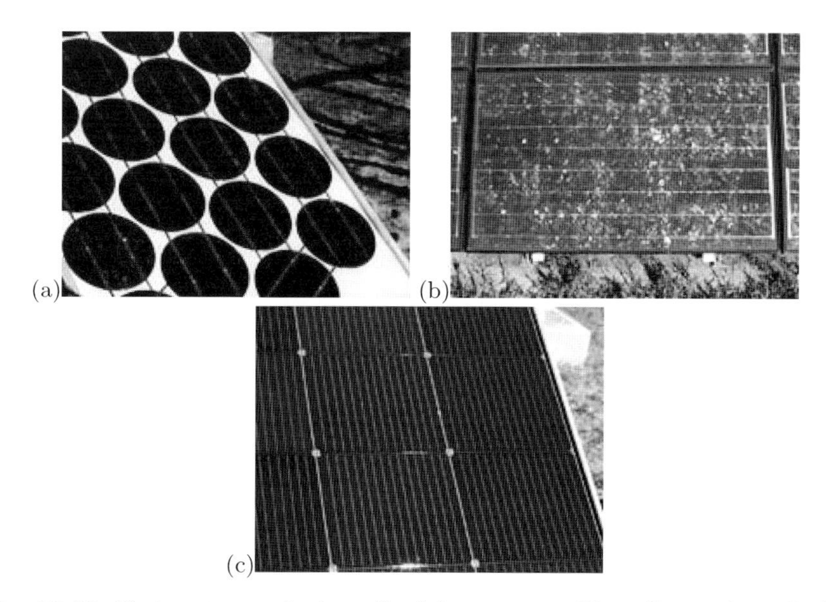

Fig. 19.59. Various types of solar cells: (**a**) monocrystalline silicon solar cell, (**b**) polycrystalline solar cell, (**c**) amorphous silicon solar cell. From [639]

In order to cover a large area and supply certain values of output voltage and current, several solar cells are connected into modules. Arrays are built up from several modules (Fig. 19.61). If solar cells are connected in parallel, the total current increases; if they are connected in series, the output voltage increases.

'Third-generation' photovoltaics attempt to go beyond the 30% limit and comprise of multijunction solar cells, concentration of sunlight, use of hot-carrier excess energy as discussed above and possibly other concepts including photon conversion [641, 642].

[8]Kapton® is a polyimide and a product and registered trademark of DuPont.

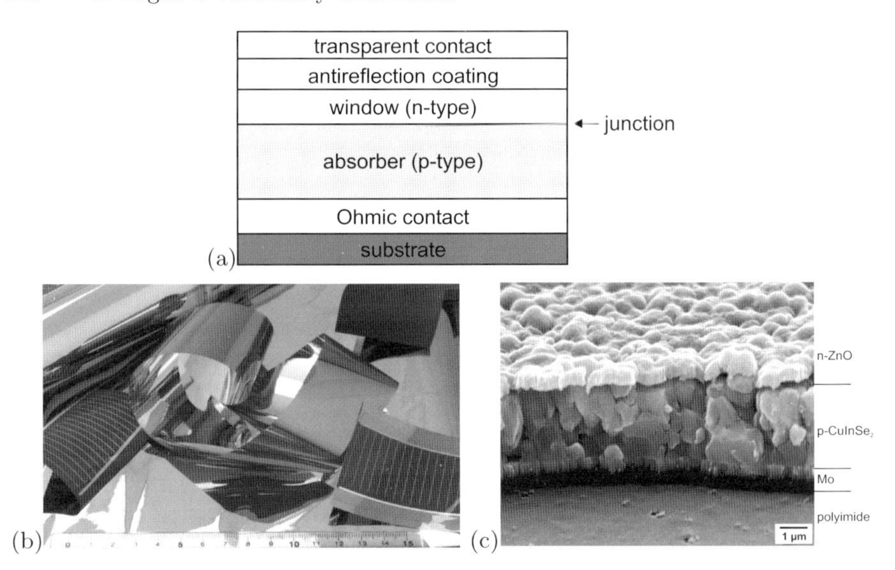

Fig. 19.60. (**a**) Schematic cross section of a polycrystalline thin-film solar cell. (**b**) Rolled sheets of CIS thin film solar cell on flexible Kapton foil. (**c**) SEM cross section of CIS thin-film solar cell. Parts (**b**) and (**c**) reprinted with permission from [640].

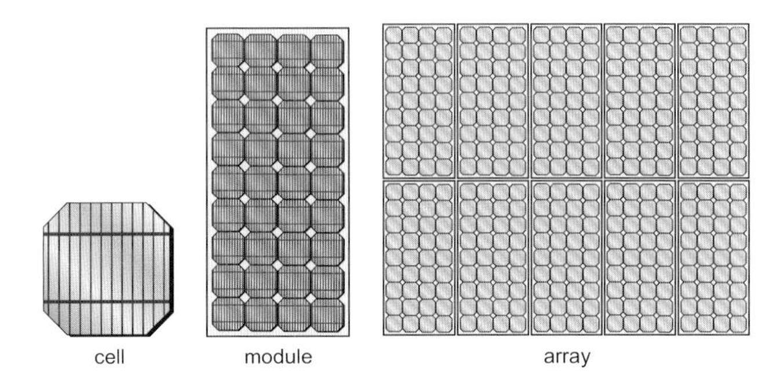

Fig. 19.61. Schematic drawing of a solar cell (with contact grid), a module (36 cells) and an array of ten modules

19.4.6 Commercial Issues

The cost[9] of producing photovoltaic (PV) modules, in constant dollars, has fallen from as much as \$50 per peak watt in 1980 to as little as \$3 per peak watt in 2004. A projected cost of 0.2 €/kWh in 2020, a third of the current cost, is realistic and competitive in many applications. In 2002, photovoltaic power of 560 MW was installed worldwide. By the end of 2003, a total photovoltaic power of about 350 MW was installed in Germany. Thus, the PV power is 0.33% of the total installed electric power of 106 GW in Germany.

[9]The following information is taken from [643] and [644].

The current market growth of 30% (worldwide) is driven by crystalline silicon cells (95% in 2002).

The energy payback period is also dropping rapidly. For example, it takes today's typical crystalline silicon module about 4 years to generate more energy than went into making the module in the first place. The next generation of silicon modules, which will employ a different grade of silicon and use thinner layers of semiconductor material, will have an energy payback of about 2 years. And thin-film modules will soon bring the payback down to one year or less. However, market growth of thin-film modules is currently slow. This means that these modules will produce 'free' and clean energy for the remaining 29 years of their expected life.

PV technology can meet electricity demand on any scale. The solar energy resource in a 100 mile-square area of Nevada could supply the United States with all its electricity (about 800 GW), using modestly efficient (10%) commercial PV modules. A more realistic scenario involves distributing these same PV systems throughout the 50 states. Currently available sites, such as vacant land, parking lots, and rooftops, could be used. The land requirement to produce 800 GW would average out to be about 17×17 miles per state. Alternatively, PV systems built in the 'brownfields', the estimated 5 million acres of abandoned industrial sites in the US, could supply 90% of America's current electricity. Solar power is expected to contribute 10% of the US energy need in 2030. For Germany, more than 2% in 2020 is probable.

In 2001, PV module shipments in the US approached the 400 MW mark, representing a $2.5 to $3 billion market. The US-based industry itself is now approaching $1 billion per year and provides 25 000 jobs. It is expected to grow to the $10–$15 billion level in the next 20 years, providing 300 000 jobs by 2025.

20 Electricity-to-Light Conversion

20.1 Radiometric and Photometric Quantities

20.1.1 Radiometric Quantities

The radiometric quantities are derived from the radiant flux (power) Φ_{e} (or usually simply Φ) that is the total power (energy per time) emitted by a source, measured in Watts. The radiant intensity I_{e} is the radiant flux emitted by a point source into a solid angle,[1] measured in Watts per steradian (or W/sr). The irradiance E_{e} is the radiant flux per area incident on a given plane, measured in $\mathrm{W/m^2}$. The radiance L_{e} is the radiant flux per area and solid angle as, e.g., emitted by an extended source, measured in $\mathrm{W/(m^2\,sr)}$.

20.1.2 Photometric Quantities

The photometric quantities are related to the visual impression and are derived from the radiometric quantities by weighting them with the $V(\lambda)$ curve.

The luminous flux (luminosity or visible brightness) Φ_{v} of a source with the radiant flux (spectral power distribution) $\Phi(\lambda)$ is given by

$$\Phi_{\mathrm{v}} = K_{\mathrm{m}} \int_0^{\infty} \Phi(\lambda) V(\lambda) \mathrm{d}\lambda \,, \tag{20.1}$$

with $K_{\mathrm{m}} = 683\,\mathrm{lm/W}$. This formula is also the definition of the unit 'lumen'. In Fig. 20.1b, the conversion function[2] $V(\lambda)$ is shown for light and dark adapted vision.[3]

[1] A solid angle Ω is the ratio of the spherical surface area A and the square of the sphere's radius r, i.e. $\Omega = A/r^2$.

[2] The $V(\lambda)$ curve has been experimentally determined by letting several observers adjust (decrease) the perceived brightness of a monochromatic light source at 555 nm to that of light sources of the same absolute radiation power at other wavelengths with so-called heterochromatic flicker photometry. The 'relative sensitivity curve for the CIE Standard Observer' was determined in 1924. The 'standard observer' is neither a real observer nor an average human observer. The curve has shortcomings, e.g., due to the used spectral band width (20–30 nm) of the light sources and the comparison of spectral power instead of the photon flux.

[3] While photopic vision is due to cones, the scotopic (dark-adapted) vision is due to rods. Rods are more than one thousand times as sensitive as the cones and

Further derived photometric quantities are luminous intensity (luminous flux per solid angle), measured in candela (cd), the illuminance (luminous flux per area), measured in lux (lx), and the luminance (luminous flux per area and solid angle). The latter is particularly important if the radiation enters an optical system, e.g. for refocusing. The radiometric and photometric quantities are summarized in Table 20.1.

Table 20.1. Radiometric and photometric quantities and units. The photometric units are lumen (lm), lux (lx=lm/m^2) and candela (cd=lm/sr)

Radiometric			Photometric		
quantity	symbol	unit	quantity	symbol	unit
radiant flux	Φ_e	W	luminous flux	Φ_v	lm
radiant intensity	I_e	W/sr	luminous intensity	I_v	cd
irradiance	E_e	W/m^2	illuminance	E_v	lx
radiance	L_e	W/m^2/sr	luminance	L_v	lm/m^2/sr

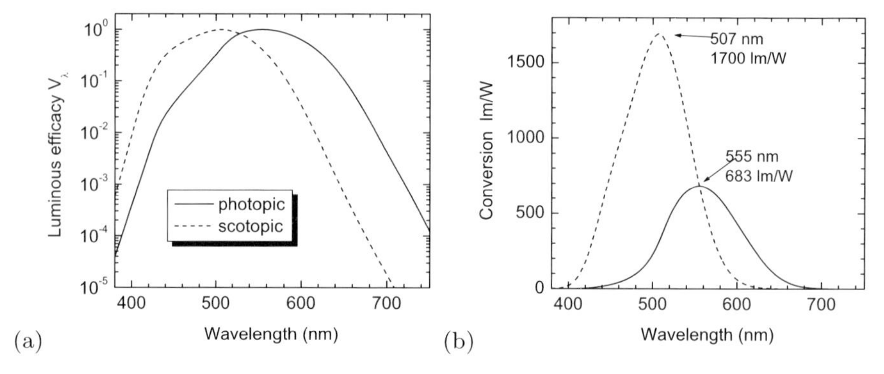

Fig. 20.1. (**a**) Relative eye sensitivity curves for photopic (light adapted, *solid line*) and (dark adapted, *dashed line*) vision. (**b**) Conversion of lumen to Watt for light-(*solid line*) and dark-adapted (*dashed line*) vision

20.2 Scintillators

A scintillator (or phosphor) is a material that converts impacting high-energy radiation into photons [645]. Besides a high conversion efficiency, the spectrum and decay time constant of the scintillator are important for display

can reportedly be triggered by individual photons under optimal conditions. Rods predominate in the peripheral vision and are not color sensitive.

applications. For display purposes, the photons are directly used for forming the image for the observer. For radiation detection, the photons are fed to a photomultiplier tube and counted.

The most prominent applications, involving the conversion of electrons, are the screens of cathode ray tubes (CRT) (acceleration voltage $> 10\,\mathrm{kV}$) and of flat panel devices, such as field-effect displays (using a low voltage for excitation, typically $< 1\,\mathrm{kV}$) or plasma displays (using the UV light from the discharge of a plasma placed between two electrodes for excitation). Further details on electroluminescent displays can be found in [646]. Other forms of radiation detected with scintillators are α-, β-, and γ-radiation, X-rays and neutrons [647]. Different excitation conditions require different phosphors for optimal performance.

20.2.1 CIE Chromaticity Diagram

The CIE[4] procedure converts the spectral power distribution (SPD) of light from an object into a brightness parameter Y and two chromaticity coordinates x and y. The chromaticity coordinates map the color[5] with respect to hue and saturation on the two-dimensional CIE chromaticity diagram. The procedure for obtaining the chromaticity coordinates for a given colored object involves determination of its spectral power distribution $P(\lambda)$ at each wavelength, multiplication by each of the three color-matching functions $\bar{x}(\lambda)$, $\bar{y}(\lambda)$, and $\bar{z}(\lambda)$ (Fig. 20.2a) and integration (or summation) of the three tristimulus values X, Y, Z

$$X = \int_{380\,\mathrm{nm}}^{780\,\mathrm{nm}} P(\lambda)\bar{x}(\lambda)\mathrm{d}\lambda \tag{20.2a}$$

$$Y = \int_{380\,\mathrm{nm}}^{780\,\mathrm{nm}} P(\lambda)\bar{y}(\lambda)\mathrm{d}\lambda \tag{20.2b}$$

$$Z = \int_{380\,\mathrm{nm}}^{780\,\mathrm{nm}} P(\lambda)\bar{z}(\lambda)\mathrm{d}\lambda\,. \tag{20.2c}$$

Y gives the brightness. The tristimulus values are normalized to yield the chromaticity coordinates, e.g. $x = X/(X+Y+Z)$. x and y obtained in this

---

[4]Commission Internationale de l'Éclairage. The color space can be described by different coordinate systems, and the three most widely used color systems, Munsell, Ostwald, and CIE, describe the color space with different parameters. The Munsell system uses hue, value, and chroma and the Ostwald system uses dominant wavelength, purity, and luminance. The more precise CIE system uses a parameter Y to measure brightness and parameters x and y to specify the chromaticity that covers the properties hue and saturation on a two-dimensional chromaticity diagram.

[5]This definition is motivated by the color vision of the eye. Two light sources will have the same color, even if they have different SPDs, when they evoke the same color impression to the human eye.

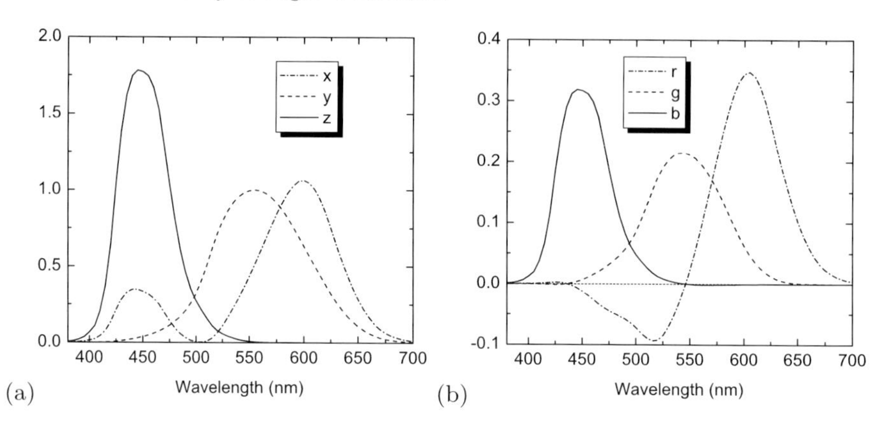

Fig. 20.2. (**a**) Color-matching functions $\bar{x}$, $\bar{y}$, and $\bar{z}$ for the calculation of the CIE chromaticity, (**b**) color-matching functions $\bar{r}$, $\bar{g}$, and $\bar{b}$ for the calculation of RGB values

way are the chromaticity coordinates. The third coordinate $z = 1 - x - y$ offers no additional information and is redundant. Therefore, the color is represented in a two-dimensional diagram, the CIE chromaticity diagram[6] as shown in Fig. 20.3a. White is represented by $x = y = z = 1/3$. In order to relate the differences between colors as perceived by the human eye more closely to the geometrical distance in the chart, a revision was made (Fig. 20.3b) with new coordinates

$$u' = 4x/(-2x + 12y + 3) \tag{20.3a}$$

$$v' = 9y/(-2x + 12y + 3) \,. \tag{20.3b}$$

For CRTs the red-green-blue (RGB) color space is used.[7] The color matching functions for RGB values are shown in Fig. 20.2b. The RGB values are related to the XYZ values according to

$$\begin{pmatrix} R \\ G \\ B \end{pmatrix} = \begin{pmatrix} 2.36461 & -0.89654 & -0.46807 \\ -0.51517 & 1.42641 & 0.08876 \\ 0.00520 & -0.01441 & 1.00920 \end{pmatrix} \begin{pmatrix} X \\ Y \\ Z \end{pmatrix} \,. \tag{20.4}$$

[6]The coloring of the chart is provided for an understanding of color relationships. CRT monitors and printed materials cannot reproduce the full gamut of the color spectrum as perceived in human vision. The color areas that are shown only depict rough categories and are not precise statements of color.

[7]RGB is an additive color system. However, printing devices use a subtractive color system. This means that the ink absorbs a particular color, and the visible impression stems from what is reflected (not absorbed). When inks are combined, they absorb a combination of colors, and hence the reflected colors are reduced, or subtracted. The subtractive primaries are cyan, magenta and yellow (CMY) and are related to RGB via $(C, M, Y) = (1 - R, 1 - G, 1 - B)$.

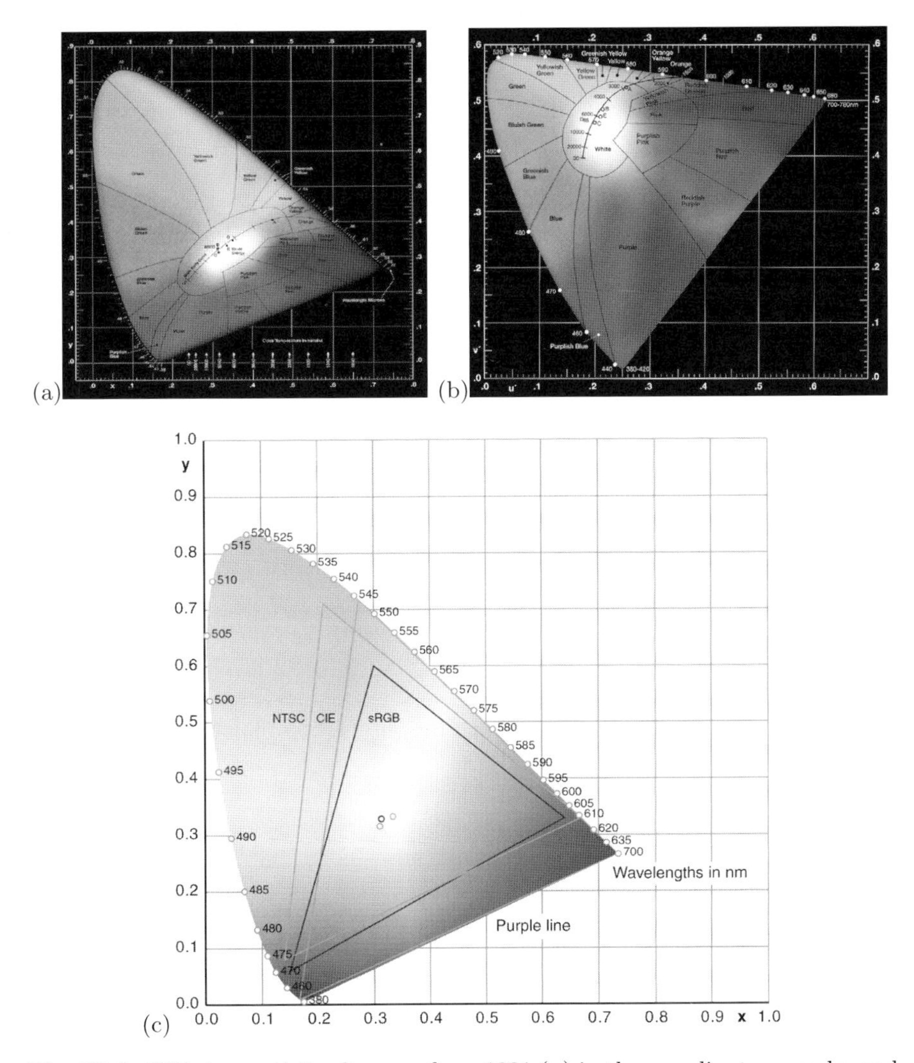

Fig. 20.3. CIE chromaticity diagram from 1931 (**a**) in the coordinates x and y and from 1976 (**b**) in the coordinates u' and v' (20.3b). The curved upper boundary is called the 'spectrum locus' and contains monochromatic colors, the straight line at the lower left is termed the 'purple boundary'. In the graph also the color of blackbody radiation is given, $T = 5440\,\mathrm{K}$ corresponds to $x = y = 1/3$. 'A', 'B', 'C', and 'E' are standard illuminants, 'D65' denotes daylight with color temperature $T = 6500\,\mathrm{K}$. (**c**) CIE chart with the color ranges of sRGB, CIE and NTSC. Part (**c**) adapted from [648]

The CIE RGB primaries from 1931 are at 700, 546.1, and 435.8 nm with the relative intensities 1.0, 4.5907, and 0.0601. A display device using three phosphors can only display colors in the triangular area of the CIE chart between the three chromaticity coordinates. For sRGB[8], the 1931 CIE primaries and the NTSC[9] norm the coordinates are given in Table 20.2 and visualized in Fig. 20.3c. An optimal coverage of the CIE chart involves monochromatic sources (for laser TV or LED displays) at about 680, 520 and 440 nm.

Table 20.2. Primaries and white points for sRGB, CIE and NTSC

primary	red		green		blue		white	
CIE	0.73467	0.26533	0.27376	0.71741	0.16658	0.00886	0.33333	0.33333
NTSC	0.6700	0.3300	0.2100	0.7100	0.1400	0.0800	0.3100	0.3160
sRGB	0.6400	0.3300	0.3000	0.6000	0.1500	0.0600	0.3127	0.3290

20.2.2 Display Applications

The once ubiquitous amber-colored monochrome displays are mostly fabricated using ZnS:Mn [646], having broad emission (540–680 nm) with its spectral peak at 585 nm ($x = 0.50$, $y = 0.50$) with an efficiency of 2–4 lm/W. In color television (and similar applications such as color computer monitors, tubes for aviation use, projection television) the image is reproduced by selective and time-multiplexed cathode excitation of three phosphors (blue, green and red) deposited on the internal face of the screen. The chromaticity coordinates of the standard CRT phosphors P-22B, P-22G and P-22R are given in Table 20.3. They cover about the color range labeled 'sRGB' in Fig. 20.3c. For blue and green ZnS:Ag ($x = 0.157$, $y = 0.069$), ZnS:Ag,Cl, ZnS:Ag,Al and ZnS:Cu,Al ($x = 0.312$, $y = 0.597$), ZnS:Cu,Au,Al are used as phosphors, respectively. Y_2O_2S:Eu ($x = 0.624$, $y = 0.337$) activated with trivalent europium (Eu^{3+}) facilitated such a gain in the brilliance of red over ZnS:Ag (more than doubled it) that it has totally replaced it at about one fifth of the cost. For reproducible image quality, precise grain-size control (median size for CRT phosphors is about 8 µm), dispersion control and surface treatment are necessary. Flat-panel displays with their lower excitation voltage require different phosphors for optimal efficiency.

20.2.3 Radiation Detection

The most commonly used scintillation detector for alpha measurements is ZnS activated with silver, ZnS:Ag. This material is not very transparent to light and is usually prepared as a large number of crystals with sub-mm size attached with an adhesive to a flat piece of plastic or other material. The

[8]Standard RGB color space as defined mainly by Hewlett-Packard and Microsoft, almost identical to PAL/SECAM European television phosphors.

[9]National television standard colors, US norm.

Table 20.3. CIE color coordinates, peak emission wavelength and decay time (10%) of standard CRT phosphors

phosphor	x	y	λ_p (nm)	decay time
P-22B	0.148	0.062	440	~20 µs
P-22G	0.310	0.594	540	~60 µs
P-22R	0.661	0.332	625	1 ms

flat screen is optically coupled to a photomultiplier tube that is attached to associated electronics. The voltage and discriminator levels are selected so that the detector is sensitive to the rather large pulses from alpha interactions but insensitive to beta- or gamma-induced pulses. The alpha particles deposit all of their energies in a small thickness of material compared to beta and gamma radiations.

Scintillation detectors for beta radiation are often made from organic materials. In an organic scintillator, the light emission occurs as a result of fluorescence when a molecule relaxes from an excited level following excitation by energy absorption from ionizing radiation. Molecules such as anthracene, trans-stilbene, para-terphenyl, and phenyl oxazole derivatives are among the many organic species that have useful scintillation properties. The organic molecules are dissolved in organic solvents and used as liquid scintillators. A classic application is in the measurement of low-energy beta radiation from, e.g. tritium, ^{14}C, or ^{35}S. In such cases, the sample containing the radioactive beta emitter is dissolved in, or in some cases suspended in, the liquid scintillation solution. The emitted beta radiation transfers energy through the solvent to the scintillator molecule that emits light, subsequently detected by photomultiplier tubes. Organic scintillator molecules can also be dissolved in an organic monomer that can then be polymerized to produce a plastic scintillator in a wide variety of shapes and sizes. Very thin scintillators have been used for alpha detection, somewhat thicker scintillators for beta detection. Large-volume plastic scintillators have been used in gamma detection, particularly for dose-related measurements.

Other inorganic crystalline scintillators, especially sodium iodide activated with thallium, NaI:Tl, have been used for gamma-ray energy measurements. Such detectors can be grown as large single crystals that have a reasonably high efficiency for absorbing all of the energy from incident gamma rays. There exists a rather large number of inorganic scintillators; some examples of these include cesium iodide activated with thallium, CsI:Tl, bismuth germanate, $Bi_4Ge_3O_{12}$, and barium fluoride, BaF_2. These are mostly used for gamma measurements but can also be prepared with thin windows and have been used for charged particle (e.g. alpha and beta) counting.

In Table 20.4, the peak emission wavelength and the characteristic decay time are listed for a variety of scintillator materials. Direct semiconductors,

Table 20.4. Emission peak wavelength and decay time of various scintillator materials

material	λ_p (nm)	decay time
Zn_2SiO_4:Mn	525	24 ms
ZnS:Cu	543	35–100 µs
$CdWO_4$	475	5 µs
CsI:Tl	540	1 µs
CsI:Na	425	630 ns
$Y_3Al_5O_{12}$:Ce	550	65 ns
Lu_2SiO_5:Ce	400	40 ns
$YAlO_3$:Ce	365	30 ns
ZnO:Ga	385	2 ns

although not offering the highest efficiency, are particularly useful for high time resolution in, e.g., time-of-flight measurements or fast scanning electron microscopy.

20.2.4 Luminescence Mechanisms

Self-Trapped Excitons

In a strongly ionic crystal, such as NaI, a hole becomes localized to an atomic site via the polaron effect. A spatially diffuse electron is attracted, and a self-trapped exciton is formed that can recombine radiatively.

Self-Activated Scintillator

In such material, the luminescent species is a constituent of the crystal. The emission involves an intraionic transition, e.g. 6p→6s in Bi^{3+} of $Bi_4Ge_3O_{12}$, or a charge-transfer transition in the case of $(WO_4)^{2-}$ in $CaWO_4$. At room temperature, nonradiative competing processes limit the efficiency.

Activator Ions

For dopant ions such as Eu^{2+} in YO_2S:Eu, Ce^{3+} in $YAlO_3$:Ce or Tl^+ in NaI:Tl, the hole and electron excited by the radiation are sequentially trapped by the same ion that then undergoes a radiative transition, in the case of Eu and Ce[10] 5d→4f, for Tl $^3P_{0,1}$→S_0. CsI:Tl has one of the highest efficiencies of 64.8 photons/keV [649].

[10] This transition is dipole allowed for Ce and partially forbidden for Eu.

Core–Valence Luminescence

In some materials, e.g., BaF_2, CsF, $BaLu_2F_8$ the energy gap between the valence band and the top core band is less than the fundamental bandgap. A radiative transition occurs when an electron from the valence band fills a hole in the top core band that has been created by the radiation. The light yield is limited to about 2 photons/keV.

Semiconductor Recombination Processes

Free excitons or excitons bound to impurities can recombine radiatively. This process is most efficient at low temperatures. At room temperature, the emission is typically much weaker ($\gtrsim 10\times$) since excitons become unbound or dissociated. Highly doped n-type semiconductors, e.g. CdS:In, exhibit recombination between donor-band electrons and holes. ZnO:Ga has an efficiency of about 15 photons/keV and a fast response (with 2.4 photons/keV emitted in the first 100 ps). Luminescence can also stem from donor–acceptor pair transitions, e.g. in PbI_2 with an efficiency of 3 photons/keV at 10 K. Isoelectronic impurities such as nitrogen in GaP:N and tellurium in CdS:Te attract an electron and subsequently a hole. In ZnS:Ag and ZnS:Cu (conduction) band to trap recombination is dominant. In a codoping scheme like CdS:In,Te, In supplies electrons in an impurity band that can recombine with holes trapped at Te.

20.3 Light-Emitting Diodes

20.3.1 Introduction

Light-emitting diodes (LEDs) are semiconductor devices in which injected carriers recombine radiatively. The recombination process leading to light emission can be of intrinsic nature, i.e. band–band recombination, or extrinsic, e.g. impurity-bound excitons. Impurity-related luminescence can also be excited via impact excitation. For an extensive treatment of LEDs see [650].

20.3.2 Spectral Ranges

Applications for LEDs can be sorted by the color of emission. In Fig. 20.4a, the standard sensitivity $V(\lambda)$ of the human eye is shown (cf. Fig. 20.1a). In the visible spectral region (about 400–750 nm) the perceived brightness of the LED depends on the eye sensitivity. It is largest in the green (at 555 nm) and drops strongly towards the red and blue.

The most important spectral regions and applications are:

- infrared ($\lambda > 800$ nm): remote controls, optocouplers, low-cost data transmission, IR interface

- visible: indicator LED, lighting[11] (room, buildings, cars), white LED for broad spectrum
- ultraviolet ($\lambda < 400$ nm): pumping of phosphors for white LEDs, biotechnology

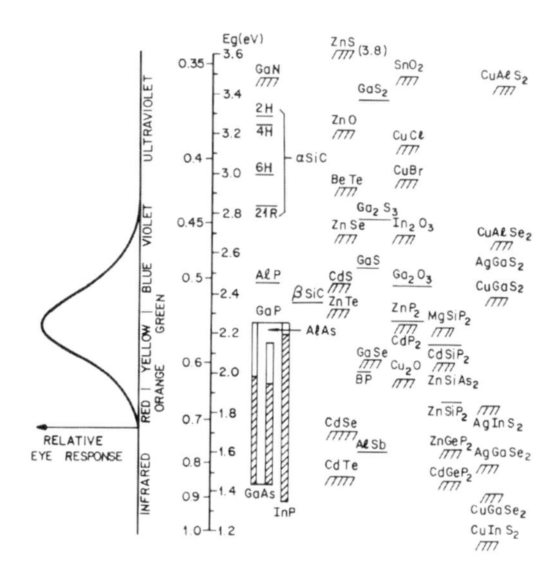

Fig. 20.4. Spectral coverage by various semiconductor materials. Reprinted with permission from [183], ©1981 Wiley

In Fig. 20.4, potentially useful semiconductors for the various spectral regions are shown. The semiconductors that are currently used for the various colors of the visible spectrum are

- red–yellow: GaAsP/GaAs, now AlInGaP/GaP
- yellow–green: GaP:N
- green–blue: SiC, now GaN, InGaN
- violet: GaN
- ultraviolet: AlGaN

20.3.3 Quantum Efficiency

The total quantum efficiency η is the number of photons emitted from the device per injected electron–hole pair. It is given by the product of the *internal* quantum efficiency η_{int} and the *external* quantum efficiency η_{ex}:

[11] Penetration of white LEDs into the general lighting market could translate (globally) into cost savings of \$$10^{11}$ or a reduction of power generation capacity of 120 GW.

$$\eta = \eta_{\mathrm{int}} \eta_{\mathrm{ex}} \ . \tag{20.5}$$

The internal quantum efficiency is the number of photons generated (inside the semiconductor) per injected electron–hole pair. High material quality, low defect density and low trap concentration are important for a large value of η_{int}. The recombination current in the pn-diode has been already given in (18.109). The external quantum efficiency is the number of photons leaving the device divided by the total number of generated photons. The geometry of the LED is of prime importance to optimize η_{ex}.

Due to the large index of refraction of the semiconductors ($n_{\mathrm{s}} \sim 2.5$–3.5), light can leave the semiconductor only under a small angle θ_{c} from the surface normal due to total reflection (cf. (9.3) and see right part of Fig. 9.3). Against air ($n_1 \approx 1$) the critical angle is

$$\theta_{\mathrm{c}} = \sin^{-1}(1/n_{\mathrm{s}}) \ . \tag{20.6}$$

The critical angle for total reflection is 16° for GaAs and 17° for GaP. Additionally, a portion of the photons that do not suffer total reflection is reflected back from the surface with the reflectivity R given by (cf. (9.7))

$$R = \left(\frac{n_{\mathrm{s}} - 1}{n_{\mathrm{s}} + 1} \right)^2 \ . \tag{20.7}$$

We note that the above formula is valid strictly for vertical incidence. For the GaAs/air interface, the surface reflectivity (for normal incidence) is about 30%. Thus, the external quantum efficiency for a LED is given by (1–R) and the critical angle by

$$\eta_{\mathrm{ex}} \cong \frac{4 n_1 n_{\mathrm{s}}}{(n_1 + n_{\mathrm{s}})^2} (1 - \cos\theta_{\mathrm{c}}) \approx \frac{4 n_{\mathrm{s}}}{(n_{\mathrm{s}} + 1)^2} (1 - \cos\theta_{\mathrm{c}}) \ . \tag{20.8}$$

The latter approximation is valid when the outer medium is air. For GaAs, the external quantum efficiency is $0.7 \times 4\% \approx 2.7\%$. Thus, typically only a small fraction of generated photons can leave the device and contribute to the LED emission.

20.3.4 Device Design

The following strategies have allowed significant improvement of LED efficiency:

Nonplanar Surfaces

With curved surfaces, the problem of total reflection can be (partially) circumvented. Spherically polished chips are feasible, but, very expensive. The epoxy seal of the standard LED case (Fig.20.6a) and its shape play a similar role, however, with a smaller index of refraction than the semiconductor, and are important for the beam shape.

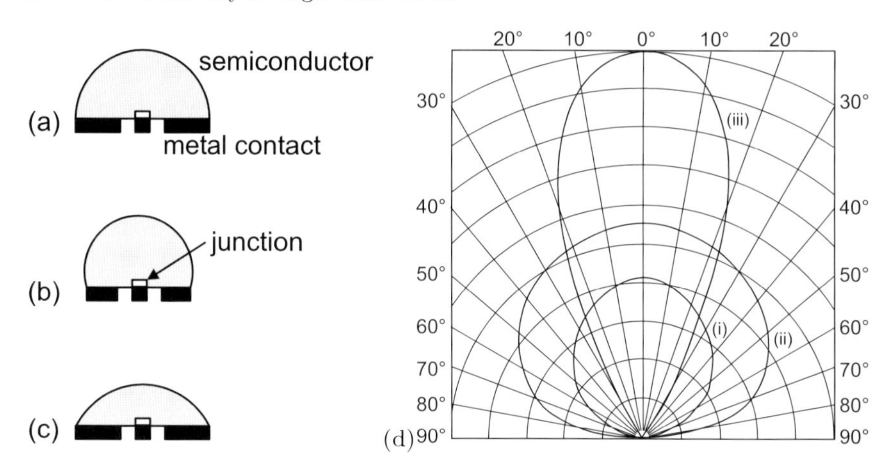

Fig. 20.5. Form of various LED casings with (**a**) hemispherical, (**b**) truncated sphere and (**c**) parabolic geometry. Adapted from [651]. (**d**) Emission characteristics for rectangular (**i**), hemispheric (**ii**) and parabolic (**iii**) geometry. Adapted from [652]

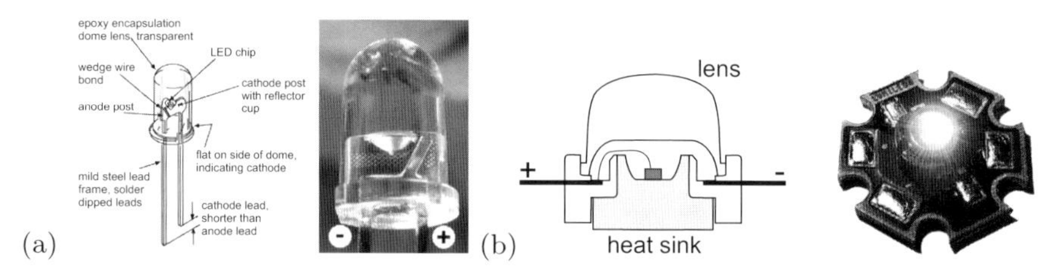

Fig. 20.6. (**a**) Standard LED casing (schematic drawing and macrophoto), (**b**) high-power mounting (schematic drawing and image of Luxeon® LED)

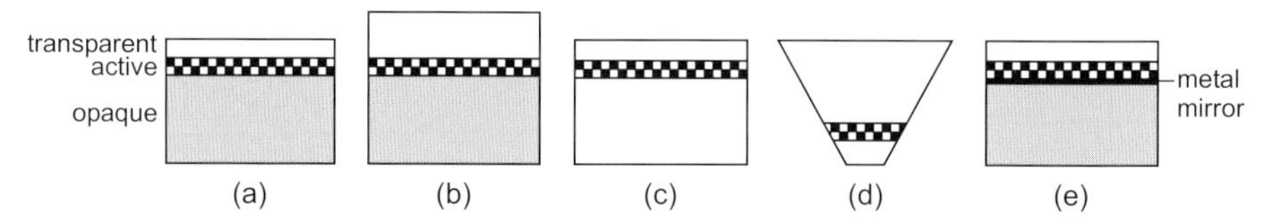

Fig. 20.7. (**a**) Standard LED layer sequence with opaque substrate (*grey*), active layer (*checkered*) and transparent top, (**b**) thick window design with thick top layer (50–70 µm). (**c**) Transparent substrate (by rebonding, see Fig. 20.8, (**d**) chip shaping (see also Fig. 20.10). (**e**) Thin-film LED with metal mirror (*black*) and rebonding (see also Fig. 20.11)

Thick-Window Chip Geometry

An increase in quantum efficiency to about 10–12% can be achieved if the top layer is fabricated with a much larger thickness (Fig. 20.7b) of 50–70 µm instead of a few µm.

Transparent Substrate

Reflection of photons is not so detrimental if they are not lost later due to absorption in the substrate. In Fig. 20.7, the evolution of LED chip design is shown schematically. In Fig. 20.8, the light path is compared for opaque and transparent substrates. The latter provides higher external quantum efficiency due to the 'photon recycling' effect. Efficiencies of 20–25% are possible. In Fig. 20.9, the technological steps are shown to fabricate a GaP LED with an AlGaInP active layer. The active layer is initially grown on GaAs due to lattice-match conditions.

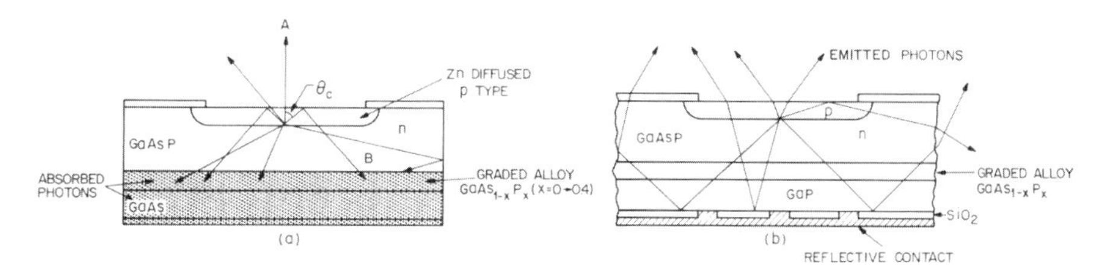

Fig. 20.8. Comparison of light path in a LED with (**a**) opaque and (**b**) transparent substrate. Reprinted with permission from [183], ©1981 Wiley

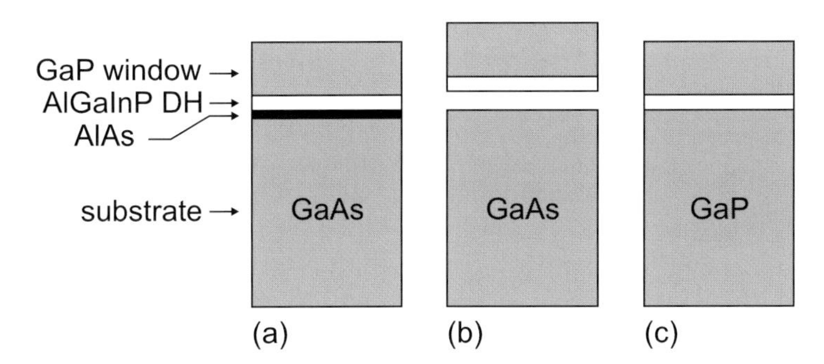

Fig. 20.9. Scheme of fabrication for red high brightness LED: (**a**) AlGaInP double heterostructure (DH) with GaP window on GaAs substrate (growth with MOCVD). (**b**) Lift-off using HF etch of sacrificial AlAs layer. (**c**) Wafer bonding on GaP (transparent for red light)

Nonrectangular Chip Geometry

If the chip is made with an inverted structure and mounted on a mirror, a very high external efficiency (> 50%) can be achieved. Typical commercial designs are shown in Fig. 20.10.

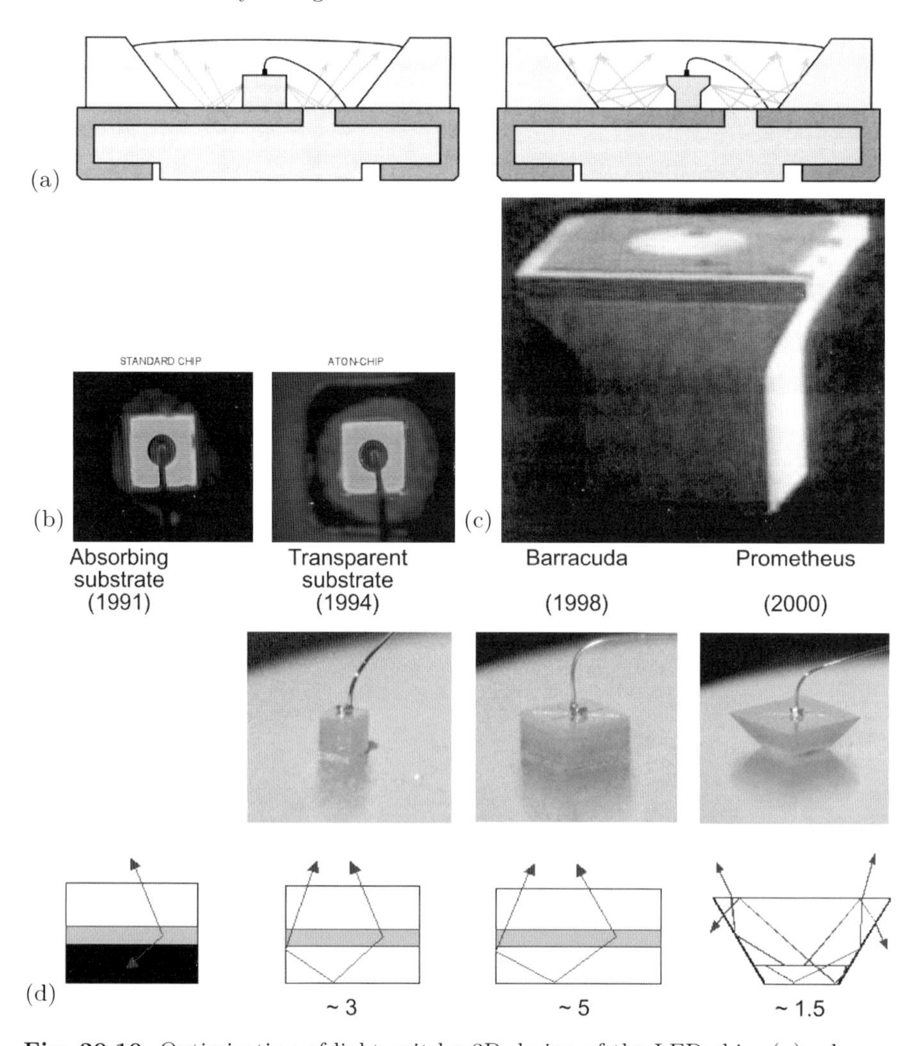

Fig. 20.10. Optimization of light exit by 3D design of the LED chip, (**a**) scheme, (**b**) emission pattern comparison and (**c**) SEM image of the ATON chip. Reprinted with permission from [653]. (**d**) Development stages towards the truncated inverted pyramid (Prometheus) chip. The lower row indicates the improvement of photon flux. Adapted from [654]

The increase in quantum efficiency allows the devices to run on much higher output power. While initially LEDs delivered power only in the mW regime, now output power in the $\sim 1\,\mathrm{W}$ regime is possible (high brightness LEDs). The higher currents made a redesign of the LED mount towards better heat sinks necessary (Fig. 20.6b). While the standard case has a thermal resistance of 220 K/W (chip size $(0.25\,\mathrm{mm})^2$ for 0.05–0.1 W and 0.2–2 lm), the high-power case has 15 K/W (chip size (0.5 mm)^2 for 0.5–2 W and 10–100 lm). An epoxy-free technique for encapsulation also enhances the color uniformity and maintains the brightness.

Thin-Film LED

In the thin-film LED design, as schematically shown in Fig. 20.11a, a metal mirror is evaporated on the LED layers. Subsequently, the metal side is wafer bonded to another metallized substrate and the original substrate is removed. Additionally, the LED surface can be patterned (before bonding) into an (hexagonal) array of (hexagonal) microprism mesas with an insulating (e.g. silicon nitride) layer with openings in order to optimize the current path. The microprisms are optimized to allow efficient reflection of light towards the emitting surface. This technology is scalable to large areas without loss in efficiency.

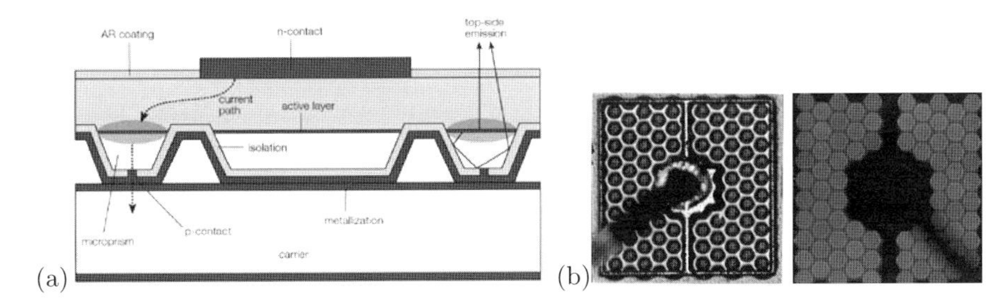

Fig. 20.11. (**a**) Scheme of thin-film LED with microprisms. (**b**) image and luminescence image of thin film AlInGaP LED (chip length: 320 μm). Reprinted with permission from [655]

White LEDs

There are different possibilities to generate white light with an LED as shown schematically in Fig. 20.12. The highest color gamut and a tunable white point can be achieved by combining a red, a green and a blue LED (Fig. 20.12a). Using a blue LED and a yellow phosphor (Figs. 20.12b and 20.13a,b), a white spectrum can be achieved that is, however, not very close to a blackbody spectrum (Fig. 20.13c). A better color rendering can be obtained with the combination of two phosphors. With an UV LED that is itself invisible (and must be shielded so no UV radiation leaves the LED), phosphors with various colors can be pumped (Fig. 20.12c). The mix of phosphors determines the white point.

Using a blue-emitting LED based on InGaN material, phosphors (similar to those used in fluorescence lamps) can be pumped. Blue light is converted into green, yellow or red light such that the resulting total spectrum appears white to the human eye. Also, a broad range of other colors can be designed (color on demand), e.g. pink or particular corporate colors. LumiLeds predicts that its Luxeon® LEDs (60 lumen/W, 5 W) will retain upwards of 90% of their light output after 9000 h and 70% through 50.000 h. Nichia targets 50% after 30 000 h.

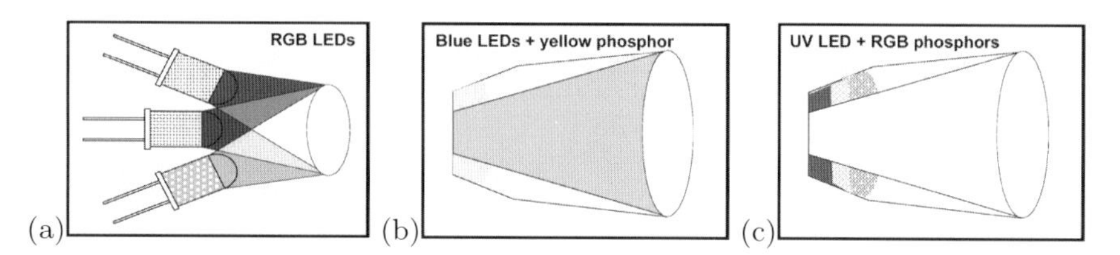

Fig. 20.12. Different strategies to generate white light with LEDs. (**a**) Addition of R, G, and B LEDs, (**b**) blue LED and yellow phosphor, (**c**) UV LED (invisible) and R, G, and B phosphors. From [654]

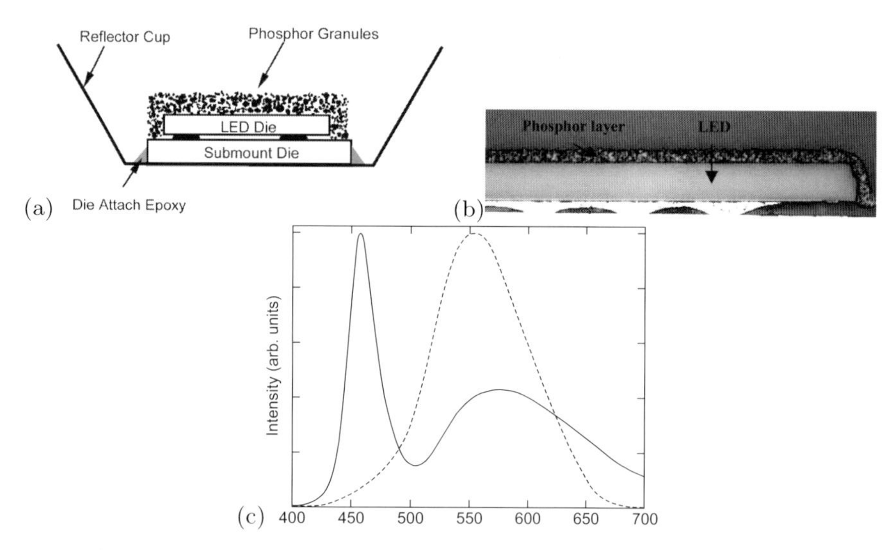

Fig. 20.13. (**a**) Scheme and (**b**) image of color conversion Luxeon® LED. From [654]. (**c**) Spectrum (*solid line*) of white LED with blue LED pumping a yellow phosphor together with eye-sensitivity curve $V(\lambda)$ (*dashed line*). Adapted from [655]

The color of a white LED depends on the operation conditions. In Fig. 20.14a the intensity vs. dc driving current characteristic of a white LED is shown. In Fig. 20.14b the chromaticity coordinates are shown for various dc currents. In order to avoid this effect, the LED is driven with pulses of a fixed amplitude and a repetition frequency that is high enough to provide a flicker-free image to the human eye, e.g. 100 Hz. The intensity of the LED is modulated via the pulse width, i.e. between 0–10 ms in this case (PWM, pulsewidth modulation).

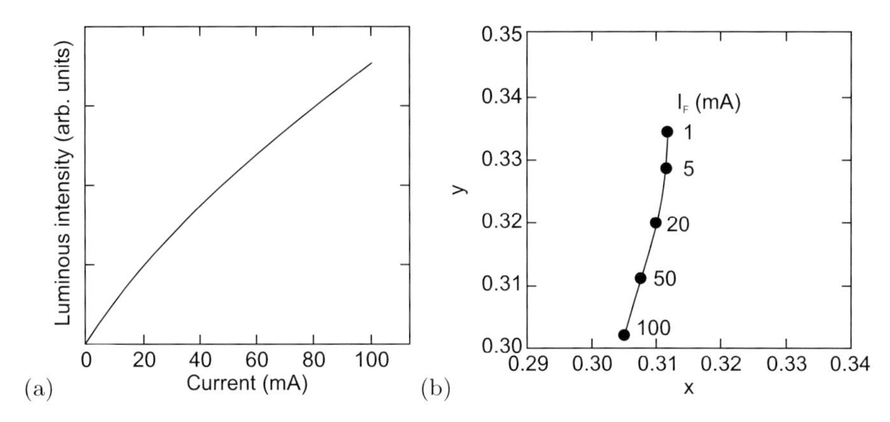

Fig. 20.14. (**a**) Luminous intensity of white LED (NSCW215) vs. dc forward current. (**b**) CIE chromaticity coordinates for various dc driving conditions as labeled. Data taken from [656]

Historic Development

In Fig. 20.15, the historic development of the LED efficiency is shown for various material systems. While the luminosity has increased by a factor of 20 per decade in the last 40 years, the price has decreased by a factor of ten per decade (Fig. 20.16). Currently, there is a need for the development of efficient LEDs in the green spectral range since their luminosity is small compared to devices for the blue and red spectral regions.

20.4 Lasers

20.4.1 Introduction

Semiconductor lasers[12] [658, 659] contain a zone (mostly called the *active layer*) that has gain if sufficiently pumped and that overlaps with an optical

[12]The term 'laser' is an acronym for 'light amplification by stimulated emission of radiation'. The amplification relies on stimulated emission, theoretically predicted by Einstein in 1917. The laser concept was first explored in the microwave wavelength region (1954, MASER using ammonia, Ch.H. Townes, Nobel prize 1964). The first optical laser (1958, US patent No. 2,929,922 awarded 1960, A.L. Schawlow, Ch.H. Townes) was the ruby laser developed in 1960 by Th. Maiman.

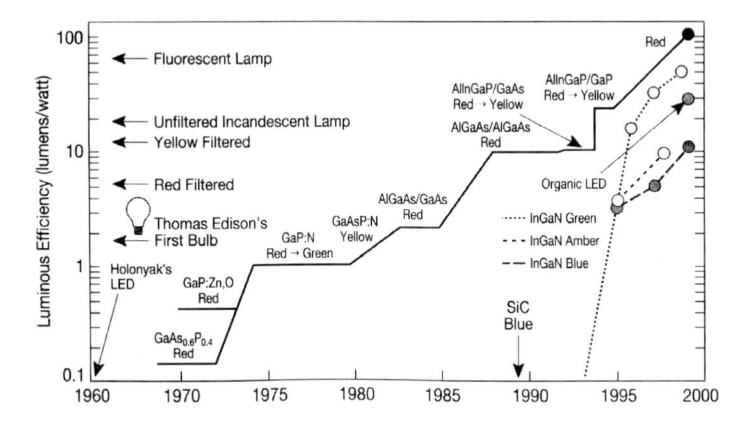

Fig. 20.15. Historic development of the efficiency of semiconductor LEDs. Reproduced from [657] by permission of the MRS Bulletin

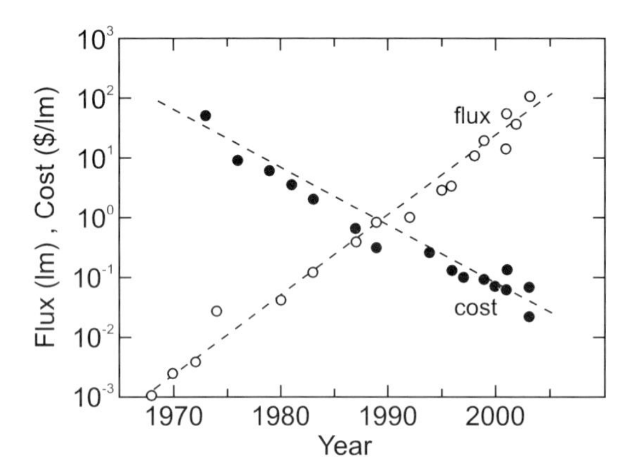

Fig. 20.16. Historic development of the flux (in lumen) and cost (in \$/lm) for semiconductor LEDs. Data from [654]

wave. The wave bounces back and forth in an optical cavity that leads to optical feedback. The part of the wave that exits the semiconductor forms the laser beam. Some of the first semiconductor lasers and a mounting design are shown in Fig. 20.18.

Generally, two main geometrical laser types, *edge* emitters (Fig. 20.19a) and *surface* emitters (Fig. 20.19b), are distinguished. The emission of the edge emitter exits through cleaved $\{110\}$ side facets[13] ($\approx 30\%$ reflectivity), of which an opposite pair acts as a Fabry–Perot optical cavity. The surface emission is directed along (001), since this is the (standard) growth direction of the heterostructure sequence making up the laser. The mirrors in a vertical-cavity surface-emitting laser (VCSEL) are made from dielectric Bragg mirrors

[13]or etched facets in possibly any direction.

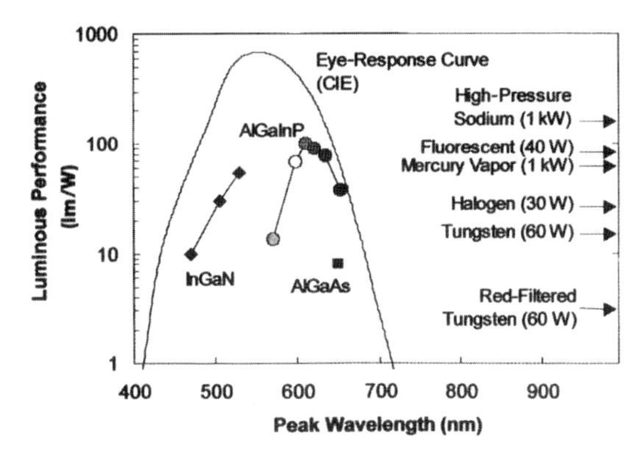

Fig. 20.17. Luminous performance of various LED materials. Adapted from [657], reprinted by permission of the MRS Bulletin

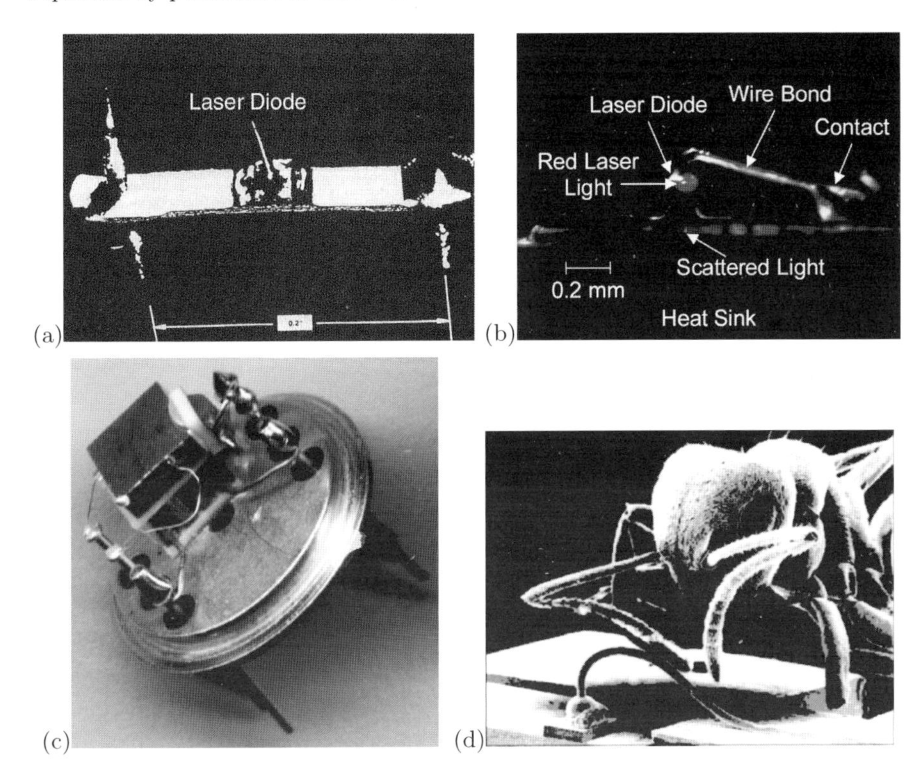

Fig. 20.18. Images of the first semiconductor lasers, 1962: (**a**) GaAs laser, Lincoln Laboratories and (**b**) GaInP laser, N. Holonyak and S.F. Bevacqua, Urbana Champaign. (**c**) Laser (at the end of gold bond wire) mounted on Peltier heat sink and a TO chip, Universität Leipzig. (**d**) Size comparison of an ant with a laser chip (underneath the bond wire)

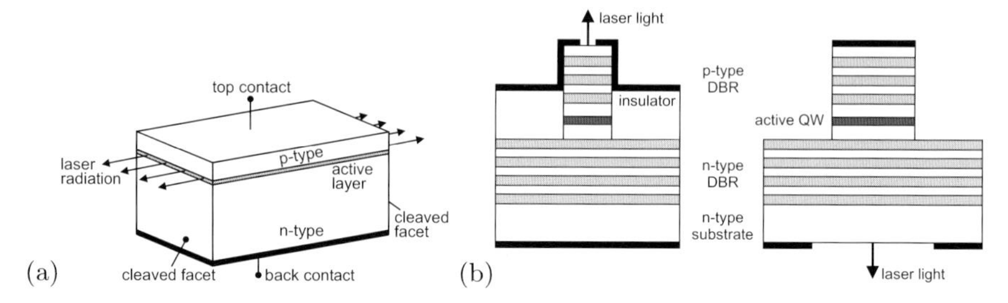

Fig. 20.19. (**a**) Schematic drawing of edge-emitting semiconductor laser. (**b**) Schematic drawings of vertical-cavity surface-emitting lasers with top emission (*left*) and emission through the substrate (*right*). *Black areas* are metal contacts

(see Sect. 17.1.4) with typically $R > 99.6\%$. Using antireflection coating on one facet, semiconductor lasers can be set up with an external cavity.[14] If both facets are antireflection coated, feedback is missing and the chip can be used as an optical amplifier (see Sect. 20.5).

Most lasers are pn-diodes and are then called laser diodes. They rely on the gain of interband transitions and the emission wavelength is determined and (more or less) given by the bandgap of the semiconductor. The cascade laser [660] (Sect. 20.4.16) is a unipolar structure with a superlattice as active layer. Here, the intersubband transitions (mostly in the conduction band but also in the valence band) carry the gain. The emission wavelength depends on the subband separation and lies typically in the far- and mid-infrared. Extensions to the THz regime and also to shorter wavelengths are possible. A third type of laser is the 'hot-hole' laser (Sect. 20.4.17), typically fabricated with p-doped Ge, which can be viewed as unipolar and functions only in a magnetic field; its emission is in the THz regime.

20.4.2 Applications

In Fig. 20.20, the revenue in the worldwide diode laser market is shown. The drop after 2000 is due to the burst of the 'internet bubble'. Nondiode laser (gas, ruby, excimer, ...) revenue is currently stable at around 2 billion US$, thus semiconductor lasers account for the largest share of all laser types sold.

The following applications rely on semiconductor lasers:

- most semiconductor lasers are fabricated for optical communication as senders
- optical information storage and retrieval (CD, DVD) with as short of a wavelength as possible, as shown in Fig. 20.21, currently about 400 nm.
- pumping of solid-state laser, typically 910 or 940 nm for pumping Nd:YAG

[14]Such external cavities can be used for manipulation of the laser properties such as wavelength tuning.

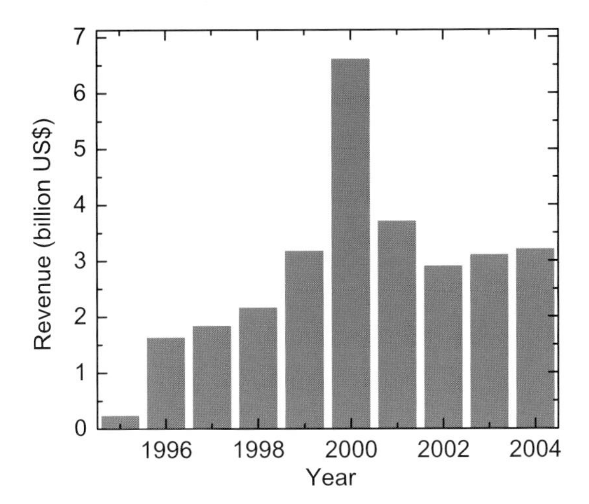

Fig. 20.20. Revenue in worldwide diode laser market. Based on numbers from [661]

- laser TV for which efficient red, green and blue lasers will be needed. A mass product can only be developed if cheap laser diodes are available as emitters.
- laser pointers, see Fig. 20.22. A red laser pointer simply uses the collimated red emission of a GaAs-based diode. In a green laser pointer, an infrared diode pumps a Nd:YAG or Nd:YVO_4 crystal. The emitted beam is then frequency doubled, typically with a $KTiOPO_4$ (KTP) crystal.
- medical instruments with a variety of wavelengths.
- remote control, position detection, distance measurement.

The market for photonic devices is much more dynamic than the electronics market. An example is the rapid change of dominating laser applications. For diode lasers, the most prominent single application moved from telecommunication (77% market share in 2000, 25% in 2003) to optical data storage (17% market share in 2000, 60% in 2003).

20.4.3 Gain

Due to current injection,[15] a nonequilibrium carrier distribution is created. After fast thermalization processes (phonon scattering), it can mostly described by quasi-Fermi levels. Sufficiently strong pumping leads to inversion, i.e. conduction-band states are more strongly populated with electrons than valence-band states (Fig. 20.23). In this case, the stimulated emission rate is stronger than the absorption rate (see Sect. 10.2.6). The thermodynamic

[15] or due to optical pumping. If electrical contacts are not available, the laser action can be invoked by supplying a high-intensity light beam, possibly in a stripe-like shape. For optically pumped semiconductor lasers see Sect. 20.4.15.

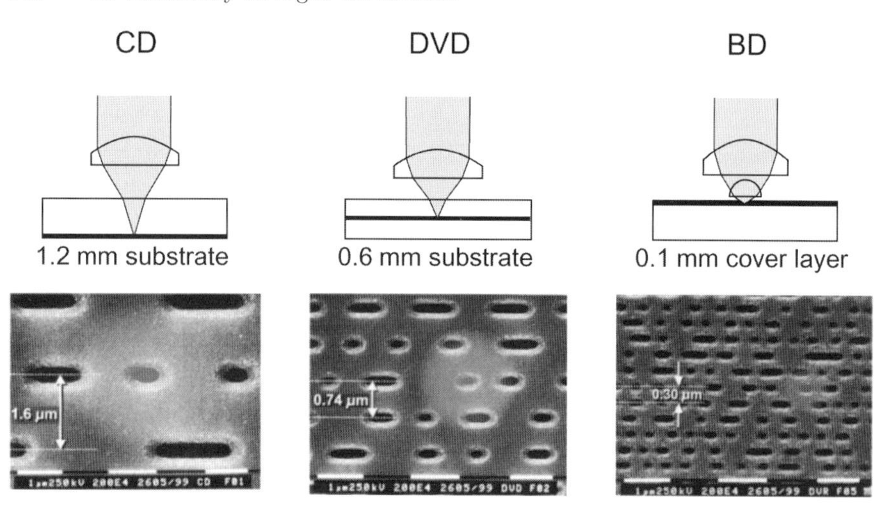

Fig. 20.21. Evolution of optical data storage technology, 'CD': compact disk (laser: 780 nm, pitch: 1.6 µm, capacity: 0.7 GB), 'DVD': digital versatile disk (laser: 635–650 nm, pitch: 0.74 µm, capacity: 4.7 GB for one layer), 'BD': 'Blu-ray' disk (laser: 405 nm, pitch: 0.32 µm, capacity: 27 GB for one layer)

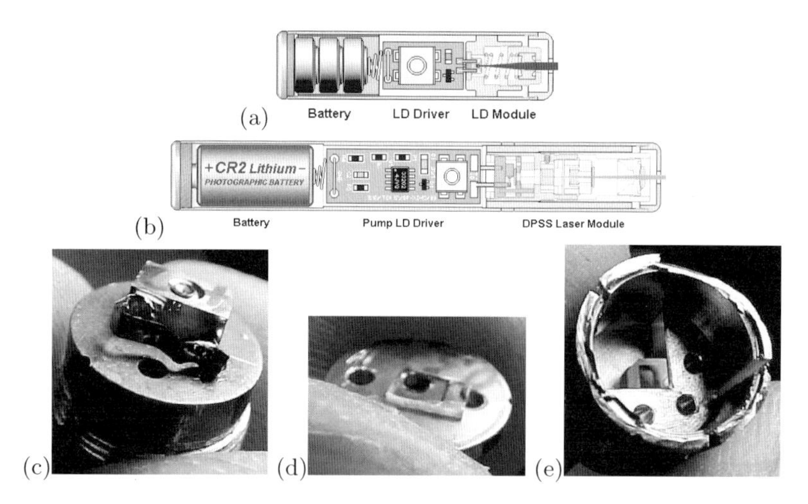

Fig. 20.22. (**a**) Scheme of red laser pointer, (**b**) scheme of green laser pointer. Parts of a green laser pointer: (**c**) pump laser diode, (**d**) YVO_4 crystal, (**e**) KTP doubler

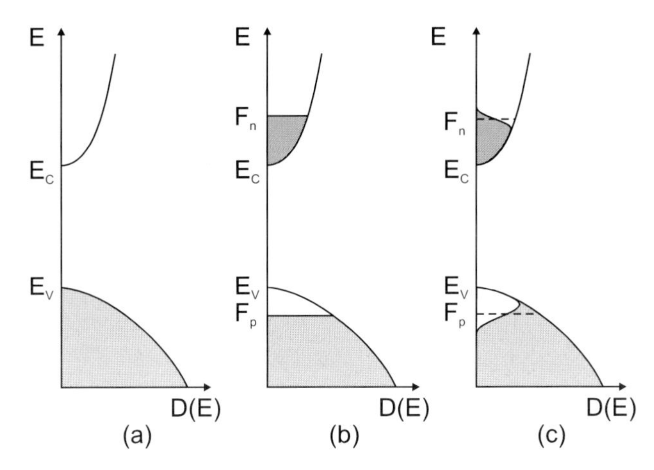

Fig. 20.23. Population (**a**) in thermodynamic equilibrium $T = 0\,\mathrm{K}$, (**b**) under inversion for $T = 0\,\mathrm{K}$, (**c**) under inversion for $T > 0\,\mathrm{K}$. *Shaded areas* are populated with electrons

laser condition (cf. (10.23)) requires the difference of the quasi-Fermi levels to be larger than the bandgap.

$$F_\mathrm{n} - F_\mathrm{p} > E_\mathrm{g} \tag{20.9}$$

The gain is defined as the (frequency-dependent) coefficient $g(\hbar\omega)$ that describes the light intensity along a path L according to

$$I(L) = I(0)\exp(gL) \; . \tag{20.10}$$

The gain spectrum as a function of the photon energy $\hbar\omega$ is given for non-k-conserving recombination by (cf. (10.5) and (10.6))

$$g(\hbar\omega) = \int_0^{\hbar\omega - E_\mathrm{g}} D_\mathrm{e}(E) D_\mathrm{h}(E') \left[f_\mathrm{e}(E) f_\mathrm{h}(E') - (1 - f_\mathrm{e}(E))(1 - f_\mathrm{h}(E')) \right] \mathrm{d}E \; , \tag{20.11}$$

with $E' = \hbar\omega - E_\mathrm{g} - E$. The gain is positive for those photon energies for which light is amplified and negative for those that are absorbed. In Fig. 20.24a, the electron and hole concentrations are shown for GaAs as a function of the quasi-Fermi energies. In Fig. 20.24b, the difference of the quasi-Fermi energies is shown as a function of the carrier density (for neutrality $n = p$). The gain spectrum is shown in Fig. 20.24c for a two-band model.[16] In the case of inversion, the gain is positive for energies between E_g and $F_\mathrm{n} - F_\mathrm{p}$. At $\hbar\omega = F_\mathrm{n} - F_\mathrm{p}$, the gain is zero (transparency) and for larger energies negative (positive absorption coefficient).

[16]One electron and one hole band are considered; the heavy and light hole bands are taken into account via the mass according to (7.14).

For a given fixed energy, the gain increases with increasing pumping and increasing carrier density n (Fig. 20.24d). For very small density, it is given as $g(n \to 0) = -\alpha$. The gain rises around transparency approximately linearly with the pumping intensity. At transparency carrier density n_{tr}, the gain is zero. Therefore, the relation $g(n)$ can be approximated as (linear gain model)

$$g(n) \approx \hat{\alpha} \frac{n - n_{\mathrm{tr}}}{n_{\mathrm{tr}}} . \tag{20.12}$$

For large carrier density, the gain saturates (at a value similar to α).

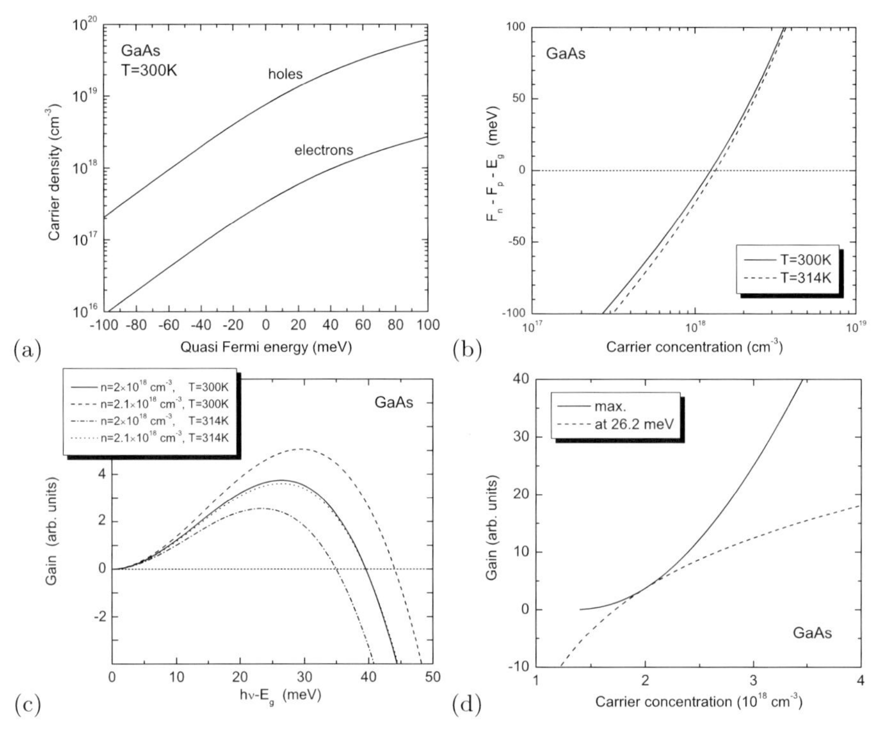

Fig. 20.24. Gain in the two-band model for GaAs. (**a**) Electron and hole concentrations at $T = 300\,\mathrm{K}$ as a function of the quasi-Fermi energies counted relative to the band edges, i.e. $F_{\mathrm{n}} - E_{\mathrm{C}}$ and $E_{\mathrm{V}} - F_{\mathrm{p}}$. (**b**) Difference of quasi-Fermi levels as a function of carrier concentration ($n = p$) for GaAs at two different temperatures. (**c**) Gain spectra according to (20.11) for $n = 2 \times 10^{18}$ and $T = 300\,\mathrm{K}$ (*solid line*), increased carrier density $n = 2.1 \times 10^{18}$ and $T = 300\,\mathrm{K}$ (*dashed line*), higher temperature $n = 2 \times 10^{18}$ and $T = 314\,\mathrm{K}$ (*dash-dotted line*) and same difference of the quasi-Fermi levels as for the *solid line*, $n = 2.1 \times 10^{18}$ and $T = 314\,\mathrm{K}$ (*dotted line*). (**d**) Maximum gain (*solid line*) and gain at a particular energy (*dashed line*), for photon energy $E_{\mathrm{g}}+ 26.2\,\mathrm{meV}$ for which the gain is maximal for $n = 2 \times 10^{18}$ and $T = 300\,\mathrm{K}$, see *solid line* in part (**c**)

20.4.4 Optical Mode

The light wave that is amplified must be guided in the laser. An optical cavity is needed to provide optical feedback such that the photons travel several times through the gain medium and contribute to amplification. We explain the light-wave management for the edge emitter first:

Vertical mode guiding

In the course of the historical development of the semiconductor laser, the most significant improvements (reduction of lasing threshold current) have

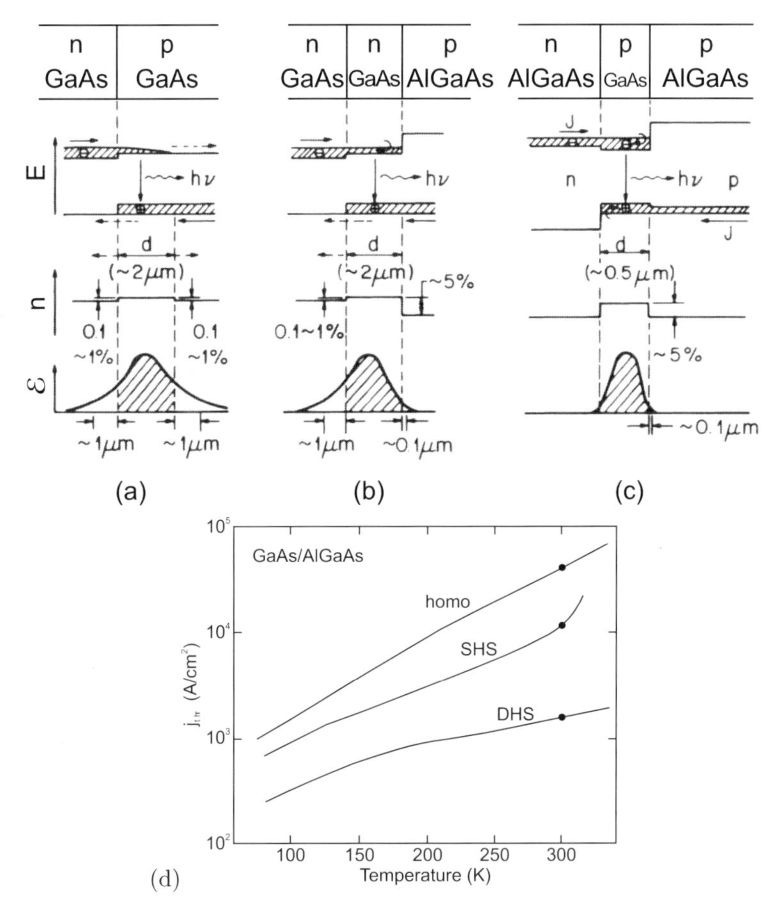

Fig. 20.25. Laser with (**a**) homojunction, (**b**) single heterostructure (SHS), (**c**) double heterostructure ('DHS'), (**d**) reduction of threshold current with design progress ('SHS': d =2 µm, 'DHS': d =0.5 µm). Adapted from [662]

been achieved through the improvement of the overlap of the optical wave with the gain medium, as shown in Fig. 20.25. From homojunctions over the single heterojunction, eventually the double heterostructure (DHS) design could reduce the laser threshold current density to the $1\,\mathrm{kA/cm^2}$ level.

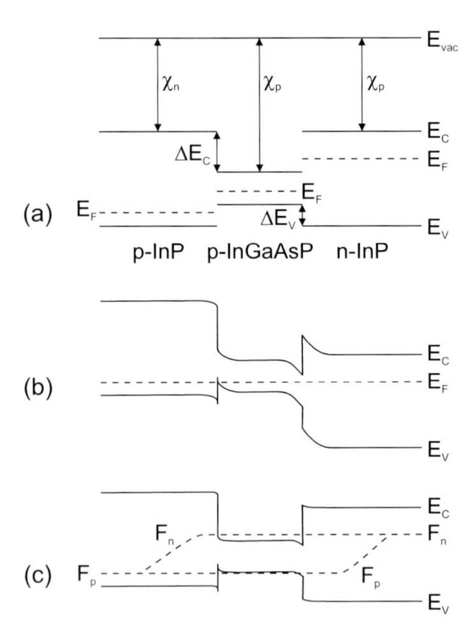

Fig. 20.26. Schematic band diagram of a pn double heterostructure (DHS) diode (InP/InGaAsP/InP) (**a**) before contact of the materials, (**b**) in thermodynamic equilibrium (zero bias, *dashed line* is Fermi level E_F = const.), (**c**) with forward bias close to flat-band conditions, *dashed lines* are quasi-Fermi levels

The band diagram of a double heterostructure is shown in Fig. 20.26 for zero and forward bias. In the DHS, the optical mode is guided by total reflection within the low-bandgap center layer, which has a larger index of refraction than the outer, large-bandgap layer.[17] When the layer thickness is in the range of λ/n_r, the form of the optical mode must be determined from the (one-dimensional) wave equation (Helmholtz equation)

$$\frac{\partial^2 \mathcal{E}_\mathrm{z}}{\partial z^2} + \omega^2 \mu\, \epsilon(z) \mathcal{E}_\mathrm{z} = 0\,. \tag{20.13}$$

In Fig. 20.27a, the shape of the optical mode for GaAs/$\mathrm{Al_{0.3}Ga_{0.7}As}$ DHS with different GaAs thickness is shown.

The optical confinement factor Γ is the part of the wave that has geometrical overlap with the gain medium, i.e. is subject to amplification. It is

[17] A smaller bandgap coincides for many cases with a larger index of refraction.

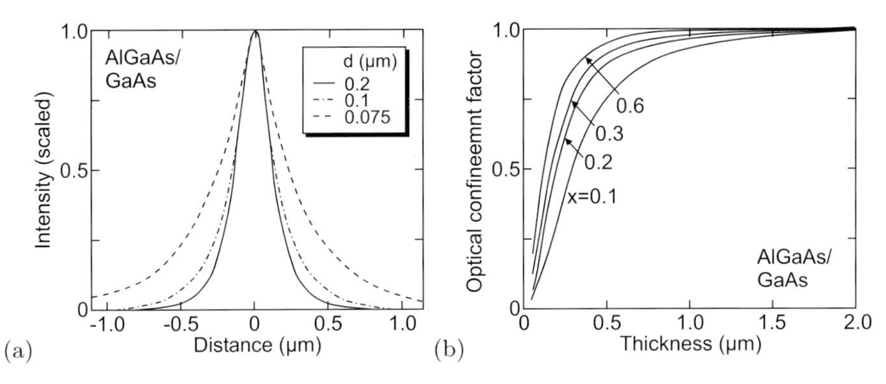

Fig. 20.27. (**a**) Optical mode (relative intensity) for various values of the thickness d of the active layer as labeled of a GaAs/$Al_{0.3}Ga_{0.7}As$ DHS laser, (**b**) Optical confinement factor Γ as a function of the thickness of the active layer and the Al concentration x of the barrier as labeled in a GaAs/$Al_xGa_{1-x}As$ DHS laser. Adapted from [658]

shown for GaAs/$Al_xGa_{1-x}As$ DHS with different GaAs thickness and different Al concentration in Fig. 20.27b. The modal gain g_{mod} that is responsible for light amplification in the cavity consists of the material gain g_{mat} due to inversion and the optical confinement factor.

$$g_{\mathrm{mod}} = \Gamma g_{\mathrm{mat}} \; . \tag{20.14}$$

In order to allow simultaneous optimization of the light mode and the carrier confinement, the separate confinement heterostructure (SCH) has been designed. Here, a single or multiple quantum well of a third material with even smaller bandgap is the active medium (Fig. 20.28a,b,d). A single quantum well has an optical confinement factor of a few per cent only. It offers, however, efficient carrier capture and efficient radiative recombination. An increase in the carrier capture efficiency can be achieved using a graded index in the barrier (GRINSCH, Fig. 20.28c).

The thin waveguiding layer leads to large divergence of the laser beam along the vertical direction, typically about 90°. The strong confinement of light also limits the maximum achievable output power due to catastrophic optical damage (COD). Several ideas have been realized to overcome this problem and achieve much smaller divergence of about 18°. The waveguide can be designed to be very thick (large optical cavity, LOC) that leads to an increase of threshold. Other schemes are insertion of a low-index layer into the confinement layer, insertion of a high-index layer into the cladding layer or the use of high-index quarter-wavelength reflecting layers [663].

Lateral Mode Guiding

Lateral waveguiding can be achieved with gain guiding and index guiding (or a mixture of the two). In the gain-guiding scheme (Fig. 20.29), the current

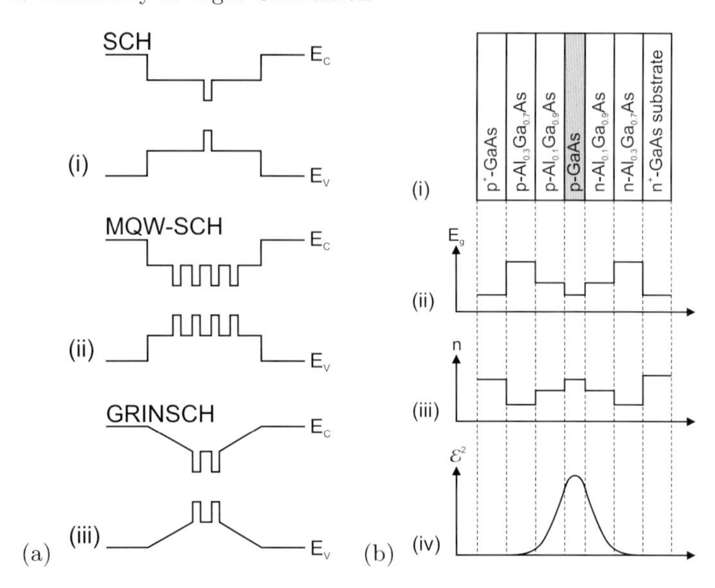

Fig. 20.28. (**a**) Various geometries of the active layer of a DHS laser with quantum wells as active medium, (**i**) single QW (separate confinement heterostructure, SCH), (**ii**) multiple QW SCH, and (**c**) GRINSCH (graded-index SCH) structure. (**b**) Layer sequence for a separate confinement heterostructure laser

path that is defined by the stripe contact and the current spreading underneath it, defines the gain region and therefore the volume of amplification that guides the optical wave. Since a high carrier density reduces the index of refraction, a competing antiguiding effect can occur. For index guiding, the lateral light confinement is caused by a lateral increase of the index of refraction. This index modulation can be achieved by using a mesa-like contact stripe (Fig. 20.30a). A shallow mesa reaches down into the upper cladding,

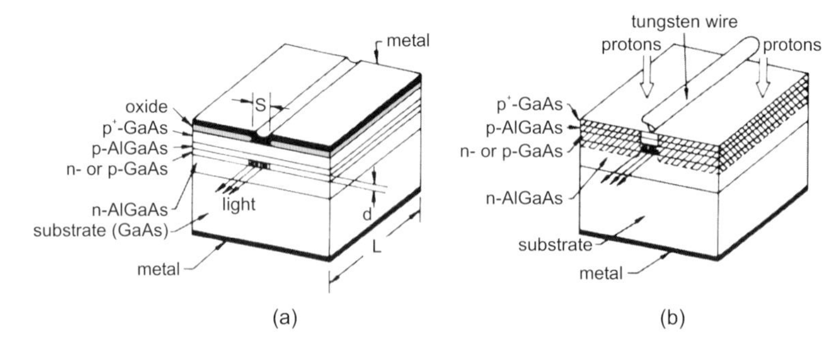

Fig. 20.29. Scheme of gain-guided lasers with stripe contact: (**a**) oxide stripe, (**b**) proton implanted with shadow mask from tungsten wire ($\sim 10\,\mu$m). Adapted from [664]

a deep mesa reaches down into or through the active layer. Possible problems with surface recombination can be avoided by regrowth of the structure (Fig. 20.30b) with a wide-bandgap material (compared to the active layer). Optimization of regrowth is targeted to achieve a well-defined surface for subsequent contact processing. A lateral pn-diode can be incorporated that avoids current spreading in the upper part of the structure.

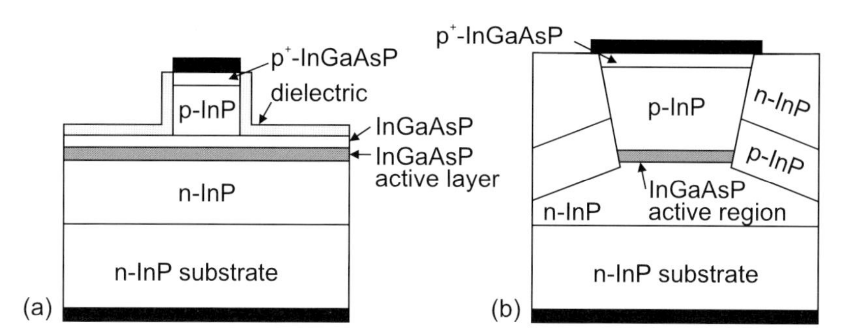

Fig. 20.30. Schematic cross section of index-guided lasers: (**a**) shallow ridge, (**b**) deep etch and regrowth. *Black areas* are metal contacts

Depending on the width of the lateral mode, it can be monomode or multimode (Fig. 20.31a). For laterally monomode lasers, the stripe width may only be a few µm. In particular for such lasers, the current spreading must be controlled. Problems can arise for wide stripe widths due to current filamentation and inhomogeneous laser emission from the facet. Since the optical mode is typically more confined in the growth direction than in the lateral direction, the far field is asymmetric (Fig. 20.31c). The vertical axis has the higher divergence and is called the fast axis. The lateral axis is called the slow axis. The asymmetric beam shape is detrimental when the laser needs to be coupled into an optical fiber or a symmetric beam profile is needed for subsequent optics. The beam can be made symmetric using special optic components such as anamorphic prisms (Fig. 20.31d) and graded-index lenses. The beam from a laterally monomode laser is diffraction limited and can therefore generally be refocused efficiently (beam quality $M^2 \gtrsim 1$).

Longitudinal Modes

The spectral positions of the laser modes for a cavity with length L is given by the condition (cf. (17.34))

$$L = \frac{m\,\lambda}{2n_\mathrm{r}(\lambda)} \,, \tag{20.15}$$

where m is a natural number and $n(\lambda)$ is the dispersion of the index of refraction. The distance of neighboring modes is given by (for large m)

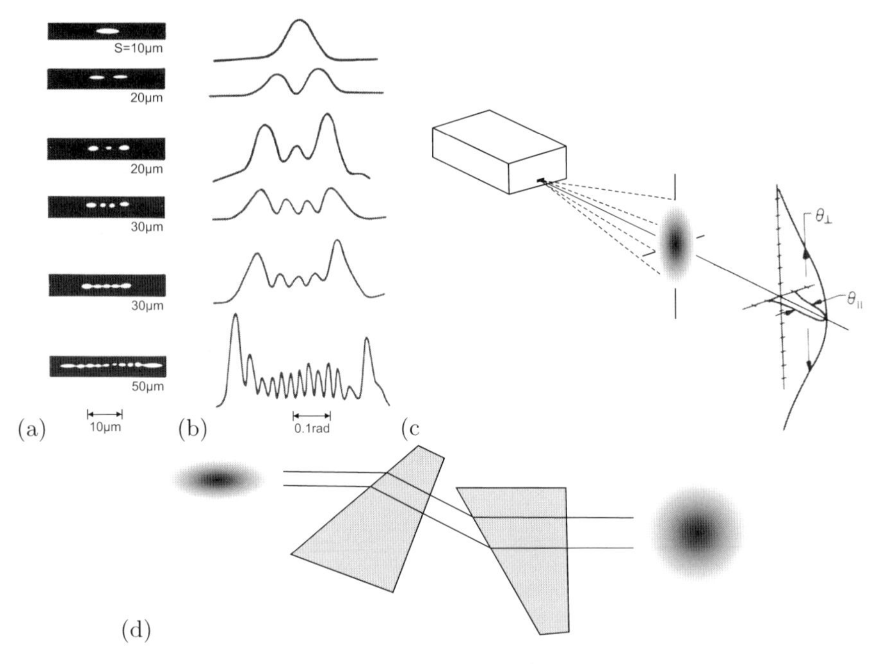

Fig. 20.31. Lateral near field (**a**) and far field (**b**) of lasers with various width S of the injection stripe as labeled. Adapted from [665]. (**c**) Typical asymmetric far field of an edge emitter. Adapted from [183]. (**d**) Correction of asymmetric far field with a pair of anamorphic prisms

$$\Delta\lambda = \frac{\lambda^2}{2n_\mathrm{r}L\left(1 - \frac{\lambda}{n_\mathrm{r}}\frac{\mathrm{d}n_\mathrm{r}}{\mathrm{d}\lambda}\right)} . \tag{20.16}$$

The dispersion $\mathrm{d}n_\mathrm{r}/\mathrm{d}\lambda$ can sometimes be neglected.

The facets of edge-emitting lasers are typically cleaved. Cleaving bears the danger of mechanical breakage and tends to have poor reproducibility, low yield and therefore high cost. Etched facets are another possibility to form the cavity mirror. The etch process, typically reactive ion dry etching, must yield sufficiently smooth surfaces to avoid scattering losses. A highly efficient distributed Bragg mirror (see Sect. 17.1.4) with only a few periods can be created by using the large index contrast between the semiconductor and air. As shown in Fig. 20.32, slabs can be etched that make a Bragg mirror with the air gaps [666]. In this way, very short longitudinal cavities can be made ($\approx 10\,\mu$m).

20.4.5 Loss Mechanisms

While the light travels through the cavity, it is not only amplified but it also suffers losses. The internal loss α_i and the mirror loss α_m contribute to the

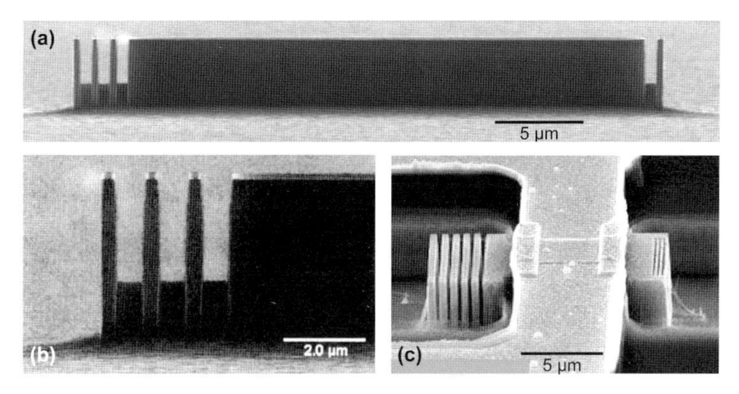

Fig. 20.32. SEM image of (**a**) an InP microlaser with third-order Bragg mirrors, (**b**) magnified view of the front facet with three slabs, (**c**) a 12-µm long microlaser with five third-order mirrors on the rear side and three first-order mirrors on the front side with top contact. From [667]; part (**b**) reprinted with permission from [666], ©2001 AIP

total loss α_{tot}

$$\alpha_{\mathrm{tot}} = \alpha_{\mathrm{i}} + \alpha_{\mathrm{m}} \, . \tag{20.17}$$

The internal loss is due to absorption in the cladding, scattering at waveguide inhomogeneities and possibly other processes. It can be written as

$$\alpha_{\mathrm{i}} = \alpha_0 \Gamma + \alpha_{\mathrm{g}}(1-\Gamma) \, , \tag{20.18}$$

where α_0 is the loss coefficient in the active medium and α_{g} is the loss coefficient outside the active medium.

The mirror loss is due to the incomplete reflection of the optical wave at the laser facets. This condition is necessary, however, to observe a laser beam outside the cavity. R_1 and R_2 are the reflectivities of the two facets, respectively. An as-cleaved facet has a reflectivity of about 30% (cf. (20.7)). Using dielectric layers on the facets, the reflectivity can be increased (high reflection, HR-coating) or decreased (antireflection, AR-coating). One round-trip through the cavity of length L has the length $2L$. The intensity loss due to reflection at the facets is expressed via $\exp(-2\alpha_{\mathrm{m}} L) = R_1 \, R_2$

$$\alpha_{\mathrm{m}} = \frac{1}{2L} \ln\left(\frac{1}{R_1 R_2}\right) \, . \tag{20.19}$$

If both mirrors have the same reflectivity R, we have $\alpha_{\mathrm{m}} = -L^{-1} \ln R$. For $R = 0.3$ a 1-mm cavity has a loss of $12\,\mathrm{cm}^{-1}$. For the internal loss a typical value is $10\,\mathrm{cm}^{-1}$, very good waveguides go down to 1–$2\,\mathrm{cm}^{-1}$.

Lasing is only possible if the gain overcomes all losses (at least for one wavelength), i.e.

$$g_{\mathrm{mod}} = g_{\mathrm{mat}} \Gamma \geq \alpha_{\mathrm{tot}} \, . \tag{20.20}$$

20.4.6 Threshold

When the laser reaches threshold, the (material) gain is pinned at the threshold value

$$g_{\mathrm{thr}} = \frac{\alpha_{\mathrm{i}} + \alpha_{\mathrm{m}}}{\Gamma} \ . \tag{20.21}$$

Since $g \propto n$, the carrier density is also pinned at its threshold value and does not increase further with increasing injection current. Instead, additional carriers are quickly converted into photons by stimulated emission. The threshold carrier density is (using the linear gain model, cf. (20.12))

$$n_{\mathrm{thr}} = n_{\mathrm{tr}} + \frac{\alpha_{\mathrm{i}} + \alpha_{\mathrm{m}}}{\hat{\alpha}\Gamma} \ . \tag{20.22}$$

For an active layer of thickness d, the threshold current density is

$$j_{\mathrm{thr}} \cong \frac{edn_{\mathrm{thr}}}{\tau(n_{\mathrm{thr}})} \ , \tag{20.23}$$

where $\tau_{\mathrm{e}}(n_{\mathrm{thr}})$ is the minority carrier lifetime at the threshold carrier density:

$$\tau(n_{\mathrm{thr}}) = \frac{1}{A + Bn_{\mathrm{thr}} + Cn_{\mathrm{thr}}^2} \ . \tag{20.24}$$

Using (20.22), we can write (for $R = R_1 = R_2$)

$$j_{\mathrm{thr}} = j_{\mathrm{tr}} + \frac{ed}{\tau\hat{\alpha}\Gamma}\left(\alpha_{\mathrm{i}} - \ln R \frac{1}{L}\right) \ , \tag{20.25}$$

where the transparency current density is $j_{\mathrm{tr}} = edn_{\mathrm{tr}}/\tau$. Thus, the plot of j_{thr} vs. $1/L$ (or the optical loss) should be linear and can be extrapolated towards the transparency current density (see Fig. 20.33a).

Any additional increase of the current j leads to stimulated emission with the rate

$$r_{\mathrm{st}} = edv_{\mathrm{g}}g_{\mathrm{thr}}N_{\mathrm{ph}} \ , \tag{20.26}$$

where v_{g} is the group velocity (mostly c_0/n_{r}) and N_{ph} is the photon density (per length) in the cavity. The photon density increases linearly beyond the threshold

$$N_{\mathrm{ph}} = \frac{1}{dv_{\mathrm{g}}g_{\mathrm{thr}}}\left(j - j_{\mathrm{thr}}\right) \ . \tag{20.27}$$

The photon lifetime

$$\frac{1}{\tau_{\mathrm{ph}}} = v_{\mathrm{g}}\left(\alpha_{\mathrm{i}} + \alpha_{\mathrm{m}}\right) = v_{\mathrm{g}}\Gamma g_{\mathrm{thr}} \tag{20.28}$$

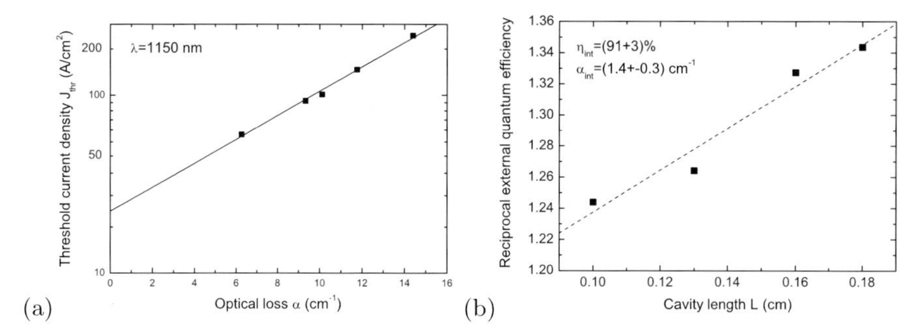

Fig. 20.33. (**a**) Threshold current density for (three-fold InGaAs/GaAs QD stack) laser ($\lambda = 1150$ nm) at 10°C with different cavity length vs. the optical loss ($\propto 1/L$). The extrapolated transparency current density is $21.5 \pm 0.9\,\mathrm{A/cm^2}$. (**b**) Inverse external quantum efficiency for the $L = 1$ mm QD laser vs. cavity length. The internal quantum efficiency determined from the plot is 91% and the internal loss is $1.4\,\mathrm{cm}^{-1}$

is introduced that describes the loss rate of photons. $v_g\alpha_m$ describes the escape rate of photons from the cavity into the laser beam(s). Therefore,

$$N_{ph} = \frac{\tau_{ph}\Gamma}{ed}\left(j - j_{thr}\right) . \tag{20.29}$$

Since the threshold depends on the carrier density, it is advantageous to reduce the active volume further and further. In this way, the same *amount* of injected carriers creates a larger carrier *density*. Figure 20.34 shows the historic development of laser threshold due to design improvements.

20.4.7 Spontaneous Emission Factor

The spontaneous emission factor β is the fraction of spontaneous emission (emitted into all angles) that is emitted into laser modes. For Fabry–Perot lasers, β is typically in the order of 10^4–10^5. The design of a microcavity can increase β drastically by several orders of magnitude to ≈ 0.1 [668] or above and thus reduce the threshold current. The photon number as a function of the pump current can be calculated from the laser rate equations and is depicted in Fig. 20.35. For $\beta = 1$, all emitted power goes into the laser mode regardless of whether emission is spontaneous or stimulated. The definition of threshold in such 'nonclassical' lasers with large β is discussed in detail in [669].

20.4.8 Output Power

The output power is given by the product of photon energy, the photon density in the cavity, the effective mode volume and the escape rate:

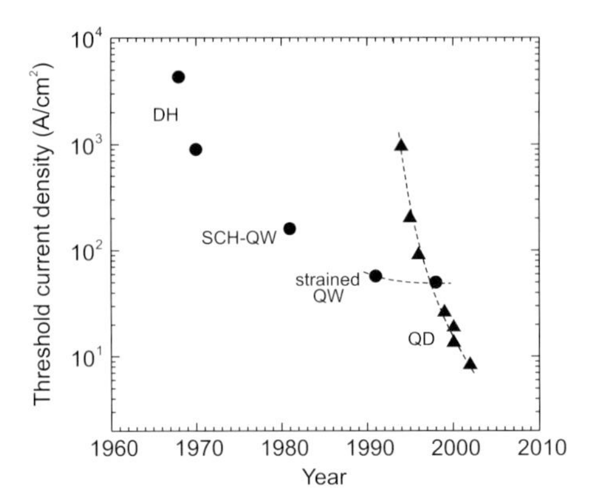

Fig. 20.34. Historic development of threshold current density (at room temperature, extrapolated for infinite cavity length and injection stripe width) for various laser designs, 'DH': double heterostructure, 'SCH–QW': separate confinement heterostructure with quantum wells. 'QD': quantum dots. *Dashed lines* are guides to the eye

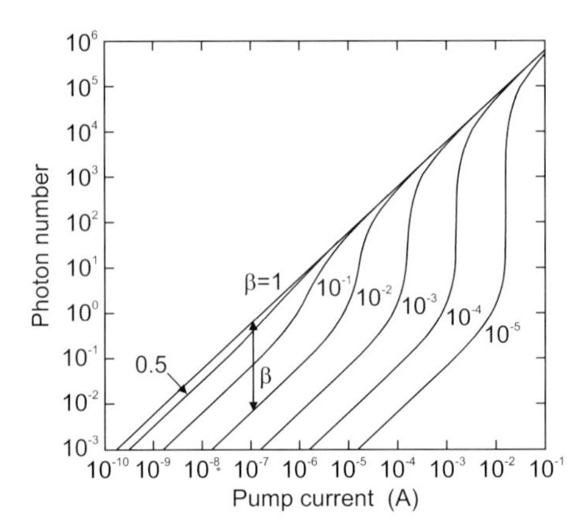

Fig. 20.35. Photon number vs. pump current for a model laser. Adapted from [670]

$$P_{\mathrm{out}} = \hbar\omega N_{\mathrm{ph}} \frac{Lwd}{\Gamma} v_{\mathrm{g}} \alpha_{\mathrm{m}} . \tag{20.30}$$

Thus, it is given by

$$P_{\mathrm{out}} = \hbar\omega v_{\mathrm{g}} \alpha_{\mathrm{m}} \frac{\tau_{\mathrm{ph}}}{e} Lw \left(j - j_{\mathrm{thr}}\right) = \frac{\hbar\omega}{e} \frac{\alpha_{\mathrm{m}}}{\alpha_{\mathrm{m}} + \alpha_{\mathrm{i}}} \left(I - I_{\mathrm{thr}}\right) . \tag{20.31}$$

To this equation, the factor η_{int} must be added. The internal quantum efficiency describes the efficiency of the conversion of electron–hole pairs into photons:

$$\eta_{\mathrm{int}} = \frac{Bn^2 + v_{\mathrm{g}} g_{\mathrm{thr}} N_{\mathrm{ph}}}{An + Bn^2 + Cn^3 + v_{\mathrm{g}} g_{\mathrm{thr}} N_{\mathrm{ph}}} . \tag{20.32}$$

All in all, now (see Fig. 20.36a)

$$P_{\mathrm{out}} = \frac{\hbar\omega}{e} \frac{\alpha_{\mathrm{m}}}{\alpha_{\mathrm{m}} + \alpha_{\mathrm{i}}} \eta_{\mathrm{int}} \left(I - I_{\mathrm{thr}}\right) . \tag{20.33}$$

The differential (or slope) quantum efficiency, also called the external quantum efficiency η_{ext}, is the slope of the P_{out} curve vs. the current in the lasing regime. It is given by

$$\eta_{\mathrm{ext}} = \frac{\mathrm{d}P_{\mathrm{out}}/\mathrm{d}I}{\hbar\omega/e} = \eta_{\mathrm{int}} \frac{\alpha_{\mathrm{m}}}{\alpha_{\mathrm{m}} + \alpha_{\mathrm{i}}} . \tag{20.34}$$

The external quantum efficiency can also be written as

$$\frac{1}{\eta_{\mathrm{ext}}} = \frac{1}{\eta_{\mathrm{int}}} \left(1 + \frac{\alpha_{\mathrm{i}}}{\alpha_{\mathrm{m}}}\right) = \frac{1}{\eta_{\mathrm{int}}} \left[1 - 2\alpha_{\mathrm{i}} L \ln\left(R_1 R_2\right)\right] . \tag{20.35}$$

Therefore, if η_{ext}^{-1} is plotted for similar lasers with different cavity length (see Fig. 20.33b), a straight line should arise from which the internal quantum efficiency (extrapolation to $L \to 0$) and the internal loss ($\propto$ slope) can be determined experimentally.

The threshold current for a given laser is determined from the P–I characteristic via extrapolation of the linear regime as shown in Fig. 20.36a. Record values for the threshold current density are often given for the limit $L \to \infty$. Due to current spreading, the threshold current density also depends on the width of the injection stripe. Record low thresholds are therefore often given for the limit $w \to \infty$.

The total quantum efficiency is given by

$$\eta_{\mathrm{tot}} = \frac{P_{\mathrm{out}}/I}{\hbar\omega/e} . \tag{20.36}$$

This quantity is shown in Fig. 20.36b for a laser as a function of the current. For a linear P–I lasing characteristic, the total quantum efficiency converges towards the external quantum efficiency for high currents because the low quantum efficiency subthreshold regime no longer plays any role. Another important figure of merit is the wall-plug efficiency η_{w} that describes the power conversion:

$$\eta_{\mathrm{w}} = \frac{P_{\mathrm{out}}}{U\,I} . \tag{20.37}$$

Additionally to the current balance discussed so far, typically a leakage current exists that flows without contributing to recombination or lasing. Carriers not captured into or escaping from the active layers can contribute to this current. The present (2004) record for wall-plug efficiency of high-power laser diodes is 65%, employing careful control of band alignment (graded

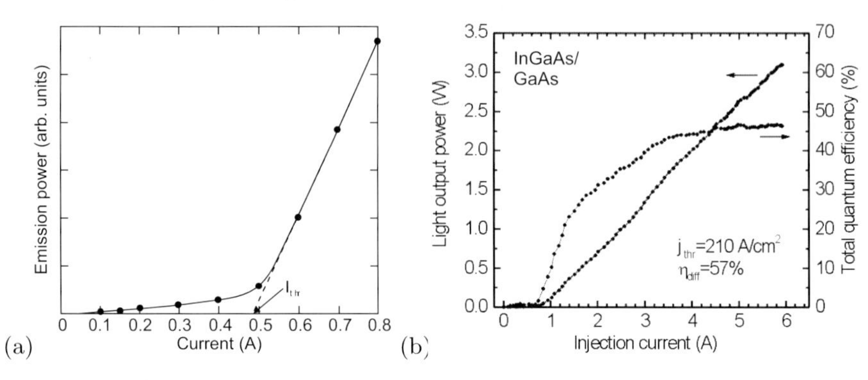

Fig. 20.36. (**a**) Typical P–I characteristic of a semiconductor laser. Adapted from [183]. (**b**) Output power and total quantum efficiency of a quantum dot laser (3 stacks of InGaAs/GaAs QDs, $L = 2\,\mathrm{mm}$, $w = 200\,\mu\mathrm{m}$, $\lambda = 1100\,\mathrm{nm}$, $T = 293\,\mathrm{K}$) vs. injection current

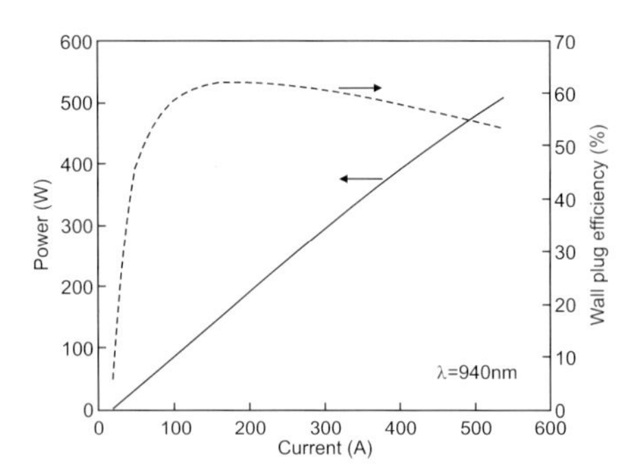

Fig. 20.37. Output power (*left scale*) and wall-plug efficiency (*right scale*) vs. driving current for a 10 mm laser bar. Adapted from [673]

junctions, avoiding voltage barriers), optical losses, Joule heating, spontaneous emission and carrier leakage. It seems possible to achieve η_{w} of 80%.

The P–I characteristic is not linear to arbitrary high currents. Generally, the output power will saturate or even decrease for increasing current. These effects can be due to increasing leakage current, increasing internal loss at high current or temperature effects, e.g. an increase of threshold with increasing temperature (see Sect. 20.4.9) and therefore a reduction of total efficiency. All nonradiative losses will eventually show up as heat in the laser that must be managed with a heat sink.

A radical effect is catastrophical optical damage (COD) at which the laser facet is irreversible (partially) destructed. Antioxidation or protective layers can increase the damage threshold to $> 20\,\mathrm{MW/cm^2}$. The record power from a single edge emitter is $\sim 12\,\mathrm{W}$ (200 µm stripe width). For a lateral monomode laser, cw power of about 1.2 W has been reached from a 1480-nm

InGaAsP/InP double quantum-well lasers with 3–5 µm stripes and 3 mm cavity length [671]. About 500 mW can be coupled into a single-mode fiber [672].

An array of edge emitters is called a laser bar. The current power record is 509 W (at 540 A) for a 10 mm wide bar (Fig. 20.37).

20.4.9 Temperature Dependence

The threshold of a laser typically increases with increasing temperature as shown in Fig. 20.38a. Empirically, in the vicinity of a temperature T_1 the threshold follows an exponential law (see Fig. 20.38b)

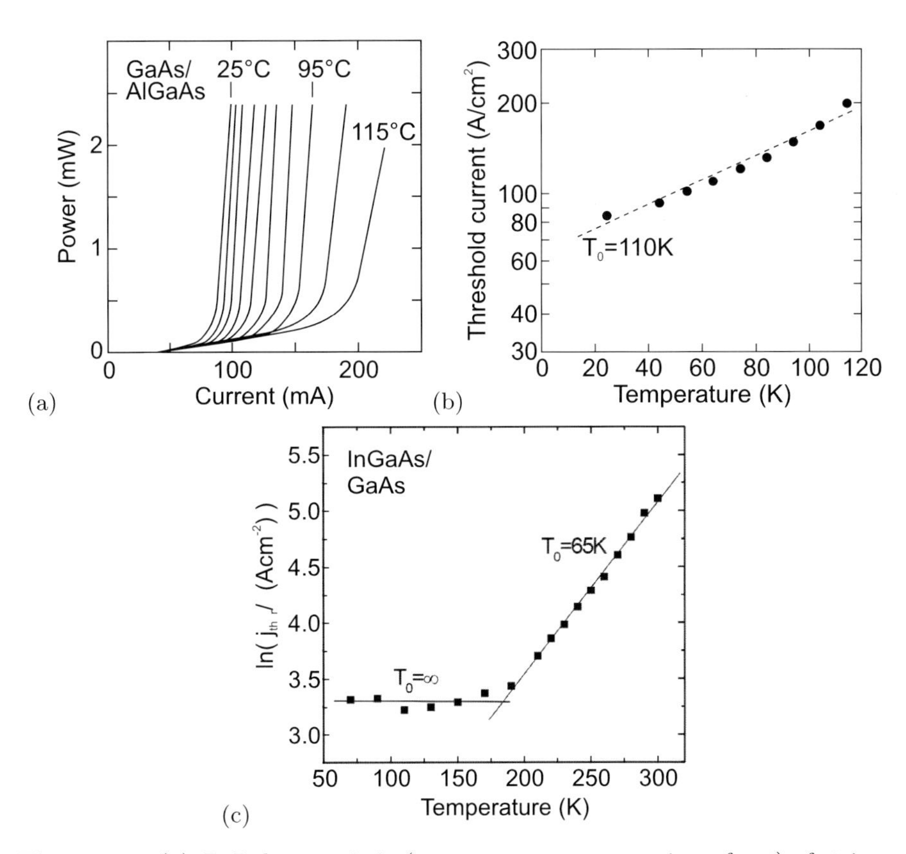

Fig. 20.38. (**a**) *P–I* characteristic (cw output power per mirror facet) of stripe-buried heterostructure laser at various temperatures of the heat sink between 25 °C and 115 °C in steps of 10 K. (**b**) Threshold current (in logarithmic scale) of this laser as a function of heat-sink temperature and exponential fit (*dashed line*) with $T_0 = 110$ K. Parts (**a**) and (**b**) adapted from [674]. (**c**) Temperature dependence of the threshold current density of a quantum dot laser (3 stacks of InGaAs/GaAs QDs, $\lambda = 1150$ nm) with T_0 (*solid lines* are fits) given in the figure

$$j_{\mathrm{thr}}(T) = j_{\mathrm{thr}}(T_1) \exp\left(\frac{T - T_1}{T_0}\right) \propto \exp\left(\frac{T}{T_0}\right) , \qquad (20.38)$$

with T_0 being the so-called characteristic temperature.[18] It is the inverse logarithmic slope, $T_0^{-1} = \mathrm{d} \ln j_{\mathrm{thr}}/\mathrm{d}T$.

T_0 summarizes the temperature-dependent loss and the carrier redistribution in **k**-space due to the change of the Fermi distribution with temperature. With increasing temperature, populated states below the quasi-Fermi level become unpopulated and nonlasing states become populated. Therefore, the gain decreases with increasing temperature. This redistribution must be compensated by an increase of the quasi-Fermi energy, i.e. stronger pumping. This effect is present for (even ideal) bulk, quantum well and quantum wire lasers. Only for quantum dots with a δ-like density of states is the change of Fermi distribution irrelevant as long as excited states are energetically well separated from the (lasing) ground state. In Fig. 20.38c, the threshold of a quantum dot laser [675] is indeed temperature independent ($T_0 = \infty$) as long as excited states are not thermally populated (for $T < 170\,\mathrm{K}$ for the present laser).

20.4.10 Mode Spectrum

In Fig. 20.39a, the mode spectrum of a typical edge-emitting laser is shown. Below threshold, the amplified spontaneous emission (ASE) spectrum exhibits a comb-like structure due to the Fabry–Perot modes. Above threshold, some modes grow much faster than others, possibly resulting in single longitudinal mode operation at high injection. The relative strength of the strongest side mode is expressed through the side-mode suppression ratio (SSR) in dB

$$SSR = 10 \log\left(\frac{I_{\mathrm{mm}}}{I_{\mathrm{sm}}}\right) , \qquad (20.39)$$

where I_{mm} (I_{sm}) is the intensity of the maximum (strongest side) mode in the lasing spectrum.

20.4.11 Longitudinal Single-Mode Lasers

In order to achieve a high SSR or single longitudinal mode lasing, the feedback must offer a higher wavelength selectivity than a simple mirror. A preferential feedback for certain modes can be obtained using a periodic dielectric structure that 'fits' to a particular wavelength, similar to a Bragg mirror. The periodic modulation of the refractive index can be made within the cavity (distributed feedback, DFB, Fig. 20.40a) or at the mirror (distributed reflection, DBR, Fig. 20.40b). In this way, monochromatic lasers with SSR $\gg 30\,\mathrm{dB}$ are possible (Fig. 20.39b).

[18] Since T_0 has the dimension of a temperature difference, it can be expressed in °C. For the sake of unambiguity it should be given, however, in K.

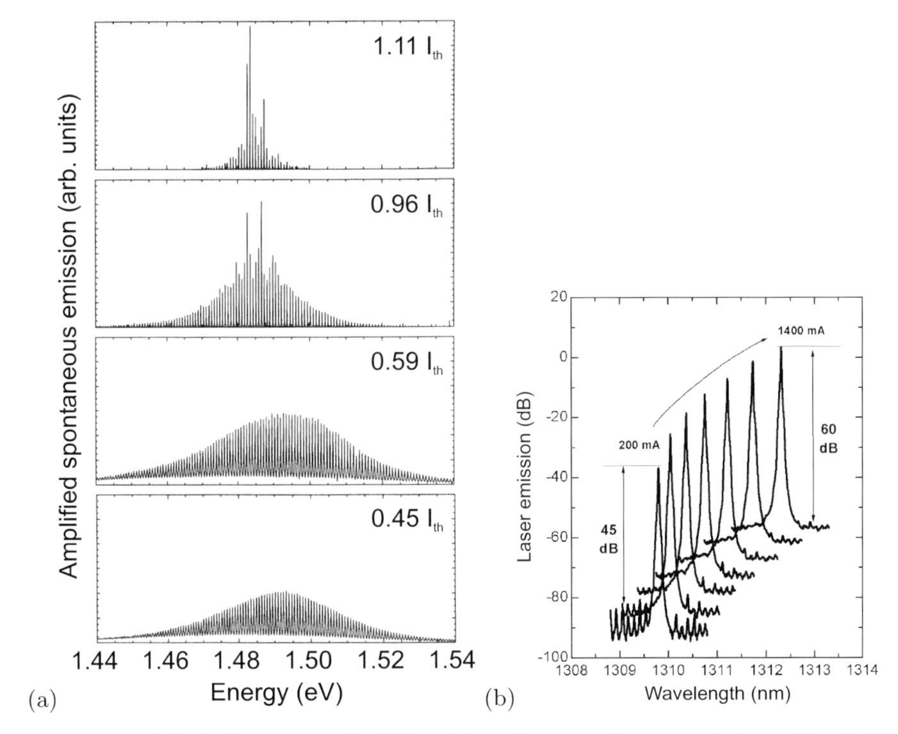

Fig. 20.39. (**a**) Mode spectra of a Fabry–Perot laser, under, at and above threshold ($I_{\mathrm{thr}} = 13.5\,\mathrm{mA}$). Adapted from [388]. (**b**) Mode spectra of a cw DFB InGaAs/InP laser with 2 mm cavity length at various currents of 200, 400, ..., 1400 mA ($I_{\mathrm{thr}} = 65\,\mathrm{mA}$), SSR>40 dB at $T = 293\,\mathrm{K}$. Adapted from [676], reproduced with permission from SPIE

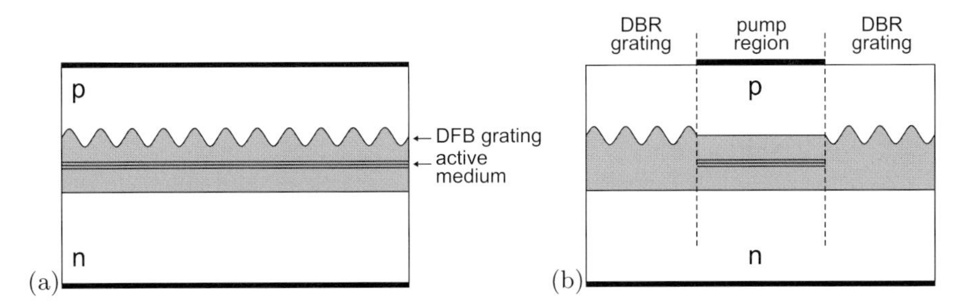

Fig. 20.40. Schematic drawing of (**a**) DFB (distributed feedback) and (**b**) DBR (distributed Bragg reflection) lasers. The active medium is schematically shown as a triple quantum well, the waveguide is shown as a *grey area*

It is possible to couple several hundred mW optical power of a laterally and spectrally monomode laser into a monomode optical fiber [677] (Fig. 20.41).

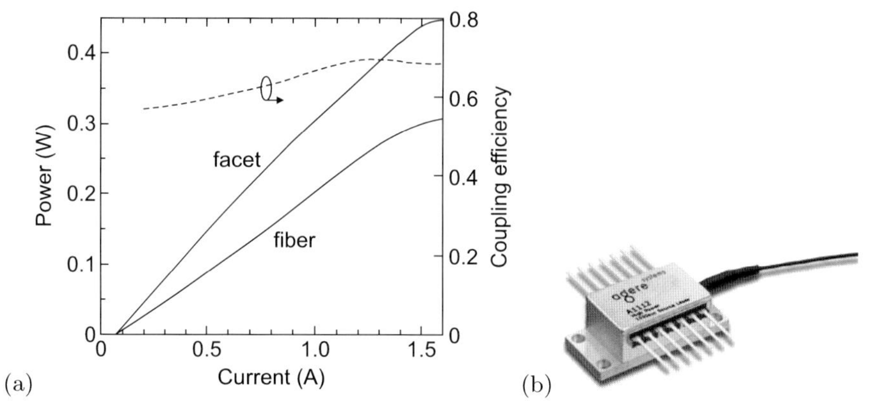

Fig. 20.41. (**a**) Output power of an InGaAsP/InP cw single-mode DFB laser at 1427 nm with 2 mm cavity length from the facet and coupled to a single-mode fiber vs. driving current ($T = 293$ K). The *dashed* line represents the coupling efficiency to the fiber (right scale). Adapted from [677]. (**b**) Package with pigtail of fiber coupled 1550 nm DFB laser with 40 mW output power in the fiber. From [678]

20.4.12 Tunability

The tunability of the emission wavelength [679] is important for several applications such as wavelength division multiplexing[19] and spectroscopy.

The simplest possibility to tune a laser is to vary its temperature and thus its bandgap. This method is particularly used for lead salt lasers in the mid-infrared region,[20] as shown in Fig. 20.42.

For monomode lasers, mode hopping, i.e. the discontinuous shift of lasing wavelength (or gain maximum) from one mode to the next, poses a problem for continuous tuning, as shown in Fig. 20.43. The continuous shift of emission wavelength is due to the temperature dependence of the index of refraction and subsequently the longitudinal modes. The index of refraction increases with increasing temperature at typically $\sim 3 \times 10^{-4}\,\mathrm{K}^{-1}$. Generally, a redshift is the consequence.

Another possibility to vary the index of refraction (and thus the optical path length) is a variation of the carrier density. The coefficient $\mathrm{d}n_\mathrm{r}/\mathrm{d}n$ is about $-10^{-20}\,\mathrm{cm}^3$. In a two-section laser, separate regions (with separately controlled currents) for gain and tuning are present. The regions are separated with deep-etched trenches to avoid crosstalk. The tuning range is limited to about 10 nm. For a mode-hopping free tuning, the control of the phase in the cavity is important and requires an additional section for the phase control. Such a three-section laser has separate regions (and current control) for the reflection, phase and amplification (or gain) region.

[19] In order to make better use of the high bandwidth of the optical fiber several information channels with closely lying wavelengths are transmitted.

[20] Note the anomalous positive coefficient $\mathrm{d}E_\mathrm{g}/\mathrm{d}T$ as discussed in Sect. 6.10.

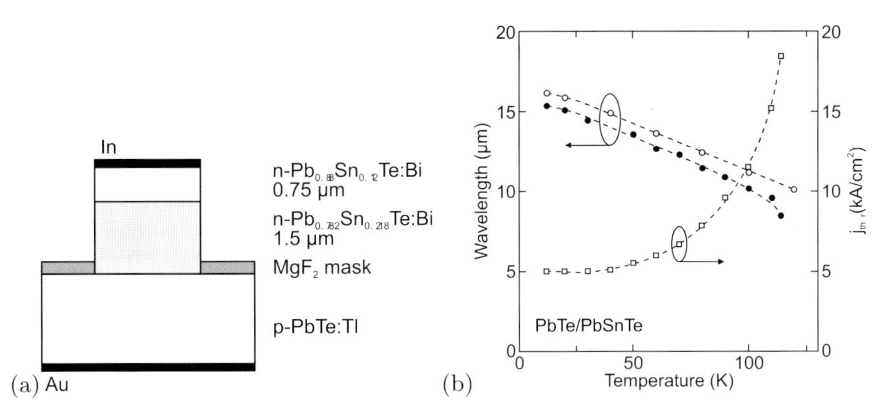

Fig. 20.42. (**a**) Schematic drawing of PbTe lead salt laser. (**b**) Tuning characteristics of such laser: Emission wavelength (*left scale*, *filled circles*: emission wavelength at cw threshold, *empty circles*: emission maximum under pulsed operation) and cw threshold current density (*right scale*) as a function of the heat-sink temperature. *Symbols* are experimental data, *dashed lines* are guides to the eye. Adapted from [680]

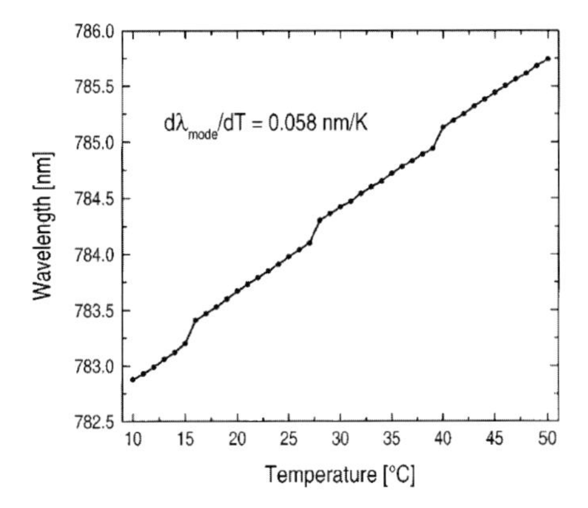

Fig. 20.43. Wavelength as a function of temperature (with mode hopping) for a GaAs-based DFB laser

Using sampled gratings, the tuning range can be strongly increased to about 100 nm. A sampled grating is a nonperiodic lattice that has several (∼ 10) reflection peaks. The laser structure has four sections (Fig. 20.44) with two mirrors that have slightly different sampled gratings. Via the carrier densities in the two mirror sections, different maxima can be brought to overlap (Vernier effect) and the position of the selected maximum can be tuned over a wide spectral range.

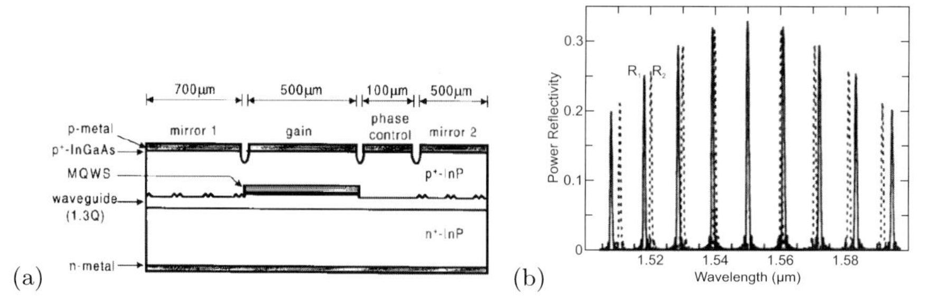

Fig. 20.44. (**a**) Schematic representation of SGDBR (sampled-grating DBR) laser with four sections. Reprinted from [681] with permission, ©1999 IEEE. (**b**) Reflectivity of two sampled gratings DBR mirrors (R_1: *Solid line*, R_2: *dashed line*)

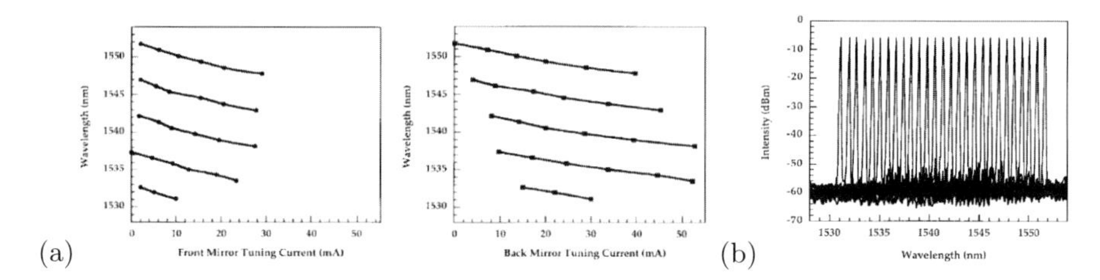

Fig. 20.45. (**a**) Tuning curves of two sampled gratings DBR mirrors for the front and back mirror current. (**b**) 27 wavelength channels (1531.12 to 1551.72 nm) with a channel separation of 1 nm. Reprinted from [682] with permission, ©1999 IEEE

20.4.13 Modulation

For transmission of information, the laser intensity must be modulated. This can be accomplished by direct modulation, i.e. modulation of the injection current, or external modulators, e.g. using, e.g., the voltage-induced shift of the absorption spectrum due to QCSE (see Sect. 12.1.2). For direct modulation, small- and large-signal modulation are distinguished.

Large-Signal Modulation

If a current pulse is fed to the laser, the laser radiation is emitted with a short time delay, the so-called turn-on delay (TOD) time. This time is needed to build up the carrier density for inversion. The time dependence of the density is (neglecting the density dependence of the lifetime)

$$n(t) = \frac{I\tau}{eAd}\left[1 - \exp\left(-\frac{t}{\tau}\right)\right] . \tag{20.40}$$

The TOD time to reach the threshold density (using (20.23)) is

$$\tau_{\mathrm{TOD}} = \tau \ln\left(\frac{I}{I - I_{\mathrm{thr}}}\right) . \tag{20.41}$$

We note that $\tau_{\mathrm{TOD}} > 0$ for $I > I_{\mathrm{thr}}$. Such a dependence is found experimentally (Fig. 20.46). The turn-on delay time decreases with increasing pump current but typically is at least 1 ns. In order to circumvent this limitation for more than about 1 GHz pulse repetition rate, the laser is biased slightly below threshold.

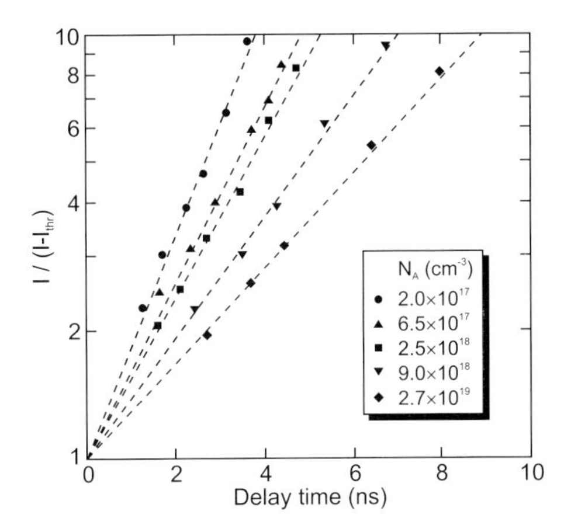

Fig. 20.46. Variation of turn-on delay time with the injected current for a laser at room temperature. Adapted from [683]

In Fig. 20.47a, the response (light emission) of a LED to a short current pulse is shown schematically. The monotonously decreasing transient (that is more or less exponential) corresponds to the carrier recombination dynamics. When a laser is excited with a steep (long) current pulse, the response exhibits so-called relaxation oscillations (RO) before the steady-state (cw) intensity level is reached (Fig. 20.47b).

In the laser, first the carrier density is built up. It surpasses the threshold density that leads to a build-up of the photon density. The laser pulse depletes the carrier density faster below threshold than the current can supply further

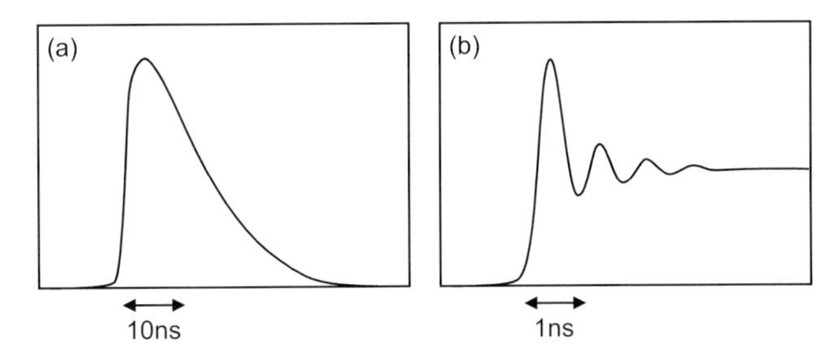

Fig. 20.47. Schematic response of (**a**) LED to current pulse and (**b**) of laser to current step

carriers. Therefore, the laser intensity drops below the cw level. From the coupled rate equations for the electron and photon densities n and N_{ph}, the relaxation oscillations have a frequency of

$$f_{\mathrm{RO}} = \frac{1}{2\pi}\left(\frac{v_{\mathrm{g}} g' S_0}{\tau_{\mathrm{ph}}}\right)^{1/2} , \tag{20.42}$$

where g' is the differential gain $g' = \mathrm{d}g/\mathrm{d}n$ $(g(n) = g_0 + g'(n - n_0))$ and S_0 is the photon density per volume that is proportional to the laser intensity. The dependence $f_{\mathrm{RO}} \propto S_0^{1/2}$ is also found experimentally (Fig. 20.48).

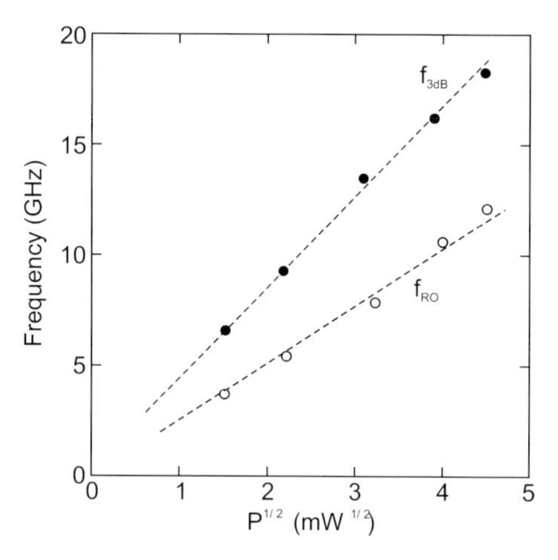

Fig. 20.48. Dependence of small-signal 3 dB cutoff frequency $f_{3\mathrm{dB}}$ (*filled symbols*) and relaxation oscillation frequency f_{RO} (*empty symbols*) on the square root of the output power P for a DFB-laser. Adapted from [684]

For digital data transmission, the laser is driven with pulse sequences. The response to a random bit pattern is called an 'eye pattern' and is shown in Fig. 20.49. The pattern consists of traces of the type shown in Fig. 20.47b. A clear distinction with well-defined trigger thresholds between 'on'- and 'off'-states can only be made if the eye formed by the overlay of all possible traces remains open. From the eye patterns in Fig. 20.49, it can be seen that the RO overshoot can be suppressed by driving the laser with a dc offset current well above threshold.

Small-Signal Modulation

In small-signal modulation, the injection current I is varied periodically by a small amount δI with $\delta I \ll I$ in the lasing regime. The current modulation leads to a corresponding variation of the output intensity. The frequency response is limited by the differential gain and the gain compression coefficient

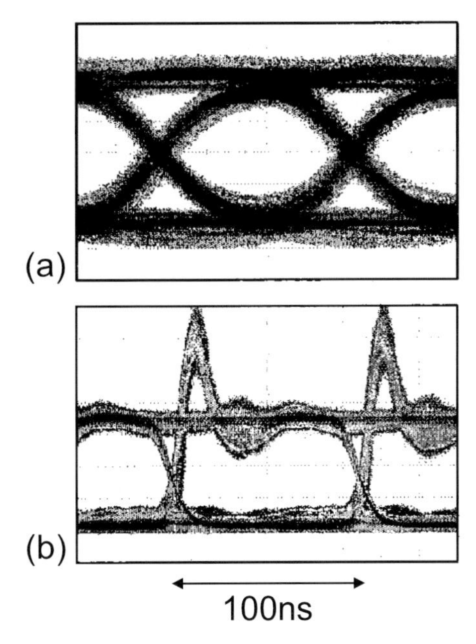

Fig. 20.49. Eye pattern of a single-mode VCSEL in response to a 10 Gb/s random bit pattern. The patterns are measured (**a**) with an offset current well above threshold and (**b**) with an offset current above but close to threshold. Adapted from [685], reprinted with permission, ©2002 IEEE

ϵ. The latter describes the saturation of gain with increasing photon density S_0 according to

$$g(n) = g_0 + \frac{g'(n - n_0)}{1 + \epsilon S_0} . \tag{20.43}$$

The frequency response shifts to higher frequency with increasing laser power as shown in Fig. 20.50b.

α factor

Another important quantity is the α factor, also called the linewidth enhancement factor [686, 687]. Due to the coupling of amplitude and phase fluctuations in the laser, the linewidth Δf is larger than expected.

$$\Delta f = \frac{\hbar\omega v_{\mathrm{g}} R_{\mathrm{spont}} \ln R}{8\pi P_{\mathrm{out}} L}(1 + \alpha^2) . \tag{20.44}$$

The linewidth enhancement is described via $(1 + \alpha^2)$ with

$$\alpha = \frac{\mathrm{d}n_{\mathrm{r}}/\mathrm{d}n}{\mathrm{d}\kappa/\mathrm{d}n} = -\frac{4\pi}{\lambda}\frac{\mathrm{d}n_{\mathrm{r}}/\mathrm{d}n}{g'} , \tag{20.45}$$

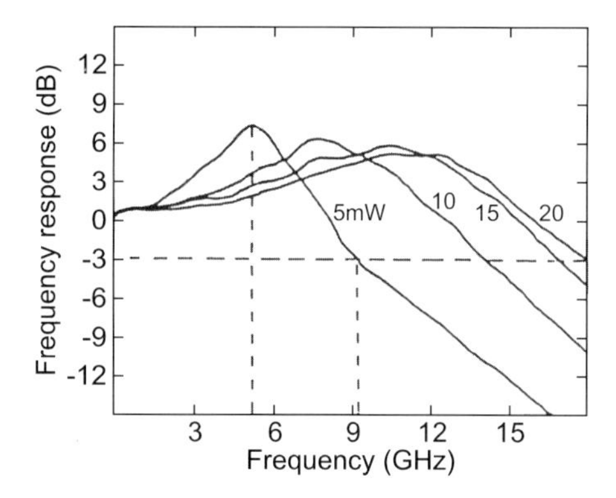

Fig. 20.50. Frequency response of a DFB-laser for various output powers as labeled. Adapted from [684]

where κ denotes the imaginary part of the index of refraction ($n^* = n_r + i\kappa$). Typical values for α are between 1 and 10. The linewidth is inversely proportional to the output power (Fig. 20.51).

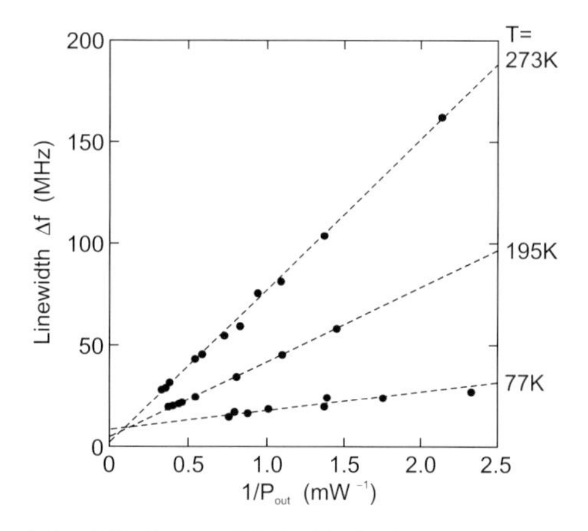

Fig. 20.51. Linewidth Δf of a cw GaAs/AlGaAs semiconductor laser at various temperatures as a function of the inverse output power P_{out}^{-1}. At room temperature $\alpha \approx 5$. Adapted from [688]

20.4.14 Surface-emitting Lasers

Surface-emitting lasers emit their beam normal to the surface. They can be fabricated from horizontal (edge-) emitters by reflecting the beam with a suitable mirror into the surface direction. This technology requires tilted facets or micro-optical components but allows for high power per area. In

Fig. 20.52, a schematic cross section of a horizontal-cavity surface-emitting laser (HCSEL) and the light emission from an array of 220 such lasers are shown. The laser contains a 45° mirror that steers the light through the substrate and a Bragg mirror to provide the cavity mirror. The facet can also be fabricated such that the emission is to the top surface (Fig. 20.53). Another possibility to couple the beam out of a horizontal cavity is a surface grating.

Now, surface-emitting lasers with vertical-cavity (VCSEL), as shown in Fig. 20.19b, will be discussed. A detailed treatment can be found in [692]. VCSELs are of increasing importance after many issues regarding their technology and fabrication have been solved. VCSEL fabrication is essentially a planar technology and VCSELs can be fabricated as arrays (Fig. 20.55).

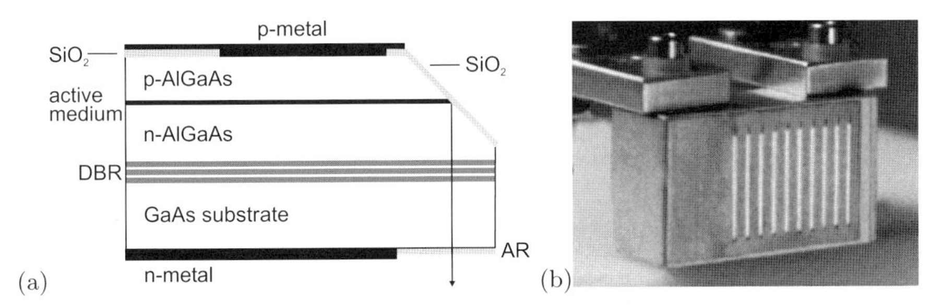

Fig. 20.52. (**a**) Principle of a surface-emitting laser. Light generated in the active region is internally reflected by the 45° angled mirror and directed through the substrate; 'AR': antireflection coating, 'DBR': epitaxial Bragg mirror. (**b**) Light emission from a 10 × 22 surface-emitting diode array. The light emission appears as stripes due to the broad beam divergence in the vertical direction. Part (**b**) reprinted with permission from [689]

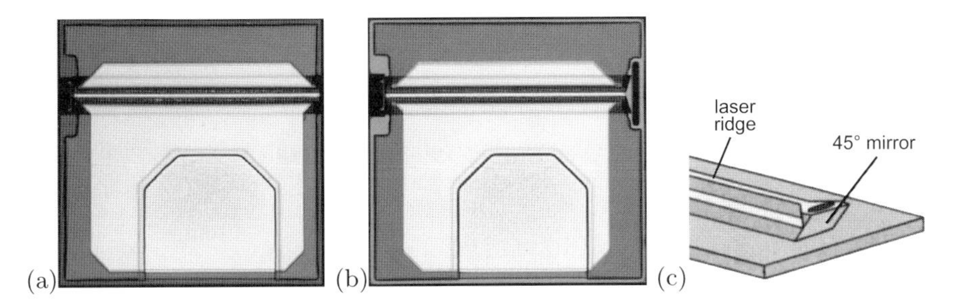

Fig. 20.53. (**a**) Horizontal Fabry–Perot cavity InP-based laser with 1310 nm emission length and 10 mW output power for modulation at 2.5 GB/s. The right facet is formed as DBR, emission is to the left. The trapezoid area in the center bottom of the image is the bond pad for the top contact. (**b**) Horizontal-cavity surface-emitting laser. Compared to (**a**), the right facet is replaced with a 45° mirror, leading to surface emission. (**c**) Schematic drawing of the tilted facet. Parts (**a**) and (**b**) from [690]

An on-wafer test of their properties is possible. They offer a symmetrical (or possibly a controlled asymmetrical) beam profile (Fig. 20.54) with possible polarization control or fixation.

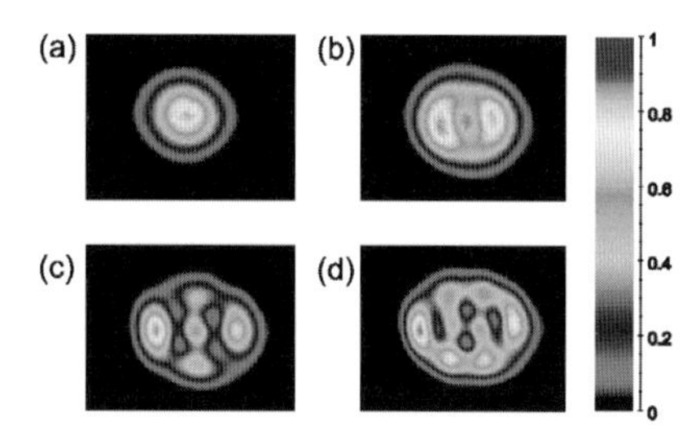

Fig. 20.54. In-plane near field of a VCSEL with 6 μm oxide aperture at various currents, (**a**) 3.0 mA, (**b**), 6.2 mA, (**c**) 14.7 mA, (**d**) 18 mA

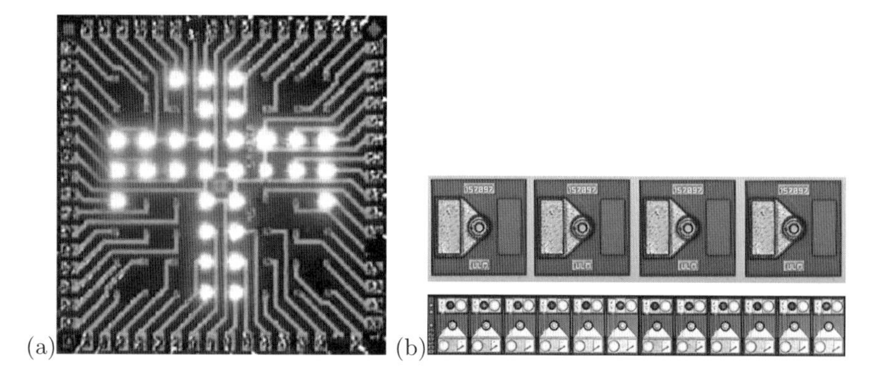

Fig. 20.55. (**a**), (**b**) VCSEL arrays. Part (**a**) reprinted from [693] with permission, part (**b**) reprinted from [694] with permission

The cavity is formed by two highly reflecting Bragg mirrors with a distance of $\lambda/2$ or $3\lambda/2$ forming a microcavity (see Sect. 17.1.6). A high index contrast can be obtained from GaAs/AlAs Bragg mirrors in which the AlAs layers have been selectively oxidized in a hot moist atmosphere. Pure semiconductor Bragg mirrors suffer typically from small index contrast and require many pairs. This poses a problem, e.g. for InP-based VCSELs. In Fig. 20.56, the distribution of light intensity along a $3\lambda/2$ cavity is shown. In the stop band of the mirrors, there is only one optical mode, the cavity mode, that can propagate along the vertical (z) direction.

The current path through the active layer can be defined with an oxide aperture. This aperture is fabricated by selective oxidation of an AlAs layer, leaving a circular opening in the center of the VCSEL pillar as shown in

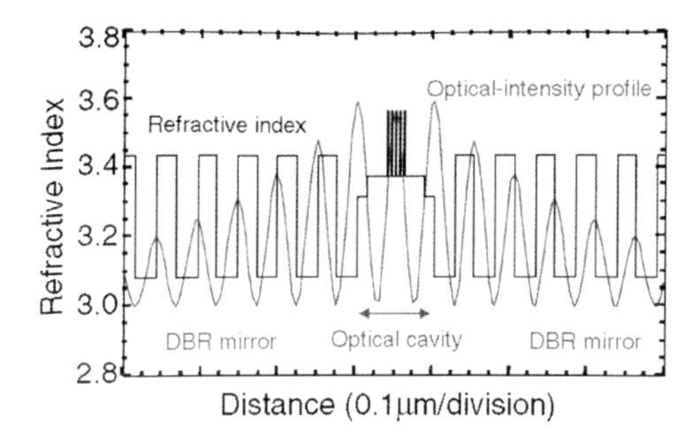

Fig. 20.56. Simulation of the longitudinal distribution of the optical field in a VCSEL structure. The active medium are five quantum wells in the center. Reproduced from [691] by permission from the MRS Bulletin

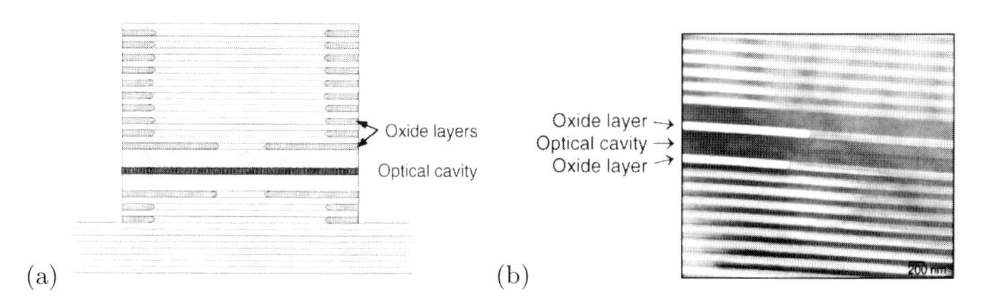

Fig. 20.57. (**a**) Schematic cross section of VCSEL with oxide aperture, (**b**) TEM image of cross section. Reproduced from [691] by permission from the MRS Bulletin

Fig. 20.57. The current can be injected through the mirrors if they are doped. Alternatively, the current can be directly fed to the active layer by so-called intracavity contacts.

The emission wavelength of a VCSEL can be shifted via a variation of temperature or pump power. Tuning of the VCSEL emission can also be accomplished by leaving an air gap between the cavity and the upper mirror [695]. Applying a voltage to the lever arm with the top mirror, the width of the air gap can be varied. This variation leads to a shift of the cavity mode and therefore of the laser emission wavelength. A VCSEL with air gap and particularly a high contrast Bragg mirror is achieved with InP/air as shown in Fig. 20.59.

20.4.15 Optically Pumped Semiconductor Lasers

An easy way to pump semiconductor lasers is optical pumping. This technique is similar to diode-pumped solid-state lasers (DPSS). A (semiconductor) pump diode illuminates a suitable semiconductor structure (Fig. 20.60). The resonator is built between the bottom Bragg mirror of the semiconductor and the output coupler. The semiconductor structure contains suitable

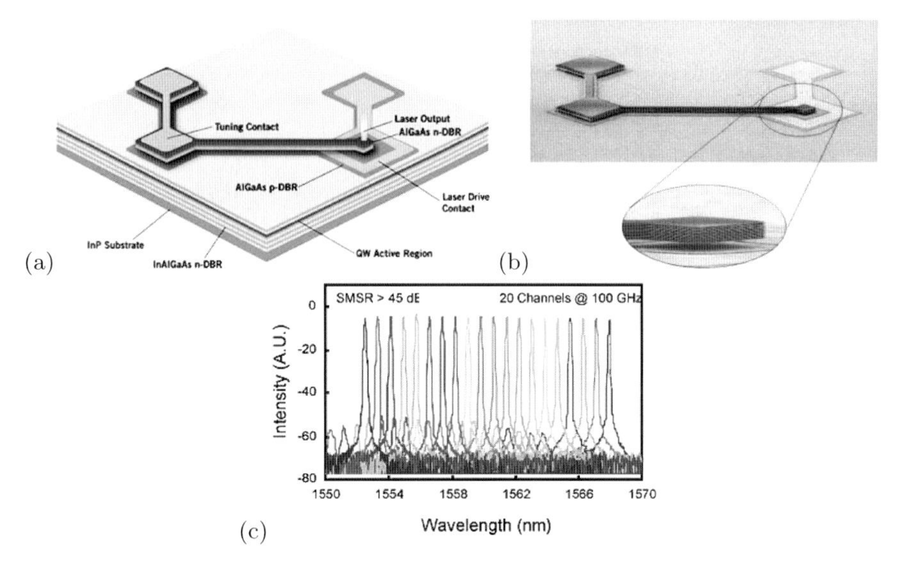

Fig. 20.58. (**a**) Schematic setup and (**b**) SEM image of VCSEL with air gap between active layer and top Bragg mirror, (**c**) spectra for different tuning conditions (width of air gap). From [696].

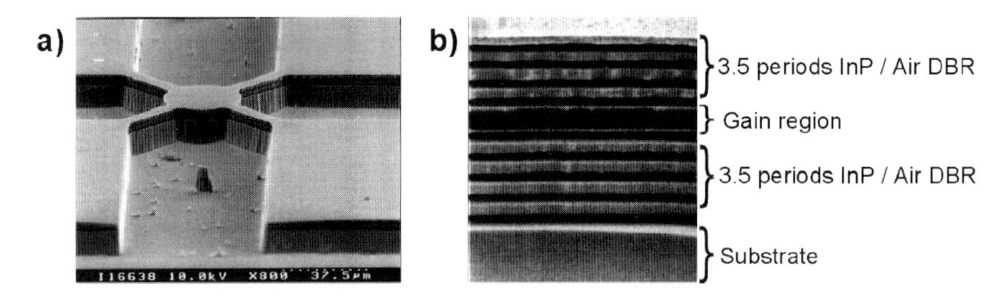

Fig. 20.59. (**a**) VCSEL with air gap and (**b**) Bragg mirror with high dielectric contrast InP/air interfaces. Reprinted with permission from [697], ©2002 IEEE

absorption layers (barriers) that absorb the pump light and quantum wells that emit laser radiation. This radiation is intracavity frequency doubled. In order to reach, e.g., a 488-nm output laser beam, a standard 808-nm pump diode is employed. The InGaAs/GaAs quantum wells are designed to emit at 976 nm. Other design wavelengths of the quantum wells allow for other output wavelengths. This technology allows compact lasers with little heat dissipation [698]. The optically pumped semiconductor laser (OPSL) is also known as a semiconductor disc laser.

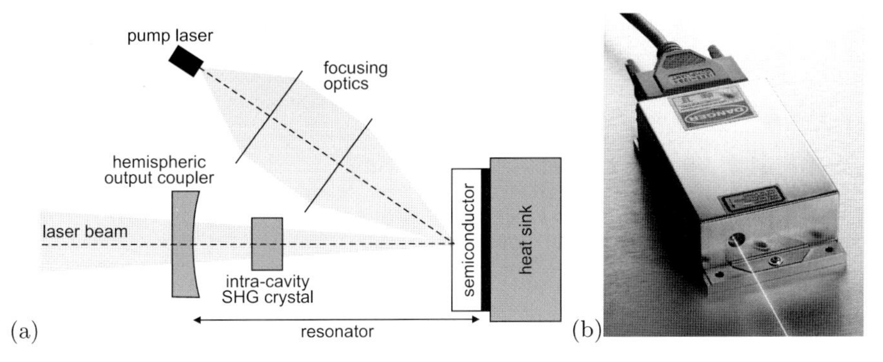

Fig. 20.60. (**a**) Schematic setup of optically pumped semiconductor laser (OPSL). The semiconductor chip consists of a Bragg mirror on the bottom, multiple quantum wells and an antireflection coating on the top. Adapted from [698]. (**b**) OPSL source (488 nm, 20 mW, footprint: $125 \times 70\,\mathrm{mm}^2$). Reprinted with permission from [699]

20.4.16 Quantum Cascade Lasers

In a quantum cascade laser (QCL), the gain stems from an intersubband transition. The concept was conceived in 1971 [700, 701] and realized in 1994 [660]. In Fig. 20.61a, the schematic conduction-band structure at operation is shown. The injector supplies electrons into the active region. The electron is removed quickly from the lower level in order to allow inversion. The electron is then extracted into the next injector. The laser medium consists of several such units as shown in Fig. 20.61b. Since every unit can deliver a photon per electron (with efficiency η_1), the total quantum efficiency of N units $\eta = N\,\eta_1$ can be larger than 1.

The emission wavelength is in the far- or mid-infrared, depending only on the designed layer thicknesses and *not* on the bandgap of the material (Fig. 20.61d). In the mid-infrared, room-temperature operation has been achieved while operation in the far-infrared requires cooling so far. Extensions to the THz-range and the infrared spectral region (telecommunication wavelengths of 1.3 and 1.55 μm) seem feasible. The cascade laser concept can also be combined with the DFB technology to create single-mode laser emission (Fig. 20.61d).

20.4.17 Hot-Hole Lasers

The hot-hole laser, which is mostly realized with p-doped Ge, is based on a population inversion between the light- and heavy-hole valence subbands. The laser operates with crossed electric and magnetic fields (Voigt configuration, typically E = 0.5–3 kV/cm, B = 0.3–2 T) and at cryogenic temperatures (T = 4–40 K) [704–706].

A significant scattering process of hot carriers is interaction with optical phonons, mainly optical phonon emission. This process has a threshold in

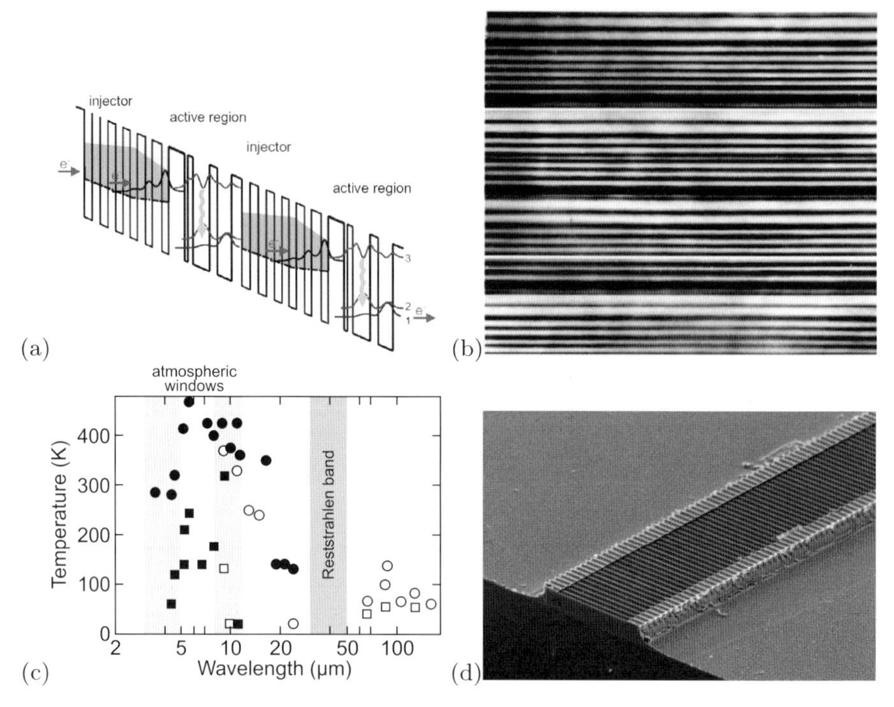

Fig. 20.61. (**a**) Schematic band diagram of quantum cascade laser. (**b**) Cross sectional TEM of cascade layer sequence. The periodicity of the vertical layer sequence is 45 nm. From [702]. (**c**) Laser emission wavelengths and operation temperatures for various realized quantum cascade lasers (*squares*: cw, *circles*: pulsed operation, *solid symbols*: InP-, *empty symbols*: GaAs-based). Data from [703]. (**d**) SEM image of a quantum cascade DFB laser (grating period: 1.6 μm). From [703]

carrier energy given by the optical phonon energy. For sufficiently high electric fields and at low temperature, hot carriers accelerate without acoustical phonon interaction (ballistic transport) along the crystallographic direction in which the electric field is applied. These hot carriers reach the optical phonon energy and lose all their energy due to emission of an optical phonon. They accelerate again, repeating this directional motion in momentum space. This motion is called *streaming motion*.

For $|E/B|$ ratios of about 1.5 kV cm^{-1}T^{-1}, the heavy holes are accelerated up to energies above the optical phonon energy (37 meV in germanium) and consequently are scattered strongly by these phonons. Under these conditions, a few per cent of the heavy holes are scattered into the light-hole band. The light holes remain at much lower energies and are accumulated at the bottom of the light-hole band below the optical phonon energy as sketched in Fig. 20.62. The continuous pumping of heavy holes into the light-hole band can lead to a population inversion. Consequently, laser radiation is emitted from optical (radiative) intervalence-band transitions (see Sect. 9.4.5). The emission wavelength is in the far-infrared around 100 µm. Typical p-Ge lasers span the frequency range 1–4 THz (300–70 µm) [707] and deliver 1–10 W peak output power for 1 cm^3 typical active volume.

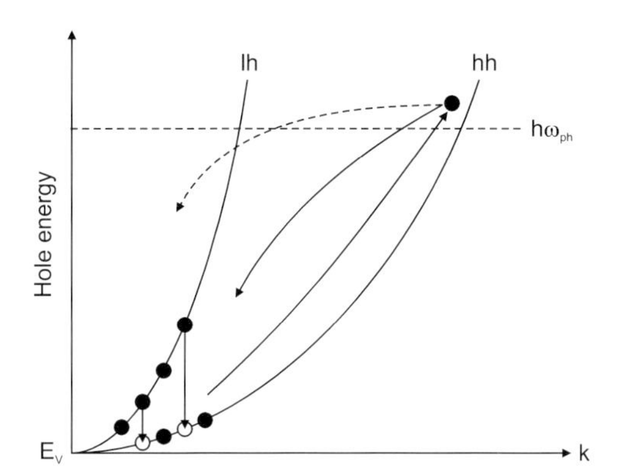

Fig. 20.62. Schematic cycle of hole motion in a hot-hole laser. *Filled* (*empty*) circles represent populated (unpopulated) hole states. The *solid lines* represent streaming motion of heavy hole, the *dashed line* represents scattering into the light hole band. *Arrows* denote radiative intervalence-band transitions

Since the applied electric field causes considerable heating, the temperature of the laser crystal rises quickly, within a few µs, up to 40 K. Then the laser action stops. Thus the duration of the electric-field excitation is limited to 1–5 µs (limiting the emission power) and the repetition frequency is only a few Hertz due to the necessary cooling. Research is underway towards high duty cycle (possibly cw) operation by using smaller volumes and planar vertical-cavities [708, 709].

20.5 Semiconductor Optical Amplifiers

If the facets of a laser cavity are antireflection coated, a laser gain medium can be used as a semiconductor optical amplifier (SOA).

A tapered amplifier geometry, as shown in Fig. 20.63a, allows for laterally monomode input and a preservation of the lateral beam quality during the propagation of the optical wave through the gain medium. The active medium is an 8-nm compressively strained InGaAs quantum well. A typical taper angle is 5–10°. The input aperture is between 5 and 7-µm. The amplifier length is 2040 µm. More than 20 dB optical amplification can be obtained (Fig. 20.63b). The self-oscillation is suppressed for currents up to 2 A by AR facet coating of 10^{-4} in a 70-nm band. The wall-plug efficiency of the discussed amplifier is up to over 40%. If such an amplifier is arranged together with a seed laser diode (master oscillator), the setup is called MOPA (master oscillator power amplifier), as shown in Fig. 20.64. A modulated input also leads to a modulated output.

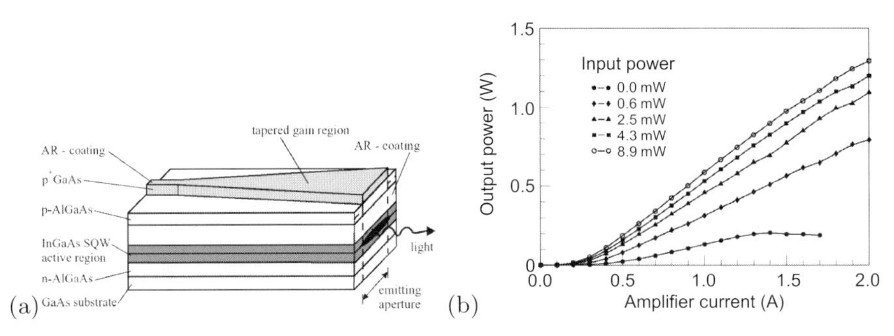

Fig. 20.63. (**a**) Schematic geometry of tapered semiconductor amplifier. (**b**) Optical output power vs. amplifier current for various values of the optical input power, taper angle was 5°. For zero input power only spontaneous and amplified spontaneous emission is observed. Reprinted with permission from [710].

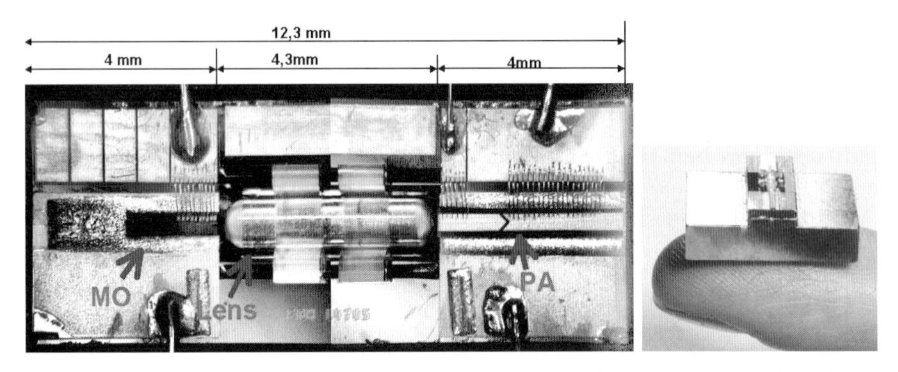

Fig. 20.64. Photographs of a MOPA arrangement of a laser (master oscillator, 'MO'), glass lens and tapered amplifier (power amplifier, 'PA') on a silicon micro-optical 'bench'. Reprinted with permission from [711]

21 Transistors

21.1 Introduction

Transistors[1] are the key elements for electronic circuits such as amplifiers, memories and microprocessors. Transistors can be realized in bipolar technology (Sect. 21.2) or as unipolar devices using the field effect (Sect. 21.3). The equivalent in vacuum-tube technology to the transistor is the triode (Fig. 21.1a). Transistors can be optimized for their properties in analog circuits such as linearity and frequency response or their properties in digital circuits such as switching speed and power consumption. Early commercial models are shown in Fig. 21.2.

21.2 Bipolar Transistors

Bipolar transistors consist of a pnp or npn sequence (Fig. 21.3). The layers (or parts) are named emitter (highly doped), base (thin, highly doped) and

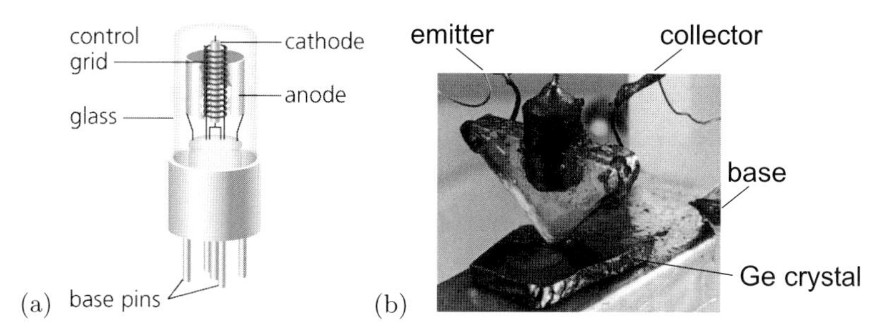

Fig. 21.1. (**a**) Schematic image of a vacuum triode. The electron current flows from the heated cathode to the anode when the latter is at a positive potential. The flow of electrons is controlled with the grid voltage. (**b**) Bell Laboratories' first (experimental) transistor, 1947

[1]The term 'transistor' was coined from the combination of 'transconductance' or 'transfer' and 'varistor' after initially such devices were termed 'semiconductor triodes'. The major breakthrough was achieved in 1947 when the first transistor was realized that showed gain (Figs. 1.4 and 21.1b).

Fig. 21.2. (**a**) First commercial, developmental (point contact) transistor from BTL (Bell Telephone Laboratories) with access holes for adjustment of the whiskers pressing on a piece of Ge, diameter 7/32"=5 mm, 1948. (**b**) First high-performance silicon transistor (npn mesa technology), model 2N697 from Fairchild Semiconductor, 1958 (at \$200, in 1960 \$28.50). The product number is still in use (now \$0.95)

collector (normal doping level). The transistor can be considered to consist of two diodes (emitter–base and base–collector) back to back. However, the important point is that the base is sufficiently thin (in relation to its minority carrier diffusion length) and carriers from the emitter (which are minority carriers in the base) can dominantly reach the collector by diffusion.

In Fig. 21.4, the three basic circuits with a transistor are shown. They are classified by the common contact for the input and output circuit. The space charges and band diagram for a pnp transistor in the base circuit configuration are depicted in Fig. 21.5. The emitter–base diode is switched in the forward direction to inject electrons into the base. The base–collector diode is switched in the reverse direction. The electrons that diffuse through the base and reach the neutral region of the collector are transported by the high drift field away from the base.

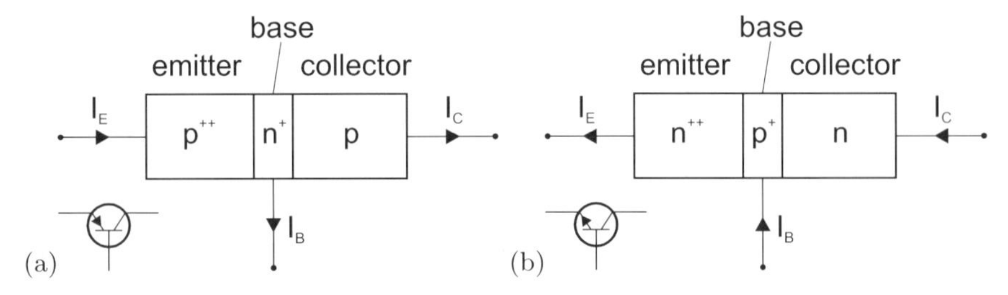

Fig. 21.3. Schematic structure and circuit symbol for (**a**) pnp and (**b**) npn transistors

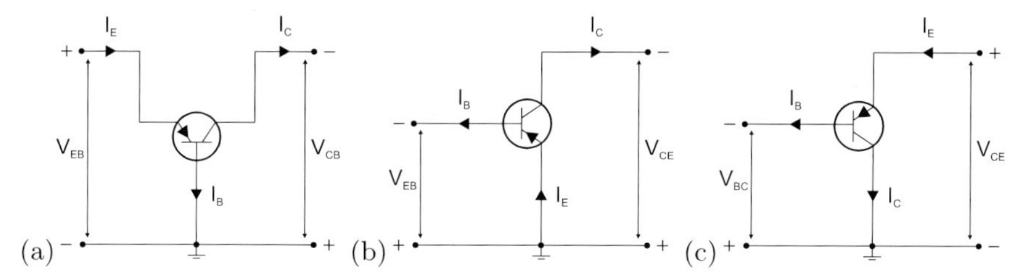

Fig. 21.4. Basic transistor circuits, named after the common contact: (**a**) Common base circuit, (**b**) common emitter circuit and (**c**) common collector circuit

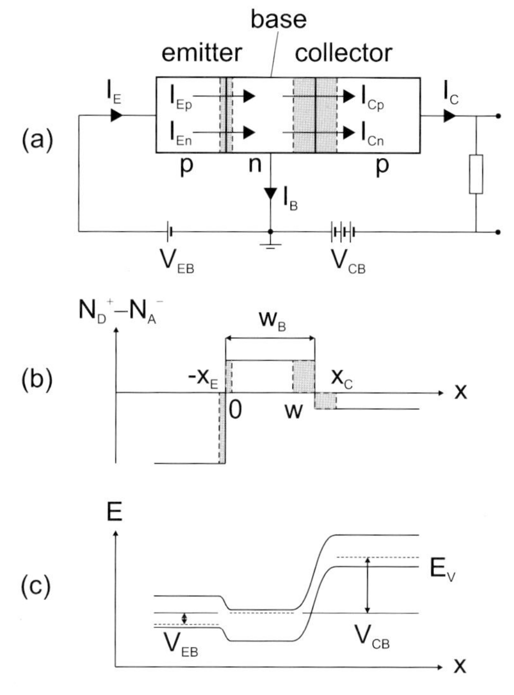

Fig. 21.5. pnp transistor in (**a**) base circuit. (**b**) Doping profile and space charges (abrupt approximation) and (**c**) band diagram for typical operation conditions

21.2.1 Carrier Density and Currents

The modeling of transistors is a complex topic. We treat the transistor on the level of the abrupt junction. As an approximation, we assume that all voltages drop at the junctions. Series resistances, capacities and stray capacities and other parasitic impedances are neglected at this point.

The major result is that the emitter–base current from the forward-biased emitter–base diode will be transferred to the collector. The current flowing from the base contact is small compared to the collector current. This explains the most prominent property of the transistor, the current amplification.

For the neutral part of the base region of a pnp transistor, the stationary equations for diffusion and continuity are

$$0 = D_{\mathrm{B}} \frac{\partial^2 p}{\partial x^2} - \frac{p - p_{\mathrm{B}}}{\tau_{\mathrm{B}}} \tag{21.1a}$$

$$j_{\mathrm{p}} = -eD_{\mathrm{B}} \frac{\partial p}{\partial x} \tag{21.1b}$$

$$j_{\mathrm{tot}} = j_{\mathrm{n}} + j_{\mathrm{p}} \,, \tag{21.1c}$$

where p_{B} is the equilibrium minority carrier density in the base. From the discussion of the pn-diode, we know that at the boundary of the depletion layer the minority carrier density is increased by $\exp(eV/kT)$ (cf. (18.76a,b)). At the boundaries of the emitter–base diode (for geometry see Fig. 21.5a)

$$\delta p(0) = p(0) - p_{\mathrm{B}} = p_{\mathrm{B}} \left[\exp\left(\frac{eV_{\mathrm{EB}}}{kT} \right) - 1 \right] \tag{21.2a}$$

$$\delta n(-x_{\mathrm{E}}) = n(-x_{\mathrm{E}}) - n_{\mathrm{E}} = n_{\mathrm{E}} \left[\exp\left(\frac{eV_{\mathrm{EB}}}{kT} \right) - 1 \right] , \tag{21.2b}$$

where n_{E} and p_{B} are the equilibrium minority-carrier densities in the emitter and base, respectively. Accordingly, at the boundaries of the base–collector diode we have

$$\delta p(w) = p(w) - p_{\mathrm{B}} = p_{\mathrm{B}} \left[\exp\left(\frac{eV_{\mathrm{CB}}}{kT} \right) - 1 \right] \tag{21.3a}$$

$$\delta n(x_{\mathrm{C}}) = n(x_{\mathrm{C}}) - n_{\mathrm{C}} = n_{\mathrm{C}} \left[\exp\left(\frac{eV_{\mathrm{CB}}}{kT} \right) - 1 \right] . \tag{21.3b}$$

These are the boundary conditions for the diffusion equations in the p-doped layers and in the neutral region of the n-doped base. For the p-layers (with infinitely long contacts), the solution is (similar to (18.101)) for $x < -x_{\mathrm{E}}$ and $x > -x_{\mathrm{C}}$, respectively

$$n(-x_{\mathrm{E}}) = n_{\mathrm{E}} + \delta n(-x_{\mathrm{E}}) \exp\left(\frac{x + x_{\mathrm{E}}}{L_{\mathrm{E}}} \right) \tag{21.4a}$$

$$n(x_{\mathrm{C}}) = n_{\mathrm{C}} + \delta n(-x_{\mathrm{C}}) \exp\left(-\frac{x - x_{\mathrm{C}}}{L_{\mathrm{C}}} \right) . \tag{21.4b}$$

L_{E} and L_{C} are the minority carrier (electron) diffusion lengths in the emitter and collector, respectively. The solution for the hole density in the neutral region in the base ($0 < x < w$) is

$$\begin{aligned} p(x) = p_{\mathrm{B}} &+ \left[\frac{\delta p(w) - \delta p(0) \exp\left(-w/L_{\mathrm{B}}\right)}{2 \sinh\left(w/L_{\mathrm{B}}\right)} \right] \exp\left(\frac{x}{L_{\mathrm{B}}} \right) \\ &- \left[\frac{\delta p(w) - \delta p(0) \exp\left(w/L_{\mathrm{B}}\right)}{2 \sinh\left(w/L_{\mathrm{B}}\right)} \right] \exp\left(-\frac{x}{L_{\mathrm{B}}} \right) . \end{aligned} \tag{21.5}$$

We shall denote the excess hole density at $x = 0$ and $x = w$ as $\delta p_{\rm E} = \delta p(0)$ and $\delta p_{\rm C} = \delta p(w)$, respectively. Typical ('normal') operation condition in the common base circuit is that $\delta p_{\rm C} = 0$ (Fig. 21.8a). In the 'inverted' configuration, the role of emitter and collector are reversed and $\delta p_{\rm E} = 0$. We can write (21.5) also as

$$p(x) = p_{\rm B} + \delta p_{\rm E} \frac{\sinh\left[(w-x)/L_{\rm B})\right]}{\sinh\left[w/L_{\rm B}\right]} + \delta p_{\rm C} \frac{\sinh\left[x/L_{\rm B})\right]}{\sinh\left[w/L_{\rm B}\right]} \,. \tag{21.6}$$

If the base is thick, i.e. $w \to \infty$, or at least large compared to the diffusion length ($w/L_{\rm B} \gg 1$), the carrier concentration is given by

$$p(x) = p_{\rm B} + \delta p(0) \exp\left(\frac{-x}{L_{\rm B}}\right) \tag{21.7}$$

and does *not* depend on the collector. In this case there is no transistor effect. A 'coupling' between emitter and collector currents that are given by the derivative $\partial p/\partial x$ at 0 and w, respectively, is only present for a sufficiently thin base.

From (21.6), the hole current densities at $x = 0$ and $x = w$ are given as[2]

$$j_{\rm Ep} = j_{\rm p}(0) = e\frac{D_{\rm B}}{L_{\rm B}} \left[\delta p_{\rm E} \coth\left(\frac{w}{L_{\rm B}}\right) - \delta p_{\rm C} \,\mathrm{csch}\left(\frac{w}{L_{\rm B}}\right)\right] \tag{21.8a}$$

$$j_{\rm Cp} = j_{\rm p}(w) = e\frac{D_{\rm B}}{L_{\rm B}} \left[\delta p_{\rm E} \,\mathrm{csch}\left(\frac{w}{L_{\rm B}}\right) - \delta p_{\rm C} \coth\left(\frac{w}{L_{\rm B}}\right)\right] \,. \tag{21.8b}$$

From (21.4a,b), the electron current densities at $x = -x_{\rm E}$ and $x = x_{\rm C}$ are given (with $\delta n_{\rm E} = \delta n(-x_{\rm E})$ and $\delta n_{\rm C} = \delta n(x_{\rm C})$) by

$$j_{\rm En} = j_{\rm n}(-x_{\rm E}) = e\frac{D_{\rm E}}{L_{\rm E}}\,\delta n_{\rm E} \tag{21.9a}$$

$$j_{\rm Cn} = j_{\rm n}(x_{\rm C}) = -e\frac{D_{\rm C}}{L_{\rm C}}\,\delta n_{\rm C} \,. \tag{21.9b}$$

The emitter current density is (similar to (18.104))

$$j_{\rm E} = j_{\rm p}(0) + j_{\rm n}(-x_{\rm E}) = e\frac{D_{\rm B}}{L_{\rm B}} \left[\delta p_{\rm E} \coth\left(\frac{w}{L_{\rm B}}\right) - \delta p_{\rm C} \,\mathrm{csch}\left(\frac{w}{L_{\rm B}}\right)\right] + e\frac{D_{\rm E}}{L_{\rm E}}\,\delta n_{\rm E} \,. \tag{21.10}$$

The collector current density is given as

$$j_{\rm C} = j_{\rm p}(w) + j_{\rm n}(x_{\rm C}) = e\frac{D_{\rm B}}{L_{\rm B}} \left[\delta p_{\rm E} \,\mathrm{csch}\left(\frac{w}{L_{\rm B}}\right) - \delta p_{\rm C} \coth\left(\frac{w}{L_{\rm B}}\right)\right] - e\frac{D_{\rm C}}{L_{\rm C}}\,\delta n_{\rm C} \,. \tag{21.11}$$

---

[2] $\coth x \equiv \sinh x/\cosh x$, $\mathrm{csch}\, x \equiv 1/\sinh x$.

In these equations, only the diffusion currents are considered. Additionally, the recombination currents in the depletion layers must be considered, in particular at small junction voltages.

21.2.2 Current Amplification

The emitter current consists of two parts, the hole current I_{pE} injected from the base and the electron current I_{nE} that flows from the emitter to the base (Fig. 21.5a). Similarly, the collector current is made up from the hole and electron currents I_{pC} and I_{pC}, respectively.

The total emitter current splits into the base and collector currents

$$I_{\mathrm{E}} = I_{\mathrm{B}} + I_{\mathrm{C}} \,. \tag{21.12}$$

The amplification (gain) in common base circuits

$$\alpha_0 = h_{\mathrm{FB}} = \frac{\partial I_{\mathrm{C}}}{\partial I_{\mathrm{E}}} = \frac{\partial I_{\mathrm{pE}}}{\partial I_{\mathrm{E}}} \frac{\partial I_{\mathrm{pC}}}{\partial I_{\mathrm{pE}}} \frac{\partial I_{\mathrm{C}}}{\partial I_{\mathrm{pC}}} = \gamma \alpha_{\mathrm{T}} M \,, \tag{21.13}$$

where γ is the emitter efficiency, α_{T} the base transport factor and M the collector multiplication factor. Since the collector is normally operated below the threshold for avalanche multiplication, $M = 1$. The current amplification in the common emitter circuit is

$$\beta_0 = h_{\mathrm{FE}} = \frac{\partial I_{\mathrm{C}}}{\partial I_{\mathrm{B}}} \,. \tag{21.14}$$

Using (21.12), we find

$$\beta_0 = \frac{\partial I_{\mathrm{E}}}{\partial I_{\mathrm{B}}} - 1 = \frac{\partial I_{\mathrm{E}}}{\partial I_{\mathrm{C}}} \frac{\partial I_{\mathrm{C}}}{\partial I_{\mathrm{B}}} - 1 = \frac{1}{\alpha_0} \beta_0 - 1 = \frac{\alpha_0}{1 - \alpha_0} \,. \tag{21.15}$$

Since α_0 is close to 1 for a well-designed transistor, β_0 is a large number, e.g. $\beta_0 = 99$ for $\alpha_0 = 0.99$.

The emitter efficiency is (A denotes the device area)

$$\gamma = A \left. \frac{\partial j_{\mathrm{p}}}{\partial I_{\mathrm{E}}} \right|_{x=0} = \left[1 + \frac{n_{\mathrm{E}}}{p_{\mathrm{B}}} \frac{D_{\mathrm{E}}}{D_{\mathrm{B}}} \frac{L_{\mathrm{B}}}{L_{\mathrm{E}}} \tanh\left(\frac{w}{L_{\mathrm{B}}} \right) \right]^{-1} \,. \tag{21.16}$$

The base transport factor, i.e. the ratio of minority carriers reaching the collector and the total number of injected minority carriers, is

$$\alpha_{\mathrm{T}} = \frac{j_{\mathrm{p}}(w)}{j_{\mathrm{p}}(0)} = \frac{\partial p / \partial x|_{x=w}}{\partial p / \partial x|_{x=0}} = \frac{1}{\cosh\left(w / L_{\mathrm{B}} \right)} \approx 1 - \frac{w^2}{2 L_{\mathrm{B}}^2} \,. \tag{21.17}$$

The approximation is valid if α_{T} is close to 1. If the base length is a tenth of the diffusion length, the base transport factor is $\alpha_{\mathrm{T}} > 0.995$. Then

$$\beta_0 = h_{\mathrm{FE}} \cong \frac{\gamma}{1-\gamma} \propto \frac{N_{\mathrm{E}}}{w\,N_{\mathrm{B}}}\,, \tag{21.18}$$

with N_{E} and N_{B} being the doping levels in the emitter and base, respectively. The base and collector current are shown in Fig. 21.6 as a function of the emitter–base voltage, i.e. the voltage at the injection diode. The collector current is close to the emitter–base diode current and displays a dependence $\propto \exp(eV_{\mathrm{EB}}/kT)$. The base current shows a similar slope but is orders of magnitude smaller in amplitude. For small forward voltages of the emitter–base diode, the current is typically dominated by a nonradiative recombination current that flows through the base contact and has an ideality factor (m in Fig. 21.6) close to 2.

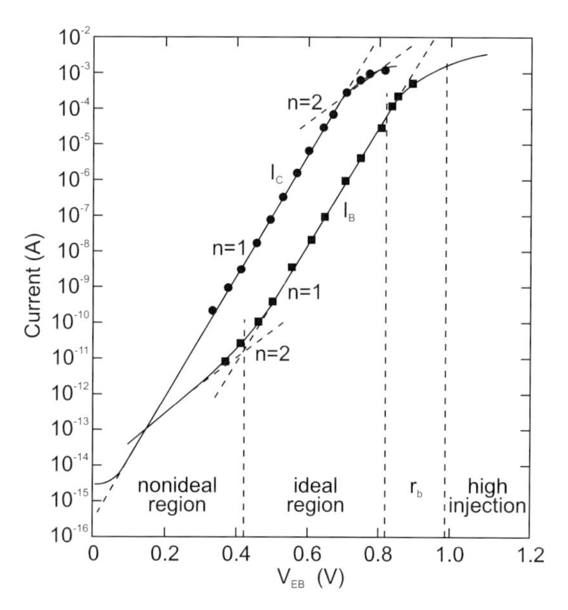

Fig. 21.6. Collector current I_{C} and base current I_{B} as a function of the emitter–base voltage V_{EB}. Adapted from [712]

21.2.3 Ebers–Moll Model

The Ebers–Moll model (Fig. 21.7) was developed in 1954 and is a relatively simple transistor model that needs, at its simplest level (Fig. 21.7a) just three parameters. It can (and must) be refined (Figs. 21.7b,c). The model considers two ideal diodes ('F' (forward) and 'R' (reverse)) back to back, each feeding a current source. The F diode represents the emitter–base diode and and the R diode the collector–base diode. The currents are

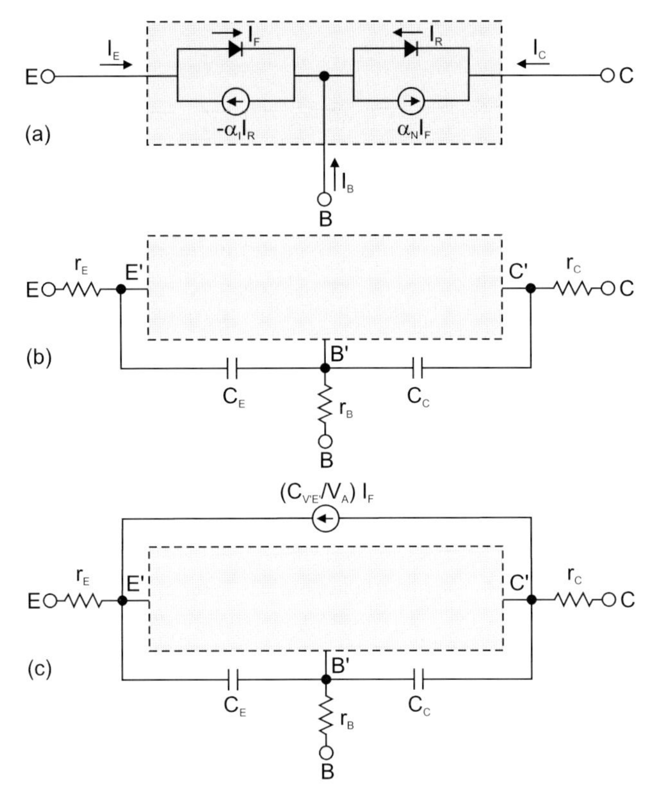

Fig. 21.7. Ebers–Moll model of a transistor, 'E': emitter, 'C': collector and 'B': base. (**a**) Basic model (*grey area* in (**b**,**c**)), (**b**) model with series resistances and depletion-layer capacitances, (**c**) model additionally including the Early effect (V_A: Early voltage)

$$I_F = I_{F0}\left[\exp\left(\frac{eV_{EB}}{kT}\right) - 1\right] \tag{21.19a}$$

$$I_R = I_{R0}\left[\exp\left(\frac{eV_{CB}}{kT}\right) - 1\right] . \tag{21.19b}$$

Using (21.8a,b)–(21.11), the emitter and collector currents are

$$I_E = \hat{a}_{11}\left[\exp\left(\frac{eV_{EB}}{kT}\right) - 1\right] + \hat{a}_{12}\left[\exp\left(\frac{eV_{CB}}{kT}\right) - 1\right]$$

$$I_C = \hat{a}_{21}\left[\exp\left(\frac{eV_{EB}}{kT}\right) - 1\right] + \hat{a}_{22}\left[\exp\left(\frac{eV_{CB}}{kT}\right) - 1\right] ,$$

with

$$\hat{a}_{11} = eA\left[p_\mathrm{B}\frac{D_\mathrm{B}}{L_\mathrm{B}}\coth\left(\frac{w}{L_\mathrm{B}}\right) + n_\mathrm{E}\frac{D_\mathrm{E}}{L_\mathrm{E}}\right] \tag{21.21a}$$

$$\hat{a}_{12} = -eAp_\mathrm{B}\frac{D_\mathrm{B}}{L_\mathrm{B}}\operatorname{csch}\left(\frac{w}{L_\mathrm{B}}\right) \tag{21.21b}$$

$$\hat{a}_{21} = eAp_\mathrm{B}\frac{D_\mathrm{B}}{L_\mathrm{B}}\operatorname{csch}\left(\frac{w}{L_\mathrm{B}}\right) = -\hat{a}_{12} \tag{21.21c}$$

$$\hat{a}_{22} = -eA\left[p_\mathrm{B}\frac{D_\mathrm{B}}{L_\mathrm{B}}\operatorname{csch}\left(\frac{w}{L_\mathrm{B}}\right) + n_\mathrm{C}\frac{D_\mathrm{C}}{L_\mathrm{C}}\right] . \tag{21.21d}$$

The currents at the three contacts are

$$I_\mathrm{E} = I_\mathrm{F} - \alpha_\mathrm{I} I_\mathrm{R} \tag{21.22a}$$

$$I_\mathrm{C} = \alpha_\mathrm{N} I_\mathrm{F} - I_\mathrm{R} \tag{21.22b}$$

$$I_\mathrm{B} = (1-\alpha_\mathrm{N})I_\mathrm{F} + (1-\alpha_\mathrm{I})I_\mathrm{R} . \tag{21.22c}$$

The last equation is obtained from (21.22a,b) using (21.12). By comparison with (21.19a,b) and (21.21a–d) we find

$$I_\mathrm{F0} = \hat{a}_{11} \tag{21.23a}$$

$$I_\mathrm{R0} = -\hat{a}_{22} \tag{21.23b}$$

$$\alpha_\mathrm{I} = \hat{a}_{12}/I_\mathrm{R0} \tag{21.23c}$$

$$\alpha_\mathrm{N} = \hat{a}_{21}/I_\mathrm{F0} = -\hat{a}_{12}/I_\mathrm{F0} = -\alpha_\mathrm{I} I_\mathrm{R0}/I_\mathrm{F0} . \tag{21.23d}$$

The constants α_I and α_N are the forward and reverse gains in the common base circuit, respectively. Both constants are larger than zero. Typically, $\alpha_\mathrm{I} \lessapprox 1$ and $\alpha_\mathrm{I} < \alpha_\mathrm{N}$. The model has three independent parameters. Equations (21.22a,b) can be rewritten as

$$I_\mathrm{E} = \alpha_\mathrm{I} I_\mathrm{C} - (1-\alpha_\mathrm{I}\alpha_\mathrm{N})I_\mathrm{F} \tag{21.24a}$$

$$I_\mathrm{C} = \alpha_\mathrm{N} I_\mathrm{E} - (1-\alpha_\mathrm{I}\alpha_\mathrm{N})I_\mathrm{R} . \tag{21.24b}$$

The model can be refined and made more realistic by including the effect of series resistances and depletion-layer capacitances, increasing the number of parameters to eight. The Early effect (see p. 587) can be included by adding a further current source. This level is the 'standard' Ebers–Moll model with a total of nine parameters. Further parameters can be added. However, as is always the case with simulations, there is a tradeoff between the simplicity of the model and to what detail a real situation is approximated.

21.2.4 Current–Voltage Characteristics

In Fig. 21.8, the hole density in the base (of a pnp transistor) is shown for various voltage conditions. In Fig. 21.9, the I–V characteristics of a bipolar transistor in common base and common collector circuit are shown. In the

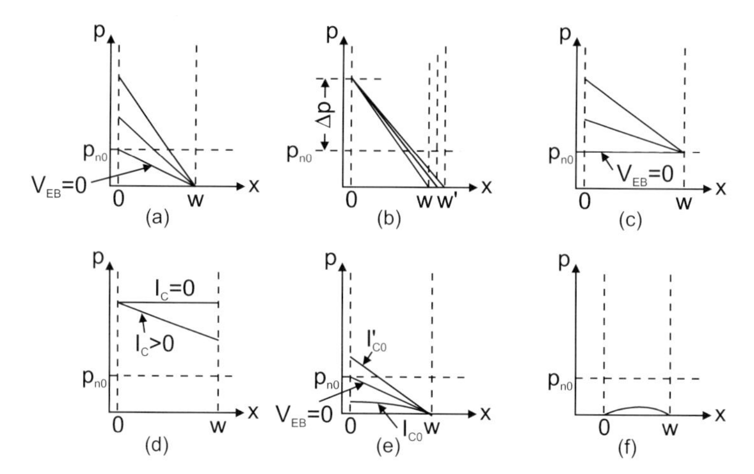

Fig. 21.8. Hole density (linear scale) in the base region (the neutral part of the base ranges from 0 to w) of a pnp transistor for various voltages. (**a**) normal voltages, $V_{\rm CB}$ = const. and various $V_{\rm EB}$ (in forward direction). (**b**) $V_{\rm EB}$ = const. and various values of $V_{\rm CB}$. (**c**) Various values of $V_{\rm EB} > 0$, $V_{\rm CB} = 0$. (**d**) Both pn-junctions in forward direction. (**e**) Conditions for $I_{\rm C0}$ and $I'_{\rm C0}$. (**f**) both junctions in reverse direction. Adapted from [183]

common base circuit (Fig. 21.9a), the collector current is practically equal to the emitter current and is almost independent of the collector–base voltage. From (21.24b), the dependence of the collector current on the collector–base voltage is given (within the Ebers–Moll model) as

$$I_{\rm C} = \alpha_{\rm N} I_{\rm E} - (1 - \alpha_{\rm I}\alpha_{\rm N}) I_{\rm R0} \left[\exp\left(\frac{eV_{\rm CB}}{kT} \right) - 1 \right] . \tag{21.25}$$

$V_{\rm CB}$ is in the reverse direction. Therefore, the second term is zero for normal operating conditions. Since $\alpha_{\rm N} \lessapprox 1$, the collector current is almost equal to the emitter current.

Even at $V_{\rm CB} = 0$ (the case of (Fig. 21.8c), holes are extracted from the base since $\partial p/\partial x|_{x=w} > 0$. A small forward voltage must be applied to the collector–base diode in order to make the current zero, i.e. $\partial p/\partial x|_{x=w} = 0$ (Fig. 21.8d). The collector saturation current $I_{\rm C0}$ is measured with an open emitter side. This current is smaller than the saturation current of the CB diode, since at the emitter side of the basis a vanishing gradient of the hole density is present (Fig. 21.8e). This reduces the gradient (and thus the current) at the collector side. The current $I_{\rm C0}$ is therefore smaller than the collector current for shorted emitter–base contact ($V_{\rm EB} = 0$). At high collector voltage, the current increases rapidly at $BV_{\rm CB0}$ due to breakdown of the collector–base diode. It can also occur that the width of the neutral base region w becomes zero (punch-through). In this case, the emitter and collector are short-circuited.

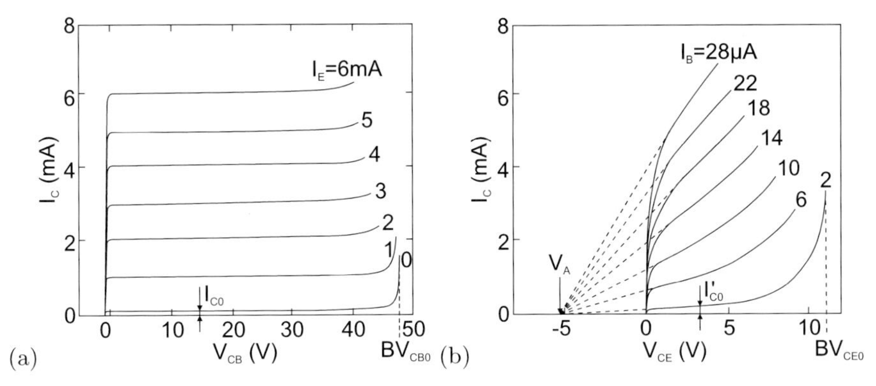

Fig. 21.9. Characteristics (I_{C} vs. V_{CB}) of a pnp transistor in (**a**) common base (CB) circuit (Fig. 21.4a) for various values of the emitter current as labeled. Adapted from [713]. (**b**) Characteristics in common emitter (CE) circuit (Fig. 21.4b). Adapted from [714]

In the common emitter circuit (Fig. 21.9b), there is a high current amplification $I_{\mathrm{C}}/I_{\mathrm{B}}$. Note that the collector current is given in mA and the base current in μA. The current increases with increasing V_{CE} because the base width w decreases and β_0 increases. There is no saturation of the I–V characteristics (Early effect). Instead, the I–V curves look as if they start at a negative collector–emitter voltage, the so-called Early voltage V_{A}. In the linear regime, the characteristic can be approximated by

$$I_{\mathrm{C}} = \left(1 + \frac{V_{\mathrm{CE}}}{V_{\mathrm{A}}}\right) \beta_0 I_{\mathrm{B}} \; . \tag{21.26}$$

The physical reason for the increase of the collector current with increasing V_{CE} is the increasing reverse voltage at the collector–base diode that causes a so-called base-width modulation, as shown in Fig. 21.8b. This causes an expansion of the CB depletion layer and subsequently a reduction of the neutral base width w. w will be smaller and smaller compared to the geometrical base width w_{B}. When w is reduced, the base transfer factor α_{T} (21.17) becomes closer to 1 and the current gain β_0 (21.15) increases. Therefore, the collector current increases for a given (fixed) base current. For transistors for which the geometrical base width is much larger than the width of the depletion layers, the Early voltage is

$$V_{\mathrm{A}} \approx \frac{e}{\epsilon_{\mathrm{s}}} N_{\mathrm{B}} w_{\mathrm{B}}^2 \; . \tag{21.27}$$

For small collector–emitter voltage, the current quickly drops to zero. V_{CE} is typically split in such a way that the emitter–base diode is well biased forward and the CB diode has a high reverse voltage. If V_{CE} drops below a certain value ($\approx 1\,\mathrm{V}$ for silicon transistors), there is no longer any bias at the

CB diode. A further reduction of V_{CE} biases the CB diode in the forward direction and quickly brings the collector current down to zero.

21.2.5 Basic Circuits

Common Base Circuit

In the common base configuration, there is no current amplification since the currents flowing through emitter and base are almost the same. However, there is voltage gain since the collector current causes a large voltage drop across the load resistor.

Common Emitter Circuit

The input resistance of the common emitter circuit (Fig. 21.10a) depends on the emitter–base diode and varies between a value of the order of 100 kΩ at small current and a few Ω at larger current and high V_{EB}. The voltage gain is

$$r_{\mathrm{V}} = \frac{V_{\mathrm{CE}}}{V_{\mathrm{EB}}} = \frac{I_{\mathrm{C}}}{V_{\mathrm{EB}}} R_{\mathrm{L}} \,, \tag{21.28}$$

where R_{L} is the load resistance in the output circuit (see Fig. 21.4b). The ratio $g_{\mathrm{m}} = I_{\mathrm{C}}/V_{\mathrm{EB}}$ is called the forward transconductance. Also, the differential transconductance $g'_{\mathrm{m}} = \partial I_{\mathrm{C}}/\partial V_{\mathrm{EB}}$ is used. The voltage gain of the common emitter circuit is typically 10^2–10^3. Since current *and* voltage are amplified, this circuit has the highest power gain.

If the input voltage V_{EB} (U_1 in Fig. 21.10a) is increased, the collector current rises. This increase causes an increase of the voltage drop across the load resistance R_{L} and a decrease of the output voltage U_2. Therefore, the phase of the input signal is reversed and the amplifier is inverting.

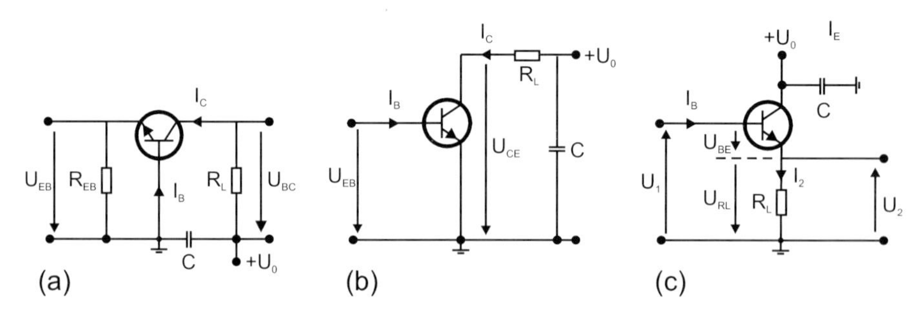

Fig. 21.10. (**a**) Common base, (**b**) common emitter and (**c**) common collector circuits with external loads

Common Collector Circuit

In Fig. 21.10c, the collector is connected to ground for alternating currents. Input and output current flow through the load resistance at which part of the input voltage drops. The input voltage is divided between the load resistor R_L and the emitter–base diode. At the transistor, the voltage $V_{BE} = V_1 - V_{RL}$ is applied. If the input voltage is increased, I_2 increases. This leads to a larger voltage drop at the load resistor and therefore to a decrease of V_{BE}, working against the original increase. The input resistance R_1 is large despite a small load resistance, $R_1 \approx \beta_0 R_L$. The input voltage is larger than V_{RL}, thus no voltage gain occurs (actually it is a little smaller than 1). The current amplification is $(\beta + 1)$. The output resistance R_2 is small, $R_2 = U_2/I_2 \approx R_L/\beta_0$. Therefore, this circuit is also called an impedance amplifier that allows high-impedance sources to be connected low-impedance loads. Since an increase of the input voltage leads to an increase of the output voltage that is present at the emitter, this circuit is a direct amplifier and is also called an emitter follower.

21.2.6 High-Frequency Properties

Transistors for amplification of high-frequency signals are typically chosen as npn transistors since electrons, the minority carriers in the base, have higher mobility than holes. The active area and parasitic capacities must be minimized. The emitter is formed in the shape of a stripe, nowadays in the 100 nm regime. The base width is in the 10 nm range. High p-doping of GaAs with low diffusion of the dopant is accomplished with carbon. Defects that would short emitter and collector at such thin base width must be avoided.

An important figure of merit is the cutoff frequency f_T for which h_{FE} is unity in the common emitter configuration. The cutoff frequency is related to the emitter–collector delay time τ_{EC} by

$$f_T = \frac{1}{2\pi\tau_{EC}} . \tag{21.29}$$

The delay time is determined by the charging time of the emitter–base depletion layer, the base capacity, and the transport through the base–collector depletion layer. It is favorable if all times are short and similar. It does not help to minimize only one or two of the three processes since the longest time determines the transistor performance.

Another important figure of merit is the maximum frequency with which the transistor can oscillate in a feedback circuit with zero loss. This frequency is denoted by f_{max}. Approximately,

$$f_{max} \simeq \left(\frac{f_T}{8\pi R_B C_{CB}} \right)^{1/2} , \tag{21.30}$$

where R_B is the base resistance and C_{CB} is the collector–base capacity. f_{max} is larger than f_T, by a factor of the order of three.

21.2.7 Heterobipolar Transistors

In a heterojunction bipolar transistor (HBT), the emitter–base diode is formed with a heterostructure diode. The desired functionality is obtained when the emitter is made from the higher-bandgap material and the base from the lower-bandgap material. The schematic band diagram is shown in Fig. 21.11 (cf. Fig. 18.64c for the emitter–base diode).

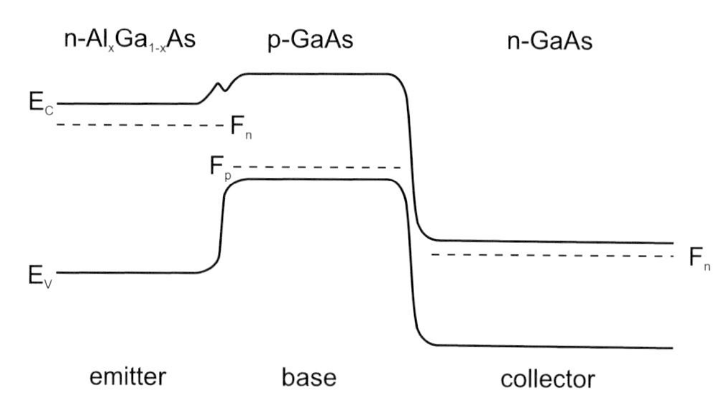

Fig. 21.11. Schematic band diagram of a heterojunction bipolar transistor

The higher discontinuity in the valence band, compared to a homojunction with the base material, provides a higher barrier for hole transport from the base to the emitter. Thus, the emitter efficiency is increased. Another advantage is the possibility for higher doping of the base without loss of emitter efficiency. This reduces the base series resistance and leads to better high-frequency behavior due to higher current gain and a smaller RC time constant. Also, operation at higher temperature is possible when the emitter has a larger bandgap. Current InP/InGaAs-based HBTs have cutoff frequencies beyond 30 GHz, SiGe-HBTs beyond 80 GHz. The high-frequency performance is influenced by the velocity-overshoot effect (cf. Sect. 8.5.2) [715].

In Fig. 21.12, an InAlAs/InGaAs HBT is shown [717]. The cutoff frequency is 90 GHz. For the layer design, a fairly thick collector with low doping was chosen. This design allows a broad depletion layer with fairly small maximum electric field and thus a high breakdown voltage of $BV_{\mathrm{CE0}} > 8.5\,\mathrm{V}$. The base is not too thin (80 instead of maybe 60 nm) to reduce the series resistance. A graded region between emitter and base was chosen to avoid a spike occurring in the conduction band (Fig. 18.64b) and keep the turn-on voltage low.

21.2.8 Light-emitting Transistors

The base current has two components. One is the recombination current in the neutral region of the emitter; this current can be suppressed in the HBT.

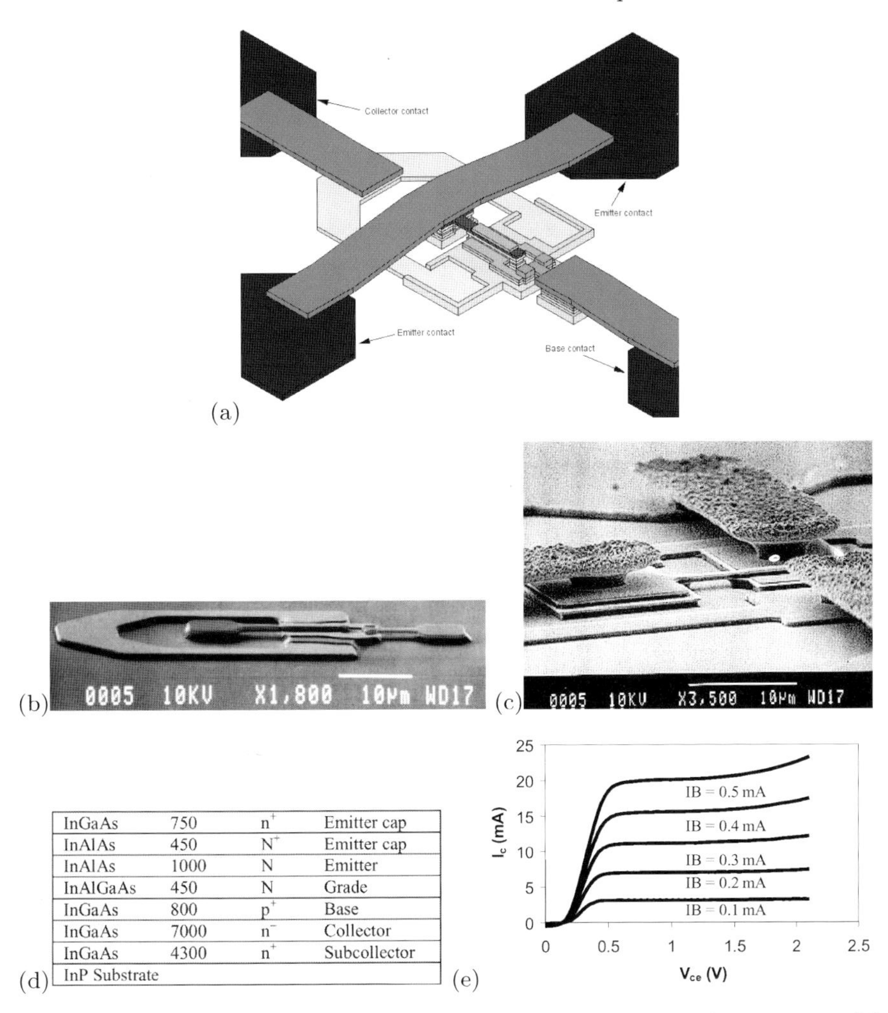

InGaAs	750	n^+	Emitter cap
InAlAs	450	N^+	Emitter cap
InAlAs	1000	N	Emitter
InAlGaAs	450	N	Grade
InGaAs	800	p^+	Base
InGaAs	7000	n^-	Collector
InGaAs	4300	n^+	Subcollector
InP Substrate			

Fig. 21.12. (**a**) Schematic layout of a high-frequency HBT and SEM images (**b**) without and (**c**) with contacts. (**d**) Epitaxial layer sequence and (**e**) static performance data. Parts (**a**,**b**) from [716], parts (**d**,**e**) from [717]

The other is the recombination in the base region itself.[3] If quantum wells are introduced into the base region, this recombination can occur radiatively between electrons and holes captured into the quantum well. The spectrum exhibits two peaks from the QWs and the GaAs barrier.

[3]Also, a recombination current in the emitter–base depletion region is possible. However, since in normal operating conditions this diode is forward biased, the depletion layer is short and the associated recombination current is small, cf. p. 452.

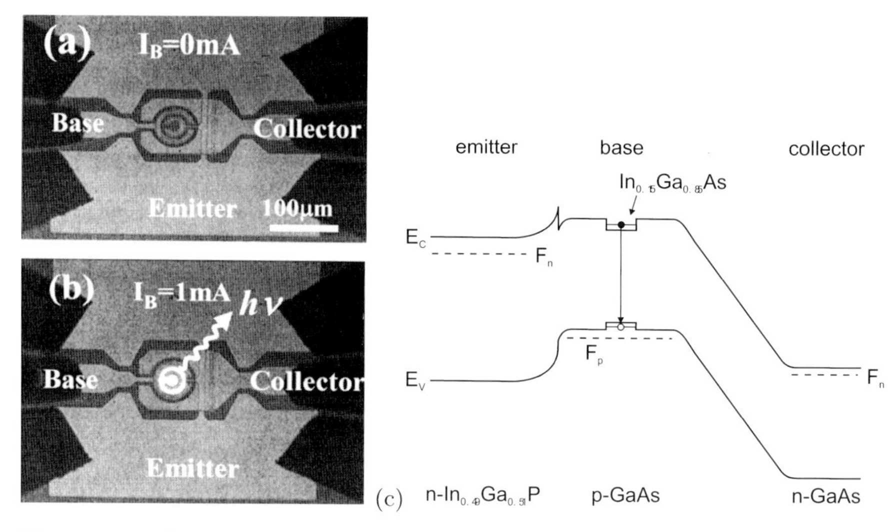

Fig. 21.13. Microscopic image of an InGaP/GaAs HBT with two 5-nm InGaAs/GaAs QWs in the 30-nm wide base at (**a**) zero base current and (**b**) at 1 mA base current in the common emitter configuration with Si CCD image of light emission. (**c**) Schematic band diagram of a HBT with a single InGaAs/GaAs quantum well in the base. Parts (**a,b**) from [718], part (**c**) adapted from [718]

21.3 Field-Effect Transistors

Next to the bipolar transistors, the field-effect transistors (FET) are another large class of transistors. FETs were conceptualized first but due to technological difficulties with semiconductor surfaces, realized second. The principle is fairly simple: A current flows through a channel from source to drain. The current is varied via the channel conductivity upon the change of the gate voltage. The gate needs to make a nonohmic contact to the semiconductor. Since the conductivity in the channel is a property related to the majority charge carriers, FETs are called unipolar transistors. FETs feature a higher input impedance than bipolar transistors, a good linearity, and a negative temperature coefficient and thus a more homogeneous temperature distribution. According to the structure of the gate diode we distinguish JFETs, MESFETs and MOSFETs, as discussed in the following:

In the junction FET (JFET), the variation of channel conductivity is accomplished via the extension of the depletion layer of the pn-junction formed by the gate and the channel material (Fig. 21.14a). The JFET was analyzed by Schottky in 1952 [25] and realized by Dacey and Ross in 1953 [26].

In a MESFET, a metal–semiconductor diode (Schottky diode) is used as rectifying contact instead of a pn-diode. Otherwise, the principle is the same as that of the JFET. After the proposal by Mead in 1966 [35], the first (epitaxial) GaAs MESFET was realized by Hooper and Lehrer in 1967 [36].

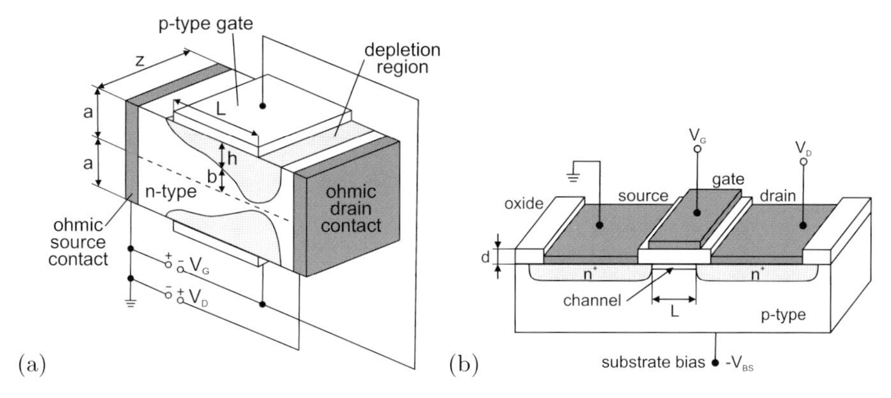

Fig. 21.14. (**a**) Shockley's model of a JFET. The *dashed line* represents the middle of the symmetric channel of total thickness $2a$. The *light grey area* is the depletion layer with thickness h. The gate length is L. The *dark grey* areas are ohmic metal contacts. Based on [26]. (**b**) Scheme of a MOSFET with channel length L and oxide thickness d. The *dark grey* areas are ohmic metal contacts. Adapted from [183]

The MESFET offers some advantages, such as the fabrication of the metal gate at lower temperature than necessary for the (diffusion or epitaxy of the) pn-diode, lower resistance, good thermal contact. The JFET can be made with a heterostructure gate to improve the frequency response.

In a MISFET, the gate diode is a metal–insulator–semiconductor diode (Fig. 21.14b). If the insulator is an oxide, the related FET is a MOSFET. When the gate is put at a positive voltage (for a p-channel), an inversion layer is formed close to the insulator–semiconductor interface. This layer is an n-conducting channel allowing conduction between the two oppositely biased pn-diodes. It can carry a high current. The MOSFET was theoretically envisioned early by Lilienfeld in 1925 [12] and realized only in 1960 by Kahng and Atalla [30].

FETs come in 'n' and 'p' flavors, depending on the conductivity type of the channel. For high-frequency applications, typically an n-channel is used due to the higher mobility or drift velocity. In CMOS (complementary MOS) technology, both n-FETs and p-FETs are integrated in high density, allowing the effective realization of logic gates with minimized power consumption.

21.4 JFET and MESFET

21.4.1 General Principle

The principal characteristic of a JFET is shown in Fig. 21.15. At $V_{\mathrm{D}} = 0$ and $V_{\mathrm{G}} = 0$, the transistor is in thermodynamic equilibrium and there are no net currents. Underneath the gate diode, a depletion layer is present. If for zero gate voltage the source–drain voltage is applied to the channel, the

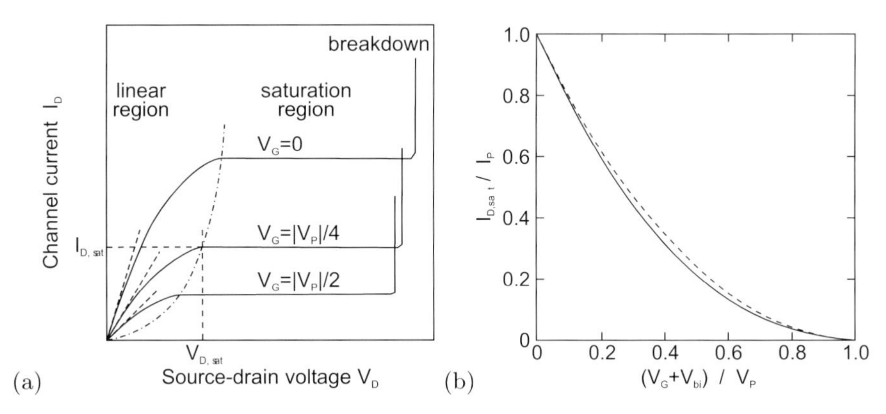

Fig. 21.15. (**a**) Principal characteristics of a JFET. The channel current $I_{\rm D}$ is shown as a function of the source–drain voltage $V_{\rm D}$ for three different values of the (absolute value of the) gate voltage $V_{\rm G}$. The saturation values $V_{\rm D,sat}$ and $I_{\rm D,sat}$ are indicated for one curve. The intersections with the *dash-dotted* line yield the saturation voltage. Adapted from [183] (**b**) Transfer behavior of a JFET for two different carrier distributions. The *inset* depicts the geometry of the upper channel half. Adapted from [183] after [719, 720]

current increases linearly. The positive voltage at the drain contact causes the expansion of the depletion layer of the (reversely biased) gate–drain pn-diode. When the two (the upper and the lower) depletion regions meet (pinch-off), the current saturates. The respective source–drain voltage is denoted as $V_{\rm D,sat}$. For high gate–drain (reverse) voltage $V_{\rm D}$, breakdown occurs with a strong increase of the source–drain current. A variation of the gate voltage $V_{\rm G}$ leads to a variation of the source–drain current. A reverse voltage leads to a reduction of the saturation current and saturation at lower source–drain voltage. For a certain gate voltage $V_{\rm P}$, the pinch-off voltage, no current can flow in the channel any longer since pinch-off exists even for $V_{\rm D} = 0$.

21.4.2 Static Characteristics

Here, we will calculate the general static behavior outlined in the previous section. We assume a long channel ($L \gg a$), the abrupt approximation for the depletion layer, the gradual channel approximation, i.e. the depletion layer depth changes slowly along x, and a field-independent, constant mobility. In this case, the two-dimensional Poisson equation for the potential distribution V can be used by solving it along the y direction (channel depth) for all x-positions (adiabatic approximation),

$$\frac{\partial^2 V}{\partial y^2} = -\frac{\rho(y)}{\epsilon_{\rm s}} \,. \tag{21.31}$$

The geometry is shown in the inset of Fig. 21.15b.

The depth h of the depletion layer in the abrupt approximation is given by (cf. (18.84), reverse voltages are counted as positive here)

$$h = \left[\frac{2\epsilon_s}{eN_D} (V_{bi} + V_G + V(x)) \right]^{1/2} . \tag{21.32}$$

Here, we have assumed homogeneous doping, i.e. N_D does not depend on y (or x). The built-in voltage (for a p*n gate diode) is given by $V_{bi} = \frac{kT}{e} \ln\left(\frac{N_D}{n_i}\right)$ (18.75a). The voltage V is the applied source–drain voltage in relation to the source. The depth of the depletion layer at $x = 0$ (source) and $x = L$ (drain) is given by

$$y_1 = h(0) = \left[\frac{2\epsilon_s}{eN_D} (V_{bi} + V_G) \right]^{1/2} \tag{21.33a}$$

$$y_2 = h(L) = \left[\frac{2\epsilon_s}{eN_D} (V_{bi} + V_G + V_D) \right]^{1/2} . \tag{21.33b}$$

The maximum value of h is a. Therefore, the pinch-off voltage V_P, at which $V_P = V_{bi} + V_G + V_D$ is such that $h = a$, is given by

$$V_P = \frac{ea^2 N_D}{2\epsilon_s} . \tag{21.34}$$

The (drift) current density along x is given by (cf. (8.41a))

$$j_x = -eN_D\mu_n\mathcal{E}_x = eN_D\mu_n \frac{\partial V}{\partial x} \tag{21.35}$$

for the neutral part of the semiconductor. Therefore, the current in the upper half of the channel is given by

$$I_D = eN_D\mu_n \frac{\partial V(x)}{\partial x} [a - h(x)] Z , \tag{21.36}$$

where Z is the width of the channel (Fig. 21.14a). Although it seems that I_D depends on x, it is of course constant along the channel due to Kirchhoff's law.[4] Using the triviality $\int_0^L I_D \mathrm{d}x = LI_D$ and $\frac{\partial V}{\partial x} = \frac{\partial V}{\partial h}\frac{\partial h}{\partial x}$ with $\frac{\partial V}{\partial h} = eN_D h/\epsilon_s$ from (21.32), we find from (21.36)

$$I_D = \frac{e^2\mu_n N_D Z a^3}{6\epsilon_s L} \left[\frac{3}{a^2}\left(y_2^2 - y_1^2\right) - \frac{2}{a^3}\left(y_1^3 - y_2^3\right) \right] . \tag{21.37}$$

This equation can also be written, using (21.34) and

---

[4] We neglect recombination, in particular since the current is a majority-carrier current.

$$I_{\mathrm{P}} = \frac{e^2 \mu_{\mathrm{n}} N_{\mathrm{D}} Z a^3}{6\epsilon_{\mathrm{s}} L} , \tag{21.38}$$

as

$$I_{\mathrm{D}} = I_{\mathrm{P}} \left[\frac{3V_{\mathrm{D}}}{V_{\mathrm{P}}} - 2\frac{(V_{\mathrm{bi}}+V_{\mathrm{G}}+V_{\mathrm{D}})^{3/2} - (V_{\mathrm{bi}}+V_{\mathrm{G}})^{3/2}}{V_{\mathrm{P}}^{3/2}} \right] . \tag{21.39}$$

The saturation current is reached for $y_2 = a$ or $V_{\mathrm{bi}} + V_{\mathrm{G}} + V_{\mathrm{D}} = V_{\mathrm{P}}$ and is given by

$$I_{\mathrm{D,sat}} = I_{\mathrm{P}} \left[1 - 3\frac{V_{\mathrm{bi}}+V_{\mathrm{G}}}{V_{\mathrm{P}}} + 2\left(\frac{V_{\mathrm{bi}}+V_{\mathrm{G}}}{V_{\mathrm{P}}}\right)^{3/2} \right] . \tag{21.40}$$

The dependence of the saturation current on $(V_{\mathrm{G}}+V_{\mathrm{bi}})/V_{\mathrm{P}}$ is depicted in Fig. 21.15b. We note that for $V_{\mathrm{bi}} + V_{\mathrm{G}} = V_{\mathrm{P}}$, the saturation current is zero since then $V_{\mathrm{D}} = 0$. The gate voltage at which a certain saturation current is reached can be determined graphically from Fig. 21.15b or numerically from (21.40). Letting γ be the saturation current in units of I_{P}, i.e. $\gamma = I_{\mathrm{D,sat}}/I_{\mathrm{P}}$, the gate voltage at the saturation point $V_{\mathrm{G,sat}}$ is given by[5]

$$V_{\mathrm{G,sat}} = V_{\mathrm{P}} \left[\frac{3}{4} - \frac{8^{-1}\mathrm{i}(\sqrt{3}-\mathrm{i})(1+8\gamma)}{\hat{\gamma}} + 8^{-1}\mathrm{i}(\sqrt{3}+\mathrm{i})\hat{\gamma} \right] - V_{\mathrm{bi}} \tag{21.41}$$

$$\hat{\gamma} = \left[-1 + 8(\gamma-1)^{3/2} + 20\gamma + 8\gamma^2 \right]^{1/3} .$$

The source–drain voltage at the saturation point decreases with decreasing saturation current as $V_{\mathrm{D,sat}} = V_{\mathrm{P}} - V_{\mathrm{bi}} - V_{\mathrm{G,sat}}$ (dashed parabola-like line in Fig. 21.15a).

If the charge-carrier distribution differs from the homogeneous distribution assumed so far, a change of transistor properties arises, as shown in Fig. 21.15b for a δ-like carrier distribution. The I–V characteristic is slightly less curved, but not linear. A linear characteristic is only achievable in the drift velocity saturation regime (see Sect. 21.4.4).

For high source–drain voltage $V_{\mathrm{D}} > V_{\mathrm{P}} - V_{\mathrm{bi}} - V_{\mathrm{G}}$, the current remains essentially at its saturation value. For very high source–drain voltage, breakdown in the gate–drain diode can occur, when the maximum voltage, which is given by $V_{\mathrm{G}} + V_{\mathrm{D}}$ at the end of the channel, is equal to the breakdown voltage V_{B}.

The forward transconductance g_{m} and the drain transconductance g_{D} are given by

$$g_{\mathrm{m}} = \frac{\partial I_{\mathrm{D}}}{\partial V_{\mathrm{G}}} = g_{\max} \left[\left(\frac{V_{\mathrm{bi}}+V_{\mathrm{G}}}{V_{\mathrm{P}}}\right)^{1/2} - \left(\frac{V_{\mathrm{bi}}+V_{\mathrm{G}}+V_{\mathrm{D}}}{V_{\mathrm{P}}}\right)^{1/2} \right] \tag{21.42}$$

$$g_{\mathrm{D}} = \frac{\partial I_{\mathrm{D}}}{\partial V_{\mathrm{D}}} = g_{\max} \left[1 - \left(\frac{V_{\mathrm{bi}}+V_{\mathrm{G}}+V_{\mathrm{D}}}{V_{\mathrm{P}}}\right)^{1/2} \right] , \tag{21.43}$$

[5] Although the arguments are complex, the result is real for $0 \leq \gamma \leq 1$.

where

$$g_{\max} = \frac{3I_{\mathrm{P}}}{V_{\mathrm{P}}} = \frac{eN_{\mathrm{D}}a\mu Z}{L} \,. \tag{21.44}$$

The drain transconductance for $V_{\mathrm{D}} \to 0$ (linear regime, dashed straight lines in Fig. 21.15a) is given by

$$g_{\mathrm{D0}} = g_{\max} \left[1 - \left(\frac{V_{\mathrm{bi}} + V_{\mathrm{G}}}{V_{\mathrm{P}}} \right)^{1/2} \right] = g_{\mathrm{m,sat}} \,, \tag{21.45}$$

which is equal[6] to the forward transconductance in the saturation regime $g_{\mathrm{m,sat}} = \partial I_{\mathrm{D,sat}}/\partial V_{\mathrm{G}}$.

21.4.3 Normally On and Normally Off FETs

The JFET discussed so far had an n-conductive channel and was conductive at $V_{\mathrm{G}} = 0$. It is termed an 'n-type, normally on' (or depletion) FET. If the channel is p-conductive, the FET is called 'p-type'. A FET that has a nonconductive channel at $V_{\mathrm{G}} = 0$ is called 'normally off' (or accumulation)

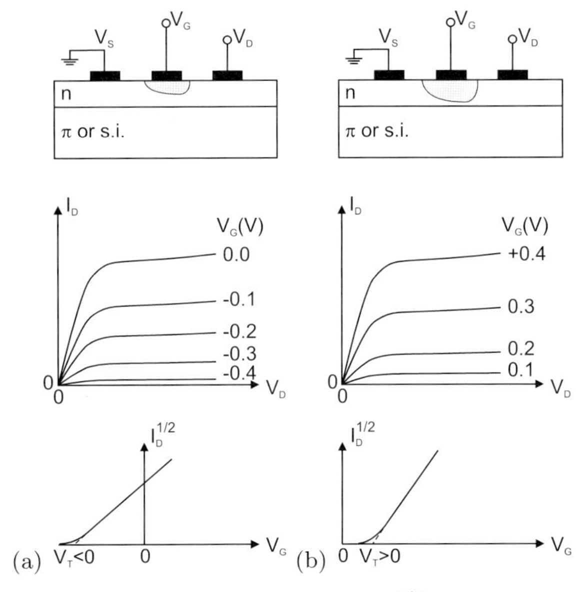

Fig. 21.16. Scheme (*top*), I_{D} vs. V_{D} (*center*) and $I_{\mathrm{D}}^{1/2}$ vs. V_{G} (*bottom*) I–V characteristics for (**a**) normally on (depletion) and (**b**) normally off (accumulation) n-type JFET. Adapted from [183]

[6]Technically, here $g_{\mathrm{D0}} = -g_{\mathrm{m,sat}}$, however, we had counted V_{G} positive for the reverse direction.

FET. In this case, the built-in voltage must be large enough to cause pinch-off. For a positive gate voltage (in the forward direction of the gate–drain diode), current begins to flow. The I–V characteristics of the four FET-types are depicted in Fig. 21.16. The circuit symbols for the four different FET types are shown in Fig. 21.17.

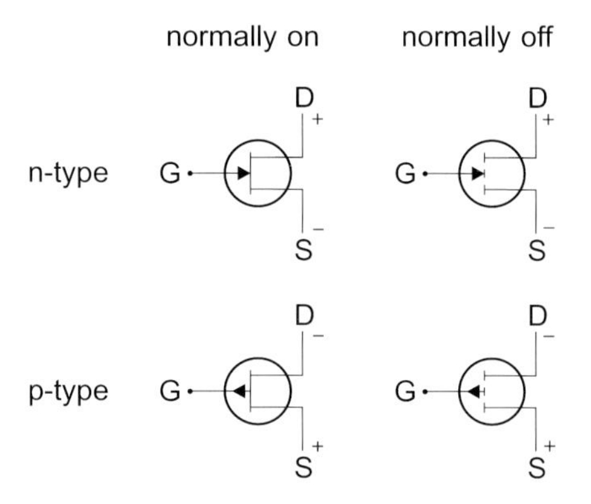

Fig. 21.17. Circuit symbols for various types of FETs

21.4.4 Field-Dependent Mobility

So far, we have considered FETs with long channels ($L \gg a$). This situation is often not the case, in particular for high-integration or high-frequency applications. For short channels, the I–V characteristics exhibit changes. The theory needs to be modified to take into account, among other effects, the electric-field dependence of the mobility (Fig. 8.9) that was discussed in Sect. 8.5.1.

Drift-Velocity Saturation

A material without negative differential mobility, such as Si or Ge, can be described with a drift-velocity model

$$v_{\mathrm{d}} = \mu \mathcal{E} \frac{1}{1 + \mu \mathcal{E} / v_{\mathrm{s}}} . \tag{21.46}$$

In this model, μ denotes the low-field (ohmic) mobility and v_{s} the drift-saturation velocity reached for $\mathcal{E} \gg v_{\mathrm{s}}/\mu$. The fraction in (21.46) describes the drift-velocity saturation.

By inserting (21.46) into (21.36), we obtain

$$I_{\mathrm{D}} = e N_{\mathrm{D}} \mu \mathcal{E}(x) \frac{1}{1 + \mu \mathcal{E}(x) / v_{\mathrm{s}}} \left[a - h(x) \right] Z , \tag{21.47}$$

and after a short calculation the drain current is given by (cf. (21.39))

$$I_D = I_P \left(1 + \frac{\mu V_G}{v_s L}\right)^{-1} \left[\frac{3V_D}{V_P} - 2\frac{(V_{bi} + V_G + V_D)^{3/2} - (V_{bi} + V_G)^{3/2}}{V_P^{3/2}}\right]. \quad (21.48)$$

The factor $1/(1+z)$ with $z = \mu V_G/v_s L$ reduces the channel current due to the drift saturation effect. The effect of the parameter z is depicted in Fig. 21.18 in comparison to $z = 0$, i.e. without the drift saturation effect (or $v_s \to \infty$). The forward conductance $g_{m,sat}$ decreases with increasing z, as shown in Fig. 21.19.

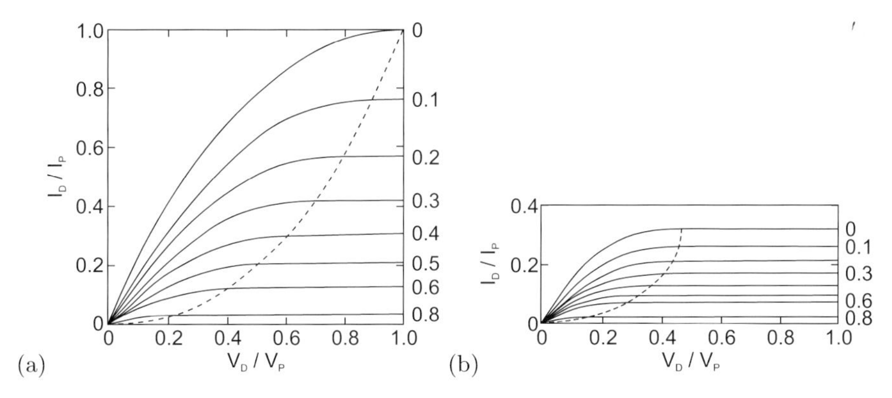

Fig. 21.18. I–V characteristic (**a**) without consideration of drift saturation ($z = 0$) and (**b**) with drift saturation ($z = 3$) for various values of $(V_G+V_{bi})/V_P$ =0, 0.1, 0.2, 0.3, 0.4, 0.5, 0.6, 0.8 as indicated at the right side. The intersections of the *dashed line* and the *solid lines* indicate the beginning of saturation. Adapted from [721]

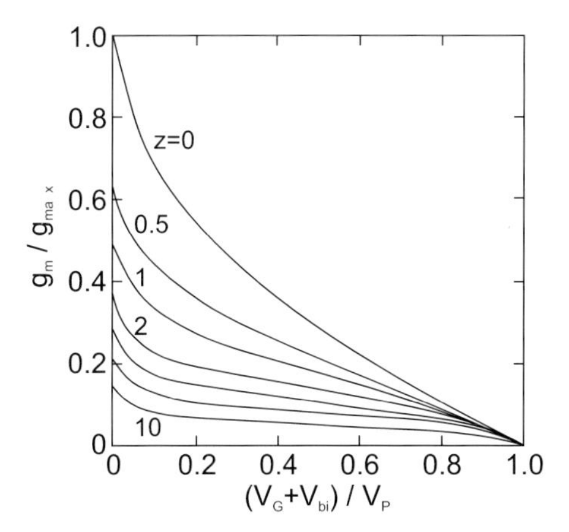

Fig. 21.19. Decrease of (saturated) forward conductance with gate voltage (according to (21.45)) and parametric dependence on z for $z = 0$, 0.5, 1, 2, 3, 5 and 10. Adapted from [721]

Two-Region Model

In order to model the GaAs drift velocity vs. field characteristic, a two-region model is used. In the front region of the channel (region I), the field is small enough and a constant mobility μ is used. In the back region of the channel (region II) is the high-field region where a constant drift velocity v_{s} is used. With increasing source–drain voltage, the region II (I) increases (decreases) in length. The relative width of region II is also increased with decreasing channel length.

Saturated-Drift Model

Here, the drift velocity is taken everywhere as v_{s}, i.e. complete drift saturation. This is a good approximation for short channels (high fields) that are in current saturation. In this case, the current is given by

$$I_{\mathrm{D}} = eN_{\mathrm{D}}v_{\mathrm{s}}\left[a - h(x)\right]Z \; . \tag{21.49}$$

Equation (21.49) is valid for homogeneous doping. For other doping profiles, the current is given by

$$I_{\mathrm{D}} = v_{\mathrm{s}}Z\int_h^a \rho(y)\mathrm{d}y \; . \tag{21.50}$$

The forward conductance is given by

$$g_{\mathrm{m}} = \frac{v_{\mathrm{s}}Z\epsilon_{\mathrm{s}}}{h(V_{\mathrm{G}})} \; . \tag{21.51}$$

The transistor is more linear if the depletion-layer depth only weakly depends on the gate voltage. This can be accomplished with a doping profile with increasing doping with depth. An increase with a power law and a stepwise or exponential increase lead to a more linear $I(V)$-dependence. In the limit of δ-like doping, a linear $I_{\mathrm{D,sat}}$ vs. V_{G} relation develops. Indeed, FETs with graded or stepped doping profiles exhibit improved linearity and are used for analog circuits.

Nonequilibrium Velocity

Below the electric field for which the drift velocity in GaAs peaks, the carriers can be considered to be in equilibrium. If the field is higher, velocity overshoot (Sect. 21.4.4 and Fig. 8.12) occurs. The carriers have a higher velocity (and ballistic transport) before they relax to the lower equilibrium (or steady-state) velocity after intervalley scattering. This effect will shorten the transit time in short-channel FETs.

21.4.5 High-Frequency Properties

Two factors limit the high-frequency performance of a FET: The transit time and the RC time constant. The transit time t_r is the time that the carrier needs to go from source to drain. For the case of constant mobility (long channel) and constant drift velocity (short channel), the transit time is given by (21.52a and b), respectively.

$$t_\mathrm{r} = \frac{L}{\mu \mathcal{E}} \approx \frac{L^2}{\mu V_\mathrm{G}} \tag{21.52a}$$

$$t_\mathrm{r} = \frac{L}{v_\mathrm{s}} \; . \tag{21.52b}$$

For a 1-µm long gate in a GaAs FET, the transit time is of the order of 10 ps. This time is typically small compared to the RC time constant due to the capacitance C_GS and transconductance. The cutoff frequency is given by

$$f_\mathrm{T} = \frac{g_\mathrm{m}}{2\pi C_\mathrm{GS}} \; . \tag{21.53}$$

21.5 MOSFETs

The MOSFET has four terminals. In Fig. 21.14b, two n-type regions (source and drain) are within a p-type substrate. The n-type channel (length L) forms underneath a MIS diode. A forth electrode sets the substrate bias. The source electrode is considered to be at zero potential. The important parameters are the substrate doping N_A, the insulator thickness d and the depth r_j of the n-type regions. Around the MOSFET structure is an oxide to insulate the transistor from neighboring devices.

21.5.1 Operation Principle

When there is no applied gate voltage, only the saturation current of the pn-diode(s) between source and drain flows. In thermodynamic equilibrium (Fig. 21.20c), the necessary surface potential for inversion at the MIS diode is $\Psi_\mathrm{s}^\mathrm{inv} \approx 2\Psi_\mathrm{B}$. If there is a finite drain voltage, a current flows and there is no longer equilibrium. In this case, the quasi-Fermi level of the electrons (or generally of the minority carriers) is lowered and a higher gate voltage is needed to create inversion. The situation at the drain is depicted in Fig. 21.21.

In nonequilibrium, the depletion layer width is a function of the drain voltage V_D. In order to reach strong inversion at the drain, the surface potential must be at least $\Psi_\mathrm{s}^\mathrm{inv} \approx V_\mathrm{D} + 2\Psi_\mathrm{B}$.

If the gate voltage is such that an inversion channel is present from source to drain, a current will flow for a small drain voltage (Fig. 21.22a). Initially,

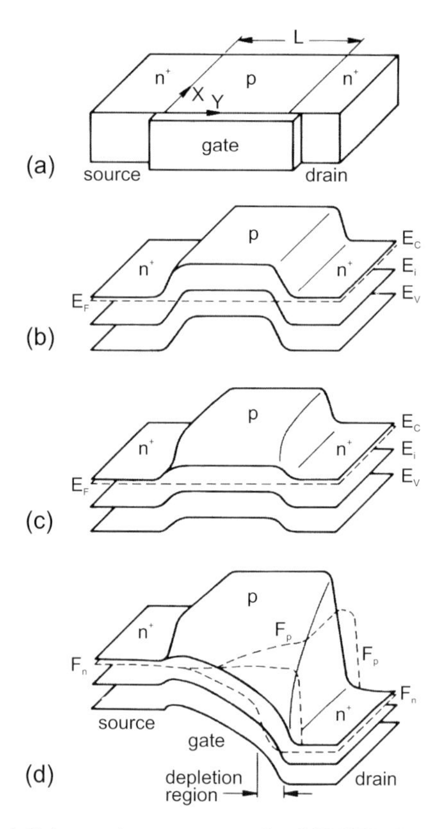

Fig. 21.20. (**a**) Schematic geometry of a MOSFET and its band diagram for (**b**) flat-band conditions for zero gate voltage (and $V_D = 0$), (**c**) thermodynamic equilibrium with reverse gate voltage (weak inversion, still $V_D = 0$) and (**d**) nonequilibrium with nonzero drain voltage and gate voltages (with most of the channel being inverted, the depletion region is indicated). Adapted from [722]

the current will increase linearly with V_D, depending on the conductivity of the channel. With increasing drain voltage, the quasi-Fermi level of the electrons is lowered until, finally at $V_D = V_{D,sat}$, the inversion channel depth becomes zero (pinch-off at point 'Y' in Fig. 21.22b). The current at this condition is denoted as $I_{D,sat}$. For a further increase of V_D, the pinch-off point moves closer to the source contact and the channel length (inverted region) is shortened (Fig. 21.22c). The voltage at the pinch-off point remains $V_{D,sat}$ and thus the current in the channel remains constant at $I_{D,sat}$.

21.5.2 Current–Voltage Characteristics

We assume now that the potential $V(y)$ varies along the channel from $V = 0$ at $y = 0$ to $V = V_D$ at $y = L$. In the gradual-channel approximation, the voltage drop V_i across the oxide is

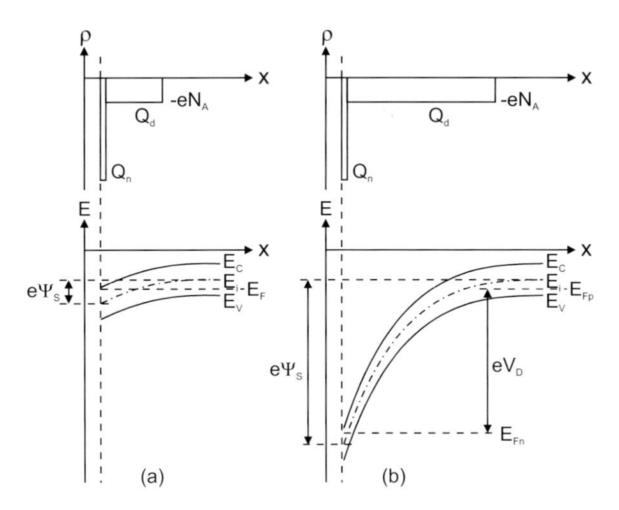

Fig. 21.21. Charge-carrier distribution (*top*) and band diagram (*bottom*) at the inverted p-region of a MOSFET for (**a**) thermodynamic equilibrium ($V_D = 0$) and (**b**) nonequilibrium at drain

$$V_i(y) = V_G - \Psi_s(y) , \tag{21.54}$$

where Ψ_s is the surface potential in the semiconductor (cf. Fig. 18.28). The total charge induced in the semiconductor (per unit area) is, using (18.64), given by

$$Q_s(y) = -[V_G - \Psi_s(y)]\, C_i , \tag{21.55}$$

with C_i being the insulator capacity (per unit area), as given in (18.65).

The inversion surface potential can be approximated by $\Psi_s(y) \approx 2\Psi_B + V(y)$ (see Fig. 21.21). Using (18.63) and (18.69), the depletion-layer charge is

$$Q_d(y) = -eN_A w_m = -\left(2\epsilon_s e N_A\, [2\Psi_B + V(y)]\right)^{1/2} , \tag{21.56}$$

such that, using (21.55), the inversion layer charge is

$$\begin{aligned} Q_n(y) &= Q_s(y) - Q_d(y) \\ &= -[V_G - V(y) - 2\Psi_B]\, C_i + \left(2\epsilon_s e N_A\, [2\Psi_B + V(y)]\right)^{1/2} . \end{aligned} \tag{21.57}$$

For the calculation of the drain current, we consider the increase of channel resistance $\mathrm{d}R(y)$ along a line element $\mathrm{d}y$ of the channel. The integral of the conductivity over the cross section A of the channel (width Z) is

$$\iint_A \sigma(x,z)\mathrm{d}x\,\mathrm{d}z = -e\mu_n \iint_A n(x,z)\mathrm{d}x\,\mathrm{d}z = Z\mu_n |Q_n(y)| . \tag{21.58}$$

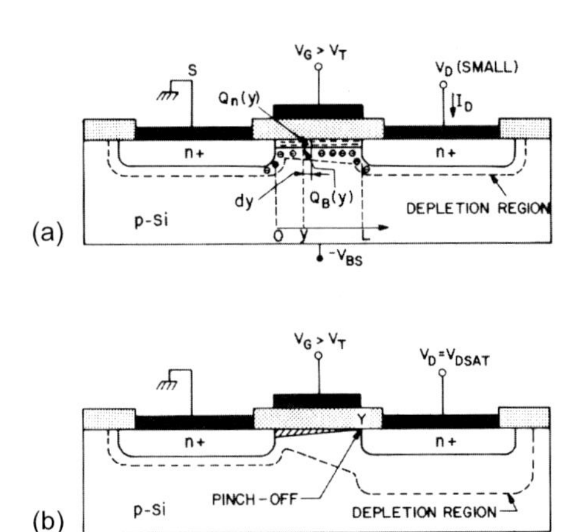

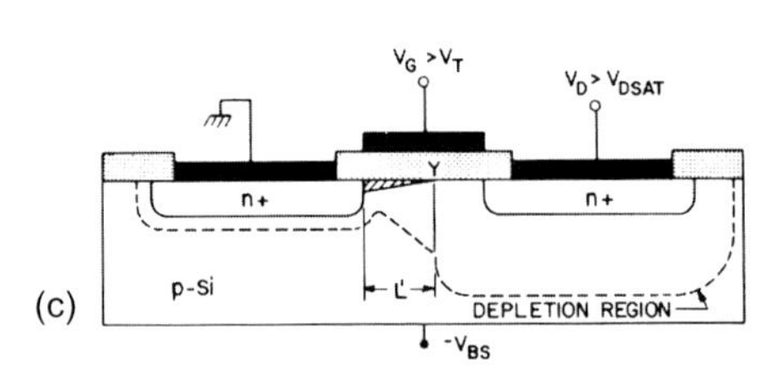

Fig. 21.22. (**a**) MOSFET with inverted channel and small source–drain voltage in linear regime, (**b**) at the start of saturation at pinch-off, (**c**) in the saturation regime with reduced channel length L'. The pinch-off point is denoted in (**b**) and (**c**) by 'Y'. The *dashed lines* denote the extension of the depletion region. Reprinted with permission from [183], ©1981 Wiley

Therefore,

$$\mathrm{d}R(y) = \mathrm{d}y \frac{1}{Z\mu_\mathrm{n}|Q_\mathrm{n}(y)|} \,. \tag{21.59}$$

Here we have assumed that the mobility is constant along the channel, i.e. not field dependent. The change of voltage across the line element $\mathrm{d}x$ is

$$\mathrm{d}V(y) = I_\mathrm{D}\,\mathrm{d}R = \frac{I_\mathrm{D}\,\mathrm{d}y}{Z\mu_\mathrm{n}|Q_\mathrm{n}(y)|} \,. \tag{21.60}$$

We note the drain current is independent of x. Using (21.57) and performing the integral of (21.60) from $V(y=0)=0$ to $V(y=L)=V_\mathrm{D}$, we find

$$I_\mathrm{D} = \mu_\mathrm{n} C_\mathrm{i} \frac{Z}{L} \left\{ \left(V_\mathrm{G} - 2\Psi_\mathrm{B} - \frac{V_\mathrm{D}}{2} \right) - \frac{2}{3} \frac{(2e\epsilon_\mathrm{s} N_\mathrm{A})^{1/2}}{C_\mathrm{i}} \left[(V_\mathrm{D} + 2\Psi_\mathrm{B})^{3/2} - (2\Psi_\mathrm{B})^{3/2} \right] \right\} . \tag{21.61}$$

In the linear regime (small drain voltage, $V_D \ll (V_G - V_T)$), the drain current is given by

$$I_D \cong \mu_n C_i \frac{Z}{L} (V_G - V_T) V_D \ . \qquad (21.62)$$

The threshold voltage V_T, i.e. the gate voltage for which the channel is opened and a current can flow, is given for small drain voltage (linear regime) by

$$V_T = 2\Psi_B + \frac{(4e\epsilon_s N_A \Psi_B)^{1/2}}{C_i} \ . \qquad (21.63)$$

The transconductances in the linear regime are easily obtained as

$$g_m = \mu_n C_i \frac{Z}{L} V_D \qquad (21.64a)$$

$$g_D = \mu_n C_i \frac{Z}{L} (V_G - V_T) \ . \qquad (21.64b)$$

The saturation current (for constant mobility) is approximately

$$I_{D,sat} \cong \mu_n C_i \frac{mZ}{L} (V_G - V_T)^2 \ , \qquad (21.65)$$

where m depends on the doping concentration and is about 0.5 for low doping. For low p-doping of the substrate, the threshold voltage in (21.65) for the saturation regime is also given by (21.63). At higher doping, the threshold voltage becomes dependent on the gate voltage. C_i denotes the insulator capacitance, $C_i = \epsilon_i/d_i$.

The forward transconductance in the saturation regime is

$$g_{m,sat} = \mu_n C_i \frac{2mZ}{L} (V_G - V_T) \ . \qquad (21.66)$$

For constant drift velocity, the saturation current is given by

$$I_{D,sat} = Z C_i v_s (V_G - V_T) \ , \qquad (21.67)$$

and the forward transconductance in the saturation regime is

$$g_{m,sat} = Z C_i v_s \ . \qquad (21.68)$$

The threshold voltage can be changed by the substrate bias V_{BS} as ($\beta = e/kT$)

$$\Delta V_T = \frac{a}{\sqrt{\beta}} \left[(2\Psi_B + V_{BS})^{1/2} - (2\Psi_B)^{1/2} \right] \ , \qquad (21.69)$$

with (L_D being the Debye length (cf. 18.58b))

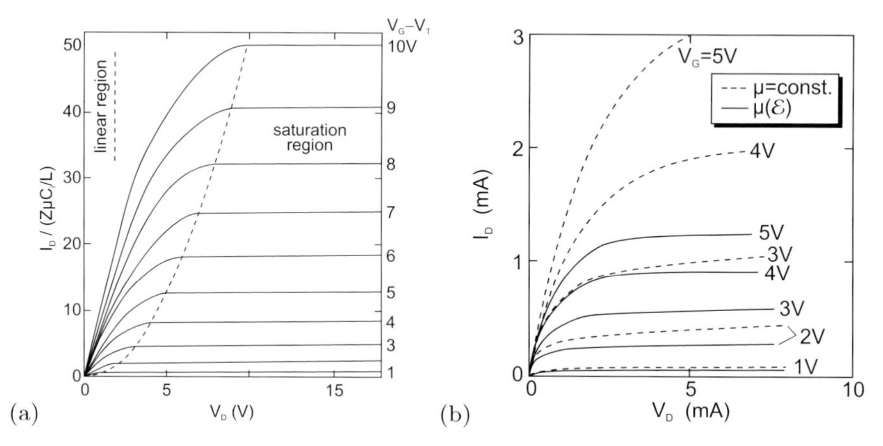

Fig. 21.23. (**a**) Idealized I–V characteristics for a MOSFET with constant mobility. The *dashed line* visualizes the drain (saturation) voltage for which the current is equal to $I_{\mathrm{D,sat}}$. The *solid lines* are for various values of the gate voltage $V_{\mathrm{G}} - V_{\mathrm{T}} = 1$–10 V. Adapted from [183] (**b**) I–V characteristics taking into account the effect of field-dependent mobility (*solid lines*) in comparison to the constant-mobility model (*dashed lines*) for various gate voltages as labeled. Adapted from [723]

$$a = 2\frac{\epsilon_{\mathrm{s}}}{\epsilon_{\mathrm{i}}}\frac{d}{L_{\mathrm{D}}} \,. \tag{21.70}$$

Experimental data are shown in Fig. 21.24. For a Si/SiO$_2$ gate diode, $a = 1$ for, e.g., $d_{\mathrm{i}} = 10\,\mathrm{nm}$ and $N_{\mathrm{A}} = 10^{16}\,\mathrm{cm}^{-3}$. For gate voltages below V_{T}, the current is given by the diffusion current, similar to a npn transistor. This regime is important for low-voltage, low-power conditions. The related drain current is termed the *subthreshold* current and is given by

$$I_{\mathrm{D}} = \mu_{\mathrm{n}}\frac{ZaC_{\mathrm{i}}n_{\mathrm{i}}^2}{2L\beta^2N_{\mathrm{A}}^2}\left[1 - \exp\left(-\beta V_{\mathrm{D}}\right)\right]\exp\left(-\beta\Psi_{\mathrm{s}}\right)\left(\beta\Psi_{\mathrm{s}}\right)^{-1/2} \,. \tag{21.71}$$

The drain current therefore increases exponentially with V_{G}, as shown in Fig. 21.24. V_{G} is roughly proportional to Ψ_{B}:

$$\Psi_{\mathrm{s}} = \left(V_{\mathrm{G}} - V_{\mathrm{FB}}\right) - \frac{a^2}{2\beta}\left\{\left[1 + \frac{4}{a^2}\left(\beta V_{\mathrm{G}} - \beta V_{\mathrm{FB}} - 1\right)\right]^{1/2} - 1\right\} \,, \tag{21.72}$$

where V_{FB} is the flat-band voltage of the gate MIS diode. The drain current is independent of V_{D} for $V_{\mathrm{D}} \gtrsim 3kT/e$.

21.5.3 MOSFET Types

MOSFETs can have an n-type channel (on a p-substrate) or a p-channel (on an n-type substrate). So far, we have discussed the normally off MOSFET.

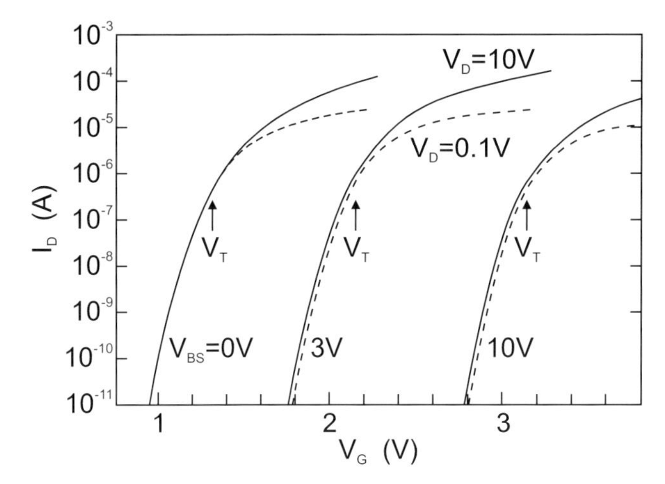

Fig. 21.24. Experimental subthreshold *I–V* characteristic of a MOSFET device with long channel (15.5 µm). *Solid lines* for $V_\mathrm{D} = 10\,\mathrm{V}$, *dashed lines* for $V_\mathrm{D} = 0.1\,\mathrm{V}$. Adapted from [724]

If there is a conductive channel even without a gate voltage, the MOSFET is normally on. Here, a negative gate voltage must be applied to close the channel. Therefore, similar to the JFET, a total of four different types of MOSFET exist, see Fig. 21.25.

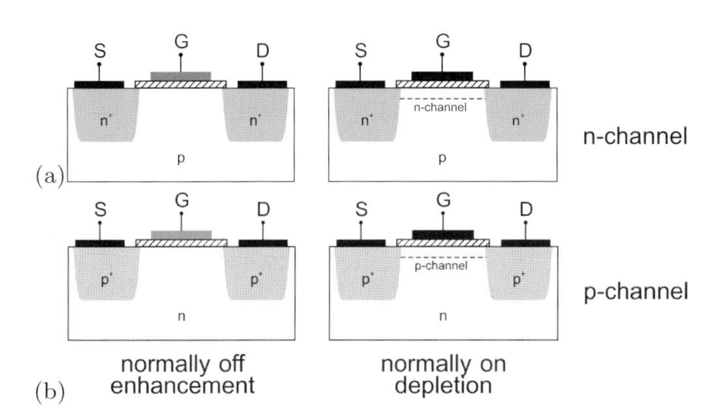

Fig. 21.25. The four MOSFET types. (**a**) Enhancement and (**b**) depletion type with n-channel (*top row*) and p-channel (*bottom row*).

21.5.4 Complementary MOS

Complementary metal–oxide–semiconductor technology (CMOS) is the dominating technology for highly integrated circuits. In such devices, MOSFETs with n-channel (NMOS) and p-channel (PMOS) are used on the same chip. The basic structure of logic circuits, the inverter, can be realized with a pair of NMOS and PMOS transistors, as shown in Fig. 21.26a with two normally off transistors. The load capacitor represents the capacitance of the following elements.

If the input voltage is $V_{\mathrm{in}} = 0$, the NMOS transistor is nonconductive ('off'). The (positive) voltage V_{DD} is at the PMOS transistor source, thus the gate is negative in relation to the source and the transistor is conductive ('on') since $-V_{\mathrm{DD}} = V_{\mathrm{Gp}} < V_{\mathrm{Tp}} < 0$ (see Fig. 21.25). The current flows through the capacitor that becomes charged to $V_{\mathrm{out}} = V_{\mathrm{DD}}$. The current then subsides, since V_{D} at the PMOS becomes zero. If the input voltage is set to V_{DD}, the NMOS transistor has a positive gate–source voltage larger than the threshold $V_{\mathrm{Tn}} < V_{\mathrm{Gn}} = V_{\mathrm{DD}}$ and becomes conductive. The charge from the capacitor flows over the NMOS to ground. The PMOS transistor has zero gate–source voltage and is in the 'off' state. In this case, the voltage V_{DD} drops entirely across the PMOS and the capacitor is uncharged with $V_{\mathrm{out}} = 0$.

In both its logic states, the CMOS inverter does not consume power. No current[7] flows in either of the two steady states since one of the two transistors is in both cases in the 'off' state. Current flows only during the switching operation. Therefore, the CMOS scheme allows for low power consumption.

The middle voltage for which $V_{\mathrm{in}} = V_{\mathrm{out}}$ can be calculated from the MOSFET characteristics. Both are, for this condition, in saturation and the currents are given by (cf. (21.65))

$$I_{\mathrm{Dn}} = \mu_{\mathrm{n}} C_{\mathrm{ox}} \frac{Z_{\mathrm{n}}}{2L_{\mathrm{n}}} \left(V_{\mathrm{M}} - V_{\mathrm{Tn}}\right)^2 \tag{21.73a}$$

$$I_{\mathrm{Dp}} = \mu_{\mathrm{p}} C_{\mathrm{ox}} \frac{Z_{\mathrm{p}}}{2L_{\mathrm{p}}} \left(V_{\mathrm{DD}} - V_{\mathrm{M}} - V_{\mathrm{Tp}}\right)^2 . \tag{21.73b}$$

With $\gamma = \frac{Z_{\mathrm{p}}}{Z_{\mathrm{n}}} \frac{L_{\mathrm{n}}}{L_{\mathrm{p}}} \frac{\mu_{\mathrm{p}}}{(-\mu_{\mathrm{n}})}$, we find from $I_{\mathrm{Dn}} = -I_{\mathrm{Dp}}$,

$$V_{\mathrm{M}} = \frac{V_{\mathrm{Tn}} + \gamma \left(V_{\mathrm{DD}} + V_{\mathrm{Tp}}\right)}{1+\gamma} . \tag{21.74}$$

As gate material, often polycrystalline silicon (poly-Si) is used (cf. Fig. 18.23). It is used instead of metals because its work function matches that of silicon closely. Also, poly-Si is more resistant to temperature. Despite its high doping, the resistance of poly-Si is two orders of magnitude larger

[7]except for the subthreshold current and other leakage currents. These need to be reduced further since the dissipated power limits chip performance (speed and device density) and battery lifetime in handheld applications.

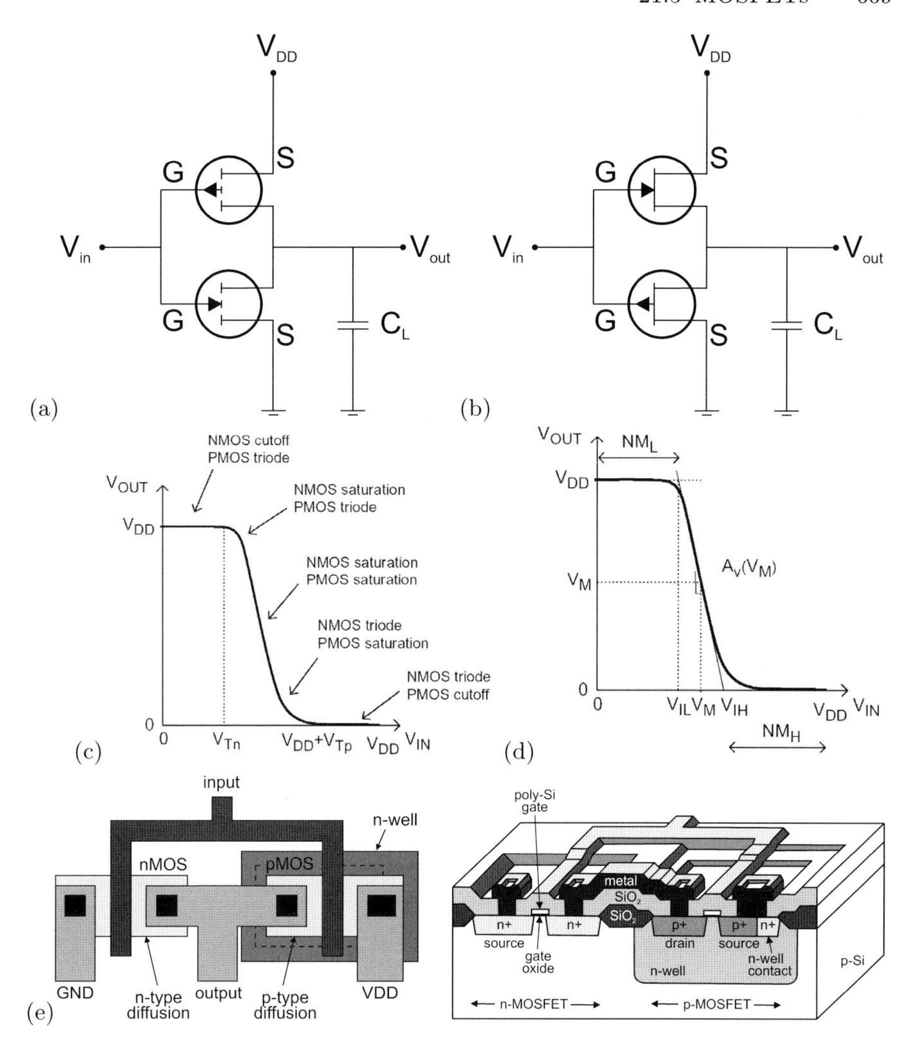

Fig. 21.26. Circuit diagram of (**a**) inverter with n-type (bottom) and p-type (normally off, enhancement mode) FETs and (**b**) inverter with p-type (bottom) and n-type (normally on, depletion mode) FETs. (**c**) Inverter characteristic with the transistor thresholds indicated, (**d**) inverter characteristic with middle voltage V_{M} indicated. $\mathrm{NM}_{\mathrm{L,H}}$ denotes the low- and high-noise margins, respectively, i.e. the voltage by which the input voltage can fluctuate without leading to switching. (**e**) Composite layout (*left panel*) and cross-sectional view (*right panel*) of CMOS inverter. Part (**e**) adapted from [725]

than that of metals. Since it is easily oxidized, it cannot be used with high-k oxide dielectrics.[8]

For optimized ohmic contacts on the n- and p-Si, different metals are used to create a small barrier height (Fig. 18.19a) and low contact resistance (cf. Sect. 18.2.6). Figure 21.27 visualizes the band edges of silicon in relation to the work functions of various metals (see Table 18.1). For example, the work function of titanium matches the electron affinity of n-Si closely. However, a direct deposition of Ti on Si results in a Schottky barrier of 0.5 eV [549]. A surface passivation with a group-VI element such as Se can help reduce this value to 0.19 eV [727].

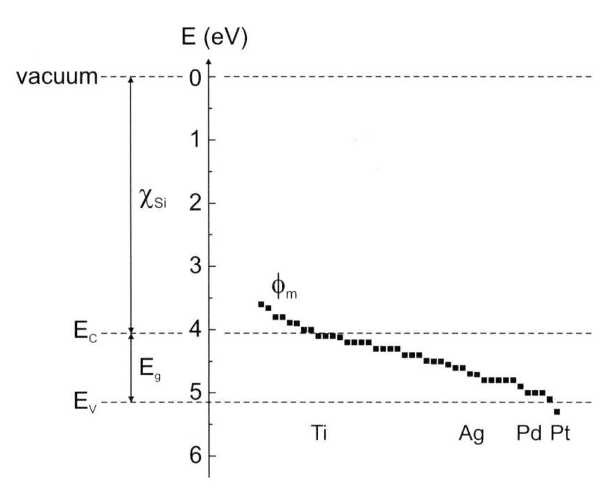

Fig. 21.27. Silicon band edges in relation to different metals and their work functions

In the latest generation of CMOS ICs the PMOS (NMOS) device has a built-in compressive (tensile) channel strain for modifying the effective mass (cf. Sect. 6.14.2), both allowing higher drive current due to higher mobility.

21.5.5 Large-Scale Integration

Compared to the first computers on the basis of vacuum tubes (triodes), e.g. ENIAC (Fig. 21.28), today's devices are extremely miniaturized and need many orders of magnitude less power per operation. ENIAC needed 174 kW of power. A comparable computing power was reached in 1971 with the few cm^2 large Intel 4004 microprocessor (Fig. 21.29b) consuming only several Watts. Also, memory chips started to become highly integrated (Fig. 21.29a).

The development of electronic circuit integration is empirically described by Moore's[9] 'law' that has been valid since the 1970s. According to this law,

[8] The term 'high-k dielectric' means a dielectric material with large dielectric constant ϵ.

[9] Gordon Moore, co-founder of INTEL.

Fig. 21.28. ENIAC, the first electronic computer (J.P. Eckert, J.W. Mauchly, 1944/5). The images show only a small part of the 18 000 vacuum tubes

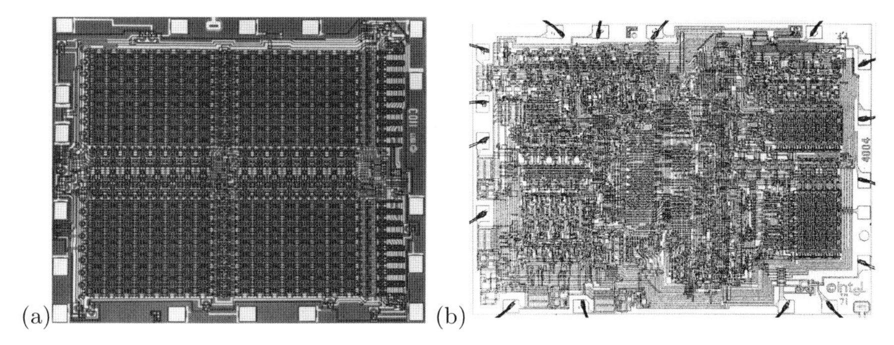

Fig. 21.29. (**a**) Intel^{TM} 1103 1 KByte (1024 memory cells) dynamic random access memory (RAM), arranged in four grids with 32 rows and columns (1970), chip size: $2.9 \times 3.5\,\mathrm{mm}^2$. (**b**) Intel^{TM} 4004 microprocessor (1971), chip size: $2.8 \times 3.8\,\mathrm{mm}^2$, circuit lines: $10\,\mu\mathrm{m}$, 2,300 MOS transistors, clock speed: 108 kHz

the number of transistors doubles every 20 months (Fig. 21.31). At the same time, the performance has been improved by an increase of the clock speed.

Moore's second law says that the cost of production also doubles for each new chip generation and is currently (2004) in the multi-billion US$ range. Most of the cost saved by integration is due to efficient *wiring* (interconnects) of the components, currently in eight layers above the active elements (transistors and capacitors) (Fig. 21.32).

Using planar technologies, LSI (large-scale integration), VLSI (very large-scale integration), ULSI (ultra large-scale integration) and further generations of devices have been conceived, driven by high-density electronic memory devices. The down-scaling of the MOSFET size requires the scaling of many other parameters such as, e.g., the operation voltage in order to keep electric fields small enough. Since also the clock speed is increased with higher integration, power consumption, being proportional to the operation voltage, the saturation current and the clock speed need to be kept constant.

Fig. 21.30. The Intel^{TM} Pentium 4 microprocessor (2000), circuit lines: 0.18 µm, 42 million transistors, clock speed: 1.5 GHz

The electronics industry is based on silicon as the material for transistors. However, many other materials are incorporated in the technology. Some recent progress was made with copper interconnects (IBM, 1997), replacing aluminum. The better electrical and heat conductivity could previously not be used since Cu is a deep level in Si (cf. Fig. 7.6). The key to success was an improved barrier technology to prevent the indiffusion of Cu into the Si. The first chip from series production, incorporating the Cu technology, was the PowerPC 750 (400 MHz) in 1998.

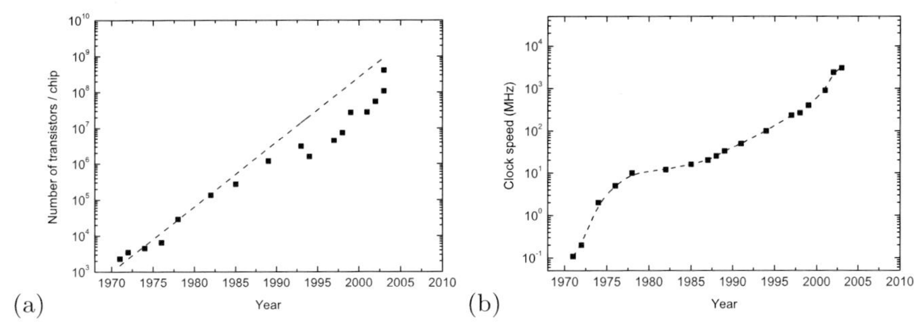

Fig. 21.31. (**a**) Moore's law on the exponential increase of transistors per chip (for INTEL^{TM} processor chips). *Dashed line* corresponds to doubling in 20 months. (**b**) Historical increase of clock speed, *dashed line* is guide to the eye. Note the almost constant rate of 10 MHz from the mid-1970s to the mid-1980s

Fig. 21.32. Cross section through a memory chip with eight layers of interconnects above the active elements

The end of the miniaturization has been theoretically predicted many times and for various feature sizes. Today, only fundamental limits such as the size of an atom seems to limit circuit design.[10] Such limits (and the effects in nanostructures in the few-nm regime) will be reached beyond 2010, projected at about 2020. Up to then, it is probable that at least a few companies will follow the road map for further miniaturization, as laid out by the Semiconductor Industry Association[11] (SIA).

[10]Only commercial profit, rather than testing physical limits, drives the miniaturization. Insufficient economic advantages or low yield of further chip generations possibly can limit or slow down large-scale integration.

[11]http://www.semichips.org

21.5.6 Nonvolatile Memories

When the gate electrode of a MOSFET is modified in such a way that a (semi-)permanent charge can be stored in the gate, a nonvolatile electronic memory can be fabricated. In the floating-gate structure (Fig. 21.33a), an insulator–metal–insulator structure is used where charge is stored in the metal and cannot escape through the insulating barriers. The 'metal' is often realized by poly-Si. In the MIOS structure (Fig. 21.33b), the insulator–oxide interface is charged. The charge can be removed by UV light (EPROM, erasable programmable read-only memory) or by a sufficient voltage across the oxide at which the charge carriers can tunnel out (Fowler–Nordheim tunneling) (EEPROM, E^2PROM, electrically erasable programmable read-only memory).

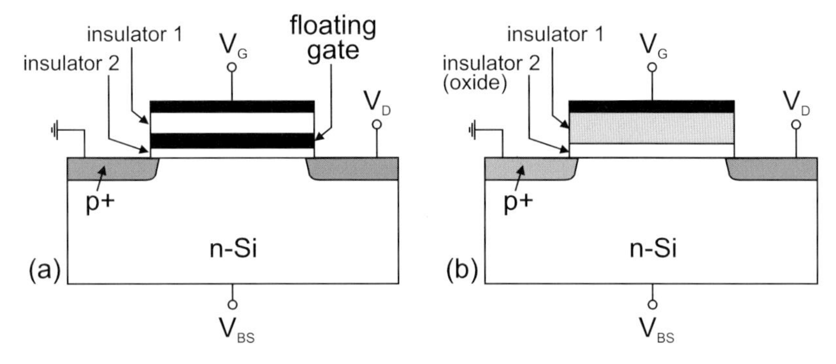

Fig. 21.33. MOSFET with (**a**) floating gate and (**b**) MIOS structure

Nowadays, a special type of EEPROM is used for the so-called *flash* memories. The stored gate charge causes a change in the MOSFET threshold voltage and is designed to switch between the on and off state. The storage time of the charge can be of the order of 100 years. Since tunneling limits the charge retention, the oxide must be sufficiently thick. Typical endurance is at least 10^6 program–erase cycles. The ultimate limit, explored currently, is to use a single electron charge to cause such an effect in the single-electron transistor (SET).

Other memory concepts include information storage via the static polarization in a ferroelectric material (either crystalline or polymer) (FeRAM), via the phase change between amorphous or polycrystalline in a chalcogenide layer (typically $Ge_2Sb_2Te_5$) upon local heating (similar to a rewritable DVD) and the related resistance change, or via a molecular configuration change (redox reaction) between crossed wire lines, not requiring diode structures at all.

21.5.7 Heterojunction FETs

Several types of field-effect transistors have been devised that use heterojunctions (HJFET).

HIGFET

As conducting channel, the two-dimensional electron gas at an undoped heterointerface is used. Such a transistor is called a heterojunction insulating gate FET (HIGFET). With forward or backward gate voltage, an electron or hole gas can be created (channel enhancement mode), as visualized in Fig. 21.34. Thus, a complementary logic can be realized. However, the p-channel suffers from low hole mobility.

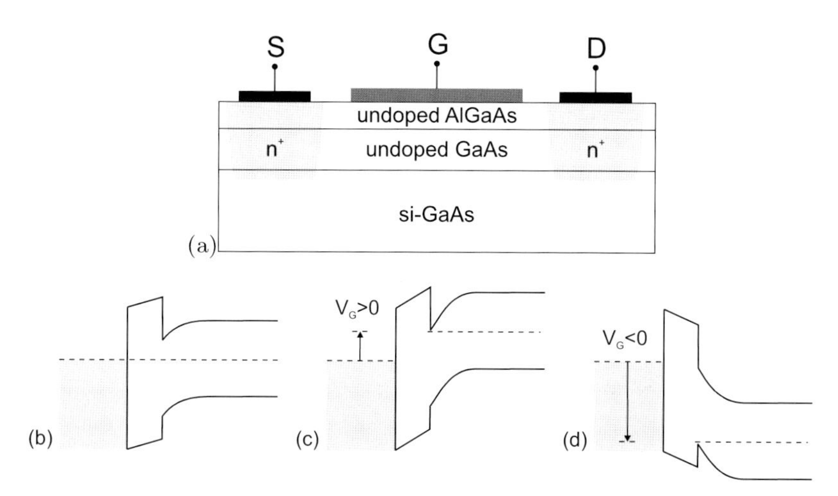

Fig. 21.34. (**a**) Scheme of a HIGFET structure with metal gate and undoped AlGaAs/GaAs heterointerface on semi-insulating GaAs. The source and drain contacts are n-doped such that this structure can be used as an n-HIGFET (see part (**c**)). (**b**) Band diagram for zero gate voltage. (**c**) Band diagram for positive gate voltage and n-channel, (**d**) for negative gate voltage and p-channel

HEMT

If the top wide-bandgap layer is n-doped, a modulation-doped FET (MODFET) is made (see Sect. 11.5.3). This structure is also called a HEMT (high electron mobility transistor) or TEGFET (two-dimensional electron gas FET). A thin undoped AlGaAs spacer layer is introduced between the doped AlGaAs and the undoped GaAs to reduce impurity scattering from carriers that tunnel into the barrier. With increasing gate voltage, a parallel conduction channel in the AlGaAs is opened. The natural idea would be to

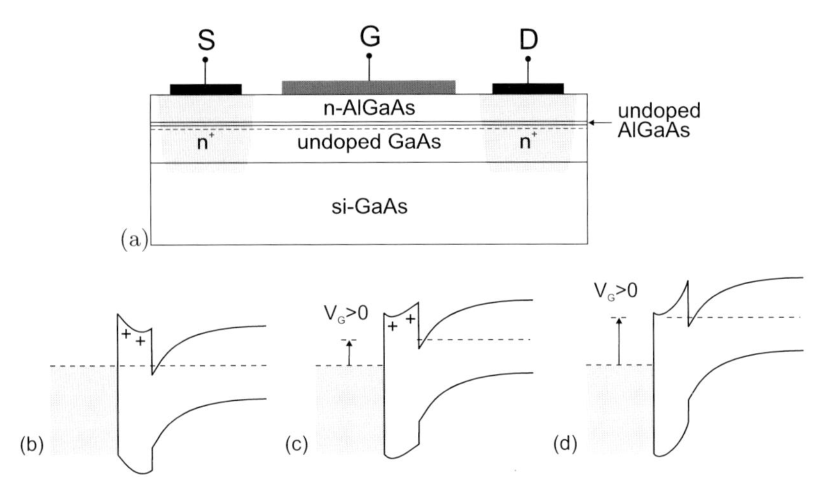

Fig. 21.35. (**a**) Scheme of a HEMT structure with n-AlGaAs/GaAs heterointerface on semi-insulating GaAs. The source and drain contacts are n-doped such that this structure can be used as an n-channel (normally-on) HEMT. The *horizontal dashed line* represents schematically the position of the 2DEG at the heterointerface on the GaAs side. (**b**) Band diagram at zero gate voltage. (**c**) Band diagram at positive gate voltage, increase of channel carrier concentration. (**d**) Band diagram at even larger positive gate voltage, formation of conducting channel in the AlGaAs layer

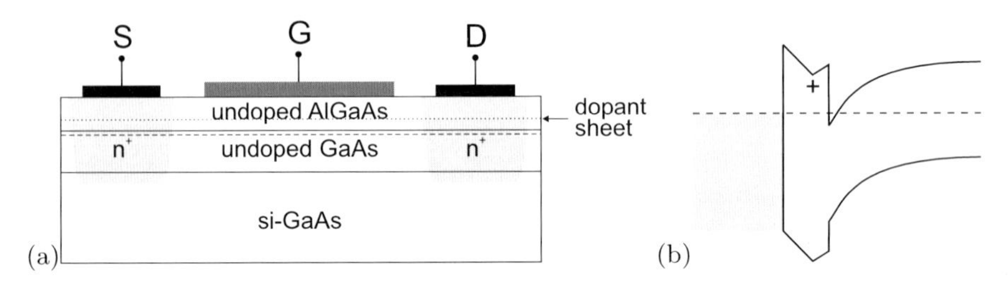

Fig. 21.36. (**a**) Scheme of a δ-doped HEMT structure with AlGaAs/InGaAs heterointerface on semi-insulating GaAs. The source and drain contacts are n-doped such that this structure can be used as an n-channel HEMT. The *horizontal dashed line* represents schematically the position of the 2DEG in the InGaAs layer. (**b**) Band diagram at zero gate voltage

increase the Al fraction in the AlGaAs to increase the quantum-well barrier height. Unfortunately, the barrier height is limited to 160 meV for an aluminum concentration of about 20%. For Al content higher than about 22%, the DX center (cf. Sect. 7.7.4) forms a deep level such that the apparent ionization energy increases drastically and no shallow donors can be used for modulation doping. An improvement for the barrier conduction problem is the use of δ-doping, i.e. the introduction of a highly doped thin (mono-)layer (Fig. 21.36), which results in higher channel carrier concentration.

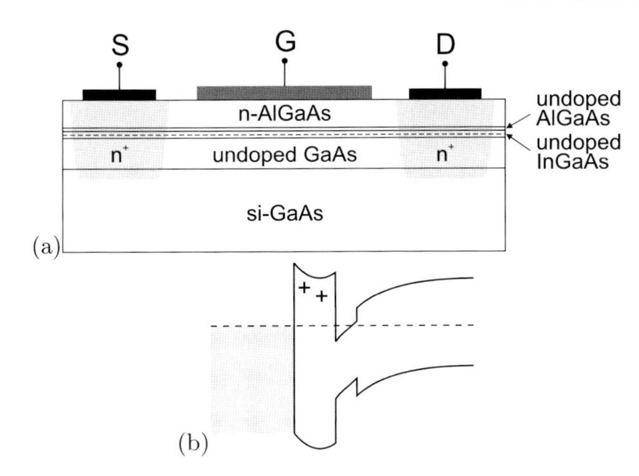

Fig. 21.37. (**a**) Scheme of a PHEMT structure with n-AlGaAs/InGaAs heterointerface on semi-insulating GaAs. The source and drain contacts are n-doped such that this structure can be used as an n-channel HEMT. The *horizontal dashed line* represents schematically the position of the 2DEG in the InGaAs layer. (**b**) Band diagram at zero gate voltage

Pseudomorphic HEMTs

Instead of increasing the height of the barrier, the depth of the well can be increased by using a low-bandgap material. On GaAs substrate, InGaAs is used. However, strain is introduced in this case and the InGaAs layer thickness is limited by the onset of dislocation formation (cf. Sect. 5.3.6) (which reduces the channel mobility and the device reliability). For $In_{0.15}Ga_{0.85}As$ (thickness about 10–20 nm), a total barrier height of about 400 meV can be obtained. A barrier height of 500 meV can be reached with an InAlAs/InGaAs structure on InP (Fig. 21.38). The InAlAs does not suffer from the problem related to DX centers. The channel indium concentration is typically 50%. The mobil-

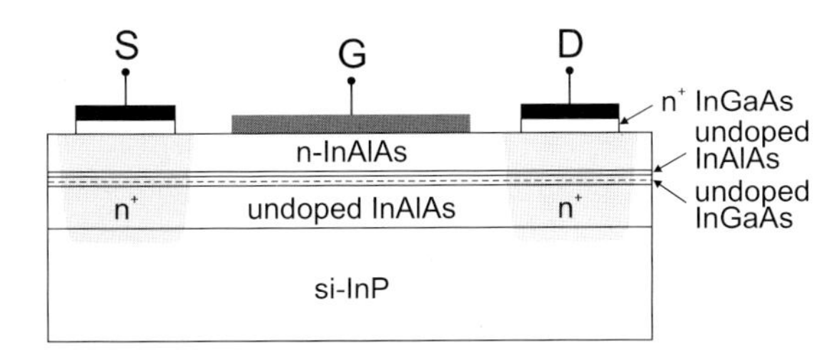

Fig. 21.38. Scheme of a PHEMT structure with n-AlInAs/InGaAs/InAlAs structure on semi-insulating InP. The source and drain contacts (with a highly doped InGaAs contact layer) are an n-doped such that this structure can be used as an n-channel HEMT. The *horizontal dashed line* represents schematically the position of the 2DEG in the InGaAs layer.

ity increases with increasing indium concentration. This InP-based HEMT structure is widely used in satellite receivers for its excellent high-speed and low-noise properties in the 100–500 GHz range and beyond.

However, the InP technology is economically less favorable than GaAs due to smaller available substrate size and higher cost (2001: 4" InP substrate: $1000, 6" GaAs substrate: $450).

Metamorphic HEMTs

A unification of the InAlAs/InGaAs structure with the best figure of merit and the GaAs substrate is achieved with the metamorphic HEMT (MHEMT). Here, a relaxed buffer is used to bring the in-plane lattice constant from that of GaAs to about that of InP. It is key that the defects occurring are confined to the relaxed buffer and do not enter the active device structure (see Fig. 21.39). The relaxed buffer is typically about 1 μm thick. It can be grown, e.g., with a graded $\mathrm{In}_x(\mathrm{Ga,Al})_{1-x}\mathrm{As}$ layer with x =0–42% or with a stepped structure with piecewise constant indium concentration in each layer. It is important that a smooth interface of the channel is achieved in order to avoid additional scattering mechanisms. For high-frequency operation, the fabrication of a small gate length is important, as shown in Fig. 21.40 for a 70-

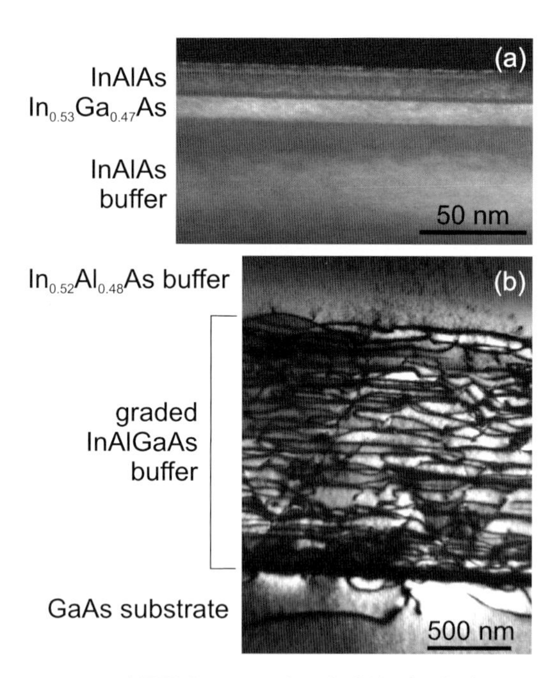

Fig. 21.39. Cross-sectional TEM image of an InAlAs/InGaAs MHEMT: (**a**) Active layer with rms surface roughness of 2.0 nm (from AFM), (**b**) graded InGaAlAs buffer layer (1.5 μm) on GaAs substrate. Reprinted with permission from [726], ©2001 AVS

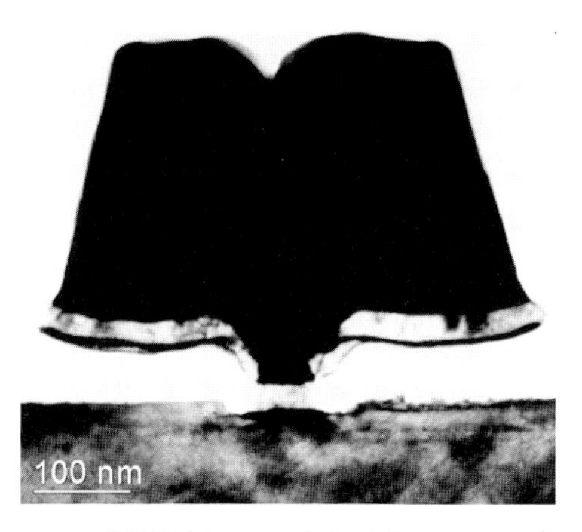

Fig. 21.40. Cross-sectional TEM image of the 70-nm gate of an InAlAs MHEMT on GaAs substrate. From [728]

nm gate of a f_{T} =293 GHz, f_{max} = 337 GHz transistor [728]. SiGe channels, providing higher mobility than pure Si, can be fabricated using graded or stepped SiGe buffer layers on Si substrate. With such Si-based MHEMTs frequencies up to 100 GHz can be achieved.

21.6 Thin-Film Transistors

Thin-film transistors (TFTs) are field-effect transistors typically fabricated as large-area arrays from thin layers of polycrystalline or amorphous silicon on cheap substrates such as glass. Their most prominent use is driving pixels in active-matrix displays such as electroluminescence (EL) displays or twisted nematic liquid crystal displays (LCD).

A schematic cross section of a TFT is shown in Fig. 21.41. Carriers in amorphous silicon have a low mobility of 0.3–0.7 cm^2/V s. Polycrystalline silicon has a mobility of about 6 cm^2/V s. With the use of laser irradiation,

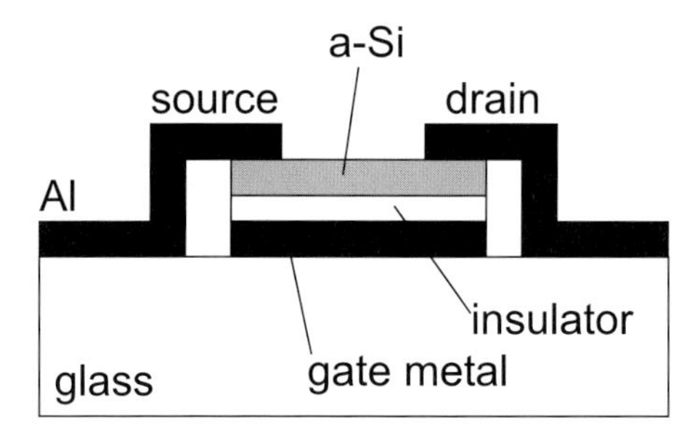

Fig. 21.41. Schematic cross section of an amorphous silicon (a-Si) thin-film transistor on glass substrate

amorphous or small-grain polycrystalline silicon layers can be recrystallized, increasing the mobility to about $330\,\mathrm{cm^2/V\,s}$ and thus improving the transistor performance.

The main optimization criteria for thin-film transistors are high on-off ratio, long-term stability, good uniformity and reproducibility, and low cost. Recently, flexible (on polymer substrate) and transparent (from polycrystalline ZnO) TFTs are being investigated for advanced applications such as all-transparent electronics and displays [729, 730].

Part III

Appendices

A Tensors

Introduction

A physical quantity $T_{ij\dots m}$ with a total of k indices that is independent of translations of the coordinate system and transforms with respect to all indices like a vector is called a tensor of rank k.

Often, Einstein's sum convention is used; a sum is built over indices with the same symbol. For example, $x'_i = D_{ij}x_j$ shall be read as $x'_i = \sum_{j=1}^{3} D_{ij}x_j$.

Rotation of Coordinate System

A rotation of the coordinate system is a transformation $\mathbf{x} \to \mathbf{x}'$ that is written in components as

$$x'_i = D_{ij}x_j \ . \tag{A.1}$$

$\mathbf{D}$ is called the rotation matrix. The inverse of the rotation matrix is $\mathbf{D}^{-1}$ with

$$D^{-1}_{kl} = D_{lk} \ , \tag{A.2}$$

i.e. it is the transposed matrix $\mathbf{D}$. The inverse transformation is $x_j = D_{ij}x'_i$. Thus,

$$D_{ij}D_{kj} = \delta_{ij} \ . \tag{A.3}$$

A simple example is the azimuthal rotation around the z-axis by an angle ϕ (in the mathematically positive direction)

$$\mathbf{D} = \begin{pmatrix} \cos\phi & -\sin\phi & 0 \\ \sin\phi & \cos\phi & 0 \\ 0 & 0 & 1 \end{pmatrix} \ . \tag{A.4}$$

For the description of a general rotation $(x, y, z) \to (X, Y, Z)$, generally three angles are necessary. Typically, the Euler angles (ϕ, θ, ψ) are used (Fig. A.1). First, the system is rotated by ϕ around the z-axis. The y-axis

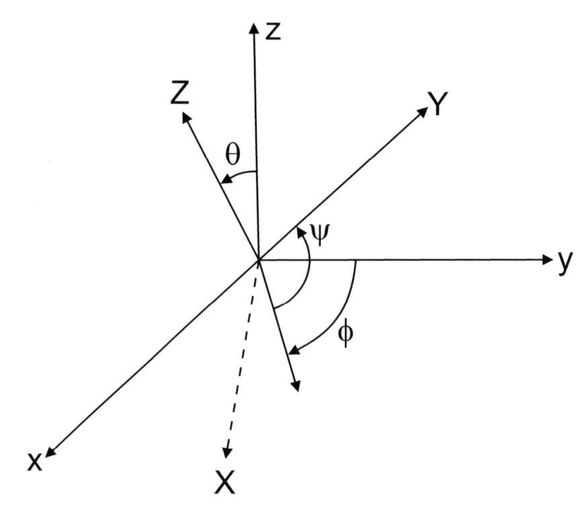

Fig. A.1. Rotation of a coordinate system (x, y, z) by the Euler angles (ϕ, θ, ψ) into the system (X, Y, Z)

becomes the u-axis. Then, the system is tilted by θ around the u-axis and the z-axis becomes the Z-axis. Finally, the system is rotated by ψ around the Z-axis.

The matrix for the general rotation by the Euler angles is

$$\begin{pmatrix} \cos\psi\cos\theta\cos\phi - \sin\psi\sin\phi & -\sin\psi\cos\theta\cos\phi - \cos\psi\sin\phi & \sin\theta\cos\phi \\ \cos\psi\cos\theta\sin\phi + \sin\psi\cos\phi & -\sin\psi\cos\theta\sin\phi + \cos\psi\cos\phi & \sin\theta\sin\phi \\ -\cos\psi\sin\theta & \sin\psi\sin\theta & \cos\theta \end{pmatrix} . \tag{A.5}$$

Rank-n Tensors

Rank-0 Tensors

A tensor of rank 0 is also called a scalar. For example, the length $v_1^2 + v_2^2 + v_3^2$ of the vector $\mathbf{v} = (v_1, v_2, v_3)$ is a scalar since it is invariant under rotation of the coordinate system. However, ‘scalar’ is not equivalent to ‘number’ since, e.g. the number $v_1^2 + v_2^2$ is not rotationally invariant.

Rank-1 Tensors

A tensor of rank 1 is a vector. It transforms under coordinate rotation $\mathbf{D}$ as

$$v_i' = D_{ij} v_j \; . \tag{A.6}$$

Rank-2 Tensors

A tensor of rank 2 is also called a dyade and is a 3×3 matrix $\mathbf{T}$ that transforms under coordinate rotation as

$$T'_{ij} = D_{ik} D_{jl} T_{kl} \ . \tag{A.7}$$

The physical meaning is the following: Two vectors $\mathbf{s}$ and $\mathbf{r}$ shall be related to each other via $s_i = T_{ij} r_j$. This could be, e.g., the current $\mathbf{j}$ and the electric field $\mathbf{E}$ that are connected via the tensor of conductivity σ, i.e. $j_i = \sigma_{ij} E_j$.

Such a physical equation only makes sense if it is also valid in a rotated coordination system. The tensor $\mathbf{T}'$ in the rotated coordinate system must fulfill $s'_i = T'_{ij} r'_j$. This implies the transformation law (A.7). $s'_k = D_{ki} s_i = D_{ki} T_{ij} r_j$ and also $s'_k = T'_{km} r'_m = T'_{km} D_{mj} r_j$. Thus, $T'_{km} D_{mj} = D_{ki} T_{ij}$ since the previous relations are valid for arbitrary $\mathbf{r}$. Multiplication by D_{lj} yields $T'_{km} D_{mj} D_{lj} = T'_{km} \delta_{ml} = T'_{kl} = D_{ki} D_{lj} T_{ij}$.

The trace of a rank-2 tensor is defined as $\mathrm{tr}\mathbf{T} = T_{ii} = T_1 1 + T_2 2 + T_3 3$. It is a scalar, i.e. invariant under coordinate rotation, since $T'_{kk} = D_{ki} D_{kj} T_{ij} = \delta_{ij} T_{ij} = T_{ii}$.

A rank-2 tensor can be separated into a *symmetric* part $\mathbf{T}^S$ and an *antisymmetric* part $\mathbf{T}^A$, i.e. $T^S_{ji} = T^S_{ij}$ and $T^A_{ji} = -T^A_{ij}$ with

$$\mathbf{T} = \mathbf{T}^S + \mathbf{T}^A \tag{A.8a}$$

$$T^S_{ij} = \frac{T_{ij} + T_{ji}}{2} \tag{A.8b}$$

$$T^A_{ij} = \frac{T_{ij} - T_{ji}}{2} \ . \tag{A.8c}$$

A rank-2 tensor can also be separated into an *isotropic* (spherical) part $\mathbf{T}^I$ and a *deviatoric* part $\mathbf{T}^D$. The isotropic part is invariant under coordinate rotation.

$$\mathbf{T} = \mathbf{T}^I + \mathbf{T}^D \tag{A.9a}$$

$$T^I_{ij} = \delta_{ij} \frac{\mathrm{tr}\mathbf{T}}{3} \tag{A.9b}$$

$$T^D_{ij} = T_{ij} - \delta_{ij} \frac{\mathrm{tr}\mathbf{T}}{3} \ . \tag{A.9c}$$

The trace of $\mathbf{T}$ is the same as that of $\mathbf{T}^I$. The trace of $\mathbf{T}^D$ is zero.

Rank-3 Tensors

A tensor of rank 3 transforms according to

$$T'_{ijk} = D_{il} D_{jm} D_{kn} T_{lmn} \ . \tag{A.10}$$

An example is the tensor $\mathbf{e}$ of piezoelectric constants that relates the rank-2 tensor of the strains ϵ with the polarization vector $\mathbf{P}$, i.e. $P_i = e_{ijk} \epsilon_{jk}$.

Rank-4 Tensors

A tensor of rank 4 transforms according to

$$T'_{ijkl} = D_{im} D_{jn} D_{ko} D_{lp} T_{mnop} \ . \tag{A.11}$$

An example is the tensor **C** of elastic constants that relates the rank 2 tensors ϵ and σ of the elastic strains and stresses, i.e. $\sigma_{ij} = C_{ijkl}\epsilon_{kl}$.

B Kramers–Kronig Relations

The Kramers–Kronig relations (KKR) are relations between the real and imaginary part of the dielectric function. They are of a general nature and are based on the properties of a complex, analytical response function $f(\omega) = f_1(\omega) + \mathrm{i} f_2(\omega)$ fulfilling the following conditions:[1]

- The poles of $f(\omega)$ are below the real axis.
- The integral of $f(\omega)/\omega$ along a semicircle with infinite radius in the upper half of the complex plane vanishes.
- The function $f_1(\omega)$ is even and the function $f_2(\omega)$ is odd for real values of the argument.

The integral of $f(s)/(s-\omega)\mathrm{d}s$ along the real axis and an infinite semicircle in the upper half of the complex plane is zero because the path is a closed line. The integral along a semicircle above the pole at $s = \omega$ yields $-\pi\mathrm{i} f(\omega)$, the integral over the infinite semicircle is zero. Therefore the value of $f(\omega)$ is given by[2]

$$f(\omega) = \frac{1}{\pi \mathrm{i}} \, \mathrm{Pr} \int_{-\infty}^{\infty} \frac{f(s)}{s - \omega} \mathrm{d}s \,. \tag{B.1}$$

Equating the real and imaginary parts of (B.1) yields for the real part

$$f_1(\omega) = \frac{1}{\pi} \, \mathrm{Pr} \int_{-\infty}^{\infty} \frac{f_2(s)}{s - \omega} \mathrm{d}s \,. \tag{B.2}$$

Splitting the integral into two parts $\int_0^\infty$ and $\int_{-\infty}^0$, going from s to $-s$ in the latter and using $f_2(-\omega) = -f_2(\omega)$ and $\frac{1}{s-\omega} + \frac{1}{s+\omega} = \frac{2s}{s^2-\omega^2}$ yields (B.3a)

$$f_1(\omega) = \frac{2}{\pi} \, \mathrm{Pr} \int_0^{\infty} \frac{s f_2(s)}{s^2 - \omega^2} \mathrm{d}s \tag{B.3a}$$

$$f_2(\omega) = -\frac{2}{\pi} \, \mathrm{Pr} \int_0^{\infty} \frac{f_1(s)}{s^2 - \omega^2} \mathrm{d}s \,. \tag{B.3b}$$

[1] The requirements for the function f to which the KKR apply can be interpreted as that the function must represent the Fourier transform of a linear and causal physical process.

[2] The Cauchy principal value Pr of the integral is the limit for $\delta \to 0$ of the sum of the integrals over $-\infty < s < \omega - \delta$ and $\omega + \delta < s < \infty$.

In a similar way, (B.3b) is obtained. These two relations are the Kramers–Kronig relations [731, 732]. They are most often applied to the dielectric function ϵ. In this case, they apply to the susceptibility, i.e. $f(\omega) = \chi(\omega) = \epsilon(\omega)/\epsilon_0 - 1$. The susceptibility can be interpreted as the Fourier transform of the time-dependent polarization in the semiconductor after an infinitely short pulsed electric field, i.e. the impulse response of the polarization. For the dielectric function $\epsilon = \epsilon_1 + \mathrm{i}\epsilon_2$, the following KKR relations hold:

$$\epsilon_1(\omega) = \epsilon_0 + \frac{2}{\pi}\,\mathrm{Pr}\int_0^\infty \frac{s\epsilon_2(s)}{s^2-\omega^2}\,\mathrm{d}s \tag{B.4a}$$

$$\epsilon_2(\omega) = -\frac{2\omega}{\pi}\,\mathrm{Pr}\int_0^\infty \frac{\epsilon_1(s)-\epsilon_0}{s^2-\omega^2}\,\mathrm{d}s\;. \tag{B.4b}$$

The static dielectric constant is thus given by

$$\epsilon(0) = \epsilon_0 + \frac{2}{\pi}\,\mathrm{Pr}\int_0^\infty \frac{\epsilon_2(s)}{s}\,\mathrm{d}s\;. \tag{B.5}$$

The integral does not diverge since ϵ_2 is an odd function and zero at $\omega = 0$. Generally the j–th momentum M_j of the imaginary part of the dielectric function is

$$M_j = \int_0^\infty \epsilon_2(\omega)\omega^j\,\mathrm{d}\omega\;. \tag{B.6}$$

Thus, $M_{-1} = \pi[\epsilon(0) - \epsilon_0]/2$.

Other KKRs are, e.g., the relation between the index of refraction n_r and the absorption coefficient α:

$$n_\mathrm{r}(\lambda) = \frac{1}{\pi}\,\mathrm{Pr}\int_0^\infty \frac{\alpha(s)}{1-s^2/\lambda^2}\,\mathrm{d}s\;. \tag{B.7}$$

If the imaginary (real) part of the dielectric function is known (for all frequencies), the real (imaginary) part can be calculated via the KKR. If the dependence is not known for the entire frequency range, assumptions about the dielectric function in the unknown spectral regions must be made that limit the reliability of the transformation.

C Oscillator Strength

The response of an oscillator to an electric field $\mathbf{E}$ is formulated with the dielectric function. The resulting polarization $\mathbf{P}$ is related to the electric field via

$$\mathbf{P} = \epsilon_0 \chi \mathbf{E} \; , \tag{C.1}$$

with χ being the electric susceptibility, and the displacement field $\mathbf{D}$ is given by

$$\mathbf{D} = \epsilon_0 \mathbf{E} + \mathbf{P} = \epsilon_0 \epsilon \mathbf{E} \; . \tag{C.2}$$

Thus the (relative) dielectric constant is

$$\epsilon = 1 + \chi \; . \tag{C.3}$$

We assume a harmonic oscillator model for a bound electron, i.e. an equation of motion for the amplitude $x = x_0 \exp(\mathrm{i}\omega t)$

$$m\ddot{x} = -Cx \; . \tag{C.4}$$

The resonance frequency is $\omega_0^2 = C/m$. The presence of a harmonic electric field E of frequency ω and amplitude E_0 adds a force $-eE$. Thus,

$$-m\omega^2 x = -m\omega_0^2 x - eE \; . \tag{C.5}$$

The polarization is given by

$$-ex_0 = \frac{e^2}{m}\frac{1}{\omega_0^2 - \omega^2} E_0 = \frac{e^2}{m\omega_0^2}\frac{1}{1 - \frac{\omega^2}{\omega_0^2}} E_0 \; . \tag{C.6}$$

The prefactor is called the (dimensionless) *oscillator strength* and will be denoted as

$$f = \frac{e^2}{\epsilon_0 m \omega_0^2} \tag{C.7}$$

in the following. The frequency-dependent dielectric function of the resonance is thus

$$\epsilon(\omega) = 1 + \frac{f}{1 - \frac{\omega^2}{\omega_0^2}} \,. \tag{C.8}$$

In the low-frequency limit, the dielectric function is $\epsilon(0) = 1 + f$, in the high-frequency limit (beyond the X-ray regime) $\epsilon(\infty) = 1$. The oscillator strength is the difference of ϵ for frequencies below and above the resonance.

For all systems, the high-frequency limit of ϵ is 1. This means that $\chi = 0$, i.e. there are no more oscillators to be polarized. The low-frequency limit includes all possible oscillators. If there are further oscillators between frequencies well above ω_0 and $\omega \to \infty$, these are summarized as the high-frequency dielectric constant $\epsilon_\infty > 1$. Equation (C.8) then reads

$$\epsilon(\omega) = \epsilon(\infty) + \frac{\hat{f}}{1 - \frac{\omega^2}{\omega_0^2}} \,. \tag{C.9}$$

The limit $\epsilon \to \epsilon(\infty)$ is only valid for frequencies above ω_0 but smaller than the next resonance(s) at higher frequencies. Another common form is to include the background dielectric constant via

$$\epsilon(\omega) = \epsilon(\infty) \left[1 + \frac{f}{1 - \frac{\omega^2}{\omega_0^2}}\right] \,. \tag{C.10}$$

Obviously, $f = \hat{f}/\epsilon(\infty)$, making the two forms equivalent.

In order to discuss the lineshape, not only for ϵ but also for the index of refraction $n^* = n_\mathrm{r} + \mathrm{i}\kappa = \sqrt{\epsilon}$, we introduce damping to our calculation by adding a term $-m\Gamma\dot{x}$ to the left side of (C.5). This term is something like a 'friction' and would cause the oscillation amplitude to decay exponentially with a time constant $\tau = 2/\Gamma$ without external stimulus. The dielectric constant is

$$\epsilon(\omega) = \epsilon(\infty) \left[1 + \frac{f}{1 - \frac{\omega^2 + \mathrm{i}\omega\Gamma}{\omega_0^2}}\right] = \epsilon' + \mathrm{i}\epsilon'' \,. \tag{C.11}$$

The real and imaginary part fulfill the Kramers–Kronig relations (B.3a) and (B.3b). For the oscillator strength, the regimes of large oscillator strength ($f \sim 1$) and small oscillator strength ($f \ll 1$) are distinguished. For the damping, two regimes should be distinguished: Small damping ($\Gamma \ll \omega_0$) and strong damping ($\Gamma \gtrsim \omega_0$). Typical lineshapes are shown in Figs. C.1 and C.2.

For small oscillator strength, i.e. $f \ll 1$, the index of refraction $n^* = \sqrt{\epsilon} = n_\mathrm{r} + \mathrm{i}\kappa$ is given by ($n_\infty = \sqrt{\epsilon(\infty)}$)

$$n_\mathrm{r} = n_\infty \left[1 + \frac{f}{2} \frac{\omega_0^2(\omega_0^2 - \omega^2)}{(\omega_0^2 - \omega^2)^2 + \Gamma^2\omega^2}\right] \tag{C.12a}$$

$$\kappa = n_\infty \frac{f}{2} \frac{\Gamma\omega_0(\omega_0^2 - \omega^2)}{(\omega_0^2 - \omega^2)^2 + \Gamma^2\omega^2} \,. \tag{C.12b}$$

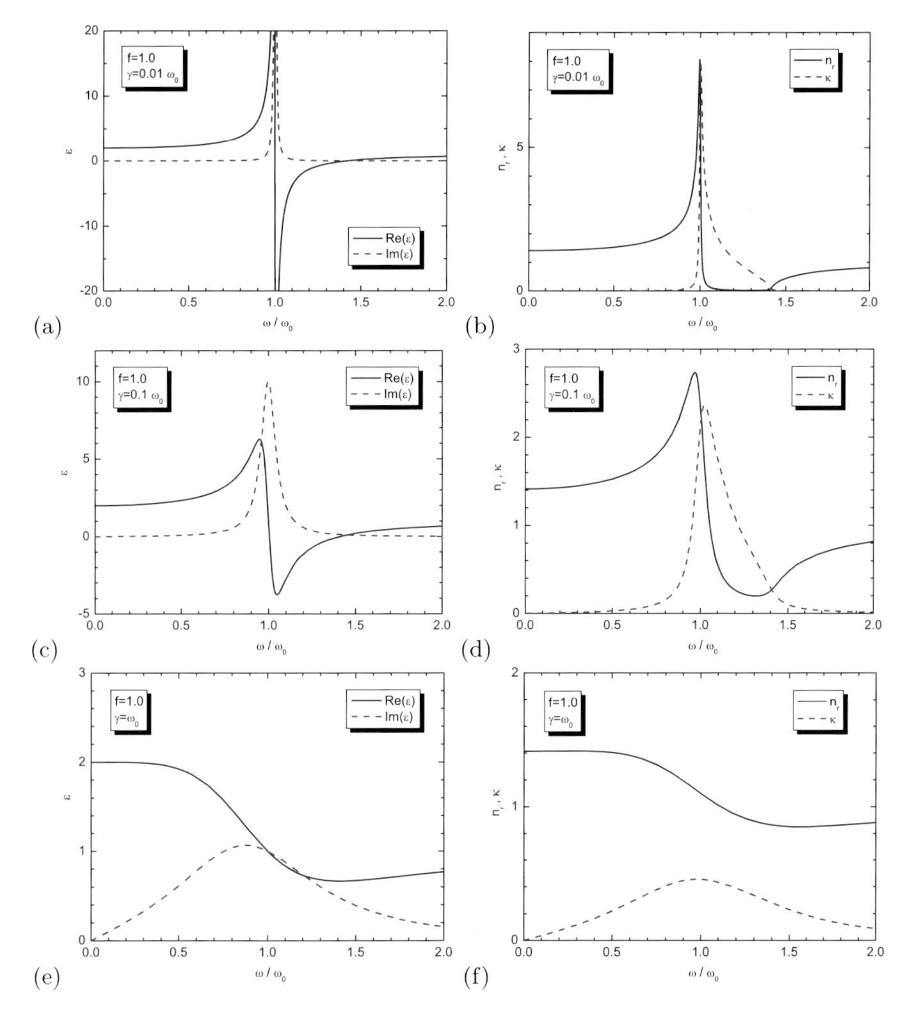

Fig. C.1. Real (*solid lines*) and imaginary (*dashed lines*) parts of the dielectric constant (**a**,**c**,**e**) and index of refraction (**b**,**d**,**f**) (C.11) for oscillator strength $f = 1$ and various values of damping: (**a**,**b**) $\Gamma = 10^{-2}\omega_0$, (**c**,**d**) $\Gamma = 10^{-1}\omega_0$, and (**e**,**f**) $\Gamma = \omega_0$

For small detuning from the resonance frequency, i.e. $\omega = \omega_0 + \delta\omega$ with $|\delta\omega|/\omega_0 \ll 1$, the index of refraction is given by

$$n_\mathrm{r} = n_\infty \left[1 - \frac{f}{4} \frac{\omega_0\, \delta\omega}{(\delta\omega)^2 + \Gamma^2/4} \right] \tag{C.13a}$$

$$\kappa = n_\infty \frac{f}{4} \frac{\omega_0\, \Gamma/2}{(\delta\omega)^2 + \Gamma^2/4} \,. \tag{C.13b}$$

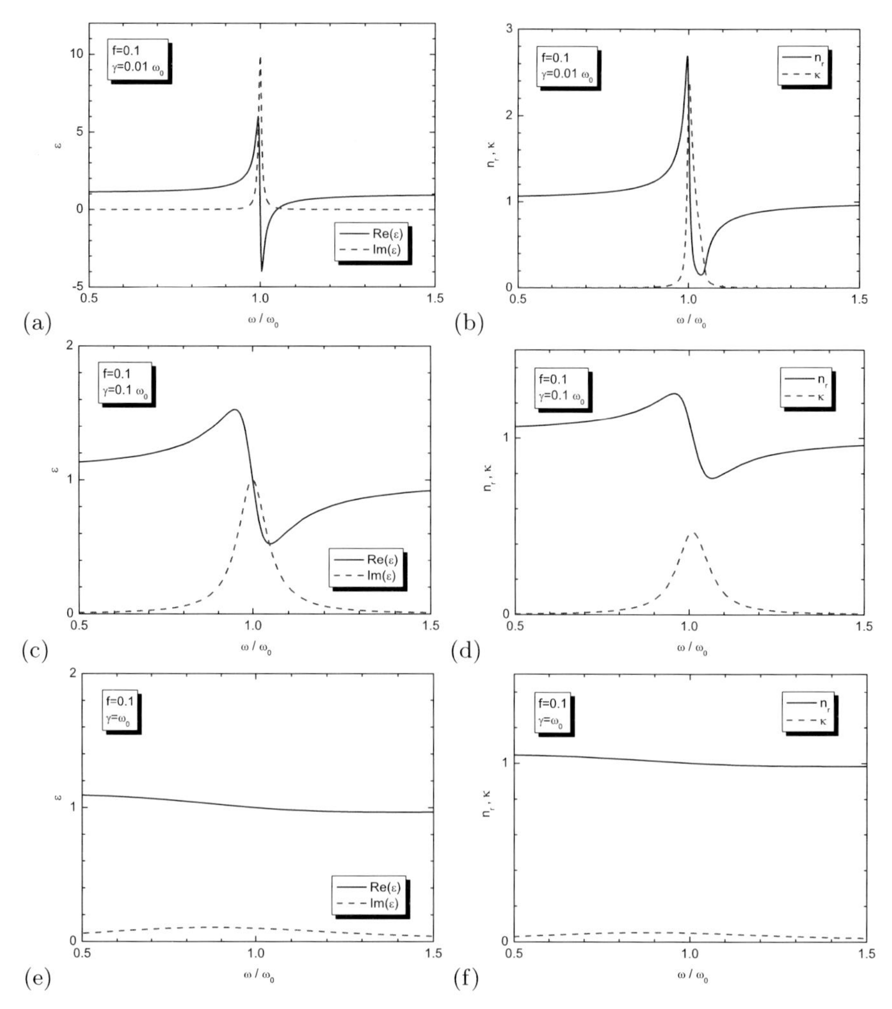

Fig. C.2. Real (*solid lines*) and imaginary (*dashed lines*) parts of the dielectric constant (**a**,**c**,**e**) and index of refraction (**b**,**d**,**f**) (C.11) for oscillator strength $f = 10^{-1}$ and various values of damping: (**a**,**b**) $\Gamma = 10^{-2}\omega_0$, (**c**,**d**) $\Gamma = 10^{-1}\omega_0$, and (**e**,**f**) $\Gamma = \omega_0$

The maximum absorption is given as

$$\alpha_{\mathrm{m}} = 2\frac{\omega_0}{c}\kappa(\omega_0) = f\,\frac{\omega_0^2}{\Gamma}\frac{n_\infty}{c}\,. \tag{C.14}$$

For zero damping, the dielectric function has a zero at

$$\omega_0' = \sqrt{1+f}\,\omega_0 \approx \omega_0\left(1+\frac{f}{2}\right)\,. \tag{C.15}$$

The latter approximation is valid for $f \ll 1$. In the region between ω_0 and ω_0', the real part of the index of refraction is very small (for the physically unrealistic case of $\Gamma \equiv 0$ it is exactly zero since $\epsilon < 0$). The reflectivity for vertical incidence $R = [(1-n^*)/(1+n^*)]^2$ in this region (width: $f\omega_0/2$) is thus very high. For larger damping (and small oscillator strength), this effect is washed out.

The frequency $\omega_{\epsilon'',\max}$ of the maximum of the imaginary part of ϵ'' of the dielectric function ($\hat{\Gamma} = \Gamma/\omega_0$) is

$$\omega^2_{\epsilon'',\max} = \frac{2 - \hat{\Gamma}^2 + \sqrt{16 - 4\hat{\Gamma}^2 + \hat{\Gamma}^4}}{6}\,\omega_0^2 \approx \omega_0^2 \left[1 - \left(\frac{\Gamma}{2\omega_0}\right)^2\right] . \tag{C.16}$$

The approximation is valid for small damping $\Gamma \ll \omega_0$. In this case, the detuned frequency of the maximum is close to ω_0 (Fig. C.3). The frequency position of the maximum of $\tan\delta = \epsilon''/\epsilon'$ is

$$\omega^2_{\tan\delta,\max} = \frac{2 + f - \hat{\Gamma}^2 + \Lambda^2}{6}\,\omega_0^2 \tag{C.17}$$

$$\Lambda^2 = \sqrt{12\,(1+f) + \left(2 + f - \hat{\Gamma}^2\right)^2} .$$

The value of $\tan\delta$ at its maximum is (Λ has the same meaning as in (C.17))

$$(\tan\delta)_{\max} = \frac{-3\sqrt{\frac{3}{2}}\, f\,\hat{\Gamma}\,\sqrt{2 + f - \hat{\Gamma}^2 + \Lambda^2}}{-8 - 8\,f + f^2 - 4\,\hat{\Gamma}^2 - 2\,f\,\hat{\Gamma}^2 + \hat{\Gamma}^4 + \left(2 + f - \hat{\Gamma}^2\right)\Lambda^2} . \tag{C.18}$$

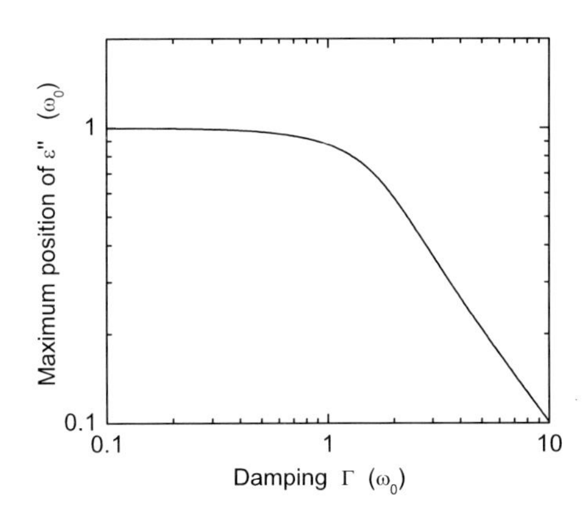

Fig. C.3. Frequency position of the maximum of ϵ'' as a function of the damping

D Quantum Statistics

Introduction

Bosons are particles with integer spin $s = n$, fermions are particles with spin $s = \frac{1}{2} + n$ with n being an integer including zero. The fundamental quantum-mechanical property of the wavefunction of a system with N such particles is that under exchange of any two particles, the wavefunction is symmetric in the case of bosons and antisymmetric in the case of fermions. For two particles, these conditions read

$$\Psi(q_1, q_2) = \Psi(q_2, q_1) \tag{D.1a}$$

$$\Psi(q_1, q_2) = -\Psi(q_2, q_1) \,, \tag{D.1b}$$

where (D.1a) holds for bosons and (D.1b) holds for fermions. The variables q_i denote the coordinates and spin of the i-th particle. The Pauli principle allows bosons to populate the same single particle state with an arbitrary number of particles (at least more than one). For fermions, the exclusion principle holds that each single particle state can only be populated once.

Partition Sum

We consider a gas of N identical particles in a volume V in equilibrium at a temperature T. The possible quantum-mechanical states of a particle is denoted as r. The energy of a particle in the state r is ϵ_r, the number of particles in the state r is n_r.

For vanishing interaction of the particles, the total energy of the gas in the state R (with n_r particles in the state r) is

$$E_R = \sum_r n_r \epsilon_r \,. \tag{D.2}$$

The sum runs over all possible states r. The total number of particles imposes the condition

$$N = \sum_r n_r \,. \tag{D.3}$$

In order to calculate the thermodynamic potentials, the partition sum Z needs to be calculated

$$Z = \sum_R \exp(-\beta E_R) \,, \tag{D.4}$$

with $\beta = 1/(kT)$. The sum runs over all possible microscopic states R of the gas, i.e. all combinations of the n_r that fulfill (D.3). The probability P_S to find the system in a particular state S is given by (canonical ensemble)

$$P_S = \frac{\exp(-\beta E_S)}{Z} \,. \tag{D.5}$$

The average number $\bar{n}_s$ of particles in a state s is given by

$$\bar{n}_s = \frac{\sum_R n_s \exp(-\beta E_R)}{Z} = -\frac{1}{\beta Z}\frac{\partial Z}{\partial \epsilon_s} = -\frac{1}{\beta}\frac{\partial \ln Z}{\partial \epsilon_s} \,. \tag{D.6}$$

We note that the average deviation $\overline{(\Delta n_s)^2} = \overline{n_s^2 - \bar{n}_s^2} = \overline{n_s^2} - \bar{n}_s^2$ is given by

$$\overline{(\Delta n_s)^2} = \frac{1}{\beta^2}\frac{\partial^2 \ln Z}{\partial \epsilon_s^2} = -\frac{1}{\beta}\frac{\partial \bar{n}_s}{\partial \epsilon_s} \,. \tag{D.7}$$

In the Bose–Einstein statistics (for bosons), the particles are fundamentally indistinguishable. Thus, a set of $(n_1, n_2, \ldots)$ uniquely describes the system. In the case of fermions, for each state n_r is either 0 or 1. In both cases, (D.3) needs to be fulfilled.

Photon Statistics

This case is the Bose–Einstein statistics (cf. Sect. D) with undefined particle number. We rewrite (D.6) as

$$\bar{n}_s = \frac{\sum_{n_s} n_s \exp(-\beta n_s \epsilon_s) \sum^{(s)}_{n_1, n_2, \ldots} \exp(-\beta(n_1\epsilon_1 + n_2\epsilon_2 + \ldots))}{\sum_{n_s} \exp(-\beta n_s \epsilon_s) \sum^{(s)}_{n_1, n_2, \ldots} \exp(-\beta(n_1\epsilon_1 + n_2\epsilon_2 + \ldots))} \,, \tag{D.8}$$

where $\sum^{(s)}$ denotes a summation that does not include the index s. In the case of photons, the values n_r can take any value (integers including zero) without restriction and therefore the sums $\sum^{(s)}$ in the numerator and denominator of (D.8) are identical. After some calculation we find

$$\bar{n}_s = -\frac{1}{\beta}\frac{\partial}{\partial \epsilon_s} \ln\left(\sum_{n_s=0}^{\infty} \exp(-\beta n_s \epsilon_s)\right) \,. \tag{D.9}$$

The argument of the logarithm is a geometrical series with the limit $[1-\exp(-\beta\epsilon_s)]^{-1}$. This leads to the so-called Planck distribution

$$\bar{n}_s = \frac{1}{\exp(\beta\epsilon_s) - 1} \,. \tag{D.10}$$

Fermi–Dirac Statistics

Now, the particle number is fixed to N. For the sum $\sum^{(s)}$ from (D.6), we introduce the term $Z_S(M)$

$$Z_s(M) = \sum_{n_1,n_2,\dots}^{(s)} \exp(-\beta(n_1\epsilon_1 + n_2\epsilon_2 + \dots)) \,, \tag{D.11}$$

when M particles are to be distributed over all states except s ($\sum_r^{(s)} n_r = M$). M is either $N-1$ if $n_s = 1$ and N if $n_s = 0$. Using Z_s, we can write

$$\bar{n}_s = \frac{1}{\frac{Z_s(N)}{Z_s(N-1)} \exp(\beta\epsilon_s) + 1} \,. \tag{D.12}$$

We evaluate $Z_s(N-1)$

$$\ln Z_s(N - \Delta N) = \ln Z_s(N) - \frac{\partial \ln Z_s}{\partial N}|_N \, \Delta N \,, \tag{D.13}$$

or

$$Z_s(N - \Delta N) = Z_s(N) \exp(-\gamma_s \Delta N) \,, \tag{D.14}$$

with

$$\gamma_s = \frac{\partial \ln Z_s}{\partial N} \,. \tag{D.15}$$

Since Z_s runs over many states, the derivative is approximately equal to

$$\gamma = \frac{\partial \ln Z}{\partial N} \,, \tag{D.16}$$

as will be shown below. Thus, we obtained so far

$$\bar{n}_s = \frac{1}{\exp(\gamma + \beta\epsilon_s) + 1} \,. \tag{D.17}$$

Equation (D.3) holds also for the average values $\bar{n}_s$, i.e.

$$N = \sum_r \bar{n}_r = \sum_r \frac{1}{\exp(\gamma + \beta\epsilon_s) + 1} \,, \tag{D.18}$$

from which the value of γ can be calculated. Given that the free energy is given as $F = -kT \ln Z$, we find that

$$\gamma = -\frac{1}{kT}\frac{\partial F}{\partial N} = -\beta\mu \,, \tag{D.19}$$

where μ is the chemical potential by definition. Therefore, the distribution function for the Fermi–Dirac statistics (also called the Fermi function) is

$$\bar{n}_s = \frac{1}{\exp[\beta(\epsilon_s - \mu)] + 1} . \tag{D.20}$$

Now, we briefly revisit the approximation $\gamma = \gamma_s$. Exactly, γ fulfills

$$\gamma = \gamma_s - n_s \frac{\partial \gamma}{\partial N} . \tag{D.21}$$

Thus, the approximation is valid if $n_s \frac{\partial \gamma}{\partial N} \ll \gamma$. Since $n_s < 1$, this means that the chemical potential does not change significantly upon addition of another particle.

Bose–Einstein Distribution

Executing (D.8) with the approximation $\gamma = \gamma_s$, the Bose–Einstein distribution is found to be

$$\bar{n}_s = \frac{1}{\exp[\beta(\epsilon_s - \mu)] - 1} . \tag{D.22}$$

E The $\mathbf{k}\cdot\mathbf{p}$ Perturbation Theory

The solutions of the Schrödinger equation (cf. Sect. 6.2)

$$H\Psi_{n\mathbf{k}}(\mathbf{r}) = \left(-\frac{\hbar^2}{2m}\nabla^2 + U(\mathbf{r})\right)\Psi_{n\mathbf{k}}(\mathbf{r}) = E_n(\mathbf{k})\Psi_{n\mathbf{k}}(\mathbf{r}) \,, \tag{E.1}$$

with a lattice periodic potential U, i.e. $U(\mathbf{r}) = U(\mathbf{r}+\mathbf{R})$ for direct lattice vectors $\mathbf{R}$, m being the free electron mass, are Bloch waves of the form

$$\Psi_{n\mathbf{k}}(\mathbf{r}) = \mathrm{e}^{\mathrm{i}\mathbf{k}\mathbf{r}} u_{n\mathbf{k}}(\mathbf{r}) \,, \tag{E.2}$$

with the lattice periodic Bloch function $u_{n\mathbf{k}}(\mathbf{r}) = u_{n\mathbf{k}}(\mathbf{r}+\mathbf{R})$.

Inserting the Bloch wave into (E.1), the following equation is obtained for the periodic Bloch function:

$$\left(-\frac{\hbar^2}{2m}\nabla^2 + U(\mathbf{r}) + \frac{\hbar}{m}\mathbf{k}\cdot\mathbf{p}\right)u_{n\mathbf{k}}(\mathbf{r}) = \left(E_n(\mathbf{k}) - \frac{\hbar^2 k^2}{2m}\right)u_{n\mathbf{k}}(\mathbf{r}) \,. \tag{E.3}$$

For simplicity, we assume a band edge $E_n(0)$ at $\mathbf{k}=0$. In its vicinity, the $\mathbf{k}\cdot\mathbf{p}$ term can be treated as a perturbation. The dispersion for a nondegenerate band[1] is given up to second order in k

$$E_n(\mathbf{k}) = E_n(0) + \sum_{i,j=1}^{3}\left(\frac{\hbar^2}{2m}\delta_{ij} + \frac{\hbar^2}{m}\sum_{l\neq n}\frac{p^i_{nl}\,p^j_{ln}}{E_n(0)-E_l(0)}\right)k_i\,k_j \,, \tag{E.4}$$

with l running over other, so-called *remote* bands. The momentum matrix element is given by $p^i_{nl} = \langle u_{n0}|p_i|u_{l0}\rangle$ (cf. (6.38)). The coefficients in front of the quadratic terms are the components of the dimensionless inverse effective-mass tensor (cf. (6.42))

$$\left(\frac{m}{m^*}\right)_{ij} = \delta_{ij} + \frac{2}{m}\sum_{l\neq n}\frac{p^i_{nl}\,p^j_{ln}}{E_n(0)-E_l(0)} \,. \tag{E.5}$$

For degenerate bands, the $p^i_{nn'}$ vanish when n and n' belong to the degenerate set and also the first-order correction is zero. In the Löwdin perturbation

[1] apart from the spin degeneracy

theory [733], the bands are separated into the close-by degenerate or nearly degenerate bands and the remote bands. The effect of the remote bands is taken into account by an effective perturbation

$$\mathbf{k} \cdot \mathbf{p} + \mathbf{k} \cdot \mathbf{p} \sum_{l \neq n} \frac{|l\rangle\langle l|}{E_n(0) - E_l(0)} \mathbf{k} \cdot \mathbf{p} \,, \tag{E.6}$$

with the index l running over all bands not being in the degenerate set. The dispersion relation is obtained by diagonalization of the Hamiltonian (E.3) in the degenerate basis but with the perturbation given by (E.6).

The spin-orbit interaction [540] adds an additional term

$$H_{\text{so}} = \frac{\hbar}{4m^2c^2} \left(\boldsymbol{\sigma} \times \nabla U\right) \mathbf{p} \tag{E.7}$$

to the Hamiltonian, where $\boldsymbol{\sigma}$ are the Pauli spin matrices and c the vacuum speed of light. In the Schrödinger equation for the Bloch functions two new terms arise:

$$\begin{aligned} &\left(-\frac{\hbar^2}{2m} \nabla^2 + U(\mathbf{r}) + \frac{\hbar}{4m^2c^2} \left(\boldsymbol{\sigma} \times \nabla U\right) \mathbf{p} + \right. \\ &\left. \frac{\hbar}{m} \mathbf{k} \left[\mathbf{p} + \frac{\hbar}{4m^2c^2} \left(\boldsymbol{\sigma} \times \nabla U\right) \right] \right) u_{n\mathbf{k}}(\mathbf{r}) = \left(E_n(\mathbf{k}) - \frac{\hbar^2 k^2}{2m} \right) u_{n\mathbf{k}}(\mathbf{r}) \,. \end{aligned} \tag{E.8}$$

The linear term in $\mathbf{k}$ is again treated as a perturbation. The first spin-orbit term in (E.8) is lattice periodic, thus the solutions at $\mathbf{k} = 0$ are still periodic Bloch functions, however, different ones from previously. If the band edge is not degenerate, the momentum operator in (E.3) is simply replaced by

$$\boldsymbol{\pi} = \mathbf{p} + \frac{\hbar}{4m^2c^2} \left(\boldsymbol{\sigma} \times \nabla U\right) \,, \tag{E.9}$$

and the band edge is still parabolic. For a degenerate band edge, the effect can be more profound, in particular it can lead to the lifting of a degeneracy.

In the 8-band Kane model [158], four bands (lowest conduction band, heavy, light and split-off hole band) are treated explicitly and the others through Löwdin perturbation theory. The basis is chosen to be diagonal in the spin-orbit interaction leaving the spin-orbit interaction Δ_0 as parameter. The band- edge Bloch functions are denoted as $|i \uparrow\rangle$, where the index $i = s, x, y, z$ labels the symmetry of the different bands. The linear combinations that diagonalize the spin-orbit interaction are given in Table E.1. The bandgap and the spin-orbit interaction are given by

$$E_{\text{g}} = E_{\Gamma_6} - E_{\Gamma_8} \tag{E.10a}$$

$$\Delta_0 = E_{\Gamma_8} - E_{\Gamma_7} \,. \tag{E.10b}$$

Table E.1. Basis set that diagonalizes the spin-orbit interaction

$\|J, m_j\rangle$	wavefunction	symmetry
$\|\frac{1}{2},\frac{1}{2}\rangle$	$\mathrm{i}\|s\uparrow\rangle$	Γ_6
$\|\frac{1}{2},-\frac{1}{2}\rangle$	$\mathrm{i}\|s\downarrow\rangle$	Γ_6
$\|\frac{3}{2},\frac{3}{2}\rangle$	$\frac{1}{\sqrt{2}}\|(x+\mathrm{i}y)\uparrow\rangle$	Γ_8
$\|\frac{3}{2},\frac{1}{2}\rangle$	$\frac{1}{\sqrt{6}}\|(x+\mathrm{i}y)\downarrow\rangle-\sqrt{\frac{2}{3}}\|z\uparrow\rangle$	Γ_8
$\|\frac{3}{2},-\frac{1}{2}\rangle$	$-\frac{1}{\sqrt{6}}\|(x-\mathrm{i}y)\uparrow\rangle-\sqrt{\frac{2}{3}}\|z\downarrow\rangle$	Γ_8
$\|\frac{3}{2},-\frac{3}{2}\rangle$	$\frac{1}{\sqrt{2}}\|(x-\mathrm{i}y)\uparrow\rangle$	Γ_8
$\|\frac{1}{2},\frac{1}{2}\rangle$	$\frac{1}{\sqrt{3}}\|(x+\mathrm{i}y)\downarrow\rangle+\sqrt{\frac{1}{3}}\|z\uparrow\rangle$	Γ_7
$\|\frac{1}{2},-\frac{1}{2}\rangle$	$-\frac{1}{\sqrt{3}}\|(x-\mathrm{i}y)\uparrow\rangle+\sqrt{\frac{1}{3}}\|z\downarrow\rangle$	Γ_7

The Hamiltonian in the basis states of Table E.1 is given by

$$\begin{bmatrix}
k^2+E_{\mathrm{g}} & 0 & \sqrt{2}Pk_+ & -\sqrt{\frac{2}{3}}Pk_z & -\sqrt{\frac{2}{3}}Pk_- & 0 & \sqrt{\frac{1}{3}}Pk_z & -\sqrt{\frac{4}{3}}Pk_- \\
0 & k^2+E_{\mathrm{g}} & 0 & \sqrt{\frac{2}{3}}Pk_+ & -\sqrt{\frac{2}{3}}Pk_z & \sqrt{2}Pk_- & \sqrt{\frac{4}{3}}Pk_+ & \sqrt{\frac{1}{3}}Pk_z \\
\sqrt{2}Pk_- & 0 & k^2 & 0 & 0 & 0 & 0 & 0 \\
-\sqrt{\frac{2}{3}}Pk_z & \sqrt{\frac{2}{3}}Pk_- & 0 & k^2 & 0 & 0 & 0 & 0 \\
-\sqrt{\frac{2}{3}}Pk_+ & -\sqrt{\frac{2}{3}}Pk_z & 0 & 0 & k^2 & 0 & 0 & 0 \\
0 & \sqrt{2}Pk_+ & 0 & 0 & 0 & k^2 & 0 & 0 \\
\sqrt{\frac{1}{3}}Pk_z & \sqrt{\frac{4}{3}}Pk_- & 0 & 0 & 0 & 0 & k^2-\Delta_0 & 0 \\
-\sqrt{\frac{4}{3}}Pk_+ & \sqrt{\frac{1}{3}}Pk_z & 0 & 0 & 0 & 0 & 0 & k^2-\Delta_0
\end{bmatrix} \tag{E.11}$$

with the energy measured from the valence-band edge in units of $\hbar^2/(2m)$ and

$$\frac{1}{2}\mathrm{i}\hbar P = \langle s|\pi_x|x\rangle = \langle s|\pi_y|y\rangle = \langle s|\pi_z|z\rangle \tag{E.12a}$$

$$k_\pm = k_x \pm \mathrm{i}k_y\ . \tag{E.12b}$$

The inclusion of remote bands renormalizes the above Hamiltonian to

$$\begin{bmatrix} Dk^2+E_\mathrm{g} & 0 & \sqrt{2}Pk_+ & -\sqrt{\frac{2}{3}}Pk_z & -\sqrt{\frac{2}{3}}Pk_- & 0 & \sqrt{\frac{1}{3}}Pk_z & -\sqrt{\frac{4}{3}}Pk_- \\ 0 & Dk^2+E_\mathrm{g} & 0 & \sqrt{\frac{2}{3}}Pk_+ & -\sqrt{\frac{2}{3}}Pk_z & \sqrt{2}Pk_- & \sqrt{\frac{4}{3}}Pk_+ & \sqrt{\frac{1}{3}}Pk_z \\ \sqrt{2}Pk_- & 0 & H_\mathrm{h} & R & S & 0 & \frac{\mathrm{i}}{\sqrt{2}}R & -\mathrm{i}\sqrt{2}S \\ -\sqrt{\frac{2}{3}}Pk_z & \sqrt{\frac{2}{3}}Pk_- & R^* & H_\mathrm{l} & 0 & S & \frac{H_\mathrm{h}-H_\mathrm{l}}{\sqrt{2}\mathrm{i}} & \mathrm{i}\sqrt{\frac{3}{2}}R \\ -\sqrt{\frac{2}{3}}Pk_+ & -\sqrt{\frac{2}{3}}Pk_z & S^* & 0 & H_\mathrm{l} & -R & -\mathrm{i}\sqrt{\frac{3}{2}}R^* & \frac{H_\mathrm{h}-H_\mathrm{l}}{\sqrt{2}\mathrm{i}} \\ 0 & \sqrt{2}Pk_+ & 0 & S^* & -R^* & H_\mathrm{h} & -\mathrm{i}\sqrt{2}S^* & -\frac{\mathrm{i}}{\sqrt{2}}R^* \\ \sqrt{\frac{1}{3}}Pk_z & \sqrt{\frac{4}{3}}Pk_- & -\frac{\mathrm{i}}{\sqrt{2}}R^* & -\frac{H_\mathrm{h}-H_\mathrm{l}}{\sqrt{2}\mathrm{i}} & \mathrm{i}\sqrt{\frac{3}{2}}R & \mathrm{i}\sqrt{2}S & \frac{H_\mathrm{h}+H_\mathrm{l}}{\sqrt{2}}-\Delta_0 & 0 \\ -\sqrt{\frac{4}{3}}Pk_+ & \sqrt{\frac{1}{3}}Pk_z & \mathrm{i}\sqrt{2}S^* & -\mathrm{i}\sqrt{\frac{3}{2}}R^* & -\frac{H_\mathrm{h}-H_\mathrm{l}}{\sqrt{2}\mathrm{i}} & \frac{\mathrm{i}}{\sqrt{2}}R & 0 & \frac{H_\mathrm{h}+H_\mathrm{l}}{\sqrt{2}}-\Delta_0 \end{bmatrix} \tag{E.13}$$

with

$$D = 1 + \frac{2}{m}\sum_{l\neq n}\frac{|\langle s|\pi_x|l\rangle|^2}{E_\mathrm{g}-E_l(0)} \tag{E.14a}$$

$$\gamma_1' = \left[1+\frac{2}{m}\sum_{l\neq n}\frac{|p^x_{xl}|^2}{E_n(0)-E_l(0)}\right] - \frac{2P^2}{3E_\mathrm{g}} \tag{E.14b}$$

$$\gamma_2' = \left[1+\frac{2}{m}\sum_{l\neq n}\frac{|p^y_{xl}|^2}{E_n(0)-E_l(0)}\right] - \frac{P^2}{3E_\mathrm{g}} \tag{E.14c}$$

$$\gamma_3' = \left[\frac{2}{m}\sum_{l\neq n}\frac{p^x_{xl}p^y_{ly}+p^y_{xl}p^x_{ly}}{E_n(0)-E_l(0)}\right] - \frac{P^2}{3E_\mathrm{g}} \tag{E.14d}$$

$$H_\mathrm{h} = (\gamma_1'+\gamma_2')(k_x^2+k_y^2) + (\gamma_1'-2\gamma_2')k_z^2 \tag{E.14e}$$

$$H_\mathrm{l} = (\gamma_1'-\gamma_2')(k_x^2+k_y^2) + (\gamma_1'+2\gamma_2')k_z^2 \tag{E.14f}$$

$$R = -2\sqrt{3}\gamma_3' k_- k_z \tag{E.14g}$$

$$S = \sqrt{3}\gamma_2'(k_x^2-k_y^2) + 2\sqrt{3}\gamma_3'\mathrm{i}k_x k_y\,. \tag{E.14h}$$

The Hamiltonian in the presence of inhomogeneous strain is given in Ref. [172]. The hole bands decouple from the conduction band for $E_\mathrm{g}\to\infty$ (six-band model [409]). The heavy and light holes can be treated separately for $\Delta_0\to\infty$ (Luttinger Hamiltonian). For the Γ_8 states, the Hamiltonian is then given by

$$\begin{bmatrix} H_\mathrm{h} & R & S & 0 \\ R^* & H_\mathrm{l} & 0 & S \\ S^* & 0 & H_\mathrm{l} & -R \\ 0 & S^* & -R^* & H_\mathrm{h} \end{bmatrix}\,. \tag{E.15}$$

F Effective-Mass Theory

The effective-mass theory or approximation (EMA), also termed the envelope function approximation, is widely used for calculating the electronic properties of carriers in potentials in an otherwise periodic crystal. The strength of the method is that the complexities of the periodic potential are hidden in the effective-mass tensor m^*_{ij}. The effective-mass theory is a useful approximation for the treatment of shallow impurities (Sect. 7.5) or quantum wells (Sect. 11.5.1) with a potential that is slowly varying with respect to the scale of the lattice constant.

For the lattice-periodic potential, the Schrödinger equation

$$H_0 \Psi_{n\mathbf{k}} = E_n(\mathbf{k}) \Psi_{\mathrm{n}\mathbf{k}} \tag{F.1}$$

is solved by the Bloch wave $\Psi_{n\mathbf{k}}$. With a perturbing potential V, the Schrödinger equation reads

$$(H_0 + V)\, \Psi_{n\mathbf{k}} = E_n(\mathbf{k}) \Psi_{\mathrm{n}\mathbf{k}} \; . \tag{F.2}$$

According to Wannier's theorem [734], the solution is approximated by the solution of the equation

$$[E_n(-\mathrm{i}\nabla) + V]\, \Phi_n = E\, \Phi_n \; . \tag{F.3}$$

The dispersion relation is expanded to second order as described in Appendix E. The function Φ_n is termed the *envelope function* since it varies slowly compared to the lattice constant and the exact wavefunction is approximated (in lowest order) by

$$\Psi(\mathbf{r}) = \Phi_n(\mathbf{r}) \, \exp\left(\mathrm{i}\mathbf{k}\mathbf{r}\right) u_{n0}(\mathbf{r}) \; . \tag{F.4}$$

References

1. J.M. Bishop, *How to Win the Nobel Prize* (Harvard University Press, Cambridge, 2003)
2. T.J. Seebeck, *Magnetische Polarisation der Metalle und Erze durch Temperaturdifferenz*, Abhandl. Deut. Akad. Wiss. Berlin, 265–373 (1822)
3. M. Faraday, Experimental Researches in Electricity, Series IV, 433 (1833)
4. W. Smith, J. Soc. Telegraph Engrs, 2, 31 (1873), Nature (issue of 20 February 1873), 303 (1873)
5. Ch. Fritts, Am. J. Sci. 26, 465 (1883)
6. F. Braun, *Über die Stromleitung durch Schwefelmetalle*, J.C. Poggendorf ed., Ann. Phy. Chem. 153, 556 (1874)
7. W.G. Adams, R.E. Day, Proc. Roy. Soc. London Ser. A 25, 113 (1876)
8. H.J. Round, Electron World 19, 309 (1907)
9. J. Königsberger, J. Weiss, Ann. Phys. 35, 1 (1911)
10. Th. Thomas, New Scientist (issue of March 29 1997), p. 55 (1997)
11. C. Kleint, Progr. Surf. Sci. 57, 253–327 (1998)
12. J.J. Lilienfeld, US patent 1,745,175, awarded 1930
13. R. de L. Kronig, W.G. Penney, Proc. Roy. Soc. (London) A 130, 499 (1931)
14. A.H. Wilson, *The theory of electronic semiconductors*, Proc. Roy. Soc. Lond. Ser. A 133, 458 (1931)
15. C. Zener, Proc. Roy. Soc. London A 130, 499 (1931)
16. J. Frenkel, Phys. Z. Sowjetunion 9, 158 (1936)
17. B. Davydov, C.R. (Dokl.) Acad. Sci. USS, 20, 283 (1938)
18. W. Schottky, Naturwissenschaften, 26, 843 (1938)
19. N.F. Mott, Proc. Camb. Philos. Soc. 34, 568 (1938)
20. R.S. Ohl, US patent 2,402,661, awarded 1946
21. W.B. Shockley, *Circuit Element Utilizing Semiconductor Material*, US patent 2,569,347 (1948)
22. J. Bardeen, W.H. Brittain, *The transistor, a semiconductor triode*, Phys. Rev. 74, 230 (1948)
23. W.F. Brinkman, D.E. Haggan, W.W. Troutman, IEEE J. Solid-State Circ. 32, 1858–1865 (1997)
24. H. Welker, Z. Naturf. 7a, 744 (1952)
25. W. Shockley, Proc. IRE, 40, 1365 (1952)
26. G.C. Dacey, I.M. Ross, Proc. IRE, 41, 970 (1953)
27. D.M. Chapin, C.S. Fuller, G.L. Pearson, J. Appl. Phys. 25, 676 (1954)
28. R.N. Hall, G.E. Fenner, J.D. Kingsley, T.J. Soltys, R.O. Carlson, Phys. Rev. Lett. 9, 366 (1962)

29. J.A. Hoerni, *Planar silicon diodes and transistors*, IRE Trans. Electron Devices, Mar. 8, 1961, p. 178; also presented at Professional Group on Electron Devices Meeting, Washington, D.C., Oct. 1960
30. D. Kahng, M.M. Atalla, *Silicon-silicon dioxide field induced surface device*, presented at Solid State Device Research Conf., Pittsburgh, PA, June 1960
31. M.I. Nathan, W.P. Dumke, G. Burns, F.H. Dill Jr., G. Lasher, Appl. Phys. Lett. 1, 62 (1962)
32. Zh.I. Alferov, R.F. Kasarinov, Inventor's Certificate No. 181737 [in Russian], Application No. 950840 (1963)
33. H. Kroemer, Proc. IEEE 51, 1782 (1963)
34. Zh.I. Alferov, V.I. Korol'kov, V.I. Maslov, A.V. Litina M.K. Krukan, A.A. Yakovenko, Fizika i Tekn. Poluprovodn. 1, 260 (1967)
35. C.A. Mead, Proc. IEEE 54, 307 (1966)
36. W.W. Hooper, W.I. Lehrer, Proc. IEEE 55, 1237 (1967)
37. Zh.I. Alferov, V.M. Andreev, E.L. Portnoi, M.K. Trukan, Fizika i. Tekn. Poluprovodn. 3, 1328 (1969)
38. I. Hayashi, IEEE Trans. Electron Devices, ED-31, 1630 (1969)
39. M. Dayah, www.dayah.com
40. P. Evers, *Die wundersame Welt der Atomis: 10 Jahre in den Physikalischen Blättern* (Wiley-VCH, Weinheim, 2002), ISBN 3-527-40359-0
41. M.T. Yin, M.L. Cohen, Phys. Rev. Lett. 45, 1004 (1980)
42. J. Bardeen, W. Shockley, Phys. Rev. 80, 72 (1950)
43. Royston M. Roberts, *Serendipidty, Accidental Discoveries in Science* (John Wiley & Sons, New York, NY, 1989), pp. 75–81
44. T. Soma, T. Umenai, phys. stat. sol. (b) 78, 229 (1976)
45. H.J.F. Jansen, A.J. Freeman, Phys. Rev. B 33, 8629 (1986)
46. J.C. Phillips, *Bonds and Bands in Semiconductors* (Academic Press, New York, 1973)
47. J.A. Van Vechten, Phys. Rev. 182, 891 (1969)
48. J.C. Phillips, J.A. Van Vechten, Phys. Rev. Lett. 22, 705 (1969)
49. G. Burns, A.M. Glazer, *Space Groups for Solid State Scientists* (Academic Press, San Diego, 1990)
50. R.E. Proano, D.G. Ast, J. Appl. Phys. 66, 2189 (1989)
51. *Amorphous Semiconductors*, M.H. Brodsky ed., Topics in Applied Physics, Vol. 36 (Springer, Berlin, 1979)
52. B. Kramer, *Electronic properties of amorphous solids*, Adv. Solid State Phys. (Festkörperprobleme) 12, 133 (1972)
53. J.R. Chelikowski, Phys. Rev. B 34, 5295 (1986)
54. J.M. Besson, J.P. Itié, A. Polian, G. Weill, J.L. Mansot, J. Gonzalez, Phys. Rev. B 44, 4214 (1991)
55. P. Lawaetz, Phys. Rev. B 5, 4039 (1972)
56. J.L. Shay, J.H. Wernick, *Ternary Chalcopyrites* (Pergamon, Oxford, 1975)
57. A. Zunger, Appl. Phys. Lett. 50, 164 (1987)
58. J.E. Jaffe, A. Zunger, Phys. Rev. B 29, 1882 (1984)
59. X. Nie, S.-H. Wei, S.B. Zhang, Phys. Rev. Lett. 88, 066405 (2002)
60. A. Guivarc'h, R. Guérin, J. Caulet, A. Poudoulec, J. Appl. Phys. 66, 2129 (1989)
61. Y. Lifshitz, X.F. Duan, N.G. Shang, Q. Li, L. Wan, I. Bello, S.T. Lee, Nature 412, 404 (2001)

62. S.-H. Wei, L.G. Ferreira, A. Zunger, Phys. Rev. B 41, 8240 (1990)
63. I.-H. Ho, G.B. Stringfellow, Appl. Phys. Lett. 69, 2701 (1996)
64. C. Bundesmann, M. Schubert, D. Spemann, A. Rahm, H. Hochmuth, M. Lorenz, M. Grundmann, Appl. Phys. Lett. 85, 905 (2004)
65. M.M. Kreitman, D.L. Barnett, J. Chem. Phys. 43, 364 (1965)
66. O. Okada, J. Phys. Soc. Jpn. 48, 391 (1980)
67. J.C. Mikkelsen Jr., J.B. Boyce, Phys. Rev. Lett. 49, 1412 (1982)
68. J.C. Mikkelsen Jr., J.B. Boyce, Phys. Rev. B 28, 7130 (1983)
69. O. Ambacher, M. Eickhoff, A. Link, M. Hermann, M. Stutzmann, F. Bernardini, V. Fiorentini, Y. Smorchkova, J. Speck, U. Mishra, W. Schaff, V. Tilak, L.F. Eastman, phys. stat. sol. (c) 0, 1878 (2003)
70. A. Zunger, S. Mahajan, *Atomic Ordering and Phase Separation in III–V Alloys* in Handbook of Semiconductors, Vol. 3 (Elsevier, Amsterdam, 1994), p. 1399–1513
71. H.S. Lee, J.Y. Lee, T.W. Kim, D.U. Lee, D.C. Choo, H.L. Park, Appl. Phys. Lett. 79, 1637 (2001)
72. M. Lannoo, J. Bourgoin, *Point Defects in Semiconductors I* (Springer, Berlin, 1981)
73. C.S. Nichols, C.G. Van der Walle, S.T. Pantelides, Phys. Rev. B 40, 5484 (1989)
74. J.-W. Jeong, A. Oshiyama, Phys. Rev. B 64, 235204 (2001)
75. H. Bracht, N.A. Stolwijk, H. Mehrer, Phys. Rev. B 52, 16542 (1995)
76. J. Gebauer, M. Lausmann, F. Redmann, R. Krause-Rehberg, H.S. Leipner, E.R. Weber, Ph. Ebert, Phys. Rev. B 67, 235207 (2003)
77. S. Kret, Pawel Dłużewski, Piotr Dłużewski, J.-Y. Laval, Philos. Mag. 83, 231 (2003). Also at http://info.ifpan.edu.pl/SL-1/sl14sub/dys1.htm
78. S.N.G. Chu, W.T. Tsang, T.H. Chiu, A.T. Macrander, J. Appl. Phys. 66, 521 (1989)
79. A.R. Smith, V. Ramachandran, R.M. Feenstra, D.W. Greve, M.-S. Shin, M. Skowronski, J. Neugebauer, J.E. Northrup, J. Vac. Sci. Technol. A 16, 1641 (1998)
80. V. Smaminathan, A.S. Jordan, *Dislocations in III/V Compounds* Semicond. Semimet. 38, 294 (1993)
81. T. Kamejima, J. Matsui, Y. Seki, H. Watanabe, J. Appl. Phys. 50, 3312 (1979)
82. Z. Liliental-Weber, H. Sohn, J. Washburn, *Structural Defects in Epitaxial III/V Layers* Semicond. Semimet. 38, 397 (1993)
83. S. Mader, A.E. Blakeslee, Appl. Phys. Lett. 25, 365 (1974)
84. S. Amelinckx, W. Dekeyser, *The structure and properties of grain boundaries*, Solid State Phys. 8, 325 (1959)
85. C.R.M. Grovenor, J. Phys. C: Solid State Phys. 18, 4079–4119 (1985)
86. NREL, http://www.nrel.gov/measurements/trans.html
87. F.L. Vogel, W.G. Pfann, H.E. Corey, E.E. Thomas, Phys. Rev. 90, 489 (1953)
88. H. Föll, http://www.tf.uni-kiel.de/matwis/amat/def_en/index.html
89. H. Föll, D. Ast, Philos. Mag. A 40, 589 (1979)
90. M. Grundmann, A. Krost, D. Bimberg, J. Cryst. Growth 107, 494 (1991)
91. F. Wolf, W. Mader, Optik 110, Suppl. 8, 1999
92. W. Mader, private communication
93. Landolt-Börnstein, *Numerical Data and Functional Relationships in Science and Technology*, New Series, Vol. 17, *Semiconductors*, O. Madelung, M. Schulz, H. Weiss, eds. (Springer, Berlin, 1982)

94. J.L.T. Waugh, G. Dolling, Phys. Rev. 132, 2410 (1963)
95. A. Göbel, T. Ruf, A. Fischer, K. Eberl, M. Cardona, J.P. Silveira, F. Briones, Phys. Rev. B 59, 12612 (1999)
96. Q.-Y. Tong, U. Gösele, *Semiconductor Wafer Bonding* (John Wiley & Sons, New York, 1998), ISBN 0-471-57481-3
97. *Wafer Bonding*, M. Alexe, U. Gösele, eds. (Springer, Berlin, 2004), ISBN 3-540-21049-0
98. A. Reznicek, R. Scholz, S. Senz, U. Gösele, Mater. Chem. Phys. 81, 277 (2003)
99. P. Kopperschmidt, S. Senz, G. Kästner, D. Hesse, U.M. Gösele, Appl. Phys. Lett. 72, 3181 (1998)
100. M. Schubert, Universität Leipzig, private communication
101. A.S. Barker, A.J. Sievers, Rev. Mod. Phys. 47, Suppl. No. 2, S1 (1975)
102. R.C. Newman, Semicond. Semimet. 38, 118 (1993)
103. M. Stavola, Semicond. Semimet. 51B, 153 (1999)
104. M.D. McCluskey, J. Appl. Phys. 87, 3593 (2000)
105. L. Hoffmann, J.C. Bach, B. Bech Nielsen, P. Leary, R. Jones, S. Öberg, Phys. Rev. B 55, 11167 (1997)
106. F. Thompson, R.C. Newman, J. Phys. C 4, 3249 (1971)
107. I.F. Chang, S.S. Mitra, Phys. Rev. 172, 924 (1968)
108. L.I. Deych, A. Yamilov, A.A. Lisyansky, Phys. Rev. B 62, 6301 (2000)
109. M. Born, I. Huang, *Dynamical Theory of Crystal Lattices* (Clarendon Press, Oxford, 1960)
110. A.S. Saada, *Elasticity, Theory and Applications* (Pergamon, New York, 1974)
111. P.N. Keating, Phys. Rev. 145, 637 (1966)
112. A.-B. Chen, A. Sher, W.T. Yost, Semicond. Semimet. 37, 1 (1992)
113. R.M. Martin, Phys. Rev. B 1, 4005 (1970)
114. R.M. Martin, Phys. Rev. B 6, 4546 (1972)
115. Ch.G. Van de Walle, Phys. Rev. B 39, 1871 (1989)
116. J.D. Eshelby, Proc. Roy. Soc. London Ser. A 241, 376 (1957)
117. M. Grundmann, O. Stier, D. Bimberg, Phys. Rev. B 52, 11969 (1995)
118. L.B. Freund, J.A. Floro, E. Chason, Appl. Phys. Lett. 74, 1987 (1999)
119. M. Grundmann, Appl. Phys. Lett. 83, 2444 (2003)
120. G.G. Stoney, Proc. R. Soc. London, Ser. A 82, 172 (1909)
121. R. Beresford, J. Yin, K. Tetz, E. Chason, J. Vac. Sci. Technol. B 18, 1431 (2000)
122. V.Y. Prinz, V.A. Seleznev, A.K. Gutakovsky, Proc. 24th Int. Conf. Semicond. Physics, Jerusalem, Israel, 1998 (World Scientific, Singapore, 1998), Th3-D5
123. O.G. Schmidt, N. Schmarje, C. Deneke, C. Müller, N.-Y. Jin-Phillipp, Adv. Mater. 13, 756 (2001)
124. J.C. Bean, L.C. Feldman, A.T. Fiory, S. Nakahara, I.K. Robinson, J. Vac. Sci. Technol. A 2, 436 (1984)
125. S. Mendach, University of Hamburg, private communication
126. R. Hull, J.C. Bean, J. Vac. Sci. Technol. A 7, 2580 (1989)
127. J.W. Matthews, A.E. Blakeslee, J. Cryst. Growth 27, 118 (1974)
128. R. People, J.C. Bean, Appl. Phys. Lett. 47, 322 (1985)
129. F.C. Frank, J. van der Merwe, Proc. R. Soc. A 198, 216 (1949); ibid 198, 2205
130. J.H. van der Merwe, J. Appl. Phys. 34, 123 (1962)
131. J.R. Willis, S.C. Jain, R. Bullough, Philos. Mag. A 62, 115 (1990)
132. B.W. Dodson, J.Y. Tsao, Appl. Phys. Lett. 51, 1325 (1987)

133. K.L. Kavanagh, M.A. Capano, L.W. Hobbs, J.C. Barbour, P.M.J. Maree, W. Schaff, J.W. Mayer, D. Pettit, J.M. Woodall, J.A. Stroscio, R.M. Feenstra, J. Appl. Phys. 64, 4843 (1988)
134. M. Grundmann, U. Lienert, D. Bimberg, A. Fischer-Colbrie, J.N. Miller, Appl. Phys. Lett. 55, 1765 (1989)
135. E. Kasper, H.J. Herzog, H. Kibbel, Appl. Phys. 8, 199 (1975)
136. P. Quadbeck, Ph. Ebert, K. Urban, J. Gebauer, R. Krause-Rehberg, Appl. Phys. Lett. 76, 300 (2000)
137. Ch. Kittel, *Quantum Theory of Solids* (John Wiley & Sons, New York, 1963)
138. J.R. Chelikowsky, M.L. Cohen, Phys. Rev. B 14, 556 (1976)
139. R. Dalven, *Electronic structure of PbS, PbSe, and PbTe*, Solid State Phys. 28, 179 (1973)
140. J.E. Jaffe, A. Zunger, Phys. Rev. B 28, 5822 (1983)
141. S. Limpijumnong, S.N. Raskkeev, W.R.L. Lambrecht, MRS Internet J. Nitride Semicond. Res. 4S1, G6.11 (1999)
142. R. Ahuja, O. Eriksson, B. Johansson, J. Appl. Phys. 90, 1854 (2001)
143. J.E. Bernard, A. Zunger, Phys. Rev. B 26, 3199 (1987)
144. E.W. Williams, V. Rehn, Phys. Rev. 172, 798 (1968)
145. K.-R. Schulze, H. Neumann, K. Unger, phys. stat. sol. (b) 75, 493 (1976)
146. R. Braunstein, A.R. Moore, F. Herman, Phys. Rev. 109, 695 (1958)
147. D.J. Wolford, W.Y. Hsu, J.D. Dow, B.G. Streetman, J. Lumin. 18/19, 863 (1978)
148. R. Schmidt, B. Rheinländer, M. Schubert, D. Spemann, T. Butz, J. Lenzner, E.M. Kaidashev, M. Lorenz, M. Grundmann, Appl. Phys. Lett. 82, 2260 (2003)
149. R. Schmidt-Grund, D. Fritsch, M. Schubert, B. Rheinländer, H. Schmidt, H. Hochmuth, M. Lorenz, D. Spemann, C.M. Herzinger, M. Grundmann, AIP Conf. Proc. 772, 201 (2005)
150. B. Kramer, phys. stat. sol. 41, 649 (1970)
151. B. Kramer, phys. stat. sol. (b) 47, 501 (1971)
152. B. Seraphin, Z. Naturf. 9a, 450 (1954)
153. Y.W. Tsang, M.L. Cohen, Phys. Rev. B 3, 1254 (1971)
154. J. Hartung, L.Å. Hansson, J. Weber, Proc. of the 20th Int. Conf. on the Physics of Semiconductors, Thessaloniki, Greece, E.M. Anastassakis, J.D. Joannopoulos, eds. (World Scientific, Singapore, 1990), p. 1875
155. Y. Varshni, Physica 34, 149 (1967)
156. K.P. O'Donnell, X. Chen, Appl. Phys. Lett. 58, 2924 (1991)
157. R. Pässler, Phys. Rev. B 66, 085201 (2002)
158. E.O. Kane, J. Phys. Chem. Solids 1, 249 (1957)
159. C. Hermann, C. Weisbuch, Phys. Rev. B 15, 823 (1977)
160. W. Shockley, Phys. Rev. 90, 491 (1953)
161. G. Dresselhaus, A.F. Kip, C. Kittel, Phys. Rev. 98, 368 (1955)
162. R.P. Feynman, Phys. Rev. B 97, 600 (1955)
163. G.D. Mahan, *Many-Particle Physics* (Plenum Press, New York, 1981)
164. P. Pfeffer, W. Zawadzki, Phys. Rev. B 41, 1561 (1990)
165. D. Bimberg, private communication, original authorship unknown
166. M. Cardona, F.H. Pollak, Phys. Rev. 142, 530 (1966)
167. J.J. Hopfield, J. Phys. Chem. Solids 15, 97 (1960)
168. J.L. Shay, B. Tell, L.M. Schiavone, H.M. Kasper, F. Thiel, Phys. Rev. B 9, 1719 (1974)

169. G.L. Bir, G.E. Pikus, *Symmetry and Strain-induced Effects in Semiconductors* (John Wiley & Sons, New York, 1974)
170. G.E. Pikus, G.L. Bir, Fiz. Tverd. Tela 1, 1642 (1956) [Sov. Phys. Solid State 1, 1502 (1959)]
171. N.E. Christensen, Phys. Rev. B 30, 5753 (1984)
172. Y. Zhang, Phys. Rev. B 49, 14352 (1994)
173. S.L. Chuang, C.S. Chang, Phys. Rev. B 54, 2491 (1996)
174. M. Kumagai, S.L. Chuang, H. Ando, Phys. Rev. B 57, 15303 (1998)
175. C. Herring, E. Vogt, Phys. Rev. 101, 944 (1956)
176. D. Aspnes, M. Cardona, Phys. Rev. B 17, 726 (1978)
177. J. Chelikowsky, D.J. Chadi, M.L. Cohen, Phys. Rev. B 8, 2786 (1973)
178. W. Pauli in a letter to R. Peierls, cf. G. Busch, Condens. Matter News 2, 15 (1993)
179. J. McDougall, E.C. Stoner, Philos. Trans. Roy. Soc. London 237, 67 (1938)
180. A.J. MacLeod, ACM Trans. Math. Softw. 24, 1 (1998)
181. A.B. Sproul, M.A. Green, J. Zhao, Appl. Phys. Lett. 57, 255 (1990)
182. A.B. Sproul, M.A. Green, J. Appl. Phys. 73, 1214 (1993)
183. S.M. Sze, *Physics of Semiconductor Devices*, 2nd edition (John Wiley & Sons, New York, 1981)
184. E.F. Schubert, *Doping in III–V Semiconductors* (Cambridge University Press, Cambridge, 1993)
185. W. Kohn, J.M. Luttinger, Phys. Rev. 98, 915 (1955)
186. V.A. Karasyuk, D.G.S. Beckett, M.K. Nissen, A. Villemarie, T.W. Steiner, M.L.W. Thewalt, Phys. Rev. B 49, 16381 (1994)
187. U. Kaufmann, J. Schneider, Adv. Electron. Electr. Phys. 58, 81 (1982)
188. W. Götz, N.M. Johnson, C. Chen, H. Liu, C. Kuo, W. Imler, Appl. Phys. Lett. 68, 3144 (1996)
189. A.J. Ptak, L.J. Holbert, L. Ting, C.H. Swartz, M. Moldovan, N.C. Giles, T.H. Myersa, P. Van Lierde, C. Tian, R.A. Hockett, S. Mitha, A.E. Wickenden, D.D. Koleske, R.L. Henry, Appl. Phys. Lett. 79, 2740 (2001)
190. O. Breitenstein, M. Langenkamp, *Lock-in Thermography, Basics and Use for Functional Diagnostics of Electronic Components*, Springer Series in Advanced Microelectronics, Vol. 10 (Springer, Heidelberg, 2003), ISBN 3-540-43439-9
191. R.M. Dickstein, S.L. Titcomb, R.L. Anderson, J. Appl. Phys. 66, 2437 (1989)
192. G.W. Brown, H. Grube, M.E. Hawley, Phys. Rev. B 70, 121301 (2004)
193. H. Alves, *Defects, Doping and Compensation in Wide Bandgap Semiconductors*, PhD Thesis, Universität Giessen (2003)
194. M.B. Johnson, O. Albrektsen, R.M. Feenstra, H.W.M. Salemink, Appl. Phys. Lett. 63, 2923 (1993) and Erratum Appl. Phys. Lett. 64, 1454 (1994)
195. A.M. Yakunin, A.Yu. Silov, P.M. Koenraad, J.H. Wolter, W. Van Roy, J. De Boeck, J.-M. Tang, M.E. Flatté, Phys. Rev. Lett. 92, 216806 (2004)
196. J.-M. Tang, M. Flatté, Phys. Rev. Lett. 92, 047201 (2004)
197. N.B. Hannay (ed.), *Semiconductors* (Reinhold Publ. Corp., New York, 1959)
198. G. Leibiger, $A^{\mathrm{III}}B^{\mathrm{V}}$*-Mischkristallbildung mit Stickstoff und Bor*, PhD Thesis, Universität Leipzig (2003)
199. M. Tao, J. Appl. Phys. 87, 3554 (2000)
200. R. Noufi, R. Axton, C. Herrington, S.K. Deb, Appl. Phys. Lett. 45, 668 (1984)
201. N.F. Mott, *Metal-insulator Transitions* (Taylor & Francis, London, 1990)

202. P.P. Debye, E.M. Conwell, Phys. Rev. 93, 693 (1954)
203. G.L. Pearson, J. Bardeen, Phys. Rev. 75, 865 (1949)
204. G.E. Stillman, L.W. Cook, T.J. Roth, T.S. Low, B.J. Skromme, in *GaInAsP Alloy Semiconductors*, T.P. Pearsall, ed. (John Wiley & Sons, New York, 1982), p. 121
205. H.C. Casey, Jr., F. Ermanis, K.B. Wolfstirn, J. Appl. Phys. 40, 2945 (1969)
206. K. Pakula, M. Wojdak, M. Palczewska, B. Suchanek, J.M. Baranowski, MRS Internet J. Nitride Semicond. Res. 3, 34 (1998)
207. M.C. Wu, Y.K. Su, K.Y. Cheng, C.Y. Chang, Solid-State Electron. 31, 251 (1988)
208. M. Ogawa, T. Baba, Jpn. J. Appl. Phys. 24, L572 (1985)
209. T. Yamada, E. Tokumitsu, K. Saito, T. Akatsuka, M. Miyauchi, M. Konagai, K. Takahashi, J. Cryst. Growth 95, 145 (1989)
210. J.L. Lievin, F. Alexandre, C. Dubon-Chevaillier, in: *Properties of Impurity States in Superlattice Semiconductors*, C.Y. Fong, I.P. Batra, S. Ciraci, eds. (Plenum, New York, 1988), p. 19
211. J. Bourgoin, M. Lannoo, *Point Defects in Semiconductors II* (Springer, Berlin, 1983)
212. *Deep Centers in Semiconductors*, S.T. Pantelides, ed. (Gordon & Breach, New York, 1986)
213. M.D. Sturge, *The Jahn-Teller Effect in Solids*, Solid State Phys. 20, 92 (1967)
214. G.A. Baraff, E.O. Kane, M. Schlüter, Phys. Rev. B 21, 5662 (1980)
215. P.M. Mooney, N.S. Caswell, S.L. Wright, J. Appl. Phys. 62, 4786 (1987)
216. P.M. Mooney, J. Appl. Phys. 67, R1 (1990)
217. *Physics of DX Centers in GaAs Alloys*, J.C. Bourgoin, ed. (Sci-Tech, Lake Isabella, CA, 1990)
218. D.V. Lang, R.A. Logan, Phys. Rev. Lett. 39, 635 (1977)
219. J. Mäkinen, T. Laine, K. Saarinen, P. Hautojärvi, C. Corbel, V.M. Araksinen, J. Nagle, Phys. Rev. B 52, 4870 (1995)
220. G.M. Martin, Appl. Phys. Lett. 39, 747 (1981)
221. J. Dabrowski, M. Scheffler, Phys. Rev. B 40, 10391 (1989)
222. J.S. Blakemore, J. Appl. Phys. 53(10), R123-R181 (1982)
223. A. Rohatgi, R.H. Hopkins, J.R. Davis, R.B. Campbell, H.C. Mollenkopf, J.R. McCormick, Solid-State Electron. 23, 1185 (1980)
224. D.C. Look, J. Appl. Phys. 48, 5141 (1977)
225. O. Mizuno, H. Watanabe, Electron. Lett. 11, 118 (1975)
226. Y. Toudic, B. Lambert, R. Coquille, G. Grandpierre, M. Gauneau, Semicond. Sci. Technol. 3 464 (1988)
227. R.P. Tapster, M.S. Skolnick, R.G. Humphreys, P.J. Dean, B. Cockayne, W.T. MacEwan, J. Phys. C: Solid State Phys. 14, 5069 (1981)
228. A. Juhl, A. Hoffmann, D. Bimberg, H.J. Schulz, Appl. Phys. Lett. 50, 1292 (1987)
229. E.S. Koteles, W.R. Datars, Can. J. Phys. 54, 1676 (1976)
230. A.M. Hennel, *Transition Metals in III/V Compounds*, Semicond. Semimet. 38, 189 (1993)
231. H.J. Schulz, J. Cryst. Growth 59, 65 (1982)
232. J. Cheng, S.R. Forrest, B. Tell, D. Wilt, B. Schwartz, P.D. Wright, J. Appl. Phys. 58, 1780 (1985)
233. D.G. Knight, B. Watt, R. Bruce, D.A. Clark, in: *Semi-Insulating III–V Materials*, (Bristol, 1990), A. Milnes, C.J. Miner, eds., p. 83 (1990)

234. A. Dadgar, O. Stenzel, A. Näser, M. Zafar Iqbal, D. Bimberg, H. Schumann, Appl. Phys. Lett. 73, 3878 (1998)
235. D. Söderström, G. Fornuto, A. Buccieri, Proc. 10th European Workshop on MOVPE, Lecce (Italy) 8–11 June 2003, PS.IV.01
236. W. Mönch, *Semiconductor Surfaces and Interfaces* (Springer, Berlin, 2001), ISBN 354067902-2
237. Y. Rosenwaks, R. Shikler, Th. Glatzel, S. Sadewasser, Phys. Rev. B 70, 085320 (2004)
238. E. Mollwo, Z. Phys. 138, 478 (1954)
239. *Hydrogen in Semiconductors*, Semicond. Semimet. 34, J.I. Pankove, N.M. Johnson, eds. (1991)
240. S.J. Pearton, J.W. Corbett, M. Stavola, *Hydrogen in Crystalline Semiconductors* (Springer, Berlin, 1992)
241. T. Sakurai, H.D. Hagstrom, J. Vac. Sci. Technol. 13, 807 (1976)
242. M.H. Brodsky, M. Cardona, J.J. Cuomo, Phys. Rev. B 16, 3556 (1977)
243. J.I. Pankove, Appl. Phys. Lett. 32, 812 (1978)
244. C.P. Herrero, M. Stutzmann, Phys. Rev. B 38, 12668 (1988)
245. M. Stavola, K. Bergmann, S.J. Pearton, J. Lopata, Phys. Rev. Lett. 61, 2786 (1988)
246. P.J.H. Denteneer, C.G. Van de Walle, S.T. Pantelides, Phys. Rev. B 39, 10809 (1989)
247. V.P. Markevich, L.I. Murin, M. Suezawa, J.L. Lindström, J. Coutinho, R. Jones, P.R. Briddon, S. Öberg, Phys. Rev. B 61, 12964 (2000)
248. A.S. Yapsir, P. Deak, R.K. Singh, L.C. Snyder, J.W. Corbett, T.-M. Lu, Phys. Rev. B 38, 9936 (1988)
249. V.P. Markevich, A.R. Peaker, J. Coutinho, R. Jones, V.J.B. Torres, S.Öberg, P.R. Briddon, L.I. Murin, L. Dobaczewski, N.V. Abrosimov, Phys. Rev. B 69, 125218 (2004)
250. E.M. Conwell, V. Weisskopf, Phys. Rev. 77, 388 (1950)
251. H. Brooks, Phys. Rev. 83 879 (1951)
252. B.K. Ridley, J. Phys. C: Solid State Phys. 10 1589 (1977)
253. D.C. Chattopadhyay, H.J. Queisser, Revs. Mod. Phys. 53, 745 (1981)
254. B.K. Ridley, *Quantum Processes in Semiconductors*, 2nd edn. (Clarendon, Oxford, 1988)
255. C.M. Wolfe, G.E. Stillman, W.T. Lindley, J. Appl. Phys. 41, 3088 (1970)
256. J.Y.W. Seto, J. Appl. Phys. 46, 5247 (1975)
257. E.H. Hall, Amer. J. Math. 2, 287 (1879)
258. L.J. van der Pauw, Philips Res. Repts. 13, 1 (1958)
259. L.J. van der Pauw, Philips Tech. Rev. 20, 220 (1958)
260. D.S. Perloff, *Four-Point Sheet Resistance Correction Factors for Thin Rectangular Samples*, Solid-State Electron. 20, 681 (1977)
261. O. Madelung, H. Weiss, Z. Naturf. 9a, 527 (1954)
262. E.M. Conwell, *High Field Transport in Semiconductors* (Academic Press, New York, 1967)
263. C. Jacoboni, C. Canali, G. Ottaviani, A. A. Quaranta, Solid-State Electron. 20, 77 (1977)
264. C. Jacoboni, F. Nava, C. Canali, G. Ottaviani, Phys. Rev. B 24, 1014 (1981)
265. T. Gonzalez Sanchez, J.E. Velazquez Perez, P.M. Gutierrez Conde, D. Pardo, Semicond. Sci. Technol. 7, 31 (1992)

266. V. Balynas, A. Krotkus, A. Stalnionis, A.T. Gorelionok, N.M. Shmidt, J.A. Tellefsen, Appl. Phys. A 51, 357 (1990)
267. J.D. Albrecht, P.P. Ruden, S. Limpijumnong, W.R. Lambrecht, K.F. Brennan, J. Appl. Phys. 86, 6864 (1999)
268. J. Pozhela, A. Reklaitis, Solid-State Electron. 23, 9 (1980) 927-933.
269. B. Kramer, A. Mircea, Appl. Phys. Lett. 26, 623 (1975)
270. H.S. Carslaw, J.C. Jaeger, *Conduction of Heat in Solids* (Clarendon Press, Oxford, 1959)
271. G.O. Mahan, *Good Thermoelectrics*, Solid State Phys. 51, 81 (1997)
272. T.H. Geballe, G.W. Hull, Phys. Rev. 110, 773 (1958)
273. W.S. Capinski, H.J. Maris, E. Bauser, I. Silier, M. Asen-Palmer, T. Ruf, M. Cardona, E. Gmelin, Appl. Phys. Lett. 71, 2109 (1997)
274. T. Ruf, R.W. Henn, M. Asen-Palmer, E. Gmelin, M. Cardona, H.J. Pohl, G.G. Devyatykh, P.G. Sennikov, Solid State Commun. 115, 243 (2000)
275. T.H. Geballe, G.W. Hull, Phys. Rev. 98, 940 (1955)
276. B. Jogai, Solid State Commun. 116, 153 (2000)
277. M. Cardona, *Modulation Spectroscopy*, Solid State Phys. Suppl. 11 (1969)
278. Ch.M. Wolfe, N. Holonyak, Jr., G.E. Stillman, *Physical Properties of Semiconductors* (Prentice Hall, Englewood Cliffs, NJ, 1989)
279. M.D. Sturge, Phys. Rev. 127, 768 (1962)
280. R.G. Ulbrich, Adv. Solid State Phys. (Festkörperprobleme) 25, 299 (1985)
281. T.P. McLean, Prog. Semicond. 5, 87 (1960)
282. P.J. Dean, D.G. Thomas, Phys. Rev. 150, 690 (1966)
283. G.A. Cox, G.G. Roberts, R.H. Tredgold, Br. J. Appl. Phys. 17, 743 (1966)
284. F. Urbach, Phys. Rev. 92, 1324 (1953)
285. S.R. Johnson, T. Tiedje, J. Appl. Phys. 78, 5609 (1995).
286. M. Beaudoin, A.J.G. DeVries, S.R. Johnson, H. Laman, T. Tiedje, Appl. Phys. Lett. 70, 3540 (1997)
287. T.S. Moss, T.D.F. Hawking, Infrared Phys. 1, 111 (1961)
288. R. Braunstein, E.O. Kane, J. Phys. Chem. Solids 23, 1423 (1962)
289. J. Struke, J. Non-Cryst. Solids 4, 1 (1970)
290. A. Baldereschi, N.O. Lipari, Phys. Rev. B 3, 439 (1971)
291. N.O. Lipari, M. Altarelli, Phys. Rev. B 15, 4883 (1977)
292. N.O. Lipari, Phys. Rev. B 4, 4535 (1971)
293. E.F. Gross, Usp. Fiz. Nauk 76, 433 (1962) [Sov. Phys.-Usp. 5, 195 (1962)]
294. Ch. Uihlein, D. Fröhlich, R. Kenklies, Phys. Rev. B 23, 2731 (1981)
295. R.J. Elliott, Phys. Rev. 108, 1384 (1957)
296. A. Shikanai, T. Azuhata, T. Sota, S. Chichibu, A. Kuramata, K. Horino, S. Nakamura, J. Appl. Phys. 81, 417 (1997)
297. J.J. Hopfield, Phys. Rev. 182, 945 (1969)
298. Y. Toyozawa, Prog. Theor. Phys., Suppl. 12, 111 (1959)
299. U. Heim, P. Wiesner, Phys. Rev. Lett. 24, 1205 (1973)
300. B. Gil, Phys. Rev. B 64, 201310 (2001)
301. T. Soma, H.-M. Kagaya, phys. stat. sol. (b) 118, 245 (1983)
302. A. Göldner, PhD Thesis, Technische Universität Berlin (2000) (Wissenschaft und Technik Verlag, Berlin, 2000), ISBN 3-89685-346-5
303. A.A. Maradudin, D.L. Mills, Phys. Rev. B 7, 2787 (1973)
304. I. Broser, M. Rosenzweig, Phys. Rev. B 22, 2000 (1980)
305. M. Rosenzweig, PhD Thesis, Technische Universität Berlin (1982)

306. G. Blattner, G. Kurtze, G. Schmieder, C. Klingshirn, Phys. Rev. B 25, 7413 (1982)
307. B. Gil, A. Hoffmann, S. Clur, L. Eckey, O. Briot, R.-L. Aulombard, J. Cryst. Growth 189/190, 639 (1998)
308. C. Weisbuch, R. Ulbrich, J. Lumin. 18/19, 27 (1979)
309. D.G. Thomas, J.J. Hopfield, Phys. Rev. 150, 680 (1966)
310. E. Burstein, Phys. Rev. 93, 632 (1954)
311. T.S. Moss, Proc. Phys. Soc. London B 76, 775 (1954)
312. V. Vashishta, R.K. Kalia, Phys. Rev. B 25, 6492 (1982)
313. R. Zimmermann, phys. stat. sol. (b) 146, 371 (1988)
314. H.-E. Swoboda, M. Sence, F.A. Majumder, M. Rinker, J.-Y. Bigot, J.B. Grun, C. Klingshirn, Phys. Rev. B 39, 11019 (1989)
315. J.P. Löwenau, S. Schmitt-Rink, H. Haug, Phys. Rev. Lett. 49, 1511 (1982)
316. L.V. Keldysh, Proc. of the 9th Int. Conf. on the Physics of Semiconductors, Moscow (Nauka, Leningrad, 1968), p. 1303
317. W.F. Brinkman, T.M. Rice, P.W. Anderson, S.T. Chui, Phys. Rev. Lett. 28, 961 (1972)
318. G.A. Thomas, T.M. Rice, J.C. Hensel, Phys. Rev. Lett. 33, 219 (1974)
319. T.L. Reinecke, S.C. Ying, Phys. Rev. Lett. 35, 311 (1975)
320. R.S. Markienwicz, J.P. Wolfe, C.D. Jeffries, Phys. Rev. B 15, 1988 (1977)
321. L.V. Butov, C.W. Lai, A.L. Ivanov, A.C. Gossard, D.S. Chemla, Nature 417, 47 (2002)
322. L.V. Butov, A.C. Gossard, D.S. Chemla, Nature 418, 751 (2002)
323. K.E. O.Hara, L.O. Suilleabhain, J.P. Wolfe, Phys. Rev. B 60, 10565 (1999)
324. M. Skolnick, A.I. Tartakovskii, R. Butté, R.M. Stevenson, J.J. Baumberg, D.M. Whittaker, *High Occupancy Effects and Condensation Phenomena in Semiconductor Microcavities and Bulk Semiconductors* in *Nano-Optoelectronics, Concepts, Physics and Devices*, M. Grundmann, ed. (Springer, Berlin, 2002), p. 273
325. H. Mahr, in *Quantum Electronics*, H. Rabin, C.L. Tang, eds. (Academic, New York, 1975), Vol. IA, p. 285
326. H.J. Fossum, D.B. Chang, Phys. Rev. B 8, 2842 (1973)
327. J.P. van der Ziel, Phys. Rev. B 16, 2775 (1977)
328. C.J. Summers, R. Dingle, D.E. Hill, Phys. Rev. B 1, 1603 (1970)
329. Sh.M. Kogan, T.M. Lifshits, phys. stat. sol. (a) 39, 11 (1977)
330. R.A. Cooke, R.A. Hoult, R.F. Kirkman, R.A. Stradling, J. Phys. D 11, 945 (1978)
331. B.L. Cardozo, E.E. Haller, L.A. Reichertz, J.W. Beemann, Appl. Phys. Lett. 83, 3990 (2003)
332. G. Göbel, Appl. Phys. Lett. 24, 492 (1974)
333. R.N. Hall, Phys. Rev. 87, 387 (1952)
334. W. Shockley, W.T. Read, Phys. Rev. 87, 835 (1952)
335. V.K. Malyutenko, Physica E 20, 553 (2004)
336. J.F. Muth, J.H. Lee, I.K. Shmagin, R.M. Kolbas, H.C. Casey, Jr., B.P. Keller, U.K. Mishra, S.P. DenBaars, Appl. Phys. Lett. 71 2572 (1997)
337. W. Gerlach, H. Schlangenotto, H. Maeder, phys. stat. sol. (a) 13, 277 (1972)
338. A. Galeskas, J. LSiCros, V. Grivickas, U. Lindefelt, C. Hallin, Proc. of the 7th International Conference on SiC, III-Nitrides and Related Materials, 1997, Stockholm, Sweden 1997, p. 533–536

339. V. Palankovski, *Simulation of Heterojunction Bipolar Transistors*, PhD Thesis, Technische Universität Wien (2000)
340. J.R. Haynes, Phys. Rev. Lett. 4, 361 (1960)
341. P.J. Dean, *Luminescence of Crystals, Molecules and Solutions*, F. Williams, ed. (Plenum, New York, 1973), p. 523
342. B. K. Meyer, H. Alves, D.M. Hofmann, W. Kriegseis, D. Forster, F. Bertram, J. Christen, A. Hoffmann, M. Straßburg, M. Dworzak, U. Haboeck, A.V. Rodina, phys. stat. sol. (b) 241, 231 (2004)
343. R.G. Ulbrich, Solid-State Electron. 21, 51 (1978)
344. P.J. Dean, M. Skolnick, J. Appl. Phys. 54, 346 (1983)
345. F.A.J.M. Driessen, H.G.M. Lochs, S.M. Olsthoorn, L.J. Giling, J. Appl. Phys. 69, 906 (1991)
346. D. Karaiskaj, M.L.W Thewalt, T. Ruf, M. Cardona, M. Konuma, Phys. Rev. Lett. 86, 6010 (2001)
347. D. Karaiskaj, M.L.W Thewalt, T. Ruf, M. Cardona, M. Konuma, Phys. Rev. Lett. 89, 016401 (2002)
348. V.A. Karasyuk, M.L.W Thewalt, S. An, E.C. Lightowlers, A.S. Kaminskii, Phys. Rev. B 54, 10 543 (1996)
349. E.F. Schubert, E.O. Göbel, Y. Horikoshi, K. Ploog, H.J. Queisser, Phys. Rev. B 30, 813 (1984)
350. J. Conradi, R.R. Haering, Phys. Rev. Lett. 20, 1344 (1968)
351. Y.S. Park, J.R. Schneider, Phys. Rev. Lett. 21, 798 (1968)
352. D. Kovalev, B. Averboukh, D. Volm, B.K. Meyer, H. Amano, I. Akasaki, Phys. Rev. B 54, 2518 (1996)
353. S. Permogorov, *Optical Emission due to Exciton Scattering by LO Phonons in Semiconductors* in: *Excitons*, E.I. Rashba, M.D. Sturge, eds. (North-Holland, 1982)
354. R. Dingle, Phys. Rev. Lett. 23, 579 (1969)
355. Th. Agne, *Identifikation und Untersuchung von Defekten in ZnO Einkristallen*, PhD Thesis, Universität des Saarlandes, Saarbrücken (2004)
356. K. Huang, A. Rhys, Proc. Roy. Soc. A 204, 406 (1950)
357. J.J. Hopfield, J. Phys. Chem. Solids 10, 110
358. M. Lax, J. Chem. Phys. 20, 1752 (1952) (1959)
359. P.J. Dean, C.H. Henry, C.J. Frosch, Phys. Rev. 168, 812 (1968)
360. W.H. Koschel, U. Kaufmann, S.G. Bishop, Solid State Commun. 21, 1069 (1977)
361. A. Juhl, *Calorimetrische Absorptionsspektroskopie (CAS) – Eine neue Methode zur Charakterisierung der optischen Eigenschaften von Halbleitersystemen*, PhD Thesis, Technische Universität Berlin (1987)
362. P.C. Findlay, C.R. Pidgeon, H. Pellemans, R. Kotitschke, B.N. Murdin, T. Ashley, A.D. Johnson, A.M. White, C.T. Elliott, Semicond. Sci. Technol. 14,1026 (1999)
363. D.H. Auston, Semicond. Semimet. 28, 85 (1990)
364. D.J. Fitzgerald, A.S. Grove, Surf. Sci. 9, 347 (1968)
365. S. Bothra, S. Tyagi, S. K. Chandhi, J. M. Borrego, Solid-State Electron. 34, 47 (1991)
366. K. Kurita, T. Shingyouji, Jpn. J. Appl. Phys. (Part 1) 38, 5710 (1999)
367. M.J. Kerr, J. Schmidt, A. Cuevas, J.H. Bultman, J. Appl. Phys. 89, 3821 (2001)

368. O. Hahneiser, M. Kunst, J. Appl. Phys. 85, 7741 (1999)
369. D.E. Aspnes, Surf. Sci. 132, 406 (1983)
370. G.K. Teal, J.B. Little, Phys. Rev. 78, 647 (1950)
371. *Crystal Growth Technology*, H.J. Scheel, T. Fukuda, eds. (John Wiley & Sons, New York, 2004), ISBN 0-471-49059-8
372. Siltronic AG
373. B.A. Joyce, J.H. Neave, P.J. Dobson, P.K. Larsen, Phys. Rev. B 29, 814 (1984)
374. J.-T. Zettler, W. Richter, K. Ploska, M. Zorn, J. Rumberg, C. Meyne, M. Pristovsek, *Real Time Diagnostics of Semiconductor Surface Modifications by Reflectance Anisotropy Spectroscopy in Semiconductor Characterization - Present Status and Future Needs*, W. Bullis, D. Seiler, A. Diebold, eds. (AIP Press, Woodbury, New York, 1996), p. 537
375. J.-T. Zettler, Prog. Cryst. Growth Charact. Mater. 35, 27 (1997)
376. C. Weisbuch, *Fundamental Properties of III–V Semiconductor Two-Dimensional Quantized Structures: The Basis for Optical and Electronic Device Applications*, R. Dingle, ed., Semicond. Semimet. 24, 1 (1987)
377. E. Ohshima, H. Ogino, I. Niikura, K. Maeda, M. Sato, M. Ito, T. Fukuda, J. Cryst. Growth 260, 16 (2004)
378. G.F. Kuznetsov, S.A. Aitkhozhin, Crystallogr. Rep. 47, 514 (2002)
379. J. Ohta, H. Fujioka, M. Oshima, K. Fujiwara, A. Ishii, Appl. Phys. Lett. 83, 3075 (2003)
380. P.J. Schuck, M.D. Mason, R.D. Grober, O. Ambacher, A.P. Lima, C. Miskys, R. Dimitrov, M. Stutzmann, Appl. Phys. Lett. 79, 952 (2001)
381. B.J. Rodriguez, A. Gruveman, A.I. Kingon, R.J. Nemanich, O. Ambacher, Appl. Phys. Lett. 80, 4166 (2002)
382. A. Strittmatter, S. Rodt, L. Reißmann, D. Bimberg, H. Schröder, E. Obermeier, T. Riemann, J. Christen, A. Krost, Appl. Phys. Lett. 78, 727 (2001)
383. A. Dadgar, J. Bläsing, A. Diez, A. Alam, M. Heuken, A. Krost, Jpn. J. Appl. Phys. 39, L1183 (2000)
384. J. Bläsing, A. Reiher, A. Dadgar, A. Dietz, A. Krost, Appl. Phys. Lett. 81, 2722 (2002)
385. Ch. G. van der Walle, J. Neugebauer, Nature 423, 626 (2003)
386. V. Gottschalch, G. Wagner, University of Leipzig, private communication
387. Y.C. Chang, J.N. Schulman, Phys. Rev. B 31, 2069 (1985)
388. S.L. Chuang, *Physics of Optoelectronic Devices* (John Wiley & Sons, New York, 1995)
389. R.C. Miller, D.A. Kleinman, A.C. Gossard, Phys. Rev. B 29, 7085 (1984)
390. R.C. Miller, A.C. Gossard, W.T. Tsang, O. Munteanu, Phys. Rev. B 25, 3871 (1982)
391. W.T. Masselink, Y.-Ch. Chang, H. Morkoç, Phys. Rev. B 28, 7373 (1983)
392. D.B. Tran Thoai, R. Zimmermann, M. Grundmann, D. Bimberg, Phys. Rev. B 42, 5906 (1990)
393. M.J.L.S. Haines, N. Ahmed, S.J.A. Adams, K. Mitchell, I.R. Agool, C.R. Pidgeon, B.C. Cavenett, E.P. O'Reilly, A. Ghiti, M.T. Emeny, Phys. Rev. B 43, 11944 (1991)
394. K.J. Moore, G. Duggan, K. Woodbridge, C. Roberts, Phys. Rev. B 41, 1090 (1990)
395. L. Esaki in *Recent Topics in Semiconductor Physics*, H. Kamimura, Y. Toyozawa, eds. (World Scientific, Singapore, 1983), pp. 1–71

396. R. Dingle, A.C. Gossard, W. Wiegmann, Phys. Rev. Lett. 34, 1327 (1975)
397. A.L. Efros, F.G. Pikus, G.G. Samsonidze, Phys. Rev. B 41, 8295 (1990)
398. F. Stern, S. Das Sarma, Phys. Rev. B 30, 840 (1984)
399. L. Pfeiffer, K.W. West, Physica E 20, 57 (2003)
400. J. Christen, D. Bimberg, Phys. Rev. B 42, 7213 (1990)
401. M.S. Skolnick, J.M. Rorison, K.J. Nash, D.J. Mowbray, P.R. Tapster, S.J. Bass, A.D. Pitt, Phys. Rev. Lett. 58, 2130 (1987)
402. E. Runge, R. Zimmermann, Adv. Solid State Phys. (Festkörperprobleme) 38, 251 (1998)
403. E. Runge, Solid State Phys. 57, 149 (2002)
404. M.S. Skolnick, E.G. Scott, B. Wakefield, G.J. Davies, Semicond. Sci. Technol. 3, 365 (1988)
405. M. Grassi Alessi, F. Fragano, A. Patané, M. Capizzi, E. Runge, R. Zimmermann, Phys. Rev. B 61, 10985 (2000)
406. J. Hegarty, L. Goldner, M.D. Sturge, Phys. Rev. B 30, 7346 (1984)
407. N.F. Mott, E.A. Davies, *Electronic Processes in Noncrystalline Materials*, 2nd edn. (Oxford University Press, New York, 1979)
408. J. Spitzer, T. Ruf, M. Cardona, W. Dondl, R. Schorer, G. Abstreiter, E.E. Haller, Phys. Rev. Lett. 72, 1565 (1994)
409. J.M. Luttinger, Phys. Rev. 102, 1030 (1956)
410. A. Jager, *Exzitonen und Franz–Keldysh-Effekt im quaternären Halbleiter InGaAsP/InP*, PhD Thesis, Phillips-Universität Marburg (1997)
411. M. Oestreich, S. Hallstein, A.P. Heberle, K. Eberl, E. Bauser, W.W. Rühle, Phys. Rev. B 53, 7911 (1996)
412. M.J. Snelling, E. Blackwood, C.J. McDonagh, R.T. Harley, C.T.B. Foxon, Phys. Rev. B 45, 3922 (1992)
413. M. Schubert, *Infrared Ellipsometry on Semiconductor Layer Structures: Phonons, Plasmons and Polaritons* (Springer, Heidelberg, 2004)
414. M. Schubert, T. Hofmann, C.M. Herzinger, J. Opt. Soc. Am. A 20, 347 (2003)
415. W. Zawadski, R. Lassnig, Solid State Commun. 50, 537 (1984)
416. H.L. Störmer, R. Dingle, A.C. Gossard, W. Wiegmann, R.A. Logan, Conf. Ser. – Inst. Phys. 43, 557 (1979)
417. A.B. Fowler, F.F. Fang, W.E. Howard, P.J. Stiles, Phys. Rev. Lett. 16, 901 (1966)
418. K. v. Klitzing, G. Dorda, M. Pepper, Phys. Rev. Lett. 45, 494 (1980)
419. G. Landwehr, Physica E 20, 1 (2003)
420. H. Aoki, T. Ando, Solid State Commun. 38, 1079 (1981)
421. M.A. Paalanen, D.C. Tsui, A.C. Gossard, Phys. Rev. B 25, 5566 (1982)
422. H. Bachmair, E.O. Göbel, G. Hein, J. Melcher, B. Schumacher, J. Schurr, L. Schweitzer, P. Warnecke, Physica E 20, 14 (2003)
423. Physikalisch-Technische Bundesanstalt
424. J.P. Eisenstein, H.L. Störmer, Science 248, 1461 (1990)
425. R.E. Prange, Phys. Rev. B 23, 4802 (1981)
426. M. Büttiker, Phys. Rev. B 38, 9375 (1988)
427. E. Ahlswede, P. Weitz, J. Weis, K. v. Klitzing, K. Eberl, Physica E 298, 562 (2001)
428. K. Lier, R.R. Gerhardts, Phys. Rev. B 50, 7757 (1994)
429. R.B. Laughlin, Phys. Rev. B 23, 5632 (1981)
430. J.K. Jain, Phys. Rev. Lett. 63, 199 (1983)

431. J.K. Jain, Phys. Rev. B 41, 7653 (1990)
432. H.L. Störmer, D. Tsui, A.C. Gossard, Rev. Mod. Phys. 71, S298 (1999)
433. D. Weiss, M.L. Roukes, A. Menschig, P. Frambow, K. v. Klitzing, G. Weimann, Phys. Rev. Lett. 66, 2790 (1991)
434. R.P. Feynman, *There's Plenty of Room at the Bottom*, After-dinner speech on December 29th 1959 at the annual meeting of the American Physical Society at the California Institute of Technology
435. D. Bimberg, M. Grundmann, N.N. Ledentsov, *Quantum Dot Heterostructures* (John Wiley & Sons, Chichester, 1999)
436. *Nano-Optoelectronics, Concepts, Physics and Devices*, M. Grundmann, ed. (Springer, Heidelberg, 2002)
437. E. Kapon, M. Walther, J. Christen, M. Grundmann, C. Caneau, D.M. Hwang, E. Colas, R. Bhat, G.H. Song, D. Bimberg, Superlatt. Microstruct. 12, 491 (1992)
438. M. Grundmann, J. Christen, M. Joschko, O. Stier, D. Bimberg, E. Kapon, Semicond. Sci. Technol. 9, 1939 (1994)
439. L. Pfeiffer, K.W. West, H.L. Störmer, J.P. Eisenstein, K.W. Baldwin, D. Gershoni, J. Spector, Appl. Phys. Lett. 56, 1697 (1990)
440. M. Grundmann, D. Bimberg, Phys. Rev. B 55, 4054 (1997)
441. W. Wegscheider, G. Schedelbeck, G. Abstreiter, M. Rother, M. Bichler, Phys. Rev. Lett. 79, 1917 (1997)
442. *Whisker Technology* A.P. Levitt (ed.) (John Wiley & Sons, New York, 1970)
443. M.T. Björk, B.J. Ohlsen, T. Sass, A.I. Perrson, C. Thelander, M.H. Magnusson, K. Deppert, L.R. Wallenberg, L. Samuelson, Appl. Phys. Lett. 80, 1058 (2002)
444. M.H. Huang, S. Mao, H. Feick, H. Yan, Y. Wu, H. Kind, E. Weber, R. Russo, P. Yang, Science 292, 1897 (2001)
445. X. Duan, Y. Huang, R. Agarwal, C. Lieber, Nature 421, 241 (2003)
446. M. Grundmann, E. Kapon, J. Christen, D. Bimberg, *Electronic and Optical Properties of Quasi One-dimensional Carriers in Quantum Wires*, J. Nonlinear Opt. Phys. and Mater. 4, 99 (1995)
447. M. Grundmann, O. Stier, D. Bimberg, Phys. Rev. B 58, 10557 (1998)
448. O. Stier, M. Grundmann, D. Bimberg, Phys. Rev. B 59, 5688 (1999)
449. O. Stier, *Theory of the optical properties of InGaAs/GaAs quantum dots* in: Nano-Optoelectronics, Concepts, Physics, Devices, M. Grundmann, ed. (Springer, Berlin, 2002)
450. R. Santoprete, B. Koiller, R.B. Capaz, P. Kratzer, Q.K.K. Liu, M. Scheffler, Phys. Rev. B 68, 235311 (2003)
451. L.P. Kouwenhoven, N.C. van der Vaart, A.T. Johnson, W. Kool, C.J.P.M. Harmans, J.G. Williamson, A.A.M. Staring, C.T. Foxon, Z. Phys. B 85, 367-373 (1991)
452. S. Tarucha, D.G. Austing, T. Honda, R.J. van der Hage, L.P. Kouwenhoven, Phys. Rev. Lett. 77, 3613 (1996)
453. N. Horiguchi, T. Futatsugi, Y. Nakata, N. Yokoyama, Appl. Phys. Lett. 70, 2294 (1997)
454. A. Forchel, R. Steffen, M. Michel, A. Pecher, T.L. Reinecke, Proc. 23rd Int. Conf. on the Physics of Semiconductors, Berlin, Germany, M. Scheffler, R. Zimmermann, eds. (World Scientific, Singapore, 1996), p. 1285
455. K.C. Rajkumar, K. Kaviani, J. Chen, P. Chen, A. Madhukar, D. Rich, Mater. Res. Soc. Symp. Proc. 263, 163 (1992)

456. B. Urbaszek, R.J. Warburton, K. Karrai, B.D. Gerardot, P.M. Petroff, J.M. Garcia, Phys. Rev. Lett. 90, 247403 (2003)
457. R.J. Warburton, C. Schäflein, D. Haft, F. Bickel, A. Lorke, K. Karrai, J.M. Garcia, W. Schoenfeld, P.M. Petroff, Nature 405, 926 (2000)
458. W. Weller, private communication
459. M. Grundmann, Adv. Solid State Phys. (Festkörperprobleme) 35, 123 (1996)
460. V.A. Shchukin, N.N. Ledentsov, D. Bimberg, *Epitaxy of Nanostructures*, (Springer, Heidelberg, 2004), ISBN 3-540-67817-4
461. D. Leonard, M. Krishnamurthy, C. M. Reaves, S. P. Denbaars, P. M. Petroff, Appl. Phys. Lett. 63, 3203 (1993)
462. D.M. Bruls, P.M. Koenraad, H.W.M. Salemink, J.H. Wolter, M. Hopkinson, M.S. Skolnick, Appl. Phys. Lett. 82, 3758 (2003)
463. Q. Xie, A. Madhukar, P. Chen, N. Kobayashi, Phys. Rev. Lett. 75, 2542 (1995)
464. F. Findeis, A. Zrenner, G. Böhm, G. Abstreiter, Solid State Commun. 114, 227 (2000)
465. D. Braun, A. Heeger, Appl. Phys. Lett. 58, 1982 (1991)
466. A. Curioni, W. Andreoni, R. Treusch, F.J. Himpsel, E. Haskal, P. Seidler, C. Heske, S. Kakar, T. van Buuren, L.J. Terminello, Appl. Phys. Lett. 72, 1575 (1998)
467. Z. Vardeny, E. Ehrenfreund, J. Shinar, F. Wudl, Phys. Rev. B 35, 2498 (1987)
468. J.-W. van der Horst, P.A. Bobbert, M.A.J. Michels, Phys. Rev. B 66, 035206 (2002)
469. J.M. Bendickson, J.P. Dowling, M. Scalora, Phys. Rev. E 53, 4107 (1996)
470. B. Rauschenbach, Leibniz-Institut für Oberflächenmodifizierung, Leipzig, private communication
471. N. Kaiser, Fraunhofer-Institute for Applied Optics, Jena, Germany
472. *Photonic Crystals and Light Localization in the 21st Century*, C.M. Soukoulis, ed., NATO Science Series C Vol. 563 (Kluwer Academic Publishers, Dordrecht, 1996)
473. J.D. Joannopoulos, R.D. Meade, J.N. Winn, *Photonic Crystals* (Princeton University Press, Princeton, NJ, 1995)
474. E.-X. Ping, J. Appl. Phys. 76, 7188, (1994)
475. K.M. Ho, C.T. Chan, C.M. Soukoulis, Phys. Rev. Lett. 65, 3152 (1990)
476. O. Toader, S. John, Science 292, 1133 (2001)
477. S.R. Kennedy, M.J. Brett, O. Toader, S. John, Nano Lett. 2, 59 (2002)
478. K. Busch, S. John, Phys. Rev. E 58, 3896 (1998)
479. F. Garcia-Santamaria, C. López, F. Meseguer, F. López-Tejeira, J. Sánchez-Dehesa, H.T. Miyazaki, Appl. Phys. Lett. 79, 2309 (2001)
480. A. Blanco, E. Chomski, S. Grabtchak, M. Ibisate, S. John, S.W. Leonard, C. Lopez, F. Meseguer, H. Miguez, J.P. Mondia, G.A. Ozin, O. Toader, H.M. van Driel, Nature 405, 437 (2000)
481. M. Lončar, D. Nedeljković, T. Doll, J. Vučović, A. Scherer, T.P. Pearsall, Appl. Phys. Lett. 77, 1937 (2000)
482. Y. Zhu, Q. Wu, S. Morin, T.W. Mossberg, Phys. Rev. Lett. 65, 1200 (1990)
483. C. Weisbuch, M. Nishioka, A. Ishikawa, Y. Arakawa, Phys. Rev. Lett. 69, 3314 (1992)
484. V. Savona, Z. Hadril, A. Quattropani, P. Schwendimann, Phys. Rev. B 49, 8774 (1994)

485. G. Khitrova, H.M. Gibbs, F. Jahnke, M. Kira, S.W. Koch, Rev. Mod. Phys. 71, 1591 (1999)
486. S.L. McCall, A.F.J. Levi, R.E. Slusher, S.J. Pearton, R.A. Logan, Appl. Phys. Lett. 60, 289 (1992)
487. R.P. Wang, M.-M. Dumitrescu, J. Appl. Phys. 81, 3391 (1997)
488. A.F.J. Levi, R.E. Slusher, S.L. McCall, J.L. Glass, S.J. Pearton, R.A. Logan, Appl. Phys. Lett. 62, 561 (1993)
489. S.-K. Kim, S.-H. Kim, G.-H. Kim, H.-G. Park, D.-J. Shin, Y.-H. Lee, Appl. Phys. Lett. 84, 861 (2004)
490. E.M. Purcell, Phys. Rev. 69, 681 (1946)
491. J.M. Gérard, B. Sermage, B. Gayral, B. Legrand, E. Costard, V. Thierry-Mieg, Phys. Rev. Lett. 81, 1110 (1998)
492. A. Kiraz, P. Michler, C. Becher, B. Gayral, A. Imamoğlu, Lidong Zhang, E. Hu, W.V. Schoenfeld, P.M. Petroff, Appl. Phys. Lett. 78, 3932 (2001)
493. J.U. Nöckel, A.D. Stone, Nature 385, 45 (1997)
494. C. Gmachl, F. Capasso, E.E. Narimanov, J.U. Nöckel, A.D. Stone, J. Faist, D.L. Sivco, A.Y. Cho, Science 280, 1556 (1998)
495. G.D. Chern, H. E. Tureci, A. Douglas Stone, R. K. Chang, M. Kneissl, N. M. Johnson, Appl. Phys. Lett. 83, 1710 (2003)
496. M. Kneissl, M. Teepe, N. Miyashita, N.M. Johnson, G.D. Chern, R.K. Chang, Appl. Phys. Lett. 84, 2485 (2004)
497. S.-Y. Lee, S. Rim, J.-W. Ryu, T.-Y. Kwon, M. Choi, C.-M. Kim, Phys. Rev. Lett. 93, 164102 (2004)
498. F. Leiter, H. Zhou, F. Henecker, A. Hofstaetter, D.M. Hofmann, B.K. Meyer, Physica B 308–310, 908 (2001)
499. T. Nobis, E.M. Kaidashev, A. Rahm, M. Lorenz, M. Grundmann, Phys. Rev. Lett. 93, 103903 (2004)
500. T. Nobis, M. Grundmann, Phys. Rev. A-72, 063806 (2005)
501. T. Nobis, A. Rahm, M. Lorenz, M. Grundmann, Proc. SPIE 6122, 93 (2006)
502. J. Wiersig, Phys. Rev. A 67, 023807 (2003)
503. Nils Ashcroft, Cornell Univerity, private communication
504. V.M. Fridkin, *Ferroelectric Semiconductors* (translated from Russian) (Plenum, New York, 1980)
505. Y. Xu, *Ferroelectric Materials and their Applications* (North Holland, Amsterdam, 1991)
506. G. Shirane, S. Hoshino, J. Phys. Soc. Jpn. 6, 265 (1951)
507. W.J. Merz, Phys. Rev. 76, 1221 (1949)
508. D.K. Agrawal, C.H. Perry, Phys. Rev. B 4, 1893 (1971)
509. G. Rupprecht, R.O. Bell, Phys. Rev. 135, A748 (1964)
510. M.E. Lines, Phys. Rev. 177, 819 (1969)
511. W.J. Merz, Phys. Rev. 91, 513 (1953)
512. P.W. Forsbergh, Jr., Phys. Rev. 76, 1187 (1949)
513. C. Gähwiller, Phys. Condens. Mater. 6, 269 (1967)
514. D.L. Smith, C. Mailhiot, J. Appl. Phys. 63, 2717 (1988)
515. O. Ambacher, TU Ilmenau, private communication
516. M. Grundmann, O. Stier, D. Bimberg, Phys. Rev. B 50, 14187 (1994)
517. A.D. Andreev, E.P. O'Reilly, Phys. Rev. B 62, 15851 (2000)
518. *Diluted Magnetic Semiconductors*, J.K. Furdyna, J. Kossut, eds. Semicond. Semimet. 25 (1988)

519. *Semiconductor Spintronics and Quantum Computation*, Nanoscience and Technology, D. Awshalom, D. Loss, N. Samarth, eds. (Springer, Berlin, 2002)
520. S.J. Pearton, C.R. Abernathy, M.E. Overberg, G.T. Thaler, D.P. Norton, N. Theodoropoulou, A.F. Hebard, Y.D. Park, F. Ren, J. Kim, L.A. Boatner, J. Appl. Phys. 93, 1 (2003)
521. I. Tsubokawa, J. Phys. Soc. Jpn. 15, 1664 (1960)
522. B.T. Matthias, R.M. Bozorth, J.H. van Vleck, Phys. Rev. Lett. 7, 160 (1961)
523. P.G. Steenecken, PhD Thesis, Rijksuniversiteit Groningen (1974), ISBN 90-367-1695-0
524. P. Wachter, in *Handbook on the Physics and Chemistry of Rare Earths*, K.A. Gschneider, Jr., L. Eyring, eds., Chap. 19, 507 (1979)
525. T. Fukumura, H. Toyosaki, Y. Yamada, Semicond. Sci. Technol. 20, S103 (2005)
526. W. Giriat, J.K. Furdyna, *Crystal Structure, Composition, and Materials Preparation of Diluted Magnetic Semiconductors* in Ref. [518], p. 1
527. C. Rigaux, *Magnetooptics in Narrow Gap Diluted Magnetic Semiconductors* in Ref. [518], p. 229
528. C. Domb, N.W. Dalton, Proc. Phys. Soc. (London) 89, 859 (1966)
529. C. Zener, Phys. Rev. 81, 440 (1951)
530. H. Akai, Phys. Rev. Lett. 81, 3002 (1998)
531. T. Dietl, H. Ohno, F. Matsukura, J. Cibert, D. Ferrand, Science 287, 1019 (2000)
532. H. Munekata, H. Ohno, S. von Molnár, A. Segmüller, L.L. Chang, L. Esaki, Phys. Rev. Lett. 63, 1849 (1989)
533. H. Ohno, M. Munekata, T. Penney, S. von Molnár, L.L. Chang, Phys. Rev. Lett. 68, 2664 (1992)
534. H. Ohno, A. Shen, F. Matsukura, A. Oiwa, A. Endo, S. Katsumoto, Y. Iye, Appl. Phys. Lett. 69, 363 (1996)
535. Y.-J. Zhao, P. Mahadevan, A. Zunger, Appl. Phys. Lett. 84, 3753 (2004)
536. H. Ohno, D. Chiba, F. Matsukura, T. Omiya, E. Abe, T. Dietl, Y. Ohno, K. Ohtani, Nature 408, 944 (2000)
537. S.T.B. Goennenwein, Th.A. Wassner, H. Huebl, M.S. Brandt , J.B. Philipp, M. Opel, R. Gross, A. Koeder, W. Schoch, A. Waag, Phys. Rev. Lett. 92, 227202 (2004)
538. P. Sharma, A. Gupta, K. V. Rao, F.J. Owens, R. Sharma, R. Ahuja, J.M. Osorio Guillen, B. Johansson, G.A. Gehring, Nature Mater. 2, 673–677 (2003)
539. H. Schmidt, M. Grundmann et al., unpublished
540. E.I. Rashba, Physica E 20, 189 (2004)
541. M.I. Dyakonov, V.I. Perel, Sov. Phys. Solid State 13, 3023 (1972)
542. S. Datta, B. Das, Appl. Phys. Lett. 56, 665 (1990)
543. A.T. Kanbicki, O.M.J. van 't Erve, R. Magno, G. Kioseoglou, C.H. Li, B.T. Jonker, Appl. Phys. Lett. 82, 4092 (2003)
544. B.T. Jonker, S.C. Erwin, A. Petrou, A.G. Petukhov, MRS Bull. 28(10), 740 (2003)
545. H.K. Henisch, *Rectifying Semiconductor Contacts* (Oxford University Press (Clarendon), London, 1957)
546. J.R. Macdonald, Solid-State Electron. 5, 11 (1962)
547. F.A. Padovani, *The Voltage–Current Characteristic of Metal–Semiconductor Contacts*, Semicond. Semimet. 7, 75 (1971)

548. E.H. Rhoderick, Inst. Phys. Conf. Ser. 22, 3 (1974)
549. A.M. Cowley, S.M. Sze, J. Appl. Phys. 36, 3212 (1965)
550. C.A. Mead, *Ohmic Contacts to Semiconductors*, B. Schwartz, ed. (Electrochemical Society, New York, 1969), p. 3–16
551. S. Kurtin, T.C. McGill, C.A. Mead, Phys. Rev. Lett. 22, 1433 (1969)
552. S.M. Sze, C.R. Crowell, D. Khang, J. Appl. Phys. 35, 2534 (1964)
553. A.M. Goodman, J. Appl. Phys. 34, 329 (1963)
554. C. Wagner, Phys. Z 32, 641 (1931)
555. W. Schottky, E. Spenke, Wiss. Veröff. Siemens Werke 18, 225 (1939)
556. H.A. Bethe, MIT Radiation Lab. Report 43-12 (1942)
557. J.H. Werner, H.H. Güttler, J. Appl. Phys. 69, 1522 (1991)
558. H. v. Wenckstern, G. Biehne, R. Abdel Rahman, H. Hochmuth, M. Lorenz, M. Grundmann, Appl. Phys. Lett. 88, (2006)
559. J.H. Werner, Appl. Phys. A 47, 291 (1988)
560. F.A. Padovani, G.G. Summer, J. Appl. Phys. 36, 3744 (1965)
561. F.A. Padovani, R. Stratton, Solid-State Electron. 9, 695 (1966)
562. C.R. Crowell, M. Beguwala, Solid-State Electron. 14, 1149 (1971)
563. C.R. Crowell, S.M. Sze, Solid-State Electron. 9, 1035 (1966)
564. J.M. Andrews, M.P. Lepselter, Solid-State Electron. 13, 1011 (1970)
565. A.Y.C. Yu, Solid-State Electron. 13, 239 (1970)
566. C.Y. Chang, Y.K. Fang, S.M. Sze, Solid-State Electron. 14, 541 (1971)
567. T. Sanada, O. Wada, Jpn. J. Appl. Phys. 49, L491 (1980)
568. N. Braslau, J.B. Gunn, J.L. Staples, Solid-State Electron. 10, 381 (1967)
569. C.L. Chen, L.J. Mahoney, M.C. Finn, R.C. Brooks, A. Chu, J.G. Mavroides, Appl. Phys. Lett. 48, 535 (1986)
570. V.L. Rideout, Solid-State Electron. 18, 541 (1975)
571. B.C. Sharma, *Ohmic Contacts to III–V Compound Semiconductors*, Semicond. Semimet. 15, 1–38 (1981)
572. G.Y. Robinson, Solid-State Electron. 18, 331 (1975)
573. I.H. Campbell, T.W. Hagler, D.L. Smith, J.P Ferraris, Phys. Rev. Lett. 76, 1900 (1996)
574. I.H. Campbell, D.L. Smith, *Physics of Organic Electronic Devices*, Solid State Phys. 55, 1 (2001)
575. C.M. Heller, I.H. Campbell, D.L. Smith, N.N. Barashkov, J.P. Ferraris, J. Appl. Phys. 81, 3227 (1997)
576. Frieder Baumann, Lucent Technologies' Bell Labs.
577. C.G.B. Garrett, W.H. Brattain, Phys. Rev. 99, 376 (1955)
578. A.S. Grove, B.E. Deal, E.H. Snow, C.T. Shah, Solid-State Electron. 8, 145 (1965)
579. B.E. Deal, E.H. Snow, C.A. Mead, J. Phys. Chem. Solids 27, 1873 (1966)
580. W.M. Werner, Solid-State Electron. 17, 769 (1974)
581. W.F. O'Hearn, Y.F. Chang, Solid-State Electron. 13, 473 (1970)
582. M. Grundmann, Solid-State Electron. 49, 1446 (2005)
583. H.K. Gummel, Solid-State Electron. 10, 209 (1967)
584. J.L. Moll, *Physics of Semiconductors* (McGraw-Hill, New York, 1964)
585. M.J.O. Strutt, *Semiconductor Devices* Vol. 1 (Academic, New York, 1966), Chap. 2
586. S.M. Sze, G. Gibbons, Appl. Phys. Lett. 8, 111 (1966)
587. M.H. Lee, S.M. Sze, Solid-State Electron. 23, 1007 (1980)

588. A. Goetzberger, B. McDonald, R.H. Haitz, R.M. Scarlet, J. Appl. Phys. 34, 1591 (1963)
589. Virginia Diodes, Inc., www.virginiadiodes.com
590. A. Dearn, L. Devlin, Plextek Ltd., UK.
591. *Tuning Varactors*, Application Note, MicroMetrics Inc., Londonderry, NH, USA
592. *BY329 Product Specifications*, Philips Semiconductors (1998)
593. *Step Recovery Diodes*, Application Note, MicroMetrics Inc., Londonderry, NH, USA
594. M.R.T. Tan, S.Y. Yang, D.E. Mars, J.L. Moll, Hewlett Packard, Palo Alto, CA, USA (1991)
595. H.J.A. Bluyssen, L.J. van Ruyven, F. Williams, Solid-State Electron. 22, 573 (1979)
596. A.L. Linsebigler, G. Lu, J.T. Yates, Chem. Rev. 95, 735–758 (1995)
597. D. Duonghong, E. Borgarello, M. Grätzel, J. Am. Chem. Soc. 103, 4685 (1981)
598. K.Y. Jung, Y.Ch. Kang, S.B. Park, J. Mater. Sci. Lett. 16, 1848 (1997)
599. J.L. Yang, S.J. An, W.I. Park, G.-Ch. Yi, W. Choi, Adv. Mater. 16, 1661 (2004)
600. Canon Inc., www.canon.com/technology
601. B.F. Levine, J. Appl. Phys. 74, R1 (1993)
602. H.C. Liu, R. Dudek, A. Shen, E. Dupont, C.Y. Song, Z.R. Wasilewski, M. Buchanan, Appl. Phys. Lett. 79, 4237 (2001)
603. Fraunhofer-Institut für Angewandte Festkörperphysik, Freiburg, www.iaf.fraunhofer
604. M.D. Petroff, M.G. Stapelbroek, US Patent No. 4-568-960 (1986)
605. N.M. Haegel, Proc. SPIE 4999, 182 (2003)
606. F. Szmulowicz, F.L. Madarsz, J. Appl. Phys. 62, 2533 (1987)
607. H. Melchior, *Demodulation and Photodetection Techniques*, Laser Handbook, F.T. Arecchi, E.O. Schulz-Dubois, eds., Vol. 1 (North-Holland, Amsterdam, 1972), p. 725–835
608. G.E. Stillman, C.M. Wolfe, Semicond. Semimet. 12, 291 (1977)
609. Datasheet *Position Sensitive Photodiodes, DL-100-7-KER pin* (2002), Silicon Sensor GmbH, Berlin, Germany, www.silicon-sensor.com
610. S.M. Sze, D.J. Coleman, A. Loya, Solid-State Electron. 14, 1209 (1971)
611. D. Kuhl, PhD Thesis, Technische Universität Berlin (1992)
612. S.Y. Chou, M.Y. Liu, IEEE J. Quantum Electr. QE-28, 2358 (1992)
613. R.J. McIntyre, IEEE Trans. Electron Devices ED-13, 164 (1966)
614. R.D. Baertsch, IEEE Trans. Electron Devices ED-13, 987 (1966)
615. K.S. Giboney, M.J.W. Rodwell, J.E. Bowers, IEEE Phot. Technol. Lett. 4, 1363 (1992)
616. J.-W. Shi, K.-G. Gan, Y.-J. Chiu, Y.-H. Chen, C.-K. Sun, Y.-J. Yang, J.E. Bowers, IEEE Phot. Technol. Lett. 16, 623 (2001)
617. W.S. Boyle, G.E. Smith, Bell Syst. Tech. J., 49, 587 (1970)
618. *Charge-coupled Devices and their Applications*, J.D.E. Beynon, D.R. Lamb, eds., (McGraw-Hill, Maidenhead, 1977)
619. A. Goetzberger, E.H. Nicollian, Bell Syst. Tech. J. 46, 513 (1967)
620. J.D.E. Beynon, Microelectron. 7, 7 (1975)
621. *Charge-coupled Device (CCD) Image Sensors*, MTD/PS-0218, Rev. 1 (2001), Eastman Kodak Company, Rochester, NY, www.kodak.com/go/ccd.

622. J.E. Carnes, W.F. Kosonocky, E.G. Ramberg, IEEE Trans. Electron Devices ED-19, 798 (1972)
623. C.H. Sequin, M.F. Tompsett, *Charge Transfer Devices* (Academic, New York, 1975)
624. D.J. Burt, Int. Conf. Technol. Appl. CCD, University of Edinburgh, p. 1 (1974)
625. SONY Corporation, www.sony.net.
626. Datasheet *CMOS Linear Image Sensor, S9226* (2003), www.hamamatsu.com
627. Foveon Inc., www.foveon.com
628. R.B. Merrill, US Patent 5,965,875, awarded 1999
629. Datasheet *Avalanche Photodiode Array, AD-LA-16-9-DIL 18* (2002), Silicon Sensor GmbH, Berlin, Germany, www.silicon-sensor.com
630. Datasheet *InGaAs Linear Photodiode Array, SU1024LE-1.7* (2003), Sensors Unlimited, Inc., www.sensorsinc.com
631. Datasheet *Quadrant Photodiode with Position Sensing, QD50-0-SD* (2004), Centrovision, OSI Systems, Inc., Newbury Park, CA, USA, www.centrovision.com
632. R. Leemans, W. Cramer, *The IIASA database for mean monthly values of temperature, precipitation and cloudiness on a global terrestrial grid*, Research Report RR-91-18. International Institute of Applied Systems Analyses, Laxenburg, pp. 61 (1991)
633. W. Shockley, H.-J. Queisser, J. Appl. Phys. 32, 510 (1961)
634. J.H. Werner, S. Kolodinski, H.-J. Queisser, Phys. Rev. Lett. 72, 3851 (1994)
635. C.H. Henry, J. Appl. Phys. 51, 4494 (1980)
636. M.B. Prince, J. Appl. Phys. 26, 534 (1955)
637. J. Mandelkorn, J.H. Lamneck, Conf. Rec. 9th IEEE Photovoltaic Spec. Conf. (IEEE, New York, 1972), p. 83
638. R.A. Arndt, J.F. Allison, J.G. Haynos, A. Meulenberg, Jr., Conf. Rec. 11th IEEE Photovoltaic Spec. Conf. (IEEE, New York, 1975), p. 40
639. Australian CRC for Renewable Energy Ltd. (ACRE), acre.murdoch.edu.au.
640. Solarion GmbH, Leipzig, Germany, www.solarion.de
641. T. Trupke, M.A. Green, P. Würfel, J. Appl. Phys. 92, 1668 (2002)
642. T. Trupke, M.A. Green, P. Würfel, J. Appl. Phys. 92, 4117 (2002)
643. US Department of Energy, http://www.eere.energy.gov/solar.
644. J.H. Werner, Adv. Solid State Phys. (Festkörperprobleme) 44, 51 (2004)
645. S.E. Derenzo, M.J. Weber, E. Bourret-Courchesne, M.K. Klintenberg, Nucl. Instrum Methods A505, 111-117 (2003)
646. Y.A. Ono, *Electroluminescent Displays*, (World Scientific, Singapore, 1995)
647. Glenn F. Knoll, *Radiation Detection and Measurement*, 3rd edn. (John Wiley & Sons, New York, 2000)
648. G. Hoffmann, F.H. Emden.
649. S.E. Derenzo, W.W. Moses, M.J. Weber, A.C. West, Mater. Res. Soc. Symp. 348, 39 (1994)
650. E.F. Schubert, *Light-emitting Diodes* (Cambridge University Press, 2003)
651. W.N. Carr, Infrared Phys. 6, 1 (1966)
652. S.V. Galginaitis, J. Appl. Phys. 36, 460 (1965)
653. OSRAM Opto Semiconductors GmbH (2001), Regensburg, Germany, www.osram-os.com
654. LumiLeds Lighting, www.lumileds.com

655. OSRAM Opto Semiconductors GmbH (2003), Regensburg, Germany, www.osram-os.com
656. Application note STS-KSE3692, Nichia Corp. (2004)
657. M.G. Craford, MRS Bull. 25(10), 27 (2000)
658. H.C. Casey, M.B Panish, *Heterostructure Lasers* (Academic Press, 1st edn., 1978), ISBN 0-12-163101-X (Two volumes. Part A: Fundamental Principles and Part B: Materials and Operating Characteristics.)
659. H. Kressel, J. Butler, *Semiconductor Lasers and LEDs* (Academic Press, 1st edn., 1977), ISBN 0-12-426250-3
660. J. Faist, F. Capasso, D.L. Sivco, C. Sirtori, A.L. Hutchinson, A.Y. Cho, Science 264, 553 (1994)
661. Lasers and Photonics Marketplace Seminar 2005
662. M.B. Panish, I. Hayashi, S. Sumski, Appl. Phys. Lett. 16, 326 (1970)
663. H. Wenzel, F. Bugge, G. Erbert, R. Hülsewede, R. Staske, G. Tränkle, Electron. Lett. 37, 351 (2001)
664. L.A. D'Asaro, J. Lumin. 7, 310 (1973)
665. H. Yonezu, I. Sakuma, K. Kobayashi, T. Kamejima, M. Ueno, Y. Nannichi, Jpn. J. Appl. Phys. 12, 1585 (1973)
666. M. Kamp, J. Hofmann, A. Forchel, S. Lourdudoss, Appl. Phys. Lett. 78, 4074 (2001)
667. M. Kamp, private communication
668. M. Fujite, R. Ushigome, T. Baba, IEEE Phot. Technol. Lett. 13, 403 (2001)
669. G. Björk, A. Karlsson, Y. Yamamoto, Phys. Rev. A 50, 1675 (1994)
670. Y. Yamamoto, S. Machida, G. Björk, Phys. Rev. A 44, 657 (1991)
671. D. Garbuzov, I. Kudryashov, A. Komissarov, M. Maiorov, W. Roff, J. Connolly, Optical Fiber Communication Conference, OSA Technical Digest Series, WD1, (Optical Society of America, Washington, D.C., 2003)
672. T. Kimura, M. Nakae, J. Yoshida, S. Iizuka, A. Sato, Optical Fiber Digest Series, ThN5 485-486 (Optical Society of America, Washington, D.C., 2002)
673. JENOPTIK Laserdiode GmbH, Pressemitteilung 19.1.2006
674. W.T. Tsang, R.A. Logan, J.P. Van der Ziel, Appl. Phys. Lett. 34, 644 (1979)
675. N. Kirstaedter, N.N. Ledentsov, M. Grundmann, D. Bimberg, V.M. Ustinov, S.S. Ruvimov, M.V. Maximov, P.S. Kop'ev, Zh.I. Alferov, U. Richter, P. Werner, U. Gösele, J. Heydenreich, Electron. Lett. 30, 1416 (1994)
676. D. Garbuzov, M. Maiorov, R. Menna, A. Komissarov, V. Khalfin, I. Kudryashov, A. Lunev, L. DiMarco, J. Connolly, Proc. SPIE 4651, 92 (2002)
677. D. Garbuzov, I. Kudryashov, A. Tsekoun, A. Komissarov, W. Roff, M. Maiorov, R. Menna, A. Lunev, J. Connolly, Optical Fiber Communication Conference 2002, Technical Digest: ThN6, (2002)
678. Datasheet *High-Power 1550 nm DFB Source Lasers, A1112*, Agere Systems (2001), www.agere.com
679. M.-C. Amann, J. Buus, *Tunable Laser Diodes* (Artech House, Boston, 1998), ISBN 0-89006-963-8
680. J.N. Walpole, A.R. Calawa, T.C. Harman, S.H. Groves, Appl. Phys. Lett. 28, 552 (1976)
681. S.-L. Lee, I.-F. Jang, C.-T. Pien, C.-Y. Wang, T.-T. Shih, IEEE Phot. Technol. Lett. 11, 955 (1999)
682. B. Mason, G.A. Fish, S.P. DenBaars, L.A. Coldren, IEEE Phot. Technol. Lett. 10, 1211 (1998)

683. C.J. Hwang, J.C. Dyment, J. Appl. Phys. 44, 3240 (1973)
684. N.K. Dutta, S.J. Wang, A.B. Piccirilli, R.F. Karlicek, Jr., R.L. Brown, M. Washington, U.K. Chakrabarti, A. Gnauck, J. Appl. Phys. 66, 4640 (1989)
685. J.S. Gustavsson, Å. Haglund, J. Bengtsson, A. Larsson, IEEE J. Quantum Electron. QE-38, 1089 (2002)
686. C.H. Henry, IEEE J. Quantum Electron. QE-18, 259 (1982)
687. Y. Yamamoto, H.A. Haus, Phys. Rev. A 41, 5164 (1990)
688. D. Welford, A. Mooradian, Appl. Phys. Lett. 40, 865 (1982); Appl. Phys. Lett. 41, 1007 (1982) (erratum)
689. Quintessence Photonics Corporation, http://www.qpc.cc
690. BinOptics Corporation, Ithaca, NY, www.binoptics.com
691. K.D. Choquette, MRS Bull. 27(7), 507 (2002)
692. K. Iga, *Vertical-cavity Surface-emitting Laser Devices* (Springer, Berlin, 2003), ISBN 3-540-67851-4
693. Sandia National Laboratories, www.sandia.gov
694. www.ulm-photonics.de
695. C.J. Chang-Hasnain, *Tunable VCSELs*, IEEE J. Sel. Topics Quantum Electron. 6, 978 (2000)
696. Bandwidth 9, www.bw9.com
697. F. Römer, C. Prott, J. Daleiden, S. Irmer, M. Strassner, A. Tarraf, H. Hillmer, IEEE LEOS International Semiconductor Laser Conference, Garmisch/Germany (2002)
698. M. Schulze, J.-M. Pelaprat, Photonics Spectra 5 (2001)
699. Datasheet *Sapphire*™ *488-20 laser* (2004), Coherent Inc., www.coherent.com
700. R. F. Kazarinov, R. A. Suris, Fiz. Tekh. Poluprovodn, 5, 797 (1971)
701. F. Capasso, K. Mohammed, A.Y. Cho, IEEE J. Quantum Electron. 22, 1853 (1986)
702. F. Capasso, www.bell-labs.com
703. J. Faist, www.unine.ch/phys/meso
704. H. Krömer, Phys. Rev. 109, 1856 (1958)
705. A.A. Andronov, I.V. Zverev, V.A. Kozlov, Yu.N. Nozdrin, S.A. Pavlov, V. N. Shastin, Pis'ma Zh. Eksp. Teor. Fiz. 40, 69 (1984) [JETP Lett. 40, 804 (1984)]
706. Opt. Quantum Electron. 23, Special Issue *Far-infrared Semiconductor Lasers*, E. Gornik, A.A. Andronov, eds. (Chapman and Hall, London, 1991)
707. E. Bründermann, *Widely Tunable Far Infrared Hot Hole Semiconductor Lasers* in *Long-wavelength Infrared Semiconductor Lasers*, H.K. Choi, ed. (John Wiley & Sons, New York, 2004), pp. 279–350, ISBN 0-471-39200-6
708. E. Bründermann, A.M. Linhart, H.P. Röser, O.D. Dubon, W.L. Hansen, E.E. Haller, Appl. Phys. Lett. 68, 1359 (1996)
709. E. Bründermann, D.R. Chamberlin, E.E. Haller, Appl. Phys. Lett. 76, 2991 (2000)
710. G. Jost, University of Ulm, Department of Optoelectronics, Annual Report 1998, p. 64
711. Ferdinand-Braun-Institut für Höchstfrequenztechnik, Berlin, www.fbh-berlin.de
712. P.G.A Jespers, *Measurements for Bipolar Devives* in *Process and Device Modelling for Integrated Circuit Design*, F. van de Wiele, W.L. Engl, P.G. Jespers, eds. (Noordhoff, Leyden, 1977)

713. M.J. Morant, *Introduction to Semiconductor Devices* (Addison-Wesley, Reading, MA, 1964)
714. H.K. Gummel, H.C. Poon, Bell Syst. Tech. J. 49, 827 (1970)
715. D.C. Herbert, Semicond. Sci. Technol. 10, 682 (1995)
716. Solid State Electronics Laboratory, University of Michigan, www.eecs.umich.edu
717. D. Cui, D. Sawdai, D. Pavlidis, S.H. Hsu, P. Chin, T. Block, Proc. of the 12th Int. Conf. on Indium Phosphide and Related Materials, p. 473 (2000)
718. M. Feng, N. Holonyak, Jr., R. Chan, Appl. Phys. Lett. 84, 1952 (2004)
719. R.R. Bockemuehl, IEEE Trans. Electron Devices ED-10, 31 (1963)
720. R.D. Middlebrook, I. Richer, Solid-State Electron. 6, 542 (1963)
721. K. Lehovec, R. Zuleeg, Solid-State Electron. 13, 1415 (1970)
722. H.C. Pao, C.T. Sah, Solid-State Electron. 9, 927 (1966)
723. K. Yamaguchi, IEEE Trans. Electron Devices ED-26, 1068 (1979)
724. R.R. Troutman, IEEE J. Solid State Circuits SC-9, 55 (1974)
725. W. Maly, *Atlas of IC technologies - An Introduction to VLSI Processes* (Benjamin/Cummings Publishing Company, San Francisco, 1987)
726. D. Lubyshev, W.K. Liu, T.R. Stewart, A.B. Cornfeld, X.M. Fang, X. Xu, P. Specht, C. Kisielowski, M. Naidenkova, M.S. Goorsky, C.S. Whelan, W.E. Hoke, P.F. Marsh, J.M. Millunchick, S.P. Svensson, J. Vac. Sci. Technol. B 19, 1510 (2001)
727. D. Udeshi, E. Maldonado, Y. Xu, M. Tao, W.P. Kirk, J. Appl. Phys. 95, 4219 (2004)
728. M. Schlechtweg, A. Tessmann, A. Leuther, C. Schwörer, M. Lang, U. Nowotny, O. Kappeler, Proc. of The European Gallium Arsenide and other Compound Semiconductors Application Symposium (GAAS 2003) (Horizon House, London, 2003), pp. 465-468
729. J.F. Wager, Science 300, 1245 (2003)
730. K. Nomura, H. Ohta, K. Ueda, T. Kamiya, M. Hirano, H. Hosono, Science 300, 1269 (2003)
731. H.A. Kramers, Nature 117, 775 (1926)
732. R. de L. Kronig, J. Opt. Soc. Am. 12, 547 (1926)
733. P.O. Löwdin, J. Chem Phys. 19, 1396 (1951)
734. G.H. Wannier, Phys. Rev. 52, 191 (1937)

Index